T0142302

Studies in Computational Intelligence

Volume 771

Series editor

Janusz Kacprzyk, Polish Academy of Sciences, Warsaw, Poland
e-mail: kacprzyk@ibspan.waw.pl

The series "Studies in Computational Intelligence" (SCI) publishes new developments and advances in the various areas of computational intelligence—quickly and with a high quality. The intent is to cover the theory, applications, and design methods of computational intelligence, as embedded in the fields of engineering, computer science, physics and life sciences, as well as the methodologies behind them. The series contains monographs, lecture notes and edited volumes in computational intelligence spanning the areas of neural networks, connectionist systems, genetic algorithms, evolutionary computation, artificial intelligence, cellular automata, self-organizing systems, soft computing, fuzzy systems, and hybrid intelligent systems. Of particular value to both the contributors and the readership are the short publication timeframe and the world-wide distribution, which enable both wide and rapid dissemination of research output.

More information about this series at http://www.springer.com/series/7092

A. N. Krishna · K. C. Srikantaiah
C. Naveena
Editors

Integrated Intelligent Computing, Communication and Security

Editors
A. N. Krishna
Department of Computer Science
and Engineering
SJB Institute of Technology
Bangalore, Karnataka
India

C. Naveena
Department of Computer Science
and Engineering
SJB Institute of Technology
Bangalore, Karnataka
India

K. C. Srikantaiah
SJB Institute of Technology
Bangalore, Karnataka
India

ISSN 1860-949X ISSN 1860-9503 (electronic)
Studies in Computational Intelligence
ISBN 978-981-13-4242-4 ISBN 978-981-10-8797-4 (eBook)
https://doi.org/10.1007/978-981-10-8797-4

This Springer imprint is published by the registered company Springer Nature Singapore Pte Ltd.
The registered company address is: 152 Beach Road, #21-01/04 Gateway East, Singapore 189721, Singapore

Contents

Improved k-Anonymity Privacy-Preserving Algorithm Using Madhya Pradesh State Election Commission Big Data

Priyank Jain, Manasi Gyanchandani and Nilay Khare

Abstract Modern technology produces a large number of public and private data sets that make the task of securing personal data unavoidable. Initially, priority was given to securing data for organizations and companies, but nowadays it is also necessary to provide security for personal data. Therefore, to achieve information security, the protection of individual data is critical. Data anonymization is a technique for preserving privacy in data publishing, which enables the publication of practically useful information for data mining while preserving the confidentiality of the information of the individual. This chapter presents the implementation of data anonymization using a proposed improved k-anonymity algorithm applied to a large candidate election data set acquired from the Madhya Pradesh (MP, India) State Election Commission. Along with greater privacy, the algorithm is executed in less time than the traditional k-anonymity algorithm, and as such, it is able to satisfy the data protection needs of the current big data environment.

Keywords Big data · Anonymization · Generalization · Suppression l-Diversity · t-Closeness · Differential privacy

1 Introduction

A majority of the work in data protection has concentrated on issues involving the nature and utility of security safeguards, and there has been little focus on the issue of versatility in data protection strategies. The reason is that most of the aspects of anonymization have been viewed as a cluster and one-time handling with regard to information sharing. Be that as it may, in recent years, the size of information collections has grown exponentially, to the point that the use of the present calculations is becoming increasingly problematic [1, 2]. Moreover, the recent growth

P. Jain (✉) · M. Gyanchandani · N. Khare
Maulana Azad National Institute of Technology, Bhopal, MP, India
e-mail: priyankjain1984@gmail.com

A. N. Krishna et al. (eds.), *Integrated Intelligent Computing, Communication and Security*, Studies in Computational Intelligence 771,
https://doi.org/10.1007/978-981-10-8797-4_1

of transient information indexes has given rise to a need to reexamine and adapt such strategies for the more current information indexes which have been gathered.

Client protection may be compromised under the following conditions:

1. Personal data, when consolidated with public data sets, may enable the retrieval of new information about the clients. Those certainties might be shrouded and should not be uncovered to others.
2. Personal data is in some cases collected and used to increase the value of business. For instance, a person's shopping habits may reveal a considerable measure of individual data.
3. The sensitive data are stored and processed in a location not secured properly and data leakage may occur during storage and processing phases [3].

Traditional data preservation methods consisted of cryptography that refers to set of techniques and algorithms for the protection of data. Cryptography alone can't guarantee the security required for normally distributed computing and big data administration. This is based on the fact that big data differs from conventional large data collections based on properties known as the three V's (velocity, variety and volume). It is these elements that make big data engineering unique [4]. These differences in design and the mind-boggling nature of big data make conventional encryption strategies inadequate for its security needs. The challenge with cryptography is all or nothing retrieval policy of encrypted data. The less sensitive information, which can be helpful in big data analytics, is likewise encrypted, and the client is not permitted access. This makes information inaccessible to individuals without an unscrambling key. Additionally, security may be breached if information is stolen before encryption, or if cryptographic keys are abused.

The Madhya Pradesh (MP) election candidate data set is a large volume of data, i.e., "big data". The data set consists of sensitive information regarding candidates. This chapter presents the implementation of an improved *k*-anonymization algorithm compared to traditional *k*-anonymization [5, 6] in a de-identification process using an election candidate data set on a distributed environment using Apache Hadoop to secure sensitive candidate information. The improved *k*-anonymity algorithm provides superior time performance compared to the traditional *k*-anonymity algorithm.

2 Related Work

k-Anonymization, along with *l*-diversity and *t*-closeness, are recently developed techniques to increase privacy [3, 7, 8].

(a) *k*-Anonymity: *k*-Anonymity is a privacy model which requires each record (belonging to one individual) to be indistinguishable from at least $(k - 1)$ other records within the published data. Therefore, an attacker who knows the values of the quasi-identifier attributes of a victim is unable to distinguish their record from at least ($k - 1$) other data records [9]. This is referred to as identity attack prevention. Records having a similar incentive for their semi-identifiers constitute a proportionality class. The first information is expected [8, 9] to be sensitive and private, and comprise numerous records. Each record may comprise the accompanying four characteristics. k-Anonymity suffers from the following problems: (i) sensitive values in an equivalence class lack diversity, and (ii) the attacker has background knowledge.

(b) *l*-Diversity [3]: Two noteworthy attacks referred to as homogeneity attack and background knowledge attack have prompted the development of another procedure, l-diversity, which is a progression of the k-anonymity display where it ensures security despite the fact that the information proprietor is unaware of any data that the interloper holds. This technique is derived from k-anonymization and shows where k records in the data set will coordinate with ($k - 1$) other information in the records with a reduction in the scale or level of detail to frame an l-diverse data set. Sensitive characteristics within an equality class must vary within each quasi-identifier equivalence class.

(c) *t*-Closeness [7, 10]: Each data set is said to satisfy t-closeness if each equivalence class has t-closeness. The model of t-closeness is an enhancement over the l-diversity variety display. A vital element of the l-diversity variety display is that it takes all given property values as the same, independent of their allotment in the information. Distribution of sensitive attributes within each quasi-identifier group should be "close" to their distribution in the entire original database.

(d) Differential privacy [11]: In Eq. 1, let D1 and D2 be two data sets. D1 and D2 are said to be neighbors on the off chance that they differ in at most one information section. A calculation M is ε-differentially private if for all neighboring data sets D1 and D2, and all yields x,

$$\Pr[\mathrm{M}(\mathrm{D1}) = \mathrm{x}] \leq \exp(\varepsilon) \Pr[\mathrm{M}(\mathrm{D2}) = \mathrm{x}] \tag{1}$$

In other words, given the yield of the calculation, one can't tell if a particular data point was utilized as a major aspect of the information in light of the fact that the likelihood of creating this yield would have been the same even without that data point. Not having the capacity to tell whether the data point was utilized at all in the calculation prevents discerning any valuable data about it from the calculation's yield alone. The computation of the function must be randomized to achieve privacy [12]. This mechanism can be used for privacy with one of the strongest mathematical guarantees [13]; however, it is yet to be applied to real-world data sets at a large scale. Therefore, further differential privacy can be explored along with machine learning for greater efficiency.

3 Preserving Privacy in Big Data

Various techniques have been developed recently with the goal of guaranteeing big data privacy and security [14–17]. These mechanisms can be grouped by stages of the big data life cycle: data generation, storage and processing. In the information era, for the assurance of protection, methods of confinement and misrepresenting information are utilized. While limitation systems attempt to constrain access to people's private information, information adulteration methods modify the preliminary information before it is discharged to a non-trusted party. Encryption systems are one way to provide security assurances during the information-stockpiling stage [3, 7]. Encryption-based systems can be additionally separated into attribute-based encryption (ABE), identity-based encryption (IBE) and storage encryption. In addition, hybrid cloud solutions can be utilized to ensure the security of sensitive data, where such data are stored in a private cloud [18, 19]. The information-preparation stage incorporates privacy-preserving data publication (PPDP) and learning extraction from the information. In PPDP, anonymization systems—for example, speculation and concealment—are utilized to secure the protection of information. Guaranteeing the utility of the information while protecting privacy is an incredible challenge in PPDP [6]. Anonymization refers to irreversibly severing a data set from the identity of the data contributor in a study to prevent any future re-identification, even by the study organizers under any condition. Before distribution, the preliminary information is altered by the predefined security prerequisites. To safeguard security, one of the accompanying anonymization operations is connected to the information [7, 20].

4 Proposed Work

The proposed work deals with the election commission data set and aims at providing privacy on a large collection of data. To this end, the Hadoop MapReduce paradigm is used.

4.1 k-*Anonymization Using MapReduce*

The first step in designing the MapReduce routines for *k*-anonymization is to define and handle the input and output of the implementation. The input is given as a <key, value> pair, where "key" represents a tuple and "value" represents 1. A comma-separated value (CSV) file format is the prerequisite for implementing the mapping and reduction operations. Once the initial key and value pair is defined and the data sequentially organized on the basis of "age" attribute, the improved *k*-anonymization algorithm is applied after MapReduce routines.

4.2 *Improved* k-*Anonymity Algorithm*

The improved *k*-anonymity algorithm provides better time performance as compared to the traditional *k*-anonymity algorithm. The following is the pseudo-code in Fig. 1 of the improved *k*-anonymity algorithm.

5 Data Set Description

The data set was collected from the Madhya Pradesh, India, State Election Commission (MP-SEC). It is a candidate data set consisting of 34 attributes and 35,000 instances. The useful attributes extracted after pre-processing include age, district code, candidate name, gender, category, mobile phone number, candidate designation, ward number, votes, marital status, auto identification and occupation. Interesting patterns can be observed from the MP-SEC data set. For example, if it is combined with demographic data, the percentage of people who were eligible and voted in the elections can be calculated. The correlation and dependency between voter caste and the winning candidate can also be found. The relationship between female candidates and their occupations may be detected, and the percentage of female candidates among total candidates can be calculated. If the MP-SEC data set is combined with Aadhar card data—which is India's sophisticated biometric identification system—the voting patterns between reserved and unreserved voter constituencies can be determined. The percentage of different age groups standing for election can also be deduced, revealing whether the proportion of younger or older candidates is greater.

Fig. 1 Improved *k*-anonymity algorithm

1) Select the AUTOID as the pivoting field, i.e. on the basis of its median; we are splitting the entire dataset into 2 partitions.
2) Recursively apply this on the reduced dataset until each partition contains at most 3 tuples.
3) Now, start the process of generalization over the age attribute of the 3-tupleset.
4) After the generalization is over, we suppress the gender or marital status or both if required, to form the equivalence class of at least 2-anonymity.
5) Finally, update the original data

6 Implementation and Results

The working environment is Ubuntu 16.x software. The end-user may utilize any operating system, as Java is platform-independent. To use IntelliJ IDEA with the Java SE Development Kit, the end-user needs to have JRE installed [21]. The Apache Hadoop framework Hadoop 2.x [20] is required, which provides Hadoop Distributed File System (HDFS) capabilities and is used to run the MapReduce paradigm and YARN resource manager [22]. The required hardware configuration consists of an Intel® Core™ i5 CPU, 8 GB of RAM and a maximum of 10 GB of storage space [23].

6.1 Performance Measures

Different performance metrics used to test the effectiveness of the proposed approach:

(1) Information loss:
This is a widely used metric that captures the entire amount of information lost due to generalization [9]. For an anonymized data set with *n* tuples and *m* attributes, the information loss, *I*, is computed as follows:

$$I = \sum_{i=1}^{n} \sum_{j=1}^{m} \frac{|upper_{ij} - lower_{ij}|}{n \cdot m \cdot |max_j - min_j|} \quad (2)$$

In Eq. 2, $lower_{ij}$ and $upper_{ij}$ represent the lower and upper bounds of attribute *j* in tuple *i* after generalization, respectively, and min_j and max_j represent the minimum and maximum values, respectively, taken by attribute *j* over all records.

(2) Running time:
The running time is measured in terms of clock time (ms), which provides good overall scalability of the method.

6.2 Problems with the Traditional Technique

A few problems are encountered when using *k*-anonymity [1, 3, 8, 9] if the equivalence class lacks diversity and the quasi-identifiers are not similar to one another. The *k*-anonymity approach does not preserve privacy if:

1. Sensitive values in an equivalence class lack diversity.
2. The attacker has background knowledge.

The proposed k-anonymity algorithm deals with these problems and provides better results with respect to time, while reducing information loss, compared with the existing algorithm.

6.3 Results

The graphs shown in Figs. 2 and 3 depict time versus data size, with the x-coordinate representing data size and the y-coordinate representing overall time (in ms). Figure 2 represents the improved k-anonymity algorithm; a total of 35,000 instances were taken from the MP-SEC candidate data sets. The experiment started with 10,000 instances and took 21 ms to complete. Further, upon increasing the number of instances to 20,000, the time taken increased in a straight line, whereas it decreased when the instances increase from 25,000 to 35,000 (high volume of data), as shown in Fig. 2. As our experimental results show, information loss using the improved k-anonymity in generalization (age attribute) = 17.566%. The proposed algorithm is thus able to satisfy industrial needs because it provides greater privacy in optimal time, with less loss of information. Figure 3 depicts the existing k-anonymity algorithm [24]. In this algorithm, time is directly proportional to data size, which means it is not an appropriate algorithm for dealing with a high volume of data in optimal time. Figure 4 illustrates the final output using the improved k-anonymity algorithm on MP-SEC candidate big data.

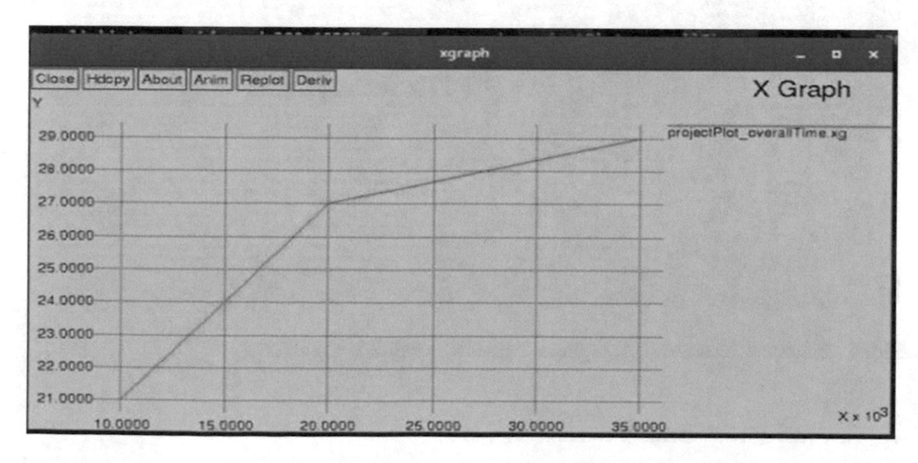

Fig. 2 Overall time versus data size in proposed improved k-anonymity algorithm

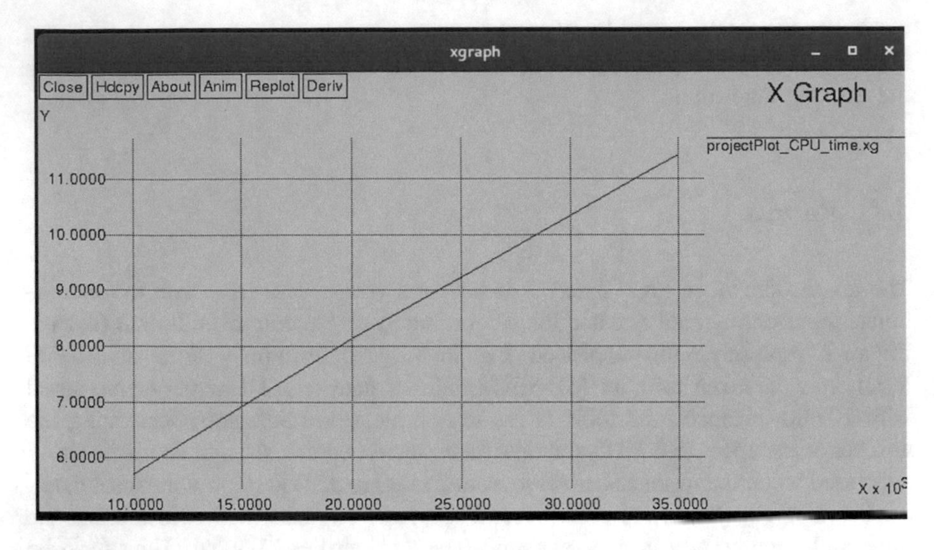

Fig. 3 Overall time versus data size in the existing *k*-anonymity algorithm

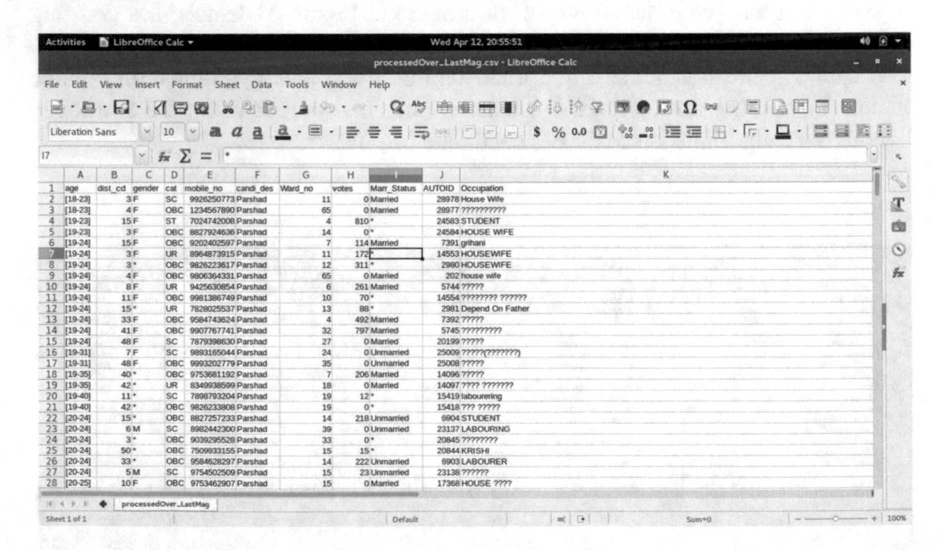

Fig. 4 Improved *k*-anonymity output on MP-SEC candidate big data

7 Conclusion and Future Work

The objective of this work is the application of a practical privacy-preserving framework that ensures the privacy of the individual while maintaining the value of the anonymized data for the researcher. The core value of this work is to promote data sharing for knowledge mining. The improved *k*-anonymity provides a privacy guarantee and data utility for data miners. Generalization and suppression are used to achieve anonymization. The improved algorithm provides better privacy in less time and with less loss of information than the traditional *k*-anonymity algorithm, making it suitable for use on a commercial scale. Moreover, since the real-world MP candidate election data set is being used, the privacy mechanism is useful for solving privacy problems on a practical level. Future work includes the implementation of more complex privacy mechanisms.

Acknowledgements We are grateful to the Madhya Pradesh State Election Commission, India, for their ardent and constant support and for providing us with the real-time candidate big data needed for the research paper.

References

1. Han, Y., B. Jiang, B. Zhou, Y. Tao, J. Pei, and Y. Jia. 2009. Continuous privacy preserving publishing of data streams. In *EDBT*.
2. https://ico.org.uk/for-organisations/guide-to-data-protection/anonymisation/.
3. Hessam Zakerdah, C.C., and K.B. Aggarwal. 2015. *Privacy-preserving big data publishing*. La Jolla: ACM.
4. G. Acampora, et al. 2015. Data analytics for pervasive health. In: *Healthcare data analytics*, 533–576.
5. https://en.wikipedia.org/wiki/K-anonymity.
6. Srikant, R., and R. Agrawal. 2000. Privacy-preserving data mining. In *SIGMOD*.
7. Gyanchandani, Manasi, Priyank Jain, Nilay Khare. Big data privacy: A technological perspective and review, *Journal of Big Data*, 3: 2016. ISSN 2196-1115.
8. Ton, A, M. Saravanan. Ericsson research (Online). http://www.ericsson.com/research-blog/data-knowledge/big-data-privacy-preservation/2015.
9. Nergiz, M., C. Clifton, and A. Nergiz. 2009. Multirelational k-anonymity. *IEEE Transactions on Knowledge and Data Engineering* 21: 1104–1117.
10. Jain, P., A.S. Umesh. 2013. Privacy preserving processing of data decision tree based on Singular Value Decomposition and sample selection. In *2013 9th international conference on information assurance and security (IAS)*, 91–95, Gammarth.
11. Dwork, Cynthia, and Aaron Roth. 2014. *The algorithmic foundations of differential privacy*.
12. Wang, Yadong, and Guang Li. 2011. Privacy-preserving data mining based on sample selection and singular value decomposition. In *IEEE international conference on information services and internet computing*, 298–301.
13. Huy, Xueyang, Jianguo Yaoy, Yu Dengy, Lei Chenz, Mingxuan Yuan, Qiang Yangz, et al. Differential privacy in telco big data platform. In *Proceedings of the VLDB Endowment*, Vol. 8, No. 12. Copyright 2015 VLDB Endowment 21508097/15/08.
14. Sokolova, M., and S. Matwin. 2015. *Personal privacy protection in time of big data*. Berlin: Springer.
15. Lu, R., H. Zhu, X. Liu, J.K. Liu, and J. Shao. 2014. Toward efficient and privacy-preserving computing in big data era. *IEEE Network* 28: 46–50.
16. Mohammadian, E., M. Noferesti, R. Jalili. 2014. FAST: Fast anonymization of big data streams. In *ACM proceedings of the 2014 international conference on big data science and computing*, article 1.
17. Zhang, X., T. Yang, C. Liu, and J. Chen. 2014. A scalable two-phase top-down specialization approach for data anonymization using systems, in MapReduce on cloud. *IEEE Transactions on Parallel and Distributed Systems* 25 (2): 363–373.

18. Xiao, Z., and Y. Xiao. 2013. Security and privacy in cloud computing. *IEEE Transaction on Communications Surveys and Tutorials* 15 (2): 843–859.
19. Wu, X. 2014. Data mining with big data. *IEEE Transactions on Knowledge and Data Engineering* 26 (1): 97–107.
20. Gyanchandani, Manasi, Priyank Jain, Nilay Khare Direndra Pratap Singh, and Lokini Rajesh. 2017. A survey on big data privacy using hadoop architecture. *International Journal of Computer Science and Network Security* 17(2).
21. http://www.javatpoint.com/.
22. https://hadoop.apache.org/docs/current/hadoop-mapreduce-client/hadoop-mapreduce-client-core/MapReduceTutorial.html.
23. https://docs.oracle.com/javase/7/docs/api/.
24. Kifle Russom, Yohannes. 2013. Privacy preserving for big data analysis. Master thesis, University of Stavanger.

Prediction of Different Types of Wine Using Nonlinear and Probabilistic Classifiers

Satyabrata Aich, Mangal Sain and Jin-Han Yoon

Abstract In the past few years, machine-learning techniques have garnered much attention across disciplines. Most of these techniques are capable of producing highly accurate results that compel a majority of scientists to implement the approach in cases of predictive analytics. Few works related to wine data have been undertaken using different classifiers, and thus far, no studies have compared the performance metrics of the different classifiers with different feature sets for the prediction of quality among types of wine. In this chapter, an intelligent approach is proposed by considering a recursive feature elimination (RFE) algorithm for feature selection, as well as nonlinear and probabilistic classifiers. Performance metrics including accuracy, sensitivity, specificity, positive predictive value (PPV), and negative predictive value (NPV) are compared by implementing different classifiers with original feature sets (OFS) as well as reduced feature sets (RFS). The results show accuracy ranging from 97.61 to 99.69% among the different feature sets. This analysis will aid wine experts in differentiating various wines according to their features.

Keywords Machine learning · Feature selection · Classifiers
Performance metrics · Prediction

S. Aich (✉)
Department of Computer Engineering, Inje University, Gimhae, South Korea
e-mail: satyabrataaich@gmail.com

M. Sain
Department of Computer Engineering, Dongseo University, Busan, South Korea
e-mail: mangalsain1@gmail.com

J.-H. Yoon
Daedong College, Busan, South Korea
e-mail: Yoonjh911@daedong.ac.kr

A. N. Krishna et al. (eds.), *Integrated Intelligent Computing, Communication and Security*, Studies in Computational Intelligence 771,
https://doi.org/10.1007/978-981-10-8797-4_2

1 Introduction

Recent years have seen a trend toward more moderate wine consumption, as studies have reported a positive correlation between wine drinking and heart rate variability [1]. Given these more modest consumption rates, the wine industry is seeking alternatives for the production of good-quality wine at less cost. Different wines have different purposes. Although analyses have shown that chemicals in different types of wine are mostly the same, the chemical concentrations vary. These days, it is critical to classify different wine for quality assurance [2]. In the past, due to lack of technological resources, it was difficult for most of the wine industry to classify wines based on chemical analysis, as it is time- and cost-intensive. These days, with the advent of machine-learning techniques, it is possible to classify wines and to determine the importance of each chemical analysis parameter in the wine and which ones to ignore in order to reduce cost. A performance comparison with different feature sets will also help to classify wine in a more distinctive way. In this chapter, an intelligent approach is proposed by considering a recursive feature elimination (RFE) algorithm for feature selection, in addition to considering the nonlinear and probabilistic classifiers.

The remainder of the chapter is organized as follows: Sect. 2 presents past research related to this field, while Sect. 3 describes the methodologies used in the current work. Section 4 discusses the results of both feature selection and classification. Section 5 puts forth the conclusions and describes future work.

2 Related Works

In the past, few attempts have been made to apply different machine-learning approaches and feature-selection techniques to wine datasets. Er and Atasoy proposed a method for assessing the quality of wines using different classifiers including support vector machines, random forest and k-nearest neighbors. They used principal component analysis (PCA) for feature selection and obtained good results using the random forest algorithm [3]. Cortez et al. proposed a method for predicting human wine taste preferences using machine-learning approaches. They used a support vector machine, a neural network and a multiple regression technique and found that the support vector machine performed better than the other classifiers [4].

Appalasamy et al. proposed a method for predicting wine quality based on physiochemical test data using classification approaches including Iterative Dichotomiser 3 (ID3) and naïve Bayes. They concluded that ID3 outperformed naïve Bayes in the case of red wine, and noted that the classification approach helped to improve the quality of wine during production [5]. Beltrán et al. proposed an approach for wine classification based on aroma chromatograms. They used PCA for dimensionality reduction and wavelet transform for feature extraction, with classifiers such as neural networks, linear discriminant analysis (LDA) and support vector machine, and found that support vector machine with wavelet transform performed better than the other classifiers [6].

Chen et al. proposed a method for predicting wine grade based on consumer sensory review. Using a hierarchical clustering approach and an association rule algorithm, they reported that the reviews demonstrated accuracy of 85.25% in predicting the grade [7]. The above-referenced works motivated us to try a different feature selection algorithm as well as different classifiers to compare performance metrics. In this chapter, we propose an RFE algorithm for feature selection, non-linear classifiers including partitioning decision tree (PART), recursive PART (RPART), bagging and C5.0, and probabilistic classifiers LDA and naïve Bayes.

3 Methodology

A flow chart of the proposed methodology is shown in Fig. 1.

A. Data collection

The wine data set is publicly available in the University of California Irvine (UCI) Machine Learning Repository database. These data were obtained from the

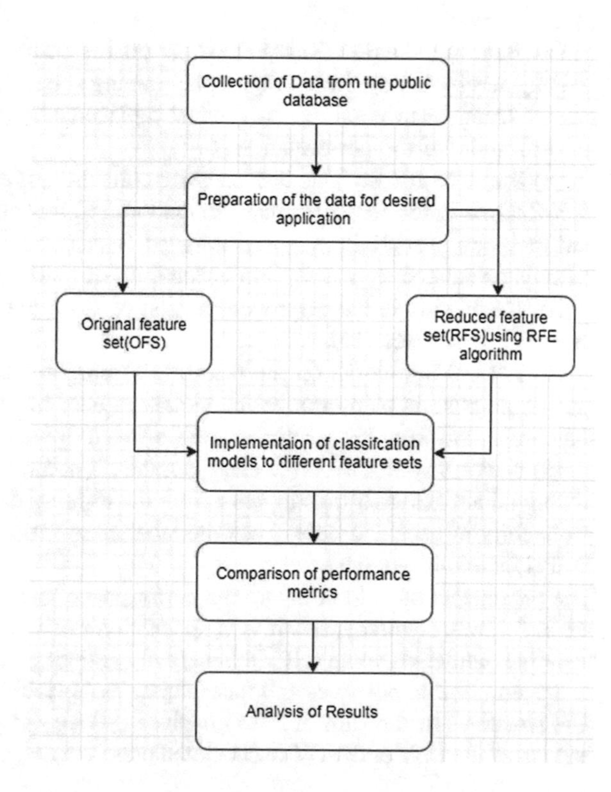

Fig. 1 Flow chart of the proposed method

chemical analysis of wines grown in the same region of Italy. The data set contains three classes of wine, each with 13 attributes [8].

B. Feature reduction

The original features collected from the data sets are referred to as the original feature set (OFS), and the reduced feature set (RFS) is obtained from the RFE algorithm. The RFE selection method is basically a recursive process that ranks features according to some measure of their importance [9].

C. Classification model

In this work, we have used a nonlinear classifier with a decision tree for classification of groups as follows: PART, RPART, C4.5, classification and regression tree with bagging (bagging CART), random forest and boosted C5.0, and probabilistic classifiers LDA and naïve Bayes.

(1) RPART: The recursive partition-based classifier basically works on the principle of the splitting technique. It is called recursive because it continues splitting until some stopping criterion is reached [10].

(2) C4.5: This is basically an improvement over the ID3 algorithm. It works on the principle of information entropy. It is also referred to as a statistical classifier [11].

(3) PART: This is basically a rule-based system that produces some pruned decision tree based on C4.5 and then tries to derive some rule and remove the instances covered by the rule. It continues until all the instances have been covered by the derived rule [12].

(4) Bagging CART: Bagging works on the principle of manipulation of the training sets by taking the average of all training sets. Basically, the training sets are created from the original sets by random replacements [13]. CART with bagging is used to improve the performance [14].

(5) Random forest: This is basically a collection of random decision trees. Each decision tree in the forest learns from random training sets and random feature sets, and then one probability score is assigned to each of them; finally, overall probability is calculated by taking into account all decision trees [15, 16].

(6) C5.0: This is the improved version of C4.5 and has numerous advantages over its predecessor [17].

(7) LDA: This works on the principle of probabilistic parametric classification and its purpose is to minimize the variances between each category by means of data projection from higher dimensional space to lower dimensional space [18, 19].

(8) Naïve Bayes: Given the class variable, it assumes conditional independence among all attributes. It tries to learn from the training data based on the conditional probability of each attribute given the label class [20].

D. Performance metrics

The parameters used to compare the performance and validations of the classifiers are as follows: accuracy, sensitivity, specificity, positive predictive value (PPV) and negative predictive value (NPV). Sensitivity is defined as the ratio of true positives to the sum of true positives and false negatives. Specificity is defined as the ratio of true negatives to the sum of false positives and true negatives. In our research, we have used the PPV and NPV to check the presence and absence of one type of wine.

Thus the PPV is the probability that the one type of wine is present, given a positive test result, and NPV is the probability that the one type of wine is absent, given a negative test result [21]. Accuracy is defined as the ratio of the number of correct predictions made to the total predictions made, and the ratio is multiplied by 100 to express it in terms of percentage.

4 Results and Discussion

We have divided the data into two groups, training and test data. We have trained each classifier based on the training data and predict the power of the classifier based on the test data. Therefore, each classifier is able to show all performance metrics—accuracy, sensitivity, specificity, PPV and NPV—based on the test data. We have applied all the classification techniques to the OFS, which includes 13 features, and the RFS, which includes 10 features, to measure the performance parameter with respect to each classifier. We separated each performance measure with respect to the OFS and RFS and plotted the column plot for better visualization. The results of each performance measure with respect to two feature sets are shown in Figs. 2, 3, 4, 5 and 6.

4.1 Comparison of Accuracy

Figure 2 shows that there is not much difference in terms of accuracy of each classifier with respect to the OFS and RFS. Random forest shows maximum accuracy of 99.69% with the OFS and also shows maximum accuracy of 99.54% with the RFS.

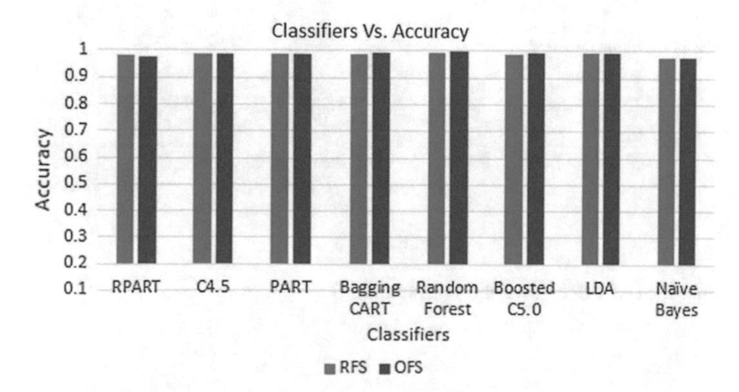

Fig. 2 Comparison of the accuracy of the OFS and RFS

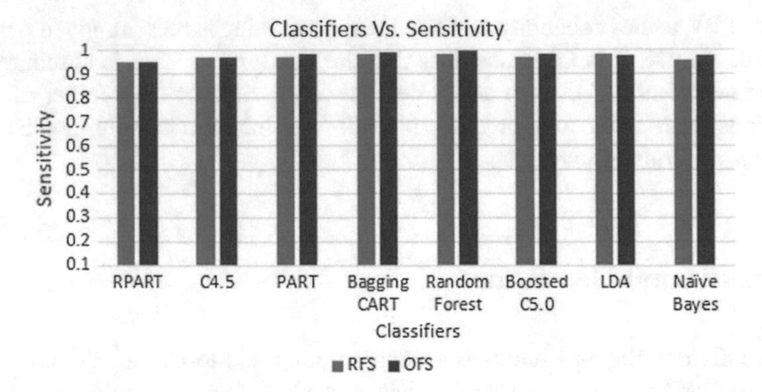

Fig. 3 Comparison of the sensitivity of the OFS and RFS

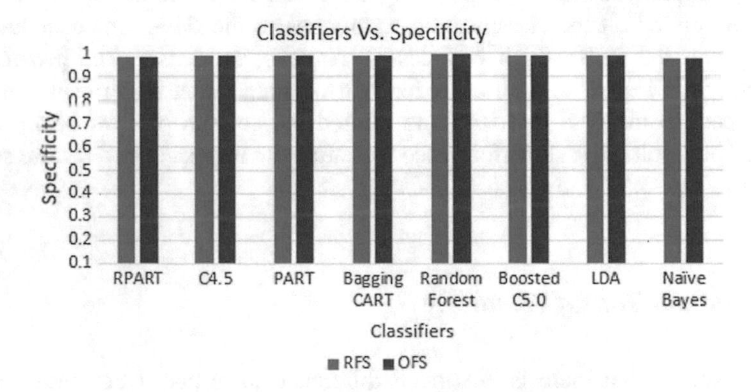

Fig. 4 Comparison of the specificity of the OFS and RFS

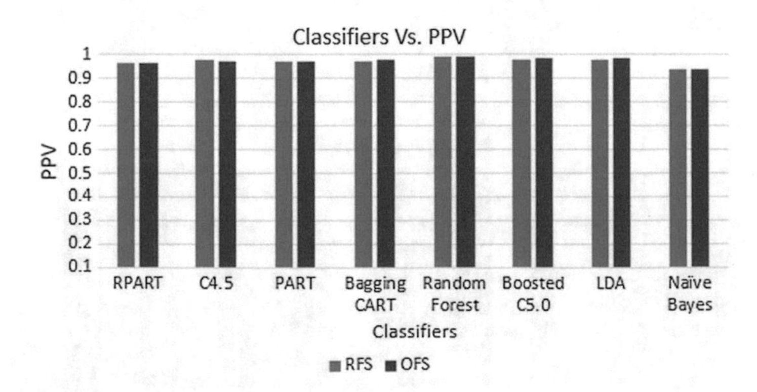

Fig. 5 Comparison of the PPV of the OFS and RFS

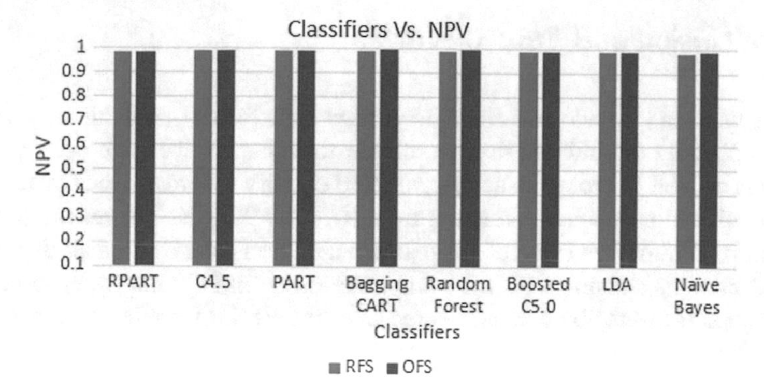

Fig. 6 Comparison of the NPV of the OFS and RFS

4.2 *Comparison of Sensitivity*

Figure 3 shows that there is not much difference in terms of the sensitivity of each classifier with respect to the OFS and RFS. Random forest shows maximum sensitivity of 0.9969 with the OFS, while bagging CART, random forest and LDA show maximum sensitivity of 0.9875 with the RFS.

4.3 *Comparison of Specificity*

Figure 4 shows that there is not much difference in terms of the specificity of each classifier with respect to the OFS and RFS. Random forest shows maximum specificity of 0.9980 with the RFS and also shows maximum specificity of 0.9969 with the OFS.

4.4 *Comparison of PPV*

Figure 5 shows that there is not much difference in terms of the PPV of each classifier with respect to the OFS and RFS. Random forest shows maximum PPV of 0.9937 with the RFS and also shows maximum PPV of 0.9907 with the OFS.

4.5 *Comparison of NPV*

Figure 6 shows that there is not much difference in terms of NPV of each classifier with respect to the OFS and RFS. Random forest shows maximum NPV of 0.9990 with the OFS, while bagging CART and random forest show maximum NPV of 0.9959 with the RFS.

5 Conclusion and Future Work

This analysis can be adopted as an important tool for the prediction of different wines depending on different feature sets. In this chapter, we have used nonlinear classifiers as well as probabilistic classifiers to classify different wines by achieving good classification accuracies ranging from 97.61 to 99.69%. The results show that using an RFS improves some of the performance measures by a small degree over the OFS in some classifiers. At the same time, using an OFS improves some of the performance measures by a small degree over the RFS. Overall, we have not seen much difference in performance measures between the RFS and OFS. However, the RFS can tell us which features should be given more importance for the prediction of difference, and it can ensure that we can achieve good performance with fewer features and less computation time. In the future, we will use other feature selection algorithms as well as other classifiers for better comparison of performance.

References

1. Janszky, I., M. Ericson, M. Blom, A. Georgiades, J.O. Magnusson, H. Alinagizadeh, and S. Ahnve. 2005. Wine drinking is associated with increased heart rate variability in women with coronary heart disease. *Heart* 91 (3): 314–318.
2. Preedy, V., M.L.R. Mendez. 2016. Wine applications with electronic noses. In *Electronic noses and tongues in food science,* 137–151. Cambridge, MA, USA: Academic Press.
3. Er, Y., and A. Atasoy. 2016. The classification of white wine and red wine according to their physicochemical qualities. *International Journal of Intelligent Systems and Applications Engineering* 4: 23–26.
4. Cortez, P., A. Cerdeira, F. Almeida, T. Matos, and J. Reis. 2009. Modeling wine preferences by data mining from physicochemical properties. *Decision Support Systems* 47 (4): 547–553.
5. Appalasamy, P., A. Mustapha, N.D. Rizal, F. Johari, and A.F. Mansor. 2012. Classification-based data mining approach for quality control in wine production. *Journal of Applied Sciences* 12 (6): 598–601.
6. Beltran, N.H., M.A. Duarte-MErmound, V.A.S. Vicencio, S.A. Salah, and M.A. Bustos. 2008. Chilean wine classification using volatile organic compounds data obtained with a fast GC analyzer. *IEEE Transactions* on *Instrumentation* and *Measurement* 57: 2421–2436.
7. Chen, B., C. Rhodes, A. Crawford, and L. Hambuchen. 2014. Wineinformatics: Applying data mining on wine sensory reviews processed by the computational wine wheel. In *IEEE international conference on data mining workshop*, 142–149, Dec. 2014.
8. Forina, M., R. Leardi, C. Armanino, and S. Lanteri. 1998. PARVUS an extendible package for data exploration, classification and correla.
9. Granitto, P.M., C. Furlanello, F. Biasioli, and F. Gasperi. 2006. Recursive feature elimination with random forest for PTR-MS analysis of agroindustrial products. *Chemometrics and Intelligent Laboratory Systems* 83 (2): 83–90.
10. https://en.wikipedia.org/wiki/Recursive_partitioning, Retrieved 19 August 2017.
11. Vijayarani, S., and M. Divya. 2011. An efficient algorithm for generating classification rules. *International Journal of Computer Science and Technology* 2 (4).

12. http://machinelearningmastery.com/non-linear-classification-in-r-with-decision-trees/, Retrieved 19 August 2017.
13. Breiman, L. 1996. Bagging predictors. *Machine Learning* 26 (2): 123–140.
14. Sun, X. 2002. Pitch accent prediction using ensemble machine learning. In *Seventh international conference on spoken language processing*.
15. Breiman, L. 2001. Random forests. *Machine Learning* 45 (1): 5–32.
16. Ellis, K., J. Kerr, S. Godbole, G. Lanckriet, D. Wing, and S. Marshall. 2014. A random forest classifier for the prediction of energy expenditure and type of physical activity from wrist and hip accelerometers. *Physiological Measurement* 35 (11): 2191–2203.
17. Is See5/C5.0 Better Than C4.5?. 2009. (Online). http://www.rulequest.com/see5-comparison.html.
18. Nilashi, M., O. bin Ibrahim, H. Ahmadi, and L. Shahmoradi. 2017. An analytical method for diseases prediction using machine learning techniques. *Computers & Chemical Engineering* 10 (6): 212–223.
19. http://www.dtreg.com Software for Predictive Modelling and Forecasting.
20. Hsu, C.C., Y.P. Huang, and K.W. Chang. 2008. Extended naive Bayes classifier for mixed data. *Expert Systems with Applications* 35 (3): 1080–1083.
21. Wong, H.B., and G.H. Lim. 2011. Measures of diagnostic accuracy: sensitivity, specificity, PPV and NPV. *Proceedings of Singapore Healthcare* 20 (4): 316–318.

Intelligent Bot Architecture for e-Advertisements

G. M. Roopa and C. R. Nirmala

Abstract Nowadays, many businesses are engaged in Internet marketing, which is an inexpensive way to reach millions of Internet users on a target webpage. It provides a mechanism for promoting products/services via the Internet and its broader space encompasses e-marketing, e-advertising, customer relationship management and promotional activities. In this work, a mobile agent framework for Internet marketing is proposed which accomplishes the task of advertiser and ad publisher by creating an environment for aggregating advertisements and parallelizing advertiser requests for ad publishing. Mobile agents migrate over the Internet with the goal of "anytime/anywhere". The use of mobile agents for publishing tasks reduce advertiser/user time, addresses the problem of limited bandwidth and high costs, provide flexible inter-coordination among the mobile agents to perform product marketing, and ensures efficient utilization of the available bandwidth. Here, various agent dispatch models are proposed to perform the task of ad publishing on various e-shops. The experiments show that in comparison to serial mobile agent dispatch models, the parallel dispatch model performs more efficiently. In the worst-case scenario, the time complexity for parallel dispatch of mobile agents is O ($\log_q$ p).

Keywords Advertisements · Agent dispatch models · Aggregation
Internet marketing · e-Shops · Mobile agents · Time complexity

G. M. Roopa (✉) · C. R. Nirmala
Department of Computer Science and Engineering, Bapuji Institute of Engineering & Technology, Davangere 577004, Karnataka, India
e-mail: roopa.rgm@bietdvg.edu

C. R. Nirmala
e-mail: crn@bietdvg.edu

A. N. Krishna et al. (eds.), *Integrated Intelligent Computing, Communication and Security*, Studies in Computational Intelligence 771,
https://doi.org/10.1007/978-981-10-8797-4_3

1 Introduction

Today, Internet marketing offers new mechanisms and business models for promoting products and services, including support for the process of buying/selling of goods over the Internet. Aggregating agent-mediated Internet marketing delivers agent-based solutions for different phases involved in publishing advertisements on various e-shops, such as identification needs, product brokering, merchant brokering, contract negotiation and agreement, payment and delivery, service and evaluation [1].

With the proliferation of Internet marketing, new leading-edge dynamic marketing solutions appear for mapping more sophisticated and efficient models for business transactions, thus facilitating e-commerce/e-advertising tasks during the contract agreement stage of the trading process. This leads to the development of marketing assistance agents with intelligent decision-making in an automated context for promoting products/services, publishing advertisements, and auction bidding processes.

Currently, the global dynamic service environment provides an integrated information framework for inter/intra-enterprise collaborations. In such a highly interdisciplinary research environment, it is necessary to design/develop dynamic services for a network economy. Thus, to be competitive in the new "e-economy", it is vital that researchers and industries are able to exploit emerging technologies that will form the basis for tomorrow's global information network. To address such a situation, in recent years, an emerging distributed-computing paradigm with mobile agent technology has been proposed for e-business, e-marketing, e-commerce and e-learning, which attempts to resolve the current problems with traditional approaches. The mobile agent paradigm, which has its foundation in artificial intelligence, is adopted to structure, design and develop software applications with complex interactions among autonomous distributed components, and has proved to be an excellent tool for modeling large-scale distributed information environments [1].

Mobile agents are independent computer programs which move according to the predefined procedures in a heterogeneous/distributed network environment asynchronously and autonomously. They are mobile, flexible, autonomous, dynamic and efficient [2]. When loaded/dispatched with an encapsulated task, they process/execute the task on a remote host machine and deliver the results back. A multi agent with the coordination mechanism is presented to help the agents interact when performing complex actions [3].

In this chapter, a mobile agent-based framework for Internet marketing is proposed which supports the real-time activities of advertisers, ad publishers and users by facilitating the parallel processing of advertiser requests for promotion/publishing ads on various e-shops and are widely accepted in distributed environment [4]. The latter task has become challenge as the number of e-shops has increased. Existing frameworks are plagued by numerous issues, including inconvenience to the customer due to wasted time and effort, high bandwidth occupancy and the increased load on the system [2]. With its optimized mobility strategy the measured processing/execution time is less for ad-publishing. Mobile agents minimize the network load when the number of e-shops for publishing increases, as they are able to adapt autonomously to the heterogeneous environment and require less bandwidth by employing their asynchronous features. With the optimized mobility strategy, the measured processing/execution time is reduced for ad publishing.

The framework is implemented as a prototype system to explore serial and parallel agent dispatch models. All the mobile agent-based procedures are implemented in Java code using the IBM Aglet toolkit and MySQL for storage.

2 Architecture Description

Figure 1 presents a working system description for mobile agent-based Internet marketing and the set of components involved for promoting/publishing the advertisements on various e-shops. The following modules are:

- Aggregator-agent-mediator: The static agent acts as the main controller for accepting/processing the advertiser request for ad publishing. For the accepted request, an agent collects and aggregates the advertisements and loads the advertisements to created mobile agents. Finally, mobile agents are dispatched to various predefined locations. It extends the task of storing/updating the registered advertiser/ad publisher details along with the product catalogue information and waits for an acknowledgement. The use of agent oriented middleware [5], aims at making e-Commerce/e-Advertising process much easier and more successful.
- E-shops: On reaching the predefined e-shops, the slave agents perform the task of posting the advertisement on various display slots based on the advertiser preference given. After the completion of the assigned task, they return to the dispatched host machine with the acknowledgement of successful ad publishing.
- Customer-agent module: The static agent acts as a personal assistance for the customer in making the decision and helps them choose the right products of their interest.

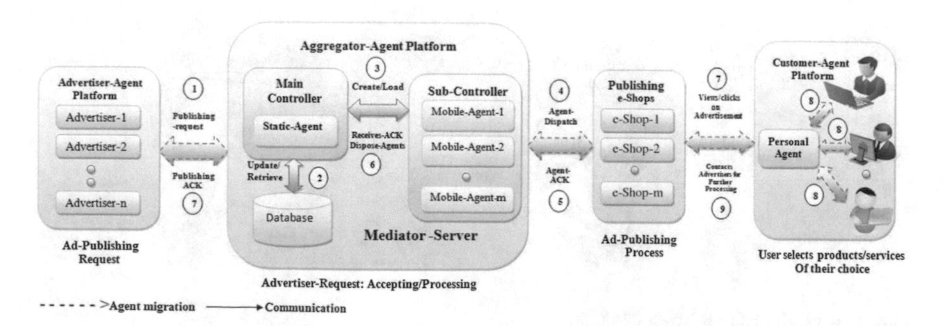

Fig. 1 Working description of mobile agent-based internet marketing

2.1 Agent Dispatch Models

For the given dispatch model, the master agent (MA) is responsible for dispatching prime working agents (PWAs) and distributing the task to these agents. Here, each PWA dispatches a cluster of PWAs/working agents (WAs). The WA then performs the assigned task of ad publishing. In this model, the MA dispatches the mobile agent based on the number of advertisers and e-shops, and trying to reduce its load significantly.

- Master agent (MA): Is a static agent which offers interfaces to the advertisers and e-shops. This agent decomposes the task, dispatches PWAs/WAs to accomplish the individual task and finally collects the results (ACK).
- Prime worker agent (PWA): These agents are created/dispatched by the MA, and they in turn are dispatched to the mediator server to dispatch WAs and distribute the task assigned by the MA. With an optimized model, the PWA further performs its own tasks in addition to its dispatching task.
- Worker agent (WA): These agents are created/dispatched by the MA/PWA to various e-shops to accomplish the task of ad publishing. After completion of the assigned task, they report to the MA to send the result back through a message [6].

To demonstrate each of the hierarchical models, dispatch tree notation is used: A dispatch tree represents a tree with the root vertex as the MA, each tree leaf ends with the WA and each non-leaf intermediate vertex functions as a PWA, if one exists. Each edge represents a directed edge denoting the dispatch process. An assumption is made regarding the height of the dispatch tree to be not less than one and should have at least two leaves. Followed by the assumption that the dispatch time for PWAs/WAs is the same.

1. Serial dispatch model (M1): In this model, the MA dispatches a cluster of mobile agents one by one, as shown in Fig. 2a. Here, the height of the dispatch tree is one. This model is in its simplest form and undoubtedly the slowest. Its time complexity is given by O(n), where n represents the number of dispatched WAs.

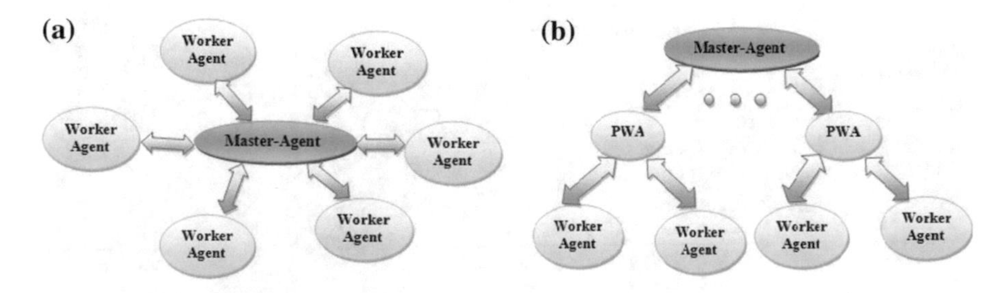

Fig. 2 **a** Model M1. **b** Model H2

2. Hierarchical dispatch model:

   - Dispatch model H2: In this model, the height of the dispatch tree is fixed at two. Figure 2b shows the layout of model H2, where the MA dispatches a group of PWAs and each PWA in turn dispatches a cluster of WAs to accomplish the assigned task of ad publishing. Initially, the MA divides all the WAs into various groups and distributes the assigned tasks along with its own task and provides the address of the PWA to return back the results (ACK).
   - Dispatch model (H_M): The above-discussed models M1 and H2 have a fixed tree height. In this model, the dispatch tree is formed with Q branches and the number is fixed. Here, MA must dispatch m PWAs if it includes more than Q WAs in its cluster, and all the WAs should be distributed with equal tasks. Figure 2c shows the layout for model H_M with 64 WAs when $Q = 4$. For the above case, let the time taken to dispatch the PWAs/WAs be Δt; then the total time taken to dispatch 64 WAs will be $12\Delta t$, which is less when compared to that of model M1 and H2.
   - Dispatch model (H_{M+}): In this model, the task of PWAs is to dispatch only the WAs. If there are p WAs to be dispatched to p e-shops,

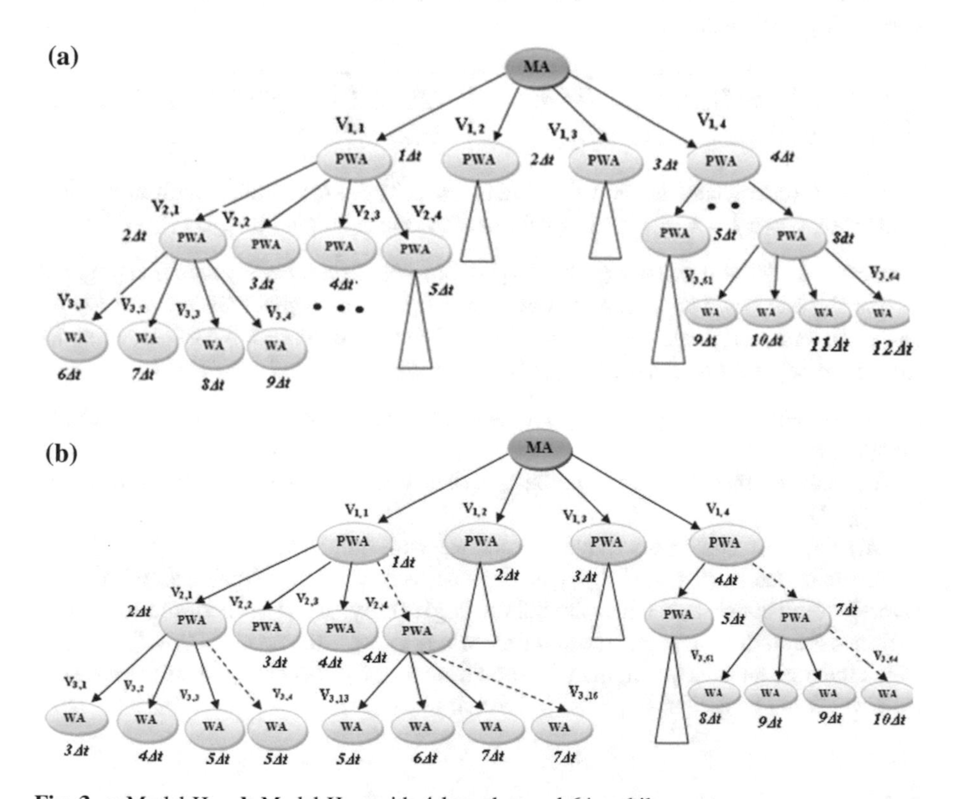

Fig. 3 **a** Model H_{M}. **b** Model-H_{m+} with 4 branches and 64 mobile agents

then q ($q < p$) PWAs must be dispatched to q e-shops for dispatching p WAs. Therefore, there are $(q + p)$ mobile agents to be dispatched in all. From Fig. 3b, it is proved that the time taken by this model is reduced when compared to that of the above-mentioned models.

3 Theoretical Proof

From the theoretical analysis, the dispatch time and the time complexity for the above-discussed models can be formulated as:

Theorem 1 *If there exists $p = q^2$ of WAs to be dispatched by model H2 and there are j PWAs and each PWA needs to dispatch i WAs, where $j = i = q^{1/2}$, then the time to dispatch all the worker agents is minimum.*

Proof Assume that the dispatch time for mobile agent is Δt. With regard to Fig. 2b, if the master agent (*vertex-V_0*) to dispatch j PWAs, then the time required to dispatch all j PWAs is given by $(j \cdot \Delta t)$. Then, if it has i WAs to dispatch, the required time to finish it is $(i \cdot \Delta t)$. Finally, the time to dispatch all the WAs with vertex ($V_{2,ji}$) is given by:

$T = (j + i) \cdot \Delta t \; where \; j \cdot i = p$	⇨	$Let \; y(x) = (x + p/x) \; and \; y'(x) = (1 - p/x^2)$
$T = (i + p/i) \cdot \Delta t$		$when \; y'(x) = 0 \; x = p^{1/2}$

Here, it is formulated to show that when $x = p^{1/2}$, $y - (x)$ obtain a minimum value. Therefore, when $j = i = (p^{1/2} \cdot T)$ a minimum value is obtained. $T_{min} = 2 \cdot p^{1/2} \cdot \Delta t$.

Theorem 2 *If there exists p ($p \geq 2$) of WAs to be dispatched by model H_M with q as the branch number of the decision tree ($q \geq 2$), height iven by $h = \log_q p$ ($p \geq 1$), and Δt represents the time required for dispatching PWAs/WAs, then the total time required to dispatch all p WAs can be formulated by: ($T = h \cdot q \cdot \Delta t$).*

Proof (i) With h = 1, the true hypothesis defines that the MA must use $(q \cdot \Delta t)$ to dispatch q WAs.

(ii) Assume that the true hypothesis exists with all value less than the height h ($h \geq 1$).

(iii) Let the dispatch tree have q branches with height h.

Assume that vertex $-(V_{h-1,\ k})$ is the last vertex to be dispatched within the same-level dispatch tree when the height is given by $(h - 1)$. Then, the total time includes the starting time and the end time when $V_{h-1,k}$ is dispatched $((h - 1) \cdot q \cdot \Delta t)$. Since the time taken is $q \cdot \Delta t$, then the total time taken to dispatch p WAs is given by:

$T = ((h - 1) \cdot q \cdot \Delta t + q \cdot \Delta t) \; h \cdot q \cdot \Delta t$

Theorem 3 *Given p ($p \geq 2$) WAs to be dispatched by model H_{M+} with q ($q \geq 2$) as the branch number of the dispatch tree, height $h = \log_q p$, and Δt = time taken to dispatch one mobile agent, then the total time required to dispatch all the WAs is given by:*

$$T = (h \cdot q - h + 1) \cdot \Delta t$$

- From Theorems 2 and 3, we can draw the following corollaries:

Corollary 1 *For the same number of WAs and same number of branches, model H_{M+} saves $(h - 1) \cdot \Delta t$ of time when compared to that of model H_M.*

Corollary 2 *Models H_M and H_{M+} exhibit the same time complexity of $O(\log_q p)$.*

Corollary 3 *When all p-WAs are dispatched using the binary dispatch tree for model H_M, the dispatch time has a minimum value of:*

$$T_{Min} = (\log_2 p + 1) \cdot \Delta t.$$

4 Performance Analysis

The performance issue is another important consideration for adopting the mobile agent approach [7] when building e-Advertising Applications. The results presented are observed from four independent executions, in which the Tahiti server is rebooted for each execution to prevent the collision of class-cache on the processing time. The set of parameters adopted to compute the performance evaluation for various defined mobile agent dispatch models are as follows: execution time, size of the ad record, number of e-shops, number of advertiser requests, dispatch time, size of dispatched agents, response time. The details of these parameters are given in Table 1.

Table 1 Details of parameter values used for conducting the experiments

Parameters	Values	Parameters	Values
Number of e-shops (m)	2–100	Size of mobile agents	50 kB to 10 MB
Number of advertiser requests (*n*)	2–64	Size of ad records	50 kB to 2 MB
Number of agent dispatch models	04	Agent dispatch time	0–50 s
Execution time	0–50 s	Request processing time	Negligible
Agent dispatch models			
Model M1: Serial dispatch (existing logic)	**Model H2**: Serial dispatch (existing logic)	**Model H_M**: Virtual parallel dispatch (partial proposed logic)	**Model H_{M+}**: Parallel dispatch (proposed logic)

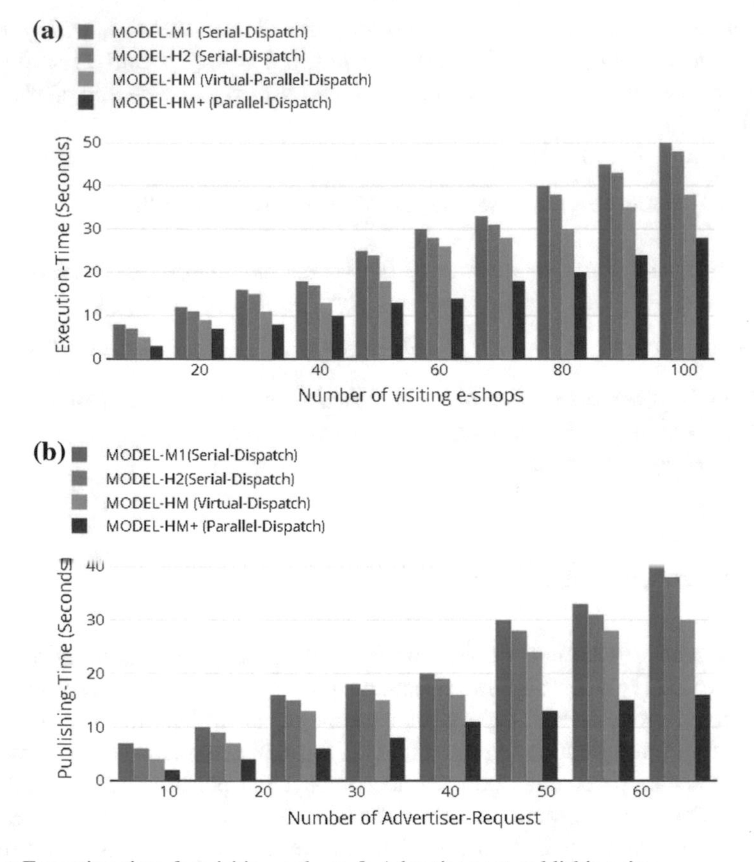

Fig. 4 **a** Execution time for visiting e-shops. **b** Advertisement publishing time

Figure 4a shows the results for publishing various ad records of varying size from 50 kB to 2 MB. Here, the performance of the serial-agent dispatch model, one by one (M1 and H2), is inferior to that of the parallel-agent models (H_M and H_{M+}) with regard to the execution time required for processing advertiser requests.

Figure 4b shows the time required for publishing the advertisements on various e-shops. The proposed framework adheres to the H_{M+} dispatch model, and the results show that the time taken by the parallel dispatch model (H_{M+}) is less when compared to that of serial dispatched agent models. It is clear from the graph that there is huge variation in performance between the serial and parallel dispatch models as the number of e-shops increases.

Figure 5a shows the corresponding graphs for the response time with regard to various mobile agent dispatch models. The important notable feature observed from this graph is that parallel mobile agents exhibit faster response times as the number of e-shops to be visited increases. Serial mobile agents suffer from performance degradation when the number of e-shops for publishing increases.

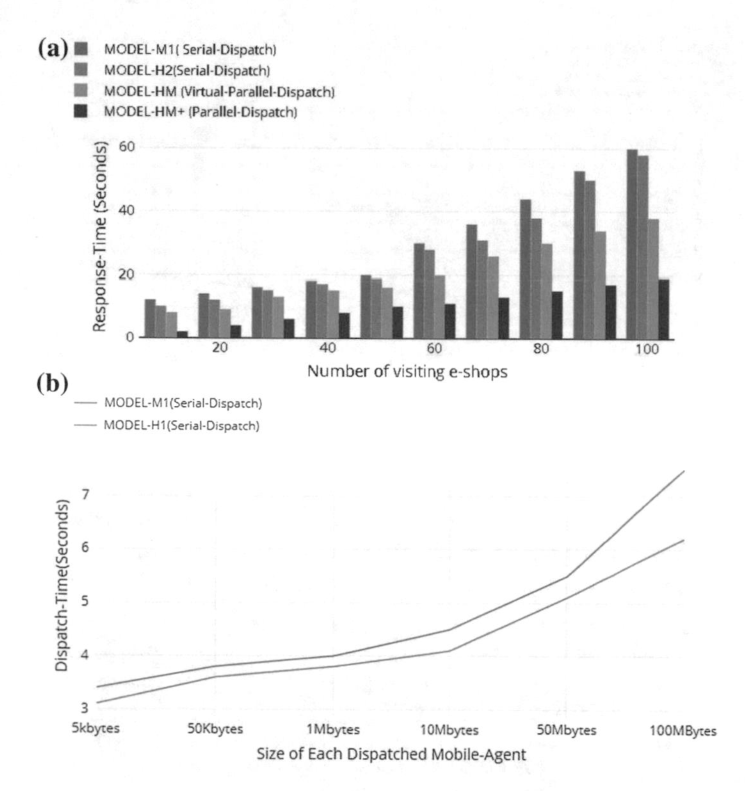

Fig. 5 **a** Response time for returning the ACK. **b** Results for dispatching four WAs

Figure 5b shows the variation in performance for various agent sizes (PWAs/WAs) based on the loaded advertisement record size. With smaller mobile agent models M1, H1 and H_M, there was no significant difference in dispatch time. But, as the number of e-shops and the size of PWAs/WAs increase, the difference widens.

From Fig. 6a, b, the number of mobile agents dispatched are 16 and 64 and the performance differences vary largely. However, the variations between the parallel models H_M and H_{M+} are smaller. For model H_{M+}, with the same number and size of WAs, the performance is improved with a smaller value of p, and it can achieve minimum dispatch time when $p = 2$. In a real-time scenario, the estimated average time becomes longer than in the local area network (LAN), but the major performance differences remain the same.

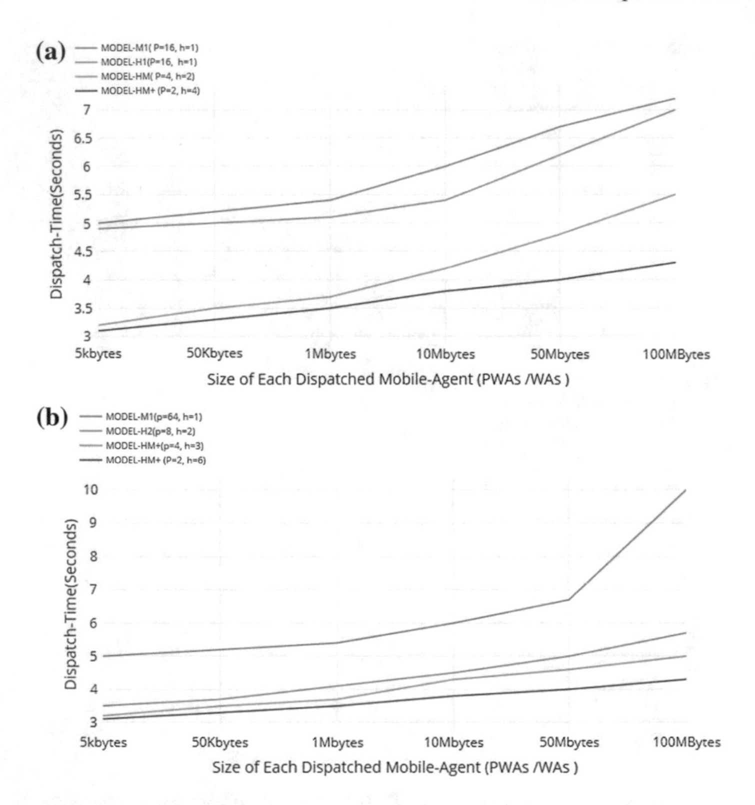

Fig. 6 **a** Results for dispatching 16 WAs. **b** Results for dispatching 64 WAs

5 Conclusion

Here, an attempt is made to implement a prototype to support Internet marketing assistance by adopting the mobile agent paradigm based on artificial intelligence, which offers the real-time activities of advertisers/e-shops/customers. This works puts forward various mobile agent dispatch models using the control strategy of a master agent responsible for creating a pool of WAs to fulfill the advertiser request in parallel. Exploiting parallel processing provides a significant benefit, helping to create an Internet marketing system with higher efficiency and providing a valuable support for customers with a "best-buy" strategy.

This work supports the agent migration process offered by dispatch model H_{M+}. The experiments show significantly improved performance with model H_{M+}, and further provide an automated update of product/service information in a real-time scenario. The improvements are achieved mainly by a platform-independent feature and try to mask the complexity that arises both at the advertiser and e-shops. With the huge advances in hardware, software and the networking environment, the individual mobile agent processing/response/execution time is significantly reduced. It is proved that the parallel dispatch model H_{M+} offers the minimum dispatch time both theoretically and practically. This work addresses the drawbacks of the existing logic and is suggested as a replacement for the client/server model. Thus, with the support of advanced standards and protocols, the mobile agent paradigm provides an efficient and effective process in the field of e-commerce, e-marketing, e-advertising and e-business.

References

1. Kowalezyk, Ryszard, Mihaela Ulieru, and Rainer Unland. Integrating mobile and intelligent agents in advanced e-commerce: A survey. *CSIRO Mathematical and Information Sciences*.
2. Che, Lei, and Xiao-Ping Yang. 2014. Research and application of mobile-agent in e-commerce system. *Applied Mechanics and Materials* 519–520: 160–161, © 2014 Trans Tech Publications, Switzerland.
3. Minand, Maw, and Nyein Oo. Mobile agent-based information retrieval for shopping Assistant. In *Proceedings of 2015 international conference on future computation technologies (ICFCT' 2015)*, 29–30 March, 2015, 211–217, Singapore.
4. Singh, Yashpal, Kapil Gulati and S. Niranjan. 2012. Dimensions and issues of mobile agent technology. *International Journal of Artificial Intelligence and Applications (IJAIA)* 3 (5).
5. Meher, Kalpana N., Prof. Sanjay Jadhav, Prof. P.S. Lokhande. 2013. Implementation of mobile agent architecture for e-commerce application for mobile devices. *International Journal of Advanced Research in Computer Engineering & Technology (IJARCET)* 2 (12).
6. Wang, Yan, Kian-Lee Tan, and Jian Ren. 2002. A study of building internet marketplaces on the basis of mobile agents for parallel processing. *World Wide Web: Internet and Web Information System* 5: 41–66, © 2002 Kluwer Academin Publishers, Netherlands.
7. Wang, Yan, and Jian Ren. Building internet marketplaces on the basis of mobile agents for parallel processing. In *Proceedings of the third international conference on mobile data management (MDM'02)*, © 2002 IEEE.

Predicting Survival of Communities in a Social Network

Shemeema Hashim, Greeshma N. Gopal and Binsu C. Kovoor

Abstract The structure of a social network changes over time. Structural change is usually studied by observing interactions within the network. The evolution of a community depends upon the changes in activity and communication patterns of individuals in the network. The major events and transitions that occur in a community are birth, death, merging, splitting, reform, expansion and shrinkage. Here, we focus on tracking and analyzing various events of a community which change over time. This chapter predicts the survival of communities based on events by extracting their most influential features.

Keywords Social networks · Community survival · Network analysis Prediction

1 Introduction

Communities are an important element in social network analysis. They are formed from the intensive interactions of a certain group of individuals in the network. Social networks contain community-like structures, in which nodes within the community are densely connected to each other and sparsely connected to nodes outside that community. Our community detection methods [1] are focused on determining communities in the social network. Social network communities are subject to change over the course of time. New members can join a community and existing members can leave the community, depending upon the changing interests

---

S. Hashim (✉)
College of Engineering, Cherthala, Cherthala, India
e-mail: shemeemahashim786@gmail.com

G. N. Gopal · B. C. Kovoor
School of Engineering, CUSAT, Kochi, India
e-mail: greeshmang@gmail.com

B. C. Kovoor
e-mail: binsukovoor@gmail.com

A. N. Krishna et al. (eds.), *Integrated Intelligent Computing, Communication and Security*, Studies in Computational Intelligence 771,
https://doi.org/10.1007/978-981-10-8797-4_4

of the entities in the network. The major events and transitions that occur in a community are birth, death, merging, splitting, expansion, reform and shrinkage [2]. In this chapter, we focus on tracking and analyzing the events in a community which change over time, and on predicting the survival of the community in the next time frame. The chapter focuses on finding the most influential features rather than on selecting all features for prediction.

2 Problem Formulation

Social network analysis and event prediction play an important role in recommendation systems and viral marketing schemes. Event detection and prediction involves the study of the features of communities for a particular chain of temporal snapshots of the community. Finding and concentrating on relevant features out of several structural attributes of a community is very significant in obtaining maximum accuracy in prediction. Relationships between these influential features and a particular event can be obtained using regression methods. Temporal chain length refers to the number of contiguous time slots selected for feature extraction. As we increase the temporal chain length, the accuracy of the prediction also increases, but a large chain length creates an enormous amount of data and demands high processing cost and time. Therefore, an optimal chain length for accurately predicting events should be considered in this method. Figure 1 shows the system model for community evolution prediction.

Temporal snapshots are considered here and communities are detected on the basis of modularity. Later, by analyzing the changes in the communities, the most influential features are identified. The relationships between these features and future events are then analyzed.

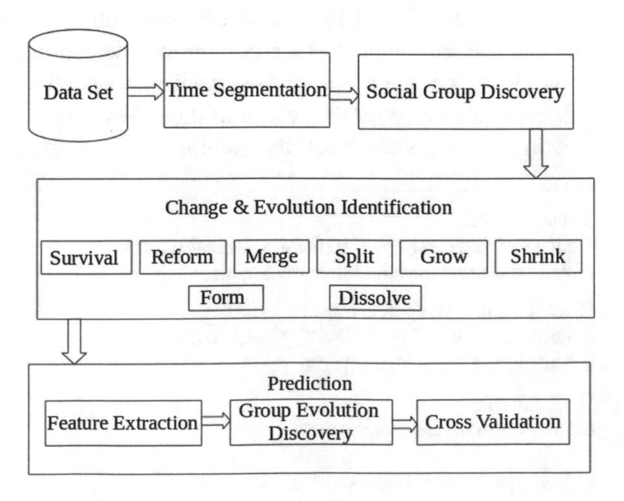

Fig. 1 System model for community evolution prediction

2.1 Events and Transitions

When a community survives into the next time frame, it may also experience different events and transitions. It expands if the number of members increases, or shrinks if the number of members declines. The major events that we have analyzed in this experiment are survival, birth, reform, death and growth [3, 4].

2.2 Features and Communities

The extraction of community features and measurement of their structural and temporal aspects is very important in tracking the evolution of communities [5]. To estimate the structural properties of a community, our community evolution analysis has considered its *node number, average betweenness, average of external links, edge number, average degree, average of internal links* and *density*.

3 Experiments and Results

The dataset for the experiment used 13 *graphml* files, which indicate networks of a code website. The file includes graphical representation of the network from April 2012 to April 2013. Using these *graphml* files, we created edge lists and vertex lists. These are both directed and undirected graphs, and are multidimensional since there are both human (coders) and non-human (repositories and their constituent parts) vertices. Out of several available community detection algorithms, the *fast-greedy* community detection method was used to find social groups in the network. Figure 2 shows the detected communities after applying the fastgreedy community detection method. Here, four different communities are illustrated.

3.1 Event Identification

The community events discussed in this work are shrinkage, expansion, birth, death and reform.

- ***Expansion and Shrinkage Events***: Figure 3a shows the expansion and shrinkage events of three communities over four timestamps. Community-1 expands during the second timestamp and starts shrinking during the third and fourth timestamps. Community-2 and Community-3 start shrinking at the second and third timestamps and expand at the fourth timestamp.
- ***Birth and Death Events***: Figure 3b shows the birth and death events of three communities over four timestamps. Birth of Community-1 and Community-2

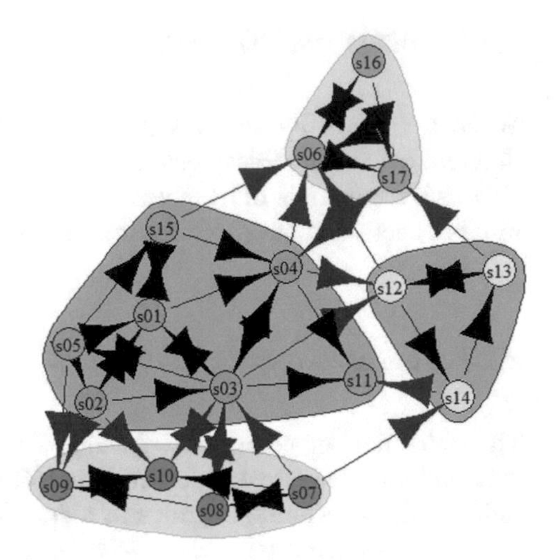

Fig. 2 Detected communities in network

occurred at the second timestamp, and Death of Community-1 and Community-3 occurred at the fourth timestamp.

- ***Reform Event***: Figure 3c shows the reform events of two communities over four timestamps. Reform of Community-1 occurs at the third timestamp, whereas the reform event of Community-2 occurs during the fourth timestamp.

3.2 Community Feature Extraction and Prediction

Here, features such as degree centrality, size, centrality, modularity and density are extracted. The relevance of features corresponding to each event are then determined using logistic regression. Logistic regression [6] is used with situations in which the observed outcome for a dependent variable can have only two possible values, "0" and "1". It makes use of one or more predictor variables that may be either categorical or continuous. *Logistic regression* is used for predicting binary dependent variables. A *generalized linear model* is used in logistic regression for predicting response variables. The *glm* function used for predicting relevant features for each event is

$$glm(formula = Response\ variable \sim Extracted\ features,\ data) \quad (1)$$

The *glm* provides coefficient values for each feature corresponding to each event as shown in Table 1. The positively valued coefficients are considered relevant features, while the negatively valued coefficients are considered irrelevant. The features of communities that are found to be relevant are then used for cross validation, by applying logistic regression. A prediction model for each event is then built to predict the occurrence of the events and transitions (Fig. 4).

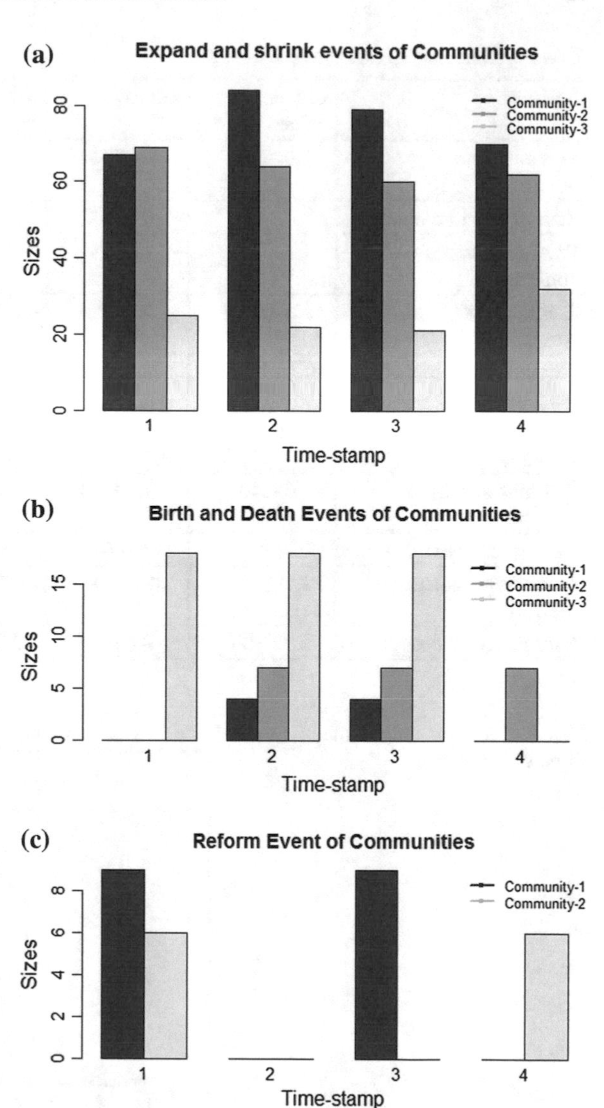

Fig. 3 **a** Expand and shrink events of three communities, **b** birth and death events of three communities, **c** reform event of two communities

Figure 5 shows the gradual increase in the value of the F-measure of a survival event for the three communities. When the F-measure reaches an optimum value, the performance of the survival event remains constant. Figure 6 shows the gradual increase in the F-measure value of reform events for three communities. When it reaches an optimum value, the F-measure of a reform event decreases gradually. Finally, the accuracy of the prediction is evaluated and its summary is shown in Table 2.

Table 1 Relevant features for each event

Features	Coefficient values	Survive	Reform	Birth	Death
Previous size of community	0.04539	−0.07644	−45.657	1.486	0.0205
Size of the community	−0.07210	0.03148	3.440	−4.298	0.0089
Previous centrality of community	7.23079	−14.9011	−1162.9	−6.253	0.5885
Centrality of community	3.51008	5.59333	−2164.8	−9.560	−1.7267
Previous cohesion of the network	−1.07950	7.61229	3.825	−7.673	−0.0172
Cohesion of the network	−1.00173	2.71478	−16.351	9.507	0.0284
Previous density of network	−1.08840	1.33089	89.652	9.668	−0.2204
Density of network	0.71177	−2.16037	−281.28	−1.983	0.1662
Previous modularity of network	2.59875	12.3591	−777.67	−7.549	−0.3053
Modularity of network	2.93197	−8.02856	2700.11	−1.150	−0.9251

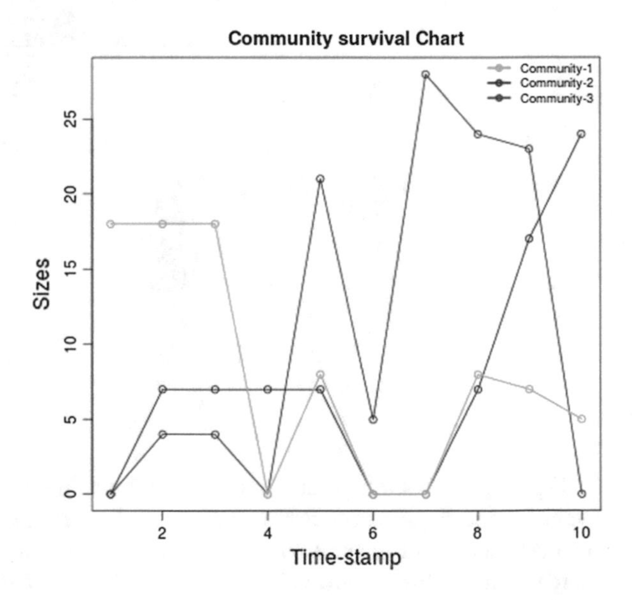

Fig. 4 Survival analysis of three communities

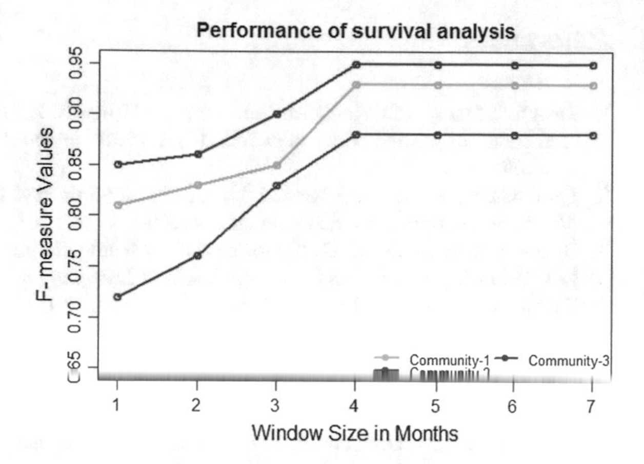

Fig. 5 Precision, recall and F-measure of survival event

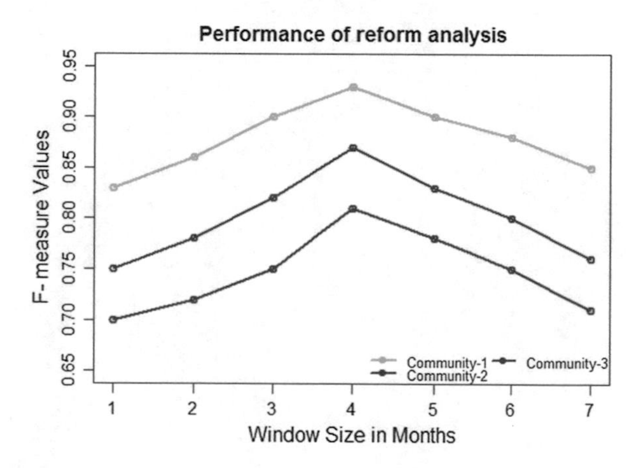

Fig. 6 Precision, recall and F-measure of reform event

Table 2 Performance analysis

Events	Precision	Recall	F-measure
Survive	0.80	0.66	0.95
Reform	0.81	0.69	0.93
Birth	0.70	0.58	0.63
Death	0.78	0.83	0.80
Growth	0.60	0.85	0.70

4 Conclusion

This work has proposed a method based on logistic regression to predict the future of a community. The influential features and their proportionality are extracted in this framework. Analysis has been performed to find the window size for each time frame. Future work might include determining algorithmic methods for finding suitable window sizes.

References

1. Takaffoli, Mansoureh, Reihaneh Rabbany, and Osmar R. Zaiane. 2014. Community evolution prediction in dynamic social networks. In *ACM international conference on web search and data mining*.
2. Spiliopoulou, Myra, Irene Ntoutsi, Yannis Theodoridis, and Rene Schult. 2006. MONIC. In *SIAM international conference on data mining*.
3. Huang, S., and D. Lee. 2011. Exploring activity features in predicting social network evolution. In IEEE *International Conference on Machine Learning and Applications*.
4. Takaffoli, Mansoureh, Justin Fagnan, Farzad Sangi, and Osmar R. Zaiane. 2012. Tracking changes in dynamic information networks. *Elsevier Journal*.
5. Goldberg, M.K., M. Magdon-Ismail, and J. Thompson. 2012. Identifying long lived social communities using structural properties. In *International conference on advances in social networks analysis and mining*.
6. Landwehr, N., M. Hall, and E. Frank. 2005. Logistic model trees. *Machine Learning*.

Leveraging Trust Behaviour of Users for Group Recommender Systems in Social Networks

Nirmal Choudhary and K. K. Bharadwaj

Abstract Group recommender systems (GRSs) provide recommendations to groups, i.e., they take all individual group member preferences into account and satisfy them optimally with a sequence of items. Few researchers have considered the behaviour of the users for group recommendation. In our work, we have exploited the trust factor among users to make group recommendations so as to satisfy the preferences of all the users. We present a novel approach to GRSs that takes into account the similarity- and knowledge-based trust between group members to enhance the quality of GRSs. The effectiveness of trust-based GRSs is compared with a baseline technique, the least-misery strategy, and it is observed that the results of computational experiments establish the superiority of our proposed model over the baseline GRSs technique.

Keywords Group recommender systems · Similarity · Knowledge
Trust

1 Introduction

Most research on recommender systems (RSs) focusses on making recommendations for individual users by filtering and presenting the users with information, products, services, and so on, according to their preferences [1]. In certain areas, however, the items to be recommended are not intended for individual usage but for a group of users—for example, watching a movie, planning a holiday, watching a TV program. Group recommender systems (GRSs) consider the preferences of all

---

N. Choudhary (✉) · K. K. Bharadwaj
School of Computer and Systems Sciences, Jawaharlal Nehru University,
New Delhi 110067, India
e-mail: nirmal32_scs@jnu.ac.in

K. K. Bharadwaj
e-mail: kbharadwaj@gmail.com

A. N. Krishna et al. (eds.), *Integrated Intelligent Computing, Communication and Security*, Studies in Computational Intelligence 771,
https://doi.org/10.1007/978-981-10-8797-4_5

individual group members and try to satisfy all of them with the best possible sequence of items.

Most GRS research considers individual group member preferences in the same manner. But, according to influential theory, a weak person is influenced by a stronger one; the same occurs with friends, in that a greater number of trusted friends on one side can influence the other with regard to a particular task, demanding agreement on that. In our work, we have exploited the trust factor among the users to make group recommendations so as to satisfy the preferences of all the users instead of an individual user. We have introduced a similarity- and knowledge-based trust model for creating a group profile. All users may not be fully satisfied, but there will be an optimal trade-off among all users of the group (trade-off of the group recommendation sequence will simply be the trade-off among the preferences of the users based on the trust between the users).

The next section includes existing group recommendation methods. The trust model used to enhance the existing group recommendation is detailed in Sect. 3. Experimental results are presented in Sects. 4 and 5 provides a summary of our work and discusses future work.

2 Related Work

The primary task of a GRS is to specify each user's preferences and then determine a compromise point on which all the group members agree equally. The idea behind group recommendation is to generate and aggregate the preferences of individual users. As explained in previous research [4], the major approaches to generating preference aggregation include (a) merging of the individual recommendations, (b) aggregation of individual rankings and (c) construction of a group preference model. The majority of GRSs use least-misery and average strategies to generate a group profile [7].

GRSs are intended to work in diverse domains, including music [9], movies [11], TV programs [12] and restaurants [8]. PolyLens [11], an extension of MovieLens, is used for recommending movies for a group of users by using the least-misery strategy. Another notable content-based recommender system is the Pocket Restaurant Finder [8], which is used for recommending restaurants to a group of users. Another example is the Collaborative Advisory Travel System (CATS) [10], which considers the behaviour of other group members for planning skiing holidays. However, a majority of the GRSs aggregate individual preferences without considering trust among group members.

Baltrunas et al. [2] noted that the effectiveness of group recommendation is influenced by several characteristics, including group size and inner group similarity. Hence, we have used trust among users based on similarity and knowledge factors to enhance the quality of group recommendation.

3 Proposed Model

This section depicts the inclusion of a trust factor based on similarity and knowledge when making recommendations for a group of users. The similarity factor depends on the common interest among users, while the knowledge factor relies on repetitive communications [3, 5]. Let G be a group of n users, $U = \{u_1, u_2, \ldots, u_n\}$ and m items $I = \{i_1, i_2, \ldots, i_m\}$ where the users express their ratings for items. An $m \times n$ user item rating matrix M is used to store the user ratings of items.

3.1 Similarity-Based Trust Model for GRSs

The rating given by the ith user to the jth item (r_{u_i, i_j}) specifies the significance of the ith user with respect to the jth item. The order of the jth item by the ith user $o(i_j)$ based on the similarity factor is computed in terms of rating difference [5]:

$$O(i_j) = \begin{cases} 0 & r_{u_i, i_j} = min \\ \frac{r_{u_i, i_j} - min}{max - min} & min < r_{u_i, i_j} < max \\ 1 & r_{u_i, i_j} = max \end{cases} \quad (1)$$

We classify each item in the group to be liked (L_Item), unliked (UL_Item) or neutral (N_Item) as follows [5]:

$$L\_Item = \{i_j : o(i) > 0.5\} \quad (2)$$

$$UL\_Item = \{i : o(i_j) < 0.5\} \quad (3)$$

$$N\_Item = \{i_j : o(i_j) = 0.5\} \quad (4)$$

The trust between user u_i and user u_j is computed as [5]:

$$Trust = \frac{1}{2}\left[\left[\frac{|L\_Item_{u_i} \cap L\_Item_{u_j}|}{|L\_Item_{u_i}|} + \frac{|UL\_Item_{u_i} \cap UL\_Item_{u_j}|}{|UL\_Item_{u_i}|}\right]\right] \quad (5)$$

where $|L\_Item_{u_i}|$ and $|UL\_Item_{u_i}|$ denote the degrees of liked and unliked items of user u_i.

Now, to calculate similarity-based trust user item ratings (S_M'), multiply the trust values by the initial user item rating matrix (M) as follows:

$$S\_M' = Trust * M \quad (6)$$

We then employ the group recommendation strategy, i.e., the least-misery strategy, to generate the similarity-based recommendation for the whole group.

3.2 Knowledge-Based Trust Model for GRSs

The trust calculated in the previous section does not have enough trained items, which results in reduced effectiveness of the group recommendation. There are some other factors that affect the computation of trust between users. The knowledge factor depends upon the frequent communication between users in the group that plays a vital role when computing trust between users.

According to [3, 5], an encounter occurs between two users when they rate the same item. As discussed earlier, the rating provided by the ith user for the jth item (r_{u_i, i_j}) specifies the importance of the jth user in the jth item. Afterward, the rating given by user i to user j at the end of the encounter is given by [5]:

$$r_{u_i}^{u_j}(e) = \begin{cases} 5 & 0.0 \le \left| r_{u_i, i_j} - r_{u_j, i_j} \right| \le 0.5 \\ 4 & 0.5 < \left| r_{u_i, i_j} - r_{u_j, i_j} \right| \le 1.0 \\ 3 & 1.0 < \left| r_{u_i, i_j} - r_{u_j, i_j} \right| \le 2.0 \\ 2 & 2.0 < \left| r_{u_i, i_j} - r_{u_j, i_j} \right| \le 3.0 \\ 1 & otherwise \end{cases} \tag{7}$$

After completion of all encounters, the collection of ratings of user u_i to another user u_j is $R_{u_i}(u_j)$, and the set of all ratings given by user u_i to the rest of the users in the group is denoted as S_i. The collection of ratings for liked, unliked and neutral encounters is represented as [5]:

$$Liked_{rat}^{u_i u_j}(e) = \left\{ r_{u_i}^{u_j}(e) | r_{u_i}^{u_j}(e) R_{u_i}(u_j) \; and \; r_{u_i}^{u_j}(e) > 3 \right\} \tag{8}$$

$$Unliked_{rat}^{u_i u_j}(e) = \left\{ r_{u_i}^{u_j}(e) | r_{u_i}^{u_j}(e) R_{u_i}(u_j) \; and \; r_{u_i}^{u_j}(e) < 3 \right\} \tag{9}$$

$$Neutral_{rat}^{u_i u_j}(e) = \left\{ r_{u_i}^{u_j}(e) | r_{u_i}^{u_j}(e) R_{u_i}(u_j) \; and \; r_{u_i}^{u_j}(e) = 3 \right\} \tag{10}$$

The trust based on the knowledge factor for each user in the group is calculated as [5]:

$$K\_Trust = \begin{cases} 0 & r_{u_i}^{u_j}(e) = 1 \\ & \frac{r_{u_i}^{u_j}(e) - 1}{4} \qquad 1 < r_{u_i}^{u_j}(e) < 5 | r_{u_i}^{u_j}(e) S_i \\ 1 & r_{u_i}^{u_j}(e) = 5 \end{cases} \tag{11}$$

Now, to compute knowledge-based trusted user item ratings (K_M'), we multiply the trust values by the initial user item rating matrix (M) as follows:

$$K\_M' = K\_Trust * M \tag{12}$$

Then we apply the least-misery strategy to generate the knowledge-based recommendation for the whole group.

4 Experiments

4.1 Data Set Description

To evaluate the performance of the proposed trust-enhanced GRS, we conducted several experiments on a Movie Lens data set consist of 100,000 ratings on 182 movies provided by 943 users. The ratings are in the range of 1–5, where 1 and 5 indicate a poor and excellent rating, respectively. Every user rated a minimum of 20 movies in the data set. In our experimental setup, we randomly selected 20 users from the data set and divided them into 60% training data and 40% test data to compute individual rating prediction.

4.2 Experiments and Results

A standard information retrieval (IR) measure, i.e., normalized discounted cumulative gain ($nDCG$) [2], was employed to examine the performance of the proposed trust-based GRS. The group recommendation techniques generate a ranked list of items, i.e., $p_1, p_2, \ldots, p_k$. The actual rating of user u for item p_i is given by r_{upi}. $IDCG$ refers to the highest achievable gain value for user u. (DCG) and nDCG at a kth rank are calculated, respectively, as follows:

$$DCG_k^u = r_{up_1} + \sum_{i=2}^{k} \frac{r_{up_i}}{log_2(i)} \qquad nDCG = \frac{DCG_k^u}{IDCG_k^u} \tag{13}$$

We performed an experiment to compare the effectiveness of our proposed model against the least-misery strategy [2] according to varying group size. We created four groups of sizes equal to 2, 4, 6 and 8. The results shown in Fig. 1 clearly demonstrate that the proposed scheme performs better than the least-misery strategy. Also, the average effectiveness (nDCG) for the knowledge-based GRS is higher than that for the similarity-based GRS.

Figure 1a shows that increasing the group size results in decreasing the average effectiveness of the GRS, whereas Fig. 1b illustrates the results for groups with high inner similarity and shows that similar groups have strong trust between users, which results in the most effective group recommendation. It is clear from Fig. 1 that increasing the group size may not always result in lower average effectiveness.

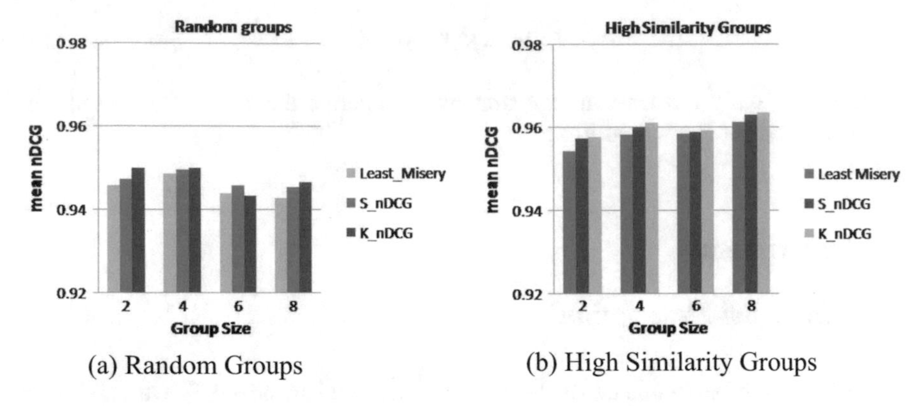

(a) Random Groups (b) High Similarity Groups

Fig. 1 Effectiveness of trust-enhanced group recommendation with the least-misery strategy

5 Conclusion and Future Directions

Here, we have proposed a new approach for group recommendation that considers similarity- and knowledge-based trust factors among users. The results presented provide experimental evidence of the efficiency of the proposed method in terms of enhanced accuracy.

As for further work, we would like to enhance the performance of our proposed trust-based model using a negotiation mechanism [6].

References

1. Adomavicius, G., and A. Tuzhilin. 2005. Toward the next generation of recommender systems: A survey of the state-of-the-art and possible extensions. *IEEE Transactions on Knowledge and Data Engineering* 17 (6): 734–749.
2. Baltrunas, L., T. Makcinskas, and F. Ricci. 2010. Group recommendations with rank aggregation and collaborative filtering. In *Proceedings of the fourth ACM conference on recommender systems*, 119–126.
3. Bharadwaj, K.K., and M.Y.H. Al-Shamri. 2009. Fuzzy computational models for trust and reputation systems. *Electronic Commerce Research and Applications* 8 (1): 37–47.
4. Jameson, A., and B. Smyth. 2007. Recommendation to groups. In *The adaptive web*, 596–627.
5. Kant, V., and K.K. Bharadwaj. 2013. Fuzzy computational models of trust and distrust for enhanced recommendations. *International Journal of Intelligent Systems* 28 (4): 332–365.
6. Kong, D.T., R.B. Lount Jr., M. Olekalns, and D.L. Ferrin. 2017. Advancing the scientific understanding of trust in the contexts of negotiations and repeated bargaining. *Journal of Trust Research* 7 (1): 15–21.
7. Masthoff, J. 2004. Group modeling: Selecting a sequence of television items to suit a group of viewers. In *Personalized digital television*, 93–141.
8. McCarthy, J.F. 2002. Pocket restaurant finder: A situated recommender system for groups. In *Workshop on mobile Ad-Hoc communication at ACM conference on human factors in computer systems*.

9. McCarthy, J.F., and T.D. Anagnost. 1998. MusicFX: An arbiter of group preferences for computer supported collaborative workouts. In *Proceedings of ACM conference on computer supported cooperative work*, 363–372.
10. McCarthy, K., L. McGinty, B. Smyth, and M. Salamó. 2006. The needs of the many: A case-based group recommender system. In *ECCBR*, *4106*, 196–210.
11. O'connor, M., D. Cosley, J.A. Konstan, and J. Riedl. 2001. PolyLens: A recommender system for groups of users. In *ECSCW*, 199–218.
12. Yu, Z., X. Zhou, Y. Hao, and J. Gu. 2006. TV program recommendation for multiple viewers based on user profile merging. *User Modeling and User-Adapted Interaction* 16 (1): 63–82.

Social Status Computation for Nodes of Overlapping Communities in Directed Signed Social Networks

Nancy Girdhar and K. K. Bharadwaj

Abstract The exponential growth in signed social networks in recent years has garnered the interest of numerous researchers in the field. Social balance theory and status theory are the two most prevalent theories of signed social networks and are used for the same purpose. Many researchers have incorporated the concept of social balance theory into their work with community detection problems in order to gain a better understanding of these networks. Social balance theory is suitable for undirected signed social networks; however, it does not consider the direction of the ties formed among users. When dealing with directed signed social networks, researchers simply ignore the direction of ties, which diminishes the significance of the tie direction information. To overcome this, in this chapter we present a mathematical formulation for computing the social status of nodes based on status theory, termed the status factor, which is well suited for directed signed social networks. The status factor is used to quantify social status for each node of overlapping communities in a directed signed social network, and the feasibility of the proposed algorithm for this metric is well illustrated through an example.

Keywords Status theory · Social balance theory · Overlapping communities
Directed signed social networks

1 Introduction

The unprecedented growth of social networks is the outcome of rich human interactions based on the strength and types of their relationships (ties) which tend to grow and evolve with time. The type of relationship (positive or negative)

N. Girdhar (✉) · K. K. Bharadwaj
School of Computer and Systems Sciences, Jawaharlal Nehru University,
110067 New Delhi, India
e-mail: nancy.gr1991@gmail.com

K. K. Bharadwaj
e-mail: kbharadwaj@gmail.com

A. N. Krishna et al. (eds.), *Integrated Intelligent Computing, Communication and Security*, Studies in Computational Intelligence 771,
https://doi.org/10.1007/978-981-10-8797-4_6

between two users depends on multiple factors (e.g., common interests, a person's status, social influence, user positivity) [10]. Complex networks with information about the types of links in addition to the links themselves are referred to as signed social networks (SSNs) [5]. Analyses of these SSNs in the literature commonly employ two theories, social balance theory [5, 6] and status theory [8]. Grounded in psychology, social balance theory is centered on the *friend-of-a-friend* (FOAF) concept. Based on this theory, a social balance factor metric is proposed to calculate the balance in the network which is the ratio of balanced triads to the total number of triads in the network. Leskovec [8] developed a theory based on the status of a node, known as status theory, which considers the fact that relationships or links are not always developed in the context of friendship; a person's status (social, economic, etc.) may also be contributor [10]. According to this theory, a positive link (i, j) indicates that user i thinks highly of user j and a negative link (i, j) indicates that user i thinks poorly of user j. Social balance theory is suitable for undirected signed social networks (UDSSNs), whereas status theory is more suitable for directed signed social networks (DSSNs).

To enable us to make sense of these complex SSNs, researchers have often used community detection [2], based on the fundamental social balance theory, which is suitable for UDSSNs. However, to deal with DSSNs, the signs of the ties are simply ignored, which dilutes the significance of the direction of the ties. To this end, we propose a new metric, which is a first attempt to consider the *direction of the links* along with the *positive and negative nature* of the nodes to compute the social *status for nodes* of overlapping communities in a *directed signed social network.*

The main contribution of our work is summarized as follows:

- Most of the work in the field of SSNs has considered metrics such as signed modularity, frustration, or social balance factor [5]. However, none of these metrics adequately considers the *link density, sign link information, or the asymmetric, directed and overlapping nature of links* [3]. Here, we propose a new metric, the *status factor*, to quantify the social status for nodes of overlapping communities of DSSNs. The salient features of this metric are as follows:
  - The status factor tries to espouse the directed nature of the links in addition to the sign information and density of the links. This is not the case for the social balance factor, as that is based on social balance theory, which is suitable for undirected signed social networks.
  - Instead of a binary membership between nodes and community, the status factor provides a degree of affiliation between a node and a community.
  - When computing the social status of a node, it ensures preservation of the overlapping nature of the nodes (presence of a node in more than one community) in DSSN community structures.

The rest of the chapter is structured as follows: Sect. 2 presents a brief discussion of related work on SSNs, and a description of the proposed algorithm for formulating the status factor metric is provided in Sect. 3. Finally, Sect. 4 concludes our work and discusses future work directions.

2 Related Work

Given the rapid proliferation of signed social networks, they have captured the attention of numerous researchers. One of the fundamental theories of signed networks is the social balance theory, proposed by Heider [6]. This theory was formulated to identify the causes of tension and conflict between positive and negative relationships and was initially focused on dyads and triads in the network. Many researchers exploited the idea of social balance in order to serve information mining in undirected signed networks. Global social balance computing was done by Facchetti et al. [4]. According to this work, presently available undirected signed networks of Epinions and Slashdot are found to be structurally balanced. Balance theory only considers the contribution of triads in the network; however, it does not incorporate the contribution of an individual's local information to the balance [7]. Also, it overlooks the impact of longer cycles to the unbalance of signed networks [8, 9].

An alternative theory to social balance is status theory, developed by Leskovec et al. [8], which stems from the idea that there may exist multiple interpretations of positive and negative links depending upon the intention of the sender. According to this theory, positive ties are directed to higher status nodes, whereas negative ties are directed to lower status nodes. The theory considers the sign of the ties and respects the order in which ties were created. Thus, status theory is more suitable for directed signed social networks.

3 Proposed Algorithm to Compute Social Status for Nodes of Overlapping Communities in a Directed Signed Social Network

This section presents the details of the algorithm for the proposed metric, termed the *status factor* (SF), which computes social status for each node of overlapping communities in a directed signed social network.

Algorithm: To compute the status of each node in the set of overlapping communities in a DSSN.

Input: Adjacency matrix A of size $n * n$, where n is the number of nodes.

$$A(i,j) = \begin{cases} +1, & \text{if node } i \text{ makes positive link with node } j \\ -1, & \text{if node } i \text{ makes negative link with node } j \end{cases} \quad (1)$$

and M is a matrix of size $m * n$ which represents community structure, where m is the number of communities and n is the number of nodes.

$$M(i,j) = \begin{cases} 1, & \text{if node } j \text{ is in community } i \\ 0, & \text{otherwise} \end{cases} \quad (2)$$

Output: Status matrix of size $m * n$ consisting of the social status value of each node in each community.

The following steps are followed to compute social status for each node of overlapping communities in a DSSN:

Step 1: Calculate for each node positive incoming degree (*p_in*), positive outgoing degree (*p_out*), negative incoming degree (*n_in*), and negative outgoing degree (*n_out*).

For each node v_k, calculate $\text{p\_out}_{(k)}$, $\text{p\_in}_{(k)}$, $\text{n\_out}_{(k)}$ and $\text{n\_in}_{(k)}$ from signed adjacency matrix A

$$\text{if } \left(A_{(k,t)} = 1\right)$$

$$\text{p\_out}_{(k)} = \text{p\_out}_{(k)} + 1;$$

$$\text{p\_in}_{(t)} = \text{p\_in}_{(t)} + 1;$$

$$\text{elseif } \left(A_{(k,t)} = -1\right)$$

$$\text{n\_out}_{(k)} = \text{n\_out}_{(k)} + 1;$$

$$\text{n\_in}_{(t)} = \text{n\_in}_{(t)} + 1;$$

End

End

Then, calculate p_k^{total}, which represents the factor which contributes to the high status of the node and n_k^{total} which representing the factor contributing to the low status of the node, where

$$p_{(k)}^{total} = \text{p\_in}_{(k)} + \text{n\_out}_{(k)} \quad (3)$$

$$n_{(k)}^{total} = \text{p\_out}_{(k)} + \text{n\_in}_{(k)} \quad (4)$$

Note: $p^{total}_{(k)}$ *is the sum of number of incoming positive links and number of outgoing negative links.* $n^{total}_{(k)}$ *is the sum of number of positive outgoing links and number of negative incoming links.*

Step 2: Now, to calculate the weights for the positive and negative contributions of nodes, take two parameters, let's say ρ and σ, respectively.

$$w\_p_{(k)} = \rho * p^{total}_{(k)} \tag{5}$$

$$\widehat{w\_n}_{(k)} = \sigma * n^{total}_{(k)} \tag{6}$$

where $\widehat{w\_n}$ denotes the intermediate weight assigned to negative contribution of the node.

Step 3: Now, for each community i in the set of overlapping communities C, let's say that p_deg are the links which contribute to the high status of the node in that community, and n_deg are the links which contribute to the low status of the node in that community.

For each node v_k that belongs to community i, compute $p\_deg_{(i,k)}$, which contributes to the high-status value of a node, is the sum of the number of incoming positive links to node k and the number of outgoing negative links from node k from the i-th community. Similarly, $n\_deg_{(i,k)}$, which contributes to the low status of node k, is the sum of the number of positive outgoing links from node k and the number of negative incoming links to the node k in the i-th community.

$$p\_in_{(i,k)};$$

$$p\_out_{(i,k)};$$

$$n\_in_{(i,k)};$$

$$n\_out_{(i,k)};$$

$$p\_deg_{(i,k)} = p\_in_{(i,k)} + n\_out_{(i,k)}; \tag{7}$$

$$n\_deg_{(i,k)} = p\_out_{(i,k)} + n\_in_{(i,k)}; \tag{8}$$

Step 4: Now, we compute positive contribution, which aids the high status $H_{(i,k)}$ of node k and negative contribution $L_{(i,k)}$, which aids the low status of node k.

For each node v_k to all communities

$$H_{(i,k)} = \frac{p\_deg_{(i,k)}}{\sum \forall i \;\; p\_deg_{(i,k)}} \tag{9}$$

If $p\_deg_{(i,k)} > 0$, then calculate intermediate negative contribution as:

$$\widehat{L}_{(i,k)} = n^{total}_{(k)} - n\_deg_{(i,k)} \tag{10}$$

The final negative contribution is given as:

$$L_{(i,k)} = \frac{\widehat{L}_{(i,k)}}{\sum_{\forall i} \widehat{L}_{(i,k)}} \tag{11}$$

Step 5: The final weight for the negative contribution will be calculated as:

$$w\_n_{(k)} = \widehat{w\_n}_{(k)} * \sum_{\forall i} L_{(i,k)} \tag{12}$$

Step 6: The social status value of each node k in community i will be computed as:

$$SF_{(i,k)} = \frac{w\_p_{(k)} * H_{(i,k)} - w\_n_{(k)} * L_{(i,k)}}{|w\_p_{(k)} - w\_n_{(k)}|} \tag{13}$$

The following example will illustrate the working of our proposed algorithm to compute social status for each node of overlapping communities in a DSSN.

Figure 1 shows a directed signed social network having three overlapping communities with four nodes. The green and red directed edges denote positive and negative links, respectively.

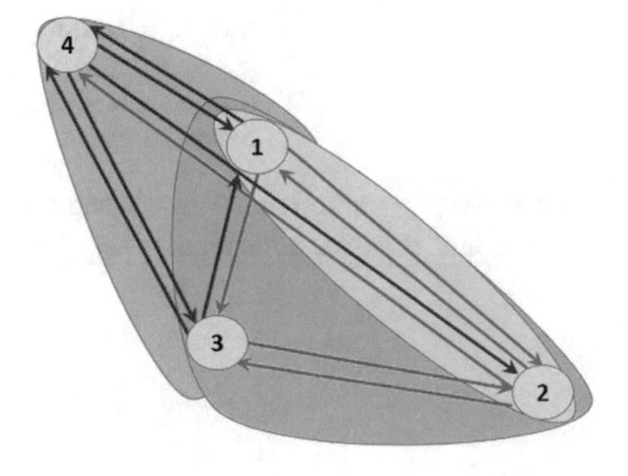

Fig. 1 A directed signed social network with three overlapping communities

From Fig. 1, the adjacency matrix A and community structure M for a given DSSN are:

$$A = \begin{bmatrix} 0 & 1 & 1 & -1 \\ 1 & 0 & 1 & 1 \\ -1 & 1 & 0 & -1 \\ -1 & -1 & -1 & 0 \end{bmatrix} \qquad M = \begin{bmatrix} 1 & 1 & 1 & 0 \\ 1 & 1 & 0 & 0 \\ 1 & 0 & 1 & 1 \end{bmatrix}$$

To compute social status for, say, node 2 of community 1, calculate $\mathrm{p\_in}_{(2)} = 2$, $\mathrm{p\_out}_{(2)} = 3, \mathrm{n\_in}_{(2)} = 1$, and $\mathrm{n\_out}_{(2)} = 0$ from adjacency matrix A.

Then, using Eqs. (3) and (4)

$$\mathrm{p}^{\mathrm{total}}_{(2)} = 2$$

$$\mathrm{n}^{\mathrm{total}}_{(2)} = 4$$

To calculate the weights for the positive and intermediate negative contribution of nodes, let us take the parameter values of $\rho = 0.8$ and $\sigma = 0.2$. Now, using Eqs. (5) and (6)

$$\mathrm{w\_p}_{(2)} = 0.8 * 2 = 1.6$$

$$\widehat{\mathrm{w\_n}}_{(2)} = 0.2 * 4 = 0.8$$

Now, calculate the links which contribute to increasing and decreasing the social status of node 2, respectively, in community 1. For this, we have to compute $\mathrm{p\_deg}_{(1,2)}$ and $\mathrm{n\_deg}_{(1,2)}$ using Eqs. (7) and (8).

$$\mathrm{p\_deg}_{(1,2)} = 2$$

$$\mathrm{n\_deg}_{(1,2)} = 2$$

The positive contribution of links which aids to high social status of node 2 in community 1 will be calculated from Eq. (9).

$$\mathrm{H}_{(1,2)} = \frac{2}{3} = 0.6667$$

By using Eqs. (10) and (11) we will calculate intermediate and final negative contribution, respectively, which aids the low social status of node 2 in community 1 as condition $\mathrm{p\_deg}_{(1,2)} > 0$ is satisfied.

$$\widehat{\mathrm{L}}_{(1,2)} = 4 - 2 = 2$$

$$L_{(1,2)} = \frac{2}{5} = 0.4$$

The final weight for negative contribution will be calculated from Eq. (12)

$$w\_n_{(2)} = 0.8 * 1 = 0.8$$

Finally, the social status value for node 2 in community 1 will be calculated from Eq. (13) as:

$$SF_{(1,2)} = \frac{1.6 * 0.6667 - 0.8 * 0.4}{|1.6 - 0.8|} = 0.9334$$

Similarly, social status for node 2 in community 2 is 0.0667. Likewise, we can compute social status for each node of overlapping communities in DSSNs that can be used to solve different leading social network problems like sparsity in DSSNs.

4 Conclusion and Future Work

In the present work, in order to compute the social status for each node of overlapping communities in directed signed social networks, we have proposed a new metric, the status factor. The proposed metric takes into account the direction of ties formed, along with the sign of ties and link density, which is well illustrated through an example. As for further work, the metric can be used to solve other significant problems of social networks, such as link prediction in directed signed social networks [5]. Another interesting area will be finding influential nodes with high status values in the network [11]. Our work provides an interesting perspective to find nodes of creation of social circles in directed signed social ego networks [1]. Furthermore, we will study the different context of links and the differences in one-way and reciprocated links formation, which will shed some light on the behavior of users of these networks [8].

References

1. Agarwal, V., and K.K. Bharadwaj. 2015. Predicting the dynamics of social circles in ego networks using pattern analysis and GA K-means clustering. *Wiley Interdisciplinary Reviews: Data Mining and Knowledge Discovery* 5 (3): 113–141.
2. Anchuri, P., and M. Magdon-Ismail. 2012. Communities and balance in signed networks: A spectral approach. In *Advances in social networks analysis and mining*, 235–242.
3. Awal, G.K., and K.K. Bharadwaj. 2017. Leveraging collective intelligence for behavioral prediction in signed social networks through evolutionary approach. *Information Systems Frontiers*, 1–23.

4. Facchetti, G., G. Iacono, and C. Altafini. 2011. Computing global structural balance in large-scale signed social networks. *Proceedings of the National Academy of Sciences* 108 (52): 20953–20958.
5. Girdhar, N., and K.K. Bharadwaj. 2016. Signed social networks: A survey. In *Proceedings of the international conferences on advances in computing and data sciences*, 326–335.
6. Heider, F. 1946. Attitudes and cognitive organization. *The Journal of Psychology* 21 (1): 107–112.
7. Kunegis, J., S. Schmidt, A. Lommatzsch, J. Lerner, E.W. De Luca, and S. Albayrak. 2010. Spectral analysis of signed graphs for clustering, prediction and visualization. In *Proceedings of the international conferences of SIAM on data mining*, 559–570.
8. Leskovec, J., D. Huttenlocher, and J. Kleinberg. 2010. Signed networks in social media. In *Proceedings of the SIGCHI conferences on human factors in computing systems*, 1361–1370.
9. Truzzi, M. 1973. An empirical examination of attitude consistency in complex cognitive structures. *Systems Research and Behavioral Science* 18 (1): 52–59.
10. Yap, J., and N. Harrigan. 2015. Why does everybody hate me? Balance, status, and homophily: The triumvirate of signed tie formation. *Social Networks* 40: 103–122.
11. Zhao, Y., S. Li, and F. Jin. 2016. Identification of influential nodes in social networks with community structure based on label propagation. *Neurocomputing* 210: 34–44.

A Genetic Algorithm Approach to Context-Aware Recommendations Based on Spatio-temporal Aspects

Sonal Linda and K. K. Bharadwaj

Abstract Context-aware recommender systems (CARS) have been extensively studied and effectively implemented over the past few years. Collaborative filtering (CF) has been established as a successful recommendation technique to provide web personalized services and products in an efficient way. In this chapter, we propose a spatio-temporal-based CF method for CARS to incorporate spatio-temporal relevance in the recommendation process. To deal with the new-user cold start problem, we exploit demographic features from the user's rating profile and incorporate this into the recommendation process. Our spatio-temporal-based CF approach provides a combined model to utilize both a spatial and temporal context in ratings simultaneously, thereby providing effective and accurate predictions. Considering a user's temporal preferences in visiting various venues to achieve better personalization, a genetic algorithm (GA) is used to learn temporal weights for each individual. Experimental results demonstrate that our proposed schemes using two benchmark real-world datasets outperform other traditional schemes.

Keywords Context-aware recommender systems · Collaborative filtering
Spatio-temporal similarity · Genetic algorithm

1 Introduction

Integrating context into user modeling for personalized recommendations has improved the performance and usefulness of recommender systems (RS). Moreover, context plays an important role in intelligent information systems, including

S. Linda (✉) · K. K. Bharadwaj
School of Computer and Systems Sciences, Jawaharlal Nehru University,
New Delhi, India
e-mail: lindasonal@gmail.com

K. K. Bharadwaj
e-mail: kbharadwaj@gmail.com

A. N. Krishna et al. (eds.), *Integrated Intelligent Computing, Communication and Security*, Studies in Computational Intelligence 771,
https://doi.org/10.1007/978-981-10-8797-4_7

web search and collaborative filtering (CF)-based RS. The recommended list of items for the active user within one contextual condition may be irrelevant in another contextual condition. For example, in the morning Alice eats breakfast at home, in the afternoon she prefers a restaurant around her office for lunch, and in the evening she loves to visit shopping malls nearest her workplace or home. Therefore, the user's changing preferences (home, restaurant, shopping malls) with changing contextual conditions (morning, afternoon, evening) adversely influence the recommended list of items and strongly affect the recommendation process. CARS tend to utilize the context in the recommendation process by exploiting the user's preferences within a particular contextual condition and suggest relevant items accordingly.

In the era of mobile internet and smartphones, spatial context and temporal context are easier to collect than other contextual information [1]. With an ever-increasing assortment of options available on the web, recommendation algorithms strive to filter out uninterested options and suggest most likely to be interested options for target users. Available options could be different venues situated in various locations (i.e. spatial), providing various services to users under various circumstances, and also time-bounded (i.e. temporal). The drifting preferences of users are captured under temporal context information, and spatial context information helps in recommending venues located close to the user [2].

The main strengths of classical CF are its cross-genre recommendation ability and that recommended items are completely independent of its machine-readable representation [3]. Towards generating personalized recommendation, context-aware collaborative filtering (CACF) is a well-established technique. However, one of the most important issues associated with CF is the cold start problem, which appears when new items or new users are introduced in the online environment. In this situation it is very difficult to give a prediction to a particular item for the new user, which requires the past user's ratings to calculate the similarities for neighborhood generation [4]. In this chapter, we combine both spatial and temporal approaches in the CACF framework by simultaneously incorporating the spatial and temporal context information into the user's rating profile. To cope with the new user cold start problem in the proposed spatio-temporal-based CACF framework, we introduce the user's demographic features into the user's rating profile. The main contributions of our proposed work are threefold:

- Incorporate a spatio-temporal approach into the CACF scheme.
- Deal with the cold start problem by incorporating demographic features of each user into the recommendation process.
- Learn temporal weights for each individual using a genetic algorithm (GA) to achieve personalized recommendation.

The rest of the chapter is structured as follows: Section 2 elaborates the work related to our proposed framework. In Sect. 3, the proposed framework is introduced. After that, the significance of temporal weights incorporated into recommendations is discussed in Sect. 4. Experiments and results are demonstrated in Sect. 5. Finally, the conclusion of our work with some future directions is mentioned in Sect. 6.

2 Related Work

In ubiquitous computing, context is a fixed set of attributes such as location, time, or different circumstances of nearby users or items. A context-dependent representation has proved to be valuable for increasing predictive capabilities of RS. A perfect context-aware recommender system (CARS) is able to reliably label each user action with an appropriate context and effectively tailor the system output to the user in that specific context [5, 6].

With the rapid growth of technological innovation, the user's activities are captured and traced with location and other contextual information, offering novel types of recommendation services. LARS considers only the spatial aspect of a user's ratings, whereas LA-LDA handles the cold start problem through a location-aware probabilistic generative model to produce top-k recommendations [7, 8]. For time-aware recommender systems (TARS), temporal context information is considered one of the most influential contextual dimensions. It facilitates tracking the evolution of user preferences to identify periodicity in user habits and interests, which lead to significant improvement of recommendation accuracy [9]. Moreover, incorporating social, temporal and spatial context information in recommendations is essential for location-based social networks, such as Foursquare, Gowalla, and BrightKite. Spatial contexts are used to restrict recommendations to locations that a user could feasibly access, whereas temporal information is utilized for accounting the change in a user's preferences over time with taking advantage of repetitive behavior [2].

2.1 Context-Aware Collaborative Filtering (CACF)

Context-aware collaborative filtering (CACF) predicts users' preferences under various contextual dimensions by using their past rating history. It is the extension of CF technique so that what other like-minded users have done in a similar context can be used to predict a user's preferences toward an activity in the current context. Adding each layer of contextual data into CF has been effective for improving predictive accuracy of RS [10].

2.2 Genetic Algorithm (GA)

GA is a heuristic solution-search or optimization technique, originally motivated by the Darwinian principle of evolution through (genetic) selection [11]. It starts with a randomly generated population of *chromosomes* (strings of binary values), and each chromosome represents a solution to a problem, which has a *fitness score*, a real number which is a measure of how good a solution it is to the particular problem. Further, it carries out recombination process of parent chromosomes using GA operators (crossover and mutation) to produce chromosomes and pass it into the successor population. This process is iterated until some stopping criterion is reached and a best solution to a given problem is evolved.

3 Proposed Spatio-temporal Recommendation for CARS

Our proposed CARS framework is based on CACF, combining predictions from spatial and temporal context similarities. Like traditional CF techniques, it follows three major steps for recommendations, namely, similarity measure, neighborhood selection, and prediction computation. In the following subsections, we will describe each step of CACF technique for the proposed framework.

3.1 Data Collection

Typically, data are collected explicitly from users where ratings for a subset of venues are given, and demographical features are added through a registration process. Formally in the CACF technique, we have a collection of users $\mathbf{U} = \{\mathbf{u_1}, \mathbf{u_2}, \ldots, \mathbf{u_n}\}$, venues $\mathbf{V} = \{\mathbf{v_1}, \mathbf{v_2}, \ldots, \mathbf{v_3}\}$, and demographic features $\mathbf{D} = \{\mathbf{d_1}, \mathbf{d_2}, \ldots, \mathbf{d_k}\}$. The rating function is defined as follows:

$$\begin{aligned}\mathbf{R}\text{: }&\mathbf{User} \times \mathbf{Venue} \times \mathbf{Spatialcontext} \times \mathbf{Temporalcontext}\\ &\times \mathbf{Demographiccontext} \rightarrow \mathbf{Rating}\end{aligned}$$

3.2 Spatio-temporal Similarity Measures

The spatio-temporal similarity (STS) measure is a combination of spatial and temporal measures. It exploits the property of spatial networks and considers time as an important factor to search similar users and similar items for a given application domain. We define spatial similarity (SS) in terms of Euclidean distance [1] $\mathbf{dist}(\mathbf{u_a}, \mathbf{v_a})$ and temporal similarity in terms of three parameters: time range of a day (TR), day range of a week (DR), and week range of a month or year (WR) [12]. $\mathbf{SS}(\mathbf{v_a}, \mathbf{v_b})$ determines spatial similarity between two venues for the active user, and $\mathbf{SS}(\mathbf{u_a}, \mathbf{u_b})$ determines spatial similarity between two users with respect to a target venue satisfying parameters $\in_1$ and $\in_2$ respectively in Eq. (1) and Eq. (2).

$$\mathbf{SS}(\mathbf{v_a}, \mathbf{v_b}) = |\mathbf{dist}(\mathbf{u_a}, \mathbf{v_a}) - \mathbf{dist}(\mathbf{u_a}, \mathbf{v_b})| \leq \in_1 \quad (1)$$

$$\mathbf{SS}(\mathbf{u_a}, \mathbf{u_b}) = |\mathbf{dist}(\mathbf{u_a}, \mathbf{v_a}) - \mathbf{dist}(\mathbf{u_b}, \mathbf{v_a})| \leq \in_2 \quad (2)$$

To compute temporal similarity, we compute the three parameters of temporal range, TR, DR, and WR, separately and then sum them to get a final value. TR considers different road conditions for running vehicles during the different time periods of a day such as morning rush hour, morning, lunchtime, afternoon, evening rush hour, late evening, and nighttime (sleeping time). DR focuses on differentiating routes of a road from working days (Monday to Friday) to weekend (Saturday and Sunday). Similarly, WR covers the scope of seasonal months in terms of weeks.

$$\mathbf{TS}(\mathbf{u_a}, \mathbf{u_b}) = \mathbf{1} - (\boldsymbol{\alpha}\mathbf{TR} + \boldsymbol{\beta}\mathbf{DR} + \boldsymbol{\gamma}\mathbf{WR}) \quad (3)$$

$$\mathbf{STS}(\mathbf{u_a}, \mathbf{u_b}) = \boldsymbol{\delta} \times \mathbf{TS}(\mathbf{u_a}, \mathbf{u_b}) + (\mathbf{1} - \boldsymbol{\delta}) \times \mathbf{SS}(\mathbf{u_a}, \mathbf{u_b}) \quad (4)$$

In Eq. (3), $\boldsymbol{\alpha}$, $\boldsymbol{\beta}$, and $\boldsymbol{\gamma}$ are temporal weights to recognize a user's temporal preference for visiting venues. In Eq. (4), $\boldsymbol{\delta}$ is a spatio-temporal parameter for utilizing the benefits of both similarities **SS** $(\mathbf{u_a}, \mathbf{u_b})$ and **TS** $(\mathbf{u_a}, \mathbf{u_b})$ in quality recommendations. When the rating data are sparse, the similarity measure that uses item similarity within the computation of user similarity has been shown to work well.

3.3 Neighborhood Selection

Neighborhood selection is one of the important steps in the CACF technique. Here, $\mathbf{STS}(\mathbf{u_a}, \mathbf{u_b})$ measure [see Eq. (4)] has been proposed for neighborhood formation to identify users with similar inclinations within a set of similar contextual information. The demographic features are the only medium to compute similarities between users when dealing with the new user cold start problem in CACF, so we have added demographic feature similarity to the $\mathbf{STS}(\mathbf{u_a}, \mathbf{u_b})$ measure. Moreover, we use both similar users and similar venues to enrich the neighbor selection process. The size of the neighborhood is based on the selection of similar users/ venues whose similarity value satisfies a certain threshold.

3.4 Prediction and Recommendation

Rating prediction is required to identify which user deliberately likes to assign what rates to which venue, and then the recommendation task is involved to select a relevant set of venues for the user from the collection of items. Some forms of recommendation may be more contextual than others, which influences the user's decision about choosing options. In such scenarios, utility of a venue in a given context may conflict with overall user preferences. The predicted value of given venue for the active user is computed by Eqs. (5)–(7).

$$\mathbf{Pred\_UB}(\mathbf{u_a},\mathbf{v_a}) = \bar{\mathbf{u}}_\mathbf{a} + \frac{\sum_{\mathbf{u_i} \in \mathbf{N_{u_a}}} (\mathbf{R_{u_i,v_a}} - \bar{\mathbf{R}}_{\mathbf{u_i}}) \times \mathbf{STS}(\mathbf{u_a},\mathbf{u_b})}{\sum_{\mathbf{u_i} \in \mathbf{N_{u_a}}} \mathbf{STS}(\mathbf{u_a},\mathbf{u_b})} \tag{5}$$

$$\mathbf{Pred\_IB}(\mathbf{u_a},\mathbf{v_a}) = \bar{\mathbf{v}}_\mathbf{a} + \frac{\sum_{\mathbf{v_i} \in \mathbf{N_{v_a}}} (\mathbf{R_{u_a,v_i}} - \bar{\mathbf{R}}_{\mathbf{v_i}}) \times \mathbf{STS}(\mathbf{v_a},\mathbf{v_b})}{\sum_{\mathbf{v_i} \in \mathbf{N_{v_a}}} \mathbf{STS}(\mathbf{v_a},\mathbf{v_b})} \tag{6}$$

$$\mathbf{Pred}(\mathbf{u_a},\mathbf{v_a}) = \boldsymbol{\mu} \times \mathbf{Pred\_UB}(\mathbf{u_a},\mathbf{v_a}) + (1-\boldsymbol{\mu}) \times \mathbf{Pred\_IB}(\mathbf{u_a},\mathbf{v_a}) \tag{7}$$

Here, the predicted rating $\mathbf{Pred}(\mathbf{u_a},\mathbf{v_a})$ [see Eq. (7)] indicates the expected likeness of the active user $\mathbf{u_a}$ to a particular venue $\mathbf{v_a}$, which is composed of user-based predicted rating $\mathbf{Pred\_UB}(\mathbf{u_a},\mathbf{v_a})$ [see Eq. (5)] and item-based predicted rating $\mathbf{Pred\_IB}(\mathbf{u_a},\mathbf{v_a})$ [see Eq. (6)] balancing with parameter $\boldsymbol{\mu}$. Finally, the recommended list of venues that have not already been visited by the active user is generated.

4 Learning Temporal Weights Using GA

Various temporal preferences are a prominent feature of recommendations, which affects a user's decision on visiting a venue. Some users might prefer to visit venues during weekend mornings in the winter season, while another prefers a Friday night in summer. We use GA to learn the optimum temporal weights using the entire set of relevance assessment. A weight vector is encoded by a string of binary digits. An example of an individual chromosome is depicted in Fig. 1.

$\longleftarrow - \alpha - \longrightarrow \longleftarrow - \beta - \longrightarrow \longleftarrow - \gamma - \longrightarrow$

w1	w2	w3	w4	w5	w6	w7	w8	w9	w10	w11	w12	w13	w14	w15
0	1	0	1	1	1	1	0	0	0	1	0	0	1	0

Fig. 1 Chromosome representation

A set of five binary values is assigned to each temporal parameter α, β and γ as temporal weights corresponding to each user indicates the user's temporal preferences, DR and WR, respectively, for visiting the venues. At the beginning, a random population is initialized to start the learning process. The selection process is based on fitness score in which chromosomes are chosen for genetic operation using the GA operators single-point crossover and bitwise mutation to produce new offspring. The average of the differences between the actual rating $ActR_i$ and predicted rating $PredR_i$ of all venues in the training sample set Tr_s is used as the fitness score (in Eq. 8) for the set of temporal weights α, β and γ. The good current generated offspring replace the previous bad offspring and form a new population to improve the next generation. GA terminates when there is no improvement in the fitness score in ten consecutive generations.

$$fitness_{score} = \frac{1}{Tr_s} \sum_{i=1}^{Tr_s} |ActR_i - PredR_i| \tag{8}$$

5 Experiments and Results

To demonstrate the effectiveness of our proposed schemes for recommendations, we conducted several experiments using two multidimensional datasets, Foursquare and ConcertTweets. From the Foursquare dataset we extracted 183 users out of 2,153,471 users, 304 venues out of 1,143,092 venues, and 518 ratings out of 2,809,581 ratings for the experiment. Similarly, we collected 69 users out of 61,803 users, 58 musical show venues out of 116,344 venues, and 218 ratings out of 250,000 ratings from the ConcertTweets dataset. Additionally, we randomly generated users' demographic features (age, gender, occupation) and incorporated these into both datasets.

In the experiments, we use leave-one-out cross validation, where datasets are used as the basis for generating two splits, the training set and testing set. According to leave-one-out cross validation, the testing set contains only one sample which provides a generalization error, and the remaining samples are considered the training set for the proposed model. We conducted an evaluation of accuracy in terms of mean absolute error (MAE). The MAE measures the deviation of predicted ratings generated by the CARS from the actual ratings specified by the user. The MAE(u_a) for active user u_a is computed by the formula:

$$MAE(u_a) = \frac{1}{n_t} \sum_{i=1}^{n_t} |ActR_{u_a,i} - PredR_{u_a,i}| \tag{9}$$

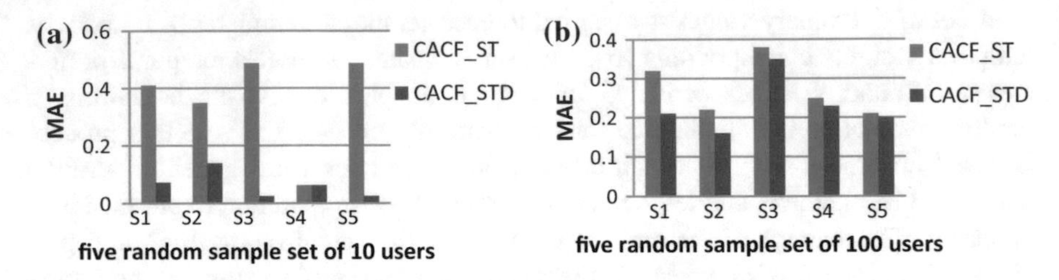

Fig. 2 Comparison between MAE of two schemes, CACF_ST and CACF_STD, using two datasets: **a** ConcertTweets and **b** Foursquare

where n_t is the cardinality of the test ratings set of user u_a. The total MAE [3] across all the active users N_T is computed as:

$$MAE = \frac{1}{N_T} \sum_{j=1}^{N_T} MAE(u_j) \tag{10}$$

5.1 Experiment 1

In the first experiment, we run the proposed spatio-temporal-based CACF (CACF_ST) scheme and spatio-temporal demographic CACF (CACF_STD) scheme. The CACF_STD is the extension of CACF_ST in which a demographic filtering approach is added for further enhancement. The comparative results in terms of MAE obtained by schemes CACF_ST and CACF_STD are depicted in Fig. 2.

5.2 Experiment 2

In the second experiment, an elitist GA is used to evolve temporal weights using GA operators with parameter values as shown in Table 1. A simple unsigned binary encoding scheme uses five bits for each of the temporal parameters α, β and γ in the

Table 1 Experimental parameters

Parameter	Values	Description
μ	0.2	Balancing the predicted ratings
α	$0 \le \alpha \le 31$	Weight for TR
β	$0 \le \beta \le 31$	Weight for DR
γ	$0 \le \gamma \le 31$	Weight for WR
δ	0.5	Balancing spatio-temporal similarities

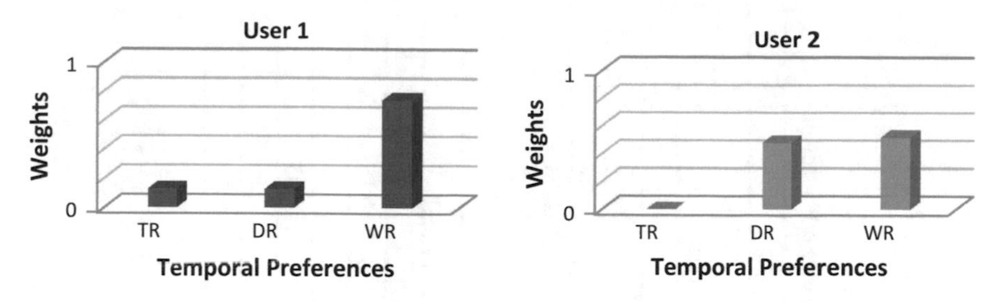

Fig. 3 Variation in temporal preferences TR, DR, and WR over the same venue for two different users

implementation process. Figure 3 shows that user 1 prefers more WR to visit a venue compared to other parameters TR and DR, whereas user 2 has equal preference of DR and WR, but no preference over TR for the same venue. The scheme CACF_STD is incorporated with different temporal weights learned by the GA, referred to as CACF_STDGA (Table 2).

In Fig. 4, a graphical representation is depicted for the MAE comparisons for the three schemes.

6 Conclusion and Future Directions

In this work, we have tackled the new-user cold start problem by incorporating a demographic filtering approach into spatio-temporal-based context-aware collaborative filtering (CACF_STD). In addition to context-based ratings, we exploited spatial, temporal, and demographic context information for recommendation. Further, to achieve personalized recommendation, we used GA to learn temporal weights/preferences of each individual user, which help in selecting appropriate venues. The results showed the effectiveness of our proposed schemes implemented using two real-world datasets: ConcertTweets and Foursquare. The scheme CACF_STDGA significantly improved the performance compared to two other schemes, CACF_ST and CACF_STD. Our future work will extend the proposed schemes by exploiting social context information [2, 13] and hybridizing the collaborative filtering approach with a reclusive method [14]. We also intend to use other types of multidimensional datasets such as Facebook and Twitter in our future work.

Table 2 Comparison of MAE of three schemes CACF_ST, CACF_STD, and CACF_STDGA using two datasets: **a** ConcertTweets **b** Foursquare

(a) MAE of various random sample sets

Sno.	10 Users			20 Users			30 Users		
	CACF_ST	CACF_STD	CACF_STDGA	CACF_ST	CACF_STD	CACF_STDGA	CACF_ST	CACF_STD	CACF_STDGA
1.	0.47	0.25	0.04	0.15	0.10	0.07	0.13	0.10	0.02
2.	0.48	0.19	0.02	0.25	0.20	0.01	0.14	0.09	0.02
3.	0.50	0.30	0.03	0.20	0.15	0.03	0.13	0.05	0.02
4.	0.41	0.28	0.07	0.17	0.14	0.05	0.12	0.03	0.03
5.	0.50	0.10	0.02	0.19	0.10	0.05	0.06	0.06	0.06

(b) MAE of various random sample sets

Sno.	50 Users			100 Users			150 Users		
	CACF_ST	CACF_STD	CACF_STDGA	CACF_ST	CACF_STD	CACF_STDGA	CACF_ST	CACF_STD	CACF_STDGA
1.	0.39	0.35	0.09	0.20	0.18	0.10	0.35	0.33	0.15
2.	0.23	0.19	0.10	0.23	0.20	0.12	0.13	0.12	0.05
3.	0.19	0.15	0.05	0.37	0.35	0.15	0.10	0.10	0.03
4.	0.33	0.23	0.12	0.30	0.20	0.13	0.15	0.12	0.10
5.	0.20	0.10	0.04	0.35	0.26	0.20	0.20	0.15	0.12

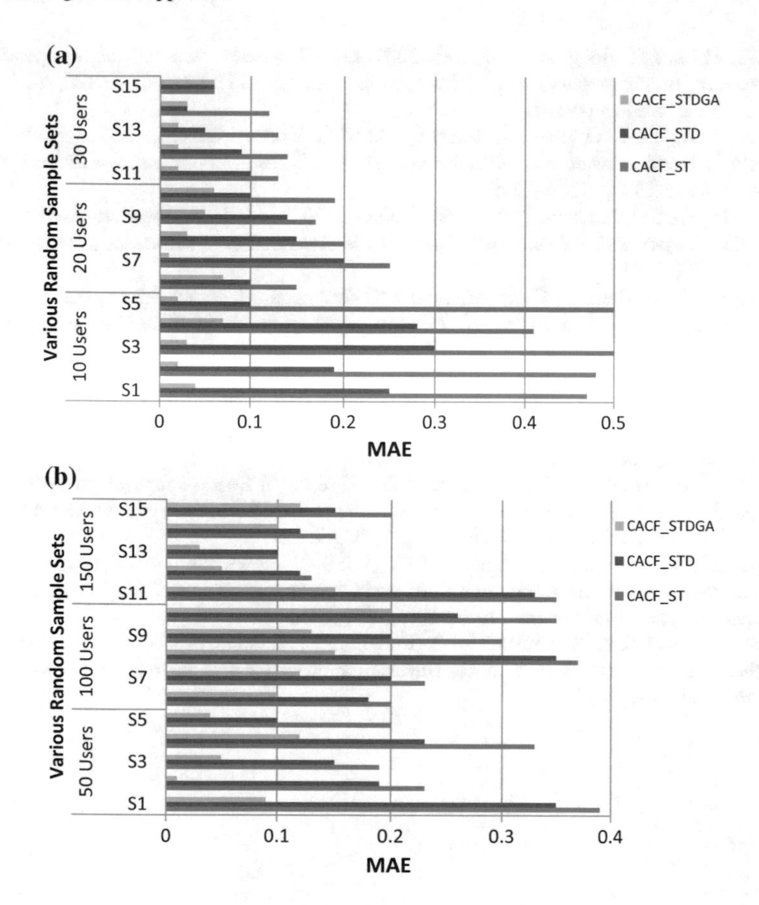

Fig. 4 Comparison of MAE for three schemes, CACF_ST, CACF_STD and CACF_STDGA, using two datasets: **a** ConcertTweets **b** Foursquare

References

1. Yin, H., and B. Cui. 2016. *Spatio-temporal recommendation in social media*. SpringerBriefs in Computer Science.
2. Stephan, T., and J.M. Morawski. 2016. Incorporating spatial, temporal, and social context in recommendations for location-based social networks. *IEEE Transactions on Computational Social Systems* 3 (4): 164–175.
3. Al-Shamri, M.Y.H., and K.K. Bharadwaj. 2009. Fuzzy-genetic approach to recommender systems based on a novel hybrid user model. *Expert Systems with Applications* 35: 1386–1399.
4. Son, L.H. 2016. Dealing with the new user cold-start problem in recommender systems: A comparative review. *Information Systems* 58: 87–104.
5. Adomavicius, G., B. Mobasher, F. Ricci, and A. Tuzhilin. 2011. Context-aware recommender systems. *AI Magazine* 32 (3): 67–80.

6. Park, M.H., J.H. Hong, and S.B. Cho. 2007. *Location-based recommendation system using Bayesian user's preference model in mobile devices*, 1130–1139., *LNCS 4611* Berlin, Heidelberg: Springer-Verlag.
7. Sarwat, M., J.J. Levandoski, A. Eldawy, and M.F. Mokbel. 2014. LARS*: An efficient and scalable location-aware recommender system. *IEEE Transactions on Knowledge and Data Engineering* 26 (6): 1384–1399.
8. Yin, H., B. Cui, L. Chen, Z. Hu, and C. Zhang. 2015. Modeling location-based user rating profiles for personalized recommendation. *ACM Transactions on Knowledge Discovery from Data* 9 (3): 1–41.
9. Campos, P.G., Díez, F., Cantador, I. (2014) Time-aware recommender systems: a comprehensive survey and analysis of existing evaluation protocols. *User Modeling and User-Adapted Interaction* 24 (1–2): 67–119.
10. Chen, A. 2005. Context-aware collaborative filtering system: Predicting the user's preference in the ubiquitous computing environment, vol. 3479, 244–253. LNCS.
11. McCall, J. 2005. Genetic algorithms for modelling and optimization. *Journal of Computational and Applied Mathematics* 184: 205–222.
12. Chang, J.W., R., Bista, Y.C., Kim, and Y.K. Kim. 2007. Spatio-temporal similarity measure algorithm for moving objects on spatial networks. In *ICCSA 2007*, eds. O. Gervasi, and M. Gavrilova, vol. 4707. LNCS. Heidelberg: Springer.
13. Agarwal, V., and K.K. Bharadwaj. 2013. A collaborative filtering framework for friends recommendation in social networks based on interaction intensity and adaptive user similarity. *Social Network Analysis and Mining* 3 (3): 359–379.
14. Kant, V., and K.K. Bharadwaj. 2013. Integrating collaborative and reclusive methods for effective recommendations: A fuzzy Bayesian approach. *International Journal of Intelligent Systems* 28 (11): 1099–1123.

Decision Tree Classifier for Classification of Proteins Using the Protein Data Bank

Babasaheb S. Satpute and Raghav Yadav

Abstract Identifying the family of an unknown protein is a challenging problem in computational biology and bioinformatics. Our aim here is to classify proteins into different families and also to identify the family of an unknown protein. For this purpose, we use the surface roughness of the proteins as a criterion. The Protein Data Bank (PDB) is the repository for protein data which contains the Cartesian coordinates of the sequences forming proteins. However, PDB coordinates give no indication of the orientation of the protein, which must be known in order to determine the surface roughness. For this purpose, we designed an invariant coordinate system (ICS) in which we took the origin as the protein center of gravity (CG). From the PDB we obtain the coordinates of all the amino acid residues which form the protein. But we are interested in the surface coordinates only in order to determine the surface similarity. Therefore, we developed a methodology to determine only the surface residues, and we recorded their coordinates. We then divided those coordinates into eight octants based on the signs of the x, y and z coordinates. For the residues in every octant, we found the standard deviation of the coordinates and created a parameter called the surface-invariant coordinate (SIC). Thus, for every protein, we obtained eight SIC values.

Keywords Protein classification · Structural classification of proteins SCOP · Protein data bank · PDB · Surface-invariant coordinate SIC · Decision tree classifier

B. S. Satpute (✉) · R. Yadav
Department of Computer Science & IT, SIET, SHUATS, Allahabad 211007, India
e-mail: satputebs@gmail.com

R. Yadav
e-mail: raghav.yadav@shiats.edu.in

A. N. Krishna et al. (eds.), *Integrated Intelligent Computing, Communication and Security*, Studies in Computational Intelligence 771,
https://doi.org/10.1007/978-981-10-8797-4_8

1 Introduction

Our aim in this work is to classify proteins into different families and also to identify the family of unknown proteins. Research has shown that the surface of a protein plays an important role in determining the protein family [1]. We have designed and used a parameter, surface-invariant coordinate (SIC), based on the surface roughness of the protein.

Proteins are important biomolecules which are formed from amino acids. There are 20 major amino acids which constitute proteins. Also, proteins are the main target when designing drugs for any disease. The surface is functionally the most important part of the protein. Though the inner residues of proteins play a major role in the formation of the backbone or the protein domain, they have much less importance from a functional perspective. For this reason, we have chosen to investigate the surface properties.

A decision tree is a classifier [2, 3] expressed as a recursive partition of the instance space. The decision tree consists of nodes that form a rooted tree, meaning it is a directed tree with a node called the "root" that has no incoming edges. All other nodes have exactly one incoming edge. A node with outgoing edges is called an internal or test node, and the other nodes are called leaves (also known as terminal or decision nodes). In a decision tree, each internal node splits the instance space into two or more sub-spaces according to a certain discrete function of the input attribute values. In the simplest and most frequent case, each test considers a single attribute, such that the instance space is partitioned according to the value of the attribute. In the case of numeric attributes, the condition refers to a range.

Each leaf is assigned to one class representing the most appropriate target value. Alternatively, the leaf may hold a probability vector indicating the probability of the target attribute having a certain value. Instances are classified by navigating them from the root of the tree down to a leaf, according to the outcome of the tests along the path.

2 Literature Survey

This section provides a brief review of some of the past work on the classification of proteins.

Datta et al. [4] utilized protein features extracted using the physicochemical properties and composition of amino acids. They employed artificial neural networks and the nearest neighbor classifier for classification purposes, achieving efficiency of 77.18%.

Bandyopadhyay [5] focused on the position of the occurrence of amino acids in each protein belonging to a specific superfamily. He used the nearest neighbor algorithm for the purpose of classification and obtained accuracy of 81.3%.

Chan et al. developed an algorithm called the unaligned protein sequence classifier (UPSEC) [6]. This is a probabilistic approach for finding the patterns in the sequences from which to classify proteins.

Angadi and Venkatesulu used an unsupervised learning technique to classify proteins using the Structural Classification of Proteins (SCOP) database. They used BLAST to create a database of p values [7].

Although the protein surface is functionally the most important, almost no work has been undertaken to classify proteins based on their surface roughness similarity.

3 Methodology for Determining Surface-Invariant Coordinates

We measure the SIC of a protein as the set of roughness values that are extracted from different indexed surface zones covered by different solid angles. We convert the Cartesian coordinate system (CCS) of a protein from the PDB to an invariant coordinate system (ICS). This ICS is necessary for structural comparison of proteins having different orientations.

3.1 Steps for Formulating the Invariant Coordinate System (ICS) [8]

I. Find the CG of the residue. For that we have taken the mean of the coordinates of the atoms. The calculated CG (say point C) was taken as the origin of the ICS.

II. Fix the mutually orthogonal axes x, y and z of the ICS.

3.2 Detection of Surface Residue from the Invariant Coordinates of the PDB Residue

We have developed a simple method for the detection of the surface residue of a protein from the ICS [8].

The following steps were adopted:

Step 1: Draw a line from the origin to the surface of the protein making the angles β with the x-axis, θ with the z-axis and γ with the y-axis.

Step 2: Detect the residue points (say the point set of n number of points $\{P\beta i\}$ $i = 1^n$) that are within a distance of 3.5 Å from line CP, making a cylinder of radius 3.5 Å.

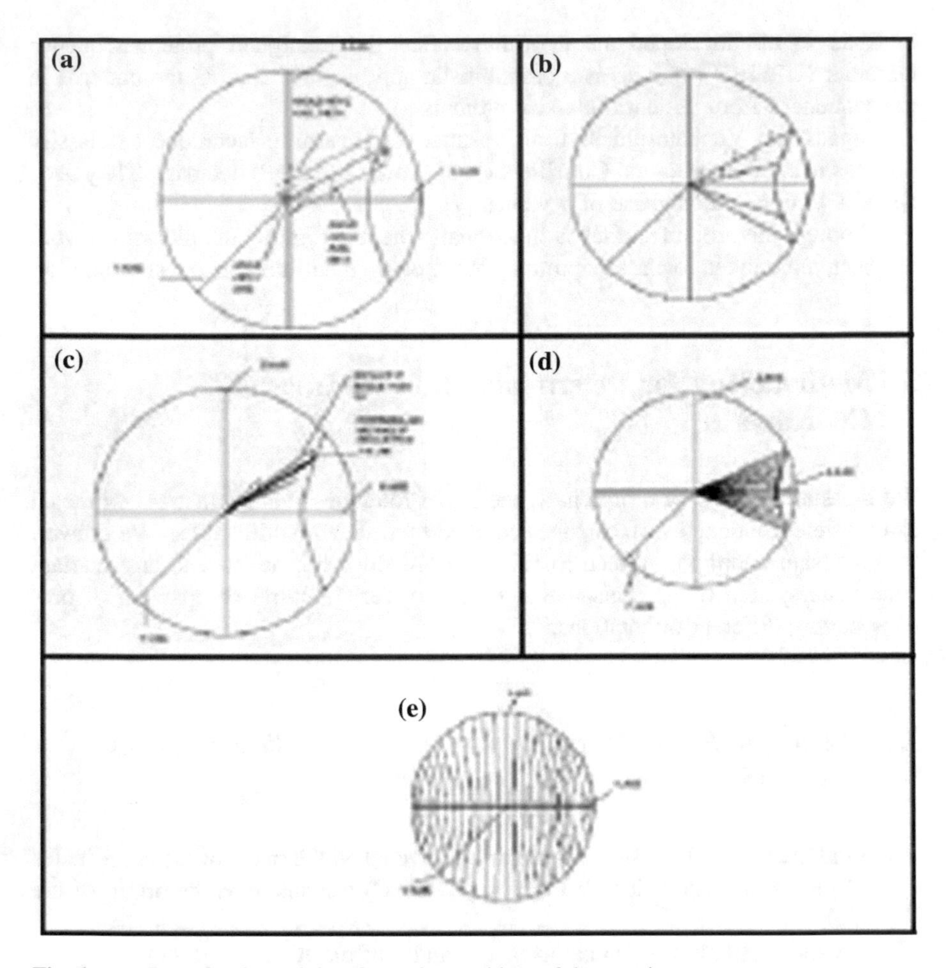

Fig. 1 **a–e** Steps for determining the surface residues of the protein

Step 3: Find the distances of all n points referenced above from C. The maximally distant point among all these points was taken as the surface residue point.

Step 4: The above was repeated by fixing the angle β and by rotating the line CP about the x-axis.

Step 5: Steps 4 and 5 were repeated by changing the angle β from 0° to 180° with discrete increase of its value covering all the residue points (Fig. 1).

3.3 Obtaining Surface-Invariant Coordinates of a Protein

I. Divide the coordinates obtained for the surface residue into eight parts for eight different octants

Our aim is to find the SIC of different octants. So, for calculation of the SIC, the coordinates for different octants should be separated.

II. Finding the SIC

We have taken the standard deviation of the distance of the residue from the center as the criteria for the measurement of the SIC.

For *N* number of solid angle partitions, **SIC S = $\{s_i\}_{i=1 \text{ to } N}$**, where

$$s_i = \sqrt{\frac{1}{M}\sum_{j=1}^{M}(x_j - \bar{x})^2} \quad (1)$$

and $\{x_j\}$ is the set of distances between the origin of the ICS and the residues in the *j*th octant, *x* bar is the mean of $\{x_j\}$ and *M* is the total number of residues.

4 Dataset/Materials

In our work, we took into consideration the proteins from the SCOP [9] database whose families are known. We took 1208 such proteins whose families are known from SCOP database. We took PDB [8] coordinates of those proteins from the PDB website. We converted those PDB coordinates into the SIC values and created a database of SIC values of the proteins family-wise. In total, for those 1208 proteins, we identified 100 families. For our work, we took into consideration only 32 families having 520 proteins belonging to these families.

The following is the sample SIC data for a few proteins (Table 1).

5 Algorithms

5.1 Training and Testing of the Decision Tree Classifier

Repeat steps 1 and 2

1. Obtain the PDB coordinates of the protein from www.rcsb.org.
2. Convert the PDB coordinates into SIC values using the SIC.

Table 1 Sample SIC dataset

O1	O2	O3	O4	O5	O6	O7	O8	Class
10.1822	9.7374	7.9482	9.2598	9.2741	8.0042	8.2156	6.5556	1
7.4712	6.7292	7.0651	7.6888	6.6028	7.9811	5.4901	6.6838	1
4.4171	5.7868	3.2321	4.0895	4.7957	3.9572	5.5311	3.8117	2

3. Prepare the database of the protein SIC values.
4. Assign each protein family a decimal class number; for example, if there are 32 classes, then the class numbers would be 1, 2, 3, …, 32.
5. Assign each protein a particular class value 1, 2, 3, …, 32.
6. Now develop the decision tree classifier program.
7. Divide the SIC database into a training dataset and test dataset in a ratio of 70:30, i.e. 70% of the protein belonging to one particular family is taken as training and 30% as testing.
8. Train the classifier with the training dataset.
9. Test the classifier performance with the test dataset.
10. Calculate the efficiency of the classifier with the following formula: Efficiency = (no. of correct predictions/total no. of test values) * 100%.
11. End.

5.2 Decision Tree for the SIC

Steps:

1. Take the octant 1 SIC value as the root of the decision tree (Fig. 2).
2. If the value is greater than 0.5, follow the right branch; otherwise, follow the left.

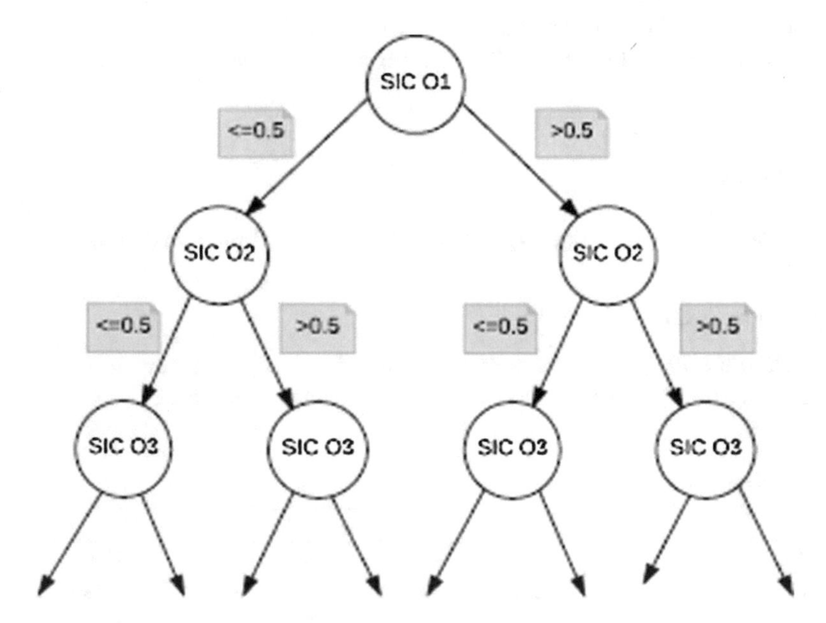

Fig. 2 Decision tree for protein classification using the SIC

Table 2 Predicted classes for class 1 and 2

O1	O2	O3	O4	O5	O6	O7	O8	Predicted class
6.6925	9.0295	9.6645	10.0935	8.9482	8.1583	9.2287	7.9731	1
6.4084	8.2107	7.9878	8.6107	8.1333	7.8973	9.355	6.2889	4
8.2812	8.4108	6.3872	10.5081	10.6953	5.9736	8.6328	6.9756	2
3.8949	3.8732	5.9014	4.8688	6.45	2.535	3.8082	5.0025	2

3. On the second level, take the octant 2 SIC values as the decision parameters.
4. If the value is greater than 0.5, follow the right branch; otherwise, follow the left.
5. Repeat steps 3 and 4 for other octant SIC values as decision parameters on the next levels (Table 2).

6 Results

I. Predicted classes for some test proteins
II. Efficiency

Total no. of proteins in the test dataset = 152.
Correct family prediction by the decision tree classifier (J48) = 90.
Therefore, efficiency = (90/152) * 100 = 59%.

7 Conclusion

Thus, by using the SIC, which is based on surface roughness of the proteins, we proved that proteins with similar surface roughness belong to the same family. The decision tree classifier for identification of the family of the protein from PDB coordinates yielded efficiency of over 59%.

References

1. Connolly, M.L. 1986. Measurement of protein surface shape by solid angles. *Journal of Molecular Graphics* 4: 3–6.
2. Richards, Joseph W., and Mark Fetherolf. 2016. *Real-world machine learning henrik brink*. ISBN 9781617291920.
3. Wang, D., and G.B. Huang. 2005. Protein sequence classification using extreme learning machine. In *Proceedings of international joint conference on neural networks (IJCNN, 2005)*, Montreal, Canada.

4. Datta, A., V. Talukdar, A. Konar, and L.C. Jain. 2009. A neural network based approach for protein structural class prediction. *Journal of Intelligent and Fuzzy Systems* 20: 61–71.
5. Bandyopadhyay, S. 2005. An efficient technique for super family classification of amino acid sequences: Feature extraction, fuzzy clustering and prototype selection. *ELSEVIER Journal of FuzzySets and Systems* 152: 5–16.
6. Ma, P.C.H., and K.C.C. Chan. 2008. UPSEC: An algorithm for classifying unaligned protein sequences into functional families. *Journal of Computational Biology* 15: 431–443. https://doi.org/10.1089/cmb.2007.0113.
7. Angadi, U.B., and M. Venkatesulu. 2012. Structural SCOP superfamily level classification using unsupervised machine learning. *IEEE/ACM Transactions on Computational Biology and Bioinformatics* 9: 601–608. https://doi.org/10.1109/tcbb.2011.114.
8. http://www.rcsb.org/pdb/.
9. http://scop.mrc-lmb.cam.ac.uk/scop/.

Logit-Based Artificial Bee Colony Optimization (LB-ABC) Approach for Dental Caries Classification Using a Back Propagation Neural Network

M. Sornam and M. Prabhakaran

Abstract Oral pain caused by bacterial infection, or caries, is an issue that can significantly impair individuals' ability to function in their day-to-day lives. Analysis of dental caries using X-ray images is tricky, and dental professionals are struggling to find a better solution in order to avoid misclassification of dental caries stages and potential false diagnosis. To avoid such classification and diagnostic inaccuracy, a method is proposed in this work that utilizes a hybrid approach combining a logit-based artificial bee colony optimization algorithm [LB-ABC] with a back-propagation neural network. This approach is implemented to boost the back-propagation algorithm for a proper training and testing process, thereby attaining the highest classification accuracy by using dental X-ray images as the numerical input generated through a gray-level co-occurrence matrix (GLCM), a texture feature extraction process. With this approach, the proposed work achieved an optimal accuracy of 99.16% and a minimized error of about 0.0033.

Keywords Logit-based artificial bee colony · Logistic function
Back-propagation neural network · Gray-level co-occurrence matrices

1 Introduction

Caries detection and classification is a challenging yet fundamental skill that all oral care experts must learn. In this chapter, a technique is proposed for determining the different stages of dental caries and provides advances to help the dental expert with this intricate assignment. Provides the assurance of trapping the caries affected tooth and treated as well. The affected caries level should be resolved, and at the same time, more serious injuries won't require a point-by-point assessment of action. Yet,

M. Sornam (✉) · M. Prabhakaran
University of Madras, Chennai 600005, Tamil Nadu, India
e-mail: madasamy.sornam@gmail.com

M. Prabhakaran
e-mail: prabhakaran_92@yahoo.com

A. N. Krishna et al. (eds.), *Integrated Intelligent Computing, Communication and Security*, Studies in Computational Intelligence 771,
https://doi.org/10.1007/978-981-10-8797-4_9

normal caries or enamel problems will require this level of examination and development. In this proposed work, a logit-based bee colony approach with a back-propagation neural network (BPNN) is used to overcome the possible flaws and problems faced in binary classification. This chapter includes an overview of the uses of bee colony algorithms for optimization problems in different fields; the approach is to use a bee colony to develop and successfully apply an image fusion technique [1]. The remainder of the chapter is organized as follows: Sect. 2 discusses related works carried out to date. Section 3 describes the proposed method, and the experimental results and analysis are presented in. Section 5 concludes the chapter and discusses future work.

2 Related Work

Previous studies in the field include the work by Mostafa et al. in which the authors introduced the concept of generating centroid values for a clustering technique to segment liver computed tomography (CT) images, and achieved acceptable accuracy [2]. Li et al. used Kapur's entropy to generate the objective function for bee colony optimization with the aim of reducing computational complexity, and showed that the modified bee colony approach yield the best results in generating the threshold value [3]. Kaur and Kaur described satellite image classification using a BPNN classifier, in which feature extraction was accomplished by means of a scale-invariant feature transform (SIFT) descriptor as an input to the BPNN, achieving a 99.9% accuracy [4]. Kavya et al. employed techniques including artificial bee colony optimization and fuzzy c-means clustering for the segmentation of retinal blood vessels to improve the accuracy of the extracted images [5]. Li et al. proposed a new technique, a modified quick artificial bee colony algorithm, for multi-level thresholding using Kapur's entropy function, which achieved a better threshold value when compared with electro-magnetism optimization [6]. Other works have included a combination of differential evaluation with a bee colony algorithm implemented for numerical dimensional problems, showing that the proposed method achieved a good global optimum [7]. A hybrid approach for satellite image segmentation using cuckoo search and an artificial bee colony algorithm achieved optimal accuracy of about 98% in comparison with other swarm intelligence approaches using the kappa coefficient [8]. Another study employed a hybridized brain tumor image segmentation technique using particle swarm optimization (PSO) and firefly algorithms to avoid unwanted objects in the image while reducing noise, achieving the highest accuracy of compressed image quality using peak signal-to-noise ratio (PSNR) and mean square error (MSE) of about 0.13 [9].

3 Proposed Method

3.1 *Caries Classification Using an LB-ABC-Based BPNN Classifier*

The main objective of the proposed work is to train the BPNN for the classification problem. Training the network to attain the best accuracy with good convergence is a challenging task. Choosing the perfect learning rate will give the proper convergence in a better way. So, this paper deals with a swarm intelligence-based heuristic technique to generate the learning rate using a logistic function called logit-based artificial bee colony (LB-ABC) algorithm for BPNN training to avoid major problems such as occurrence of local minima, to minimize testing errors and also to achieve better convergence results. The step-by-step process of the entire proposed work flow is described in Fig. 1.

Dataset: The periapical dental X-ray images are acquired from the med lab archives of 120 images in total as a data set [10].

3.2 *Segmentation*

The segmentation process is assumed for the best element extraction process for the images gathered from the panoramic dental radiography. This technique is used to lessen the noise occurring with some of the input images for the data set. The process utilizes the combined filters called convolution and averaging to diminish the noise occurring in the images taken as the input data set for classification (Fig. 2).

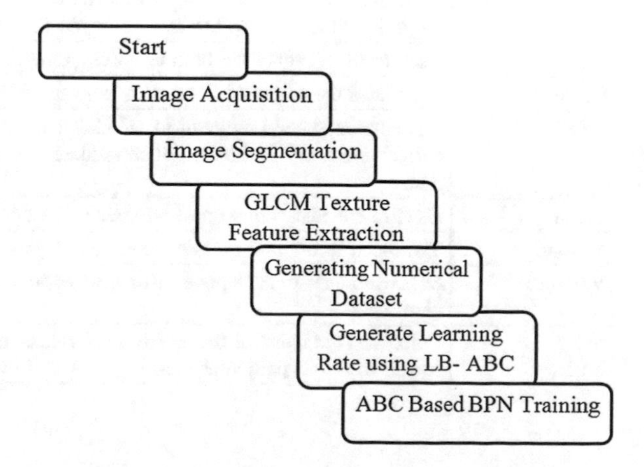

Fig. 1 Flowchart for the proposed method

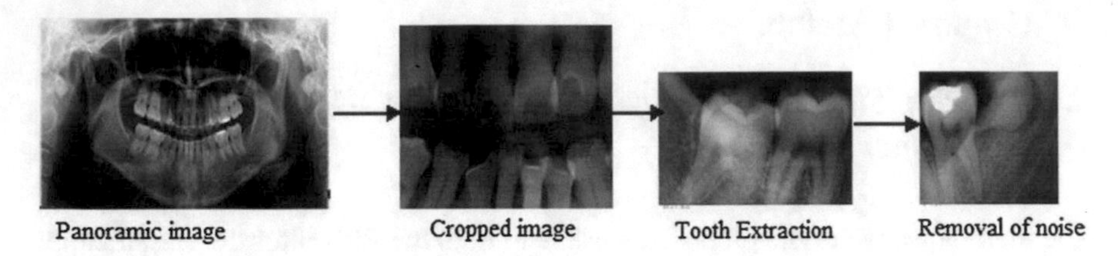

Fig. 2 X-ray image segmentation process

3.3 Feature Extraction Using a Gray-Level Co-occurrence Matrix

The gray-level co-occurrence matrix (GLCM) enables texture feature extraction, i.e., determining the internal structure of an image globally. Whole dental X-ray images were used for surface extraction of each individual image by calculating the GLCM using ten GLCM features: energy, homogeneity, contrast, correlation, dissimilarity, angular second moment (ASM), mean, median, variance, and standard deviation. The working principles of all the texture features are depicted in Table 1. Ten GLCM texture features are extracted for corresponding dental X-ray images, as shown in Table 2.

Table 1 GLCM texture feature extraction

GLCM features	Detailed texture feature extraction process
Energy	Is exactly the inverse of the entropy, which is the process of making the values linearly for the calculation of the arrangements of pixels in an image
Entropy	This term is derived from the concept of thermodynamics in physics; it is related to the aggregation of energy
Homogeneity	With the different contrasting windows, the variance and contrast values might be different; also referred to as the inverse method
Contrast	Used to differentiate the unique values generated form a single image
Correlation	Calculates the extended dominion from the different grayscale images
ASM	The angular second moment in GLCM features used to calculate the correctness which is the sum of the square of every pixel noted in a particular image
Mean	Used to calculate the sum of squares of particular image pixel values
Median	The mean of the two middle pixel values presented in a particular image
Variance	Sum of squares of each pixel differentiated in the number of terms in the given pixel values
SD	Calculates the mean of the entire pixel values, the sum of the squares and the square root of a particular value

Table 2 GLCM feature extraction for dental X-ray images

X-ray Images	Energy	Entropy	Homogeneity	Contrast	Correlation	ASM	Mea	Median	Variance	SD
	0.1805	0.6674	0.8953	0.5872	0.9227	1.779	0.00	1.5258	1.1813	2.830
	0.1805	0.6260	0.8953	0.5872	0.9227	1.765	0.00	1.5432	1.1765	3.830
	0.1805	0.6867	0.8953	0.5872	0.9227	1.742	0.00	1.4324	1.1764	2.047
	0.1805	0.6569	0.8953	0.5872	0.9227	1.167	0.00	1.3211	1.1674	3.319
	0.1805	0.6765	0.8953	0.5872	0.9227	1.214	0.000	1.2143	1.1962	1.787

3.4 Logistic Function

Initializing the learning rate in the vicinity of 0.0 and 1.0 is more difficult than between forces of two, in light of the fact that there are many forces of two incorporated into that range, and, thus one gets into the issue of drastically altered numbers. To avoid this issue and to avoid approximation of the learning rate, the LB-ABC algorithm is proposed. (The intrinsic unit is a unit of information derived from intrinsic logarithms with the power of e.) [11, 12].

The logistic function for any value α is derived as inverse-logit:

$$P(Y) = \frac{\exp(\mathrm{a} + \mathrm{b} \cdot \mathrm{X})}{1 + \exp(\mathrm{a} + \mathrm{b} \cdot \mathrm{X})} \tag{1}$$

3.5 Artificial Bee Colony

The ABC is a stochastic improvement algorithm propelled by the searching conduct of bumble bees. The algorithm speaks to arrangements in the given multi-dimensional inquiry space as food sources (nectar) and maintains a populace of three sorts of honey bee (utilized, spectator, and scout) to search for the best nourishment source (arrangement). The general strategy of the ABC is that it starts with irregular arrangements and more than once attempts to discover better arrangements via looking through the areas of the present best of their food source position.

Pseudo-code 1: Logit-based artificial bee colony algorithm

Steps:

Input: Initialize the population $x_{j,k.}$, iteration = 1, repeat until a new solution of v_{jk} is produced, where j is the neighbor of x for the employed bees (where k is a result of and a random number lying in the range [1, −1]).

Apply the greedy selection for xi and v_i to find the G-best amongst the p-best.

Find the probability using Eq. 1.

$$p = \frac{fit_i}{\sum_{i=1}^{SN} fit_i} \tag{1}$$

Calculate the fitness values of the solutions using Eq. 2.

$$fitness = \frac{1}{1+} f_i \tag{2}$$

if $(f_{i \geq 0})$ normalize the output P between [0 and 1].

Construct the new result V_I for the onlooker bees for the attained result x_i.

Apply the greedy selection method for the onlooker bees to find the global best. Regulate the unrestrained value (source), and change with randomly generated result xi for the scout bees.

$$W_{jk} = \min i_k + \text{ random } (0, 1) * (\max ij_k * \min i_k) \quad (3)$$

By repeating the above steps, finalize the best food source attained.

$$\text{Itr} = \text{itr} + 1.$$

Calculating the error based on the distance between the employed bees. Generate the learning rate by using the inverse-logit function

$$D = \frac{[(\exp(a + b * x)]}{[1 + \exp(a + b * x)/o]} \quad (4)$$

where a, b, x and o are three different bees and error (distance generated from the employed bees) as the parameters for the logistics function.

Output: The value generated is considered as the learning rate to train the BPNN.

3.6 Back-Propagation Neural Network

A BPNN is simply a process of learning based on errors; an artificial neural network (ANN) consists of an input layer, a hidden layer and an output layer. Here, the work deals with the supervised learning by providing the input with a corresponding output and enables machine learning based on the given parameters.

The flow chart in Fig. 3 elaborates the proposed LB-ABC approach using a logistic function to generate the learning rate process. The BPNN architecture shows the transformation of the image to a numerical dataset as the input for dental caries classification using GLCM texture feature extraction in Fig. 4.

4 Experiment Results and Analysis

The BPNN classifier is compared with the LB-ABC-based BPNN classifier for dental caries classification which is differentiated from the approximated and LA-ABC-generated learning rate for classification. The experimental results are presented in Table 3.

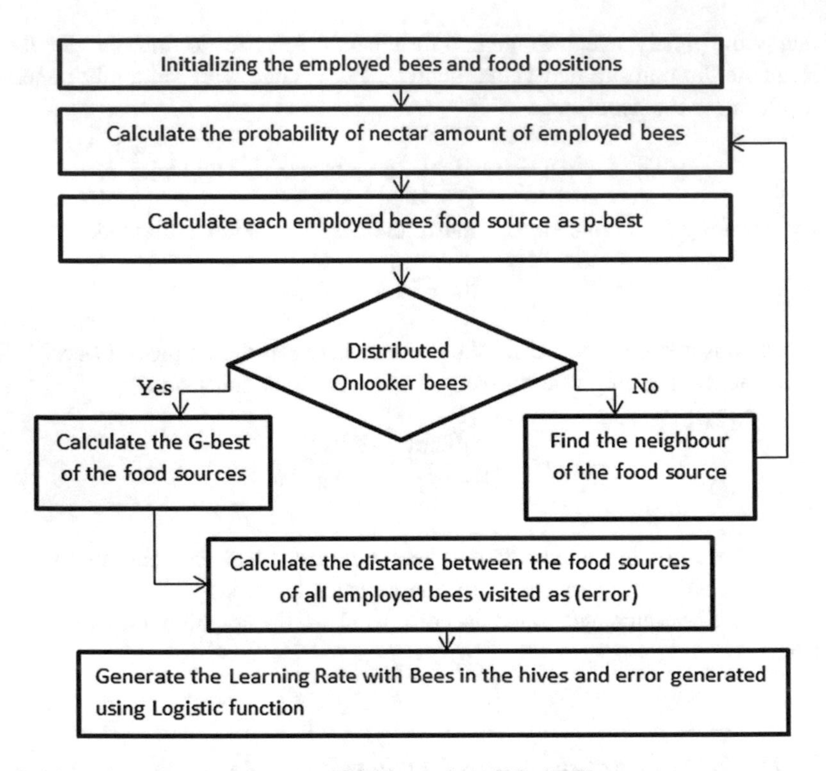

Fig. 3 Flow chart for a logit-based artificial bee colony [LB-ABC]

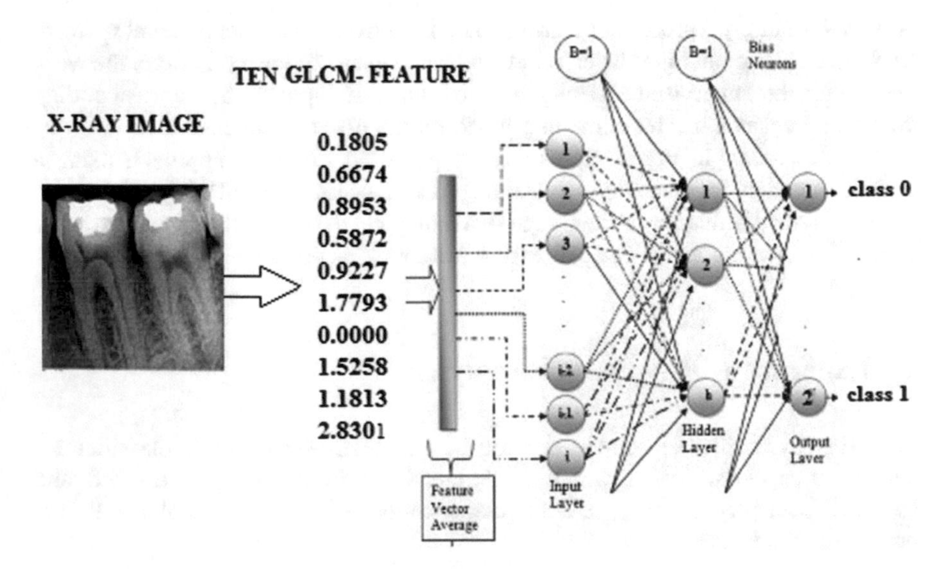

Fig. 4 Dental caries classification using a BPNN

Table 3 GLCM feature extraction for dental X-ray images

Network	Hidden nodes	Epochs	Approximate learning rate for the BPNN	Accuracy (%)	MSE
BPNN classifier	40	500	0.1	89.76	2.134
	40	1000	0.2	91.85	1.982
	40	1500	0.3	96.50	0.126
Proposed LB-ABC with BPNN	Hidden nodes	Epochs	Learning rate generated by the LB-ABC	Accuracy (%)	MSE
	20	2000	0.04	98.7	0.108
	15	4000	0.05	99.16	0.008
	10	5000	0.06	99.1667	0.00339

4.1 *Stages of Dental Caries Classification*

Diagnosing caries is a challenging task in that caries detection alone is not sufficient for the diagnosis. As such, the impact of the caries stage on a tooth is sufficient for dentists to proceed with the proper diagnosis process, shown in Table 4. The American Dental Association discussed the four stages of caries in Fig. 5. Dental caries classification using the BPNN classifier which is compared for the classification of caries stages [13], along with the predicted treatment for different stages of caries, is suggested in Table 4.

The performance of the proposed method is tested with three different activation functions (reLU, sigmoid, tanh). The tanh (Fig. 6c) activation function gives the best convergence compared with Fig. 6a, b in Fig. 6 for the parameters used for the LB-ABC with the BPNN classifier in Table 5.

The graphical representation of the LB-ABC shows that the tanh activation function delivers good computational results, as deliberately shown in Fig. 7 and

Table 4 GLCM feature extraction for dental X-ray images

X-ray images	BPNN output	Classifier	Stages	Diagnosis prediction
	0.0128882	Normal	No caries	No treatment required
	0.7397654	Abnormal	Moderate dentin caries	Root canal treatment
	0.9888456	Abnormal	Enamel with severe dentin caries	Laser treatment
	0.8897654	Abnormal	Initial enamel with moderate dentin caries	Usage of fluoride for prevention
	0.9817422	Abnormal	Severe dentin caries	Either drilling or laser treatment

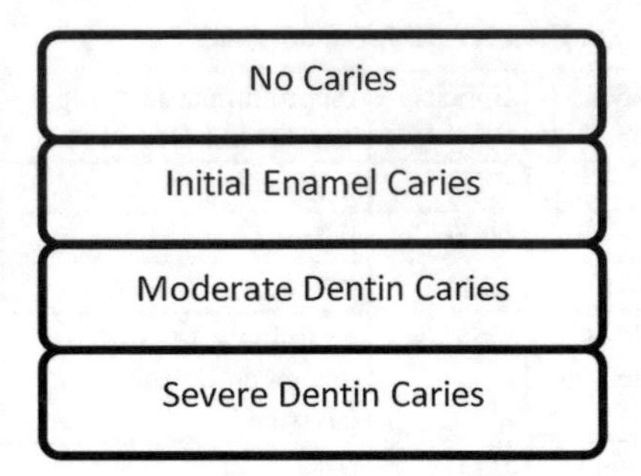

Fig. 5 Stages of dental caries

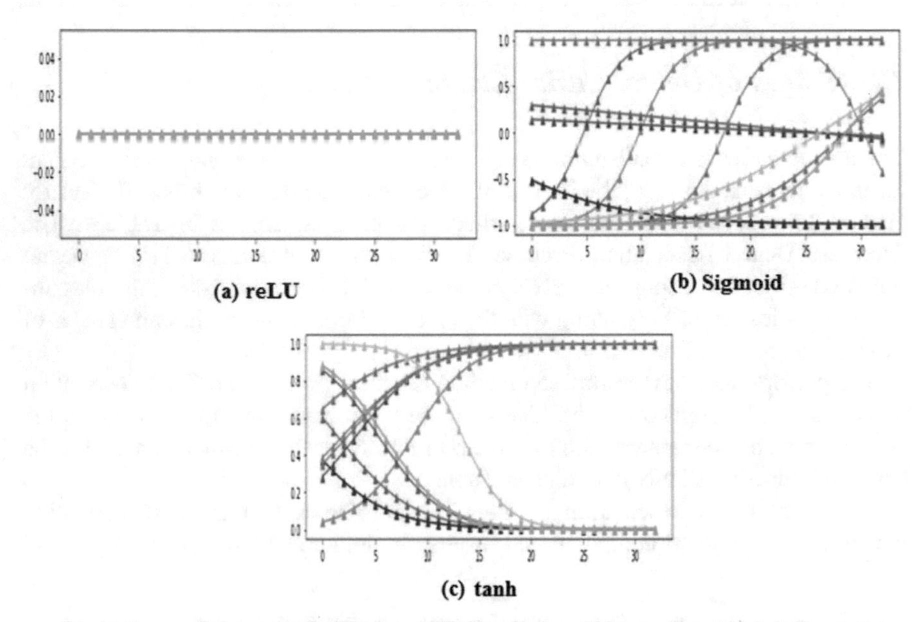

6 (a, b, c) -------Represents the Number of Hidden Nodes. (Whereas X and Y Axis Represents No of Nodes and Convergence Value).

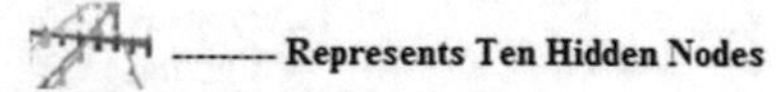

Fig. 6 Proposed LB-ABC convergence graph

Table 5 Experimental results for LB-ABC based BPNN classifier

	Proposed LB-ABC with BPNN classifier				
Activation	Epochs	Hidden modes	Learning rate	Final loss	Execution time in min.
reLU	5000	10	0.06	172.37	1.61
Sigmoid	5000	10	0.06	0.3711	1.65
tanh	5000	10	0.06	0.0033	1.38

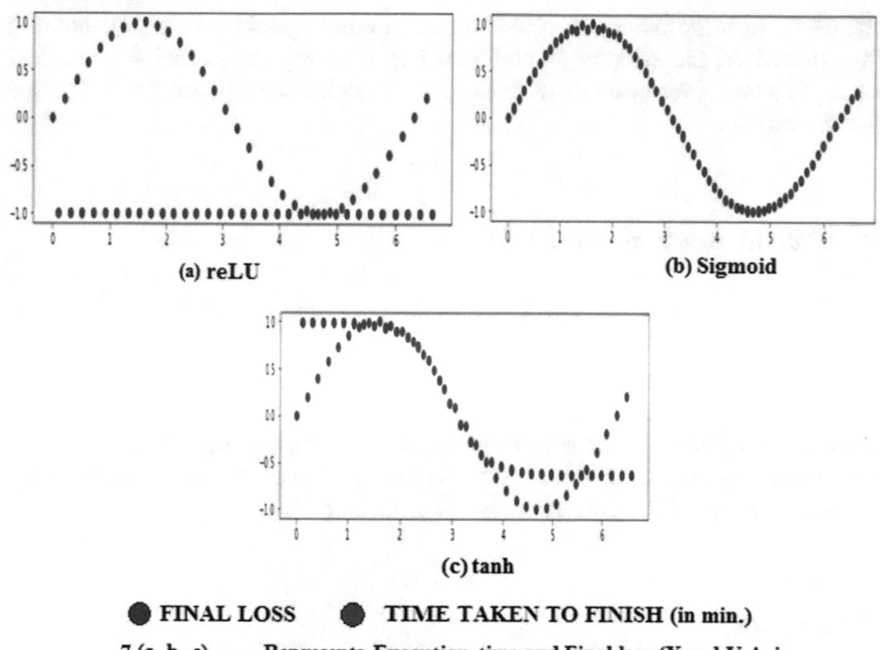

Fig. 7 Execution time and loss convergence graph

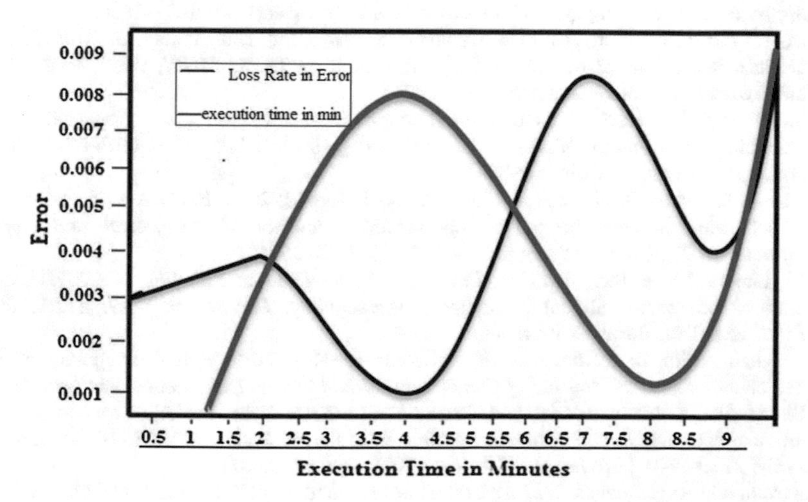

Fig. 8 Convergence graph of the tanh activation function which produces the best results from Table 5

compared with the other two activation functions (sigmoid and reLU) shown in Fig. 7b and 7c, respectively. In addition, Fig. 8 clearly shows that the tanh activation function gives better convergence results with respect to minimum error and fast execution.

5 Conclusion and Future Work

This chapter demonstrates efficient classification using a BPNN network with LB-ABC techniques for the classification of normal teeth and teeth affected by dental caries. The learning rate generated by the LB-ABC for the BPNN classifier achieved the best training and testing accuracy of 99.16%, with a reduced MSE of about 0.0033. Future work is planned to combine the proposed method with an automated classification system for multiple deadly diseases to reduce the diagnostic complexity through machine-learning techniques.

References

1. Sharma, P.K., V.S. Bhavya, K.M. Navyashree, K.S. Sunil, and P. Pavithra. 2012. Artificial bee colony and its application for image fusion. *IJ Information Technology and Computer Science* 42–49. https://doi.org/10.5815/ijitcs.2012.11.06.
2. Mostafa, Abdalla, Ahmed Fouad, Mohamed Abd Elfattah, Aboul Ella Hassanien, Hesham Hefny, Shao Ying Zhu, and Gerald Schaefer. 2015. CT liver segmentation using artificial bee colony optimisation. In *19th international conferences on knowledge based and intelligent information engineering systems. Procedia Computer Science* 15: 1622–1630.
3. Li, Linguo, Lijuan Sun, Jian Guo, Chong Han, Jian Zhou, and Shujing Li. 2017. A quick artificial bee colony algorithm for image thresholding. In *Lic MDPI*, Basel, Switz, 1–19. http://www.mdpi.com/journal/information.2017.
4. Shifali, and Gurpreet Kaur. 2016. Satellite image classification using back propagation neural network. *Indian Journal of Science and Technology* 9 (45): 1–8. ISSN: 0974-6846. https://doi.org/10.17485/ijst/2016/v9i45/97437.
5. Kavya, K., M.G. Dechamma, and B.J. Santhosh Kumar. 2016. Extraction of retinal blood vessel using artificial bee-colony optimization. *Journal of Theoretical and Applied Information Technology* 88 (3): 535–540. ISSN: 1992-8645.
6. Li, Linguo, Lijuan Sun, Jian Guo, Chong Han, Jian Zhou, and Shujing Li. 2017. A quick artificial bee colony algorithm for image thresholding. *Information* 1–19, Article 2017, Licensee MDPI, Basel, Switzerland.
7. Abraham, Ajith, Ravi Kumar Jatoth, and A. Rajasekhar. 2012. Hybrid differential artificial bee colony algorithm. *Journal of Computational and Theoretical Nanoscience* 1–9.
8. Singla, Shelja, Priyanka Jarial, and Gaurav Mittal. 2015. Hybridization of cuckoo search & artificial bee colony optimization for satellite image classification. *IJARCCE* 4 (6): 326–331. ISSN: 2319-5940. https://doi.org/10.17148/IJARCCE.2015.4671.
9. Khyati, and Amit Doegar. 2017. Hybrid nature inspired brain tumor segmentation using PSO and firefly swarm intelligence. In *Conference proceeding. CNFESMH-2017*, 29th July, 453–460. ISBN: 978-81-934083-9-1.

10. Med lab archives for periapical Dental X-ray Images from. https://mynotebook.labarchives.com/share/Vahab/MjAuOHw4NTc2Mi8xNi9UcmVlTm9kZS83NzM5OTk2MDZ8NTIuOA.
11. Gayou, Olivier, Shiva K. Das, Su-Min Zhou, Lawrence B. Marks, David S. Parda, and M. Miften. 2008. A genetic algorithm for variable selection in logistic regression analysis of radiotherapy treatment outcomes. *American Association of Physicists* 35 (12): 5426–5433. https://doi.org/10.1118/1.3005974.
12. Peterson, Leif E. 2014. Evolutionary algorithms applied to likelihood function maximization during poisson, logistic, and Cox proportional hazards regression analysis. In *2014 IEEE congress on evolutionary computation (CEC)*, 1054–1061, July, 2014.
13. Jeffery, B. 2013. A review of dental caries detection technologies. Academy of General Dentistry. ICDAS. 09-2013, 100–108. www.dentaleconomics.com.

Use of Modified Masood Score with Prediction of Dominant Features to Classify Breast Lesions

N. M. Sonali, S. K. Padma, N. M. Nandini and T. S. Rekha

Abstract In this chapter, we propose a novel method to improve the diagnostic accuracy for breast lesions based on certain cellular and nuclear features which become the criteria for final diagnosis. In the medical field, the accuracy of the diagnosis affects proper treatment of the condition. To overcome the problem of interobserver variability, a method of scoring is used to grade the lesions considered for the study. We have used the modified Masood score and designed an algorithm which classifies the various grades of breast lesions. We have used three classifiers: adaptive boosting, decision trees and random forests. These classifiers are effective tools for classifying breast lesions into benign, intermediate and malignant conditions. Principal component analysis using covariance and correlation is used to improve the diagnostic accuracy and for taking only the dominant features sufficient for classification.

Keywords Modified Masood score · Adaptive boosting · Decision tree
Random forest · Principal component analysis · Covariance
Correlation

N. M. Sonali (✉) · S. K. Padma
Department of Information Science and Engineering,
Sri Jayachamarajendra College of Engineering, Mysore, India
e-mail: sonalinandish@gmail.com

S. K. Padma
e-mail: skp@sjce.ac.in

N. M. Nandini · T. S. Rekha
Department of Pathology, JSS Medical College,
JSS University, Mysore, India
e-mail: nmnandini@jssuni.edu.in

T. S. Rekha
e-mail: tsrekha@jssuni.edu.in

A. N. Krishna et al. (eds.), *Integrated Intelligent Computing, Communication and Security*, Studies in Computational Intelligence 771,
https://doi.org/10.1007/978-981-10-8797-4_10

1 Introduction

A breast lesion is an extra growth or lump formed on the breast. It develops into cancer when there is growth of cancer cells in the breast tissue. It is necessary to know the kind of lesion so that it can be treated accordingly by an oncologist [1]. In the medical field, computer-aided diagnosis is used to enhance the identification of the type of breast lesion in order to improve classification accuracy and to enable appropriate treatment for the patient at an early stage [2].

Wolberg and Mangasrin [3] proposed that multi-surface pattern separation is useful to differentiate breast lesions defined by characteristic features. They also proposed the concept of representing the samples under consideration as a point on a vector space with a line separating the samples. The method is used to classify breast lesions into two categories, benign and malignant.

To improve the detection of breast lesion, it is also important to look into samples having an intermediate condition where the sample is neither benign nor malignant. Accurate diagnosis is necessary because statistical projections show that in India alone, the cases of breast cancer would impact 100,000 patients by the year 2020 [4]. The detection of the disease is done using fine-needle aspiration cytology (FNAC), which is a cost-effective method for cancer diagnosis [5].

Masood [5] has classified breast disease into a multi-class classification and has proposed a scoring system called the Masood score. Mridha et al. [6] modified the Masood score for their study. In [7, 8], the modified Masood score was used and proposed with validation using the histopathological correlation. In [9], the results have proved that the modified Masood score is a better scoring system than the Masood score for classification of breast lesions.

Hence, we have considered the modified Masood scoring system for our study. It is also observed that there exists inter-observer variability when classifying the lesions. Therefore, we propose an automated system to perform classification which will overcome the above limitation. To automate the process, adaptive boosting, decision tree and random forest classifiers are used for classifying our data [10–12], and studies are conducted to determine the best accuracy among these classifiers.

The remainder of the chapter is organized as follows: The proposed method is presented in Sect. 2. The data set information, experiments and results obtained are presented in Sect. 3. The chapter is concluded, along with future enhancements, in Sect. 4.

2 Proposed Methodology

The proposed methodology involves two stages: (i) principal component analysis (PCA) using the method of covariance, and (ii) prediction of dominant features.

2.1 *Data Dimensionality Reduction*

In the proposed algorithm, for classifying the breast lesion samples, the PCA method is used where a matrix X (M × D) is considered.

$M = m_1, m_2, m_3, \ldots, m_n$ is the total number of samples or the rows of the matrix. $D = d_1, d_2, d_3, \ldots, d_n$ indicates the dimension or features of the samples considered.

$$\bar{X} = \sum \frac{m_i}{M} \tag{1}$$

Here, m_i is the ith sample and $\bar{X}$ is the mean of i samples.

$$\text{Covariance (C)} = \sum_{i=1}^{M} \frac{(m_i - \bar{X})}{M - 1} \tag{2}$$

The PCA performs linear transformation of data using decomposition of the matrix into eigenvalues (λ) and eigenvectors (V) based on the covariance given by

$$\mathrm{V} = \mathrm{v}_1, \mathrm{v}_2, \ldots, \mathrm{v}_\mathrm{d} \tag{3}$$

$$\Lambda = \lambda_1, \lambda_2, \ldots, \lambda_\mathrm{d} \tag{4}$$

To obtain the eigenvalues and eigenvectors, the calculation is performed by using

$$\mathrm{XV} = \lambda \mathrm{V} \text{ or } \mathrm{XV} - \lambda \mathrm{V} = 0 \tag{5}$$

Maximum transformation of the values takes place for maximum covariance when the linear transformation between features is orthogonal to each other. To determine the maximum covariance of features, the eigenvectors are arranged as

$$\mathrm{V}_{(\mathrm{dxd})} \text{ as } \mathrm{V} = \mathrm{v}_1 > \mathrm{v}_2 > \cdots > \mathrm{v}_\mathrm{d} \text{ respectively.} \tag{6}$$

$$\text{Hence } \mathrm{X}_{(\mathrm{m} \times \mathrm{d})} \cdot \mathrm{V}_{(\mathrm{d} \times \mathrm{d})} = \mathrm{P}_{(\mathrm{m} \times \mathrm{d})} \tag{7}$$

P is the matrix with principal components.

$$\text{Accuracy} = \text{test size} - \text{error/test size} \tag{8}$$

Algorithm 1: Proposed method

Data: Breast lesion samples with d features
Result: Classification of samples into respective classes
Calculate the mean value $\bar{X} = \sum \frac{m_i}{M}$
Calculate the covariance using Eq. 2
Calculate λ and V using Eq. 5
Obtain $X_{(mxd)} \cdot V_{(dxd)} = P_{(mxd)}$
Repeat
Accuracy = test size-error/test size
Until all ratios of train : test are covered

2.2 *Prediction of Dominant Features*

The dominant features are obtained by transforming the data using PCA by means of correlation. The steps of obtaining the mean are the same as the method used in Sect. 2.1.

The standard deviation is obtained using the equation

$$\text{SD}_{\text{x}} = \sqrt{\frac{\sum_{i=1}^{M} m_i - \bar{X}}{M-1}} \tag{9}$$

The correlation between features is obtained using correlation coefficient r

$$\text{r} = \frac{1}{M-1} * \frac{\bar{X}}{SD_X} \tag{10}$$

The eigenvalues and eigenvectors are obtained using *r* instead of *C*, and the procedure is the same as in steps 5, 6 and 7 in Sect. 2.1.

The dominant features are predicted by knowing the percentage of each feature for the corresponding principal component. This is obtained by using load

$$\text{Load} = \text{V} \times \sqrt{\lambda} \tag{11}$$

3 Experimental Setup

3.1 *Data Set*

For data analysis using the modified Masood score, the samples were obtained from the Department of Pathology at JSS Hospital, Mysore. The study uses 254 such samples, with 102 classified as benign, 58 as intermediate, and 92 malignant.

3.2 Experimentation and Results

The study was conducted by choosing the adaptive boosting, decision tree and random forest classifiers to classify the breast lesions. In the first step, the samples are randomly selected and divided between train and test at ratios of 70:30 and 80:20. The redundant features are removed using the PCA by choosing only the first principal component, first two principal components and so on, until the accuracy has been measured using all six principal components. The data is analyzed using 10 iterations by choosing a different set of random samples in each iteration, and the average accuracy is calculated.

In Table 1, we can observe that the model performs the best by considering the random forest classifier when PCA-reduced data is considered with the first two principal components for both ratios of train-to-test considered.

From Table 2, we can identify the features having the highest percentage contribution to the first two principal components, which is sufficient for classification. The features are cellular arrangement, nucleoli and chromatin clumping, because they contribute 65% of the total percentage of features in P1 and P2.

Table 1 Accuracy obtained by taking whole data and the first to sixth principal components

Data/Total no. of principal components considered	Train-to-test ratio	Accuracy of classifiers (%)		
		AB	DT	RF
Whole data	70:30	81.82	87.01	92.10
	80:20	80.39	92.16	94.12
1	70:30	76.62	92.21	96.10
	80:20	86.27	92.16	96.08
2	70:30	80.52	94.81	**97.40**
	80:20	87.25	96.08	**98.40**
3	70:30	84.42	96.10	97.40
	80:20	86.27	96.00	96.08
4	70:30	81.82	92.21	93.51
	80:20	78.43	88.24	90.20
5	70:30	83.12	93.51	94.81
	80:20	80.39	90.20	94.00
6	70:30	81.82	87.01	92.10
	80:20	80.39	92.16	94.12

AB adaptive boosting; *DT* decision tree; *RF* random forest classifiers

Table 2 Percentage contribution of features to each principal component P1 to P6

Features	P1	P2	P3	P4	P5	P6
Cellular arrangement	**17.1**	12.7	04.5	20.0	36.1	08.6
Cellular pleomorphism	16.0	12.1	49.2	9.6	05.3	04.8
Myoepithelial cells	16.7	11.3	25.2	35.7	0.83	07.4
Anisonucleosis	16.9	13.5	18.2	19.7	26.9	20.7
Nucleoli	**17.1**	19.2	01.9	09.3	17.7	33.2
Chromatin clumping	16.2	**31.0**	0.92	05.5	12.5	25.3
Total percentage	100	100	100	100	100	100

4 Conclusion

In this chapter, a new automated method for classifying breast lesions based on a modified Masood score is presented. The system is chosen for the study because it is simple and cost-effective for categorizing breast lesions. The dominant features of cellular arrangement, nucleoli and chromatin clumping are confirmed to be valid by the pathologists involved in the study. The random forest classifier gives the highest accuracy of 98.4%. A system which classifies breast lesion images by considering only the dominant features is being developed.

References

1. NCI Homepage. http://www.cancer.gov/types/breast/patient/breast-treatment-pdq#sectionall. Accessed 14 Sept 2017.
2. Michael, J. 2002. Three decades of research in computer applications in health care. *Journal of the American Medical Informatics Association* 9 (2): 144–160.
3. Wolberg, W.H., and O.L. Mangasrin. 1990. Multisurface method of pattern separation for medical diagnosis applied to breast cytology. *Proceedings of the National Academy of Sciences of the United States of America* 87 (23): 9193–9196.
4. Takiar, R., D. Nadayil, and A. Nandakumar. 2010. Projections of number of cancer cases in India (2010–2020) by cancer groups. *Asian Pacific Journal of Cancer Prevention* 11 (4): 1045–1049.
5. Masood, Shahla. 2005. Cytomorphology of fibrocystic change, high-risk proliferative breast disease and premalignant breast lesions. *Clinics in Laboratory Medicine* 25 (4): 713–731.
6. Mridha, Asit Ranjan, V.K. Iyer, K. Kapila, and K. Verma. 2006. Value of scoring system in classification of proliferative breast disease on fine needle aspiration cytology. *Indian Journal* of *Pathology & Microbiology Journal* 49 (3): 334–340.
7. Nandini, N.M., T.S. Rekha, and G.V. Manjunath. 2011. Evaluation of scoring system in cytological diagnosis and management of breast lesion with review of literature. *Indian Journal of Cancer* 48 (2): 240–245.
8. Sheeba, D., and Chitrakala Sugumar. 2016. Palpable breast lesons-cytomorphological analysis and scoring system with histopathological correlation. *IOSR-JDMS* 15 (10): 25–29.

9. Cherath, Smrithi Krishna, and Savithri Moothiringode Chithrabhaum. 2017. Evaluation of Masood's and modified Masood's scoring system in the cytological diagnosis of palpable breast lump aspirates. *Journal* of *Clinical* and *Diagnostic Research* 11 (4): EC06–EC10.
10. Freund, Yoav, and E. Robert Schapire. 1999. A short introduction to boosting. *Journal of Japanese Society for Artificial Intelligence* 14 (5): 771–780.
11. Quinlan, J.R. 1986. Induction of decision trees, vol. 1, 81–86. Kluwer Academic Publishers.
12. Breiman, L. 2001. Random forests 45 (1): 5–32.

A Review of Description Logic-Based Techniques for Robot Task Planning

R. Gayathri and V. Uma

Abstract Planning and scheduling problems have become a promising area of research in the field of artificial intelligence (AI). In order to provide collision avoidance and enable successful completion of the task, the robot requires sophisticated knowledge about the environment for processing the tasks or methods. Knowledge representation and reasoning (KR & R) techniques provide such a level of knowledge to perceive the task environment. These techniques include description logic (DL) and ontology-based approaches. DL provides low-level knowledge to describe an environment. With advances in KR using ontology, high-level knowledge domain descriptions can be obtained more effectively. Knowledge representation techniques aid in temporal, ontological and spatial reasoning in a modeled environment. In this chapter, low-level KR & R techniques for task planning are summarized.

Keywords Knowledge representation · Reasoning · Robot · Ontology and linear temporal logic · Description logic

1 Introduction

In artificial intelligence, planning is concerned with organizing a sequence of actions (A) in order to achieve a certain goal (G) [7]. Task planning is used to construct a plan, and each plan has a sequence of ordered actions that enable the robot to perform efficiently [1].Knowledge representation and reasoning (KR & R)-based planning

---

R. Gayathri (✉) · V. Uma
Department of Computer Science, Pondicherry University, Puducherry 605014, India
e-mail: gayu339@yahoo.com

V. Uma
e-mail: umabskr@gmail.com

A. N. Krishna et al. (eds.), *Integrated Intelligent Computing, Communication and Security*, Studies in Computational Intelligence 771,
https://doi.org/10.1007/978-981-10-8797-4_11

techniques can represent the structural domain knowledge that aids temporal, ontological and spatial reasoning in a modeled environment, and it is used to successfully accomplish the tasks with shorter computational time [12].

Deterministic task planning problems can be represented as a tuple (I_0, A, G), where I_0 is an initial state of the domain knowledge, A is a set of actions defined by the preconditions and effects, and G is the goal state [8]. Representing and providing complete domain knowledge plays a vital role in making the robot perform actions efficiently and effectively [3, 11, 13]. This chapter provides a comprehensive survey of low-level representation techniques that may enable a robot to perform in an autonomous and intelligent manner. In addition, various deterministic task planning techniques with real-world considerations are discussed.

The remainder of this chapter is organized as follows: Sect. 2 describes how knowledge representation (KR) techniques can contribute to task planning. Section 3 presents a discussion on research contributions using low-level knowledge representation techniques, and the chapter is concluded in Sect. 4.

2 Knowledge Representation Techniques

Task planning knowledge representation techniques can be classified into two categories, high-Level and low-level domain KR [15]. Description logic is used for low-level domain representation, which can aid in accomplishing the goals in a shorter computational time [4]. The KR system is at the core of understanding knowledge in the domain, and it incorporates various types of representation techniques [15]. The taxonomy of knowledge representation techniques is illustrated in Fig. 1.

2.1 Low-Level Domain Knowledge Representation

The low-level domain involves two primary aspects, knowledge base and reasoning services, which are used to describe a KR system using description logics [8].

Description logic Description logic (DL) is a well-known family of logic-based KR formalism which is decidable in representing and reasoning the conceptual knowledge in a structured and semantically well-known manner [2]. The characterization of a DL system is accomplished using constructor, terminological assertion T (TBox), assertions about individual instances (ABox) and reasoning mechanisms of TBox and ABox [14].

In Fig. 2, Female, Human and Woman are atomic concepts, and the atomic roles are has-child and is-a. Using constructor operators, the concept "Human that is not female" is expressed as

Human $\wedge\neg$ Female

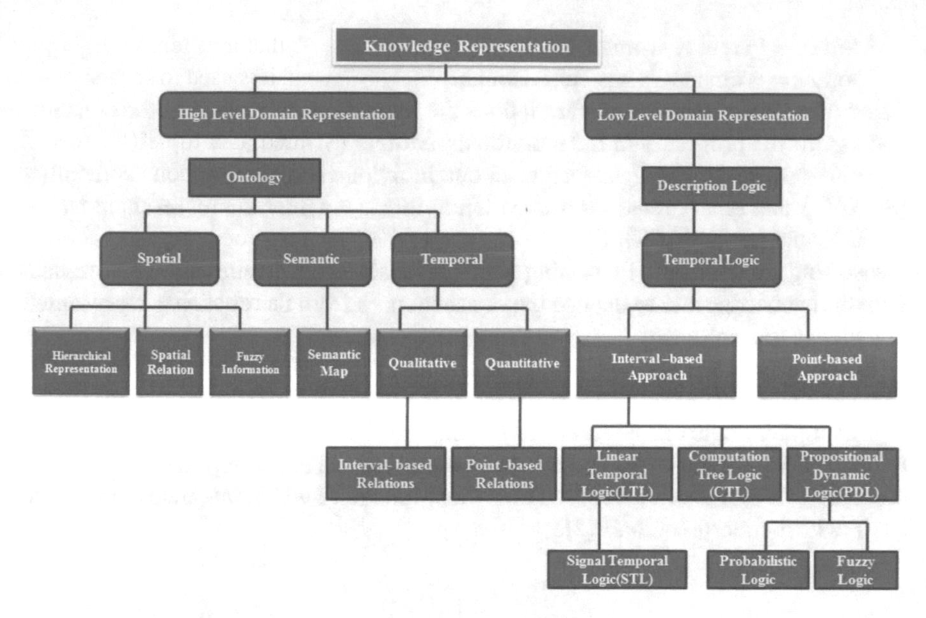

Fig. 1 Knowledge representation (KR) planning techniques

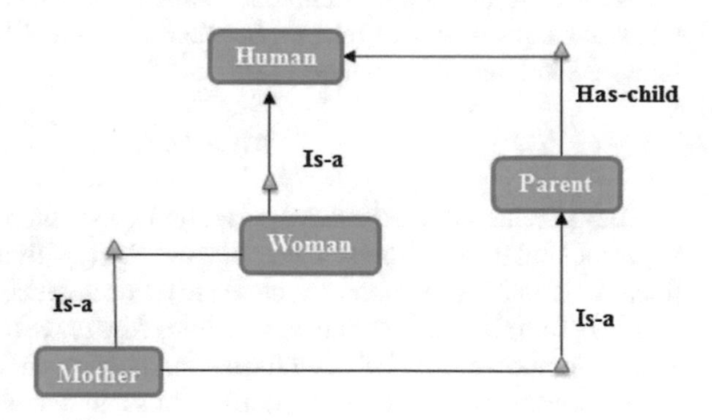

Fig. 2 Example DL system based on concepts and roles

Terminological part. The terminological assertion part, called TBox, is derived from the acyclic TBox and general TBoxes. An acyclic TBox can represent an atomic concept and is defined by

Parent ≡ Father ∨ Mother

General TBoxes can capture complex domain constraints, permitting the general concept inclusions (GCIs) which contains subsumption forms that can be expressed as C ⊆ D

where C and D represent the arbitrary concept expressions. Subsumption is used to detect whether the first concept C denotes a subset of the second concept D

Mother ⊆ Woman

Mother ⊆ Parent

In the above statement, the general inference concept of "mother" is expressed as

Mother = Parent ∧ Woman
Mother = Woman ∧ ∃ has-child.Human

Assertion part. An assertion part defines the individuals representing the concepts and asserts the properties of the individuals. Mother (SUSAN)

With respect to Fig. 2, assertion about individuals is represented by Mother (SUSAN), and binary relations between individuals via roles are represented by

Has-child (SUSAN, TOM)

Reasoning mechanism. Reasoning services enable us to represent the complete domain knowledge and to deduce the consequences from the explicitly represented knowledge. For instance,

If Mother is a Woman and Woman is a Human, then Mother is a Human can be inferred using the transitive relation.

Description logic can be extended to temporal description logic, which includes two approaches, point-based and interval-based approaches [2]. A point-based approach contains a flow of relation between time t and function I which integrates point time t ∈ T with an interpretation I [2].

$$I(t) = (\Delta^{I,t}, .^{I,t}). \tag{1}$$

An interval-based approach contributes to the semantics of a temporal model which contains a flow of relation between time t and function I, which integrates interval i with an interpretation I [2].

$$I(i) = (\Delta^{I,i}, .^{I,i}). \tag{2}$$

They provide more expressive power and good computational properties. Generally, temporal logic is based on temporal entities (i.e. time points, time intervals) and linear or branching temporal structures [6]. For instance,

Play (Tom) if $\text{Tom}^{I,t} \in \text{Play}^{I,t}$. i.e., *Tom plays at the tth moment*.

Linear temporal logic LTL is a formal language used for expressing the task planning procedure with high-level specifications and removing any ambiguity in specifications [5, 9]. The syntax of the LTL formula is inductively defined as [5],

$$\Phi = \phi \wedge \psi \mid \phi \vee \psi \mid \pi \mid \neg\pi \mid \phi \mid \phi \cup \psi \mid \Diamond\phi \mid \Box\phi$$

Computation tree logic Computation tree logic is a well-known family of branching time logic which allows the explicit quantification over the possible futures and incorporates the possible states. It is a tree-like structure, as shown in Fig. 3, which contains different set of paths, and any one of the actual paths is optimized. The syntax of the CTL formula is expressed as

$$\Phi = P \mid \neg\phi \mid \phi \wedge \phi \mid \phi \vee \phi \mid \phi \rightarrow \phi \mid AX\phi \mid EX\phi \mid A[\phi \cup \phi] \mid E[\phi \cup \phi]$$

Propositional dynamic logic Propositional dynamic logic is a formalism which can express the propositional behaviors of programs. PDL syntax is based on two sets

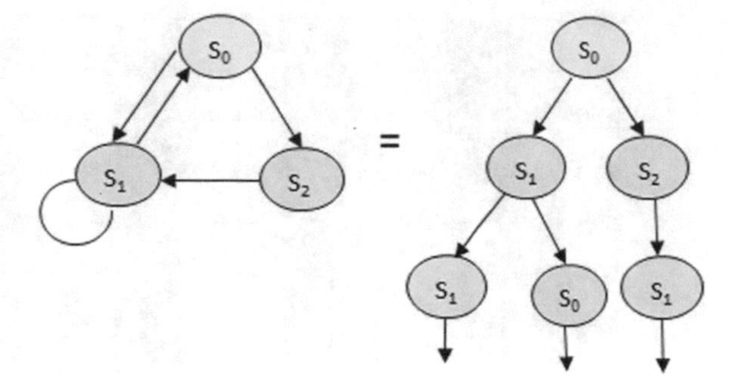

Fig. 3 Infinite computation tree structure

of symbols, namely atomic programs and atomic formulas. These two sets can be expressed with suitable operators, such as

$$\Phi, \Phi' \rightarrow \top \mid \bot \mid A \mid P \mid \phi \vee \phi' \mid \phi \wedge \phi' \mid \neg\phi \mid \phi; \phi' \mid \phi^* \mid \phi \cup \phi'$$

Unfortunately, interval-based temporal logic is undecidable in all flows of time, and it is difficult to represent the relations between more than two intervals [9].

3 Discussion

The robot task planning problems are divided into task planning and motion planning methods [10]. Robotics planning incorporates numerous tasks, and each task requires KR & R techniques. These techniques are used to describe the complete domain

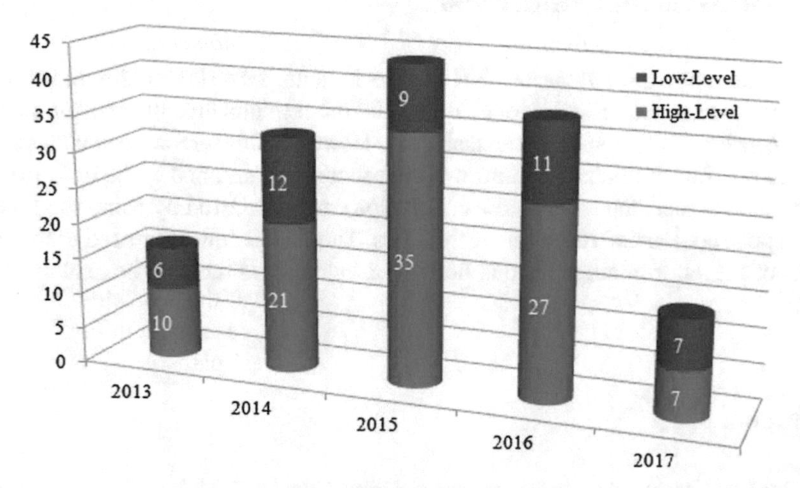

Fig. 4 Details of research contributions from studies using knowledge representation techniques since 2013

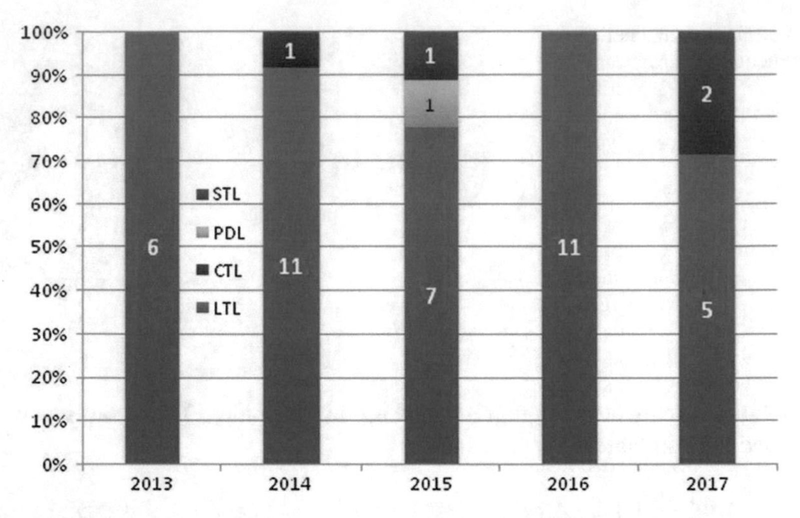

Fig. 5 Details of low-level knowledge representation techniques widely used in literature

knowledge and to infer new conclusions in the robotic domain. To understand the significance of the techniques in KR & R, a literature survey was performed. Year-wise analysis since 2013 and the details of the research works using DL and ontology are shown in Fig. 4.

Further analysis was conducted of contributions in the field of task and motion planning using various DL-based KR techniques, and the results are shown in Fig. 5.

4 Conclusion and Future work

This chapter has explored various KR & R techniques for task planning in robotics. These techniques are essential for enabling the robot to perceive the environment in order to successfully complete the task goal. However, this method can also lead to reduced run time efficiency. To improve the robot's performance of the task, precise information is represented and new conclusion can be inferred by using the knowledge representation and reasoning techniques. These techniques are used to increase the robot's level of intelligence that helping it to realize and reach the goal successfully.

References

1. Alatartsev, Sergey, Sebastian Stellmacher, and Frank Ortmeier. 2015. Robotic task sequencing problem: a survey. *Journal of intelligent & robotic systems* 80 (2): 279.

2. Baader, Franz. 2003. *The description logic handbook: theory, implementation and applications*. Cambridge: Cambridge university press.
3. Beetz, Michael, Moritz Tenorth, and Jan Winkler. Open-ease. 2015. In *2015 IEEE international conference on robotics and automation (ICRA)*,1983–1990. New York: IEEE.
4. Cho, Joonmyun , Hyun Kim, and Joochan Sohn. 2010. Implementing automated robot task planning and execution based on description logic kb. *Simulation, modeling, and programming for autonomous robots*, 217–228.
5. He, Keliang, Morteza Lahijanian, Lydia E Kavraki, and Moshe Y Vardi. Towards manipulation planning with temporal logic specifications. 2015. In *2015 IEEE international conference on robotics and automation (ICRA)*, 346–352. New York: IEEE.
6. Kress-Gazit, H, Georgios E Fainekos, and George J Pappas. 2005. Temporal logic motion planning for mobile robots. *IEEE transactions on robotics* 21(5):2020–2025.
7. Manjunath, T.C., and C. Ardil. 2008. Design of an artificial intelligence based automatic task planner or a robotic system. *International journal of computer and information science and engineering* 2: 1–6.
8. Milicic, Maja. 2008. Action, time and space in description logics.
9. Plaku, Erion, and Sertac Karaman. 2016. Motion planning with temporal-logic specifications: progress and challenges. *AI communications* 29 (1): 151–162.
10. Srivastava, Siddharth, Eugene Fang, Lorenzo Riano, Rohan Chitnis, Stuart Russell, and Pieter Abbeel. 2014. Combined task and motion planning through an extensible planner-independent interface layer. In *2014 IEEE international conference on robotics and automation (ICRA)*, 639–646. New York: IEEE.
11. Stenmark, Maj, and Jacek Malec. 2015. Knowledge-based instruction of manipulation tasks for industrial robotics. *Robotics and computer-integrated manufacturing* 33: 56–67.
12. Tenorth, Moritz, and Michael Beetz. 2017. Representations for robot knowledge in the knowrob framework. *Artificial intelligence* 247: 151–169.
13. Tenorth, Moritz, Alexander Clifford Perzylo, Reinhard Lafrenz, and Michael Beetz. 2013. Representation and exchange of knowledge about actions, objects, and environments in the roboearth framework. *IEEE transactions on automation science and engineering*, 10 (3): 643–651.
14. Wu, Xiuguo, Guangzhou Zeng, Fangxi Han, and Rui Wang. 2007. Goal representation and reasoning based on description logics (dls). In *International conference on intelligent systems and knowledge engineering, ISKE*.
15. Zhang, Shiqi, Mohan Sridharan, Michael Gelfond, and Jeremy Wyatt. 2014. Towards an architecture for knowledge representation and reasoning in robotics. In *International conference on social robotics*, 400–410. Berlin: Springer.

Classification of Proteins Using Naïve Bayes Classifier and Surface-Invariant Coordinates

Babasaheb S. Satpute and Raghav Yadav

Abstract Protein classification is one of the challenging problems in computational biology and bioinformatics. Our aim here is to classify proteins into different families using the surface roughness similarity of proteins as a criterion. Because Protein Data Bank (PDB) (http://www.rcsb.org/pdb/ [1]) coordinates give no indication of the orientation of the protein, we designed an invariant coordinate system (ICS) in which we took as the origin the protein's center of gravity (CG). From PDB we found the surface residue coordinates. We then divided those coordinates into eight octants based on the sign of x, y and z coordinates. For the residues in each octant, we found the standard deviation of the coordinates and created a parameter called the surface-invariant coordinate (SIC). Thus, for every protein we obtained eight SIC values. We also made use of the Structural Classification of Proteins (SCOP) (http://scop.mrc-lmb.cam.ac.uk/scop/ [2]) database. SCOP classifies proteins on the basis of the surface structure of the protein. As it is a classification problem, we used the naïve Bayes classifier algorithm for the classification to achieve better results.

Keywords Protein classification · Structural classification of proteins SCOP · Protein data bank · PDB · Surface-invariant coordinate SIC · Naïve Bayes classifier

1 Introduction

The surface of a protein plays an important role in determining the family of the protein [3]. We have designed a parameter, surface-invariant coordinate (SIC), which is based on the protein surface roughness.

B. S. Satpute (✉) · R. Yadav
Department of Computer Science & IT, SIET, SHUATS, Allahabad 211007, India
e-mail: satputebs@gmail.com

R. Yadav
e-mail: raghav.yadav@shiats.edu.in

A. N. Krishna et al. (eds.), *Integrated Intelligent Computing, Communication and Security*, Studies in Computational Intelligence 771,
https://doi.org/10.1007/978-981-10-8797-4_12

1.1 Why Study Protein Surfaces?

The surface is the functionally most important part of the protein. For the purpose of drug design, when we think about how the drug acts, the thing that comes to our mind is the cleft or cavity on the surface of the protein. Thus, in drug design, we are not dealing with the whole structure of the protein. We are mostly dealing with its surface region. The functional regions of the protein are largely the cavity or the cleft on the surface of the protein. Therefore, from a drug design perspective, the whole structure of the protein is not important; only the cleft or the cavities on the surface are needed. This may apply to pharmacophore drug design or structure-based drug design. The area of action of the drug is the residue and its moiety on the protein surface [3]. Although the inner residues of proteins play a major role in the formation of the backbone or the protein domain, functionally they have less importance. Thus we are studying the surface properties.

1.2 Naïve Bayes Classifier [4]

The naïve Bayes classifier classifies the dataset based on the features that are independent of each other. The naïve Bayes classifier equation is expressed as follows:

$$\mathrm{P(Class\ A|Feature1, Feature2)} = \frac{\mathrm{P(Featur1|Class\ A) \cdot P(Featur2|Class\ A) \cdot P(Class\ A)}}{\mathrm{P(Featur1) \cdot P(Featur2)}} \tag{1}$$

2 Literature Survey

In this section, we briefly review some of the past work on the classification of proteins.

Datta et al. [5] used the protein features extracted using physicochemical properties and composition of amino acids. They used artificial neural networks and nearest-neighbor classifier for classification purposes, and achieved efficiency of 77.18%.

Bandyopadhyay [1] stressed the position of occurrence of amino acids in each protein belonging to a specific superfamily. He used a nearest-neighbor algorithm for the purpose of classification, obtaining accuracy of 81.3%.

Chan and colleagues developed an algorithm which is called an unaligned protein sequence classifier (UPSEC) [2]. This is a probabilistic approach for finding patterns in the sequences with which to classify proteins.

An adaptive multiobjective genetic algorithm (AMOGA) was proposed by Vipsita et al. [6] to enhance the performance of the radial basis network used for protein classification purposes. The authors used principal component analysis for selecting features.

Most of the above and other work is focused on the sequence similarity of the proteins. Little to no work has explored the classification of proteins based on their surface roughness similarity, though surfaces are functionally most important, as stated in the introduction section of this paper. The main focus of this work is the classification of proteins based on their surface roughness similarity.

3 Methodology for Finding Surface-Invariant Coordinates

The structures for the proteins considered in our methodology are downloaded from PDB (www.rcsb.org) [7]. Positional descriptions of the atoms of a protein in PDB are given by a Cartesian coordinate system (CCS), which gives no indication of the orientation of a protein. Because of this, we convert the CCS of the PDB of a protein to an invariant coordinate system (ICS) that we designed.

3.1 Steps for Formulating the Invariant Coordinate System (ICS) [3]

I. Find the center of gravity (CG) of the residue. For this, we have taken the mean of the coordinates of the atoms. The calculated CG (say, point C) was taken as the origin of the ICS.
II. Fix the mutually orthogonal axes X, Y and Z of ICS.

Fixing the Z-axis of the ICS: We refer to the three mutually orthogonal axes of the ICS as 'X', 'Y' and 'Z'. From the PDB we have taken the coordinate of Cα only as the representative of the concerned residue for reduction of the computational cost. For detection of a point 'Z' on the Z-axis of ICS, we measured the Euclidean distance of the Cα atoms from C. The point placed at the maximum distance from C was taken as the point 'Z' and the line CZ as the Z-axis of ICS (Fig. 1).

Fixing the X-axis of the ICS: After fixing the Z-axis we fixed the X-axis. For this, first we drew a plane that is perpendicular to line CZ and passes through point C (Fig. 2). Then the perpendicular distance from the plane to the Cα atom of each residue was calculated. Then the Cα atom within a distance of 2 Å from the drawn plane was considered (shown in Fig. 3 as a lamellar strip). Among these points within the lamellar strip, the point that is placed at the maximum distance from the CG is taken as the point T. The foot of the perpendicular drawn from T to the plane, '*normal to CZ and passing through C*,' was referred as point X, where line CX was taken as the X-axis of the ICS for that protein (shown in Fig. 3).

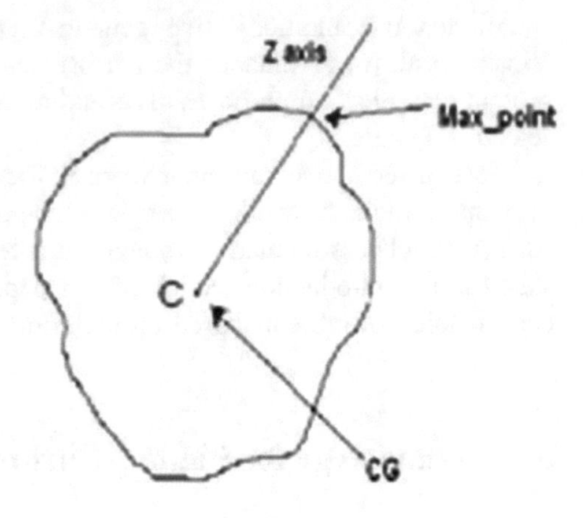

Fig. 1 Selecting the Z-axis

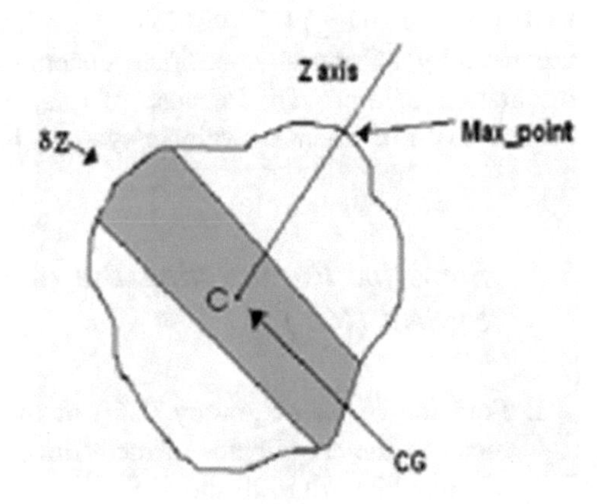

Fig. 2

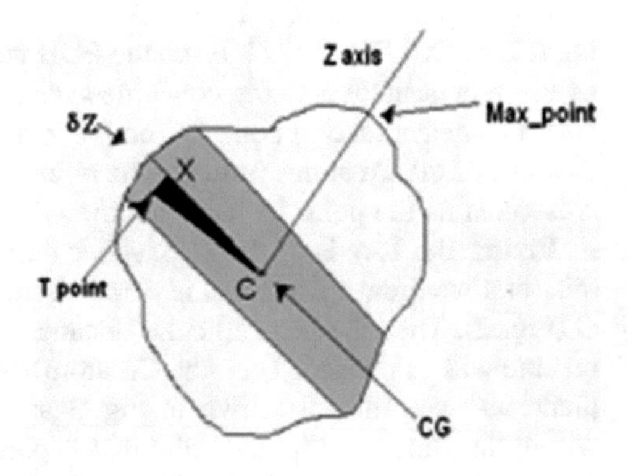

Fig. 3 Selecting the X-axis

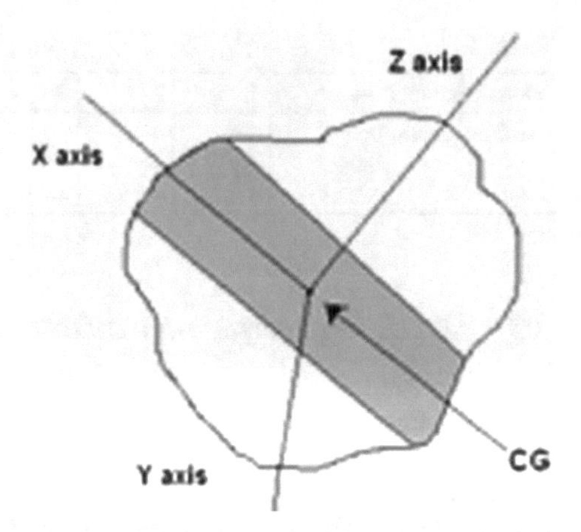

Fig. 4 Selecting the Y-axis

Fixing the Y-axis of ICS: Once the X-axis and Z-axis of ICS were fixed, the Y-axis was naturally selected as the line passing through the origin and perpendicular to both the Z-axis and X-axis of ICS. Graphically it can be represented as shown in Fig. 4.

After formulating the ICS, we transformed the coordinates of the protein from the CCS of the PDB to ICS.

3.2 *Detection of Surface Residue from the Invariant Coordinates of the PDB Residue*

We have developed a simple new method for the detection of the surface residue of a protein from the invariant coordinate system [3].

The following steps were adopted:

Step 1: Draw a line from the origin to the surface of the protein making the angles β with the X-axis, θ with the Z-axis and γ with the Y-axis.

Step 2: Detect the residue points (say, point-set of *n* number of points {Pβi} $i = 1^n$) that are within a distance of 3.5 Å from line CP, making a cylinder of radius 3.5 Å.

Step 3: Find the distances of all the n-points referenced above from C. The maximally distant point among all these points is taken as the surface residue point.

Step 4: The above is repeated by fixing the angle β and by rotating the line CP about the X-axis.

Step 5: Steps 4 and 5 are repeated by changing the angle β from 0 to 180 degrees with a discrete increase of its value covering all the residue points.

Table 1

Octant number →	1	2	3	4	5	6	7	8
Residue attribute →	+x +y +z	+x +y −z	+x −y +z	−x +y +z	+x −y −z	−x +y −z	−x −y +z	−x −y −z

3.3 Obtaining Surface-Invariant Coordinates of Proteins

I. Divide the coordinates obtained for the surface residue into eight parts for eight different octants

Our aim is to find the SIC of different octants. Thus, for the calculation of the SIC, the coordinates for different octants should be separated. The ways in which the coordinates can be divided into eight octants for different X, Y and Z values is represented in Table 1.

II. Finding the SIC

We have taken the standard deviation of the distance of the residue from the center as the criterion for the measurement of the SIC.

For 'N' number of solid angle partitions, **SIC S = $\{s_i\}_{i=1 \text{ to } N}$** where

$$S_i = \sqrt{\frac{1}{M}\sum_{j=1}^{M}(x_j - \bar{x})^2} \tag{2}$$

and $\{x_j\}$ is the set of distances between the origin of ICS and the residues in the *j*-th octant, x bar is the mean of $\{x_j\}$ and M is the total number of residues.

4 Dataset/Materials

We have taken into consideration 32 protein families [8].

We divided the database into training and testing datasets in a ratio of approximately 70:30, i.e. if the particular family has 100 proteins, we took 70 for training purposes and 30 for testing purposes. Thus we made a training set of 368 proteins and a testing set of 152 proteins.

The following is sample SIC data for a few proteins (Table 2).

Table 2 Sample SIC dataset

O1	O2	O3	O4	O5	O6	O7	O8	Class
7.0854	9.2391	8.7488	10.6253	9.9041	7.682	9.1408	7.5573	1
8.3177	9.0952	9.3581	10.3914	9.3855	8.3609	9.1254	7.8014	1
6.1429	2.7372	3.3138	3.3391	5.78	2.7364	2.8848	2.3299	2
2.951	3.3183	4.3524	3.8183	5.2429	3.4312	4.351	3.3544	2

Table 3 Predicted classes for class 1 and 2

O1	O2	O3	O4	O5	O6	O7	O8	Predicted class
6.6639	4.8488	6.7637	6.2136	8.2353	5.1398	3.6423	6.7937	1
6.4084	8.2107	7.9878	8.6107	8.1333	7.8973	9.355	6.2889	1
7.1333	7.8147	6.7865	6.9799	7.4299	8.1397	5.6747	6.6888	3
4.8608	3.9759	4.4412	3.6137	3.8416	5.4154	3.2619	6.6444	5
5.3169	6.0456	2.9878	5.2529	5.4511	3.9913	4.2371	3.9147	2

5 Algorithm

Repeat steps 1 and 2.

1. Obtain the PDB coordinates of the protein from www.rcsb.org.
2. Convert the PDB coordinates into SIC values.
3. Prepare the database of the protein SIC values.
4. Assign each protein family a decimal class number. For example, if there are 32 classes, the class numbers would be 1, 2, 3… 32.
5. Assign each protein to a particular class value 1, 2, 3… 32.
6. Now develop the naïve Bayes classifier program.
7. Divide the SIC database into a training dataset and test dataset at a ratio of 70:30, i.e. 70% of the proteins belonging to one particular family are taken as training and 30% as testing.
8. Train the classifier with the training dataset.
9. Test the classifier performance with the test dataset.
10. Calculate the efficiency of the classifier with the following formula: Efficiency = (no. of correct predictions/total no. of test values) * 100%.
11. End.

6 Results

I. The following are the predicted classes for some test proteins (Table 3).

II.Efficiency

Total number of proteins in the test dataset = 152.
Correct family prediction by Naïve Bayes classifier = 88.
Therefore, efficiency = (88/152) * 100 = 58.9%.

7 Conclusion

We used the SIC parameter very effectively to identify a family of proteins based on the surface roughness of the protein. Most of the work done in the past considers sequence similarity as a criterion to classify proteins. However, because surface roughness plays an important role in many protein functions, we considered a parameter based on the surface roughness of the proteins and used it effectively for classification.

References

1. http://www.rcsb.org/pdb/.
2. http://scop.mrc-lmb.cam.ac.uk/scop/.
3. Connolly, M.L. 1986. Measurement of protein surface shape by solid angles. *Journal of Molecular Graphics* 4: 3–6.
4. Bandyopadhyay, S. 2005. An efficient technique for superfamilyclassification of amino acid sequences: Feature extraction, fuzzy clustering and prototype selection. *ELSEVIER Jounal of FuzzySets and Systems* 152: 5–16.
5. Vipsita, S., B.K. Shee and S.K. Rath. 2010. An efficient technique for protein classification using feature extraction by artificial neural networks IEEE India conference: Green energy, computing and communication, INDICON.
6. Wang, D., and G.B. Huang. 2005. Protein sequence classification using extreme learning machine. In *Proceedings of international joint conference on neural networks* (*IJCNN, 2005*), Montreal, Canada.
7. Brink, Henrik, Joseph W. Richards, and Mark, Fetherolf. *Real-World Machine Learning*. ISBN 9781617291920.
8. Datta, A., V. Talukdar, A. Konar, and L.C. Jain. 2009. A neural network based approach for protein structural class prediction. *Journal ofIntelligent and Fuzzy Systems* 20: 61–71.

A Vision-Based On-road Vehicle Light Detection System Using Support Vector Machines

J. Arunnehru, H. Anwar Basha, Ajay Kumar, R. Sathya and M. Kalaiselvi Geetha

Abstract Vehicle light detection and recognition for collision avoidance presents a major challenge in urban driving conditions. In this chapter, an optical flow method is used to extract moving vehicles in a traffic environment, and hue-saturation-value (HSV) color space is adopted to detect vehicle brake and turn light indicators. In addition, a morphological operation is applied to obtain the precise vehicle light region. The proposed Vehicle Light Block Intensity Vector (VLBIV) feature extraction from the vehicle light region is realized by a supervised learning method known as support vector machines (SVM). Analysis is carried out on the vehicle signal recognition system which interprets the color videos taken from a front-view video camera of a car operating in traffic scenarios. This technique yields average accuracy of 98.83% in SVM (RBF) in 36 VLBIV features when compared to an SVM (polynomial) classifier.

Keywords Optical flow · HSV color space · Support vector machine
Performance measure

J. Arunnehru (✉) · H. Anwar Basha · A. Kumar
Department of CSE, SRM University (Vadapalani Campus), Chennai, India
e-mail: arunnehru.aucse@gmail.com

H. Anwar Basha
e-mail: anwar.mtech@gmail.com

A. Kumar
e-mail: ajayku87@gmail.com

R. Sathya · M. Kalaiselvi Geetha
Department of CSE, Annamalai University, Annamalainagar, Chennai, India
e-mail: rsathyamephd@gmail.com

M. Kalaiselvi Geetha
e-mail: geesiv@gmail.com

A. N. Krishna et al. (eds.), *Integrated Intelligent Computing, Communication and Security*, Studies in Computational Intelligence 771,
https://doi.org/10.1007/978-981-10-8797-4_13

1 Introduction

In recent years, advanced driver assistance systems for moving vehicles have emerged as a promising area of research in computer vision, given their significant implications for driver safety. Tracking a moving vehicle in a video sequence is essential in a variety of applications including video-based traffic surveillance, driver assistance systems and advanced vehicle safety alarms. Vehicle detection in video image sequences is an important component of driver assistance systems. Video frames contain measurement noise for a number of reasons, such as camera movement, cluttered background, vehicle movement, low visibility conditions and other obstacles in the video scene. The installation and use of the camera, and the method for detecting vehicles and vehicle lights can be either static or dynamic. In this work, the surveillance camera is fixed in the vehicle to obtain important information about the road traffic. The styles of vehicle lights are different; the most commonly used are brake lights and left and right turn indicators. In a traffic environment, the vehicle must respond to the instruction of vehicle lights in the forward visual field. Vehicle features may contain significant information about the vehicle rear end, such as plate number, vehicle logo and vehicle lights. Vehicle logos and number plates contain relevant information for vehicle identification and verification. In general, vehicle number plates are centrally positioned on the front and rear of the vehicle, whereas the position of the logo differs among various manufactures. This study is aimed at vehicle detection and vehicle light recognition. A novel vehicle detection algorithm is applied by using optical flow extraction and recognition of a vehicle light based on Vehicle Light Block Intensity Vector(VLBIV) features for collision avoidance.

1.1 Related Research

Moving vehicle detection and recognition methods have been investigated for many years. Several vision based techniques have been proposed in the literature [1]. Vehicle classification systems [2–5] comprise an essential area of research for video and traffic surveillance, intelligent transportation and security. Reference [6] proposed the use of an optical flow method for detecting and interpreting changes in the environment, from the viewpoint of a camera. References [7, 8] presents a vehicle detection method using multiple spectral vehicle features, such as Harr-like and edge features, for detecting vehicles. A recognition algorithm [9] is used to measure and to get vital information regarding turn lights for an intelligent vehicle. The vehicle tail light detection algorithm [10] is proposed, in which texture features are used to detect vehicle tail lamp area. Robust and lightweight detection of observed signals in the vehicle ahead [11] is critical, especially in autonomous vehicle applications such as turn signals and brake lights.

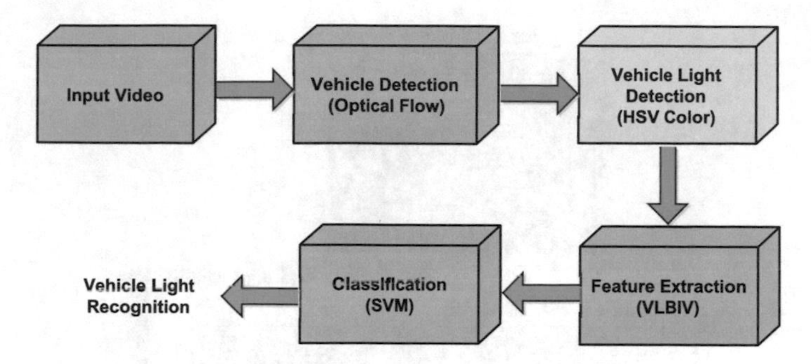

Fig. 1 Overview of the proposed approach

2 Proposed Work

This section presents a method for vehicle detection and vehicle light recognition for traffic video surveillance. The publicly available dataset for vehicle light detection is limited. In this work, datasets were recorded at 25 frames per seconds on state highways in order to evaluate our proposed method. The optical flow technique is employed to detect vehicles, and the hue-saturation-value (HSV) color space model is used to detect vehicle signal lights in order to extract the VLBIV features. The overview of the proposed approach is shown in Fig. 1. The remainder of this chapter is organized as follows. Section 3 describes the experimental setup. Finally, Sect. 4 concludes the chapter.

2.1 Vehicle Detection Using Optical Flow

Optical flow is used in the detection of objects that are currently moving in a scene. Object detection and tracking in video sequences is related to numerous problems including traffic surveillance, video surveillance, motion capture and vehicle tracking. An optical flow method computes the displacement of the pixel intensity from the current frame to the next consecutive frame. The main objective is to track the contour of a vehicle in each input frame. The video sequence is rich with flexible information. The main drawback of a computer vision based approach for vehicle detection is the fact that each frame is evaluated individually. There is no correlation between consecutive frames. A typical video sequence contains measurement noise due to traffic environments and camera calibration, so preprocessing steps are introduced to remove the noise while enhancing the computational time. In order to remove the noise, the color images are converted into grayscale images and a Gaussian-filtering algorithm is applied. The proposed work is evaluated by comparing two consecutive images. The movement of a pixel within two consecutive frames is not very significant. The calculation of the Lucas-Kanade algorithm [12,

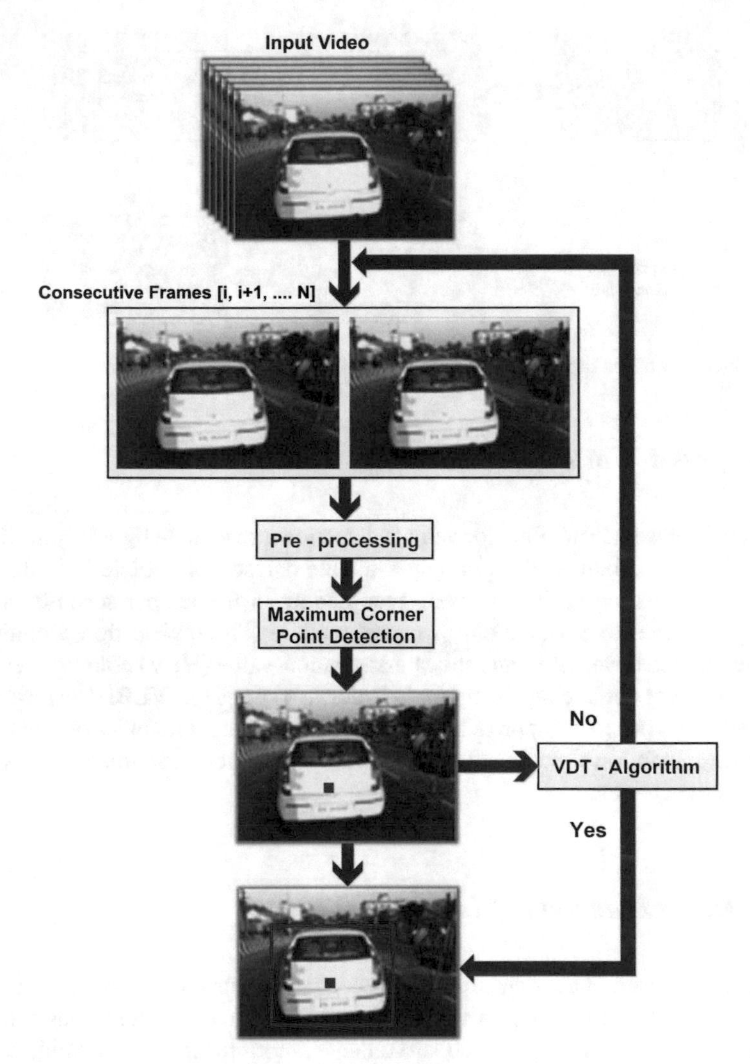

Fig. 2 Vehicle detection using the optical flow method

13] is chosen because independent investigation is found to be the most reliable algorithm for tracking. The proposed algorithm, vehicle detection and tracking (VDT) is used to locate the feature point at the *n*th frame. Figure 2 shows the flowchart for the vehicle detection method based on optical flow point features. Table 1 represents the algorithm for the vehicle detection and tracking (VDT) method. The algorithm is used to draw the bounding box around the vehicle, as shown in Fig. 2. The detected vehicle image is fixed to be of size 240×240 for future analysis without any loss of information, as described in Table 1.

Table 1 Algorithm for vehicle detection and tracking (VDT)

Algorithm 1: Vehicle Detection and Tracking (VDT)

Input: Image $IM(i, j)$, optical flow feature point height $p0.x$, optical flow feature point width $p0.y$, image height IM_h, image width IM_w, rectangle $v.r(i, j)$, rectangle height $v.r_h$, rectangle width $v.r_w$. $IM_h = 480$, $IM_w = 640$.
Output: Draw rectangle and track vehicles in video sequences.

1. Compute optical flow feature point, image $IM(i, j) = (p0.x, p0.y)$.
2. If $(p0.x \geq 30 \ \&\& \ p0.x \leq 390 \ \&\& \ p0.y < 350)$
   - Evaluate rectangle position in image $v.r(i, j) = p0.x - 30, p0.y - 80, p0.x - 110, p0.y - 130$
   - Resize of $v.r(i, j) = size(v.r_h = 240, v.r_w = 240)$
3. else if $(p0.x \geq 390 \ \&\& \ p0.x \leq 430 \ \&\& \ p0.y \leq 350)$
   - Evaluate rectangle position in image $v.r(i, j) = p0.x - 60, p0.y - 50, p0.x - 250, p0.y - 150$
   - Resize of $v.r(i, j) = size(v.r_h = 240, v.r_w = 240)$
4. else if $(p0.x \geq 480 \ \&\& \ p0.x \leq 500 \ \&\& \ p0.y \leq 350)$
   - Evaluate rectangle position in image $v.r(i, j) = p0.x - 170, p0.y - 60, p0.x - 250, p0.y - 180$
   - Resize of $v.r(i, j) = size(v.r_h = 240, v.r_w = 240)$
5. else $(p0.y \geq 350 \ \&\& \ p0.y \leq 640)$
   - Evaluate rectangle position in image $v.r(i, j) = p0.x - 130, p0.y - 180, p0.x - 200, p0.y - 230$
   - Resize of $v.r(i, j) = size(v.r_h = 240, v.r_w = 240)$
   - End if
6. Vehicle detection $v.r(i, j)$ and Extraction.
7. Go to step 2.

2.2 *Vehicle Signal Detection*

With the increasing demand for traffic rules and the prevalence of vehicle light detection in traffic surveillance, there is a great need for technologies that are able to automatically recognize vehicle signal indication types. We found that red-green-blue (RGB) values could not represent the preferred color range for the color segmentation shown in Fig. 3a. The additional natural and sensible color space for this problem is the hue-saturation-value (HSV) color space, which is a more practical and natural color space. HSV can be represented as a reversed cone, with hue as the angle, satu-

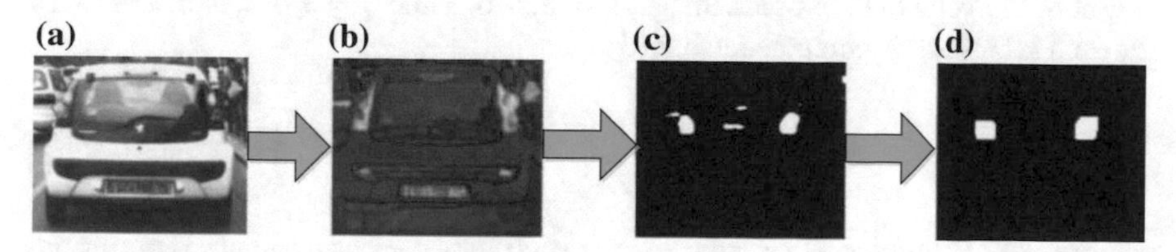

Fig. 3 Vehicle detection using HSV color space

ration as the radius and value as the height. Vehicle signal information $X_t(i,j)$ shown in Fig. 3c or vehicle light detection image is calculated using

$$X_t(i,j) = \begin{cases} 1, & \text{if } D_t(i,j) > HSV_{range}; \\ 0, & \text{otherwise}; \end{cases} \tag{1}$$

$$HSV_{range} = \begin{cases} 64 \geq H \leq 178 \\ 80 \geq S \leq 253 \\ 98 \geq V \leq 256 \end{cases} \tag{2}$$

where $X_t(i,j)$ is the intensity value of the pixel at (i,j) in the kth frame. The parameters of the HSV range color space are displayed in Eq. 2. After detecting the vehicle, light images often have distinct noise in detected vehicle areas. Morphological operations including erosion and dilation are employed to remove the isolated noise in the vehicle image shown in Fig. 3d.

2.3 Vehicle Light Block Intensity Vector (VLBIV)

Vehicle light block intensity is the most essential feature of vehicle recognition and is used to express brake and indicator information. The key idea behind the block division of the vehicle image into two equal halves is to precisely segment the vehicle light with the left turn light on the left part of vehicle image, and the right turn light on the right part. For the purpose of uniformity, vehicle image is considered to be of size 240×240 without any loss in information. The procedure for generating VLBIV is described as follows.

The $w \times h$ vehicle image is divided into two blocks, each $w/2 \times h$ in size, where w and h represent the width and height of the block image, respectively. The block is 120×240 in size, as shown in Fig. 4. Each block is further divided for analysis of the VLBIV feature extraction procedure. The $W \times H$ vehicle light block image is divided into $k \times q$, each $W/k \times H/q$ in size, where W and H represent the width and height of the vehicle light block image, $k = 3, 4, 6, 5$ and $q = 3, 4, 3, 5$. The VLBIV for each sub-block is computed using (3).

$$\frac{\sum_{i,j}^{r,c} Count \ \ (p(i,j) \geq 10)}{r \times c} \tag{3}$$

where $r = W/k$ and $c = H/q$ and r, c is the each sub-block size and $p(i,j)$ represents the pixel intensity value of each sub-block. $r \times c$ is the total number of pixels in each sub-block, where *count* is the total number of pixels whose value is higher than 10 in each block. The average intensity value of pixels in each sub-block is computed as a vehicle light block intensity vector. In this work 18, 32, 36 and 50 dimension

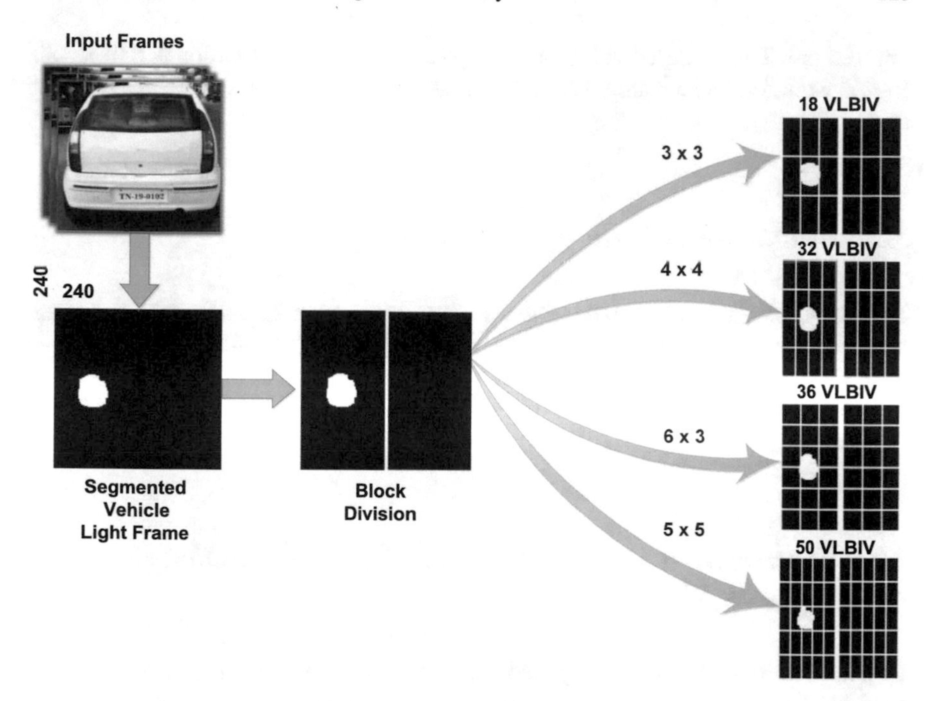

Fig. 4 Edge block intensity vector (EBIV) feature extraction

VLBIV features are extracted for carrying out the experiments and are modeled with an SVM classifier. The VLBIV feature extraction procedure is shown in Fig. 4.

3 Experimental Setup

In this section, the experiments are conducted to evaluate the performance of the proposed VLBIV features for vehicle signal indication recognition. The segmented vehicle light features are trained and tested with an SVM classifier (RBF and Polynomial) [14] for recognition.

3.1 Dataset

A large collection of vehicle signal indication videos were collected from Chidambaram urban roads, and Namakkal highways and urban roads for traffic surveillance. The videos were recorded in daytime over different time periods between 6 am to 6 pm. The recorded videos are 25 fps and the video resolution is 640×480. In this work, the videos are taken into a training set at 30 min and testing set at 45 min.

3.2 Evaluation Criteria

This work presents a systematic analysis of multi-class classification systems. Standard evaluation metrics including precision (P) $= TP/(TP + FP)$, recall (R) $= TP/(TP + FN)$, specificity (S) $= TN/(TN + FP)$ and F-measure $= 2PR/(P + R)$ are used to evaluate the performance of the proposed system. These measures provide the best perspective on a classifier's performance for classification.

3.3 Experimental Results for Vehicle Light Recognition Using an SVM Classifier

To exhibit the success of the proposed method for vehicle signal indication recognition, we conducted comprehensive experiments on the collected datasets. Figure 5i Table shows the confusion matrix for vehicle signal indication recognition using 18 VLBIV features with the SVM RBF kernel. Figure 5ii Table shows the confusion matrix for vehicle signal indication recognition using 32 VLBIV features with the SVM RBF kernel. Figure 5iii Table shows the confusion matrix for vehicle signal indication recognition using 36 VLBIV features with an SVM RBF kernel. Figure 5iv Table illustrate the confusion matrix for vehicle signal indication recognition with an SVM RBF kernel using 50 VLBIV features. Further, we consider an experiment to estimate the robustness of the proposed approach to precision, recall and specificity for vehicle signal indication recognition due to brake lights, and left and right lights of the vehicle. Table 2 indicates precision, recall and specificity values for vehicle light recognition with the SVM RBF kernel using 18, 32, 36 and 50 VLBIV features.

18	Brake	Left	Right
Brake	100	00	00
Left	17.00	75.00	08.00
Right	07.00	00	93.00

(i) 18 VLBI

32	Brake	Left	Right
Brake	100	00	00
Left	07.00	73.00	00
Right	08.50	00	91.50

(ii) 32 VLBI

36	Brake	Left	Right
Brake	100	00	00
Left	03.00	96.50	01.50
Right	00	00	100

(iii) 36 VLBI

50	Brake	Left	Right
Brake	100	00	00
Left	06.00	94.00	00
Right	25.00	00	75.00

(iv) 50 VLBI

Fig. 5 Performance comparison of vehicle signal indication recognition using different VLBIV features

Table 2 Precision, recall and specificity values for vehicle light recognition with an SVM RBF kernel

	18 VLBIV features			32 VLBIV features			36 VLBIV features			50 VLBIV features		
Vehicle signals	P (%)	R (%)	S (%)	P (%)	R (%)	S (%)	P (%)	R (%)	S (%)	P (%)	R (%)	S (%)
Brake	100	80.64	92.25	100	86.95	88	100	97.08	96.59	100	76.33	84.5
Left	75.00	100	100	93.00	100	100	96	100	100	94.00	100	100
Right	93.00	92.07	100	91.91	100	96	100	99.01	99.27	75.00	100	100

Table 3 Precision, recall and specificity for vehicle light recognition with an SVM polynomial kernel

	18 VLBIV features			32 VLBIV features			36 VLBIV features			50 VLBIV features		
Vehicle signals	P (%)	R (%)	S (%)	P (%)	R (%)	S (%)	P (%)	R (%)	S (%)	P (%)	R (%)	S (%)
Brake	89	83.7	91.5	84	81.6	90.5	93	93	96.3	87	83.7	91.3
Left	71	73.9	87.5	83	76.1	87	89.9	93.7	96.8	82	84.5	92.5
Right	78	79.6	90	74	84.1	93	94.9	91.3	95.5	85.9	86.7	93.5

Table 3 provides the precision, recall and specificity values for vehicle light recognition with an SVM polynomial kernel using 18, 32, 36 and 50 VLBIV features.

The performances of the four types of VLBIV feature dimensions 18, 32, 36 and 50 for the vehicle signal classification. The average F-measure values obtained for 18, 32, 36 and 50 VLBIV feature dimension were 79.33, 80.33, 92.61, 84.95 and 89.33, 94.88, 98.83, 89.67 for the SVM polynomial and the RBF kernel, respectively. The highest recognition rate obtained for the RBF kernel was 98.83% and for the polynomial kernel was 92.61% on 36 VLBIV features for vehicle signal indication.

4 Conclusion and Future Work

Intelligent driver assistance systems require dynamic brake and turn light signal information to make autonomous decisions. The optical flow feature points are identified as vehicle position, and the vehicles are extracted. The vehicle light detection of brake and indicators use an HSV color space. Vehicle Light Block Intensity (VLBIV) features are extracted to train and test with SVM (Polynomial and RBF) classifiers. The proposed method yields average accuracy of 98.83% in SVM (RBF) for 36 VLBIV features when compared to an SVM (polynomial) with 18, 32, 36 and 50 VLBIV features. Future work will explore the effectiveness of VLBIV features using other classification techniques for vehicle signal indication recognition.

References

1. Arunnehru, J., and M. Kalaiselvi Geetha. 2013. Motion intensity code for action recognition in video using PCA and SVM. In *Mining Intelligence and Knowledge Exploration*, 70–81. Cham: Springer.
2. Lu, Jianbo, Hassen Hammoud, Todd Clark, Otto Hofmann, Mohsen Lakehal-ayat, Shweta Farmer, Jason Shomsky, and Roland Schaefer. 2017. A System for Autonomous Braking of a Vehicle Following Collision. No. 2017–01–1581. SAE Technical Paper.
3. Zhang, Bailing, Yifan Zhou, Hao Pan, and Tammam Tillo. 2014. Hybrid model of clustering and kernel autoassociator for reliable vehicle type classification. *Machine Vision and Applications* 25 (2): 437–450.
4. Almagambetov, Akhan, Senem Velipasalar, and Mauricio Casares. 2015. Robust and computationally lightweight autonomous tracking of vehicle taillights and signal detection by embedded smart cameras. *IEEE Transactions on Industrial Electronics* 62 (6): 3732–3741.
5. Hillel, Aharon Bar, Ronen Lerner, Dan Levi, and Guy Raz. 2014. Recent progress in road and lane detection: a survey. *Machine Vision And Applications* 25 (3): 727–745.
6. Nourani-Vatani, Navid, Paulo Vinicius Koerich Borges, Jonathan M. Roberts, and Mandyam V. Srinivasan. 2014. On the use of optical flow for scene change detection and description. *Journal of Intelligent & Robotic Systems* 74 (3–4): 817–846.
7. Kim, Giseok, and Jae-Soo Cho. 2012. Vision-based vehicle detection and inter-vehicle distance estimation for driver alarm system. *Optical Review* 19 (6): 388–393.
8. Jen, Cheng-Lung, Yen-Lin Chen, and Hao-Yuan Hsiao. 2017. Robust detection and tracking of vehicle taillight signals using frequency domain feature based adaboost learning. In *IEEE International Conference on Consumer Electronics-Taiwan (ICCE-TW)*, 423–424.
9. Li, Yi, Zi-xing Cai, and Jin Tang. 2012. Recognition algorithm for turn light of front vehicle. *Journal of Central South University* 19: 522–526.
10. Tong, Jian-jun and Zou Fu-ming. 2005. Speed measurement of vehicle by video image. *Journal of Image and Graphics* 10(2): 192–196.
11. Casares, Mauricio, Akhan Almagambetov, and Senem Velipasalar. 2012. A robust algorithm for the detection of vehicle turn signals and brake lights. In *IEEE ninth international conference on advanced video and signal-based surveillance (AVSS)*, 386–391.
12. Song, Hua-jun, and Mei-li Shen. 2011. Target tracking algorithm based on optical flow method using corner detection. *Multimedia Tools and Applications* 52 (1): 121–131.
13. Enkelmann, Wilfried. 1988. Investigations of multigrid algorithms for the estimation of optical flow fields in image sequences. *Computer Vision, Graphics, and Image Processing* 43 (2): 150–177.
14. Tom, Mitchell. 1997. *Machine learning*. McGraw-Hill computer science series.

A Novel Approach for Community Detection Using the Label Propagation Technique

Jyoti Shokeen, Chhavi Rana and Harkesh Sehrawat

Abstract In this chapter, we propose a new label propagation-based approach to detect community structure in social networks. This is a multiple label propagation technique in which a node can obtain labels of different communities, which allows researchers to discover overlapping communities. One important advantage of this approach is the updating of node labels over time that makes it dynamic. Given an underlying social network, we assume that each node receives a unique label id similar to its node id in the initial phase. We allow each node to accept multiple labels from its neighbors if each of the neighbors has a high common neighbor score, which naturally encompasses the idea of *overlapping communities*.

1 Introduction

The goal of social network analysis is to investigate and analyze the relationships between individuals or groups in a network with the aim of determining how these relationships arise and decline, and the consequences that emerge [1]. The widely studied research sphere of social network analysis is to comprehend the structure of networks in terms of communities, as the detection of communities helps us gain insights into the network. In the literature, a general definition of community within social networks is lacking. Several authors have attempted to define community, but the definition varies with the algorithms. In context of social networks, a

J. Shokeen (✉) · C. Rana · H. Sehrawat
University Institute of Engineering and Technology, Rohtak, Haryana, India
e-mail: jyotishokeen12@gmail.com

C. Rana
e-mail: chhavi1jan@yahoo.com

H. Sehrawat
e-mail: sehrawat_harkesh@yahoo.com

A. N. Krishna et al. (eds.), *Integrated Intelligent Computing, Communication and Security*, Studies in Computational Intelligence 771,
https://doi.org/10.1007/978-981-10-8797-4_14

community is a group of people sharing similar interests who communicate frequently with others in the group. In broad terms, a community has many intra-links and fewer inter-links. Generally, there exist two types of communities in social networks: disjoint communities and overlapping communities. When a node in a network belongs to only a single community, it is referred to as a disjoint community. When the nodes in a network belong to more than one community, the communities are termed as overlapping communities. There is a considerable amount of literature on community detection algorithms for social networks [2–9, 11, 12, 14]. Graph partitioning, clustering, label-propagation, clique, game theory and genetic algorithms are some of the techniques that have been used detect communities in social networks. Of these, label propagation-based algorithms have been proved to be the fastest [9].

The chapter is organized as followings. The first section gives a brief overview of community definitions in social networks. The second section examines the related work in detecting community structure using label propagation algorithms. A new methodology with some definitions is described in the third section. Our conclusions are drawn in the final section.

2 Related Work

Several community detection techniques have been implemented in the past 10 years. Many of these algorithms fail to detect communities in large networks or they exhibit issues related to complexity of time and space. An algorithm recently proposed by [9] gives promising results in terms of complexity in linear time. Label propagation is a community detection technique in which each node possesses a label that denotes the community to which it belongs. A node acquires the label given to most of its neighboring nodes. Prior knowledge about communities is not needed in label propagation-based algorithms. This section briefly explains these algorithms.

Raghavan et al. [9] were the first to propose label propagation algorithm that uses only network structure to detect both disjoint and overlapping communities. The algorithm is computationally inexpensive and takes approximately linear time, $O(m)$, where m is the number of edges. Unfortunately, it has some downsides, the first of which is the random initialization of node labels that results in unpredictable performance. Secondly, it follows a greedy approach for updating labels to produce tremendously large communities, due to which small communities are wiped out. Recently, many researchers have devised algorithms to address the problem of the original label propagation algorithm. Gregory [2] proposed the Community Overlap Propagation Algorithm (COPRA), which is an extended version of the original algorithm proposed by [9], but it was the earliest label propagation-based algorithm that detected overlapping communities. It permits each node to update its coefficients by calculating the average coefficients of the neighbor nodes, synchronously at each time step. The parameter v monitors the highest number of communities to which nodes can relate. However, this parameter is node independent. When some nodes in the network belong to a small number of communities and some nodes belong to a large number of communities, then it becomes tough for the algorithm to find an appropriate value of v that complies with both types of nodes simultaneously. The time per iteration for the algorithm is $O(vmlog(vm/n))$, and time taken by the initial and last steps is $O(v(m + n) + v^3n)$, where n, m and v refer to the number of nodes, edges and maximum number of communities per node, respectively. COPRA also works well for bipartite and weighted networks. The Speaker-Listener Label Propagation Algorithm (SLPA) [12] is another advancement over LPA [9] for discovering overlapping structures in social networks. Unlike the original LPA and COPRA, where the nodes forget the knowledge acquired in the preceding iterations, SLPA gives memory to each node to store received labels. The time complexity of SLPA is $O(Tm)$, where T and m denote the maximum number of iterations and total edges, respectively. By generalizing the interaction rules, SLPA can also fit for weighted and directed networks.

3 Proposed Work

Unlike other algorithms [9] in which nodes are initially given unique labels, we use quite a different strategy to assign labels to the nodes in the initial stage. We propose an Overlapping Multiple Label Propagation Algorithm (OMLPA) that assigns multiple labels to the nodes in the network to allow the detection of overlapping communities. A community is made of densely interconnected nodes, which implies that nodes sharing connections with a large number of nodes are likely to belong to the same community. The common neighbor similarity (CNS) [10], which is based on the Jaccard index, is used to detect the prime communities. The common neighbor (CN) is extensively used in social networks to find common neighbors between any two nodes. Nonetheless, the majority of algorithms like in [11] centered on CN assess common neighbors within one hop only. As social networks are dynamic in

Table 1 Notations used in algorithm

Notation	Description
v_i	Node i
$\Gamma(v_i)$	Set of neighbors of node i
$C(v_i)$	Community of node v_i
$C(v_i)(t)$	Community of node v_i at time t
$CN_t(v_i, v_j)$	Set of common neighbors of nodes v_i and v_j
$v_i.free$	Node i not belonging to any core
$L_t(v_i, v_j)$	Closeness between neighbors v_i and v_j at time t
$D_t(v_i)$	Change in degree of node v_i from time 1 to t
$CNS_t(v_i, v_j)$	Common Neighbor Score of nodes v_i and v_j at time t

nature, and the users (often known as nodes) change degree with time. Here, we use the concept of change in degree defined by Yao et al. [13]. Within a time interval $(1, 2, \ldots, t, \ldots, T)$, the change in degree due to influence by common neighbors is defined as follows:

$$D_t(v_c) = \frac{1}{\sum_{t=2}^{T} d_{t-1,t}/\Delta T},$$

where T and ΔT indicate the present time and the time interval between 1 and T, and $d_{t-1,t}$ represents the Euclidean distance of node v_c from time $t-1$ to t. At time T, node v_c is the common neighbor of node v_i and node v_j (Table 1).

Algorithm 1 Overlapping Multiple-Label Propagation Algorithm

Require: $G(v, e)$
Ensure: Community Structure
1: Initially propagate unique label (node id) to all nodes in network

$$C_{v_1}(0) = v_1, C_{v_2}(0) = v_2, \ldots\ldots., C_{v_n}(0) = v_n$$

2: Sort all nodes in decreasing order of their degree in the node set.
3: Set t=1
4: For each v_i in the node set
5: **if** $D(v_i) \geq 3$ and $v_i.free = true$ **then**
6: find v_j with the largest degree in $\Gamma[v_i]$ **and** $v_j.free = true$
7: **end if**
8: Compute $CN_t(v_i, v_j)$ at time t
9: Find $L_t(v_i, v_j)$ between the nodes in the set $CN_t(v_i, v_j)$
10: Calculate $D_t(v_c)$ of each common neighbor v_c in $CN_t(v_i, v_j)$
11: Compute $CNS_t(v_i, v_j)$ as follows:

$$CNS_t(v_i, v_j) = L_t(v_i, v_j). \sum_{v_m \in CN_t(v_i, v_j)} D_t(v_c)$$

12: **if** two nodes have equal highest CNS **then**
13: assign them to one group.
14: **else** $\{CNS(v_j) > CNS(v_k)\}$
15: then assign label of v_j to v_k.
16: **end if**
17: **if** a node has more than one neighbor with the same highest CNS **then**
18: then the node adopts multiple labels of all such nodes.
19: **end if**
20: **repeat**
21: $\dfrac{\text{edges between X and Y}}{\text{vertices in Y}} > \mu \times \dfrac{\text{edges in X}}{\text{vertices in X}}$ {Merge communities X and Y}
22: **until** each node possesses the same label as most of its neighboring nodes
23: **if** no node changes its community further **then**
24: algorithm stops
25: **else**
26: set $t = t + 1$ and **goto** step 3
27: **end if**

The algorithm begins with assigning unique labels to every node in the network. The nodes are then arranged in decreasing order of their degree in the node set. A node v_i with degree (≥3) is chosen randomly from the network, provided that the node is free, i.e., the node is not yet assigned to any community. Then, a free node j with the largest degree in the neighborhood of node v_i is chosen. Both nodes i and j are added to the core. Then, the algorithm finds the common neighbors of nodes v_i and v_j. $L_t(v_i, v_j)$ calculates the closeness between the common neighbors of nodes v_i and v_j at time t. In each iteration, the degree $D_t(v_c)$ of each of the common neighbors of these nodes is calculated. The common neighbor score (CNS) of the nodes is calculated, which is computed as follows:

$$CNS_t(v_i, v_j) = L_t(v_i, v_j). \sum_{v_m \in CN_t(v_i, v_j)} D_t(v_c).$$

When two nodes in the network possess same highest CNS, then the nodes are assigned to the same group. But if the CNS of a node is greater than the CNS of another node, then the node with the smaller CNS attains the label of the node with the higher CNS. Also, if a node possesses more than one neighbor with the same highest CNS, then the node acquires more than one label of all such nodes. A situation may arise in a network where many small communities develop, making it more difficult to analyze the networks. In such a case, the algorithm imposes a condition on the formed communities in which smaller communities merge into larger communities, until each node acquires the label possessed by most of its neighboring nodes.

4 Conclusion and Future Work

With the revolution in analyzing social networks, it is quite intuitive to detect communities to gain insights into a network. In this chapter, we have proposed a label propagation-based community detection technique that uses an updating strategy of degree of nodes with time to fit real-world networks. The algorithm detects both disjoint and overlapping communities. Our method is a clear advance on current methods of detecting community structure. Due to the space issue, we aim to implement the algorithm on real-world networks as our future work. To further our research we plan to extend the algorithm for bipartite networks and weighted networks.

Acknowledgements The first author wishes to say thanks to the Council of Scientific and Industrial Research (CSIR) for financial assistance received in the form of JRF.

References

1. Breiger, R.L. 2004. The analysis of social networks. In *Handbook of Data Analysis*, ed. M. Hardy, and A. Bryman, 505–526. London: SAGE Publications.
2. Gregory, S. 2010. Finding overlapping communities in networks by label propagation. *New Journal of Physics* 12 (10): 103018.
3. Havvaei, E., and N. Deo. 2016. A game-theoretic approach for detection of overlapping communities in dynamic complex networks. 1: 113–324. arXiv:1603.00509.
4. Hu, W. 2013. Finding statistically significant communities in networks with weighted label propagation. *Social Network* 2: 138–146.
5. Lancichinetti, A., and S. Fortunato. 2010. Community detection algorithms: A comparative analysis. *Physical Review E* 80 (5): 056117. https://doi.org/10.1103/PhysRevE.80.056117.
6. Newman, M.E., and G. Reinert. 2016. Estimating the number of communities in a network. *Physical Review Letters* 117–125.
7. Newman, M.E., and M. Girvan. 2004. Finding and evaluating community structure in networks. *Physical Review E* 69 (2): 026113.
8. Pizzuti, C. 2009. Overlapped community detection in complex networks. In *Proceedings of the 11th Annual Conference On Genetic And Evolutionary Computation*, 859–866, ACM.
9. Raghavan, U.N., R. Albert, and S. Kumara. 2007. Near linear time algorithm to detect community structures in large-scale networks. *Physical Review E* 76: 36106.
10. Steinhaeuser, K., and N.V. Chawla. 2008. Community detection in a large real-world social network. *Social Computing, Behaviora-Cultural Modeling Prediction*, 168–175. Berlin: Springer.
11. Wu, Z.H., Y.F. Lin, S. Gregory, H.Y. Wan, and S.F. Tian. 2012. Balanced multi-label propagation for overlapping community detection in social networks. *Journal of Computer Science Technology* 27: 468–479.
12. Xie, J., and B.K. Szymanski. 2012. Towards linear time overlapping community detection in social networks. In *Advanced Techniques in Knowledge Discovery Data Mining*, ed. P.-N. Tan, S. Chawla, C.K. Ho, and J. Bailey, 25–36. Berlin: Springer.
13. Yao, L., L. Wang, L. Pan, and K. Yao. 2016. Link prediction based on common-neighbors for dynamic social network. *Procedia Computer Science* 83: 82–89.
14. Zhou, L., K. Lu, P. Yang, L. Wang, and B. Kong. 2015. An approach for overlapping and hierarchical community detection in social networks based on coalition formation game theory. *Expert Systems with Applications* 42: 9634–9646.

Learner Level and Preference Prediction of E-learners for E-learning Recommender Systems

D. Deenadayalan, A. Kangaiammal and B. K. Poornima

Abstract An effective e-learning system must identify learning content appropriate for the needs of the specific learner from among the many sources of learning content available. The recommendation system discussed here is a tool to address such competence. Identifying learner levels, and thereby identifying the appropriate learning content, is possible only if the learning content is prepared using a proven instructional strategy which covers various learner levels. Therefore, in the proposed method, the learning content is prepared using David Merrill's First Principles of Instruction, a problem-based approach that has four phases of instruction: activation, demonstration, application and integration. These four phases are used to predict learner levels for three different types of media content, namely text, video and audio, and to determine a rating. The naïve Bayes classifier assumes that the presence of a particular feature in a class is unrelated to the presence of any other feature. Learner ratings are used to predict learner preferences as to the type of content. To identify the learner's level, the learner rating and the instructional phases which they prefer most are used. The same classifier is used to identify the level and preferences of the learner. To estimate predictive accuracy, a *k*-fold cross-validation technique is used. The experimental results show that the proposed classifier yields maximum accuracy of 0.9794 and a maximum kappa statistic of 0.9687 for learning level and preference, respectively.

Keywords E-learning · Learning content · Naive bayes · David Merrill Learner preference · Learner level · *k*-Fold cross-validation

D. Deenadayalan (✉) · A. Kangaiammal · B. K. Poornima
Department of Computer Applications, Government Arts College, Salem 636007, Tamil Nadu, India
e-mail: deena.duraisamy@gmail.com

A. N. Krishna et al. (eds.), *Integrated Intelligent Computing, Communication and Security*, Studies in Computational Intelligence 771,
https://doi.org/10.1007/978-981-10-8797-4_15

1 Introduction

E-learning is facilitated through electronic devices, either standalone or over a network, using intranet, extranet, and the internet. The explosion of learning resources in the World Wide Web is increasing exponentially. Learners are facing difficulties in choosing learning resources that are suitable and relevant to their learning needs, due to information overload from the web. Accessing the right learning content for the learner is the main problem in any e-learning system.

There are educational taxonomies that are better suited for classroom teaching and learning, followed by evaluation. However, those taxonomies may not be suitable for e-learning, which differs in principle from classroom teaching and learning. E-learning systems demand a specific instructional design approach that combines the features required for addressing the above-mentioned competence. David Merrill's First Principles of Instruction (FPI) is an instructional approach that is problem-centered and object-oriented, and has been proven suitable for e-learning.

One of the most important applications of recommender systems in learning environments is material recommendation. Recommendation systems use the opinions of users to help individuals more effectively identify material and content of interest from a possibly overwhelming set of choices [1]. By using material recommender systems in learning environments, personalization and information overload are addressed. Recommender systems can overcome the information overload by filtering out irrelevant learning resources and can bring in personalization support by recommending relevant resources to the learners according to their personalized preferences.

Learning resources are the same for all learners, but different learners need different information according to their level of knowledge and learner preferences. The learner's learning level will affect the way they learn with the help of text, video, or audio support. Some learners like text content, some prefer to analyze visual content, and a few want to hear the audio content (preferable for blind learners). For identifying the learner's learning level and learning preferences, learner log data can be useful.

Classifying the content for different levels of learners and identifying learners' media preference are the most crucial elements for a successful e-learning system. The naïve Bayes classifier works with the concept of probability. Hence, the proposed system adopts this classifier to group the learners based on their level and their media preferences.

2 Review of the Literature

E-learning is an internet-based approach for enhancing knowledge and performance. With the rapid evolution of computer and internet technologies, e-learning has emerged as an important learning alternative. The learner's need for audiovisual aid and instructional material in e-learning has increased recently, and such content has been incorporated within modules to attract a learner's consideration and interest [2].

The main purposes of recommendation systems include analyzing user data and extracting useful information for further predictions. A recommender system is a software program that helps users to identify learning content that is most interesting and relevant to them from a large content selection. Recommender systems may be based on collaborative filtering (by user ratings), content-based filtering (by keywords), or hybrid filtering (by both collaborative and content-based filtering). Recommendation systems should recommend a learning task to a learner based on the tasks already performed by the learner and their successes based on the tasks [3].

The authors have worked on a recommender system for book readers. These recommendation systems help users by presenting them with the learning objects that they would be more interested in based on their known preferences. The ratings devised are as follows: not useful, poor, useful, very useful, and no value, poor value, valuable, very valuable [4]. Learners need multimedia learning content in the form of video, audio, images, or text as a learning material in e-learning. The use of multimedia-based teaching material enhances learning and increases productivity. It is therefore important that content developers not only understand the concepts behind the development of multimedia but also have a good grasp of how to implement some of the processes involved in courseware production [5].

Learners need learning aids in the form of video, audio, or text as a learning material in e-learning. Recommendation systems can produce item recommendations from a huge collection of items based on user preferences. The degree of preference is measured by a rating score. Using that rating score, learning preferences may be predicted as to whether the learner prefers text content or audiovisual content [6].

E-learning systems deliver learning content to all learners in the same way. However, different learners require different learning material according to their level of knowledge. The instructional strategy which is David Merrill's First Principles of Instruction (FPI) approach can be used for identifying the level of the learner [7, 8]. The FPI approach has four phases of instruction: activation, demonstration, application, and integration (ADAI) [9].

Activation remembers the previous knowledge or understanding and creates a learning situation for the new problem. Demonstration expresses a model of the skill required for the new problem. Application applies the skills obtained to the new problem. Integration provides the capabilities and applies the acquired skill to another new situation [10]. Any learning content can be categorized into one of those four phases. The four phases can be considered as the learning levels of the learners.

3 Proposed Work

The literature discussed above identified issues in the e-learning recommendation system such as content preparation with the use of an instructional strategy to suit different levels of learners and devising the learner preferences using learners' ratings. The proposed system architecture is presented in Fig. 1.

In Fig. 1, the learner has to understand the learning content and rate each piece of content on a scale of 1 to 5, indicating poor, below average, average, good, or very good, respectively. The content is of three different media types, namely text, video, and audio, for each of the four phases of instruction, which include activation, demonstration, application, and integration. These four phases are considered the four learning levels of the learners.

The proposed method uses the naïve Bayes classifier for predicting the learner level as well as learner preferences. For the dataset, the learners must rate the learning content according to their preferences—some prefer the textual form of content, some prefer the video form, and others prefer an audio form; thus, their rating of the content is according to their media preference. Similarly, for the learner level, some want to be at the activation level, some want demonstration, some want the application level, and some want integration; thus their rating is according to the level of learning in which they reside.

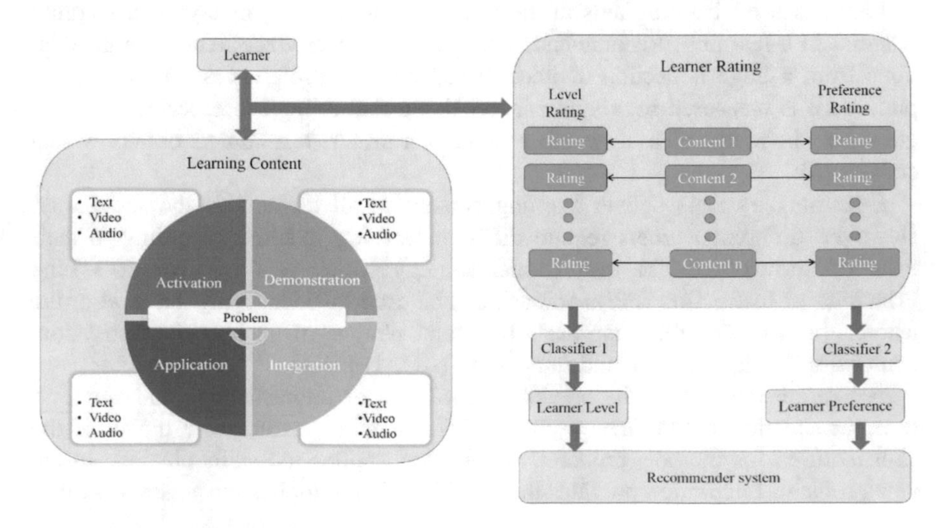

Fig. 1 Proposed system architecture

Based on the rating for level and preference given by the learners, classifiers are constructed using naïve Bayesian classification. These two classifiers for level and preference can predict the learner level and the learning content media type which will be taken as input by the recommender system. The recommender system can make use of these predictions to decide on the right learning content and learning material for the right learner.

4 Results and Discussion

The learning content is developed for 10 different topics, and content for each of the four different phases/levels, and for each phase, three different media types. In this experiment, 15 students from PG level (including three blind students) have been observed to rate the learning content.

The experimental results are observed for two different parameters, level and preference. To estimate the accuracy of classifier-1 for learning level and classifier-2 for learning preference, a *k*-fold cross- validation method is used. As cross-validation is suitable when sufficient data are available, this proposed work fixes this method. The fold parameters considered are 2, 5, 7, and 10. The maximum fold considered is 10; this is found to be adequate and accurate [11].

The cross-validation method has been executed at different resampling levels at 2, 5, 7, and 10. As shown in Table 1, the minimum accuracy in the false category is 0.8211 for 7-fold and the maximum is 0.8288 for 2-fold, whereas in the true category the minimum is 0.9261 for 2-fold and the maximum is 0.9794 for 10-fold. The kappa level for the false category minimum is 0.7260 for 7-fold and the maximum is 0.7384 for 2-fold. The true level category minimum is 0.8857 for 2-fold and the maximum is 0.9687 for both 7-fold and 10-fold. It is understood that the higher the kappa, the higher` the true positives and false negatives and thus higher accuracy. The results show that the kappa achieved is 0.9687, and hence the classifier accuracy is highly acceptable.

Table 1 Evaluation measures for learning preference

Cross-validation parameter (F = fold)	Kernel	Accuracy	Kappa
F = 2	False	0.8288	0.7384
	True	0.9261	0.8857
F = 5	False	0.8238	0.7303
	True	0.9511	0.9247
F = 7	False	0.8211	0.7260
	True	0.9794	0.9687
F = 10	False	0.8222	0.7277
	True	0.9794	0.9687

Table 2 Evaluation measures for learner level

Cross-validation parameter (F = fold)	Kernel	Accuracy	Kappa
F = 2	False	0.8216	0.7417
	True	0.9174	0.8743
F = 5	False	0.8257	0.7389
	True	0.9576	0.9189
F = 7	False	0.8289	0.7461
	True	0.9636	0.9671
F = 10	False	0.8291	0.7476
	True	0.9636	0.9671

Similar to the preference classifier, Table 2 shows that the minimum accuracy in the false category is 0.8216 and the maximum is 0.8291, whereas in the true category the minimum is 0.9174 and the maximum is 0.9636. The minimum kappa level for the false category is 0.7389 and the maximum is 0.7476, while for the true level category the minimum is 0.8743 and the maximum is 0.9671. The results indicate that the overall predictive accuracy is highly acceptable.

5 Conclusion

Learning depends mostly on the learner and the learning content, i.e. learner's learning level and the learner preference as to the form of media content. Knowing the level of the learner and understanding the type of learning of the material required are the two major tasks of any recommender system. This work has addressed the choice of these two parameters for the e-learning system. Using classification and prediction, the preferences of the learning content for the respective level of learner can be predicted. Naïve Bayes classification is used with the *k*-fold cross-validation technique. The identified learner level and preference can help any recommender system choose the learning content (simple-to-complex or lower-to-higher) to be offered with the respective learning material type such as text, audio, or video. Therefore, this work, when further refined with more specific parameters such as the nature of the problem/topic, dynamic behavior of the learner, and more, would be even more useful.

Acknowledgements This research work is supported by the University Grants Commission (UGC), New Delhi, India under Minor Research Projects Grant No. F MRP-6990/16 (SERO/UGC) Link No. 6990.

References

1. Salehi, Mojtaba, Mohammad Pourzaferani, and Seyed Amir Razavi. 2013. Hybrid attribute-based recommender system for learning material using genetic algorithm and a multidimensional information model. *Egyptian Informatics Journal* 14: 67–78.
2. Poornima, B.K., D. Deenadayalan, and A. Kangaiammal. 2017. Text preprocessing on extracted text from audio/video using R. *International Journal of Computational Intelligence and Informatics* 6 (4): 267–278.
3. Sikka, Reema, Amita Dhankhar, and Chaavi Rana. 2012. A Survey Paper on E-Learning Recommender System. *International Journal of Computer Applications* 47 (9): 0975–888.
4. Crespo, R.G., et al. 2011. Recommendation System based on user interaction data applied to intelligent electronic books. *Elsevier, Human Behavior* 27: 1445–1449.
5. Parlakkilic, and Karslioglu. 2013. Rapid multimedia content development in medical education. *Journal of Contemporary, Medical Education* 1 (4): 245–251.
6. Liu, Ying, and Jiajun Yanga. 2012. Improving ranking-based recommendation by social information and negative similarity, information technology and quantitative management (ITQM 2015). *Procedia Computer Science* 55: 732–740.
7. Poornima, B.K., D. Deenadayalan, and A. Kangaiammal. 2015. Efficient content retrieval in E-learning system using semantic web ontology. *Advances in Computer Science and Information Technology (ACSIT)* 2 (6): 503–507. ISSN: 2393-9915.
8. Poornima, B.K., D. Deenadayalan, and A. Kangaiammal. 2015. Personalization of learning objects in e-learning system using David Merrill's approach with web 3.0. *International Journal of Applied Engineering Research (IJAER)* 10 (85): 318–324.
9. Merrill, M.D., 2007. First principles of instruction: a synthesis. In *Trends and issues in instructional design and technology*, 2nd ed., 62–71.
10. Nirmala K., and M. Pushpa. 2012. Feature based text classification using application term set. *International Journal of Computer Applications* 52 (10): 0975–8887.
11. Gupta, G.K. 2006. *Introduction to data mining with case studies*. Limited: Prentice-Hall Of India Pvt.

Machine Learning-Based Model for Identification of Syndromic Autism Spectrum Disorder

V. Pream Sudha and M. S. Vijaya

Abstract Autism spectrum disorder (ASD) is characterized by a set of developmental disorders with a strong genetic origin. The genetic cause of ASD is difficult to track, as it includes a wide range of developmental disorders, a spectrum of symptoms and varied levels of disability. Mutations are key molecular players in the cause of ASD, and it is essential to develop effective therapeutic strategies that target these mutations. The development of computational tools to identify ASD originated by genetic mutations is vital to aid the development of disease-specific targeted therapies. This chapter employs supervised machine learning techniques to construct a model to identify syndromic ASD by classifying mutations that underlie these phenotypes, and supervised learning algorithms, namely support vector machines, decision trees and multilayer perceptron, are used to explore the results. It has been observed that the decision tree classifier performs better compared to other learning algorithms, with an accuracy of 94%. This model will provide accurate predictions in new cases with similar genetic background and enable the pathogenesis of ASD.

Keywords Syndromic ASD · Machine learning · Decision tree
SVM · MLP

1 Introduction

Syndromic autism spectrum disorder (ASD) represents a group of childhood neurological conditions, typically associated with chromosomal abnormalities or mutations in a single gene. It refers to conditions in which ASD occurs in conjunction with additional phenotypes and dysmorphic features. This study focuses on

V. Pream Sudha (✉) · M. S. Vijaya
PSGR Krishnammal College for Women, Coimbatore, Tamil Nadu, India
e-mail: preamsudha@psgrkc.ac.in

M. S. Vijaya
e-mail: msvijaya@psgrkc.ac.in

A. N. Krishna et al. (eds.), *Integrated Intelligent Computing, Communication and Security*, Studies in Computational Intelligence 771,
https://doi.org/10.1007/978-981-10-8797-4_16

syndromic ASD that is associated with coexisting genetic disorders including tuberous sclerosis complex (TSC), fragile X syndrome (FXS), Rett syndrome (RTT), Timothy syndrome and Phelan–McDermid syndrome (PMS).

TSC is an autosomal dominant genetic disorder characterized by benign tumors in the brain and high penetrance of ASD. It is caused by mutations in the *TSC1* or *TSC2* genes producing hamartin and tuberin, respectively. FXS is the most common monogenic cause of syndromic ASD. FXS accounts for about 2% of all ASDs and is caused by mutations in the *FMR1* gene, leading to hypermethylation of its promoter. RTT is caused by mutations in the methyl-CpG-binding protein 2 (*MECP2*) gene. Timothy syndrome is caused by mutations in the *CACNA1C* gene and is characterized by congenital heart malformations, cardiac arrhythmia, syndactyly, weakened immune system and premature death. Mutations in the *SHANK3* gene lead to PMS, characterized by epilepsy, kidney dysfunction and cardiac anomalies.

The field of machine learning holds promise for aiding humans in the analysis of ASD. A few works that have been carried out by researchers to predict or classify the genes associated with ASD are outlined below. Researchers in [1] applied several supervised learning approaches to prioritize ASD or intellectual disability (ID) disease gene candidates based on curated lists of known ASD and ID disease genes. In [2], authors explored the most important genes which are strictly associated with autism using gene expression microarrays using a two-stage ensemble system. The results combined with a genetic algorithm and support vector machine (SVM) classifier showed increased accuracy. A machine-learning approach based on a human brain-specific functional gene interaction network was used in [3] to present a genome-wide prediction of genes associated with autism. The work involved an evidence-weighted linear SVM classifier. Various other research works [4–10] have been carried out to reveal the genetics involved in ASD.

This literature survey suggests that the recognition of syndromic ASD coexisting with disorders such as RTT, FXS, TSC, PMS and Timothy syndrome can be carried out by modeling the different types of mutations that are involved in these disorders.

2 Proposed Work

The study takes into account five monogenic disorders RTT, FXS, TSC, PMS and Timothy syndrome associated with syndromic ASD, and four types of pathogenic gene mutations, namely missense, nonsense, synonymous and frameshift mutations, underlying them. The genes responsible for the syndromic ASDs under study are listed in Table 1. It is rather a complex task of classifying mutations in complex syndromic ASD genes *SHANK3*, *MECP2*, *CACNA1C*, *TSC2* and *FMR1*, taking into account the genetic variations due to synonymous and non-synonymous single-nucleotide polymorphisms underlying the autistic phenotypes. Hence, it is proposed in this research work to harness machine learning techniques to model the

Table 1 Genes associated with syndromic ASD

Gene name	Syndrome	Autism rate (%)
SHANK3	Phelan–McDermid syndrome	75
MECP2	Rett syndrome	61
CACNA1C	Timothy syndrome	60
TSC2	Tuberous sclerosis complex syndrome	36–50
FMR1	Fragile X syndrome	30–60

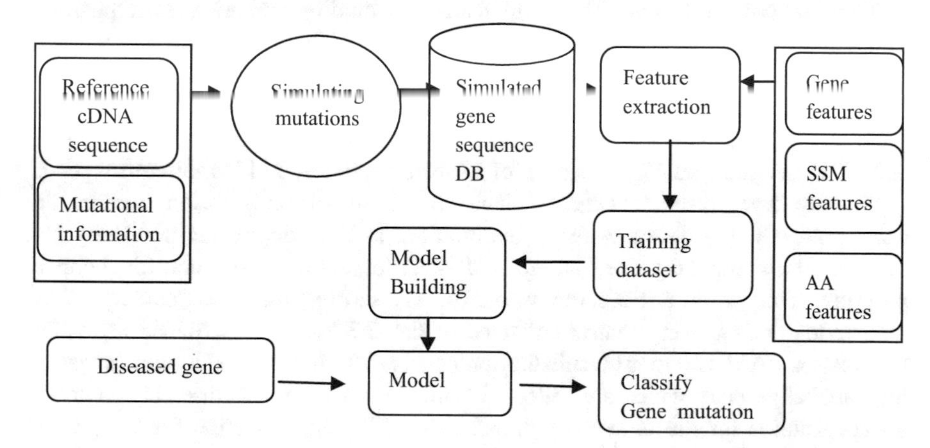

Fig. 1 Proposed framework

four types of mutations that originate the disorders associated with syndromic ASD and thereby classify ASD disease gene sequences. Initially, as shown in Fig. 1, CDNA sequences of the syndromic ASD genes *SHANK3*, *MECP2*, *CACNA1C*, *TSC2* and *FMR1* are collected from the HGMD[1] database. The mutational information about these genes is collected from the SFARI Gene[2] database.

The mutated sequences of genes are generated and then stored in a corpus, which consists of 200 sequences with 50 records for each mutation category. Consider the missense mutational information for the *SHANK3* gene such as nucleotide change 612 C > A which indicates that in the position 612 the nucleotide changes from C to A alter the protein from Asp to Glu. For example, the cDNA sequence of the *SHANK3* gene before and after nucleotide change is given below.

ATGGACGGCCCCGGGGCCAGCGCCGTGG...............GCGGCAG*C*...
ATGGACGGCCCCGGGGCCAGCGCCGTGG...............GCGGCAG*A*

[1]http://www.hgmd.cf.ac.uk.

[2]https://sfari.org.

2.1 Feature Extraction

The feature extraction step combines the gene-specific features (GS), substitution matrix features (SM) and amino acid change residues (AARC) to determine the dissimilarity between the mutations. A total of 15 attributes that describe different aspects of a mutation were investigated. These attributes can be categorized into three groups SM, GS, AARC: six features extracted from published substitution scoring matrices, five gene specific and four features related to amino acid changes.

Gene-specific features: The gene features including mutation start position, mutation end position, length of mutation, length of CDNA sequence and the type of mutational variation, i.e. substitution, insertion, deletion or duplication, are characteristics extracted. The gene-specific features for the above-mentioned *SHANK3* gene sequence will be as follows: mutation start position, 612; mutation end, 612; mutation length, 1; length of CDNA sequence, 7113; mutation type, 1.

Substitution matrix features: The values in a substitution matrix are candidate-predictive features for differentiation, as they depict the similarity and distance of a particular pair of amino acids with respect to a particular biochemical property. This work utilizes the values of six scoring matrices, namely WAC matrix, log-odds scoring matrix collected in 6.4–8.7 PAM, BLOSUM80 substitution matrix, PAM-120 matrix, substitution matrix (VTML160) and mutation matrix, for initial alignment, which are collected from the AAindex database. The mutation matrix features for the above-mentioned *SHANK3* gene sequence for the protein alteration Asp-Glu will be 2.7.

Amino acid change residues: The mutated sequences are translated to generate protein sequences that provide the amino acid observed values, and its expected value is extracted from the SFARI Gene autism database. The bigram encoding method extracts patterns of two consecutive amino acid residues in a protein sequence and counts the occurrences of the extracted residue pairs. There are 20^2 combinations of bigrams, which is huge, and so the standard deviation and the mean z-score between these values are calculated using the following formulae.

$$\text{Standard deviation} = \sqrt{\sum_{i=1\text{ to }400} (x-\mu)/(n-1)}$$

$$\text{Mean } z-\text{score} = \sum_{i=1\text{ to }400} ((x-\mu)/\sigma)/n$$

where μ is the mean value of the occurrence of the jth bigram, $12{,}009 < j < 400$, in the dataset, σ is its standard deviation, x_j is its corresponding count, and $n = 400$. The dataset with 200 records, each with 15 features, is employed for building the model.

3 Results and Discussion

The training dataset includes 200 instances of five types of syndromic ASD genes involving four types of genetic mutations. Each instance is a record with 15 features, as mentioned in Sect. 2. The standard supervised machine learning techniques decision tree induction, multilayer perceptron and SVM were applied to these data, and a tenfold cross-validation technique was used to estimate their predictive performance. The results obtained from the classifiers were analyzed through precision, recall, F-measure, accuracy, specificity and ROC, which is tabulated in Table 2. The decision tree model yields better performance, with high precision of 0.98 (Fig. 2), recall of 0.94 (Fig. 3), accuracy of 0.98 (Fig. 4) and F-measure of 0.96 (Fig. 4).

Table 3 depicts class-wise performance analysis of the decision tree classifier. The sensitivity value for class 1 is 1, which shows that all relevant instances of this class have been retrieved over total relevant instances. The specificity is high for classes 2, 3 and 4. Balanced accuracy of a classifier is defined as the average accuracy obtained on all classes. The balanced accuracy for class 1 is high compared to other classes.

A receiver operating characteristic (ROC) curve can be used to estimate classifier performance using a combination of sensitivity and specificity. The ROC curves of the three classifiers are depicted in Figs. 5, 6 and 7. The area under the ROC curve is a common metric of performance and is depicted in Table 4. The AUC of the decision tree (0.9082) is higher than that of the MLP (0.5403) and SVM (0.8414).

Table 2 Performance analysis of classifiers

Classifier	Precision	Recall	F-measure	Accuracy	Specificity
SVM	0.97	0.88	0.92	0.96	0.98
MLP	0.95	0.88	0.91	0.95	0.99
Decision tree	0.98	0.94	0.96	0.98	0.99

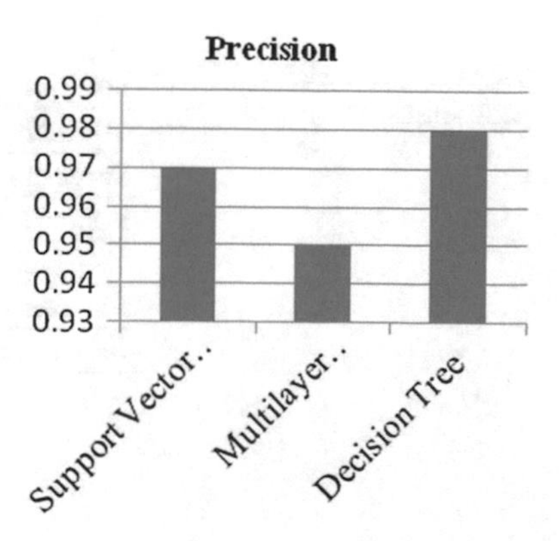

Fig. 2 Comparison of classifier precision

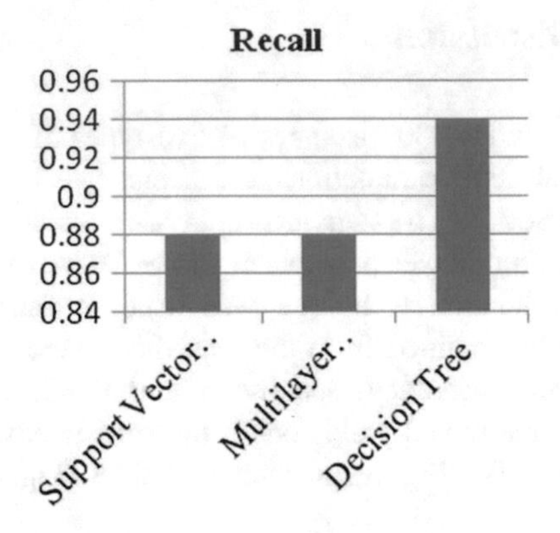

Fig. 3 Comparison of classifier recall

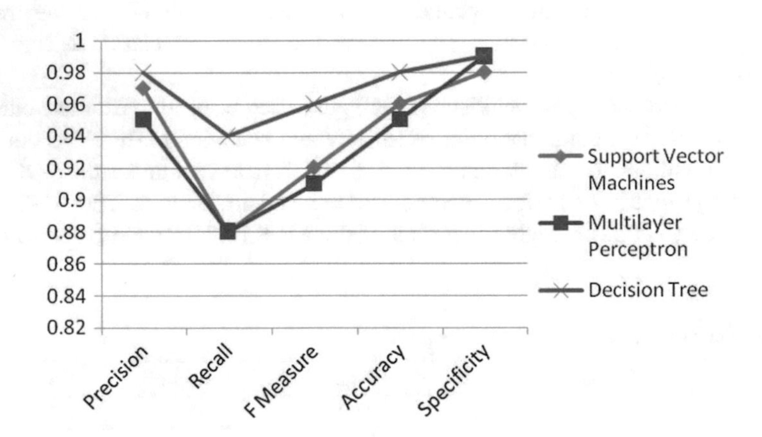

Fig. 4 Comparison of classifier performance

Table 3 Class-wise performance analysis of decision tree classifier

Statistics by class	Class 1	Class 2	Class 3	Class 4
Sensitivity	1.0000	0.7143	0.9444	0.8571
Specificity	0.8750	1.0000	1.0000	1.0000
Pos. pred. value	0.8974	1.0000	1.0000	1.0000
Neg. pred. value	1.0000	0.9677	0.9800	0.9836
Prevalence	0.5224	0.1045	0.2687	0.1045
Balanced accuracy	0.9875	0.9614	0.9742	0.9728

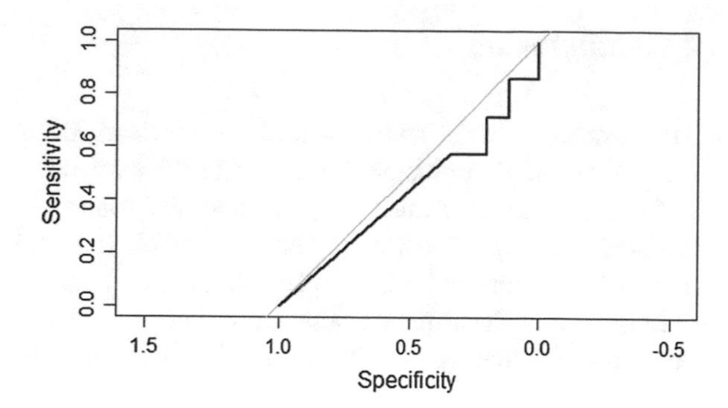

Fig. 5 ROC of multilayer perceptron

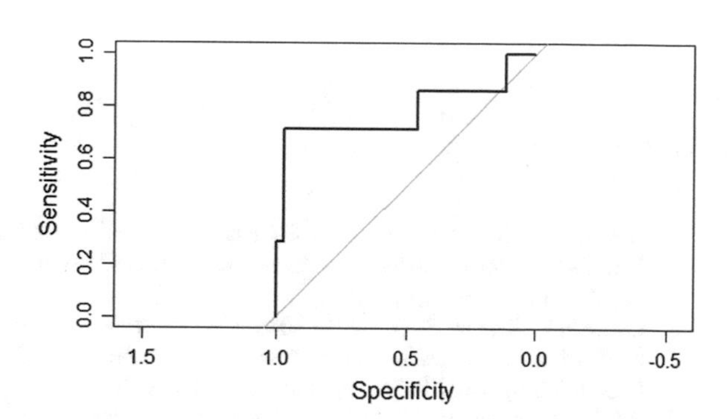

Fig. 6 ROC of support vector machine

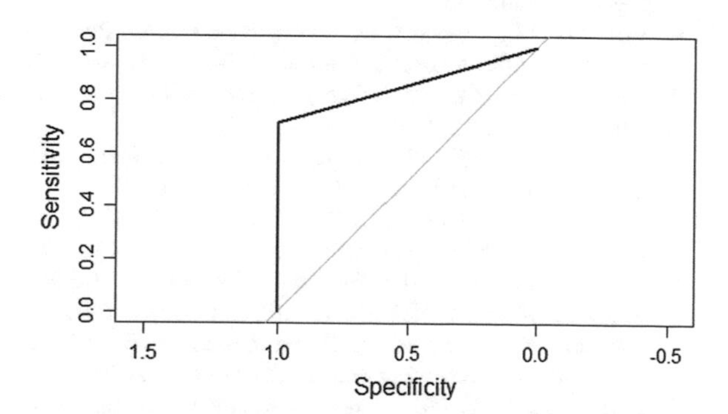

Fig. 7 ROC of decision tree

Table 4 AUC of the classifiers

Classifier	Area under ROC curve
Decision tree	0.9082
SVM	0.8414
MLP	0.5403

4 Conclusion

The research work has modeled recognition of syndromic ASD caused by nucleotide polymorphisms that underlie the autistic phenotypes TSC, RTT, FXS, PMS and Timothy syndrome. The features related to genes and mutations were extracted from gene sequences and a multiclass classification model was built by employing supervised machine learning algorithms. The predictive performance of classifiers that discriminate between mutations shows that decision trees yield the best accuracy among algorithms, at 98%. Various other genes can be included for study, and a new scoring matrix may expand the feature space, making the classifier more effective. This system finds its long-term benefit in pathogenesis studies to promote individualized risk prediction based on personal genetic profiles.

References

1. Jamal, Wasifa, et al. 2014. Classification of autism spectrum disorder using supervised learning of brain connectivity measures extracted from synchrostates. *Journal of Neural Engineering* 11 (4): 046019.
2. Latkowski, T., and S. Osowski. 2015. Computerized system for recognition of autism on the basis of gene expression microarray data. *Computers in Biology and Medicine* 56: 82–88. https://doi.org/10.1016/j.compbiomed.2014.11.004.
3. Krishnan, Arjun, et al. 2016. Genome-wide prediction and functional characterization of the genetic basis of autism spectrum disorder. *Nature Neuroscience* 19: 1454–1462. https://doi.org/10.1038/nn.4353.
4. Ronemus, M., I. Iossifov, D. Levy, and M. Wigler. 2014. The role of de novo mutations in the genetics of autism spectrum disorders. *Nature Reviews Genetics* 15: 133–141.
5. De Rubeis, S., et al. 2014. Synaptic, transcriptional and chromatin genes disrupted in autism. *Nature* 515: 209–215.
6. Sanders, S.J., et al. 2015. Insights into autism spectrum disorder genomic architecture and biology from 71 risk loci. *Neuron* 87: 1215–1233.
7. Chang, J., S.R. Gilman, A.H. Chiang, S.J. Sanders, and D. Vitkup. 2015. Genotype to phenotype relationships in autism spectrum disorders. *Nature Neuroscience* 18: 191–198.
8. Liu, L., J. Lei, and K. Roeder. 2015. Network assisted analysis to reveal the genetic basis of autism. *The Annals of Applied Statistics* 9: 1571–1600.
9. Hormozdiari, F., O. Penn, E. Borenstein, and E.E. Eichler. 2015. The discovery of integrated gene networks for autism and related disorders. *Genome Research* 25: 142–154.
10. Cotney, J., et al. 2015. The autism-associated chromatin modifier CHD8 regulates other autism risk genes during human neurodevelopment. *Nature Communications* 6: 6404.

CBHiveQL: Context-Based Hive Query Language for Large Data Analysis

N. Parimala and Gaurav Kumar

Abstract HiveQL is a 'query at a time' SQL like language where each query is independent of the previous queries. However, in the real world, queries are interrelated, and the sequence is dictated by the thought process of the user. To support interrelated queries, we define the notion of a context which specifies the domain of the data for discourse. In this paper, we propose Context-Based Hive Query Language (CBHiveQL), an extension of HiveQL, which supports the definition of a context. In CBHiveQL, every query is executed in its context created by the sequence of previous queries, and in turn provides the context for the subsequent query. Additionally, CBHiveQL provides query constructs to save context, restore context and perform the union of contexts. Our experiments show that there is a significant improvement in the execution time of successive retrieval queries; this enhancement in retrieval efficiency becomes significant when petabytes of data are handled.

Keywords Data analysis · MapReduce · Hive query language
Context · CBHiveQL

1 Introduction

Over 80% of enterprise data consists of unstructured data, which tends to grow at a much faster pace than traditional relational information. For example, weather forecasting, sensor detection, banking and telecommunication systems generate terabytes of data every hour of every day [1]. The data are diverse, high in volume and speed, and inconsistent. Complex characteristics and the huge volume of data

N. Parimala (✉) · G. Kumar
School of Computer & Systems Sciences, Jawaharlal Nehru University,
New Delhi, India
e-mail: dr.parimala.n@gmail.com

G. Kumar
e-mail: gaurav37_scs@jnu.ac.in

A. N. Krishna et al. (eds.), *Integrated Intelligent Computing, Communication and Security*, Studies in Computational Intelligence 771,
https://doi.org/10.1007/978-981-10-8797-4_17

make traditional technology inefficient and expensive, struggling to store and perform the required analytics to gain understanding from the contents of these unstructured data. Much of the data being created today is not analyzed at all, and therefore, processing such datasets efficiently has become a priority of the moment. One of the most suitable technologies is Hadoop, introduced by Doug Cutting [2], which is being used in Yahoo and Facebook, among others. It is an open-source software framework for storing and processing the massive amount of data on commodity hardware in a distributed manner. The architecture of Hadoop consists of two parts, Hadoop Distributed File System (HDFS) and MapReduce programming. HDFS [3] is a distributed, scalable, and highly fault-tolerant file system designed to store massive files ranging from terabytes to petabytes across a large scale of commodity hardware. MapReduce [4] is a simple and powerful programming model framework that splits the large datasets into independent chunks processed in a parallel manner. The MapReduce framework consists of two necessary functions for its implementation: *map* and *reduce* functions. The user needs to define only the *map* and *reduce* functions while its framework automatically parallelizes the computation across a large scale of machine clusters. This computational model takes a set of input key/value pairs and produces a set of output key/value pairs. Since it is a low-level and general purpose programming model, it is not as costly as a query language such as SQL, and as a result, the user ends up spending hours writing a program even for doing a simple analysis. Also, it is often hard to program real-life problem in terms of map-reduce function because of key-value pair identification. Therefore, various attempts have been made to create a simpler interface on top of MapReduce to provide a high-level query language interface that facilitates specification of high-level queries. These include Pig [5], SCOPE [6], Cheetah [7], and Hive [8].

The Pig system provides a new language called Pig Latin [5], which is designed to fit in a swift spot between the declarative style of SQL and the low-level procedural style of MapReduce for analyzing large-scale data. The Pig system compiles Pig Latin programs into a sequence of MapReduce jobs and executes over Hadoop, an open-source MapReduce environment. In Pig Latin, the user specifies a sequence of steps where each step specifies a single and high-level data transformation, which is statistically different from the SQL approach where the user specifies the set of declarative constraints that collectively define the result.

Cheetah [7] is a high-performance flexible and scalable custom data warehouse system built on top of MapReduce. It provides a SQL-like query language, which is designed specifically for analyzing online advertising application for various simplifications and custom optimizations. Similar to Cheetah, Hive [8] provides a generic data warehouse infrastructure for managing and querying large datasets residing in a distributed environment. It provides direct access to files stored on HDFS via a SQL-like query language called Hive Query Language (HiveQL). It supports data definition (DDL) statements to create tables with specific serialization formats and partitioning and bucketing columns. Users can load data from external sources and insert query results into Hive tables via the load and insert data manipulation (DML) statements respectively. HiveQL also supports multi-table insert, where users can perform multiple queries on the same input data using a single HiveQL statement. Hive optimizes these queries by sharing the scan of the input data.

The query interface for all the above-mentioned systems is based on a 'query-at-a-time' paradigm. That is, when the user specifies a query, it is executed and has no bearing on subsequent queries. As discussed in [9], users tend to ask a sequence of related queries, rather than a single query, which is dictated by the thought process of a user.

Poppe et al. [10] have developed a human-readable context-aware CAESAR model for event query processing. This model significantly simplifies the specification of rich event-driven application semantics by the explicit support of context windows. It also opens new multi-query optimization opportunities by associating appropriate event queries with each context. For example, consider the following scenario for a bank's client data [11], which is related to direct marketing campaigns of a Portuguese banking institution. These data are used to populate the two tables Customer (ID, age, job, marital status, education, sex, region, income) and Account (bid, account_num, account_type, balance, car_loan, house_loan, loan_amount), where the user is interested in analyzing the loans taken by the high-income (income > 10,000 and balance > 5000) female clients who belong to rural or town regions.

1. Get the female customers who have an income more than 10,000.
2. Do they have an account that has a minimum balance of 5000?
3. How many clients have taken a car loan?
4. How many of them belong to a rural or town region?

Note that all the queries are related to the same context of female clients with an income higher than 10,000. The second query and third query are centered on the financial capability of the customer. The fourth query wants to know the number of clients among them who belong to a rural or town region. The point to note is that all the queries are around a single topic of interest. Further, they evolve as we go.

To address this issue, we propose Context Based Hive Query Language (CBHiveQL), which is built over and above HiveQL. It includes the definition of a 'context' that incorporates the topic of interest of the user. Our proposed query language concept most resembles the SCOPE (Structured Computations Optimized for Parallel Execution) scripting language, which is another high-level SQL-like language developed by Microsoft [6] for web-scale data analysis. This scripting language consists of a sequence of commands where commands are data transformation operators that take one or more rowsets as input, perform some operation on the data, and output a rowset. Here, a command takes the resulting rowset of the previous command as input. It supports writing a program using traditional nested SQL expressions or as a series of simple data transformations, i.e., computation as a series of steps and implemented on a non-open-source distributed computing system called Cosmos, which is developed by Microsoft [6].

Our work is different in terms of context execution and computation implementation. In CBHiveQL, all queries are executed within a context. Instead of the 'Select From Where' format of HiveQL, we have a 'Select Where' format as the data are always fetched from the context and thus, can remain unspecified. Details of Context and its query constructs are shown in the subsequent section with the examples. The contribution of the paper is as follows:

- The notion of a context is defined to capture the data relevant to a topic of interest.
- The query language CBHiveQL is defined which permits querying within a context.

The layout of the paper is as follows. Section 2 describes Context Based HiveQL. The architecture is explained in Sect. 3. The working of the system with results is demonstrated with some examples in Sect. 4. Section 5 is the concluding section.

2 Querying in a Context

As brought out above, we support querying within a context. Before explaining the constructs of Context Based HiveQL, we first define a context.

2.1 Context

A context C consists of tables and the data corresponding to the tables. The tables are those which are added to the context as and when needed until the context is closed. All the columns of these tables are part of the context. The tuples of all tables that satisfy the conditions in a query are the only ones retained in the context. Formally, a context is defined as two tuples Context $C = <T, D>$ where $T = \{T_1, T_2, \ldots T_n\}$ is a finite set of tables; D is the data corresponding to the tables. A Context Based Query (CBQ) is executed in its context and updates the context. Let the current context be $C = <T, D>$. When a CBQ is executed, the tuples satisfying the conditions in the CBQ are used to update the context. The updated context is $C = <T, D'>$ where D' is the set of filtered tuples of these tables.

2.2 Context Based Hive Query Language

There are seven query constructs in CBHiveQL. These are Create Context, Add Table, Select, Save Context, Recall Context, Union, and Close. We explain the syntax of each of these in turn.

- **Create Context**: Create Context <context_name>; Create Context creates a context with the given name. After the execution of this statement, the current context is <null>.
- **Add Table:** Add Table <table_name> [ON <col_name$_1$ = table_name. col_name$_2$>]; Add Table adds the table to the current context. If the current context is empty, then the command is to be specified without any condition. The table is added to the current context. To add a second table to the current context, the user has to specify the join condition. The user can add only one table at a time.
- **Select**: Select col$_1$, col$_2$ … col$_n$ [Where where_condition] [Group By col_list] [Cluster By col_list | [Distribute By col_list] [Sort By col_list]] [Limit number]; the select statement follows the same syntax as in HiveQL except that the FROM clause is not specified. The select statement is executed in its context. After successful completion, the context is updated as explained above in Sect. 2.1.
- **Save Context**: Save Context <context_name>; Save Context saves the current context with the given name for later analysis. The current context is not altered in any way.
- **Recall Context**: Recall Context <context_name>; Recall Context fetches the saved context whose name is specified in the query. Once the context is recalled, it becomes the current context, and the user can now continue querying. The current context that existed at the time of executing this statement is lost unless it has been explicitly saved.
- **Union All**: Context <context_name$_1$> Union All <context_name$_2$>; The union of two saved contexts is possible only if they are union-compatible. Two contexts C_1 = <T, D> and C_2 = <T', D'> are said to be union compatible if T = T' i.e. they have the same set of tables. When this statement is executed, the union of the corresponding tables in the two contexts is performed. The tables with the new data create the newly updated context. The new context is C = <T, D"> where D" represents the data of the individual tables in T. If the contexts are not union-compatible, an error is thrown.
- **Close**: Close Context <context_name>; This query closes the current context. It releases the memory occupied by the analysis. After the execution of this statement, the current context is <null>.

2.3 An Example

To bring out the usage of the constructs of the query language, consider the following example of Sect. 1. The first requirement is that a context for querying has to be created which is specified as

Q1: Create Context A

At the end of execution of this query, the context is null. In order to query about the customers, first, the table has to be added to the context. The Add command is given below

Q2: Add Table customer

Now, context A contains customer data. But we are interested only in female customers who have income of more than 10,000. So, these are selected from the context in the following query.

Q3: Select ID, age, job, education, region, income where income > 10,000

Note, first, the absence of the WHERE clause as the data is fetched from context A. After the execution of the query, the context contains only the selected columns and only of those customers who satisfy the condition. Now, the data query (b) is in the table Account. So Account table has to be added to the current context.

Q4: Add Table Account on ID = Account.bid

The context A is updated to contain account information as well. But the user is interested in female clients who have a minimum balance of 5000, which is achieved by executing the next query.

Q5: Select age, job, region, income, balance where sex = 'female' and balance ≥ 5000

Now, the user wants to know the car_loan of each customer

Q7: Select age, job, region, income, car_loan, house_loan, balance where car_loan = 'yes' and age ≥ 25

After the execution of the above query, context A contains the data about female customers who have taken a car loan and earn more than 10,000 with a minimum balance of 5000. The last query is

Q8: Select age, job, region, income, house_loan, balance where region='rural' or region = 'town'

In the above sequence of queries, the context data is used to answer a given query. After execution, the context is updated.

2.4 *Comparison of HiveQL and CBHiveQL*

In this section, the sequence of queries for the example of Sect. 2.3 is given using HiveQL.

Qa: Select ID, age, job, education, region, income From Customer where income > 10,000

Qb: Select ID, age, job, education, region, income From Customer JOIN Account on Customer.id = Account.bid where income > 10,000 and sex = 'female' and balance ≥ 5000

Qc: Select age, job, region, income, car loan, house loan, balance From Customer JOIN Account on Customer.id = Account.bid where income > 10,000 and sex = female and balance ≥ 5000 and car_loan = 'yes'

Qd: Select age, job, income, housing, balance From Customer JOIN Account on Customer.id = Account.bid where income > 10,000 and sex = 'female' and balance ≥ 5000 and car_loan = 'yes' and region = 'rural' and region = 'town'

It can be observed from the above sequence of queries that the user has to repeat the condition of the 'where' clause of the previous query each time in the subsequent query. This is because in HiveQL each query is independent.

3 System Architecture

Figure 1 shows the overall layout of system architecture and its interaction with Hive and Hadoop. We have developed a Graphical User Interface (GUI) for context based querying using Java, which interacts with the Hive Thrift server as shown in Fig. 1.

The GUI accepts Context Based Query from the user and Hive framework executes one or more HiveQL queries corresponding to each CBHiveQL query using information stored in the Metastore. For each CBHiveQL query, a table is created, and its structural information is stored in the Metastore. It is a system catalog, which contains metadata about the tables stored in Hive. This metadata is specified during table creation and reused every time the table is referenced in HiveQL. Thrift is a framework for cross-language services, where the server written in one language can also support clients in another language. The GUI uses Java Database Connectivity (JDBC) providing an access point to Hive for Java applications. The driver manages the life cycle of the HiveQL statement during compilation, optimization, and execution. After receiving the HiveQL statement from the Thrift Server or any other interface, it creates a session handle which is later used to keep track of statistics. The driver invokes the compiler after receiving

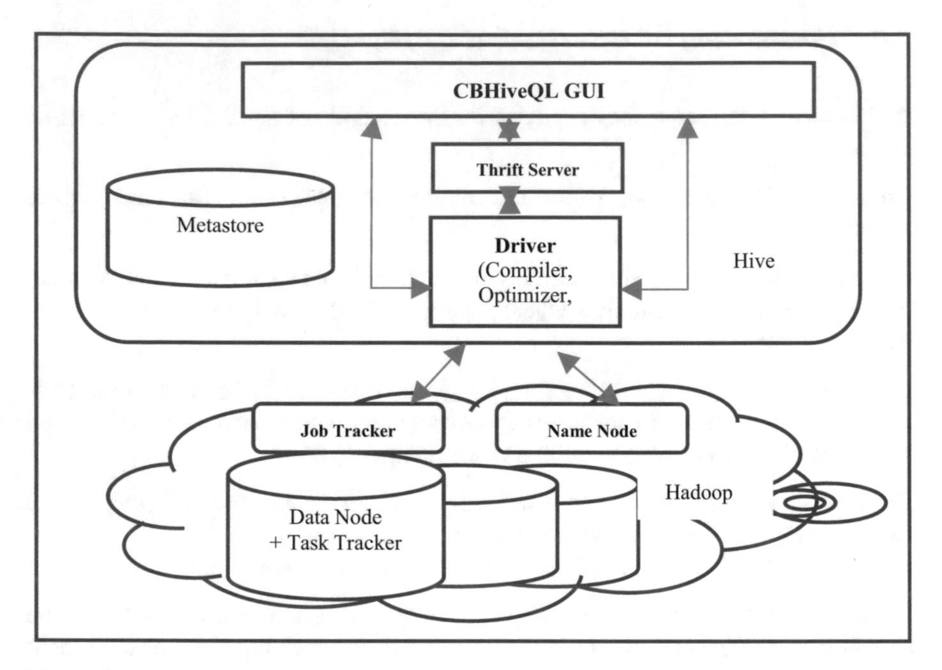

Fig. 1 CBHiveQL architecture

HiveQL statement, and it translates the HiveQL statement into Direct Acyclic Graph (DAG) of MapReduce jobs. Then Hadoop Engines execute the MapReduce jobs in topological order.

4 Implementation

Experiments were conducted on the Hadoop 2.6.0-Hive 1.2.1 cluster with an Intel i5-3470 processor, 4 GB of RAM, 360 GB hard disk and Ubuntu-14.04 Linux operating system. We used version Java 1.7 and NetBeans IDE 8 to develop our GUI interface for querying. In this experiment, datasets of Bank Marketing [12], which are related to the direct marketing campaigns of a Portuguese banking institution, were used to populate the two tables as shown in Sect. 1. Each table contains 10,48,575 tuples of customer and account. The working of the system is shown below through a sequence of queries. Because of the paucity of space, only a few screenshots are shown. The first query creates context A, which is <null>.

Query 1: Create Context A

The user can add a table to start the analysis.

Query 2: Add Table Customer

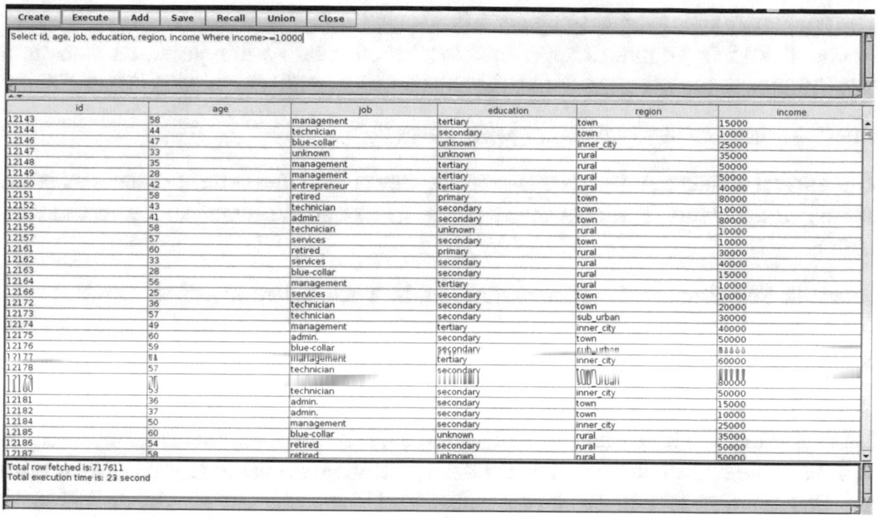

Create | Execute | Add | Save | Recall | Union | Close

Select id, age, job, education, region, income Where income>=10000

id	age	job	education	region	income
12143	58	management	tertiary	town	15000
12144	44	technician	secondary	town	10000
12146	47	blue-collar	unknown	inner_city	25000
12147	33	unknown	unknown	rural	35000
12148	35	management	tertiary	rural	50000
12149	28	management	tertiary	rural	50000
12150	42	entrepreneur	tertiary	rural	40000
12151	58	retired	primary	town	80000
12152	43	technician	secondary	town	10000
12153	41	admin.	secondary	town	80000
12156	58	technician	unknown	rural	10000
12157	57	services	secondary	town	10000
12161	60	retired	primary	rural	30000
12162	33	services	secondary	rural	40000
12163	28	blue-collar	secondary	rural	15000
12164	56	management	tertiary	rural	10000
12166	25	services	secondary	town	10000
12172	36	technician	secondary	town	20000
12173	57	technician	secondary	sub_urban	30000
12174	49	management	tertiary	inner_city	40000
12175	60	admin.	secondary	town	50000
12176	59	blue-collar	secondary	sub_urban	[illegible]
12177	[illegible]	management	tertiary	inner_city	60000
12178	57	technician	secondary	[illegible]	[illegible]
[illegible]	[illegible]	[illegible]	[illegible]	[illegible]	80000
[illegible]	53	technician	secondary	inner_city	50000
12181	36	admin.	secondary	town	15000
12182	37	admin.	secondary	town	10000
12184	50	management	secondary	inner_city	25000
12185	60	blue-collar	unknown	rural	35000
12186	54	retired	secondary	rural	50000
12187	58	retired	unknown	rural	50000

Total row fetched is:717611
Total execution time is: 23 second

Fig. 2 Select income

Table Customer is added to context A, and now, A = <T, D> where T = {Customer} and D is the data of customer table. Let the next query be

Query 3: Select ID, age, job, education, region, income where income > 10,000
After the execution of the query, the context is A = <T, D'> where D' contains only customers whose income is higher than 10,000 as shown in Fig. 2.
Let us say that the user wants to determine the customers who have savings accounts with a minimum balance of 5000. He/she needs to add 'Account' table that contains the bank detail of customers. This is shown in Query 4.

Query 4: Add Table Account on id = Account.bid

The specified condition is used to join the table bank with the current context. Now, Context A = <T', D"> where T' = {Customer, Account} and D" contains the data of D' and that of the table Account. The user can now frame Query 5.

Query 5: Select age, job, region, income, balance where account_sex = 'female' and balance ≥ 5000

Context A now contains only female customers who have a minimum balance of 5000. Continuing with this thought process, let us say that the following queries are executed.

Query 6: Select age, job, region, income, loan, housing, balance where car loan = 'yes' and age ≥ 25

The context A after executing Query 6 is A = <T'', D''' > where T'' = {Customer, Account} and D''' contains appropriately filtered data. At this point, let us assume that the user saves the current context as customer_with_cloan using Query 7.

Query 7: Save Context customer_with_cloan

The current context A is saved. It may be noted that the current context A is not altered in any way. The next query picks up the clients who belong to the rural region.

Query 8: Select age, job, income, housing, balance where region = 'rural'.
Let us assume that the user saves this context for further analysis at a later stage using Query 9.

Query 9: Save Context customer_in_rural
Let us say that the user wants to analyze the data for town region also. However, the current context does not contain this data, but context customer_with_cloan does. So, this context has to be first recalled and then queried as shown below in Query 10 and Query 11.

Query 10: Recall Context customer_with_cloan

Query 11: Select age, job, income, housing, balance where region = 'town'

Let us assume the user also saves this context for further analysis.

Query 12: Save Context customer_in_town

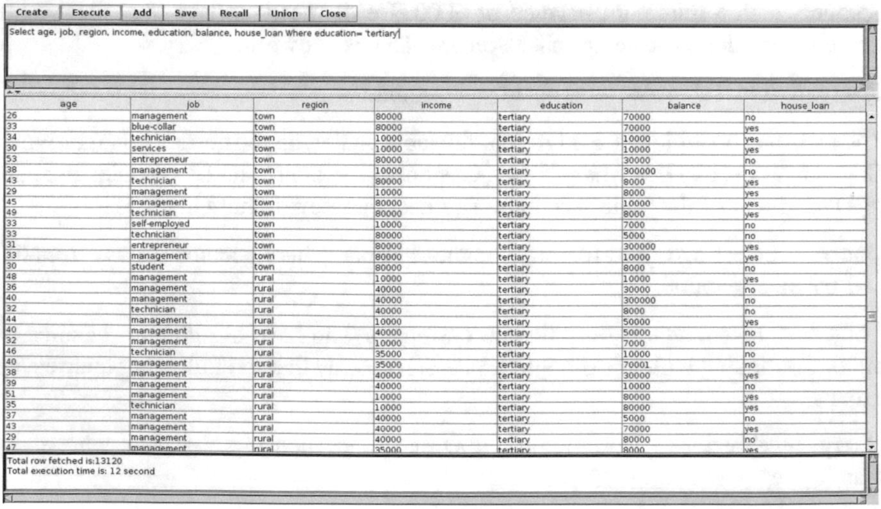

Create | Execute | Add | Save | Recall | Union | Close

Select age, job, region, income, education, balance, house_loan Where education= 'tertiary'

age	job	region	income	education	balance	house_loan
26	management	town	80000	tertiary	70000	no
33	blue-collar	town	80000	tertiary	70000	yes
34	technician	town	10000	tertiary	10000	yes
30	services	town	10000	tertiary	10000	yes
53	entrepreneur	town	80000	tertiary	30000	no
38	management	town	10000	tertiary	300000	no
43	technician	town	80000	tertiary	8000	yes
29	management	town	10000	tertiary	8000	yes
45	management	town	80000	tertiary	10000	yes
49	technician	town	80000	tertiary	8000	yes
33	self-employed	town	10000	tertiary	7000	no
33	technician	town	80000	tertiary	5000	no
31	entrepreneur	town	80000	tertiary	300000	yes
33	management	town	80000	tertiary	10000	yes
30	student	town	80000	tertiary	8000	no
48	management	rural	10000	tertiary	10000	yes
36	management	rural	40000	tertiary	30000	no
40	management	rural	40000	tertiary	300000	no
32	technician	rural	40000	tertiary	8000	no
44	management	rural	10000	tertiary	50000	yes
40	management	rural	40000	tertiary	50000	no
32	management	rural	10000	tertiary	7000	no
46	technician	rural	35000	tertiary	10000	no
40	management	rural	35000	tertiary	70001	no
38	management	rural	40000	tertiary	30000	yes
39	management	rural	40000	tertiary	10000	no
51	management	rural	10000	tertiary	80000	yes
35	technician	rural	10000	tertiary	80000	yes
37	management	rural	40000	tertiary	5000	no
43	management	rural	40000	tertiary	70000	yes
29	management	rural	40000	tertiary	80000	no
47	management	rural	35000	tertiary	8000	yes

Total row fetched is:13120
Total execution time is: 12 second

Fig. 3 Select education

If the user wants to analyze the clients who belong to the inner city region as well as town region, the union of the contexts is performed as given in Query 13 and subsequently queried (Query 14).

Query 13: Context customer_in_rural Union All customer_in_town

Query 14: Select age, job, income, region, balance, house_loan where education = 'tertiary'

The result is shown in Fig. 3.

Finally, the query given below clears current context A.

Query 15: Close Context A

5 Results

The above sequence of queries was executed on client data which consisted of approximately 10,48,575 records for each of the tables, namely Customer and Bank.

The result shown in Fig. 4 is only for the retrieval queries. It indicates that there is improvement in the execution time as long as the new table is not added. Even when a new table is added, since the new table is joined with the existing tables in the context with fewer tuples, the execution time is not higher than what it would be in the absence of context. If we execute the query on terabytes or petabytes of data, then there will be significant improvement in the execution time for each successive query.

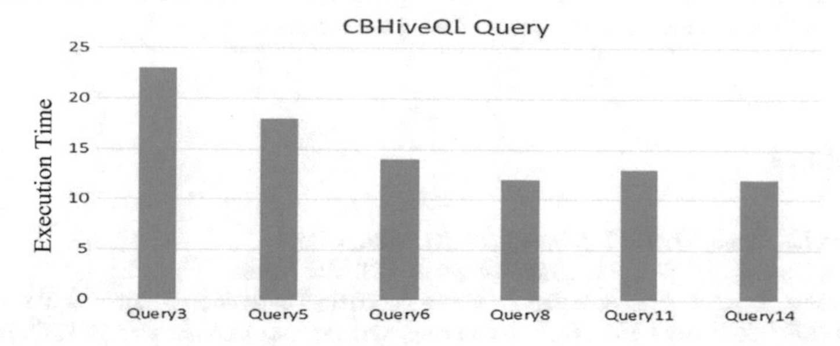

Fig. 4 Result of query execution in context

6 Conclusion

In this paper, we have proposed Context Based Hive Query Language (CBHiveQL), which is built over and above HiveQL. The purpose of this language is to be able to specify related queries in a context. The user has to specify only 'Select' and 'Where' clauses in CBQ, while its architecture is designed in such a way that each query fetches the data from its context. CBHiveQL consists of commands to create a context, query within a context, add tables to a context and delete a context. The user can also save the current context for later analysis and recall it to resume the analysis. The user can also perform union operations between two saved contexts for an in-depth analysis. It may be argued that the same effect may be achieved by creating a concept of CTAS. We believe that CBHiveQL provides a higher level of abstraction for querying. It reflects the thought process of the user rather than implementation aspects. As noted in [12], the word 'context' has been used to mean several different concepts as category restriction, query personalization, topics in a text while performing document analysis, narrowing down the search in a search engine, etc. The notion of a context defined here is similar to [13] where continuous OLAP queries are executed in a context. In this paper, it is applied to HiveQL queries.

We have shown that there is a considerable gain in the execution time when CBHiveQL is used. We are not aware of any other work that incorporates the notion of a context with HiveQL.

We have implemented the existence of more than one table in the context by taking recourse to the definition of a CTAS consisting of all the tables. We can explore if we can implement the mapping of a context to HiveQL by maintaining multiple tables and joining them only when the user uses the table in a query. As a result, the join can be postponed to the point of reference.

Acknowledgements One of the authors of this paper (G. Kumar) would like to thank CSIR (MHRD) for funding fellowship throughout during the research.

References

1. Robert D. Schneider. 2012. *Hadoop for dummies*. Wiley.
2. White, Hadoop T. 2009. *The definitive guide*. O'Reilly Media.
3. Shvachko, K., H. Kuang, S. Radia, R. Chansler. 2010. The hadoop distributed file system. In *2010 IEEE 26th symposium on mass storage systems and technologies (MSST)*, pp. 1–10. IEEE. https://doi.org/10.1109/msst.2010.5496972.
4. Dean, J., and S. Ghemawat. 2008. MapReduce : simplified data processing on large clusters. *ACM* 51 (1): 107–113. https://doi.org/10.1145/1327452.1327492.
5. Olston, C., B. Reed, U. Srivastava, R. Kumar and A. Tomkins. 2008. Pig latin: a not-so-foreign language for data processing. In *Proceedings of the 2008 ACM SIGMOD international conference on management of data*, pp. 1099–1110, ACM. https://doi.org/10.1145/1376616.1376726.

6. Chaiken, R., B. Jenkins, P.Å. Larson, B. Ramsey, D. Shakib, S. Weaver, and J. Zhou. 2008. SCOPE: easy and efficient parallel processing of massive data sets. *Proceedings of the VLDB endowment* 1 (2): 1265–1276. https://doi.org/10.14778/1454159.1454166.
7. Chen, S. 2009. Cheetah: a High performance, custom data warehouse on top of MapReduce. *Proceedings of the VLDB endowment* 3: 1459–1468. https://doi.org/10.14778/1920841.1921020.
8. Thusoo, A., J.S. Sarma, N. Jain, et al, Hive—a petabyte scale data warehouse using Hadoop. In *2010 IEEE 26th international conference on data engineering (ICDE 2010)*. https://doi.org/10.1109/icde.2010.5447738.
9. Sun, M., and J.Y. Chai. 2007. Discourse processing for context question answering based on linguistic knowledge. *Knowledge-Based Systems* 20: 511–526. https://doi.org/10.1016/j.knosys.2007.04.005.
10. Poppe, O., C. Lei, E.A. Rundensteiner, DJ Dougherty. 2016. Context-aware event stream analytics. In: EDBT, pp. 413–424. https://doi.org/10.5441/002/edbt.2016.38.
11. Moro, Sérgio, Cortez, Paulo Laureano, Raul. 2011. Using data mining for bank direct marketing: an application of the CRISP-DM methodology. In *Proceedings of the European simulation and modelling conference*.
12. Finkelstein, L., E. Gabrilovich, Y. Matias, et al. 2002. Placing search in context: the concept revisited. *ACM Transactions on Information Systems* 20: 116–131. https://doi.org/10.1145/503104.503110.
13. Parimala, N., and S. Bhawna. 2012. Continuous multiple OLAP queries for data streams. *International Journal of Cooperative Information Systems* 21: 141–164. https://doi.org/10.1142/s0218843012500037.

A Novel Private Information Retrieval Technique for Private DNS Resolution

Radhakrishna Bhat and N. R. Sunitha

Abstract Recently, advancements in various analytical techniques have enabled to encourage the unethical business to violate *user privacy* and market the analytical results. Existing user privacy-preserving techniques based on intractability assumptions have proved to offer only conditional user privacy. Thus, interest in perfect (i.e., unconditional) user privacy-preserving information retrieval techniques are receiving enormous attention. We have successfully constructed a single database perfect privacy-preserving information retrieval technique using Private Information Retrieval (PIR). We have proposed a novel perfect privacy-preserving PIR technique in a single database setting with non-trivial communication cost for private Domain Name System (DNS) resolution. We have further extended the proposed scheme to a computationally efficient scheme by varying the security parameter without losing the level of user privacy.

Keywords Private information retrieval · Quadratic residuosity · Perfect privacy Private DNS resolution · Fully qualified domain name

1 Introduction

Consider a scenario where the *user* wants to retrieve the domain name related information (i.e., the IP address of the domain name server) from a *trusted-but-curious* type *DNS server*, privately in a reasonable time where the server maintains a plain database of the domain name-related information.

R. Bhat (✉) · N. R. Sunitha
Siddaganga Institute of Technology, B H Road, Tumakuru 572103, Karnataka, India
e-mail: rsb567@gmail.com

N. R. Sunitha
e-mail: nrsunithasit@gmail.com

A. N. Krishna et al. (eds.), *Integrated Intelligent Computing, Communication and Security*, Studies in Computational Intelligence 771,
https://doi.org/10.1007/978-981-10-8797-4_18

Private information retrieval (PIR) is a protocol that hides the user's interest from the curious server. In PIR, the user retrieves the *i-th* bit privately from *n* bit server. Private Block Retrieval (PBR) is a realistic extension of PIR where an entire block is retrieved at once out of *u* blocks. Privacy preservation using PIR was first introduced by Chor et al. [6] in an information-theoretic setting. If a computationally bounded (or computationally intractable) server entity is involved, then the scheme is considered as computationally bounded PIR (cPIR), in which the user's privacy is preserved based on well-defined cryptographic intractability assumptions. If information-theoretically secure non-colluding replicated database servers entities are involved, then the scheme is considered as information-theoretic PIR (itPIR), in which the user's privacy is preserved based on information theory.

COMPUTATIONALLY BOUNDED PIR (cPIR): Initially, a scheme based on the quadratic residuosity assumption (QRA) using a single database was presented by Kushilevitz and Ostrovsky [14]. Chor and Gilboa [5] also showed that a computationally bounded privacy scheme is possible using one-way functions with minimal database replication. By introducing ϕ-hiding assumption, Cachin et al. [3] presented another single database scheme. Ishai et al. [11] showed that anonymity techniques are sufficient to construct an efficient scheme. Further, Aguilar-Melchor and Gaborit [17] presented scheme using coding theory and lattice assumptions. The first multi-query scheme was introduced by Groth et al. [10]. As an improvement to the existing schemes, Jonathan and Andy [20] presented a trapdoor group-based scheme. Further, Kushilevitz and Ostrovsky [15] presented a one-way trapdoor permutation of that scheme. Chang [4] presented a scheme using a Paillier cryptosystem. A decision subgroup problem called ϕ-hiding assumption formed the basis of the Gentry and Ramzan [8] scheme. The first keyword-based PIR search [1] was introduced to apply PIR on an existing server data structure. Using the well-known composite residuosity assumption, Lipmaa [16] introduced a new scheme in which the size of the computation mainly depends on the size of the database. Cachin et al. [2] also presented a communication efficient scheme using the ϕ-hiding assumption.

PRIVATE DNS SEARCH: DNS privacy is addressed in [19, 23], using private information retrieval as the underlying privacy preserving technique. There are several DNS privacy concerns also presented in [12, 13, 21, 22, 24].

Problem with existing schemes: Let us consider a scenario where the user wants to retrieve the *i*-th bit of *n* bit database privately from a computationally bounded server by generating *identically distributed* queries. In other words, in order to retrieve the required bit information from the server privately, the user would send an information-theoretically indistinguishable query to the server, then the server would send a response involving the whole database to the user. Since most of the existing cPIR schemes are generating intractability assumption-based references to retrieve a bit, on the assumption that the server has limited computation capacity, they clearly fail to achieve the above-mentioned scenario with the non-trivial communication $o(n)$.

Our contributions: The basic assumption is that the server has to abide by the "non-repudiation" requirement. Specifically, the server must not modify or fake or manipulate any block of the database in order to reveal privacy during or after the communication without intimating to the user. Note that the term *perfect* refers to zero percent privacy leak of the user's interest on the server, *privacy* refers to "user privacy", perfect privacy-preserving and information-theoretic are interchangeable until and unless explicitly stated. Informally, an *information-theoretically secure* or *perfect privacy-preserving* query in a single database PIR setting is a query with identically distributed input parameters and a privacy revealing probability equivalent to "random guessing" probability.

Organization: The preliminaries and notations are described in Sect. 2. The PIR-based perfect privacy-preserving information retrieval scheme is described in Sect. 3. An example of private DNS resolution (both in iterative and recursive modes) is described in Sect. 4. The open problems, along with the conclusion, are discussed in Sect. 5.

2 Preliminaries and Notations

Let $[a] \triangleq \{1, \ldots, a\}$ and $[a, b]$ denote any integer taking all the values from a to b, iteratively. Let a' denote a variable. Let $N = pq$ be an RSA composite modulus, where $p, q \equiv 3 \pmod 4$. Let Q_R be the quadratic residue set, $\overline{Q}_R$ be the quadratic non-residue set with *Jacobi Symbol* (*JS*) 1 and $\overline{Q}_R^{-1}$ be the quadratic non-residue set with *Jacobi Symbol* -1. Let k be the security parameter.

Quadratic residuosity: For all element a belonging to $\mathbb{Z}_N^*$, if there exists a square of another element b congruent to a, then a is quadratic residue, otherwise a is quadratic non-residue.

Reduced trapdoor bit function (*rltdf*) is the reduced form of the quadratic residuosity-based trapdoor function of [7]. For all $\alpha \in \mathbb{Z}_N^{+1}$, the reduced trapdoor bit function is $\mathcal{F}_{rltdf}$=($\alpha^2 \equiv \beta \pmod N$)) and the "*hx*" value of α as described in [7] is considered as as a "trapdoor bit". The sole purpose of this trapdoor bit is to serve as a trapdoor during the inverse process as $\mathcal{F}_{rltdf}^{-1}$=($\sqrt[jx=0,hx]{\beta \pmod N} \equiv \alpha \pmod N$)).

Goldwasser–Micali scheme [9]: For all plaintext bit $a \in \{0, 1\}$, $\forall \beta \xleftarrow{R} \mathbb{Z}_N^{+1}$, $\forall \alpha \in \overline{Q}_R$, the encryption is $\mathcal{F}_{gm}(a,\ \alpha,\ \beta) = (\beta^2 \cdot \alpha^a \equiv z \pmod N)$), where z would be a quadratic residue if a=0, otherwise z would be a quadratic non-residue.

Paillier encryption [18]: Let $\lambda = lcm(p-1, q-1)$. For all plaintext $M \in \mathbb{Z}_{N^2}^*$ such that $M = (m_1 + Nm_2)$, $\{\forall y \in \mathbb{Z}_{N^2}^*, (L(y^\lambda \bmod N^2))^{-1} \pmod N$ where $\forall a \in \mathbb{Z}_N^*$, $L(a) = (a-1)/2\}$, the encryption is $\mathcal{F}_{pil}(M, y, N) = (y^{m_1} \cdot (m_2)^N \equiv z \pmod{N^2})$).

Paillier tree: Let us consider a binary tree of Paillier encryption in which each *node* is a Paillier encryption $\mathcal{F}_{pil}(\cdot)$ as described above. Then, for all plaintext bits $m_i \in \mathbb{Z}_N^* : i \in [1, 2^{h+1}]$, the binary tree of Paillier encryptions with height 1 is

$$\mathcal{F}_{pil}(m_1, m_2, y, N) = (Z = z_1 + Nz_2)$$

$$\mathcal{F}_{pil}(z_1, m_3, y, N) = Z_1 \qquad \mathcal{F}_{pil}(z_2, m_4, y, N) = Z_2$$

3 Proposed Scheme

Let n bit string database $\mathcal{DB} = \{\mathcal{DB}_1, \mathcal{DB}_2, \ldots, \mathcal{DB}_u\}$, where $|\mathcal{DB}_j| = v, j \in [1, u]$. Let $\mathcal{DB}_z = S_{z,low} \cup S_{z,high}$, $z \in [u]$, be the z-th string of the database $\mathcal{DB}$, where the substring $S_{z,low} = \{\mathcal{B}_{z,j} = (b_i, b_{i+1}) : i = i + 2, i \in [1, v{-}1], j \in [1, h], h = v/2\}$, and the substring $S_{z,high} = \{\mathcal{B}'_{z,j} = (b_i, b_{i+1}) : i = i + 2,\ i \in [2, v{-}2],\ j \in [1, h{-}1], h = v/2\}$. Let the trapdoor bit sets for each string $\mathcal{DB}_z$ be $\mathcal{S}_z$ and $\mathcal{T}_z$. Note that each string $\mathcal{DB}_z$ is arranged in $S_{z,low}$ and $S_{z,high}$ in such a way that the second bit of each $\mathcal{B}_{z,i}$, $1 \leq i \leq h-1$, is the same as the first bit of $\mathcal{B}'_{z,i}$.

QRA based 2-bit encryption: This is the main building block of the proposed PIR construction. For all $a, b \in \{0, 1\}$, for all $\alpha \in \mathbb{Z}_N^{+1}$, for all public key $\mathcal{K}_1, \mathcal{K}_2 \in \mathbb{Z}_N^{+1}$, the encryption is

$$\mathcal{T}_{tb}(a, b, \alpha, N, \mathcal{K}_1, \mathcal{K}_2) = ((\alpha \cdot \mathcal{K}_j) \cdot \mathcal{K}_{j'} \equiv \beta \pmod N), \tag{1}$$

where $j, j' \in [2]$. One of the possible quadratic residuosity property combinations of input and public key (i.e, $\alpha \in Q_R$, $\mathcal{K}_1 \in Q_R$, $\mathcal{K}_2 \in \overline{Q}_R$) is given in Table 1. Other possible combinations are ($\alpha \in Q_R$, $\mathcal{K}_1 \in \overline{Q}_R$, $\mathcal{K}_2 \in Q_R$), ($\alpha \in \overline{Q}_R$, $\mathcal{K}_1 \in Q_R$, $\mathcal{K}_2 \in \overline{Q}_R$) and ($\alpha \in \overline{Q}_R$, $\mathcal{K}_1 \in \overline{Q}_R$, $\mathcal{K}_2 \in Q_R$). Any one of the above mentioned combinations can be used for encryption, and their respective inverses can be used for decryption. In order to decrypt Eq. 1, private key (p, q), the quadratic residuosity property of the ciphertext β and the second bit b must be known in advance, as shown in Table 2. Therefore, this encryption-decryption method successfully encrypts two bits and decrypts only one bit.

The proposed scheme is a 3-tuple (QG, RC, RR) protocol, in which the user privately retrieves the subset of information from the server using user-centric "public key cryptography". In order to retrieve the subset from the server, the user generates a perfect privacy-supported PIR query $\mathcal{Q}$. That is, query inputs are uniformly drawn from $\mathbb{Z}_N^{+1}$ using query generation algorithm *QG* and sent to the server. The server then generates and sends back the corresponding response $\mathcal{R}$ by involving the entire database using response creation algorithm *RC*. Finally, the user retrieves the required subset privately using response retrieval algorithm *RR*. The detailed description of the protocol is as follows.

Table 1 2-bit encryption Table for all $a, b \in \{0, 1\}, \forall\alpha \in \mathcal{Q}_R, \forall\mathcal{K}_1 \in \mathcal{Q}_R, \forall\mathcal{K}_2 \in \overline{\mathcal{Q}}_R$

Plaintext		Encryption ($\mathcal{T}_{tb}$)
a	b	
0	0	$(\alpha \cdot \mathcal{K}_1) \cdot \mathcal{K}_1 \equiv (\beta \in \mathcal{Q}_R)$
0	1	$(\alpha \cdot \mathcal{K}_1) \cdot \mathcal{K}_2 \equiv (\beta \in \overline{\mathcal{Q}}_R)$
1	0	$(\alpha \cdot \mathcal{K}_2) \cdot \mathcal{K}_1 \equiv (\beta \in \overline{\mathcal{Q}}_R)$
1	1	$(\alpha \cdot \mathcal{K}_2) \cdot \mathcal{K}_2 \equiv (\beta \in \mathcal{Q}_R)$

Table 2 Possible residuosity property combinations of a public key when the residuosity property combinations of ciphertext β, input α are given

Property (α)	Property (β)	b	a	$(\beta \cdot \mathcal{K}_{j'}^{-1}) \cdot \mathcal{K}_j^{-1} \equiv \alpha$
$\mathcal{Q}_R$	$\mathcal{Q}_R$	0	0	$(\beta \cdot \mathcal{K}_1^{-1}) \cdot \mathcal{K}_1^{-1} \equiv \alpha$
$\mathcal{Q}_R$	$\overline{\mathcal{Q}}_R$	1	0	$(\beta \cdot \mathcal{K}_2^{-1}) \cdot \mathcal{K}_1^{-1} \equiv \alpha$
$\mathcal{Q}_R$	$\overline{\mathcal{Q}}_R$	0	1	$(\beta \cdot \mathcal{K}_1^{-1}) \cdot \mathcal{K}_2^{-1} \equiv \alpha$
$\mathcal{Q}_R$	$\mathcal{Q}_R$	1	1	$(\beta \cdot \mathcal{K}_2^{-1}) \cdot \mathcal{K}_2^{-1} \equiv \alpha$

- *Query Generation (QG)*: Generates the perfect privacy-preserving query $\mathcal{Q} = (N, (\mathcal{K}_1, \mathcal{K}_2) \xleftarrow{R} \mathbb{Z}_N^{+1}, \mathcal{K}_3 \xleftarrow{R} \overline{\mathcal{Q}}_R, \alpha \xleftarrow{R} \mathbb{Z}_N^{+1})$.
- *Response Creation (RC)*: Using the query $\mathcal{Q}$ and the database $\mathcal{DB}$, generates the response $\mathcal{R}$ with the size less than the database size (in bits) as follows. For all $z \in [1, u]$,

$$(\beta_{z,1}, \mathcal{S}_z) = \mathcal{T}_{tb,i}(\mathcal{B}_{z,i} \in \mathcal{S}_{z,low}, N, \mathcal{F}_{rltdf}(\mathcal{T}_{tb,i-1}), \mathcal{K}_1, \mathcal{K}_2)$$

$$\textit{and} \tag{2}$$

$$(\beta_{z,2}, \mathcal{T}_z) = \mathcal{T}_{tb,j}(\mathcal{B}'_{z,j} \in \mathcal{S}_{z,high}, N, \mathcal{F}_{rltdf}(\mathcal{T}_{tb,j-1}), \mathcal{K}_1, \mathcal{K}_2),$$

where $2 \leq i \leq h, 2 \leq j \leq (h-1)$ and each $\mathcal{F}_{rltdf}(\cdot)$ generates one trapdoor bit. Hence, the trapdoor bit set $\mathcal{S}_z$ consists of $(v/2) - 1$ trapdoor bits, and the trapdoor bit set $\mathcal{T}_z$ consists of $(v/2) - 2$ trapdoor bits. Note that, since there are $(v-1)$ 2-bit encryption functions (each 2-bit encryption function is described in Eq. 1) executing in Eq. 2, it is intuitive that any two successive 2-bit functions generate one trapdoor bit (through $\mathcal{F}_{rltdf}(\cdot)$). Therefore, there are exactly $(v-3)$ trapdoor bits generated from Eq. 2. Since data security of each $\mathcal{T}_{tb}(\cdot)$ is covered under QRA, the server sends plain trapdoor bit sets directly. In order to send the last bit $b_{z,v}$ and the ciphertexts $\beta_{z,1}, \beta_{z,2}$ securely, the scheme encrypts $b_{z,v}$ and $\beta_{z,1}, \beta_{z,2}$ using a Paillier tree with height 1 as

$$\mathcal{F}_{pil}(\beta_{z,1}, \beta_{z,2}, \cdot) = (C = c_1 + Nc_2)$$

$$\mathcal{F}_{pil}(c_1, \mathcal{T}_{tb}(b_{z,v}, b_{z,v}, \cdot), \cdot) = C_{z,1} \qquad \mathcal{F}_{pil}(c_2, \mathcal{F}_{gm}(b_{z,v}, \cdot), \cdot) = C_{z,2}$$

Therefore, the final response $\mathcal{R} = \{(C_{z,1}, C_{z,2}, S_z, \mathcal{T}_z) : z \in [1, u]\}$.

- *Response Retrieval (RR)*: Initially, decrypts $C_{z,1}$ and $C_{z,2}$ to get the last bit $b_{z,v}$ and $(\beta_{z,1}, \beta_{z,2})$ by inverting the above Paillier tree. Further, using private key (p, q), $b_{z,v}$, $(\beta_{z,1}, \beta_{z,2})$ and $(S_z, \mathcal{T}_z)$, inverts both recursive functions described in Eq. 2 to get the required bit or the string $\mathcal{DB}_z$.

Theorem 1 *All the queries generated from query generation algorithm (QG) exhibit perfect privacy.*

Proof For any randomly selected inputs $\alpha, \beta \in \mathbb{Z}_N^{+1}$, let $\mathcal{Q}_i^{\alpha} = (\cdot,(\cdot, \cdot),\cdot, \alpha)$ and $\mathcal{Q}_j^{\beta} = (\cdot,(\cdot, \cdot),\cdot, \beta)$, $i, j \in [u]$. Since the inputs α, β are uniformly distributed over $\mathbb{Z}_N^{+1}$, $\text{PR}[\mathcal{Q}_i^{\alpha} : Ad(n, 1^k, \mathcal{Q}_i^{\alpha}) = 1] = \text{PR}[\mathcal{Q}_j^{\beta} : Ad(n, 1^k, \mathcal{Q}_j^{\beta}) = 1] = \text{PR}[\mathcal{Q}_i^{\beta} : Ad(n, 1^k, \mathcal{Q}_i^{\beta}) = 1] = \text{PR}[\mathcal{Q}_j^{\alpha} : Ad(n, 1^k, \mathcal{Q}_j^{\alpha}) = 1]$, where $\text{PR}[\cdot]$ is the privacy probability and Ad is the curious server. Therefore, all the randomly generated queries with their uniformly distributed inputs are always "identically distributed" over $\mathbb{Z}_N^{+1}$. All such identically distributed queries always exhibit perfect privacy.

Theorem 2 *The server communication in terms of total number of trapdoor bits is always non-trivial.*

Proof It is clear from Eq. 2 that for all $z \in [u]$, $|S_z| = (v/2) - 1$ and $|\mathcal{T}_z| = (v/2) - 2$. Therefore, there are $(v/2) - 1 + (v/2) - 2 = v - 3$ number of trapdoor bits generated during the encryption of each string $\mathcal{DB}_z$. In total, there are $u(v - 3)$ trapdoor bits generated from the *RC* algorithm which is always less than the database size uv.

Computation reduction: In order to achieve computational efficiency, instead of using only sufficiently large security parameter k, we can use a combination of sufficiently large and negligible security parameters in the *RC* algorithm of the proposed scheme. For any negligible security parameter $k' < k$ and $\forall \alpha \in \mathbb{Z}_N^{+1}$, $\forall i \in [u]$, $\text{PR}[\mathcal{Q}_i^{\alpha} \xleftarrow{R} \{0,1\}^k : Ad(n, 1^k, \mathcal{Q}_i^{\alpha}) = 1] = \text{PR}[\mathcal{Q}_i^{\alpha} \xleftarrow{R} \{0,1\}^{k'} : Ad(n, 1^{k'}, \mathcal{Q}_i^{\alpha}) = 1]$, since user privacy is independent of the security parameter. (Note that the data privacy is still dependent on the security parameter). The main logic here is to generate a separate set of (public, private) key pairs using both k and k' from the *QG* algorithm, encrypt $S_{z,low}$ with the public key generated from k and encrypt $S_{z,high}$ with the public key generated from k' or vice versa from the *RC* algorithm so that the overall computation time for the PIR response is reduced.

4 Private DNS Resolution Application

The greatest advantage of the proposed scheme is that it can be directly adapted to existing DNS data structures. Let us consider that a *DNS resolver* wants to privately retrieve the address of the domain name "sit.ac.in". In the iterative mode of DNS resolution through *DNS resolver*, each communicating party namely the *DNS resolver* (termed as *res*) and the respective *name server* adopt the proposed PIR scheme of Sect. 3 in which *DNS resolver* generates a perfect privacy-preserving PIR query $\mathcal{Q}_{(res)->(\cdot)}$ using query generation algorithm *QG* which it sends to the respective name server. The name server creates the PIR response $\mathcal{R}_{(\cdot)->(res)}$ using response creation algorithm *RC* and sends it back to the *DNS resolver*. The resolver retrieves the respective sub-domain privately using the response retrieval algorithm *RR*. The process continues until the *DNS resolver* gets the address of the leftmost domain name (i.e., *sit*) as shown in the left part of Fig. 1. In recursive mode, each parent node applies the proposed PIR scheme to each of its child nodes separately and each child node sends the PIR response of its own database and PIR responses of its child nodes. Ultimately, the *root* name server sends the PIR responses of all the nodes of the tree to the *DNS resolver*. Finally, *DNS resolver* retrieves the required address of the domain name as shown in Fig. 1.

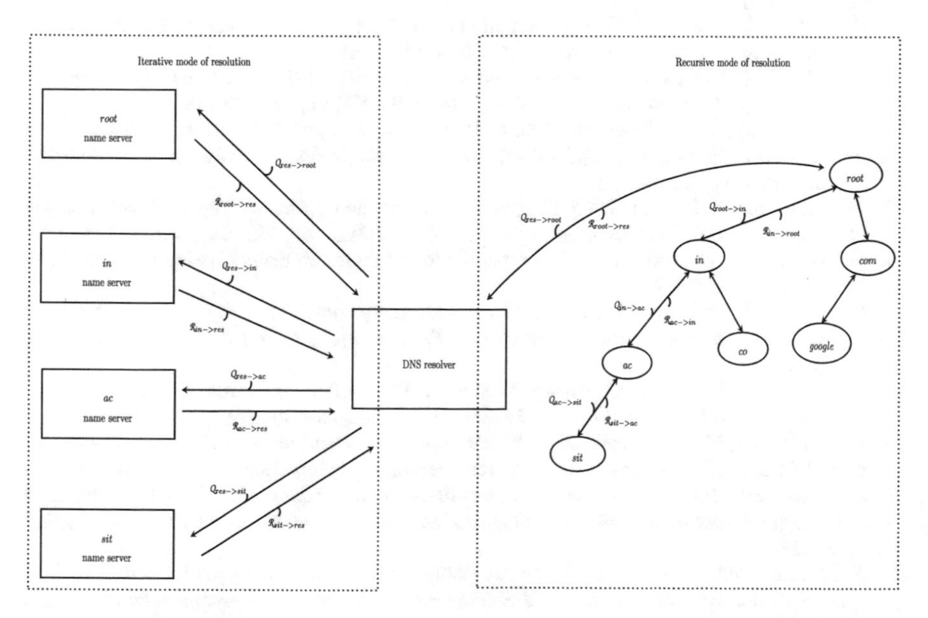

Fig. 1 Perfect privacy preserving private DNS resolution in both iterative and recursive modes through *DNS resolver*

5 Conclusion with Open Problems

We have successfully constructed a computationally efficient perfect privacy-preserving PIR scheme for private DNS resolution using the recursive 2-bit encryption function. The proposed scheme can be easily adopted in both iterative and recursive DNS resolution processes. There are several open problems such as reducing further communication, increasing robustness and concurrent process execution facility in DNS resolution that need deeper and more comprehensive understanding.

References

1. Benny, C., G. Niv, and N. Moni. 1998. Private information retrieval by keywords. Cryptology ePrint Archive, Report 1998/003. http://eprint.iacr.org/1998/003.
2. Cachin, C., S. Micali, and M. Stadler. 1999. In *Computationally private information retrieval with polylogarithmic communication*, 402–414. Berlin: Springer.
3. Cachin, C., S. Micali, M. Stadler. 1999. Computationally private information retrieval with polylogarithmic communication. In *Proceedings of 17-th theory and application of cryptographic techniques. EUROCRYPT'99*, 402–414. Berlin: Springer.
4. Chang, Y.C. 2004. Single database private information retrieval with logarithmic communication, 50–61. Berlin: Springer.
5. Chor, B., N. Gilboa. 1997. Computationally private information retrieval (extended abstract). In *Proceedings of 29-th STOC. STOC '97*, 304–313. ACM.
6. Chor, B., O. Goldreich, E. Kushilevitz, and M. Sudan. 1995. Private information retrieval. In *Proceedings of the 36-th FOCS. FOCS '95*, 41–50. IEEE Computer Society.
7. Freeman, D.M., O. Goldreich, E. Kiltz, A. Rosen, and G. Segev. 2009. More constructions of lossy and correlation-secure trapdoor functions. Cryptology ePrint Archive, Report 2009/590. http://eprint.iacr.org/2009/590.
8. Gentry, C., and Z. Ramzan. 2005. Single-database private information retrieval with constant communication rate. In *Proceedings of 32nd ICALP. ICALP'05*, 803–815. Berlin: Springer.
9. Goldwasser, S., and S. Micali. 1984. Probabilistic encryption. *Journal of computer and system sciences* 28 (2): 270–299.
10. Groth, J., A. Kiayias, and H. Lipmaa. 2010. Multi-query computationally-private information retrieval with constant communication rate. In *Proceedings of 13-th PKC. PKC'10*, 107–123. Berlin: Springer.
11. Ishai, Y., E. Kushilevitz, R. Ostrovsky, A. Sahai. 2006. Cryptography from anonymity. In *Proceedings of 47-th FOCS. FOCS '06*, 239–248. IEEE Computer Society.
12. Kaiser, D., and M. Waldvogel. 2014. Efficient privacy preserving multicast dns service discovery. In *2014 IEEE international conference on high performance computing and communications; 2014 IEEE 6th international symposium on cyberspace safety and security; 2014 IEEE 11th international conference on embedded software and system (HPCC,CSS,ICESS)*, 1229–1236.
13. Kang, A.R., and A. Mohaisen. 2016. Assessing DNS privacy under partial deployment of special-use domain names. In *2016 IEEE conference on communications and network security (CNS)*, 358–359.
14. Kushilevitz, E., and R. Ostrovsky. 1997. Replication is not needed: Single database, computationally-private information retrieval. In *Proceedings of 38-th FOCS. FOCS '97*, 364. IEEE Computer Society.
15. Kushilevitz, E., and R. Ostrovsky. 2000. One-way trapdoor permutations are sufficient for non-trivial single-server private information retrieval. In *Proceedings of 19-th Theory and Application of Cryptographic Techniques. EUROCRYPT'00*, 104–121. Berlin: Springer.

16. Lipmaa, H. 2010. First cpir protocol with data-dependent computation. In *Proceedings of 12-th information security and cryptology. ICISC'09*, 193–210. Berlin: Springer.
17. Melchor, C.A., and P. Gaborit. 2007. A lattice-based computationally-efficient private information retrieval protocol.
18. Paillier, P. 1999. In *Public-key cryptosystems based on composite degree residuosity classes*, 223–238. Berlin: Springer.
19. Radhakrishna, B., and N. Sunitha. 2014. Optar: Optional pir based trusted address resolution for dns. *IJACEN* 2 (8): 23–28.
20. Trostle, J., and A. Parrish. 2011. Efficient computationally private information retrieval from anonymity or trapdoor groups. In *Proceedings of 13-th ISC. ISC'10*,114–128. Berlin: Springer.
21. Yuchi, X., G. Geng, Z. Yan, and X. Lee. 2017. Towards tackling privacy disclosure issues in domain name service. In *2017 IFIP/IEEE symposium on integrated network and service management (IM)*, 813–816.
22. Zhao, F., Y. Hori, and K. Sakurai. 2007. Analysis of privacy disclosure in dns query. In *2007 international conference on multimedia and ubiquitous engineering (MUE'07)*, 952–957.
23. Zhao, F., Y. Hori, and K. Sakurai. 2007. Two-servers pir based dns query scheme with privacy-preserving. In *The 2007 International Conference on Intelligent Pervasive Computing (IPC 2007)*, 299–302.
24. Zhu, L., Z. Hu, J. Heidemann, D. Wessels, A. Mankin, and N. Somaiya. 2015. Connection-oriented DNS to improve privacy and security. In *2015 IEEE symposium on security and privacy*, 171–186.

Support Vector Regression for Predicting Binding Affinity in Spinocerebellar Ataxia

P. R. Asha and M. S. Vijaya

Abstract Spinocerebellar ataxia (SCA) is an inherited disorder. It arises mainly due to gene mutations which affect gray matter in the brain causing neurodegeneration. There are certain types of SCA that are caused by repeat mutation in the gene, which produces differences in the formation of protein sequence and structures. Binding affinity is very essential to know how tightly the ligand binds with the protein. In this work, a binding affinity prediction model is built using machine learning. To build the model, predictor variables and their values such as binding energy, IC_{50}, torsional energy and surface area for both ligand and protein are extracted from the complex using AutoDock, AutoDock Vina and PyMOL. A total of 17 structures and 18 drugs were used for learning the support vector regression (SVR) model. Experimental results proved that the SVR-based affinity prediction model performs better than other regression models.

Keywords Binding affinity · Docking · Ligand · Machine learning
Prediction · Protein · Protein structure

1 Introduction

Spinocerebellar ataxia is a hereditary disorder characterized by gray matter alterations affecting motor control. The disorder is due to mutations in the genes, which result in brain and spinal cord degeneration. Each SCA type features its own symptoms [1]. SCA occurs due to genetic mutations in the genes that produce differences in the formation of protein sequence and structures. Certain types of spinocerebellar ataxia are caused by repeat mutations. They are SCA1, SCA2,

P. R. Asha (✉) · M. S. Vijaya
Department of Computer Science, PSGR Krishnammal College for Women,
Coimbatore, India
e-mail: ashamscsoft@gmail.com

M. S. Vijaya
e-mail: msvijaya@psgrkc.com

A. N. Krishna et al. (eds.), *Integrated Intelligent Computing, Communication and Security*, Studies in Computational Intelligence 771,
https://doi.org/10.1007/978-981-10-8797-4_19

SCA3, SCA6, SCA7, SCA8 and SCA10 [2]. A parent possessing 39 repeats of polyglutamine may pass this to their offspring, resulting in an increase of repeats and causing mutation [3].

Molecular docking is used to forecast binding modes. Earlier study on the ligand-receptor binding procedure gave us the lock-and-key principle, whereby the protein fits into the macromolecule a bit like a lock and key. Induced-fit principles takes the lock-and-key theory an additional step, stating that the active site of the macromolecule is constantly reshaped by interactions with the ligands, since the ligands communicate with the molecule [4, 5]. There is plenty of docking software available [6]. If a structure is not available for the sequence, homology modeling can be done to get the structure.

Affinity is a measure of the strength of attraction between a receptor and its ligand. The capability of substance to create a co-ordination bond with a receptor is known as binding affinity. The binding affinity of a substance with a receptor depends upon the interaction force of attraction between the ligands and their receptor binding sites. A substance with high-affinity binding shows an extended duration at the receptor binding site compared to low-affinity binding.

A matter area unit getting to be any drug or emotional chemical which might be detected by the target cell. Some common ligands are a unit hormones, mediators, and neurotransmitters, etc. Generally, ligands are endogenous in nature as they're created naturally within the body. for each substance there's a elite receptor that they need to be complementary to every utterly completely different in their size and science.

The binding affinity of a substance to a receptor is crucial, as a number of the binding energy is employed within the receptor to cause a conformational modifications. This leads to altered behavior to a degree of the associated particle channel or the supermolecule. Ligands similar to medication also can have some specificity for the binding website on a receptor similar to a target organic compound. Therefore, the efficiency of a drug depends on its binding affinity for the binding data processor additionally as its binding ability to cause the specified effects [7].

Binding affinity prediction through machine learning is important, as much research is carried out by capturing the affinity values, which are available in database such as PDBBind, AffinDB1, AffinDB.

Many analyses are meted out with the complexes that binding affinity is thought. the aim has been studied and also the want is known from literature survey.

Xueling Li et al. proposed a method for automatic protein-protein affinity binding based on svr-ensemble. Two-layer support vector regression (TLSVR) model is used to implicitly capture binding contributions that are hard to explicitly model. The TLSVR circumvents both the descriptor compatibility problem and the need for problematic modeling assumptions. Input features for TLSVR in first layer are scores of 2209 interacting atom pairs within each distance bin. The base SVRs are combined by the second layer to infer the final affinities. This model obtains a good result [8].

Volkan Uslan, Huseyin Seker proposed a method for peptide binding affinity using hybrid fuzzy support vector regression. A hybrid system (TSK-SVR) that has helped improve the prophetical ability of TSK-FS considerably with the help of support-based vector methodology was developed and incontestable with the productive applications within the prediction of amide binding affinity being considered one among the tough modelling issues in bioinformatics. prophetical performances are improved the maximum amount as thirty fourth in comparison to the simplest performance given within the literature. The improvement gain for all tasks is found to be thirteen. 8%. aside from up the prediction accuracy, this analysis study has conjointly known aminoalkanoic acid options "Polarity," "Hydrophobicity constant," and "Zimm–Bragg parameter" being the extremely discriminating options within the amide binding affinity information sets. Therefore, these aminoalkanoic acid options is also doubtless thought of for higher style of peptides with applicable binding affinity [9].

Manoj Bhasin, G.P.S. Raghava implemented on the analysis and prediction of binding affinity of faucet binding peptides victimization cascade SVM. They projected with a dataset of 409 peptides that bind to human faucet transporter with varied affinity were analyzed to explore the property and specificity of faucet transporter. The quantitative matrix was generated on the idea of the contribution of every position and residue in binding affinity. The correlation of r = zeroes. 65 was obtained between by experimentation determined and foreseen binding affinity by employing a quantitative matrix. More a support vector machine (SVM)-based technique has been developed to model the faucet binding affinity of peptides. The correlation (r = zero.80) was obtained between the expected and experimental measured values by victimization sequence-based SVM. The responsibility of prediction was more improved by cascade SVM that uses options of amino acids along side the sequence. an especially smart correlation (r = zero.88) was obtained between measured and foreseen values, once the cascade SVM-based technique was evaluated through jackknife testing [10].

Volkan Uslan, Huseyin Seker proposed a method for the quantitative prediction of HLA-B * 2705 amide binding affinities mistreatment Support Vector Regression to achieve insights into its role for the Spondyloarthropathies. They projected the prediction of human category I MHC cistron HLA-B * 2705 binding affinities mistreatment international intelligence agency, before process, standardization and have choice were performed. The descriptors that form the amino acid composition were normalised to [0, 1] to ensure that every descriptor represented within the same range of values. Subsequently, Multi-Cluster Feature choice (MCFS) methodology was used. the prediction of human category I MHC allelomorph HLA-B * 2705 binding affinities mistreatment Sluzhba Vneshney Razvedki, before process, normalization and have choice were performed. The descriptors that form the amino acid composition were normalised to [0, 1] to ensure that every descriptor represented within the same range of values. Subsequently, Multi-Cluster Feature Selection (MCFS) method was used [11].

From the background study it absolutely was perceived that the majority of the works were supported the complex for which binding affinity was provided by the database. This emphasizes the need of more research on affinity prediction with known structures and unknown drugs by planning and explanation the effective options for generating new model. This research work is proposed to create a model from which the affinity is predicted. A total of 17 structures and 18 drugs were used to prepare the training data set.

2 Affinity Prediction Model

The binding interaction between a single biomolecule to its ligand defines the binding affinity. Affinity prediction is very essential to discover new drugs for any disorder or disease. Therefore the affinity prediction of protein and ligand is very challenging and complicated task. The main focus of this analysis is to supply associate economical machine learning approach for predicting affinity. Six types of SCA are caused by repeats mutation namely SCA1, SCA2, SCA3, SCA6, SCA7 and SCA8.

Each gene has a number of three dimensional protein structures which can be accessed from PDB [12] is a crystallographic information for the three-dimensional structural information of huge biological molecules, like proteins and nucleic acids. Three dimensional structures for each gene is obtained from OCA and proteopedia which is available in gene cards [13]. The framework of proposed model is illustrated in Fig. 1.

3 Methodology

Affinity Prediction is essential in assessing the performance of drugs. The significance of affinity prediction in the field of drug assessment will lead to produce better drugs. The research work is initiated by structure acquisition from gene cards, which contains structure that is mapped with six types of SCA. Structure is docked with the drugs to get essential features to build the model which predicts binding affinity. Training dataset is prepared from docked complexes and features are extracted. Affinity prediction models are built using support vector regression.

3.1 Docking

Docking is performed by preparing receptor (protein) and ligand (drug). Protein is manufactured by converting the protein into pdbq format that is by adding hydrogen, computing geastier charge and kolman charges are added. Protein pdbq is converted to pdbqt by not adding partial charges to the protein because the protein possesses the charge. Protein is covered by grid box and size, centre of the box are noted and it is saved as grid parameter file. Ligand is prepared by converting the ligand into pdbqt format that is by detecting root in torsion tree. In search parameters genetic algorithm is used. Population size and number of GA

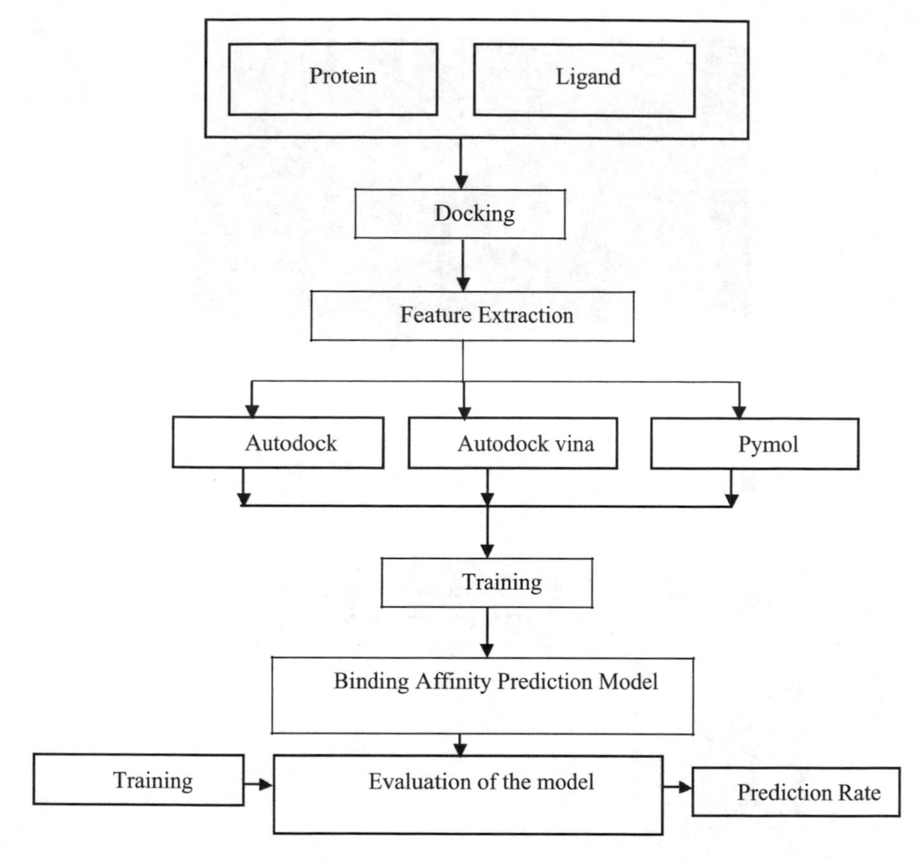

Fig. 1 Affinity prediction model

runs is given as 150 and 100 respectively. The number of runs should be given 100 as minimum to get better docking results. Each run will give one docked conformation. Output is saved as docking parameter file. The next step is to run autogrid which will generate grid log file and autodock generates docking log file. The docking log file is essential to check the docking results. The docking conformation is chosen based on which conformation is high and provide better result. The best docking conformation for a sample complex is shown in Fig. 2.

The explanatory variable and response values that are identified from auto dock, auto dock vina and PYmol plays vital role in building model that predicts affinity in an effective way. The features that are identified to build binding affinity prediction model are binding energy range, binding energy, ligand efficiency, pIC_{50}, electrostatic energy, torsional energy, internal energy, intermolecular energy, vanderwal's energy, complex RMS, ligand RMS, binding affinity, rmsd l.b, rmsd u.b, molecular weight of ligand, protein and complex, charges for protein, surface area for molecule and solvent for protein, ligand and complex. The better docking conformation with

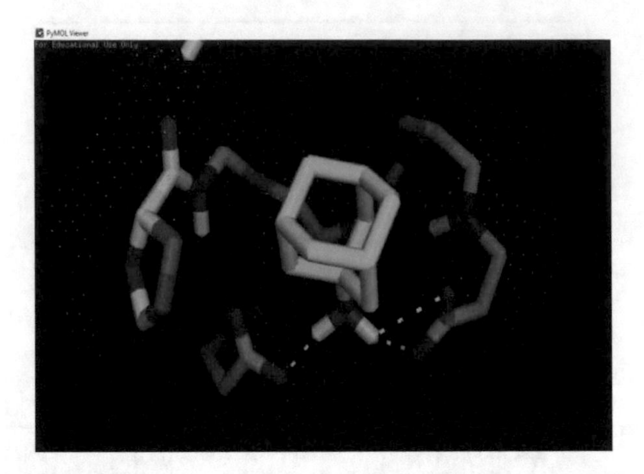

Fig. 2 Best docking conformation

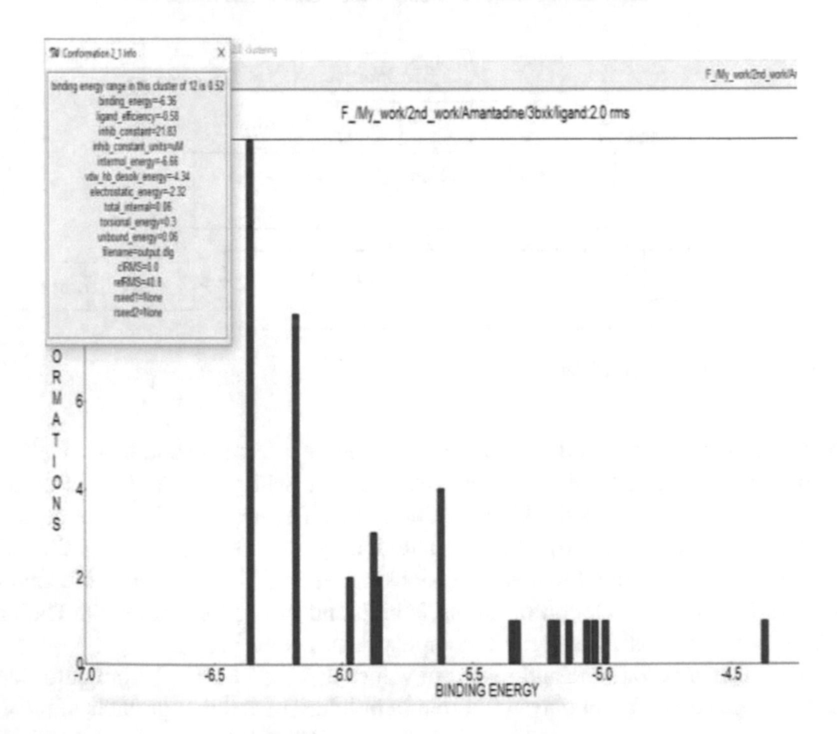

Fig. 3 Values for better conformation

values is shown in Fig. 3. The predictor variable and their values are taken from tools like autodock, autodock vina and PYmol are explained in Table 1.

Other independent values are identified from Pymol are molecular weight of protein, ligand and complex, surface area of both solvent and molecular for protein, ligand and complex, charges for protein (Fig. 4).

Table 1 Predictor variables and their description

Predictor variables	Description
Binding energy range	Range of cluster in which energy falls
Binding energy	Affinity of ligand-protein complex
Ligand efficiency	Binding energy per atom of ligand to protein
Inhibition constant	Confidence of inhibitor
Intermolecular energy	Energy between non-bounded atoms
Desolvation energy	Energy lose of the interaction between substance
Electrostatic energy	Amendment on the electricity non delimited energy of substance
Total internal energy	Change of all energetic terms
Torsional energy	Dihedral term of internal energy
clRMS	Difference between current conformation and the lowest energy conformation in its cluster
refRMS	Difference between current conformation coordinates and current reference structure
Binding affinity	Strength of attraction between a molecule and ligand
Rmsd/lb	Differing in however atoms square measure matched within the distance calculation
Rmsd/ub	Matches every atom in one conformation with itself within the alternative conformation

3.2 *Support Vector Regression*

Support vector machines tackle classification and regression issues of non-linear mapping input file into high dimensional feature areas, whereby a linear call surface is meant. Support vector regression algorithms square measure supported the results of the applied mathematics theory of learning given by vapnik, that introduces regression on the fitting of a record v to the information. SVM regression estimates the worth of w to get the operate

$$f(\bar{x}) = (\bar{w} \cdot \bar{x}) + b, \bar{w}, \bar{x} \in R^N, b \in R$$

By introducing insensitive loss function as

$$Y - f(\bar{x})/e = max\{o, |y - f(\bar{x})| - \varepsilon\}$$

It retains all the properties that characterize highest margin algorithms of support vector machines like duality, sparseness, kernel and convexity. It's become a strong

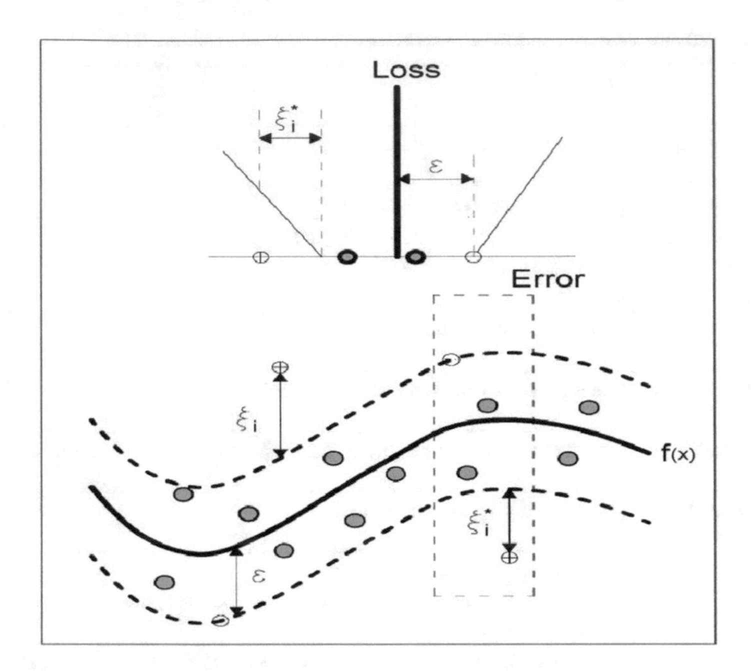

Fig. 4 Loss & error of SVR

technique for prophetical information analysis with several applications in varied areas of study like biological contexts, drug discovery, technology, topographic point frequency prediction, image pursuit, compression etc.

The model made by support vector regression (SVR) solely depends on a set of the coaching information, as a result of the price perform for building the model ignores any coaching information that ar near the model prediction.

The support vector regression uses a value perform to live the empirical risk so as to reduce the regression error. Though there are several decisions of the loss functions to calculate the value, e.g., least modulus loss perform, quadratic loss perform, etc., the ε insensitiveness loss perform is such a perform that exhibits the sparsely of the answer [14].

4 Experiment and Results

The dataset with 306 instances are prepared from docked complex. The data analysis and binding affinity prediction was allotted victimisation LIBSVM tool for support vector regression [15]. LIBSVM is easy, easy-to-use and economical computer code for classification and regression. It solves C-SVM classification, nu-SVM classification, one class-SVM, epsilon-SVR and nu-SVR. LIBSVM consists of a coaching module (svm-train) and a prediction module (svm-predict).

The performances of the three models of SVRs with linear, polynomial and RBF kernels were evaluated based on three criteria, the prediction accuracy, correlation coefficient and mean squared error.

Correlation may be a standardized variety that describes, however closely the 2 variables area unit connected. The parametric statistic continuously lies between −1 and one. A constant of one representing good direct correlation, zero represents no correlation, and −1 represents smart correlation. Statistic measures the applied mathematical correlation between the actual values ai and expected values pi exploitation the formula,

$$\text{Correlation Coefficient} = S_{PA}/\sqrt{S_P S_A}$$

$$\text{where } S_{PA} = \sum_i \frac{(p_i - \bar{p})(a_i - \bar{a})}{n-1}, \quad Sp = \sum_i \frac{(p_i - \bar{p})^2}{n-1}$$

$$\text{and } S_A = \frac{(a_i - \bar{a})^2}{n-1}$$

Error term is that the excellence between the regression line and actual info points accustomed construct the road and root mean sq. error is evaluated practice,

$$\text{RMSE} = \frac{\sqrt{(p_1 - a_1)^2 + \ldots + (p_n - a_n)^2}}{n}$$

Support vector regression with linear, polynomial, RBF kernels is enforced with parameters C-regularization parameter, d—degree of polynomial and gamma severally. Regularization parameter C is assigned totally different values within the vary of 0.1–0.5 and located that the model performs higher and reaches a stable state for the worth C = 0.5. The results of the prediction model supported nu-SVR with linear kernel is shown in Table 2.

Degree of polynomial d is assigned totally different values within the vary of one to four with c = 1. The results of the prediction model supported nu-SVR with polynomial kernel is shown in Table 3.

Gamma g is assigned in several values from zero.5 to one with the regularization parameter c and located that model performs higher and reaches stable within the

Table 2 SVR with linear kernel

Linear kernel	c = 0.1	c = 0.2	c = 0.3	c = 0.4	c = 0.5
Accuracy	60	65.3	65.3	65.5	65.5
Correlation coefficient	0.7	0.72	0.72	0.73	0.73
Mean squared error	0.1724	0.1890	0.1890	0.2450	0.2450

Table 3 SVR with polynomial kernel

D	d = 1	d = 2	d = 3	d = 4
Accuracy	80	75	60	58
Correlation coefficient	0.8	0.7	0.55	0.42
Mean squared error	0.2463	0.3654	0.5267	0.6620

Table 4 SVR with RBF kernel

G	c = 1		c = 2	
	g = 0.5	g = 1	g = 0.5	g = 1
Accuracy	80	82	82.5	83
Correlation coefficient	0.9	0.92	0.93	0.94
Mean squared error	0.0691	0.0458	0.0458	0.0005

Table 5 Performance analysis of SVR

SVR kernels	Accuracy prediction (%)
Linear	63
Polynomial	69
RBF	82

values c = 1, g = 1. The results of the prediction model supported nu-SVR with RBF kernel is shown in Table 4.

The average and comparative performance of the international intelligence agency primarily based regression models in terms of prophetic accuracy, correlation coefficient and mean squared error for binding affinity prediction is given in Table 5 and performance evaluation is given in Fig. 5.

The experiments were also carried out in our previous work with the same dataset using other regression techniques like Random forest, artificial neural

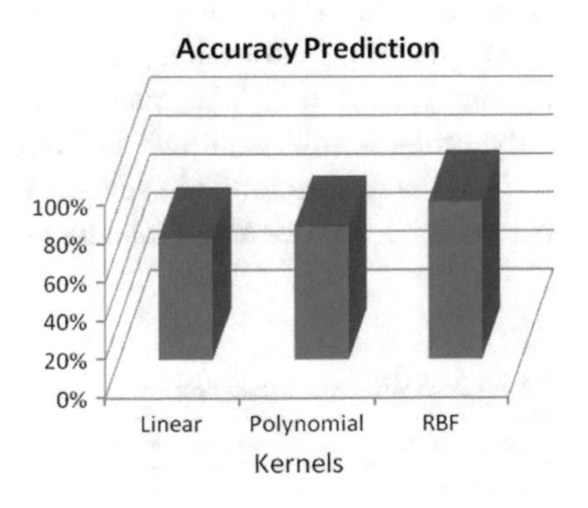

Fig. 5 Performance evaluation of SVR

Table 6 Performance analysis of all regression models

Regression algorithm	Accuracy prediction (%)
SVR	82
RF	78
ANN	60
LR	75

networks and linear regression, and the recognition rate obtained was 78%, 60% and 75% respectively. From the above results it is observed that SVR with RBF kernel provides better result when compared to three regression techniques. The results of various regression techniques are given in Table 6.

5 Conclusion

This work demonstrates the event of affinity prediction model exploitation 3 dimensional structure of macromolecule and matter. The values of AutoDock, Vina and PYmol are extracted as options and a model is constructed by using supervised machine learning algorithms including RF, ANN, simple regression and SVR. The performance of the training strategies was evaluated and supported their prognosticative accuracy. The results indicate that the SVR outperforms in prediction than Random Forest, simple regression and artificial neural network. Because the drug is crucial for the disorder, it demands additional correct prediction of binding affinity. It's ascertained that the SVR model with RBF kernel is healthier in predicting the binding affinity. The work can be further extended by adding more structures and ligands. and repeating the experiment with other regression techniques.

References

1. Thomas, C. Weiss. 2010. *Ataxia spinocerebellar: SCA facts and information.*
2. Thomas, D. Bird. 2016. *Hereditary ataxia overview.*
3. Whaley, N.R., S. Fujioka, and Z.K. Wszolek. 2011. Autosomal dominant cerebellar ataxia type I: A review of the phenotypic and genotypic characteristics. https://doi.org/10.1186/1750-1172-6-33.
4. Fischer, E. 1894. Einfluss der configuration auf die working derenzyme. *Berichte der Deutschen Chemischen Gesellschaf* 27: 2985–2993.
5. Koshland Jr., D.E. 1963. Correlation of structure and function in enzyme action. *Science* 142: 1533–1541.
6. Kuntz, I.D., J.M. Blaney, S.J. Oatley, R. Langridge, and T.E. Ferrin. 1982. A geometric approach to macromolecule-ligand interactions. *Journal of Molecular Biology* 161 (2): 269–288.
7. http://chemistry.tutorvista.com/inorganic-chemistry/binding-affinity.html.

8. Li, X., M. Zhu, X. Li, H.Q. Wang, and S. Wang. 2012. Protein-protein binding affinity prediction based on an SVR ensemble. In *Intelligent Computing Technology*, ICIC 2012, ed. D.S. Huang, C. Jiang, V. Bevilacqua, J.C. Figueroa, vol. 7389. Lecture Notes in Computer Science, Springer: Berlin, Heidelberg.
9. Volkan, Uslan, and Huseyin Seker. 2016. Quantitative prediction of peptide binding affinity by using hybrid fuzzy support vector regression. *Applied Soft Computing* 43: 210–221.
10. Bhasin, M., and G.P.S. Raghava. 2004. Analysis and prediction of affinity of TAP binding peptides using cascade SVM. *Protein Science: A Publication of the Protein Society* 13 (3): 596–607. https://doi.org/10.1110/ps.03373104.
11. Volkan, Uslan, and Huseyin Seker. 2016. Binding affinity prediction of S. Ccerevisiae 14-3-3 and GYF peptide-recognition domains using support vector regression. In *2016 IEEE 38th annual international conference of the engineering in medicine and biology society (EMBC)*, 3445–3448, ISSN 1558-4615.
12. Berman, Helen M., John Westbrook, Zukang Feng, Gary Gilliland, T.N. Bhat, Helge Weissig, Ilya, N. Shindyalov, and Philip E. Bourne. 2000. Protein data bank, *Nucleic Acids Research*, 28 (1): 235–242.
13. Rebhan, M., V. Chalifa-Caspi, J. Prilusky, and D. Lancet. 1997. GeneCards: Integrating information about genes, proteins and diseases. *Trends in Genetics* 13: 163.
14. Soman, K.P., R. Loganathan, and V. Ajay. 2009. *Machine learning with SVM and other kernel methods.*
15. LIBSVM is an open source tool. http://www.csie.ntu.edu.tw/cjlin/libsvm.

HDFS Logfile Analysis Using ElasticSearch, LogStash and Kibana

B. Purnachandra Rao and N. Nagamalleswara Rao

Abstract Hadoop is an open-source software framework for storing and processing large sets of data on a platform consisting of commodity hardware. Hadoop is mostly designed to handle large amounts of data, which can easily run into many petabytes and even exabytes. Hadoop file sizes are usually very large, ranging from gigabytes to terabytes, and large Hadoop clusters store millions of these files. Hadoop depends on large number of servers so it can parallelize work across them. Server and storage failures are to be expected, and the system is not affected by non-functioning storage units or even failed servers. Traditional databases are geared mostly for fast access to data and not for batch processing. Hadoop was originally designed for batch processing, such as the indexing of millions of web pages, and provides streaming access to datasets. Data consistency issues that may arise in an updatable database are not an issue with Hadoop file systems, because only a single writer can deal with write operation. Activity on the server will be captured by logs. There are two types of logs, and they can be generated from web servers or application servers or both. Access logs and error logs are two types of log files. An access log will have client info, whereas an error log consists of exceptions and error info. This chapter will address the log file analysis process using ElasticSearch, LogStash and Kibana. We can show the frequency of errors by the given time period using different forms such as trend graphs, bar graphs, pie charts and gauge charts.

Keywords HDFS · NameNode · DataNode · ElasticSearch
LogStash · Kibana · Index · Discover · Visualize · Dashboard
Timestamp · Error message · Log file analysis

B. Purnachandra Rao (✉) · N. Nagamalleswara Rao
Department of Computer Science & Engineering, ANU College of Engineering & Technology, Guntur, India
e-mail: pcr.bobbepalli@gmail.com

A. N. Krishna et al. (eds.), *Integrated Intelligent Computing, Communication and Security*, Studies in Computational Intelligence 771,
https://doi.org/10.1007/978-981-10-8797-4_20

1 Introduction

Apache Hadoop [1] is an open source framework for storing and processing large amounts of data in a distributed environment among clusters. Hadoop systems have a high throughput feature and a write-once-read-many-times characteristic [2, 3]. When the system has a heavy workload, there are many interactions among the NameNode and DataNode. Disk-based file systems help Hadoop [1], MapReduce [4], Dryad and HPCC frameworks to meet their exponential storage demands. The NameNode system acts as a master server, storing namespace and regulating the client's access to files. The Hadoop Distributed File System (HDFS) support common file system operations such as read and write files and creates and deletes directories. Cluster has a number of DataNodes. Data storage is done by DataNodes. Client operations are performed as per the client instructions on the DataNode. The NameNode provides the empty blocks list from the namespace. Generally, the user data are stored in the files of HDFS. HDFS stores data in HDFS files, each of which consists of a number of blocks (default size is 128 MB). The default block size is customizable using the HDFS configuration.

2 Literature Review

Hadoop processes large amounts of data where a massive quantity of information is processed using a cluster of nodes. A server log file is a text file that is written as activity is generated by the web server. Log files gather a variety of data about information requests to your web server. Server logs play the role of an incoming sign-in sheet. Server log files can give information about what pages get the most and the least traffic [5]. Hadoop distributed file systems (HDFS) [6] have the capability to store huge amounts of data. The NameNode loads the edit log and replays the edit log to update the metadata it loaded into memory in the previous step. The NameNode also updates the fsimage file with the updated HDFS state information [7, 8]. The NameNode starts running with a fresh, empty edits file. The DataNode daemon connects to NameNode and sends it block reports that list all data blocks stored by a DataNode. When the HDFS client wants to read or write some data into/from the DataNode, the client contacts the NameNode for block info. Empty blocks connect as a pipeline [9, 10]. The client writes the data to first block, which is copied to a second block, and so on until the end of the pipeline. Then the acknowledgement is transferred back until the head of the data blocks in the pipeline. When there is a deletion operation, the hard link from the block is disconnected, thus the data block remains in the same directory [11]. The main role of log file analysis is in the trouble shooting process. Suppose a couple of transactions occurred and the outcome has mixed results. Based on the outcome, we can filter the success and failure of the job events. Data visualization is the presentation of analytics output in a pictorial or graphical format to understand the analysis better and make business decisions based on the data [12]. It shows the number of users interacting with the server on a daily basis and who interacts with only read permission and who with read and write operations. In this way, we can analyze the front-end operations using the log file analysis. The log data analysis is useful to analyze the data related to machine IP address, the usernames of those interacting frequently with the HDFS, the operations they are performing, such as read, write and delete. Once the analysis of these fields is complete, we can add the results to any visualizing tool.

3 Problem Statement

The HDFS operations are recorded in a log file. Log files can be generated from web servers and application servers. A web server has access log and error log. A log file can have any content with a unique delimiter. If a couple of applications send data to each other and the user wants to debug each operation, it is required to log into each server and go through the log file. If the number of applications becomes high, then it is hard to go through each and every box log file. With a small number of boxes or applications this is not an issue. But as the number of boxes increases then it becomes cumbersome to read the log files. It is better to collect and integrate all the server log info to read it collectively.

4 Proposal

Even if the number of applications increases, we can read all the log files in one database, irrespective of the data type, followed by parsing the log file and storing it in the database with representation. ElasticSearch is a noSQL database. LogStash is used for parsing the log files, and Kibana represents the data using dashboard. We can integrate the ELK stack into Hadoop log files so that the updates from the log file are automatic. Hadoop log files are stored as events in ElasticSearch. Each event is called a document, which is in JSON format. Logstash is integrated with the server where error logs are generated. Using LogStash, we can parse any type of log file. We can customize the parsing by writing matching patterns. Parsed log content will be stored in ElasticSearch. Kibana is a user interface tool to represent the data available in the ElasticSearch in different display formats such as bar graphs, gauge representation, line charts, and pie charts.

5 Implementation

We installed an ELK stack on the Windows system and wrote the LogStash configuration file to access the Hadoop log files available on the server and parse them using matching patterns. Then they are stored in ElasticSearch. All log contents are stored in the ElasticSearch index as documents in JSON format. Index is an equivalent term to relation instance, and event is equivalent to tuple in the relational database. All events are stored in Index. I have designed a custom dashboard using the data stored in the ElasticSearch index with different display formats such as graphs, line charts, and pie charts. Figure 1 shows Hadoop, EasticSearch, LogStash, and Kibana architecture. Hadoop log file 1, Hadoop log file 2, Hadoop log file 3 are three types of log files from the HDFS. These are referenced by the LogStash config file. This parses the data and stores the parsed data in ElasticSearch. Using customized dashboards such as graph, pie chart and line chart, we can represent the data in the Kibana dashboard.

The LogStash config file has three types of data: Input, filter and output. It is necessary to specify the log file location Input, any filter pattern in filter location

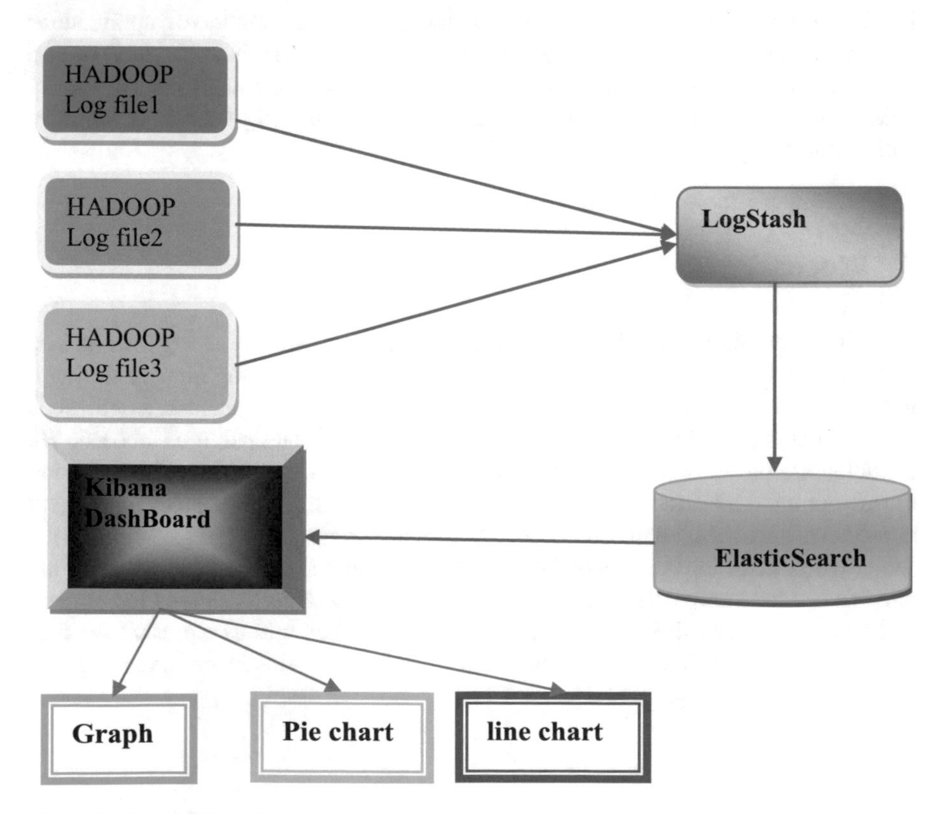

Fig. 1 Hadoop ELK architecture

```
    input {
    file {
      type => "BusinessDashboard_Image_Data"
        path => [ "C:/Logs/Hadoop/*.log" ]
        start_position => "beginning"
    }
}
    filter {
  grok {
     match => { "message" => "%{TIMESTAMP_ISO8601:timestamp} %{LOGLEVEL:loglevel} %{GREEDYDATA:message}" }
   }

}

output {

     stdout {
         codec => rubydebug
     }

     elasticsearch{
         hosts => localhost
         index => "logstash-cdsdate-%{+YYYY.MM.dd}"

     }
}
```

Fig. 2 LogStash config file format

and output is directed to ElasticSearch. Here we need to create an Index for each type of input so that the metrics for each input log file can be easily collected. Figure 2 shows the LogStash config file format. The log file data has different formats such as date (yyyy-mm-dd), hh:min:sec, and loglevel. We can parse the data using the config file and save the data according to field format.

Once we create the index in ElasticSearch, we can retrieve it from the Kibana dashboard, as shown in Fig. 3. Here we have opted for week-wise error/warn count. Using the Kibana dashboard, we can verify the error message. Kibana provides a couple of options for intervals, even as frequently as 1 s. Figure 4 shows the error frequency using a 60-s interval.

Figure 5 Dashboard with different types of graphs, including a bar graph, gauge, pie chart, area graph, and shows a trend in the error analysis.

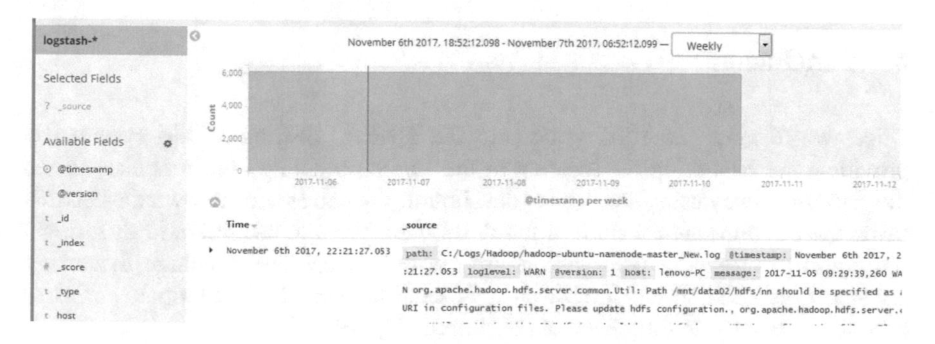

Fig. 3 ERROR/WARN with an interval of a week

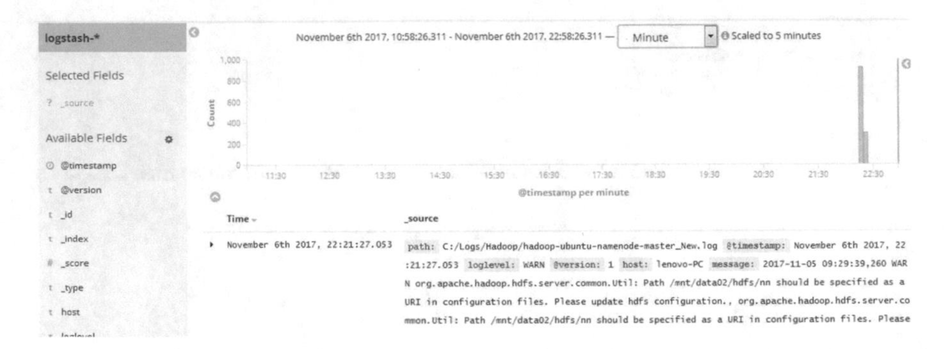

Fig. 4 Error with an interval of 60 s

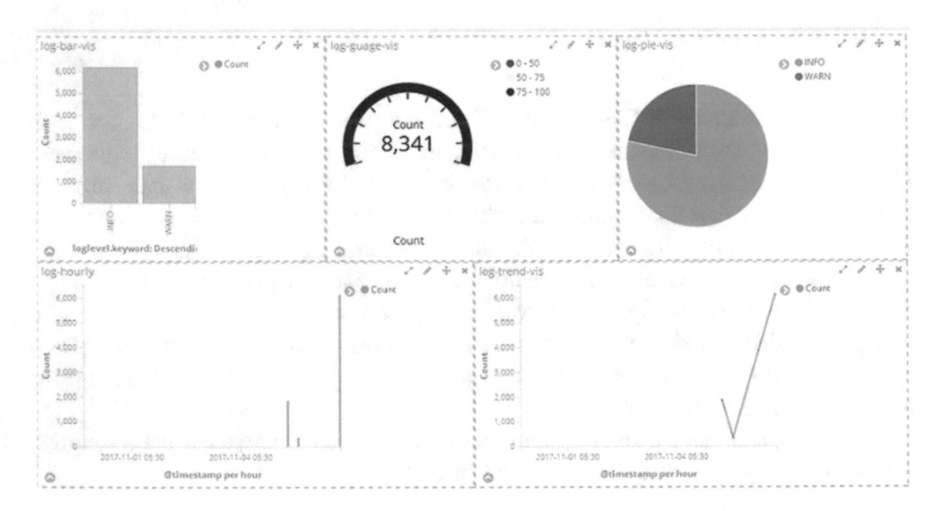

Fig. 5 Dashboard

6 Conclusion

Once we integrate the ELK stack into the Hadoop distributed file system (by providing the HDFS log file location in the LogStash config file), it is easy to see the error frequency using the Kibana dashboard. We can create our own dashboards using the customization feature in the dashboard. We can use filters to drill down the report to provide a field-wise report. With this one can conclude that using Kibana, ElasticSearch and LogStash it is easy to parse the Hadoop log files to produce a report using the Kibana dashboard (Fig. 6).

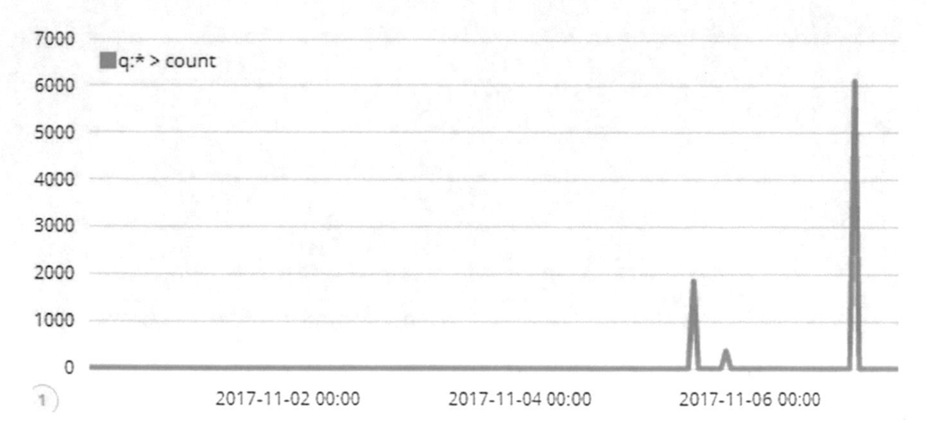

Fig. 6 Error timeline

References

1. Apache Hadoop. Available at Hadoop Apache.
2. Apache Hadoop. *Distributed file system*. Available at Hadoop Distributed File System Apache.
3. Scalability of hadoop distributed file system.
4. Kakade, Archana, and Suhas Raut. 2014. Hadoop distributed file system with cache technology. *Industrial Science* 1 (6). ISSN: 2347-5420.
5. Gavandi, Pushkar. 2016. Web server log processing using Hadoop. *International Journal for Research in Engineering Application & Management (IJREAM)*, 01 (10).
6. Dean, J., and S. Ghemawat. 2004. Mapreduce: Simplified data processing on large clusters. In *Proceedings of the 6th conference on symposium on operating systems design and implementation (OSDI'04)*, 137–150, Berkeley, CA, USA.
7. Tankel, Derek. 2010. *Scalability of Hadoop distributed file system*, Yahoo Developer Work.
8. Alapati, Sam R. *Expert Hadoop administration, managing, tuning and securing*.
9. Shafer, J., S. Rixner, and A.L. Cox. 2010. The Hadoop distributed Filesystem: Balancing portability and performance. In *Proceedings of IEEE international symposium on performance analysis of systems and software (ISPASS 2010)*, White Plains, NY.
10. Wang, Feng et al. 2009. *Hadoop high availability through metadata replication*, IBM China Research Laboratory, ACM.
11. Porter, George. 2010. Decoupling storage and computation in Hadoop with SuperDataNodes. *ACM SIGOPS Operating System Review*, 44.
12. Ankam, Venkat. *Big data analytics*. Packet Publishing Ltd. ISBN 978-1-78588-469-6.

Analysis of NSL-KDD Dataset Using K-Means and Canopy Clustering Algorithms Based on Distance Metrics

H. P. Vinutha and B. Poornima

Abstract Networks play a major role in our daily lives, and it is growing at an exponential rate. As networks grow, intruder activity also increases, so it is necessary to provide security. An intrusion detection system (IDS) is a model that can be used to analyze anomalous behavior in a network. In this chapter, the NSL-KDD dataset, which is an improved version of KDD-99 dataset, is used for this analysis. Clustering is an unsupervised learning method that divides the data into different groups based on their similarities. K-means is a straightforward algorithm with expensive computation. To overcome the limitations of the K-means of clustering, canopy clusters are used. Canopy clusters work as pre-clusters for K-means algorithms. A data mining tool called WEKA is used to analyze both algorithms. Results are shown for improved accuracy and detection rate after using canopy clustering over K-means clustering using two distance metrics, namely Euclidean and Manhattan. This reduces the number of incorrectly classified instances using K-means clustering.

Keywords NSL-KDD dataset · K-means clustering · Canopy clustering
Distance metrics

1 Introduction

Networks are growing at an exponential pace. As networks grow, providing network security has increasing importance. Nowadays, protective measures are used to monitor and protect the networks, but intruder activity is ever increasing. In order the pursuit of providing better network security, an effective and efficient dynamic

H. P. Vinutha (✉)
CS&E Department, Bapuji Institute of Education & Technology, Davangere, India
e-mail: vinuprasad.hp@gmail.com

B. Poornima
IS&E Department, Bapuji Institute of Education & Technology, Davangere, India
e-mail: poornimateju@gmail.com

A. N. Krishna et al. (eds.), *Integrated Intelligent Computing, Communication and Security*, Studies in Computational Intelligence 771,
https://doi.org/10.1007/978-981-10-8797-4_21

mechanism has been built called the intrusion detection system (IDS). Incoming and outgoing network traffic is monitored by IDS, which looks for signs of intrusion. There are two types of IDS: host-based intrusion detection systems (HIDS) and network-based intrusion detection systems (NIDS). HIDS is used to monitor the host machine of the computer system, and NIDS monitors the network traffic of the system. Attack detection is classified into two types: misuse detection and anomaly detection. Misuse detection uses a known signature to monitor the system; if the incoming pattern doesn't match with the signature, then it is considered an intruder. Anomaly detection observes the behavior of the system; if a small deviation takes place in the behavior of the system, then it is considered an intruder. Recently, most IDS are rule-based. The strength of the rule-based IDS absolutely depends on the system's developer. The main intention of developing IDS is to improve the detection rate of the system and reduce the false alarm rate generated by the system.

2 Dataset

Datasets are important to verify the effectiveness of the IDS either by using real-time data or a benchmark dataset. A benchmark dataset is a simulated dataset designed and developed by MIT Lincoln labs for the purpose of research. Most of the research work is carried out using a benchmark dataset. In 1999, DARPA 98 was developed by summarizing a network connection with 41 features per connection. Later, to obtain competence with international knowledge discovery and data mining tools, the KDD99 benchmark dataset is formed. The major drawback of the KDD99 dataset is that it contains much redundant and incomplete data. To overcome the drawbacks of KDD99, the NSL-KDD 99 dataset was designed. This dataset contains 42 features including a class type called normal or anomalous. These 42 attributes of each network connection vector are labeled as one of the following categories: Basic, Traffic, Content, Host.

The number of records in the NSL-KDD dataset is recorded as training and testing datasets with the type of normal and attacks. There are 37 actual attacks, but among those 21 are considered for the training dataset and the remaining 16 are considered for the testing dataset along with training attacks. These attacks are divided into four main categories, namely Denial of Service (DOS), Probing (probe), Remote to Local (R2L) and User to Root (U2R). Table 1 shows the list of attacks under these categories.

The original dataset is not suitable to use directly for any detection techniques. It has to be pre-processed and saved into a suitable format. The considered NSL-KDD dataset has to be pre-processed by applying cleaning and transformation steps. The cleaning step of the dataset handles the missing values and noise in the dataset. Instances of each single connection of the dataset contain 42 features including a

Table 1 Attack categories

Category	Attack types
Normal	Normal
DOS	apacha2, back, land, mailb omb, neptune, pod, processtable, smurf, trardrop, teardrop, udpstorm
U2R	buffer, overflow, httprunnel, loadmodule, perlsrootkit, solattack, xterm
R2L	ftp-write, guess_password, imap, multihop, named, phf, sendmail, snmp getattack snmpguess spy warezclient, warezmaster, worm, xlock, xsno op
Probing	Ips weep mcan, nmap, portsweep, saint, satan

Table 2 Transformation table

Type	Feature name	Numeric value
Protocol type	TCP	0
	UDP	1
	ICMP	2
Flag	OTH	3
	REJ	4
	RSTO	5
	RSTOS0	6
	RSTR	7
	S0	8
	S1	9
	S2	10
	S3	11
	SF	12
	SH	13

class feature. To make it suitable for determining it is necessary to transform the nominal features into numeric values. The assigned transformation values are given in Table 2.

3 Related Work

In [1], the author showed that clustering is a technique, which analyzes the dataset in an effective and efficient manner. They used a K-means clustering algorithm to analyze high dimensional data and divided the larger data into smaller clusters. These clusters are used to classify the dataset at the stage of classification. A modified K-means clustering algorithm is used to show the improvement in accuracy and execution time.

In [2], the author proposed a more accurate NIDS. A K-means clustering algorithm is a distance-based algorithm, which is used widely used in research.

They analyzed Euclidean and Manhattan distance matrices on a K-means algorithm using the KDD Cup 99 dataset. Finally, experimental results are used to prove that the performance of Manhattan distance is better than that of Euclidean distance metrics.

Another work [3] reviews an existing K-means clustering algorithm. According to the author, the traditional approach of K-means clustering is expensive and is more sensitive to outliers, which produce unstable results. To solve these issues, several improved methods of K-means clustering are studied with different approaches. The final conclusion is that an existing algorithm with different prospective can eliminate the limitation of the K-means clustering algorithm.

In [4], the author used canopy clustering to overcome the drawback of K-means clustering. The canopy clustering algorithm works as a pre-clustering technique for K-means clustering. By using canopies and providing an initial cluster center, the existing clustering algorithm works normally. This works well for higher dimensional datasets and minimizes the execution time.

In [5], the author proposed a double-guard system to detect and prevent attacks. The IDS containing a double guard builds models of multi-tiered requests from both ends and summarizes the alerts generated by IDS so that effective detection and prevention can be done.

In [6], the author proposed an idea that can be used for inexpensive and approximate distance measure. Canopy clustering is used, which partitions the data into overlapping subsets called canopies. These canopies are then used to measure the exact distance occurring in a common canopy.

4 Clustering

Clustering is an unsupervised learning method that plays an important role in data mining applications. It divides the data into groups, and these groups are in their features. The collection of similar objects is grouped together into one cluster. Clustering works as a pre-classifier, giving more accuracy in classification result. Clustering algorithms can be performed by several methods, each method using different principals. The different clustering types are partitioning clustering, density-based clustering, distance-based clustering, grid-based clustering and hierarchical clustering. Clustering techniques can handle unlabeled data so they are appropriate for network anomaly detection [3].

4.1 Simple K-Means Clustering

K-means clustering is an important partitioning clustering method. In this method, spherically shaped, mutually exclusive clusters are formed. K indicates the number of clusters. K-means clustering uses a distance metric to calculate the distance between the data objects into k disjoint clusters. K is a parameter specified by the user [3]. The steps of K-means algorithm are as follows [2]:

Step 1—Randomly choose the cluster center as a K-data object.

Step 2—Apply a distance metric to calculate the distance between each data object and cluster center.

Step 3—Data objects are assigned to the nearest cluster.

Step 4—Cluster center is updated.

Step 5—Repeat steps 2 to 4 until no change in the cluster center.

To calculate the distance between the data objects in K-means clustering, distance metrics are defined. Euclidean distance is a commonly used distance metric. Another distance metric called Manhattan distance metrics can be used in the K-means algorithm. Equation 1 is used to find Euclidean distance, and Eq. 2 is used to find Manhattan distance [2].

For n-attributes, consider X, Y as two data objects

$$\text{Euclid}\,(\text{X}, \text{Y}) = \left[\sum_{j=1}^{n} (x_j - y_j)^2\right]^{1/2} \tag{1}$$

$$\text{Manhat}\,(\text{X}, \text{Y}) = \left[\sum_{j=1}^{n} |x_j - y_j|\right] \tag{2}$$

K-means is a straightforward algorithm, and it is expressive in computation. Run time of the algorithm is O(k * n * i). K is the number of clusters, n is the total number of points, i is the number of iterations [6].

A major limitation of the algorithm is that it is time consuming to join the clusters. Therefore, it is necessary to reduce the run-time and to improve the efficiency.

4.2 Canopy Clustering

Canopy clustering is an unsupervised algorithm. It works as a pre-clustering algorithm for K-means clustering. This can process large datasets efficiently, and it creates canopies that pre-partition the dataset to analyze the existing slow method such as K-means clustering [6].

Steps of canopy clustering [6]:

Step 1—Define two thresholds, T1 and T2.

Step 2—Canopies are formed by selecting random points from the list.

Step 3—Update all the points in the list with approximate distance.

Step 4—Assign the points to a canopy that falls within T1.

Step 5—The points that fall within the threshold T2 are removed from the list.

Step 6–Repeat steps 2 to 5 until the list is empty.

5 Proposed Method

Figure 1 gives the proposed method. Our main intention is to reduce the time complexity and to evaluate the performance in terms of accuracy, detection rate and false positive rate of K-means clustering. In this proposed method we are implementing the K-means clustering algorithm individually and with canopies using a canopy clustering algorithm. Output of canopy clustering is an input to K-means clustering. We have assigned Euclidean and Manhattan distance metrics to reduce the number of incorrectly classified instances. Finally, results are compared for K-means with and without canopy using distance metrics.

6 Results and Analysis

To evaluate performance, NSL-KDD training dataset is used, which contains 25,192 instances and 42 attributes including class attribute. In order to reduce the dimensionality of the dataset feature selection, algorithms are used. The feature selection is done by using correlation attribute evaluation, Chi-square, gain ratio and information gain algorithms with ranker search method. All the features are ranked based on their importance, and unimportant features and irrelevant features are that are ranked least are removed from the dataset. After observing the output of feature selection algorithms some features are selected for removal: 7, 9, 11, 13, 15, 16, 17, 18, 19, 20, 21, 22. This reduced dataset is used to evaluate the performance of K-means and K-means with canopy for both Euclidean and Manhattan distance metrics. The performance is measured using the following Eqs. 3 and 4 [2]:

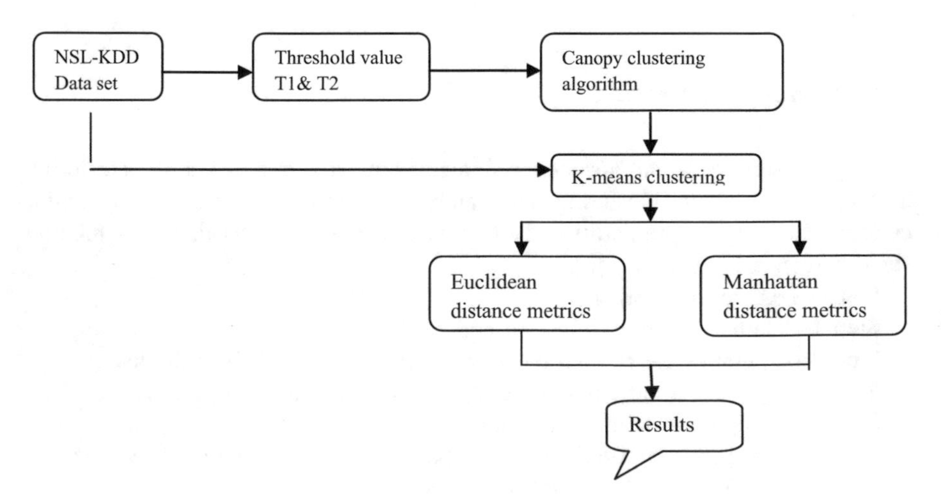

Fig. 1 Proposed method

Table 3 Performance measure of both the algorithms for both distance metrics methods

Algorithm	Method	Accuracy (%)	Detection rate (%)
K-means	Euclidean	82.35	80.68
	Manhattan	83.42	82.27
K-means with canopy	Euclidean	90.63	82.91
	Manhattan	93.51	83.12

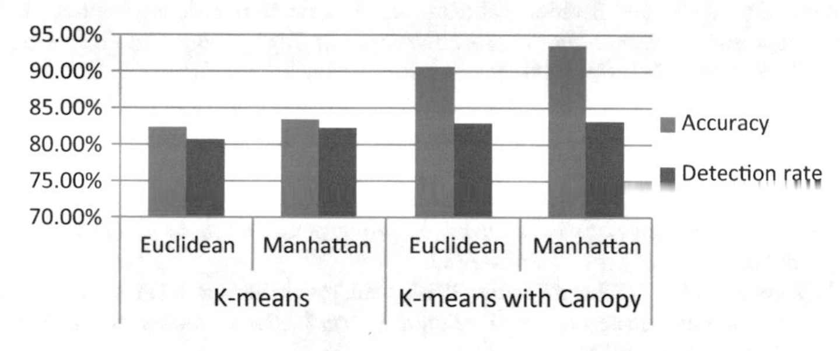

Fig. 2 Bar graph of accuracy and detection rate

$$\text{Accuracy} = \frac{(TP + TN)}{(TP + TN + FP + FN)} \tag{3}$$

$$Detection\ Rate = \frac{(TP)}{TP + FP} \tag{4}$$

True positive (TP): number of attack data correctly classified.
True negative (TN): number of normal data correctly classified.
False positive (FP): number of normal data incorrectly classified.
False negative (FN): number of data incorrectly classified [2].

Table 3 shows the performance measure of K-means and K-means with canopy algorithms using both the Euclidean and Manhattan distance metrics methods. Figure 2 shows the bar graph of accuracy and detection rate.

7 Conclusion

In this proposed method we have used an NSL-KDD dataset to achieve high accuracy and detection rate for simple K-means clustering algorithms. In order to achieve this, a canopy clustering algorithm is used, which works as a pre-cluster for K-means clustering algorithm. We have compared results for both K-means and K-means with canopy clustering algorithms. The accuracy and detection rate are improved after using K-means with canopy clustering.

References

1. Arpit, Bansal, Mayur Sharma, and Shalini Goel. 2017. Improved K-mean clustering algorithm for prediction analysis using classification technique in data mining. *International Journal of Computer Applications* 157 (6): 0975–8887.
2. Nasooti, Hadi, and Marzieh Ahmadzideh. 2015. The Impact of distance metrics on K-means clustering algorithm using in network intrusion detection data. *International Journal on Computer Network and Communication Security* 3 (5): 225–228.
3. Kavita Shiudkar, M.S., and Sachine Takmare. 2017. Review of existing methods in K-means clustering algorithm. *International Research Journal of Engineering and Technology (IRJET)* 04 (02): 2395–0056. ISSN 2395-0072.
4. Ambika, S., and G. Kavitha. 2016. Overcoming the defects of K-Means clustering by using canopy clustering algorithm. *IJSRD—International Journal for Scientific Research and Development* 4 (05). ISSN 2321-0613.
5. Bhor, Priyanka, and Rajguru Punam. 2015. Network intrusion detection and prevention using K-Means clustering. *International Journal of Advance Research in Computer Science and Management Studies* 3 (3). ISSN 2321-7782.
6. Amresh, Kumar, and S. Yashwant Ingle. 2014. Canopy clustering: a review on pre-clustering approach to K-Means clustering. *International Journal of Innovations and Advancement in Computer Science* 3 (5). ISSN 2347-8616.

A Big Data Architecture for Log Data Storage and Analysis

Swapneel Mehta, Prasanth Kothuri and Daniel Lanza Garcia

Abstract We propose an architecture for analysing database connection logs across different instances of databases within an intranet comprising over 10,000 users and associated devices. Our system uses Flume agents to send notifications to a Hadoop Distributed File System for long-term storage and ElasticSearch and Kibana for short-term visualisation, effectively creating a data lake for the extraction of log data. We adopt machine learning models with an ensemble of approaches to filter and process the indicators within the data and aim to predict anomalies or outliers using feature vectors built from this log data.

Keywords Big data analysis · Log data storage · System architecture
Anomaly detection · Unsupervised learning

1 Introduction

Large data collections are emerging as the key to carrying out impactful scientific research, especially in the areas of high-energy physics. In most cases, these communities are subject to geographical constraints and there are often a number of requirements placed on the data in terms of high-availability, low-latency, other storage and computing demands. Gorton et al. emphasise this in [1], elucidating the computational challenges and constraints placed on the operational flexibility of large-scale data management systems. Addressing a critical facet of this issue, our project proposes a scalable and secure and central repository capable of storing

S. Mehta (✉)
Dwarkadas J. Sanghvi College of Engineering, Mumbai, India
e-mail: swapneel.mehta@djsce.edu.in

P. Kothuri · D. L. Garcia
European Organisation for Nuclear Research, Meyrin, Geneva, Switzerland
e-mail: prasanth.kothuri@cern.ch

D. L. Garcia
e-mail: daniel.lanza@cern.ch

A. N. Krishna et al. (eds.), *Integrated Intelligent Computing, Communication and Security*, Studies in Computational Intelligence 771,
https://doi.org/10.1007/978-981-10-8797-4_22

consolidated audit data comprising listener, alert and OS log events generated by database instances. This platform will be used to extract data in order to filter outliers utilising machine learning approaches. The reports will provide a holistic view of activity across all oracle databases and the alerting mechanism will detect abnormal activity, including network intrusion and usage patterns, and alert the user [2]. Database connection logs are analysed to flag potentially anomalous or malicious connections to the database instances within the network of the European Organisation for Nuclear Research (CERN). We utilise this research to shed light on patterns within the network in order to understand better the temporal dependencies and implement them in the decision-making process within the CERN system. These models are trained on subsets of a data lake that comprises daily connection logs across all instances of databases on the network. The data lake comprises JavaScript Object Notation (JSON) logs that may be visualised using short-term storage, ElasticSearch, Grafana and Kibana or pushed to long-term storage on Hadoop Distributed File Storage (HDFS). We extract subsets of these data and apply models that vary based on different parameters of the data and enable us to classify the outliers among the dataset as anomalies.

1.1 Database Instances in the CERN Network

CERN follows an open access policy for most of its internal systems, which extends to a significant percentage of database instances across the organisation. While this reduces the rule-based access control load on the system, it leads to a number of novel issues including, but not limited to an unnecessarily large number of connection requests, connections accessing more resources than necessary, misused or abused credentials and more security-oriented issues such as malware causing repetitive connections from a host machine (Fig. 1).

The CERN network comprises 10,000 active users and 1500 different users [2], each with devices, including personal and officially supplied equipment. While most of the software installed on the systems is regulated and updates are managed centrally, there is considerable privilege afforded to a user to manage a device personally and consequently the loopholes for installation of user-defined software or scripts. These can end up being both inefficient and malicious in their operation, raising multiple issues within the system. Specifically, these systems might be compromised and utilised to initiate either multiple (spam) connections or use stolen credentials with malicious intent. There are numerous security issues with such a setup if there is unmonitored access to database instances. It can be used in multiple ways to throw the network into disarray due to the load on servers. We aim to use some concrete approaches in analysing these database connections by building models that can accurately fit these data and classify further connection

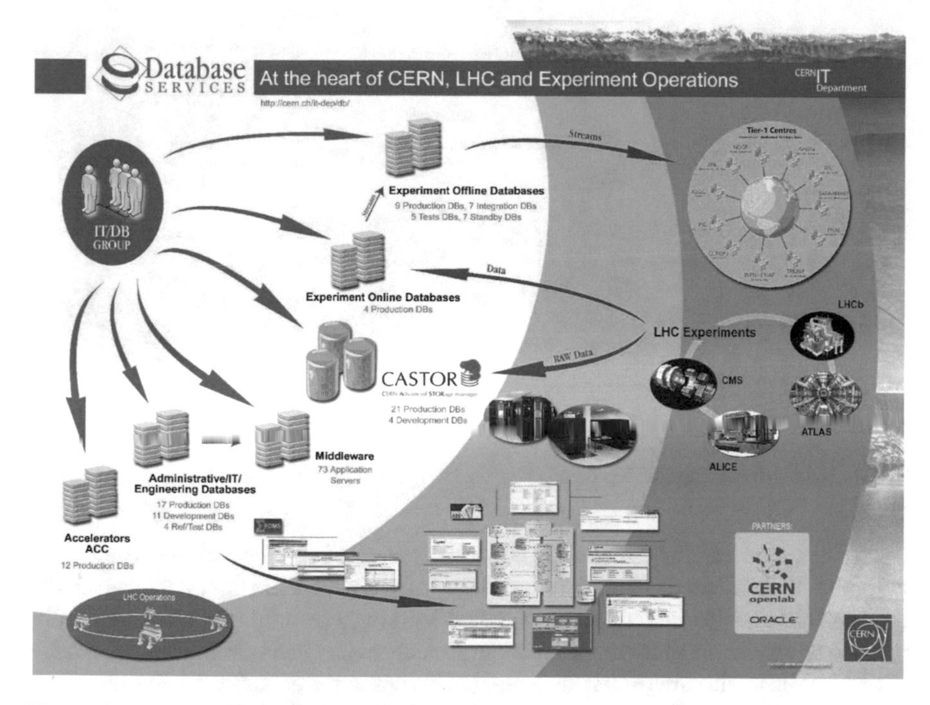

Fig. 1 Overview of the database system at CERN [3]

requests by evaluating the likelihood of being anomalous. The architecture used to store this log data needs to have a low footprint and support real-time analytical pipelines reliant on time-series and sequence-based approaches as emphasised in [4, 5].

2 Current Architecture

Various point-solutions are presented within [6–9]; however, there is no clear concept of common components within systems that can be identified in order to integrate architectures and fashion a system specific to petabyte-scale data intensive applications. Chervenak et al. [10] address this issue by proposing a 'Grid' with the focus on a uniform, policy-driven storage system, as well as management of metadata. In [11], Google adopts a 'listener-based approach' to log monitoring. These serve as models for a generalised storage architecture, which we develop to serve our use-case (Fig. 2).

Database connection logs comprise multiple parameters per successfully established or failed connection, which are then utilised in building the feature vector for analysis. The architecture comprises multiple instances of Oracle databases, each of which have an Apache Flume agent running on it. Apache Flume is a reliable and

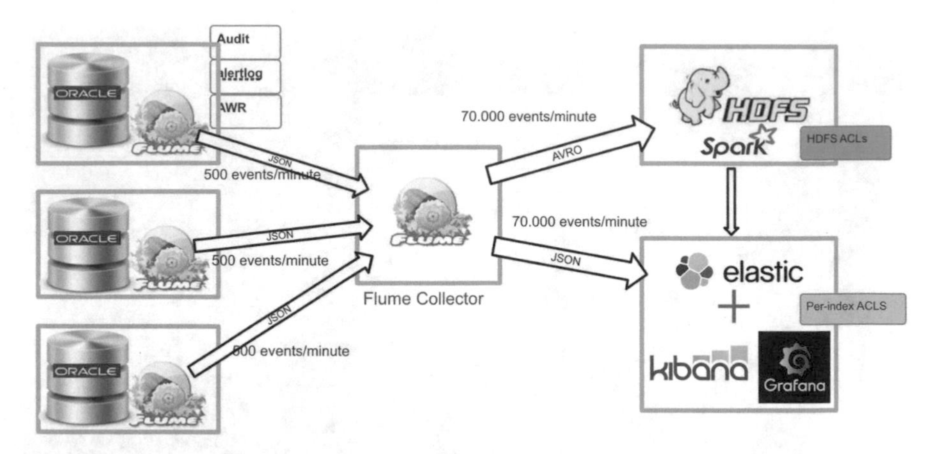

Fig. 2 System for storage of log data [3]

distributed service for efficiently collecting, clustering, and transporting large amounts of data, often comprising system logs. The system architecture for this service encapsulates the concepts of data streaming flows. Flume offers reliability due to the arsenal of failover and recovery mechanisms incorporated into a tenable set of options. The simplified data model allows for extensible online analytical implementation. It offers unprecedented scalability, reliability and performance [12]. Flume supports sources such as Avro, Thrift or logs, and a variety of sinks. It runs in the background, eliminating the need for scheduling jobs, and ensures all updates are streamed to HDFS as the configuration permits. In case of systems that do not load data directly, Flume provides an integration that allows for CSV and log data to be supported [13]. Oracle Real Application Clusters (RAC) is an Oracle Database version involving clustering of instances across the system. It provides a comprehensive, high-availability stack that can be used as the foundation of a database cloud system, as well as a shared infrastructure, ensuring high availability, scalability, and agility for any application [2, 14]. The log data are transported via a central Flume Collector to ElasticSearch and Kibana for short-term storage primarily meant for data visualisation and long-term storage on HDFS.

2.1 Data Pipelines

The current system includes a data ingestion pipeline that allows for the streaming of data from heterogeneous sources including two internally developed sources that it supports—JDBCSource and LogFileSource. These impart the ability to consume data from database tables, as well as log files (Fig. 3).

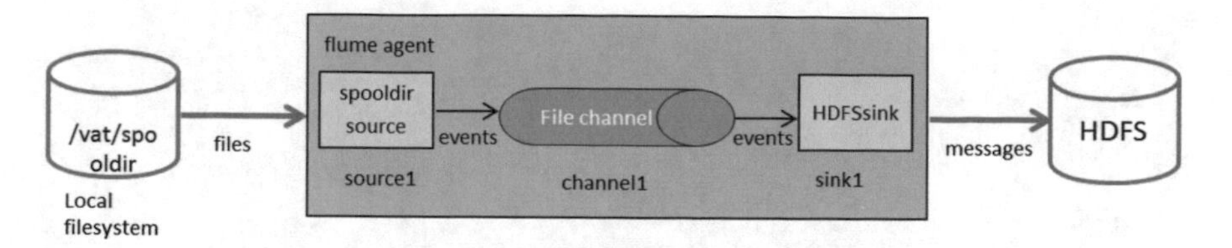

Fig. 3 High-volume ingestion of event data into hadoop

2.2 *Data Storage*

There have been comparisons and experiments performed in order to evaluate the performance of data storage using the Avro and Parquet formats and also others as in [5, 15, 16], and we adopt a similar approach in providing a comparative evaluation of different types of queries on data stored in both these formats (Figs. 4, 5, 6 and 7).

We find that Parquet serves our use-case for scan performance and analytical queries. Our results match the ones shown, which are presented in the form of a

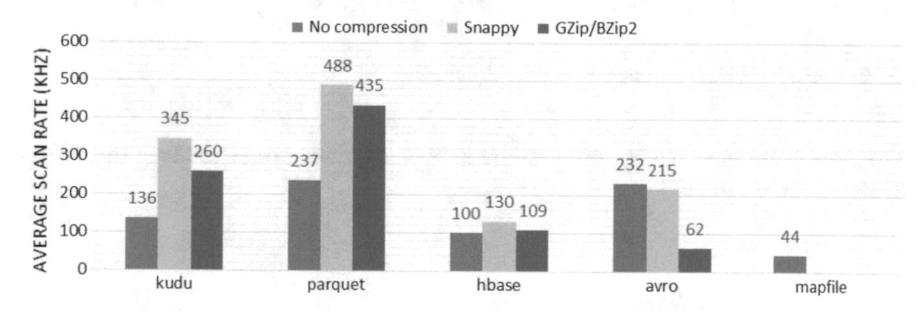

Fig. 4 Average scan speed (records per second)

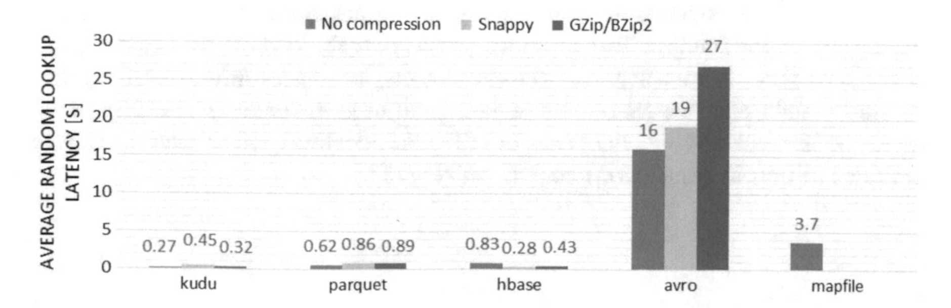

Fig. 5 Random data lookup latency (seconds)

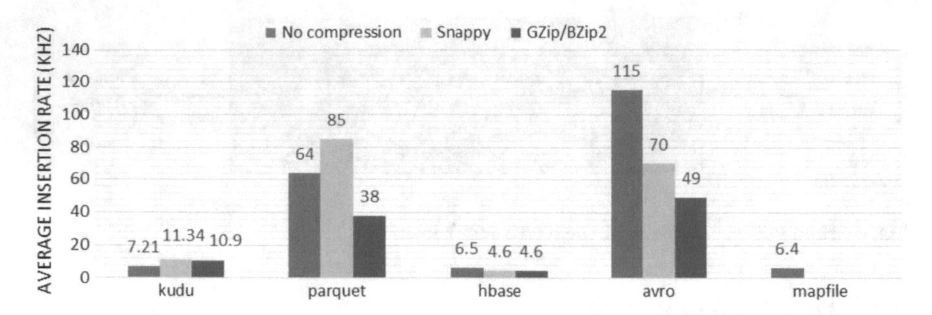

Fig. 6 Ingestion speed (1000 records per second)

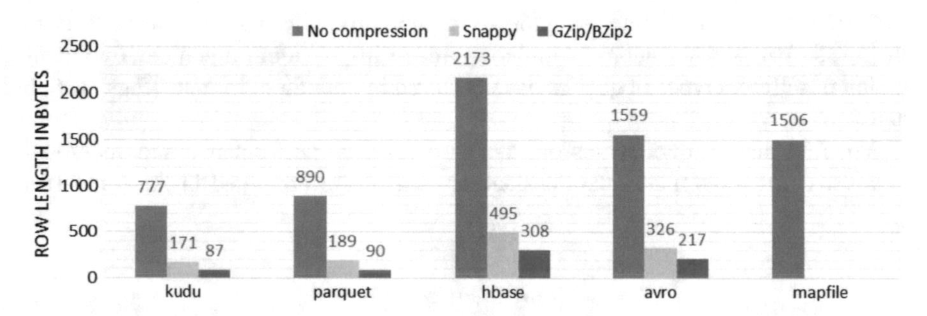

Fig. 7 Space utilisation—average row length (bytes)

comparison of data format performance across activities corresponding to experiments presented in [17].

2.3 Log Data Fields

The log consists of a number of different fields each with a value corresponding to the details of the connection. Some of the important connection details include fields such as timestamp, client_program (software that the client uses to connect), client_host (host from where the client connects; not necessarily specified), client_ip (IP address of the client), client_port (port used to connect), client_protocol (network protocol that the client uses to initiate connection), client_user (username provided to initiate connection), and so on (Fig. 8).

hour_of_the_day	day_of_the_week	client_user	client_host	client_ip	client_program	CONNECT_DATA_INST/	service_name
11.821111111111112	1	merge	pcamsj2.cern.ch	137.138.188.167	python	INT11R2	int11r.cern.ch
11.82611111111111	1						INT6R1
11.821111111111112	1	merge	pcamsj2.cern.ch	137.138.188.167	python	INT11R2	int11r.cern.ch
11.826666666666666	1						INT11R2
11.821111111111112	1	merge	pcamsj2.cern.ch	137.138.188.167	python	INT11R2	int11r.cern.ch
11.82611111111111	1	merge	pcamsj2.cern.ch	137.138.188.167	python		int11r.cern.ch
11.821111111111112	1	merge	pcamsj2.cern.ch	137.138.188.167	python	INT11R2	int11r.cern.ch
11.82611111111111	1						INT6R1
11.821111111111112	1	merge	pcamsj2.cern.ch	137.138.188.167	python	INT11R2	int11r.cern.ch
11.82638888888889	1						INT11R1
11.821111111111112	1	merge	pcamsj2.cern.ch	137.138.188.167	python	INT11R2	int11r.cern.ch
11.82638888888889	1						INT11R1
11.821111111111112	1	merge	pcamsj2.cern.ch	137.138.188.167	python	INT11R2	int11r.cern.ch
11.821388888888889	1	merge	pcamsj2.cern.ch	137.138.188.167	python	INT11R2	int11r.cern.ch

Fig. 8 Extracting user connection details to build a feature vector

3 Anomaly Detection

Anomaly detection is the process of identifying unusual or outlying data from a given dataset. It has been a subject of interest as the amount of data available for analysis rises disproportionately as compared to the amount of available tagged data. Specifically, we utilise features extracted from established connection logs in order to determine the outliers for a given subset of temporal data obtained from a data lake. The data inherently possesses some degree of contamination, which is why we set a clustering threshold as a cutoff for percentage of outliers within a data slice. Most anomaly detection models use two fundamental assumptions about the data:

1. A majority of the data are normal; the percentage of outliers is a distinct minority within the dataset.
2. The anomalies are statistically different from the normal data, usually on the basis of some metric that the model is required to detect [18] (Fig. 9).

Models employed include K-Nearest neighbours, Isolation Forests, Local Outlier Factor, and One-Class Support Vector Machines. Our experiments are of a preliminary nature and focus on the real-time streaming along with accuracy of detection. Because of the lack of 'clean data', we are forced to train the models on inherently contaminated data. This contamination refers to the percentage of original data that we assumed to be outliers since it is impossible to obtain a completely normal dataset for a given window of network activity. The pipeline involves the pre-processing of data in order to reduce high-dimensional data to a lower number of dimensions and better visualise it as well as build models to analyse it (Table 1).

The solution we arrive at involves the use of an ensemble of these methods to analyse the data points collectively and arrive at outliers that are detected by more than one model, thus filtering out many false positives in the process. The conclusion we arrive at is verified by performing a manual security audit of these connections in order to confirm their anomalous nature. We find that the system

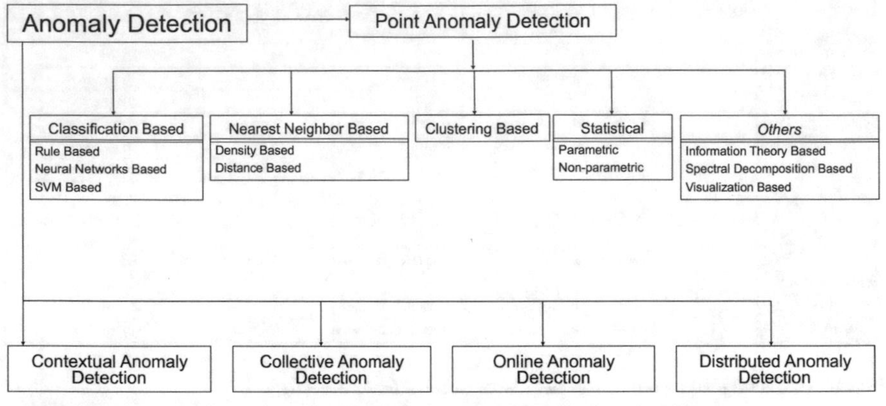

Fig. 9 An overview of strategies for outlier detection [14]

Table 1 Comparison of thresholds for outliers across models

Approach	Title	Contamination (%)
Distance	k-Nearest neighbours	2
Density	Isolation forests	3
Density	Local outlier factor	5
Classification	One-class SVM	2

filters the possible anomalies down to 10% of the original data. While there are false-positives, the true anomalies are either unusual logins, malware scans running on systems or inadvertent connection requests. We find that this approach reduces much of the monitoring load solicited originally to track undesirable activities or agents within the system.

4 Conclusion

Hadoop and its assortment of components provide a well-established solution for data warehousing and provide a reliable pipeline for the transport, storage and streaming of log data produced by control systems at CERN. Spark is a system that has proven to be scalable in streaming applications when used in tandem with Hadoop. Further, it supports querying in SQL that simplifies the learning curve for users, should such a case arise. From our tests we find that the choice of the storage engine is very important for the overall performance of the system. The system can efficiently handle data synchronisation and perform near real-time change propagation.

Acknowledgments The authors would like to acknowledge the contributions of Mr. Eric Grancher, Mr. Luca Canali, Mr. Michael Davis, Dr. Jean-Roch Vlimant, Mr. Adrian Alan Pol, and other members of the CERN IT-DB Group. They are grateful to the staff and management of the CERN Openlab Team, including Mr. Alberto Di Meglio, for their support in undertaking this project.

References

1. Gorton, I., P. Greenfield, A. Szalay, and R. Williams. 2008. Data-intensive computing in the 21st century. *Computer* 41 (4): 30–32.
2. Grancher, E., and M. Limper. 2013. Oracle at CERN. https://indico.cern.ch/event/242874/.
3. Lanza, D. 2016. Collecting heterogeneous data into a central repository. https://indico.cern.ch/event/578615/
4. Baranowski, Z., M. Grzybek, L. Canali, D.L. Garcia, and K. Surdy. 2015. Scale out databases for CERN use cases. In *Journal of physics: Conference series*, vol. 664, no. 4, 042002. IOP Publishing.
5. Kothuri, P., D. Lanza Garcia, and J. Hermans. 2016. Developing and optimizing applications in hadoop. In *22nd international conference on computing in high energy and nuclear physics, CHEP*.
6. Moore, R., C. Baru, R. Marciano, A. Rajasekar, and M. Wan. 1997. Data-intensive computing. In: *Practical digital libraries: Books, bytes, and bucks*, 105–129.
7. W. Johnston. 1997. Realtime widely distributed instrumentation systems. In: *Practical digital libraries: Books, bytes, and bucks*, 75–103.
8. Shoshani, A., L.M. Bernardo, H. Nordberg, D. Rotem, and A. Sim. 1998. Storage management for high energy physics applications. In *Proceedings of computing in high energy physics 1998 (CHEP 98)*. http://www.lbl.gov/arie/papers/proc-CHEP98.ps.
9. Foster, I., and C. Kesselman (eds.). 1999. *The grid: Blueprint for a future computing infrastructure*. Florida: Morgan Kaufmann Publishers.
10. Chervenak, A., I. Foster, C. Kesselman, C. Salisbury, and S. Tuecke. 2000. The data grid: Towards an architecture for the distributed management and analysis of large scientific datasets. *Journal of Network and Computer Applications* 23 (3): 187–200.
11. Ledain, J.E., J.A. Colgrove, and D. Koren. 1999. *Efficient virtualized mapping space for log device data storage system*. Veritas Software Corp., U.S. Patent 5,996,054.
12. Apache Flume. https://flume.apache.org/.
13. Oracle real application clusters. http://www.oracle.com/technetwork/database/options/clustering/rac-wp-12c-1896129.pdf.
14. Chandola, V., A. Banerjee, V. Kumar. 2009. Outlier detection—A survey. Technical Report TR07–17, University of Minnesota.
15. Plase, D., L. Niedrite, and R. Taranovs. 2017. A comparison of HDFS compact data formats: Avro versus Parquet. *Mokslas: Lietuvos Ateitis*, 9 (3): 267.
16. Plase, D., L. Niedrite, and R. Taranovs. November 2016. Accelerating data queries on Hadoop framework by using compact data formats. In *2016 IEEE 4th workshop on advances in information, electronic and electrical engineering (AIEEE)*, 1–7. IEEE.
17. Baranowski, Z., L. Canali, R. Toebbicke, J. Hrivnac, and D. Barberis. 2016. On behalf of the ATLAS collaboration, 2016. A study of data representation in hadoop to optimize the data storage and search performance for the ATLAS EventIndex. In *22nd international conference on computing in high energy and nuclear physics, CHEP*.
18. Denning, D.E. 1987. An intrusion-detection model. *IEEE Transactions on Software Engineering* 2: 222–232.

Hadoop Based Architecture for a Smart, Secure and Efficient Insurance Solution Using IoT

P. K. Binu, Akhil Harikrishnan and Sreejith

Abstract The "Internet of Things" (IoT) is among the most highly subsidized and promising topics in both academia and industry these days. Contemporary developments in digital technology have raised the interest of many researchers towards implementation in this area. The influence of IoT within the insurance field is vital. This chapter asserts an innovative concept of IoT pooled with an insurance application, which is beneficial for insurance companies to monitor and analyze the health of their clients continuously. Numerous insurance companies are clustered together to provide a standardized health status monitoring of clients. Since there is a large amount data generated by the system, we adopt Hadoop in the background to map the data effectively and to reduce it into a simpler format. We assimilate Sqoop tool to enable data transfer between Hadoop and RDBMS, in consort with Apache Hive for providing a database query interface to the Hadoop. By consuming the output from Hadoop MapReduce, a non-probabilistic binary linear classifier predicts the policyholder's chances of developing some health problems. Ultimately, the resultant outcomes are presented on the user's smartphones. The Apache Ranger framework interweaved with the Hadoop ecosystem aims to ensure data confidentiality. The endowments are granted to the policy holders based on the health report generated by our system. To evaluate the efficiency of the system, experiments are conducted using various policyholder's health datasets and from the results, it is observed that SVM predicts sepsis with an accuracy of approximately 86%. While testing with the medical dataset, SVM proved to be more accurate than the C4.5 algorithm.

Keywords Internet of things (IoT) · Hadoop · SVM · Big data
Hive

P. K. Binu (✉) · A. Harikrishnan · Sreejith
Department of Computer Science and Applications, Amrita School of Engineering, Amrita Vishwa Vidyapeetham, Amrita University, Amritapuri, India
e-mail: binu@am.amrita.edu

A. Harikrishnan
e-mail: akhilharikrishnan@am.students.amrita.edu

Sreejith
e-mail: sreejiths@am.students.amrita.edu

A. N. Krishna et al. (eds.), *Integrated Intelligent Computing, Communication and Security*, Studies in Computational Intelligence 771,
https://doi.org/10.1007/978-981-10-8797-4_23

1 Introduction

Every one of us may have seen or experienced an unexpected hospitalization, a sudden accident or any sort of unpredictable mishap at least once in our lives. This is the driving force for maintaining insurance. Insurance firms are not fully effective in minimizing the risk confronted by their clients. But this is gradually changing with the inception of the Internet of Things (IoT) in this field.

Previously, required customer data were not available, and information insurers could access was usually subjective or imprecise, which leads to intensified operational costs for the insurance firms. Contemporary insurance policy systems do not contain a client's monitored health data, and they are also insufficient to keep a client's information secured. In this chapter we present an advanced IoT based insurance solution using Hadoop architecture. The system proposes a very successful IoT portfolio of clients by leveraging critical benefits of big data collection and assisting insurance experts to analyze better ongoing risk profiling for reducing amplified operating costs of the organization as well [1]. This e-health monitoring IoT system is specifically designed for an insurance syndicate for monitoring and analyzing the health of their policyholders using different sensors [2]. The health data collected from clients by the system using various sensors are relocated to Arduino for necessary processing. After correlating with default readings there, the data are updated to the database. The Sqoop tool facilitates data fetching between the database and Hadoop.

This system generates large volumes of data. To map the data effectively and to simplify it, we use Hadoop. For secure storage of data in Hadoop, we adopted Apache Ranger, which is again a part of the Hadoop system. The collected readings are then mapped and reduced in the Hadoop system for efficient storage and processing.

Hive provides a SQL like interface for querying and analyzing large amounts of data stored in the Hadoop. Data processed by the MapReduce programming model is classified using an SVM classifier to predict the health state of each policyholder, their chance of developing some health problems. Ultimately, the results are presented on a user's smartphone, and if eligible, some endowments are granted to the client by the company based on their health report generated by our system.

2 Related Works

Hao Chen et al. describes an e-health monitoring (EHM) ecosystem [2]. The authors also conduct an evaluative study of the contemporary EHM systems and its associated concepts. Even though EHMs are independent of the network, they tend to be considerably dependent on various vital capabilities extracted from the specific network technologies. Some key capabilities are also proposed through their work. Based on the IoT reference model, three EHM technical models are proposed. Technical Model 1 is about device-to-device communication. Short distance device-to-device communication can be achieved with the help of Bluetooth, Zig Bee, etc. where data transfer is direct. Model 2 concerns the provided network between device and server.

Joshi [3] provides a basic introduction to the Internet of Things. The author explains the elementary concepts of IoT, the background and its challenges. IoT implementations in various fields are also evaluated. The author also reviews the IoM (Internet of Media), IoS (Internet of Service), IoE (Internet of Energy) and several IoT use cases such as environmental monitoring, infrastructure management, medical and health care systems, transport systems, industrial applications, energy management, building and home automation.

To fulfil the needs of an urban population and to manage metropolitan development more smartly and efficiently, various IoT technologies can be implemented. Anand Paul et al. proposes an IoT based smart city system [4] supported by the power of Big Data Analytics. The proposed system is designed with several sensor deployments at various locations to capture real time city data. Several stages involved in system implementation are data generation and collecting, aggregating, filtration, classification, preprocessing, computing and decision making.

There is no existence for Internet of Things without the assistance of Big Data and Cloud Computing archetypes. Improperly managed sensor data generates inappropriate results. The authors propose a comprehensive architecture [5], which uses Big Data and Cloud Computing with IoT for an improved experience of data analysis and storage. The approach is capable of overcoming certain insufficiencies of an IoT system.

IoT technologies are truly capable in successfully enhancing the smart power utilization service systems with its overall assessment, trustworthy communication and robust data processing capacities. This paper reviews [6] an electricity service system based on intelligent networking. A smart power management system consisting of various sensors for continuous power grid monitoring is evaluated. At regular time intervals, the sensor values are transferred to the cloud.

3 Methodology

To diminish the risk of being encumbered by substantial expenses for preventive care or medical bills as a result of a medical condition or a mishap, various personal health insurance policies offered by an insurance provider aids a policy holder. Our system is primarily designed to be used by an insurance syndicate for the purpose of health monitoring and analysis of their policy holders [7]. The main cornerstones of this system's architecture are the sensors embedded in a wearable widget, Arduino board and Hadoop. Additionally, the system is supported by the power of SVM, a supervised machine learning algorithm implemented with the aim of classifying data. It utilizes a procedure called the kernel trick to convert non-separable categories of a client's data into separable categories and then, based on the transformations, the algorithm finds an optimal borderline between the probable outputs or it figures out how to separate data based on the class labels or outputs defined.

In the initial phase, the health status monitoring of the clients is executed by means of a wearable wristband, which consists of three sensors for heartbeat, blood pressure and temperature. The policyholders are required to use this wearable wristband for the purpose of reading these health parameters. The readings are relocated to Arduino for necessary processing and they are correlated with the default values in Arduino for determining health status of the clients [8, 9]. It is an IoT platform that supports effective collection and storage of sensor data on to the cloud through HTTP across the web or even just through a LAN. These values are updated and temporarily stored on the database integrated as a part of the system, and the Sqoop tool facilitates fetching of necessary data based on processing requirements. MySQL is deployed as the local database server, where sensor recordings are stored.

A short time interval is set on Arduino for recording new readings, in which an average of the readings recorded so far are computed and then are updated to the MySQL database. This practice offers the benefit of avoiding the effort of updating or rewriting the sensor readings each time when a new value is received. If any of the parameters such as blood pressure, temperature or heartbeat are found to be abnormal, then the user will receive an SMS notification too, regarding their poor health conditions.

Big Data technologies are extremely significant in providing much improved and precise analysis, which assists for more solid decision-making and thus results in reduced cost, improved operational efficiencies, and decreased risk intensities for the business [10]. In order to couple the real potentials of Big Data, an infrastructure capable of managing and processing huge volumes of structured and unstructured data in real-time with enhanced efficiency is a necessity [11, 12] (Fig. 1).

We used Apache Hadoop, the open source software framework specifically designed to handle large datasets. The principal part of Hadoop consists of a storage module, as well as a processing module such as the Hadoop Distributed File System (HDFS) and a MapReduce programming model, respectively. The principal components of HDFS cluster include a NameNode for cluster metadata management and a DataNode for satisfying data storage provisions [13].

In our architecture, Hadoop processes the sensor readings fetched by the Sqoop tool from the database and also reduces the dataset to produce necessary output using MapReduce. The fundamental concept of MapReduce is that it splits a task into numerous simplified subtasks, processes each subtask in parallel and combines the results from each subtask together to produce a final output. As the programs written in the MapReduce framework are automatically parallelized, programmers are only required to write map and reduce functions while ignoring parallel processing implementation. The MapReduce algorithm encompasses two focal responsibilities, namely map and reduce. The map task is always executed before the reduce task.

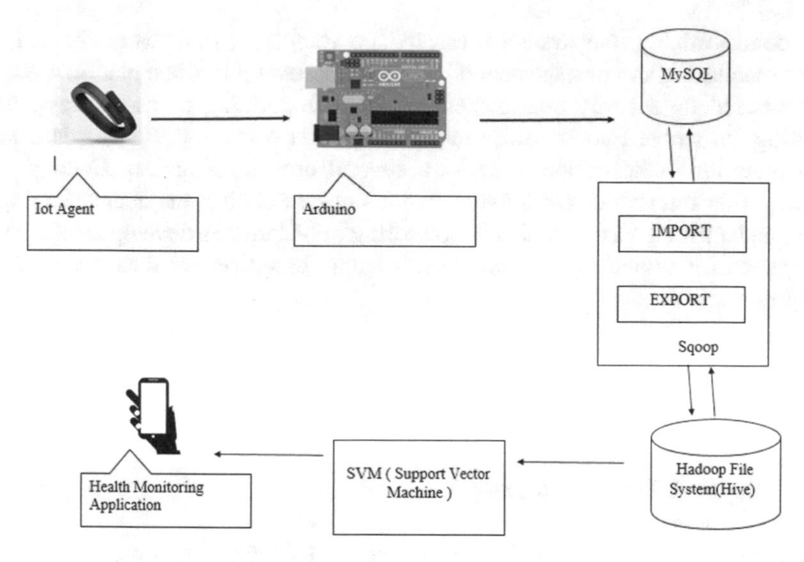

Fig. 1 Overall system architecture

The input data is stored on HDFS either as a file or directory and is processed by a mapper and creates several tiny chunks of data. The reducer accepts the output from a mapper, pools those data chunks into a smaller set of data, which is then stored on the HDFS. Sqoop is a command line interface application that enables efficient import and export of bulk data between RDBMS and Hadoop. It is capable of importing data to Hbase, Hive, HDFS and also to the database. While Sqoop transfers data, it splits the data into different partitions and maps only jobs are set for each individual partition with mappers managing transmission of dataset assigned to it.

The Map Reduce model outputs a CSV file which is classified using a support vector machine (SVM) classifier. The classifier is trained using an existing dataset. The data values of a new policyholder, such as age, sex, heartbeat, blood pressure, and body temperature, are tested and the classifier predicts whether or not the client has any medical condition. Support vector machines (SVM) are supervised learning models, which comprise a set of associated learning algorithms. The contrasting feature of SVM from other classifiers is that it is capable of finding the best separating line even for an assorted dataset. SVM consist of a training phase and a testing phase. In the first phase, the SVM model is trained with a set of training data samples. The data from the training dataset is used by the algorithm to learn rules for performing prediction or classification. In the second phase, testing dataset, which is totally unfamiliar, is used for testing the model. As the rules learned from the training dataset are also applicable to testing data, an error rate is computed [14].

The individual health data collected to be processed by the insurance application may contain various sensitive personal information about the policyholders. So, data confidentiality emerges as a significant area of concern. For accomplishing data confidentiality, our system uses Apache Ranger, a framework for enabling better monitoring and management of data security over a Hadoop platform. Ranger supports federal security management along with auditing of user access, while securing the whole Hadoop ecosystem with greater efficiency. Ranger demands a high precision authorization to perform several operations on any Hadoop component. This framework can support various access control methods such as role based and attribute based. A health monitoring application is developed as a part of the system for providing real time health status based on the data generated for every client.

3.1 Algorithms

3.1.1 Support Vector Machines

Support vector machines (SVM) introduced in COLT-92 by Boser, Guyon and Vapnik encompasses a new class of learning algorithms and is motivated by statistical learning theory, which has proved capable of providing good heuristic performance.

Let there be a set of instances Xn, where n = 1, 2, …, z, with labels Yn ϵ {1, −1}. The key task in training SVMs is to solve the following quadratic optimization problem: minα f(α) = ½αT Mα − dTα subject to 0 ≤ αn ≤ U where n = 1, 2, …, z, YTα = 0. While d is the vector of all instances, U is the upper bound of all variables; M is an l by l symmetric matrix with Mnm = YnYm K (Xn, Xm) where K (Xn, Xm) is the kernel function.

3.1.2 MapReduce Algorithm

MapReduce (Jeffrey and Sanjay 2004) is a framework for writing applications to process Big Data Analytics on a vast number of servers using a parallel distributed algorithm on a cluster. The foremost benefit of MapReduce (MR) is its adeptness in parallel processing of large cliques of data stockpiled in the Hadoop cluster. MapReduce applications are conceived and built depending on certain business scenarios. While processing large datasets, each logical record in the input data is applied with a mapping function to generate intermediate key-value pairs. The next phase called reduce is applied to data that shares common key to extract combined data, respectively.

Map:

Every input key-value pair is applied with the map function and generates a random number of intermediate key-value pairs.

The mapper function is given by:-

map (in_Key, in_Value) -> list (intermediate_Key, intermediate_Value)

The map function accepts inputs in the form of key-value pairs while a MapReduce program accepts a single file or a set of files as its input. The default value is a data record and the offset of the data record from the starting of the data file is usually selected as the key.

The mapper produces an intermediate key-value pair for each word present in a document. The task of reducer is to sum up all the counts for each word. Each line of the file is given as the mapper input, and each mapper results in a set of key value pairs, with a single word as the key and 1 as the value.

map: (K1, V1) → [(K2, V2)]

K1 is the file name and V1 is the file content. Internally, within the map function, the user may give any random key-value pair, which is represented as [K2, V2] in the list.

Reduce:

For the purpose of generating output key-value pairs, the reduce function is executed for every value possessing equivalent intermediate keys

reduce(intermediate_Key,list [intermediate_Value]) -> list (out_Key, out_Value)

The key intention of each reduce function is to generate output by processing the intermediate values associated with a specific key which is generated by a mapper function. As there exist an intermediate "group-by" procedure, the reducer function's input is a key-value pair in which the key K2 is the outcome from mapper phase and a list of values V2, which shares the common key.

reduce: (K2, [V2]) → [(K3, V3)].

3.2 Data Design

With the aim of disease prediction, the sensors incorporated in the wearable device reads necessary heath values from each policyholder. We mainly focus on three key parameters, namely blood pressure, temperature and heartbeat, as many diseases can be diagnosed with disparities in any of these parameters.

These recordings are stored on the MySQL database, and the Sqoop tool retrieves data based on necessity. When a new policyholder is included, the system will consider these parameter values, and the probability for having any health disorders are predicted based on the variations detected based on these values when compared with their corresponding normal rates. Table 1 shows a sample of patient data that the system considers for processing.

Table 1 Patient data

Name	Gender	Age	BP	Temp	Heart beat	Sepsis identified
Joe	M	39	60	26	110	T
Tom	M	23	92	44	108	F
Reid	M	35	102	36	68	F
Ben	M	35	75	41	115	T
Torre	F	28	79	42	72	F

4 Experiments and Results

In this section, we conducted an experimental analysis to evaluate the overall efficiency of our system in generating reliable an IoT portfolio for insurers. The simulation of the system is substantiated with the CSV data samples collected from clients for processing by the sensors embedded in the wearable networked wristband. After the required readings are recorded, it is repositioned to MySQL database, and with the assistance of Sqoop, our application retrieves essential data from RDBMS for executing further operational procedures [15, 16]. MapReduce processing is done internally within the cloud and the resultant outcome is transferred on to the monitoring application's interface as shown in Fig. 2.

In this system, we have focused on numerical data such as body parameters. These data are analyzed in two different ways. The first method uses Hadoop, which can handle large amounts of data, and the next method is the SVM algorithm, which is used to classify the patients with and without sepsis. The large amount of data from a patient's body parameters are reduced using Hadoop.

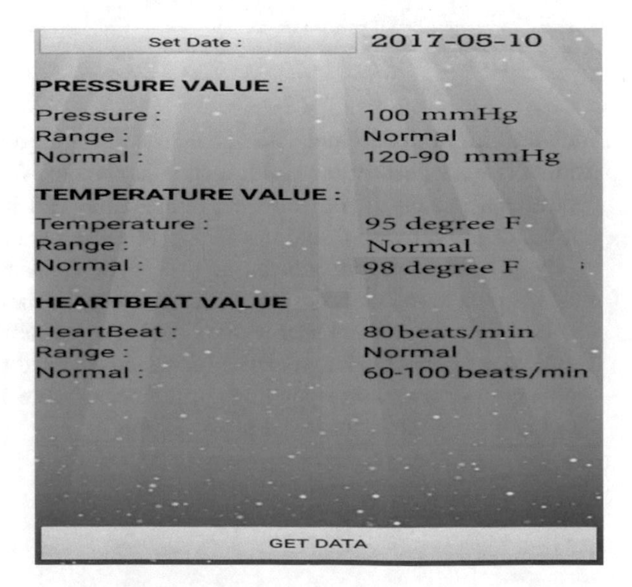

Fig. 2 Health monitoring application interface

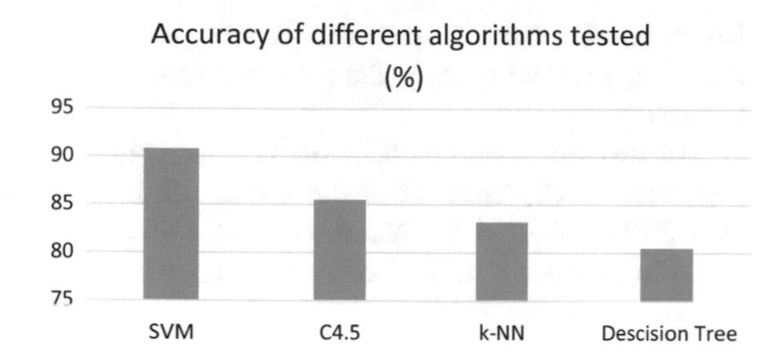

Fig. 3 Accuracy comparison of different algorithms tested (%)

The disease on which we focus to predict using this system is sepsis. Sepsis is a lethal health condition, which arises when the body's reaction to any infection results in impairment of its own cells and organs. Sepsis is detectable by variations in blood pressure, temperature and heart beat as its ordinary symptoms are fever, high heart beat and increased breathing rates.

The trained dataset includes the attributes age, name, gender, body parameters and the patient's with/without sepsis. This dataset contains attributes of 450 people with and without sepsis, collected from health centers.

From the existing data mining algorithms, SVM, C4.5, k-NN and Decision Tree were suitable for predicting sepsis. The main goal of using a classifying algorithm was to identify the category or class to which the new data belong, and in this case, the new data are the body parameters. In the study, the best classifying algorithm that can be implemented to predict sepsis was SVM [17]. From the results obtained by comparing the different algorithms in Fig. 3, we decided to use SVM.

When a new user is accommodated as a part of the insurance system, none of their health details are known to the application to conclude whether or not they are healthy. After the required parameter values are collected and inputted to the application, the interface generates a sample health status record for each client.

Our methodology depends on SVM classifier for health status prediction and also to recognize the probabilities of any threats to health status. Figure 4 shows the relation between count of records in the dataset and classifier accuracy. From the results, we infer that these two factors are highly positively correlated to each other. As the volume of data records increases, classifier accuracy also increases.

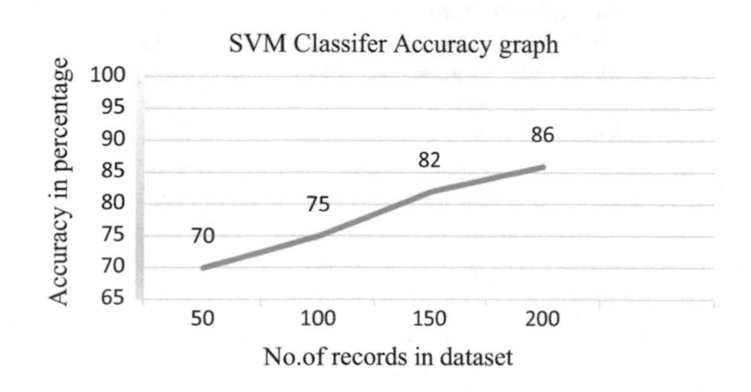

Fig. 4 Classifier accuracy graph

Likewise, classifier accuracy degrades as the number of data records reduces. For achieving greater accuracy, the classifier should be trained with massive amounts of records.

The trustworthiness of the system is analyzed by considering 200 health datasets collected for experimental analysis from different policyholders of the insurance firm. SVM classified the dataset with about 86% accuracy. The classifier accuracy is noted to improve with an increased number of records.

5 Conclusion and Future Scope

Using the real-time health data captured using smart widget, threats to health are detected and alerts are provided on time. For Big Data Analytics, the Hadoop ecosystem is incorporated as a backbone of our architecture. Client data confidentiality is ensured for improved user trust and satisfaction. When compared to the existing systems of similar interest, our system is more efficient in satisfying various requirements of the insurers, as well as their clients while simultaneously being truthful to both the parties. The system is practically implemented and tested on real data. The SVM prediction accuracy reached approximately 86% when 200 real-world health datasets were considered for experimental study.

In our future advancements, we plan to provide better authentication through biometrics. The proposed system only focuses on single disease prediction. The work is also considering extensions to multiple disease predictions. We are aiming to provide clients with smartphone alerts on health advice based on their current health status along with precautions to be followed for better prevention. The next version of this insurance application is aimed to be structured with an isometrics alarm to remind clients about workout times along with video tutorials for increasing the quality of exercise to benefit the policyholders with better health. The proposed system works locally with MySQL server and in the upcoming work, Hadoop could be redeployed to the cloud to enable its connectivity to any cloud servers.

Acknowledgments We would like to express our sincere gratitude to the faculty, Dept. of Computer Science and Applications, Amrita Vishwa Vidyapeetham, Amritapuri for their incessant support and guidance throughout this project. Our special thanks to Dr. M. R. Kaimal, Chairman, Dept. of CSA, Amrita Vishwa Vidyapeetham, Amritapuri for his support throughout the venture. We would also like to appreciate all the reviewers for their valuable opinions for improving the work.

References

1. P. Dineshkumar. 2016. Big data analytics of IoT based health care monitoring system, 55–60.
2. Jia, X., H. Chen, and F. Qi. 2012. Technical models and key technologies of e-health monitoring. In *2012 IEEE 14th International Conference e-health networking, application & services* 2012, 23–26.
3. Joshi, M. 2015. Internet of Things: A Review Introduction: IOT Framework IoS IoE IoT Use Cases IoM," no. March, 2015.
4. Rathore, M.M., A. Ahmad, and A. Paul. 2016. IoT-based smart city development using big data analytical approach. In *2016 IEEE international conference on automation*, 1–8.
5. Behera, R.K., S. Gupta, and A. Gautam. 2016. Big-data empowered cloud centric Internet of Things. In *Proceedings—2015 international conference on man machine interfacing, MAMI 2015.*
6. Duan, F., K. Li, B. Li, and S. Yang. 2015. Research of smart power utilization service system based on IoT. no. Asei, 1427–1430.
7. Pinto, S., J. Cabral, and T. Gomes. We-care : An IoT-based health care system for elderly people.
8. Luo, J., Y. Chen, K. Tang, and J. Luo. 2009. Remote monitoring information system and its applications based on the internet of things. *International Conference on Future. BioMedical Information Engineering*, 482–485.
9. Mukherjee, S., K. Dolui, and S.K. Datta. 2014. Patient health management system using e-health monitoring architecture, 400–405, 2014.
10. Berlian, M.H., et al. 2016. Design and implementation of smart environment monitoring and analytics in real-time system framework based on internet of underwater things and big data. In *2016 international electronics symposium*, 403–408.
11. Ahmad, A., M.M. Rathore, A. Paul, and S. Rho. 2016. Defining human behaviors using big data analytics in social internet of things. In *2016 IEEE 30th International Conference on Advanced Information Networking and Applications,* 1101–1107.
12. Idris, Muhammad, Shujaat Hussain, Mahmood Ahmad, Sungyoung Lee. 2015. Big data service engine (BISE): Integration of big data technologies for human centric wellness data. IEEE.
13. S. Saravanan. 2015. Design of large-scale content-based recommender system using hadoop MapReduce framework. In *2015 8th international conference on contemporary computing, IC3 2015*, 302–307.
14. Sukanya, M.V., S. Sathyadevan, U.B. Unmesha Sreevani Benchmarking support vector machines implementation suing multiple techniques. In *2015 Advances in Intelligent Systems and Computing*, vol. 320, 227–238.
15. Sharma, R., N. Kumar, N.B. Gowda, T. Srinivas. 2015. Probabilistic prediction based scheduling for delay sensitive traffic in internet of things. *Procedia Computer Science* 52 (1): 90–97.
16. Sai, A., S. Salim, P.K. Binu, and R.C. Jisha. 2015. A hadoop based architechture using recursive expectation maximization algorithm for effective and foolproof traffic anomaly detection and reporting. *International Journal of Applied Engineering Research* 10 (55): 2101–2106.
17. Viswanathan, K., K. Mayilvahanan, R. Christy Pushpaleela. 2017. Performance comparison of SVM and C4.5 algorithms for heart disease in diabetics. *International Journal of Control Theory and Applications.*

An Efficient Optimization Technique for Scheduling in Wireless Sensor Networks: A Survey

N. Mahendran and T. Mekala

Abstract Wireless sensor networks (WSNs) use a large number of tiny sensor devices for monitoring, gathering and processing data with low hardware complexity, low energy consumption, high network lifetime, scalability, and real-time support. Sensor node deployment, coverage, task allocation, and energy efficiency are the main constraints in WSNs that impact the node lifetime. Scheduling allows the platform to improve the performance of WSN. Scheduling the sensor nodes and category of sensor data minimizes the energy consumption and increases the lifetime of sensor nodes. This chapter describes the concept of optimization techniques in WSNs to extend performance. We surveyed four metaheuristic optimization approaches to enhance the scheduling performance because these approaches help to find optimal solutions quickly. Optimizing the sensor node placement through scheduling of sensor and sensor data allows a better quality of service (QoS) in WSN. In this chapter, we survey such optimization techniques as ant colony optimization (ACO), particle swarm optimization (PSO), genetic algorithm (GA), and artificial bee colony (ABC) for scheduling methods.

Keywords WSNs · Optimization techniques · Energy consumption
Network lifetime

1 Introduction

Wireless sensor networks group many tiny sensors and sinks that congregate and analyze data packets from various fields. WSN-related research topics have increased in recent years. They have been extensively used in environment, military, vehicle, medical, and other fields. However, the sensor node's data processing

N. Mahendran (✉) · T. Mekala
M. Kumarasamy College of Engineering, Karur, Tamilnadu, India
e-mail: mahe.sec@gmail.com

T. Mekala
e-mail: mekalathangavel@gmail.com

A. N. Krishna et al. (eds.), *Integrated Intelligent Computing, Communication and Security*, Studies in Computational Intelligence 771,
https://doi.org/10.1007/978-981-10-8797-4_24

capability is relatively low and its energy is high. Figure 1 shows the basic structure of WSNs [1].

The figure shows that the entire sensor node is grouped together, called a cluster. Each group contains one lead called a cluster head (CH) or router. The data generated from sensor nodes is forwarded to CH. Through different routers, the data will reach the destination. Meanwhile, the energy consumption is related to the data sensing, data processing and transmission distance. To optimize this functioning is unique to WSNs [2]. Here we discuss some scheduling methods and optimization algorithms in WSN to increase efficiency.

The remainder of the chapter is organized as follows: In Sect. 2, scheduling schemes are surveyed, and in Sect. 3 we introduce optimization techniques. An ACO algorithm is explained in Sect. 4, and PSO algorithms in Sect. 5. ABC and GA algorithms are presented in Sects. 6 and 7, respectively. Finally, Sect. 8 is dedicated to the conclusion.

2 Scheduling Scheme

A scheduling algorithm is classified as one of two types: packet scheduling and node scheduling. Node scheduling is an efficient way to manage the energy in WSNs. In this approach, not all nodes are active at the same time, which reduces energy consumption. Packet scheduling assigns data packets to the sensor nodes [3–5]. This algorithm is associated with resource selection, resource discovery, resource filtering, and resource scheduling policy. The characteristics of ant colony optimization follow in this same manner. The ants exchange information with each other in order to detect the optimal path. The idea of a parallel ant colony (PAC) is

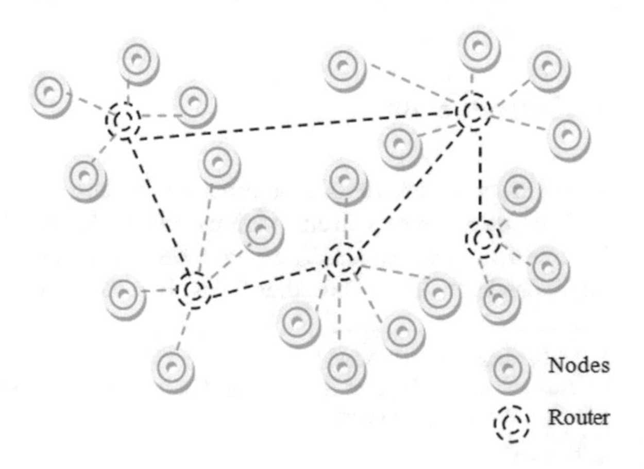

Fig. 1 Structure of WSNs

developed using parallel construction phase. In the PAC algorithm, multiple colonies are implemented simultaneously. Optimality is found by comparing all PAC with each other [6]. The only drawback with the ACO is the short distance convergence. In order to solve the problem, the honey bee algorithm (HBA) can be utilized [7].

The HBA utilizes better features of ACO along with high distance capabilities to find the optimal path. It utilizes the foraging system associated with honey bees to find an optimal path out of the available alternative path. This algorithm converges faster as compared to existing ACO.

The genetic algorithm (GA) has initialization, mutation crossover phases, and it continues until an optimal result is obtained. The genetic algorithm consumes a large amount of time for the complex problem. It terminates when a solution reaches a satisfactory level. Using this approach, ACO and HBA enhance job scheduling performance. The ACO selects the path, which is optimal in nature, but distance dependent. In order to overcome this problem, HBA is utilized; it may not always give the optimal path, but speed is enhanced and it is not distance dependent. In PAC, multiple processes can be executed at the same time and it selects the optimal path.

3 Optimization Algorithms

An efficient optimizer used in wireless sensor networks solves routing and scheduling problems to minimize energy consumption. In WSNs, the main constraint is task allocation, energy efficiency, network lifetime and node deployment. There is no distinct algorithm that is proper for improving network lifetime and energy consumption. There exist a number of optimization algorithms with bio-mimetic algorithms (pattern matrix-based methods), derivative-based algorithms, and derivative-free algorithms (classical algorithm) [8].

3.1 Derivative-Based Algorithms

This algorithm is used in discrete modeling and many scientific applications. Some examples are hill climbing and conjugate gradient Newton's method. It is also called gradient-based algorithm, which is used to resolve unrestrained optimization issues such as energy minimization, and it also uses the information of the derivative.

3.2 Derivative-Free Algorithms

This algorithm employs the importance of the objective purpose, not the information of the derivative. The Hooke-Jeeves pattern search, trust-region and Nelder-Mead downhill are the methodologies for derivative-free algorithms. The Hooke-Jeeves search is a new direction by integrating the history of iterations.

3.3 Bio-mimic Algorithms

Recently, optimization algorithms are evolving frequently from bio-mimetic and nature-inspired due to the easy fitting for universal optimization mechanisms. There exist bio-mimic optimization algorithms, including ant colony optimization (ACO), particle swarm optimization (PSO), genetic algorithm (GA), bat algorithm (BA), cuckoo search (CS) and firefly (FF). The selection of algorithms for correct solution plays a major role in any research work.

4 Ant Colony Optimization

This method is motivated by the food searching activities of ants. When ants are in search for food, they drop a pheromone on the trail, which creates a path for them. The liquid pheromone evaporates as time passes. Therefore, pheromone distribution on the path is a suggestion of probability for the path. Because of its probabilistic and dynamic nature, this algorithm is used for optimization technologies where topology modifies frequently [2].

This algorithm focuses on the next node through the comparison of distance between the residual energy and the nodes, which ensures that there is less possibility of nodes with minimum energy selected as the next hop node. Therefore, the proposed ACO algorithm improves energy balancing, the stability of WSNs and, eventually, extends the lifetime of the WSNs.

This algorithm not only helps the nodes in effectively and quickly finding the shortest path to the destination node, but can also improve the work efficiency of the WSNs. The existing protocol is called low-energy adaptive clustering hierarchy (LEACH); the CH consumes more energy due to forwarding data to the destination. The proposed ACO algorithm can help each cluster head node to find the shortest path distance to the destination node. Hence, it reduces the energy consumption of data packet transmission and also reduces the work [3]. Likewise, it schedules the sensor nodes to sleep/awake mode to increase the performance of WSN.

Algorithm 1: ACO Optimization

```
1: Input  {Si = Set Current Location}
2: Input   {Fs = Find Sink}
3: Initialize: Si, Fs, Pheromone, number of ants
4: do
5:      calculate Si
6:      Find Fs
7:      Selection of optimum values
8:      save :Pheromone Exploration and update pheromones
9: while (maximum iteration)
```

5 Particle Swarm Optimization

The PSO was introduced by Eberhert and Kennedy in 1995. It is also known as the population-based optimization approach. The performance of this technique is based on the normal social activities of bird flocking. It starts by initializing a population of arbitrary searching and looks for solutions that are more optimal by informing generations. In PSO, the "bird" refers to particle solution and the search space is called a "particle". The PSO algorithm with fitness and velocities, has a fitness value, which is calculated by fitness functions, and also has velocities to the shortest path for the particle to fly. It is calculated on the basis of local best position and global best position [9, 10].

- Current position of particle.
- Current velocity of particle.

These two parameters are used to updating the rules for finding the next position of a particle. Based on the velocity, each particle position is adjusted and the distinction between its current particle positions and the best situation is established so far. As the representation is repeated, the swarm centers more and more on an area of the search space containing the best quality solution. It is significant to note that the process engages both intelligence and social interaction so that birds learn from their own experience (local best search) and also from the knowledge of others around them (global best search). The process is initialized with a collection of random particles (solutions).

This idea uses node deployment for the purpose of increasing the WSN lifetime with minimum energy constrained sensor nodes. Ease of implementation, computational efficiency, elevated quality of solutions, and speed of junction are strengths of PSO. It is limited in continuous and real number space, thereby leading to discrete BPSO and discrete multi-valued PSO. The deployment of PSO and its variant results in a simple, fast, computationally efficient and easy to implement technique and is capable of very efficiently deploying the sensors. PSO outperforms by minimizing the energy consumption while maximizing the connectivity and coverage in the networks.

Algorithm 2: PSO Algorithm

```
1: Input  {Si   collection of sensor}
2: Input  {Ts   collection of targets monitored by sensor s}
3: Initialize Particles
4: Begin
5:    Evaluate fitness
6:    Search locally & globally
7:      if(Gen > Population) then
8:         Gen++;
9:       End if
10:           Update particle position & velocity
11:        End
```

Future research method on [7] PSO in WSNs applications expected to focus on:

- Transformation of accessible simulation mechanisms into real-world applications.
- Progress of PSO algorithm in hardware application.
- Progress of a parameter less black-box PSO.
- Cross-layer optimization mechanism through PSO algorithm.
- Assigning priority for data packets

6 Artificial Bee Colony Optimization (ABC)

Karaboga (2005) has projected the ABC scheme through swarm-based intelligence motivated by the smart foraging activities of honey bees. It is made of food sources and three bee groups, namely viewers, scouts, and employed bees. The job of employed bees is to find the path to the particular food source, take an amount of food from the source, and finally, unload the food into a food store. The bees receive the food and perform a dance called the waggle dance, which holds information about the position of the food source, food quality, direction and distance to the food. The details about all the recently found food sources are available to a viewer bee.

The new position of quality food sources is identified by the viewers and scout bees, and they change their status to employed bees to visit the position of the quality food source. The exploitation and exploration processes must be carried out concurrently in a search process. Viewers and employer bees perform the exploration process while scout bees control the exploitation process. The positions of the viewers, employers and scout bees are changed due to frequent cycles of the search processes (cycle = 1, 2, 3, 4 …, MCN), where MCN = a maximum number of the cycle per search process. Depending on the local valuable information, the employed bees change position in their memory and check the nectar quantity of the new position. The bee stores its new position and deletes the old one if the nectar amount is higher than that of the previous one or it keeps the position of the previous one in their memory. The probability P_n of picking a food source by Viewer bees is designed as follows:

$$P_n = \frac{f_n}{\sum_{n=1}^{m} f_n} \tag{1}$$

where f_n is the fitness value of a solution n, and m is number of position of target.

Algorithm 3: ABC algorithm

```
1: Input  {Si: Sensor set }
2: Input  { Ts: Collection of targets monitored by sensor node }
3: Input  {ns= Current iteration solution,
               ps =Previous iteration solution}
4: Initialize population
5: Begin
6:     Evaluate fitness
7:         do
8:             create new population
9:              Search ns
10:                   if(ns is better than ps) then
11:                      Store ns and forget the ps
12:                    else
13:                      Keep the ps as it is
14:                   End if
15:                 while(maximum iteration)
16:             End
```

7 Genetic Algorithm

A GA places the most popular algorithm based on the evolutionary method, which is supported by the concept of biological systems invented by Darwin in Holland at the 1970s. GA acts as a heuristic search evolutionary algorithm globally. GA uses random decision space through a selection process, crossover, fitness and mutation operators arranged to grow a global optimum solution. Another significant operator of GA is elitism. Its job is to accumulate an elite chromosome for the next generation.

The main pros of GA over existing optimization algorithms are the facility to use more intricate problems and parallelism implementation. Parallel implementation processes use multiple genes. In GA, the coded answers are represented as binary strings contain 1s and 0s.

GA algorithm applied as

1. Initial population arbitrarily generated in a search space and finds fitness value.
2. Population is created on the basis of these mechanisms as selection, repair, crossover, and mutation operations.

- Select parents for crossover and mutation.
- Crossover mechanism exchanges genetic materials between individuals or phenotypes to produce offspring.
- This algorithm handles population and again changes it to create a new generation of chromosomes.
- Continue this process until a maximum number of generations is achieved.

GA used in wireless sensor networks finds a number of clusters and their heads in the arranged environment. There is no communication with the global optimization process. Table 1 represents the optimization algorithm in WSNs.

Table 1 Optimization algorithm in WSNs

Parameters	ACO	PSO	ABC	GA
Representation	Undirected graph	Dimensions for vector position and speed	Solution pool, random variable	Binary form 0s and 1s, random variables
Operators	Pheromone updates and trial evaporation	Initial values updates and evaluation	Employed bees, viewers and scouts	Selection, crossover, mutation
Advantages	Inherent parallelism using dynamic environment Provides better solutions through a positive feedback approach Provides distributed computation and evades premature convergence	Easy implementation Using less parameters to adjust global search	Robust and gives good quality of solution	It uses chromosome encoding to find optimization difficulty Easily shifted to existing models and simulations
Disadvantages	Theoretical analysis is difficult Coding is not straightforward	Holds maximum memory	Poor exploitation	Need more running time

Algorithm 4: GA Algorithm

```
1: Input  {Randomly Create initial population }
2: Begin
3:    Evaluate the chromosomes fitness
4: do
5:   Select two parents from population
6:   Perform single point crossover
7:   Perform mutation
8:  while(termination)
9:     Solution Set
10:        End while
11:       End
```

8 Conclusion

In this survey, we compared the various parameters of biomimetic optimization techniques and their performance in WSNs is estimated. The performance of the ACO, PSO, GA and ABC algorithms leads to advantages in energy consumption and node lifetime in WSNs. Finally, we conclude that on tailoring these algorithms to various issues in WSN leads to better performance in the overall system evaluation.

References

1. Deepika, T., and N. Mahendran. 2015. Comparitive analysis of optimization algorithms in wireless sensor networks. *International Journal of Applied Engineering Research* 38. ISSN 0973-4562.
2. Peng Li, Huqing Nie, Lingfeng Qiu and Ruchuan Wang. 2017. Energy optimization of ant colony algorithm in wireless sensor network. *International Journal of Distributed Sensor Networks* 13 (4). https://doi.org/10.1177/1550147717704831.
3. Muralitharan Krishnan, Vishnuvarthan Rajagopal, Sakthivel Rathinasamy. 2018. Performance evaluation of sensor deployment using optimization techniques and scheduling approach for K-coverage in WSNs. https://doi.org/10.1007/s11276-016-1361-5.
4. Gomathi, R., and N. Mahendran. 2015. An efficient data packet scheduling schemes in wireless sensor networks. In *Proceeding 2015 IEEE international conference on electronics and communication systems (ICECS'15)*, 542–547. ISBN: 978-1-4799-7225-8. https://doi.org/10.1109/ecs.2015.7124966.
5. Vanithamani, S., and N. Mahendran. 2014. Performance analysis of queue based scheduling schemes in wireless sensor networks. In *Proceeding 2014 IEEE international conference on electronics and communication systems (ICECS'14)*, 1–6. ISBN: 978-1-4799-2320-5. https://doi.org/10.1109/ecs.2014.6892593.
6. Rajwinder Kaur and Sandeep Sharma. 2017. A review of various scheduling techniques considering energy efficiency in WSN. *International Journal of Computer Applications (0975–8887)* 16 (28).

7. Raghavendra V. Kulkarni, and Ganesh Kumar Venayagamoorthy. 2011. Particle swarm optimization in wireless sensor networks: A brief survey. *IEEE Transactions on Systems, Man, and Cybernetics*. https://doi.org/10.1109/tsmcc.2010.2054080.
8. Akhtaruzzaman Adnan, Md., Mohammd Abdur Razzaque, Ishtiaque Ahmed and Ismail Fauzi Isnin. 2014. Bio-mimic optimization strategies in wireless sensor networks: A survey. *Sensors* 14: 299–345. https://doi.org/10.3390/s140100299.
9. Mahendran, N., Dr. S. Shankar and T. Deepika. 2015. A survey on swarm intelligence based optimization algorithms in wireless sensor networks. *International Journal of Applied Engineering Research* 10 (20). ISSN 0973-4562.
10. Kalaiselvi, P., and N. Mahendran. 2013. An efficient resource sharing and multicast scheduling for video over wireless networks. In *Proceeding 2013 IEEE international conference on emerging trends in computing, communication and nanotechnology (ICECCN'13)*, 378–383. ISBN: 978-1-4673-5036-5. https://doi.org/10.1109/ice-ccn.2013.6528527.

A Genetic-Based Distributed Stateless Group Key Management Scheme in Mobile Ad Hoc Networks

V. S. Janani and M. S. K. Manikandan

Abstract In this chapter, we propose a stateless distributed group key management (GKM) framework, a genetic-based group key agreement (genetic-GKA) scheme, for supporting the dynamic key management mechanism in mobile ad hoc networks (MANETs). This scheme operates on polynomial computation over a finite area for group key establishment. The group key update is fixed to O (log n) with an effective GKM scheme, regardless of number of key updating information. Therefore, the key update process in this scheme has an O (log n) key update execution and storage cost, with n number of nodes in the group communication. Our proposed scheme is very efficient with implementation of hash and Lagrange polynomial interpolation. The simulation results show that the proposed GKM scheme achieves higher performance and security than the existing key management schemes. Moreover, the proposed scheme is effective in detecting both selfish and malicious attacks with low overhead.

Keywords Group key management · Genetic · MANET · Group key

1 Introduction

In many potential applications of mobile ad hoc networks (MANETs), such as on the battlefield and in rescue operations and other civilian commercial applications, providing secure group communication is significant. The most efficient method for achieving a cryptographically secure group communication is by deployment of a public key infrastructure (PKI) framework with a symmetric group key shared among the dynamic nodes for data encryption. Group key management (GKM) is a

V. S. Janani (✉) · M. S. K. Manikandan
Department of ECE, Thiagarajar College of Engineering, Madurai, India
e-mail: jananivs1987@gmail.com

A. N. Krishna et al. (eds.), *Integrated Intelligent Computing, Communication and Security*, Studies in Computational Intelligence 771,
https://doi.org/10.1007/978-981-10-8797-4_25

well-established cryptographic technique used to authorize and maintain a group key in a group communication through secured channels. It is an effective security mechanism for node-to-node unicast, broadcast or multicast communication. Establishment of the key is performed with distributed group key agreement (GKA) protocols, where the nodes collaboratively determine the group key. When considering a highly dynamic network such as MANET, the group membership in the communication changes frequently with the addition and deletion of nodes in the group. This introduces issues with updating the group key, which must be reconstructed in a secure and timely manner. The group manager or the cluster head (CH) broadcasts the messages regarding key updates to legitimate members to construct a new group key. An important security requirement during the key update operation is access control in terms of forward and backward secrecy. This access control mechanism prevents (i) new members from obtaining the past group key (backward secrecy) and (ii) revoked or evicted members from accessing the present or future group keys (forward secrecy). Therefore, providing an efficient and scalable immediate group keying procedure for dynamic groups is a tedious problem, especially in dynamic group communication with minimal size of key update messages and maximum performance benefits. Nevertheless, there are several constraints in establishing a group key update in a PKI communication system to secure the ad hoc networks. It is clear that the limitations of the long-established key updating techniques should be minimized in order to make the PKI-based security viable for secure group communication.

The proposed research work concentrated on developing an effective GKM scheme with a dynamic and stateless key update mechanism. Here, a stateless GKM scheme is introduced in order to resolve the inherent key update issues for ad hoc networks. For secure communication, each cluster members are accomplished to produce a transient subgroup. The communication nodes in the key agreement protocols collaboratively compute the session keys. The stateless schemes broadcasts Lagrange interpolation polynomials with a session key for pairwise key establishment with a one-round key update process and minimizes the computation overhead. Each key update message contains M shared nodes and authentication data for all nodes to recover the group key. In the proposed schemes, the CH is only required to distribute the updated group key to nodes. A node can obtain the lost keys after obtaining the updated message. The proposed schemes have performance advantages when compared with the existing approaches. Moreover, the schemes possess several distinguishable features, including (i) a stateless key update mechanism and (ii) a practical solution for out-of-sync problems. Moreover, the proposed scheme ensures access control with forward and backward secrecy. The proposed scheme integrates the GKA protocol with constant round efficiency in order to reduce the time for key updating in a hierarchical access control. This authenticated GKM protocol can provide implicit key authentication, session key security, and group confidentiality. In addition, this scheme resists attacks on key control, unknown key sharing, and key-compromise impersonation.

2 Related Works

The most well-known distributed group key establishment protocols in PKI systems are the key agreement schemes where all the communication members determine session keys in a collaborative fashion [1]. Many GKM methods have been considered for ID-based cryptography for several group communications in a PKI MANET system. One important KM method, the group key agreement (GKA), allows more than one party to share the group key and further makes its communication. It also provides key authentication among senders and receivers; thus it is also referred to as an authenticated key agreement (AKA) protocol. Numerous network providers apply this protocol to their present distributed system as it offers secure communication [2, 3]. The last decade has witnessed an exciting growth of research on GKA and its dynamic key updating. For the dynamic MANET environment, the conventional KM methods were found to be inactive due to its variance in topology as well as being power-constrained. The protocols employed in [4–7] have proven schemes to offer security as well as integrity over MANETs. Once a key updating procedure happens, these schemes are ineffective with a hierarchical network topology. A stateless protocol was introduced as a powerful key updating mechanism to overcome the update loss issue of the stateful scheme. The key update schemes proposed by [8–10] work on mathematical relations and bilinear pairing for key update processes. A broadcast key update scheme, a potential solution to the stateless issue, uses polynomial and linear algebra-based computations as given in [11–14]. These protocols are efficient in terms of cost and implementation. The conventional GKA schemes are vulnerable to out-of-sync problems and key-compromise impersonation attack while updating the key, whenever nodes are added or revoked in the network: (a) communication rounds grow proportionally when the number of group members is larger; (b) less key secrecy and confidentiality with lower access control (forward and backward secrecy); (c) the size of a key update message should be minimized for a larger and highly dynamic network.

It is therefore clear that the drawbacks of the long-established GKM techniques should be minimized in order to make the PKI framework viable for secured multicast communication. In this pursuit, the proposed research work concentrates on developing a distributed stateless GKA scheme for self-organized MANET environments, which quantifies nodes behaviour in the form of trust in hexagonal non-overlapping clusters. In this paper, an efficient stateless GKM scheme, namely genetic-based GKA (genetic-GKA) scheme is presented to provide public key infrastructural security in a MANET framework. A genetic algorithm is used to search for appropriate key-generating parameters (KGPs, or code segments) and efficient stateless key updating from the code segments. The proposed scheme is constructed based on polynomial broadcasting, in which each node possesses M polynomial degrees to perform a group key update by its private key. Moreover, the distributed MANET setting allows the nodes to establish a group key by themselves in a self-organized and cooperative fashion. This system design avoids the single-point-of-failure issue and network bottleneck. The proposed scheme shows more efficient performance than the existing GKM schemes.

3 Proposed Genetic-GKA Scheme

Genetic algorithms (GA) are iterative optimization techniques that represent a heuristic tool of adaptive search mechanisms which have been studied in recent years. This technique selects the best-fitted individuals (i.e., nodes) to survive any mischievous network conditions and works with a pool of binary strings called chromosomes, in an iterative process. The suitable nodes are selected by evolution of fitness function among the MANET nodes. To develop an eventually reliable population of individuals, the GA uses three processes: fitter individual selection, crossovers and mutations. The intrinsically parallel feature of the algorithm makes it well suitable to scan the large MANET environment to manage multi-optimization issues. The GA works as follows: with a pool of individuals that are encoded in binary code segments, a population is randomly chosen for evaluation. Based on the fitness value, the individuals with highest fitness are selected to generate new offspring. Mutations are performed with these offspring to produce a new population of nodes. The prime objective of any security mechanisms is to provide security to a network in terms of availability, authentication, integrity, confidentiality and non-repudiation, which can be achieved with the help of key management schemes. The proposed distributed GKM approach uses a GA in which the nodes in the group communication contribute to the group key establishment by performing a hash of the random number within each cluster partitions during the initialization stages.

The genetic-GKA model is developed over a hierarchical cluster architecture where the nodes are grouped into a non-overlapping hexagonal structure with spatial reuse benefits. It includes two types of nodes including the cluster head (CH) and member nodes monitored by a distributed certificate authority (CA). The CA sends the information monitored in the network to more powerful CH nodes with larger memory size. The CH combines the received data from the CA and distributes the aggregated results to member nodes within each cluster. The CH in the cluster-based MANET generates the key-pair and transmits it among the other header nodes securely. In the PKI framework, the private key provides communication among the cluster members and public keys distribute the key update information to the cluster members. The member nodes generate KGPs, which are used for session group key (SGK) generation in key updating operations. The KGPs are further divided into code segments that are embedded into cluster members and CHs before deployment. The CH further assembles the segments to rebuild the KGP at definite intervals, the segments of which would be harder for an attacker to know. The member node delivers m number of code segments to the CH from which the CH chooses a permutation of certain code segments and broadcasts it to the members. Each cluster member generates a new key with the permutation of code segments. Thus, the member nodes establish a common session key with a CH in each cluster. The GA analyses the appropriate key-generating code segments. The operation of genetic-GKA includes the key agreement with an appropriate code segments set. For key agreement, a population of N individuals with m code segments is maintained. Initially, the segment is encoded in a chromosome (here, CMs) as a gene. The CH chooses m segments to assign an initial CM to the segment pool, which is further increased exponentially with all member contributions. The genetic-GKA then selects a relevant CM for a simple crossover mutation to generate new offspring. The code segments of the offspring are exposed to repetitive evolution throughout the network functionality. The best among such offspring are considered as the KGPs to perform key updating in the next phase. To measure the quality of a gene from initial population, the genetic-GKA selects segments with a high rate of fitness. The rate of fitness is defined as the actual rate, which an individual determines by being tested in contribution to the subsequent generation. In the stateless key updating phase, the selected code segments are encoded randomly and distributed to the CHs and the CA.

3.1 Group Key Management

The prime goal of the proposed key management system is to generate keys to the nodes to manage keys, to encrypt/decrypt the messages and thus prevent the illegal usage of handling certified keys. That is, it facilitates providing the highest security to networks, which will manage several attacks. The code segments are encoded as CMs to make it suitable for applying the genetic mechanism. The group members generate the code segments which are then used for constructing the KGPs. The pool of code segments (considered as a gene) accumulates all the operations and operands that are operated to form the segments. With these genes, different permutations and combinations of keys are generated. For example, with k number of genes, $k!$ combinations of keys are generated. In order to measure the uncertainty of an outcome and to compute the distribution function of key values, the predictability of the code segment pool is measured. To construct an efficient quality of gene from the initial population, the GA chooses a parent code segment of a high rate of fitness. The fitness function thus conveys the feedback of the genetic-GKA scheme in managing the appropriate segment searching to improve the quality of CMs. This fitness is evaluated as in (1):

$$FR(\text{Ch}) = \sum_{j=1}^{t} \mathbb{P}_n log_2(1/\mathbb{P}_n) \qquad (1)$$

where j is the randomly selected k keys from the pool of m CMs, and $\mathbb{P}$ is the probability function of k CMs, with $\sum_{n=1}^{t} \mathbb{P}_n = 1$; $n = 1, 2, 3 \ldots k$. If the FR(Ch) value is large, the distribution key is objective or more likely to be uniform. To apply the GA, the population of possible genes is initialized and refreshed during CM evolution. The common point mutation and crossover mutation is combined in the proposed genetic-GKA scheme. From the code segment pool, a crossover point is chosen randomly by the simple crossover function with the parent CMs (Ch). The offspring are generated with two parent genes and kth gene as the point of crossover, as follows in (2) and (3):

$$Offspring_1^{j+1} = (\mathrm{Ch}_{a1}^{j+1}, \ldots, \mathrm{Ch}_{ak}^{j+1}, \mathrm{Ch}_{b(k+1)}^{j+1}, \ldots, \mathrm{Ch}_{bM}^{j+1}) \quad (2)$$

$$Offspring_2^{j+1} = (\mathrm{Ch}_{b1}^{j+1}, \ldots, \mathrm{Ch}_{bk}^{j+1}, \mathrm{Ch}_{a(k+1)}^{j+1}, \ldots, \mathrm{Ch}_{aM}^{j+1}) \quad (3)$$

where the ath CM and bth CM are the parent CMs randomly selected from the jth generation as in (4) and (5).

$$\mathrm{Ch}_a^j = (\mathrm{Ch}_{a1}, \ldots \mathrm{Ch}_{ak}, \ldots, \mathrm{Ch}_{aM}) \quad (4)$$

and

$$\mathrm{Ch}_b^j = (\mathrm{Ch}_{b1}', \ldots \mathrm{Ch}_{bk}', \ldots, \mathrm{Ch}_{bM}') \quad (5)$$

During each successive generation of offspring, a portion of the initial population is chosen to breed a new generation. The fitness process selects individual solutions with fitter genes being selected. The existing selection methodologies perform a random selection operation, which is an expensive procedure. To control the population, the relative distance (D) for each solution pair is measured in the solution pool, which is given as in (6):

$$D(s_1) = \min_{s' \in \mathbb{P}} d(s_1, s') \quad (6)$$

The population s_1 is either added or discarded from the population pool with respect to the diversity factor ω. That is, the population is added if $D(s_1) \geq \omega$ and otherwise it is discarded from the pool. Based on the size of MANETs, the proposed genetic-based model can dynamically optimize the key parameters. Therefore, intruders are not able to recognize the initial value of the key, session keys, updated timing of keys, or clustering groups. During the key-generation and communication period, the proposed genetic-based system will use various random key generators for minimizing the vulnerabilities of KM methods.

4 Performance Analysis

The simulation analysis of the proposed schemes has been performed and evaluated in the QualNet 4.5 environment along with Visual Studio 2013 (IDE), VC ++ (programming language) and NSC_XE-NETSIMCAP (SDK). The comparison among different methods is run in this simulation environment of 40 nodes. This analysis follows a random walk mobility (RWM) model with the 'T_{halt}' value of '0'. At various time intervals, each node varies its rate of mobility. In this simulation environment, all mobile nodes allocate 4 Mbps as their bandwidth or the channel capacity. The overhead of key management is also counted with respect to the average number of messages sent as well as the number of exponents. As shown in Figs. 1 and 2, this comparison analysis comprises all the key requests and responses in the existing ID-based multiple secrets key management scheme (IMKM) scheme [6] and GKM [15] with the proposed methodologies.

Figure 1 clearly reveals that the overhead is similar at all probable node mobility, making both the existing IMKM scheme and the proposed methodologies robust towards dynamic mobility. Here, the existing IMKM method necessitates higher overhead than the proposed methodology. Among these, the proposed scheme possesses the lowest overhead, at different node mobility. For various numbers of nodes, the communication overhead is counted with respect to the number of exponents that is shown the in Fig. 2. Here, the proposed genetic-GKA scheme possesses less overhead for a larger number of nodes than the other three methods. The proposed schemes are compared with other schemes, namely, the multiparty key agreement [15] scheme and [4], with respect to the computation and communication costs, as depicted in the following Table 1. The factors involved in this analysis are communication round, pairings, memory and power consumption. The analysis shows that the performance of the proposed scheme is superior to that of the two existing protocols. Therefore, the proposed schemes demonstrate an absolute improvement in the above-mentioned factors, thus minimizing both computation and communication costs.

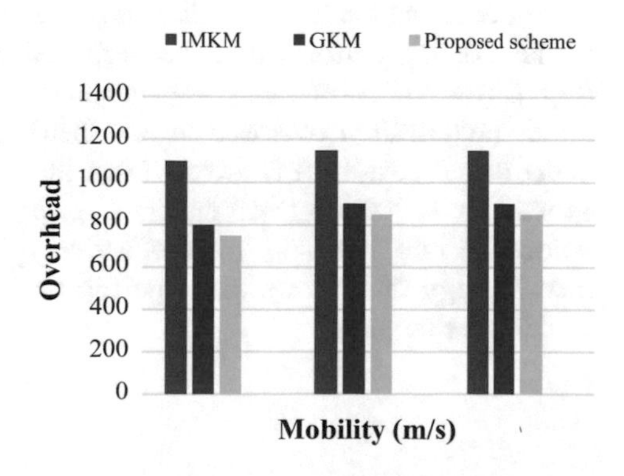

Fig. 1 Overhead versus mobility

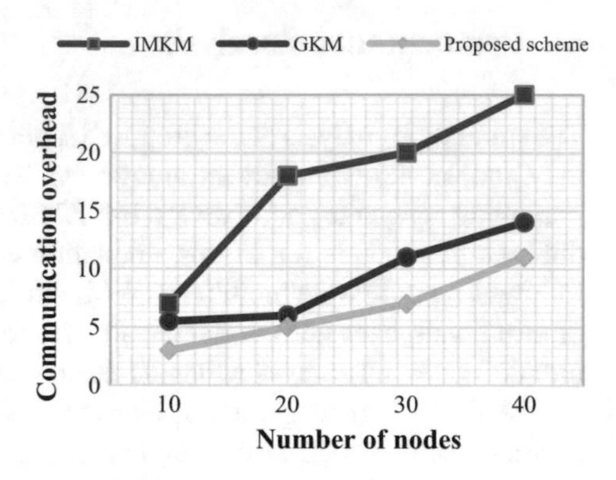

Fig. 2 Communication overhead

Table 1 Comparison of key management schemes

Factors	Schemes		
	GKA [15]	GKM [4]	Genetic-GKA
Communication rounds	2	2	1
Number of pairing computations	$2x$	$2x$	x
Memory	$x-1$	$2x-2$	$(2x/g)-2$
Power consumption (mW)	1207	1114	272
Communication complexity	Two broadcasts	One broadcast	One broadcast

5 Conclusions

In MANETs, secure key management and rekeying is more problematic than in classical networks due to the number of nodes, and the lack of infrastructure. This research addresses a secure and efficient scheme for cluster-based MANETs. Each time members leave or join the cluster, the rekeying operation is efficiently carried out for adapting the high mobility and varying link qualities of a MANET. Contrary to the existing performance, the proposed schemes are efficiently rebuilding the (log (n)) keys. A one-way key generation procedure is presented to minimize the communication overhead in key distribution and bandwidth usage. From the experimental results, it is revealed that the proposed scheme enhances the security as well as the performance factors (i.e., power consumption, overhead, rekeying, revocation rate and time) more effectively than the other schemes. The proposed methodology thus shows better performance in MANETs when compared to the existing techniques.

References

1. Boyd, C. 1997. On key agreement and conference key agreement. In *Proceedings of second Australasian conference information security and privacy (ACISP '97)*, 1270, 294–302, LNCS.
2. Lee, P.P.C., J.C.S. Lui, and D.K.Y. Yau. 2006. Distributed collaborative key agreement and authentication protocols for dynamic peer groups. *IEEE/ACM Transaction on Netwoking* 14 (2): 263–276.
3. Kim, Y., A. Perrig, and G. Tsudik. 2004. Group key agreement efficient in communication. *IEEE Transactions on Computers* 53 (7): 905–921.
4. Lin, J.-C., K.-H. Huang, F. Lai, and H.-C. Lee. 2009. Secure and efficient group key management with shared key derivation. *Computer Standards and Interfaces* 31 (1): 192–208.
5. Zhang, Z., C. Jiang, and J. Deng. 2010. A novel group key agreement protocol for wireless sensor networks. In *2010 international conference on measuring technology and mechatronics automation*, 230–233.
6. Li, L.-C., and R.-S. Liu. 2010. Securing cluster-based ad hoc networks with distributed authorities. *IEEE Transactions on Wireless Communications* 9 (10): 3072–3081.
7. Zhao, K., et al. 2013. A survey on key management of identity-based schemes in mobile ad hoc networks. *Journal of Communications* 8 (11): 768–779.
8. Jarecki, S., J. Kim, and G. Tsudik. 2011. Flexible robust group key agreement. *IEEE Transaction Parallel Distributed Systems* 22 (5): 879–886.
9. Du, X., Y. Wang, J. Ge, and Y. Wang. 2005. An id-based broadcast encryption scheme for key distribution. *IEEE Transaction on Broadcast* 51 (2): 264–266.
10. Gu, J., and Z, Xue. 2010. An efficient self-healing key distribution with resistance to the collusion attack for wireless sensor networks. In *Proceedings of the IEEE international conference on communications (ICC)*, 1–5.
11. Nabeel, M., and E. Bertino. 2014. Attribute based group key management. *Transaction on Data Privacy* 7 (3): 309–336.
12. Tang, S., et al. 2014. Provably secure group key management approach based upon hyper-sphere. *IEEE Transactions on Parallel and Distributed Systems* 25 (12): 3253–3326.
13. Yanming, S., M. Chen, A. Bacchus, and X. Lin. 2016. Towards collusion-attack-resilient group key management using one-way function tree. *Computer Networks* 104 (20): 16–26.
14. Chen, Y.-R., and W.-G. Tzeng. 2017. Group key management with efficient rekey mechanism: A Semi-stateful approach for out-of-synchronized members. *Computer Communications* 98 (15): 31–42.
15. Lin, C-H., H-H. Lin and J-C. Chang. 2006. Multiparty key agreement for secure teleconferencing. *IEEE International Conference on Systems, Man and Cybernetics* 3702–3707.

A Hybrid Approach for Traffic Delay Estimation

Pushpi Rani and Dilip Kumar Shaw

Abstract Traffic delay caused by traffic congestion is a major problem in developing countries as well as in developed countries. Delays occur due to technical factors (traffic volume, green light time, cycle time) and non-technical factors (weather conditions, road conditions, visibility). Plenty of research work has been presented for traffic delay estimation based on these factors. This chapter proposes a hybrid approach for traffic delay estimation for intersections and links on road networks. The proposed method incorporates a technical factor with non-technical factors. Simulation results reveal that both types of factors have a significant impact on delay estimation, and hence cannot be neglected.

Keywords Delay estimation · Technical factors · Non-technical factors
Traffic volume · Fuzzy sets

1 Introduction

Traffic congestion represents a significant problem that has arisen in many areas of the world, particularly in developing countries due to limited resources and infrastructure. This congestion causes unacceptable delays in traffic, resulting in increased travel time. Therefore, delay estimation will provide travellers valuable information so that they can choose an alternate route if it is available. Traffic delays are generally influenced by technical factors (e.g., green light time, traffic volume, vehicle types, intersection geometric design) and non-technical factors (e.g., weather conditions, visibility, extreme temperatures, road conditions) [1]. Technical factors pose linear or deterministic behaviour while non-technical factors exhibit non-linear and non-deterministic behaviour. Although both technical and non-technical factors have their own significance for traffic delay estimation, the former can be depicted by mathematical formalism and the latter is based on human

P. Rani (✉) · D. K. Shaw
Department of Computer Applications, NIT, Jamshedpur, India
e-mail: pushpi.05.wit@gmail.com

A. N. Krishna et al. (eds.), *Integrated Intelligent Computing, Communication and Security*, Studies in Computational Intelligence 771,
https://doi.org/10.1007/978-981-10-8797-4_26

knowledge. Plenty of research has been found in the literature for traffic delay estimation based on these factors.

Earlier, many researchers proposed mathematical models for estimation based solely on technical factors [1]. After a few decades, some authors concluded the mathematical models are not well suited for delay estimation because of the non-linear and non-deterministic behaviour of the problem [2, 3]. They analysed that the non-technical factors are significantly involved in causing delays in traffic. Here, a fuzzy logic-based model is proposed that considers non-technical factors in traffic delay estimation [4, 5]. Furthermore, all the existing models for delay estimation are either based on technical factors or non-technical factors. However, both factors have the identical impact on traffic delay; neglecting any factors results in poor estimation of traffic delay. Hence, the aim of this work is to propose a hybrid approach for traffic delay estimation considering both technical and non-technical factors.

The rest of the chapter is organized as follows. A brief summary of related work is presented in Sect. 2. Section 3 describes the proposed method in detail. A discussion of the results is presented in Sect. 4, and Sect. 5 concludes the chapter.

2 Related Works

Many researchers have paid attention to estimating delay due to traffic congestion, as it is useful in reducing traffic delays by informing travellers [1]. However, estimating traffic delays intersections and links of a road network is a challenging task.

The United States Transportation Research Board [6] has created a Highway Capacity Manual (HCM) model in the form of mathematical equations for estimation of traffic delay. However, the HCM utilises an inappropriate mathematical modelling approach with respect to the stated problem, that is, delay estimation. Murat [7] highlighted the limitations of the HCM and concluded that delay estimation is not possible in case of weather variability. Another mathematical model has been provided by Henry et al. [8] as a solution of delay estimation. Their methods considered only technical factors like traffic volume and ignored non-technical factors because these could not be modelled properly in mathematical engineering.

Qiao et al. [1] considered non-technical factors also and developed a fuzzy logic-based delay estimation method. They compared this model with other delay estimation models and found their system performed better. Another fuzzy logic-based approach has been proposed by Younas et al. [3] to estimate traffic delays. The authors promoted the usage of fuzzy logic as it is well suited for solving the problems which are difficult to model mathematically and bear fuzziness. This approach cannot estimate road link delay and is limited to delay estimation at intersections only. Tamimi and Zahoor [9] estimated link delay using a fuzzy logic model, but they included too many parameters, which made the system complex. Also, this approach only estimates link delay; delays at intersections cannot be estimated.

Further, Murat [7] integrated a fuzzy model with a neural network for traffic delay estimation at junctions. In this approach, mainly technical factors, such as traffic volume, average queue and signal cycle ratio, were included as input parameters; non-technical parameters, such as weather and road conditions, were ignored.

3 Proposed Works

While analysing traffic delay, several factors need to be considered, such as weather and road conditions, road length, visibility, traffic volume and green light time. The proposed method considers only three factors—weather condition, road condition and traffic volume—for traffic delay estimation. These factors are used as input parameters for the proposed system, and each is discussed below.

Weather Condition
The condition of the weather is an essential feature of delay estimation, having a great impact on traffic delays. For example, heavy rain floods roads or reduces traction, which reduces vehicle speed and traffic flow, resulting in traffic delay. Furthermore, travellers experience inconvenience even after the rain has stopped, until road surfaces have been cleared for efficient movement of traffic. Variability in weather affects various other factors directly or indirectly. While visibility is directly influenced by weather conditions such as heavy rain, traffic volume is affected indirectly.

Road Condition
The road condition is another inducing factor which should be considered when estimating traffic delays. Bad road conditions disturb the flow of traffic, resulting in delays [3]. Various factors, such as number of lanes and road link surface and quality, determine good or bad road conditions. Bad road conditions decrease vehicle speed, causing congestion, and ultimately delay.

Weather and road conditions are both non-technical factors [1], as they are simply terms to describe the environment. Due to the inherent fuzziness of these factors, it is inappropriate to represent them in mathematical engineering. Hence, they are represented by fuzzy sets. Three symmetric triangular and trapezoidal fuzzy sets have been defined from 0 (poor condition) to 10 (good condition) to describe the weather condition (Fig. 1). Four fuzzy sets from 0 (poor condition) to 10 (very good condition) are used to describe the road condition (Fig. 2).

In order to determine factors affecting traffic delay, the weather and road conditions are considered input parameters. The output of these parameters (Fig. 3) lies

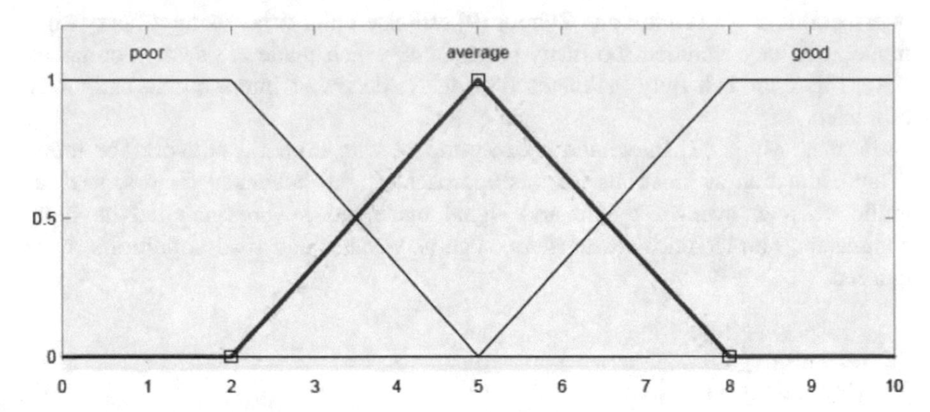

Fig. 1 Fuzzy inputs for weather condition

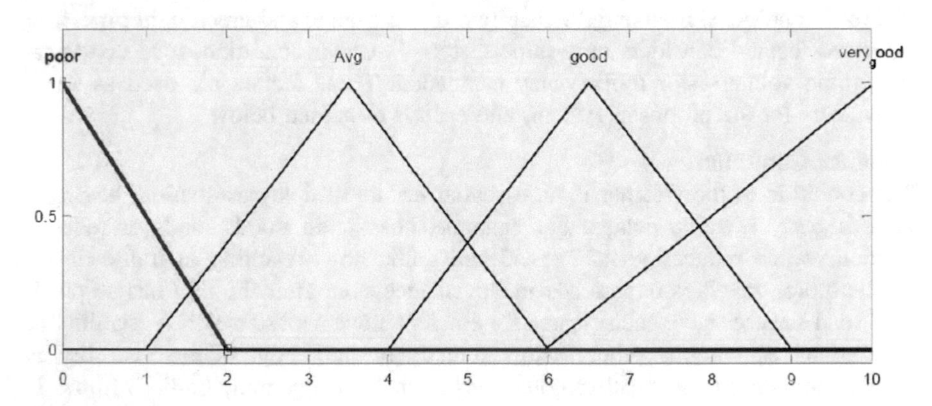

Fig. 2 Fuzzy inputs for road condition

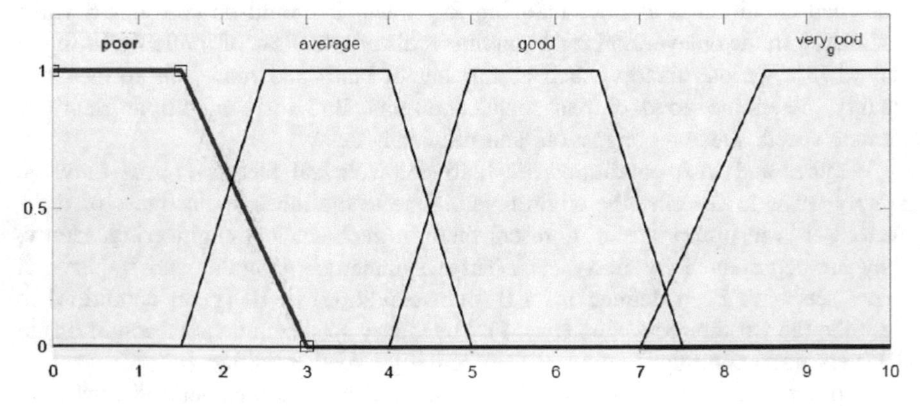

Fig. 3 Fuzzy outputs for environmental condition

between 0 and 10, and corresponds to poor, average, good and very good, representing the condition of the environment.

Traffic Volume

Traffic volume is defined as the actual number of vehicles passing a particular section of road during a particular time. For example, the traffic volume in 10 min can be obtained by counting the number of vehicles that cross a point on the roadway over a 10-min period. Although it is a tedious and complex task, several methods are available for this purpose. Furthermore, using traffic volume simplifies computing the rate of traffic flow. It is a technical factor, and hence easy to represent in mathematical engineering. It depends on several other factors including green light time and road length, and has a significant impact on traffic delay. In the proposed model, it is represented by triangular fuzzy sets from 0 (low) to 800 (higher) as shown in Fig. 4.

The proposed method estimates the delay in traffic within a road network. Figure 5 presents a schematic of the proposed hybrid approach. At the initial level, it takes two non-technical factors, weather condition and road condition, as inputs, and gives the environment condition. Environment condition works as an input for the next level and is combined with the technical factor traffic volume obtained from sensors. Further, the input values are fuzzified, which converts the crisp value into fuzzy sets. Fuzzy rules are then applied to the fuzzified inputs. The inference engine executes all rules in parallel. In the last step, defuzzification occurs. In this process, the rules obtained from the inference engine are aggregated according to the centroid method. Rule aggregation spawns a crisp output—in this case, estimated delay.

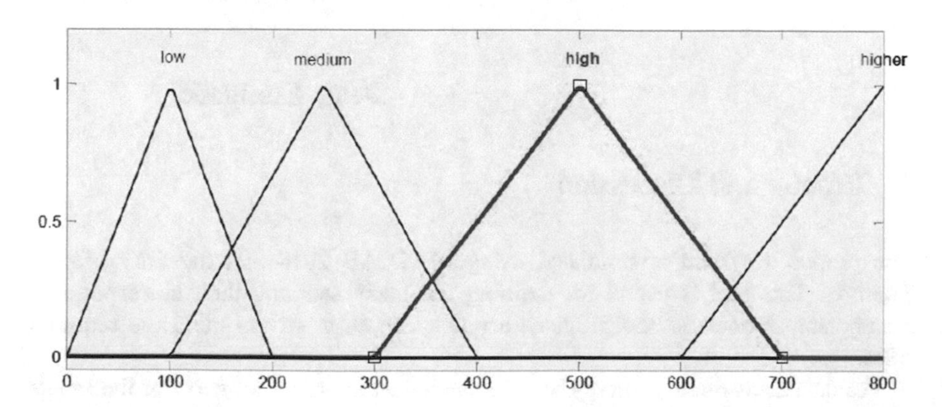

Fig. 4 Fuzzy inputs for traffic volume

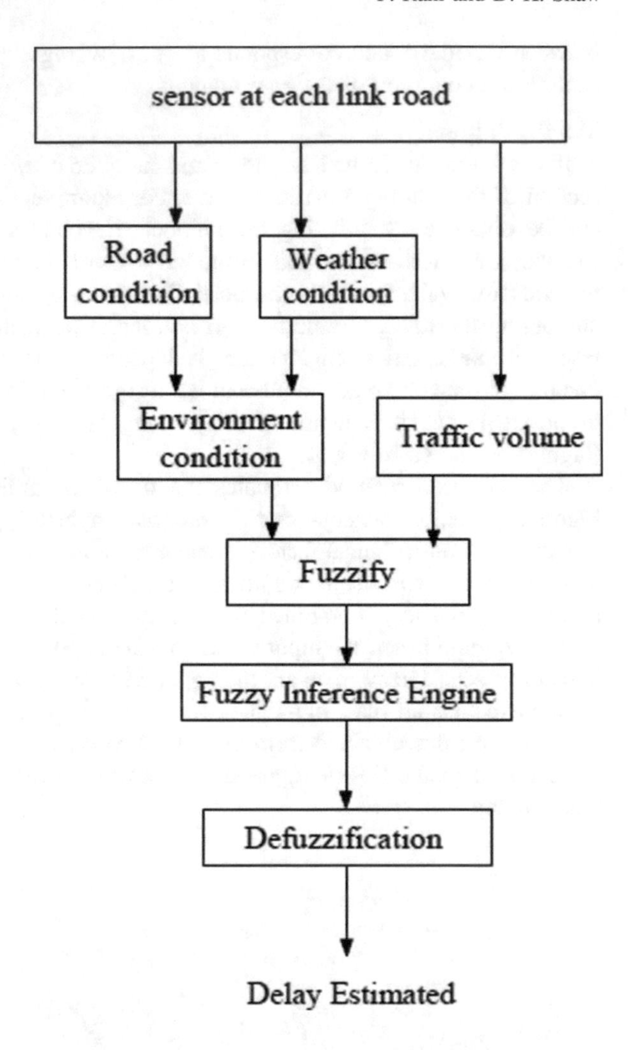

Fig. 5 Hybrid model for delay estimation

4 Results and Discussion

The proposed system is simulated using MATLAB 2014 and the Fuzzy Logic Toolbox. This tool is useful for defining the fuzzy sets and their corresponding membership functions, and provides a quick and easy way to build the required rules.

For data acquisition, a survey was conducted which provides some of the inputs to the system. Because of the considerable time required to collect the survey information, historical data and a traffic monitoring system were also used to gather information. After compiling all the information, an appropriate and complete data

Table 1 Results of fuzzy inference system

Serial no.	Traffic volume	Environmental conditions	Estimated delay
1	55	8.05	5
2	106	2.77	8.12
3	187	1.86	6.67
4	268	2.59	5.45
5	393	3.95	4
6	466	4.77	4
7	539	7.14	3.61
8	620	7.41	2.47
9	679	7.95	2
10	774	9.77	2

set among the pool of data must be selected. Out of selected dataset, 10 random values are used to generate the results. Table 1 presents the results of the fuzzy inference system. Analysis of the results reveals that traffic delay is highly affected by traffic volume; the traffic delay decreases with increasing traffic volume and vice versa. Higher traffic volume implies that large numbers of vehicles are smoothly moving on a particular section of road in a particular time; hence, the possibility of traffic delay is low. Moreover, if the traffic volume is low, the road is highly congested and traffic delay is high. But, this is not always true, as it also depends on other factors. The first value in the results shows that traffic delay is moderated while traffic volume is low and environment condition is very good. In this case, it is affected by environment condition, hence couldn't be ignored.

The proposed system incorporates fewer input parameters in comparison to existing methods [9], resulting in fewer rules generated, making the system simple and efficient. In addition, the proposed system considers both the intersection and road link while estimating the delay; this forms a system more remarkable than the existing methods.

5 Conclusions

This chapter has presented a hybrid approach for estimating traffic delay. The proposed method considers two non-technical factors (weather and road conditions) and one technical factor (traffic volume). The experimental results illustrated in the chapter exhibit the effectiveness of the proposed method. Analysis of the results indicates that both factors have a significant impact on delay estimation; hence neither technical nor non-technical factors can be neglected. This hybrid approach is well suited for intersections and links, as it includes both technical and non-technical factors, and can easily adapt to changes in traffic control and the roadway environment.

References

1. Qiao, F., P. Yi, H. Yang, and S. Devarakonda. 2002. Fuzzy logic based intersection delay estimation. *Mathematical and Computer Modeling* 1425–1434.
2. Palacharla, P.V., and P.C. Nelson. 1999. Application of fuzzy logic and neural networks for dynamic travel time estimation. In *International conference on operational research*, 145–160.
3. Younas, I., A.I. Rauf, and W.H. Syed. 2006. Towards traffic delay estimation using fuzzy logic. In *International business information management conference*, Bescia, Italy.
4. Zahoor, M., Irfan Younas, and Abad Ali Shah. 2012. A hybrid system for vehicle routing and traffic delay estimation. *International Journal of Computer Aided Engineering and Technology* 4.1: 32–48.
5. Martín, S., Manuel G. Romana, and Matilde Santos. 2016. Fuzzy model of vehicle delay to determine the level of service of two-lane roads. *Expert Systems with Applications* 54: 48–60.
6. TRB. 200. Highway capacity manual. *Transportation Research Board*, Special Report 209, National Research Council, Washington, DC, USA.
7. Murat, Y.S. 2000. Comparison of fuzzy logic and artificial neural networks approaches in vehicle delay modeling. *Transportation Research Part C: Emerging Technologies* 316–334.
8. Henry, X.L., J.S. Oh, S. Oh, L. Chau, and W. Recker. 2001. On-line traffic signal control scheme with real-time delay estimation technology. In *California partners for advanced transit and highways (PATH)*. Working Papers: Paper UCB-ITS-PWP-2001.
9. Tamimi, S., and Muhammad Zahoor. 2010. Link delay estimation using fuzzy logic. In *2010 The 2nd international conference on computer and automation engineering (ICCAE)*, vol. 2. IEEE.

Digital Privacy and Data Security in Cloud-Based Services: A Call for Action

Varun M. Deshpande and Mydhili K. Nair

Abstract The emergence of cloud-based services over the past two decades has created a paradigm shift in computing. Fueled by the universal reach of internet services, all the knowledge areas have openly embraced digitization of data storage, computation and communication. Social connectivity has made a village out of our digital world and we are more connected than ever before. With this multitude of facilities, comes the challenge of dealing with security and privacy concerns of our socially and digitally connected world. This work discusses the real-world scenario and highlights these concerns from the perspective of each stakeholder for our holistic understanding. The need for trustable solutions for secure cloud-based solutions is highlighted. Possible solutions that have been proposed include user identity protection by data masking and secure data sharing, incorporation of tools and practices for building a secure development life cycle, and developing holistic, universal privacy laws that are technically correct and auditable in real time. We call upon fellow researchers to take up research in these open research areas to defend the fortress of digital privacy and data security of cloud-based web applications.

Keywords Cloud · Digital privacy · Data security · Social networking

1 Introduction

This work is aimed at providing a real-world and practical perspective on current data security and privacy concerns in cloud-based services, specifically in the domain of social networking and e-commerce.

V. M. Deshpande (✉)
Department of C.S.E., Jain University, Bangalore, India
e-mail: varundesh@gmail.com

M. K. Nair
Department of I.S.E., M S Ramaiah Institute of Technology, Bangalore, India
e-mail: mydhili.nair@msrit.edu

A. N. Krishna et al. (eds.), *Integrated Intelligent Computing, Communication and Security*, Studies in Computational Intelligence 771,
https://doi.org/10.1007/978-981-10-8797-4_27

1.1 Era of Computing Power

Computing power has increased in leaps and bounds over the last few decades, and with the passing of each decade, is exponentially increasing the level of advancement in the electronics industry. Computing power which was available in some of the iconic space missions in the past is now available in an average smartphone at affordable cost. The golden rule of the electronics industry, propagated by Intel–Moore's Law [1]—predicted that the density of transistors on a chip would double each year, while also increasing in efficiency. This observation has stood the test of time, and we live in era of sub-14-nm transistors. All this has dramatically increased the available computing power and aided the rapid expansion of our digital capabilities. We can compute faster and handle more complex research problems. It has helped the world stay connected, effectively making the world flat and reducing it to a global village, breaking the physical barriers between countries.

1.2 Emergence of the Social Connectivity Sector

We humans are seven billion in number. Still, it is possible to reach almost any individual today with short notice. This has been possible due to advancement in technology in internet and cloud computing, telecommunications and wireless communication. Application of these technologies in the field of social connectivity in past two decades has truly changed the way we communicate and do business. Each digitally connected person is empowered with an ability to connect with any other person through the social networking platforms like Facebook, Google and many others. The era of social connectivity is maturing day by day and people have wholeheartedly embraced this with open arms and made it part of their daily life. Gone are the days when data communication to distant places was costly, difficult and challenging. This can be done with a tap of a button on one's smartphone and we can connect to other persons using video, audio or text. Such is the power of social connectivity. The possibilities of bringing people together are endless, limited only by our imagination.

1.3 Data Monetization Avenues

Knowledge and technology creation eventually leads to wealth generation. Even in the case of social connectivity, the same philosophy is applicable. Technology innovation is not sustainable unless the service providers find a way to monetize the services they are hosting and providing to all. Social networking platforms such as Facebook and Google provide a majority of their services free of cost. They serve upwards of one billion users every day. This generates a huge amount of data, especially user-generated data. It is this data that helps them sustain the business. The business model that social networking service providers incorporate intersects with the requirements of another segment of online services, *e-commerce* websites.

For survival of e-commerce websites, advertisements and conversion of potential users to paying customers is essential. For this reason, e-commerce websites advertise themselves in the social networking platforms which actively engage over a billion prospective customers every day. Service providers generate revenue by hosting these advertisements through an advertisement framework which either they maintain themselves or which is used as a metered service. Shen et al. [2], in their experiment-based research work published in a business journal, show us how a social networking platform which houses an interactive environment for consumers is an ideal environment for online advertising.

1.4 Role of Users in the Ecosystem

Users are the clear majority of the world's digitally connected population who use online services for staying connected with their peers, friends and family. They are the reason behind the success of any social networking site (SNS). Users form the backbone and the foundation of social networking ecosystem. Their acceptance or rejection of a product is reflected by the success of the service provider. However, typical of a capitalistic society, the consumers of the services are exploited in a way for selling advertisements and using user-generated data for monetization purposes. Thus, users' privacy has come under tremendous risks. Especially in the past decade or so, hackers and other malicious agents have cracked the security walls of service provider fortresses which host personally identifiable and sensitive user information. This causes users to question their trust in service providers. It seems like a side effect of a booming and connected world. However, the cost of data loss which includes personal and financial information can never be overstated. Hence, the role of users in the ecosystem needs to be re-evaluated and there is a need for building user-centric systems which uphold the respect of a user's data and their privacy.

1.5 Ensuring Privacy in a Digital World: A Lost Cause?

Privacy and social connectedness are very close to being antonyms to each other. When a user shares some personal information about oneself on a social networking platform, it becomes public information. Kissell [3], in his book, writes about the conflicts that exist in ensuring privacy of publicly shared information. He provides best practices and guidelines on what and how to share information on public platforms to ensure a reasonable amount of privacy. Tang et al. [4], in a comprehensive study of privacy and security preservation for private data in cloud-based services. In an elaborate study conducted through a survey, the authors shared details of security and privacy issues and how they have been addressed. They mark that most organizations have had to perform trade-offs between security and functionality. They have proposed a host of open areas for further research to be performed.

We believe that there is still a place for ensuring data privacy in cloud-based services. As we follow the footsteps of the previous researchers, we are close to finding trustable solutions to practical issues with respect to data security and privacy.

2 Review of the Current Real-World Scenario

As we narrow down the scope of study to social networking and e-commerce, we look at practical problems in the area from different perspectives.

2.1 Security: Why Is It Important?

Security is one of the basic requirements of any cloud-based web application. When the web application is online, it is open to attack from malicious entities and hackers. Unless service providers safeguard their service, they will surely come under attack sooner or later, exposing them to attacks such as denial of service (DoS) and data theft. Hence, ensuring web and data security of the product is a critical requirement. It is better to be paranoid about data security than to be sorry later. In addition to product security, users' digital identities need to be protected, as they are the primary users of the system. User and service provide data and privacy become the most important aspect of running a business online. Let us consider some of the specific security concerns that are associated in our domain of study.

2.2 Web Application Security Threats

It is an unfortunate fact that hackers have access to some of the sophisticated computing resources and tools in that they develop new techniques to attack cloud-based services for financial benefit. As security researchers, we are in a constant race with these malicious entities to stay ahead of the curve and ensure web applications are well protected. One of the most common ways of attacking a web application is through man-in-the-middle attacks. In this case, a hacker is able to intercept communication between a client and a server through well-crafted proxy configuration. They modify the messages being sent to and from the client and server, introducing their malicious scripts into both client and server. Ten of the most prominent web security threats have been listed by the *Open Web Application Security Project* (OWASP) [5], and they are as follows: *Injection, Broken Authentication and Session Management, Cross-Site Scripting (XSS), Insecure Direct Object References, Security Misconfiguration, Sensitive Data Exposure, Missing Function Level Access Control, Cross-Site Request Forgery (CSRF), Using Known Vulnerable Components, Un validated Redirects and Forwards.* Recent ransomware such as WannaCry and Petya is also a big cause for concern, not just for the end users, but also for service providers.

2.3 User Data Privacy Issues and Rampant Privacy Breaches

An article in WIRED magazine [6] reported that a recent data breach at Equifax, resulting in the theft of data for 143 million users (including personally identifiable information along with corresponding financial data), was preventable. A security patch which was available in March was not applied to the web application, making it vulnerable to attack. This relatively easy hack lead to data theft of mammoth proportions. This has most certainly opened a can of worms with respect to privacy breaches. This, along with several other security breaches in the recent past, such as Yahoo and MySpace in 2016, and Dropbox and LinkedIn in 2012, represents a very serious cause for concern.

2.4 Lack of Consistency in Privacy Laws

Service providers are usually globally functioning organizations. They house cloud data centers in different locations spanning countries and continents. Some of the user-generated data originating from a specific country may be stored in a data center located in a different country. The privacy laws that are applicable in one country are different from another country's laws. Nonetheless, a service provider is expected to abide by the privacy laws of each country in which it functions. This creates ambiguity in the laws and service providers have ended up in litigations because of differential privacy laws in different countries. It is also not clear regarding which laws are all-pervasive and a standard to be followed. Deshpande et al. [7] discussed in detail privacy policies in various regions such as India, USA and Europe and came to a similar conclusion. Lack of universally accepted privacy laws has created problems with respect to data jurisdictions, privacy concerns, etc.

3 Perspectives from Each Stakeholder: Where Is the Concern?

3.1 Service Provider: Here to Do Business

Service providers such as Facebook and Google provide services free for their users. In return, they benefit from the advertisers who sell their products to the users on the SNS. This is the high-level business model where service providers provide services to the users in exchange for financial payment from e-commerce sites. By sharing user data with other advertisement networks and other stakeholders, the SNS generates revenue.

In the social networking domain, there exist two major types of data sharing where user privacy is lost. One is the information that other users of an SNS can view about the person who shared the info. This can be managed by the user by updating the privacy of the posts and controlling who can see their posts on the SNS. Facebook provides such options and it educates the users from time to time about it as well. The second is the user data shared between service provider and advertisement networks and other third parties for data monetization purposes. In the first scenario, it doesn't lead to any revenue generation for the service provider. However, the second form of data sharing is the primary revenue-yielding method for SNS service providers. User privacy is lost while sharing details with advertisement providers; this is the major concern here.

3.2 Advertisement Providers: Here to Sell Ads

Advertisers invest money to reach out to their potential customers among the millions of online users. Each advertisement will not make sense for every customer. For example, an ad to sell a villa is not suitable for a student in his teens. However, it is applicable for a professional who is of age 25 or more. In this way, the advertisement networks need to find enough data points from the user's contextual data to deliver the right advertisement to the user. This ensures that there is a good chance for the user to see the advertisement and business context to be established. This is the reason they insist on getting a maximum amount of data related to users and their contextual information. While user data is being procured, there is chance of leakage of personal information. This is a cause for concern, as this personal information could be misused.

3.3 Other Third Parties: Transforming Data Through Aggregation

Service providers also share user's personal information with other third parties such as data analytics, partners, subsidiaries etc. There is a realistic chance that such data could be combined and then aggregated using the advances in Big Data. This aggregate data could be used to create a digital identity of a person and launch social engineering attacks etc.

3.4 End Users: Stay Connected with the World

End-users are people who want to use the platform that social networking provides and use it for connecting with their friends, colleagues and family. With the current business model, they can view advertisements which are targeted towards them. Their data is used to find relevant advertisement for them. Some users may not want to see the advertisements; however, they are not able to stop them.

3.5 Law Enforcement Agencies: Creating a Fair Play Platform for All

Law enforcement agencies of different countries are responsible for creating and enforcing privacy and data security laws. Recently, in a landmark judgment, the Supreme Court of India declared that, except in a few exceptional circumstances, the "Right to Privacy" is a fundamental right. This gives impetus for law enforcement agencies to expedite their duty diligently. In the case of any data breach or complaint against any organization, they investigate and prosecute the guilty as applicable by law. Currently, the laws are in a very nascent stage and not consistent among countries. Also, the real-time verifiability and accountability is not available right now. We need to consider solving these concerns over the longer run.

3.6 Solutions Need to Be Trustable and Verifiable

Solutions that arise from this churn of thoughts need to be transparent. The policies need to be technically correct and verifiable in real time. Privacy and integrity of the communication could be achieved by use of globally trusted entities like use of digital certificates, etc. This is to ensure that there is no avenue for misuse of user data.

4 Constraints and Possible Solutions

In this section, we highlight the constraints, possible solutions and research directions for finding trustable solutions to the above-mentioned concerns.

4.1 The Business Model Should not Be Disturbed

In the history of mankind, we have been able reach the highest level of connectedness in the past decade, through social connectivity. Facebook and Google have been some of the major pioneers in this venture. They are able connect people by providing services free of cost because of the business model which allows them to do it. E-commerce websites have blended into this business model as they virtually fund the service providers to keep the service going. If we take advertisers out of the equation, service providers will not be able to monetize the service, eventually leading to context collapse and demise of the business model. Hence, it is essential to keep the business model intact while we find solutions to privacy concerns.

4.2 Privacy Policies and Laws Need to Be Revisited

As discussed earlier, we need to look at privacy laws from a holistic perspective and see if it can be made universal, addressing the ambiguity of cross-boundary data privacy. We also need to make these policies technically sound and their implementation verifiable in real time through automated auditing mechanisms so any breach of security can be reported and addressed instantly.

4.3 Application Security Needs to Be Tightened

Service providers need to follow security best practices during development. A secure development life cycle needs to be formulated based on company needs and diligently followed. Web security testing tools such as Burp can be used during testing.

4.4 Secure Data-Sharing Practices

The social networking business model requires user data to be shared by the SNS with the service providers. Currently, service providers are defining their own data usage policies and sharing the data with advertisement providers and other third parties. We need to ensure that there is a trust-based framework setup to ensure that data exchange happens without affecting user privacy.

4.5 User Identity Protection Using Data Anonymization

Advertisement networks require user context details to provide the best-matching ads for the user's current context. We need to find efficient solutions to ensure that user data which is shared between an SNS and advertisement providers doesn't reveal identifiable information about users. However, all the required data points which are required for fetching high-quality advertisement matches should be shared. This would ensure that user identity is protected. This can be achieved through data anonymization techniques such as rule-based data masking.

5 Conclusion and Call for Action

The current work is a concise and humble attempt to put forth digital privacy and data security concerns regarding cloud-based services. We concentrated on the social networking and e-commerce business model to draw observations. We discussed how emergence of social connectivity has made the world a global village and some of the concerns which need to be dealt with. We reviewed the current scenario in the real world related to security, emphasizing web application security, privacy breaches and lack of consistent privacy laws. The security concerns were analyzed from the perspective of each stakeholder to give a holistic view of the security and privacy concerns. The need for finding trustable solutions was highlighted. With constraints of the business model in mind, we provided possible research areas for further work to find trustable software solutions for secure cloud-based services. They include secure data-sharing polices which incorporate identity-protecting anonymization techniques. Tools and practices to ensure a high level of web application security and which work towards holistic, technically sound and universally applicable privacy laws are proposed. We believe that these research focus areas hold the key to address some of the challenges we are currently facing in cloud-based services. With a generic framework, the solutions could be reused for other knowledge areas as well, such as health care, education and defense. We call upon the research community to take up these open research areas and work towards securing our connected world.

References

1. 50 Years of Moore's Law. https://www.intel.com/content/www/us/en/silicon-innovations/moores-law-technology.html.
2. Shen, George Chung-Chi, Jyh-Shen Chiou, Chih-Hui Hsiao, Chun-Hsien Wang, and Hsin-Ni Li. 2016. Effective marketing communication via social networking site: The moderating role of the social tie. *Journal of Business Research* 69 (6): 2265–2270, ISSN: 0148-2963.
3. Kissell, Joe. 2017. *Take control of your online privacy*, 3rd Edn. TidBITS Publishing, Inc. ISBN: 9781615424856.
4. Tang, Jun, Yong Cui, Qi Li, Kui Ren, Jiangchuan Liu, and Rajkumar Buyya. 2016. Ensuring security and privacy preservation for cloud data services. *ACM Computing Surveys* 49 (1), Article 13.
5. OWASP Top Ten Web Threats. https://www.owasp.org/index.php/Top_10_2013-Top_10.
6. Equifax has no excuse. https://www.wired.com/story/equifax-breach-no-excuse/.
7. Deshpande, Varun M., and Mydhili K. Nair. 2017. Open standards for data privacy policy framework in context of trusted social networking. *International Journal of Scientific Research in Computer Science, Engineering and Information Technology (IJSRCSEIT)* 2 (5). ISSN: 2456-3307.

Mobile Agent-Based Improved Traffic Control System in VANET

Mamata Rath, Bibudhendu Pati and Binod Kumar Pattanayak

Abstract Due to the increasing number of inhabitants in metropolitan cities, people in well-developed urban areas routinely deal with traffic congestion problems when traveling from one place to another, which results in unpredictable delays and greater risk of accidents. Excessive fuel utilization is also an issue and poor air quality conditions are created at common traffic points due to vehicle exhaust. As a strategic solution for such issues, groups of urban communities are now adopting traffic control frameworks that employ automation as a solution to these issues. The essential test lies in continuous investigation of data collected online and accurately applying it to some activity stream. In this specific situation, this article proposes an enhanced traffic control and management framework that performs traffic congestion control in an automated way using a mobile agent paradigm. Under a vehicular ad hoc network (VANET) situation, the versatile proposed executive system performs systematic control with improved efficiency.

Keywords VANET · Smart city · Traffic management · Mobile agent Sensor

M. Rath (✉)
C. V. Raman College of Engineering, Bhubaneswar, Odisha, India
e-mail: mamata.rath200@gmail.com

B. Pati
Department of CS and IT, S O A Deemed to be University, Bhubaneswar, Odisha, India
e-mail: patibibudhendu@gmail.com

B. K. Pattanayak
Department of Computer Science and Engineering, S O A Deemed to be University, Bhubaneswar, Odisha, India
e-mail: binodpattanayak@soauniversity.ac.in

A. N. Krishna et al. (eds.), *Integrated Intelligent Computing, Communication and Security*, Studies in Computational Intelligence 771,
https://doi.org/10.1007/978-981-10-8797-4_28

1 Introduction

In the age of technology, vehicular ad hoc networks (VANETs) are becoming more mainstream in traffic administration and control frameworks. Smart urban communities have recognized that problems with traffic congestion can be better addressed with the use of VANETs, as it is possible to establish communication between the vehicle and the network infrastructure. Along these lines, readily available street condition data and course data can be coordinated with VANET-connected intelligent vehicles during travel, and smart choices can be made in advance of any issues developing. The use of VANETs in smart urban areas reduces the risk of blockage, accidents, wrongdoing, stopping issues and financial overhead. The many innovations in remote technology have enabled the evolution of smart vehicles and intelligent transport systems. In addition, drivers and customary driving frameworks have likewise changed over to smart drivers with more specialized information and direction from movement controllers, understanding them and responding accordingly. VANETs bolster adaptable correspondence between vehicles and movement-controlling frameworks in both infrastructure-based and remote networks without a stationary framework. The proposed activity blockage arrangement in smart cities utilizes an enhanced, specialized approach to the issue with effective information investigation made by versatile operator progressively under a VANET scenario in a smart city.

Mobile agents are adaptable programs that are able to travel autonomously from node to node among personal computer (PC) workstations, customers and servers, and between controllers and accessories. They are self-governing and re-configurable during run time, so they can be used with any application. Mobile agents are widely utilized as a part of database applications, digital signatures and on-request organization applications. The throughput of mobile agent-based applications is determined to a great extent by the level of complexity of the required task.

As proposed in Fig. 1, our mobile agent-based improved traffic control system (MITS) has been designed with mobile agent logic embedded in the microcontroller regarding the input information of the traffic scenario. Based on congestion status, an improved congestion control mechanism is formulated under the control of the mobile agent. To prevent congestion, the vehicles are re-directed towards other routes or, depending on their choice, they may continue to move in the same traffic with some waiting point slots. In addition to redirecting routes, this system also employs security and a reliable mechanism for improved traffic management. The challenge during the design of the system lies in carefully selecting the control logic which can be smoothly implemented to communicate between platforms.

In the proposed application, the mobile agent is used to implement the congestion control algorithm in an automated traffic control system in a smart city context. The article is organized in the following manner. Section 1 focuses on the introduction. Section 2 describes an extended literature survey on the related area and similar application details. Section 3 illustrates a detailed demonstration of the

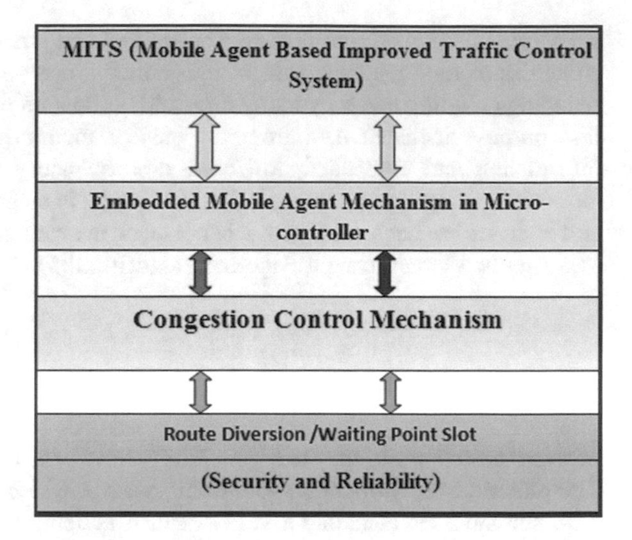

Fig. 1 Basic model of the proposed system

proposed model under a VANET, and detailed technical functionality is explained in terms of various modules used in the system. Next, Sect. 4 demonstrates the simulation and comparative analysis carried out using an ns-3 simulator, including technical configuration and results. Section 5 concludes the paper by discussing opportunities for further improvements to the model.

2 Literature Survey

This section presents some of the innovative and interesting research work in this field with the development of various applications and improved traffic control systems in the context of a smart city. Tawalbeh et al. in [1] describes a mobile cloud computing-based model and the use of big data analytics for healthcare applications. A low-cost real-time smart traffic management system is presented in [2] using the Internet of Things (IoT) and Big Data. It provides improved service by implementing traffic indicators to update the traffic-related information at every instance of events taking place in the network. In this approach, low-cost sensors capable of implementing vehicle-detecting applications are embedded in the middle of the road at intervals of 500–1000 m. IoT devices [3] are used to acquire online traffic-related information quickly and further send the data for processing at Big Data analytics centers. Intelligence-based analytical tools are used to scrutinize the traffic density and model solution strategies. Congested traffic is managed by a mobile application based on the explored traffic density and alternate solution of this. An on-road air quality monitoring and control method is proposed [4] with development of an agent-based model that includes urban road network infrastructure and assesses real time and approximate air pollution indices in different segments of the road and produces recommended strategies for road users. Such strategies include reducing vehicle usage in the most polluted road segments and reducing pollution levels by increasing vehicle flow on the roads. Data sets used for this purpose included data from air quality monitoring stations, road network information and embedded, low-cost, e-participatory pollution sensors. Mobile cloud computing supports technical development in smart cities. Comparative study and analysis has been done with a benchmark evaluation of Hama's graph package and Apache Giraph using a PageRank algorithm [5].

3 The Proposed Approach

In this research work, the intelligent MITS has been proposed as shown in Fig. 2. The planned framework deals with traffic control system MITS as the core module, with sub-modules including a video control system, traffic control system, supervisory computer control system and peripheral devices. The traffic control system manages and controls the heavy traffic during predefined rush schedule [6] on the road. It uses the video monitoring system to identify excess traffic using a video camera, and when the volume of vehicles in a particular path exceeds a predefined threshold value, it alerts the smart traffic control module of the MITS through an alarm signal indicating congestion status. A signal transmission and communication module carries out transmission of control packets.

As shown in Fig. 2, all the above-described modules provide necessary information to the mobile agent for further controlling the traffic, congestion detection and prevention. Similarly, the mobile agent also controls the video surveillance regarding security aspects of the system and incident management. In the proposed

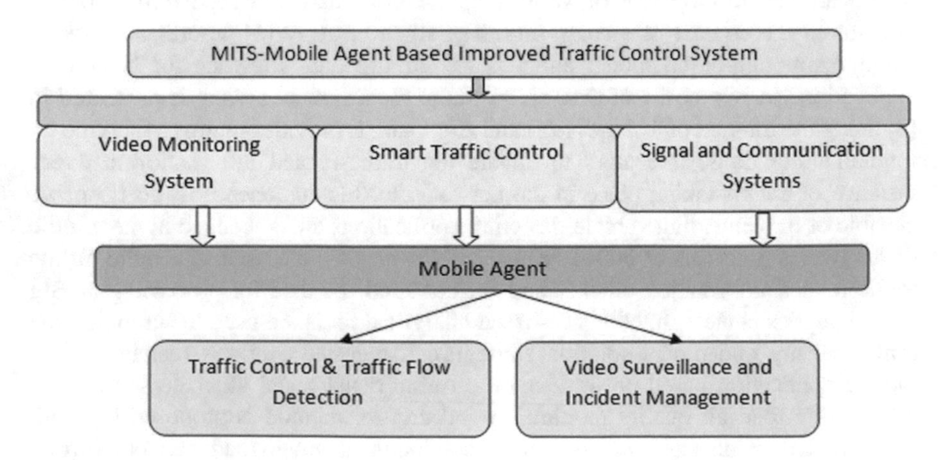

Fig. 2 Block diagram of the proposed system

system, smart data analytics are used to tactfully handle the congestion situation [7] and [8] and control the congestion with the implementation of a dynamic mobile agent.

This section explains the reason for choosing a mobile agent approach in this application. Mobile agents are utility programming systems that are part of utilization program interfaces [9]. Mobile agents are likewise occasion-driven modules with self-controlling capacity, and they can travel from one PC to another in a heterogeneous system in order to check the particular application status of their hosts [10]. Thus they represent attributes—for example, versatile, mobile, social, intelligent, responsive, objective-driven and self-ruling—and they are customized to satisfy the intended targets of their design. They transit independently, starting with one device, and then the next, in execute a series of actions. They employ security measures and caution host PCs if any security infringement is detected. During attacks, these agents have self-defense program modules by which they can counteract these attacks. Mobile agents comprise two components, the agent and the agent stage. The mobile agent contains the program code and the status data regarding the computational rationale to be completed on the agent stage. Mobile agents and their utility in systems administration frameworks during steering improves embedded layer correspondence with reduced computational system stacks on the host PCs. As the agents have the weight of operational exercises, dormancy because of handling at the hub is limited. A product agent is a program module, the goal of which is to play out a particular capacity according to the prerequisite. It can likewise be portrayed as a product protest set inside a problematic application.

Each smart vehicle considered in this article carries self-identifiable data in its smaller-scale controller chip installed in the vehicle printed circuit board, so upon entering the VANET zone of the MITS, the vehicles are controlled and observed by the mobile agent. In the busy traffic scenario, video camera connected at the traffic towers sends information to the video control system [11] and [12] and frequently sends traffic status data to the MITS controller. As this traffic demonstration was intended for a normal traffic condition, during reenactment, we have considered a traffic limit of 1000–2000 vehicles at a traffic point. During the peak time of day—for example, workplace time between 9 a.m. and 10 a.m., traffic congestion is highest.

As shown in the flowchart of the proposed approach in Fig. 3, the function of the MITS is triggered when any smart vehicle enters the VANET zone of the proposed traffic range. The MITS traffic control system uses a counter register to store the number of vehicles entering, which is sent by a roadside sensor device within a specified range. A threshold value is set as the maximum number of vehicles allowed. When the number of vehicles entering the MITS zone exceeds this threshold, the sensor device triggers an alarm, preventing further vehicle entry. This mechanism was implemented using mobile agent logic during the simulation. The sensor device automatically generates a query message to the vehicle if it opts to re-route or continue along the existing route, with some delay due to congestion.

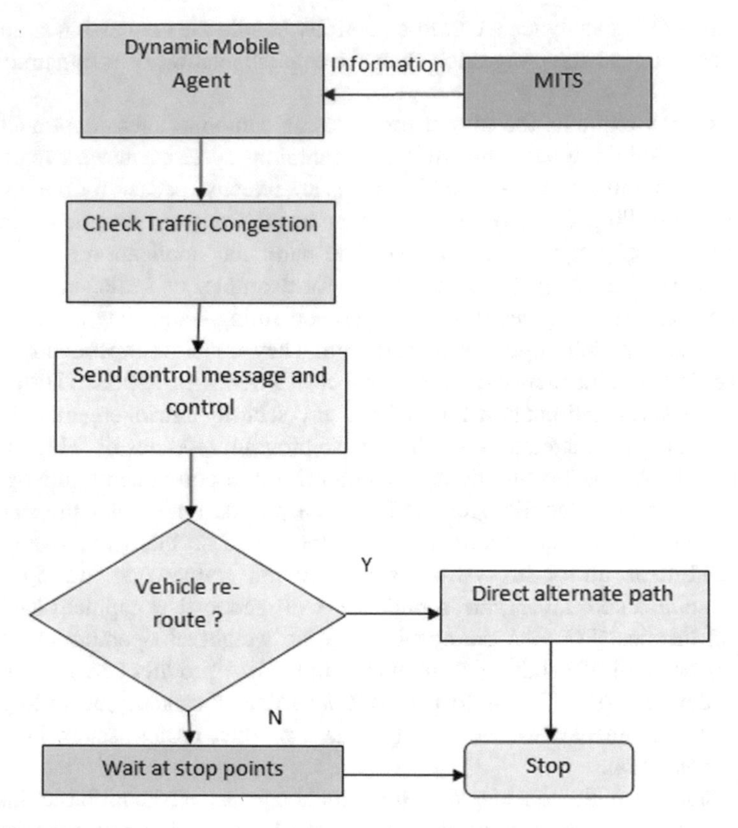

Fig. 3 Flowchart of the proposed approach

If the former option is chosen, the mobile agent automates the sensor to provide a proper signal for movement of the vehicle towards a new route. If the latter option is chosen, the intelligent logic in the mobile agent calculates the wait and stop points for the vehicle, considering the congestion status of the traffic.

4 Performance Analysis and Evaluation

The ns-3 network simulator was used for simulation of the proposed model. The nodes in a VANET can sense mobile information around them, which helps in monitoring the traffic, controlling the speed of the vehicles and many other traffic congestion parameters. In a VANET, the vehicles acting as mobile stations move within the VANET network range. Mobile stations are configured with various communication parameters. Two types of communication are possible using the VANET nodes: vehicle-to-vehicle and vehicle-to-infrastructure (V2I). In this

Table 1 Network parameters used in the simulation

Parameter	Value
Channel type	Wireless channel
Propagation	Radio propagation model
Network interface	Wireless PHY
MAC type	MAC/802_11
Interface queue type	Drop tail
Antenna model	Antenna/OmniAntenna
Simulation tool	ns-3

simulation, we used the V2I approach in which vehicles interact with and respond to the roadside sensor units through network routing devices at the road side.

Table 1 depicts the network parameters used during ns-3 simulation. There are two basic elements in MOVE [mobility model generator for vehicular networks], the map editor and the vehicle movement screen. To create the road topology, the map editor is used, and to create the vehicle movements, the vehicle movement editor is used. This section shows the comparative analysis of the proposed model with a traditional traffic system and another similar approach called a dynamic traffic management system (DTMS) [13]. To dynamically coordinate the traffic lights in a vehicle-to-infrastructure smart application, a novel system has been proposed [13]. This novel approach develops a working prototype model for flexible traffic congestion control based on traffic density, which is calculated by averaging the acquired speed rate of the vehicles. In this approach, the vehicles are required to be equipped with a central unit that contains speed data and calculates the average speed from an onboard sensor. This data is delivered to the server by the Zigbee protocol [14] using a Zigbee transmitter placed on every vehicle and receivers placed on light posts. These receivers then transmit the collected data to local servers which are placed at cluster points. Using data-mining techniques, these data are refined at cluster heads and the traffic signals are controlled using intelligent logic based on mined data. This pioneering approach to traffic control is very similar to our proposed smart traffic control approach. Therefore, we have considered this DTMS technique to compare our results for its performance.

The simulation has been carried out under three traffic scenarios with variable numbers of vehicles, and the results are obtained by extracting output values from a trace file created after the simulation. The data has been processed as required using AWK scripts. Figure 4 shows the comparative delay analysis in which our proposed system takes minimum time for processing; hence the delay is reduced compared to the DTMS and traditional system. Similarly, Fig. 5 depicts the average number of stop positions by the vehicles as determined by the mobile agent logic with varying traffic conditions. This depends on the waiting points controlled by the sensor signal point. It can be seen that in MITS, the average waiting time at stop points is less than the similar DTMS approach.

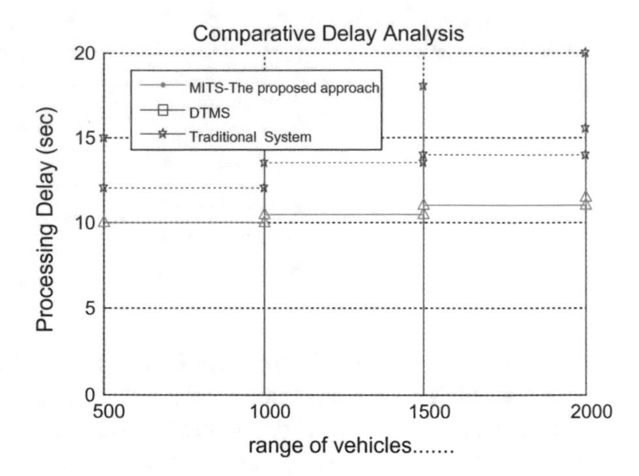

Fig. 4 Comparative delay analysis

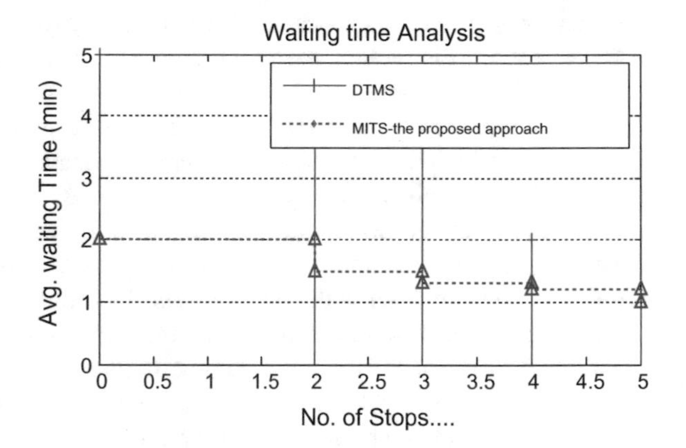

Fig. 5 Analysis of waiting time at stop points

5 Conclusion

With the development and improvement of technology, modern and instructive systems are evolving as well as there is greater chance of business and better extent of training and also explore in creating urban areas. The way of life of individuals in large metropolitan communities is similarly influenced by different application and administration frameworks. Many urban communities are undergoing transformations to smart cities by establishing mechanized frameworks in every conceivable segment. This article proposed a cutting-edge traffic control framework utilizing associated vehicle technology under a VANET arrangement with a coordinated approach for comprehending general traffic-related issues in a high-volume traffic area. Various modules, including like video monitoring, intelligent traffic control framework, signaling framework and mobile agents, are incorporated into the model. Simulation output demonstrates enhanced traffic control in traffic squares as it utilizes propelled technology of computerized vehicles, mobile agents and an improved information system.

References

1. Tawalbeh, L.A., R. Mehmood, E. Benkhlifa, and H. Song. 2016. Mobile cloud computing model and big data analysis for healthcare applications. *IEEE Access* 4: 6171–6180.
2. Rizwan, P., K. Suresh, and M.R. Babu. 2016. Real-time smart traffic management system for smart cities by using Internet of Things and big data. In *2016 international conference on emerging technological trends (ICETT)*, 1–7. Kollam.
3. Sun, Y., H. Song, A.J. Jara, and R. Bie. 2016. Internet of Things and big data analytics for smart and connected communities. *IEEE Access* 4: 766–773.
4. El Fazziki, A., D. Benslimane, A. Sadiq, J. Ouarzazi, and M. Sadgal. 2017. An agent based traffic regulation system for the roadside air quality control. *IEEE Access* 5: 13192–13201.
5. Siddique, K., Z. Akhtar, E.J. Yoon, Y.S. Jeong, D. Dasgupta, and Y. Kim. 2016. Apache Hama: An emerging bulk synchronous parallel computing framework for big data applications. *IEEE Access* 4: 8879–8887.
6. Kumar, N., A.V. Vasilakos, and J.J.P.C. Rodrigues. 2017. A multi-tenant cloud-based DC nano grid for self-sustained smart buildings in smart cities. *IEEE Communications Magazine* 55 (3): 14–21.
7. Ding, Z., B. Yang, Y. Chi, and L. Guo. 2016. Enabling smart transportation systems: A parallel spatio-temporal database approach. *IEEE Transactions on Computers* 65 (5): 1377–1391.
8. Singh, D., C. Vishnu, and C.K. Mohan. 2016. Visual big data analytics for traffic monitoring in smart city. In *2016 15th IEEE international conference on machine learning and applications (ICMLA)*, Anaheim, CA, 886–891.
9. Younes, H., O. Bouattane, M. Youssfi, and E. Illoussamen. 2017. New load balancing framework based on mobile AGENT and ant-colony optimization technique. In *2017 intelligent systems and computer vision (ISCV)*, Fez, Morocco, 1–6.
10. Cao, Jiannong, and Sajal Kumar Das. 2012. Mobile agents in mobile and wireless computing. In *Mobile agents in networking and distributed computing*, vol. 1, 450. Wiley Telecom. https://doi.org/10.1002/9781118135617.ch10.
11. Yuan, W., et al. 2015. A smart work performance measurement system for police officers. *IEEE Access* 3: 1755–1764.
12. Schleicher, J.M., M. Vögler, S. Dustdar, and C. Inzinger. 2016. Application architecture for the internet of cities: Blueprints for future smart city applications. *IEEE Internet Computing* 20 (6): 68–75.
13. Ramachandra, S.H., K.N. Reddy, V.R. Vellore, S. Karanth, and T. Kamath. 2016. A novel dynamic traffic management system using on board diagnostics and Zigbee protocol. In *2016 international conference on communication and electronics systems (ICCES)*, Coimbatore, 1–6.
14. Elahi, Ata, Adam Gschwender. 2009. Introduction to the ZigBee wireless sensor and control network. In *Zigbee wireless sensor and control network*. Pearson Publishers.

Analyzing the Applicability of Smartphone Sensors for Roadway Obstacle Identification in an Infrastructure-Free Environment Using a Soft Learning Approach

Chandra Kishor Pandey, Neeraj Kumar and Vinay Kumar Mishra

Abstract Modern-day smartphones are inbuilt with numerous sensors capable of identifying critical information from various fields. Smartphone sensors are cheap, handy, easily available and therefore useful for several purposes. Smartphone sensors can be used to identify roadway obstacles and assist vehicle drivers in handling various obstacles while driving. In this study, we analyzed the applicability of smartphone sensors to assess their usefulness in reference to roadway obstacle detection in an infrastructure-free environment using a soft learning approach. Using our approach, we found that an accelerometer, CMOS and localization sensors are the most useful and cost-effective sensors which can be used for obstacle detection and tracking in infrastructure-free environments.

Keywords Smartphone · Sensors · Obstacle detection · Accuracy
Efficiency · Driver support system

1 Introduction

In the last decade, as the number of vehicles on the road has increased rapidly, road accidents have increased proportionally, causing numerous fatalities and vehicle damage. Road accidents result from undefined road infrastructure, unmaintained roads, various obstacles situations and drivers failing to follow traffic rules while driving. Nowadays, high-range vehicles are equipped with driver support systems which help make driving safe, easy and comfortable. But these systems are costly

C. K. Pandey · N. Kumar (✉)
Department of Computer Application, Shri Ramswaroop Memorial University, Deva Road, Barabanki, Uttar Pradesh, India
e-mail: neeraj.cs@srmu.ac.in

V. K. Mishra
Department of Computer Application, Shri Ramswaroop Memorial Group of Professional Colleges, Lucknow, Uttar Pradesh, India

A. N. Krishna et al. (eds.), *Integrated Intelligent Computing, Communication and Security*, Studies in Computational Intelligence 771,
https://doi.org/10.1007/978-981-10-8797-4_29

and require special training and utilization which is not possible in low-range vehicles. Driver support systems are sensor-based. An encoder tests actual volume and converts it to radio beams which can be understood by a receiver or through a device known as a sensor [1]. Modern-day smartphones are equipped with various sensors, according to price and range, which can be utilized for developing applications to collect road traffic information and assist drivers in frequent decision making in various roadway situations to ensure a safe journey [2]. In this study, we analyzed the usefulness of different types of sensors available in modern-day smartphones. These include:

- **Accelerometer**: It is the most commonly available sensor in current-day smartphones; it determines the successive difference in velocity in 3-axis dimensions. When someone moves the smartphone in any direction, movement data from this sensor will quit change, but if left motionless, it will go on a level surface.
- **Gyroscope**: This senses angular velocity and is more advanced than an accelerometer. It is also known as a gyro sensor or angular rate sensor or angular velocity sensor. This sensor tells the rate of rotation of a device through the three sensor axes, and senses critical information from its surrounding not easily sensed by humans.
- **Magnetometer**: These are computational instruments generally used for magnetization recording of a magnetic substance or to estimate the capacity of attraction and the vector quantity of flux at any arbitrary point in space.
- **Proximity sensor**: A proximity sensor is a sensing device which collects information about nearby objects without any direct contact up to a given range. This sensor converts information on object movement into electrical signals. If any object comes within the covered area of proximity sensor, it emits an infrared beam and waits for its reflections. When it gets surety about reflections, it then pronounces an object is nearby. It makes hand-held devices more useful, responsive and functional.
- **Radiation sensor**: It is one of the earliest sensors used to identify different types of rays. It not only has the applications in the area of smart sensing, it has applications in various medical instruments and other electronic gadgets.
- **Location (GPS) sensor**: A global positioning system (GPS) sensor is an orbiter-driven steering tracking system, generally showing where an object is or has been. It is used to determine one's location. It also provides the physical position of a stationary object on earth.
- **Ambient light sensor**: An ambient light sensor automatically sets the liquid crystal display (LCD) intensity to reduce battery power consumption and extend battery life. It is typically used for backlighting control in any number of LCD display applications, providing a key benefit in mobile device applications.
- **Thermometer**: Modern smartphones have thermometer sensors to monitor the temperature inside the device and its battery. If a part inside a smartphone seems to be overheating, the smartphone automatically shuts down to prevent damage.

- **Barometer**: This is a passive sensor which measures atmospheric pressure. It easily finds the actual height of device in relation to sea level. It also enhances the GPS accuracy by giving accurate data.
- **Humidity sensor**: This is also a passive sensor which measure humidity in atmosphere. Measuring humidity is a tedious, cost-intensive task, and a critical process in determining health and safety issues. Therefore, humidity sensors play is an important role in control of industrial systems and human health.
- **Pedometer**: It is used for counting the number of steps a user has taken. A dedicated pedometer saves time and is power-efficient.
- **Fingerprint sensor**: This is a modern sensor generally available in recent smartphones. Using this sensor smartphones may be operated in safe and secure manner. A smartphone having this sensor don't require swiping of the phone screen and password pattern lock or pin lock.
- **CMOS sensor**: Complementary metal-oxide semiconductor (CMOS) sensor is used to capture the photographic data from the surroundings. It collects digital data more accurately and effectively.
- **Microphone**: A microphone sensor detects sound signals and converts the detected sound signal into an electrical signal. Generally, all of the microphone sensors employ various techniques such as light modulation, piezoelectric generation, capacitance change, and electromagnetic induction to produce an electrical voltage signal from mechanical vibration.

2 Related Work

Shah et al. evaluated various tools and techniques for optimization of mobile batteries. For this, they reviewed the latest published research articles on mobile battery optimization [1]. Engelbrecht et al. conducted an in-depth survey and compared various approaches used for identifying vehicles using intelligent transportation system (ITS) applications through mobile sensors. The purpose of this study was to analyze various techniques for assessing driving behavior and the possibility of using mobile sensors for assessing driving behavior using an ITS [2]. Bala et al. proposed an efficient learning engine approach battery for power saving in Android mobile phones [3]. Bhoraskar et al. presented a non-intrusive method that uses smartphone sensors for sensing various obstacles [4]. Sendra et al. presented a review of various power-saving and energy-optimization methods for wireless sensor networks. In their methods, they gave an efficient approach to saving power in battery used by mobile sensors [5]. Mednis et al. proposed an approach for pothole detection using smartphones equipped with off-the-shelf microphones and localization devices [6]. Mohan et al. focused basically on the sensing side, employing an algorithm that uses a smartphone's accelerometer, microphone, GSM [global system for mobile communication], and/or GPS sensors to detect potholes, bumps, braking, and honking [7]. Vo surveyed various approaches for fingerprint-based localization to sense and match different information from the roadside for location finding [8]. Dinh et al. proposed a learning framework for robust obstacle detection, recognition, and tracking preceding vehicles and pedestrians based on a deep learning approach [9].

3 Soft Learning Approach

To obtain the correct result when analyzing any raw data, we must first determine the parameters for analysis. Our proposed approach analyzes various sensors on the basis of two attributes, positional accuracy and energy efficiency (Fig. 1).

3.1 *Positional Accuracy*

For analyzing sensor attributes, accuracy is the one of the most important keys. Positional accuracy of sensors in smartphones is one of the main characteristics which affect directly identification of raw data from the environment. If a sensor inside a smartphone is positioned appropriately, then collection of raw data will be easily carried out, with less effort. It is possible that some sensors may be accurately positioned such that they are most useful for raw data collection and processing, whereas other sensors may not be positioned accurately and hence are less useful. The positional accuracy of sensors can be quantified using the measured positions and true positions based on a method described in Burrough's *Principles of Geographic Information Systems*. We used this method to test the positional accuracy of smartphone sensors to determine the applicability of various sensors. In this

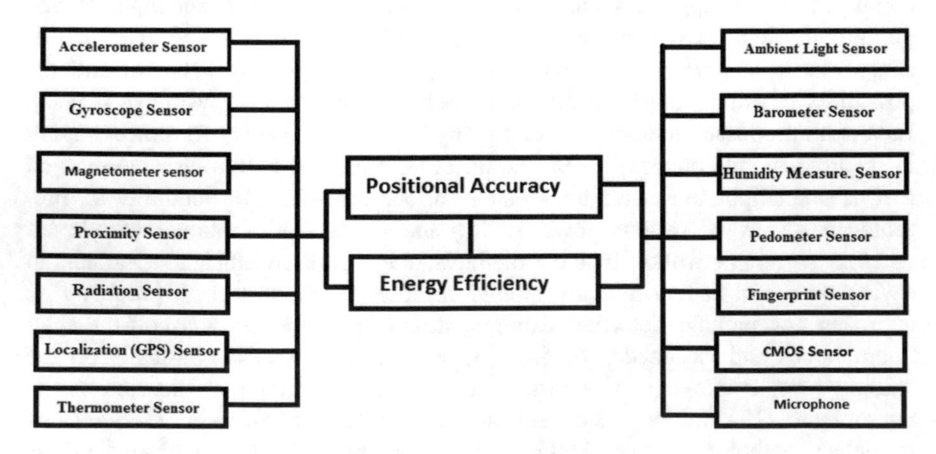

Fig. 1 Soft learning approach

Table 1 Standard deviation of distance table

X	XAverage	X − XAverage	(X − XAverage)^2
1.40	1.50	–0.10	0.01
2.10	1.50	0.60	0.36
1.10	1.50	–0.40	0.16
1.40	1.50	–0.10	0.01

Table 2 Root mean square error table

Error parameter	Value	(Error parameter)^2	Value
e1	1.40	e1^2	1.96
e2	2.10	e2^2	4.41
e3	1.10	e3^2	1.21
e4	1.40	e4^2	1.96

method, the two parameters are assumed to justify the positional accuracy of each type of sensors, and the average distance between these two parameters suggests choosing a good sensor. The two attributes for finding positional accuracy of sensor are measured position (MesP) and true position (TrueP). The difference between these two positions is termed as distance error. Different errors requiring calculation before computing the accuracy of sensors are the distance errors of every tested point, average distance error, standard deviation (SD) of distance error, and root mean square error (RMSE). This method can be used to validate the positional accuracy of data that meet the accuracy standard of sensors or not. The detail calculation is illustrated as follows (Tables 1, 2, 3 and 4):

Table 3 Positional accuracy of smartphone sensors

Sensor no.	Sensor name	Z value	Positional accuracy
1	Accelerometer sensor	3.5	1.25
2	Gyroscope sensor	3.1	0.85
3	Proximity sensor	2.3	0.05
4	Ambient light sensor	2.8	0.81
5	Magnetometer sensor	2.7	0.45
6	Localization sensor	4.0	1.75
7	Radiation sensor	3.2	1.52
8	CMOS sensor	4.5	2.25
9	Microphone sensor	3.4	1.15
10	Fingerprint sensor	2.4	0.15
11	Humidity sensor	2.5	0.25
12	Barometer sensor	2.9	0.65
13	Thermometer sensor	2.4	0.15
14	Pedometer sensor	2.5	0.25

$$\text{Distance error} = (\text{XTrueP} - \text{XMesP})^2 + (\text{YTrueP} - \text{YMesP})^2 \tag{1}$$

$$\text{XMesP} = 1.40,\ 2.1,\ 1.1,\ 1.40$$

Then, average distance error (XAverage) = (1.40 + 2.1 + 1.1 + 1.40)/4 = 1.50

$$\text{Standard deviation of distance error} = \left(\sum(\text{X} - \text{XAverage})^2/\text{n}\right) \tag{2}$$

Therefore standard deviation of distance error = 0.54/4 = 0.135

$$\text{Root Mean Square (RMS)Error} = \frac{e1^2 + e2^2 + e3^2 + \cdots + e_n^2}{\text{n}} = \sum \text{e}_\text{i}{}^2/\text{n} \tag{3}$$

Therefore, RMSE = (1.96 + 4.41 + 1.21 + 1.96)/4 = 9.54/4 = 2.39
Therefore,

$$\text{Positional Accuracy} = \text{Zvalue} + \text{Standard Deviation} - \text{RMS} \tag{4}$$

Positional accuracy of various sensors is obtained as follows.

Table 4 Energy efficiency of smartphone sensors

Sensor no.	Sensor name	W_{in}	W_{out}	Energy efficiency (%)
1	Accelerometer sensor	34	12	35
2	Gyroscope sensor	28	11	39
3	Proximity sensor	23	13	56
4	Ambient light sensor	26	14	53
5	Magnetometer sensor	26	14	53
6	Localization sensor	24	12	50
7	Radiation sensor	19	11	57
8	CMOS sensor	32	14	44
9	Microphone sensor	15	37	40
10	Fingerprint sensor	31	13	42
11	Humidity sensor	36	15	41
12	Barometer sensor	32	14	44
13	Thermometer sensor	36	15	41
14	Pedometer sensor	28	11	39

3.2 *Energy Efficiency*

A smartphone is powered by portable batteries. In smartphones, power consumption is very high because smartphones are being intensively used and applications installed in smartphones work on the basis of computer vision and image processing techniques. A smartphone user expects that his/her battery use duration should be as long as possible. The overall daily available time of battery depends on power consumption required by different onboard input/output systems and applications. Most of the input systems are sensors such that overall energy efficiency depends upon power consumption required by individual sensors. Power consumption among sensors varies, which reduces total available battery time. If a sensor consumes less power and collects information efficiently, then it will be useful for many applications.

Energy efficiency is a measure of energy consumed by a sensor in a process. While a sensor is operating, it to be assumed that it will waste battery charge. The energy efficiency of sensor can be calculated by as follows, and is expressed as a percentage.

$$\text{energy efficiency} = \frac{\text{energy output}}{\text{energy input}} \times 100 \tag{5}$$

$$\eta = \frac{W_{out}}{W_{in}} \times 100\% \tag{6}$$

where η = sensor energy efficiency, $\mathrm{W_{out}}$ = the work produced by the sensing process, and $\mathrm{W_{in}}$ = the energy input into the sensing process.

We have determined the energy efficiency of various sensors using the above formula based on sensor power consumption and raw data collection.

4 Results Analysis

We collected various properties of modern smartphone sensors and analyzed them using our proposed soft learning approach on the basis of two parameters, i.e. positional accuracy and energy efficiency. We determined based on our soft learning approach that some sensors have good positional accuracy, some sensors have good energy efficiency, and some have both good positional accuracy and good energy efficiency. Selection of appropriate sensors for developing user applications used by vehicle drivers for obstacle detection and tracking may be decided on the basis of our proposed soft learning approach. Graphical representation of the results obtained through our soft learning approach is shown below (Fig. 2).

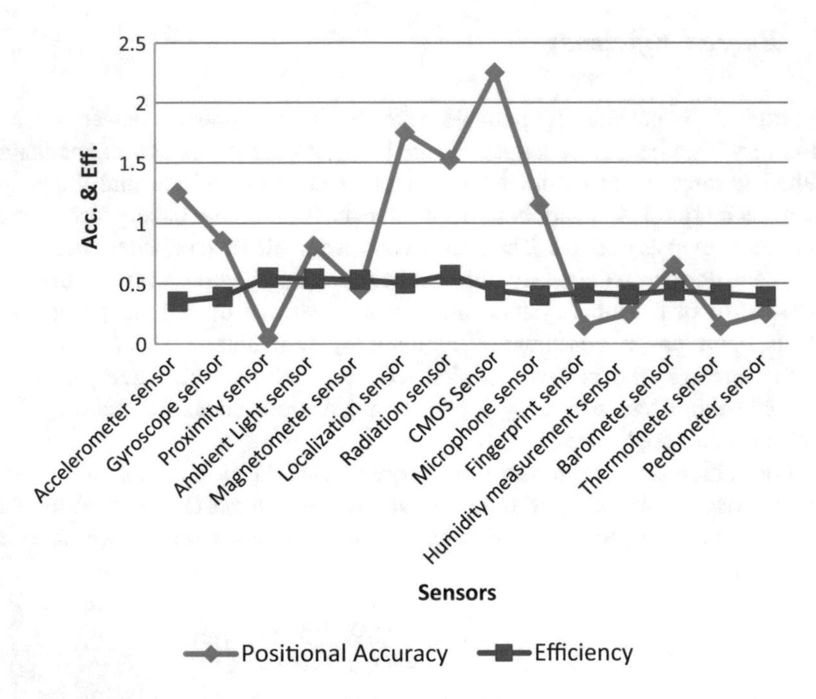

Fig. 2 Positional accuracy and energy efficiency of mobile sensors

5 Conclusion and Future Work

In this study, we analyzed the characteristics and functionality of various sensors available in modern-day smartphones. After detailed analysis, we found that accelerometer, CMOS and localization sensors are useful sensors embedded in modern-day smartphones which can be utilized for identifying roadway obstacles in infrastructure-free environments. The information retrieved via these sensors can be utilized for developing driver support systems on smartphones. In the future, mobile sensors will play an important role in developing user-friendly applications for smartphones which can be used for various purposes.

References

1. Shah, M. Ali, et al. 2015. Energy efficiency in smart phones: A survey on modern tools and techniques. In *2015 21st international conference on automation and computing (ICAC)*, IEEE.
2. Engelbrecht, Jarret, et al. 2015. Survey of smartphone-based sensing in vehicles for intelligent transportation system applications. *IET Intelligent Transport Systems* 9 (10): 924–935.
3. Bala, Rimpy, et al. 2013. Battery power saving profile with learning engine in Android phones. *International Journal of Computer Applications* 69 (13).

4. Bhoraskar, Ravi, et al. 2012. Wolverine: Traffic and road condition estimation using Smartphone sensors. In *2012 fourth international conference on communication systems and networks (COMSNETS)*, IEEE.
5. Sendra, Sandra, et al. 2011. Power saving and energy optimization techniques for wireless sensor networks. *JCM* 6 (6): 439–459.
6. Mednis, Artis, et al. 2010. RoadMic: Road surface monitoring using vehicular sensor networks with microphones. *NDT* 2.
7. Mohan, Prashanth, et al. 2008. Nericell: Rich monitoring of road and traffic conditions using mobile smartphones. In *Proceedings of the 6th ACM conference on embedded network sensor systems*. ACM.
8. Vo, Quoc Duy, et al. 2016. A survey of fingerprint-based outdoor localization. *IEEE Communications Surveys and Tutorials* 18 (1): 491–506.
9. Nguyen, Vinh Dinh, et al. 2016. Learning framework for robust obstacle detection, recognition, and tracking. *IEEE Transactions on Intelligent Transportation Systems* 18 (6): 1633–1646.
10. Oyewobi, S.S., et al. 2013. Mobile terminals energy: A survey of battery technologies and energy management techniques. *International Journal of Engineering and Technology* 3 (3): 282–286.
11. Misra, Archan, et al. 2011. Optimizing sensor data acquisition for energy-efficient Smartphone-based continuous event processing. In *2011 12th IEEE international conference on mobile data management (MDM)*, vol. 1. IEEE.
12. Eriksson, Jakob, et al. 2008. The pothole patrol: Using a mobile sensor network for road surface monitoring. In *Proceedings of the 6th international conference on mobile systems, applications, and services*. ACM.
13. Brisimi, Theodora S., et al. 2016. Sensing and classifying roadway obstacles in smart cities: The street bump system. *IEEE Access* 4: 1301–1312.
14. RadhaKrishna, Kini A. NCRB News Letter. http://ncrb.nic.in, Web. 2, August 2017.
15. Gageik, Nils, et al. 2015. Obstacle detection and collision avoidance for a UAV with complementary low-cost sensors. *IEEE Access* 3: 59.

Design of Compact Multiband Annular-Ring Slot Antenna

C. R. Prashanth and Mahesh N. Naik

Abstract An annular-ring microstrip antenna (ARMSA) with a defective ground structure for multiband operation is proposed. The proposed antenna operates at three different frequency bands, at S/C/X bands. Initially, the antenna is operated in dual-band mode from 1 to 12 GHz; then, by introducing an annular ring slot, a multiband mode antenna is realized. A microstrip feed line is used to excite the annular-ring patch antenna embedded on FR-4 substrate with a dielectric constant $\varepsilon_r = 4.4$. The designed antenna is circularly polarized with waveport excitation; it is best suited for multiband applications.

Keywords Annular ring · Microstrip antenna · Defective ground structure Multiband · ARMSA · Microstrip feed

1 Introduction

In today's digital world, the use of the Internet is ubiquitous, yet establishing a network to every corner of the world is challenging. Efficient transmission of data requires a strong network which in turn requires strong antennas. Following World War II, antennas were mainly of wired type, e.g. long wires, dipoles, helical, etc., [1] either used as a single element or in an array. Later development produced aperture types such as open-ended wave guides, slots, and horns, reflector antennas used in RADAR, space and remote sensing applications.

Later, during the 1950s, frequency-independent antennas were developed operating over 10–10,000 MHz and used in applications such as TV, point-to-point communication and feeding for reflectors. During the 1970s, microstrips or patch

---

C. R. Prashanth (✉) · M. N. Naik
Department of Telecommunication Engineering, Dr. Ambedkar Institute of Technology, Bengaluru, India
e-mail: prashanthcr.ujjani@gmail.com

M. N. Naik
e-mail: maheshnn595@gmail.com

A. N. Krishna et al. (eds.), *Integrated Intelligent Computing, Communication and Security*, Studies in Computational Intelligence 771,
https://doi.org/10.1007/978-981-10-8797-4_30

antennas conformal to the surface [2] were developed. A new antenna array design referred to as a smart antenna based on basic technology of 1970s and 1980s is sparking interest, especially for wireless applications. One of the first developments in early twentieth century was the use of commercial AM radio stations by aircraft for navigation. Radio was used to transmit pictures visible as television during the 1920s. Commercial TV transmission started in North America and Europe in the 1940s. In 1963, color TV was being broadcast commercially and the first communication satellite, Telstar, was launched. In 1947, AT&T commercialized mobile telephone service. An advanced analog mobile cell phone system developed by BELL labs and which provided much more capacity was introduced to US citizens in 1978.

Antennas are required by any radio receiver and transmitter to couple the electrical connection to an electromagnetic field. Antennas are omni-directional or directional [3]. To design an antenna with large directivity, electrical dimension must be increased with an antenna element, or multiple elements can be used in a so-called array.

The constant improvement in design of antennas in the twenty-first century and its variants have led to a new inclination towards multiband wireless communication and networking [3]. The fundamental parameters to be considered in antenna implementation are dimension, expenditure and simplicity of fabrication. A microstrip patch array is shown in Fig. 1. There are several administrative and commercial applications in wireless communication systems for reliable transcription of data where microstrip antennas play an important role. These antennas are suitable for both planar and non-planar structures, [4] and are simple to design and manufacture with suitable printing technology.

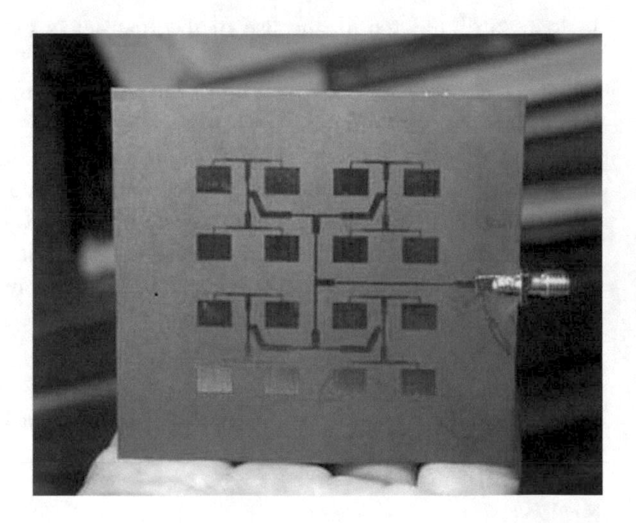

Fig. 1 Microstrip patch array

2 Literature Review

In this section, various works by authors in the field of microstrip antennas, defective ground structure, dual-band applications, multiband, annular ring antennas, circular patches and feeding techniques are discussed. Khanna et al. [5] presented a work on a multiband annular-ring microstrip antenna (ARMSA) with a deflected ground plane structure. The designed antenna operates over frequency ranges of 2.73–3.2, 5.34–5.95, 7.84–8.48 and 10.19–10.78 GHz. Also, the antenna is designed by varying the patch parameters, ground parameters and feed parameters, giving good results of gain and efficiency.

Paras and Gangwar [6] presented a work on an optimized annular-slot circular microstrip antenna. A generic algorithm was used in optimization of the circular microstrip antenna with 270 annular patches having probe feed. The generic algorithm tool provides good radiation pattern characteristics for singular or multiband operations and is best suited for wireless local area networks (WLANs) and advanced frequency RADAR or scatter meters. Ado Muhammad and Idris [7] proposed an ARMSA and performed simulation using electromagnetic simulation and an optimization simulator based on the moment of methods. The work was performed for dual-band characteristics in a frequency range of 4.5–9.2 GHz. The proposed antenna shows a return loss less than −19 dB and a voltage standing wave ratio (VSWR) less than 1.5 within an acceptable range, best suited for dual-band wireless application.

Ammann and Bao [8] studied miniaturized annular-ring patch antennas. The proposed patch had a cross-slot ground defect of compact size for desired frequency. It can be easily matched to 50 Ω. Compact circularly polarized properties and wide circularly polarized bandwidths were obtained. Wu et al. [9] presented work on planar tapered annular-slot antennas. The tapered slot provides a uniform radiation pattern and low profile, best suited for ultra-wide-band applications. Frequency domain and time domain characteristics of an antenna and the impact of an antenna's geometric variation on the performance were carefully carried out. Huang and Hsia [10] presented work on planar elliptical antennas with a ground pattern in a single substrate. It was designed for ultra-wide-band applications by etching a notch at the ground plane 90° to the microstrip line.

3 Model

The structure of a microstrip antenna consists of a circular patch on one side of the substrate and a ground plane on either side [11]. The two annular ring slots are made concentric, centered at C_1, both coupled to each other and excited by the same feed. The ground plane is made defective and excited using a microstrip feed line with waveport, which provides a planar structure [10]. The design is carried out on FR-4 substrate with dielectric constant of $\varepsilon_r = 4.4$ and thickness of $h = 1.6$ mm as shown in Fig. 2. Antenna dimensions of $35 \times 30 \times 1.6$ mm^3 are chosen.

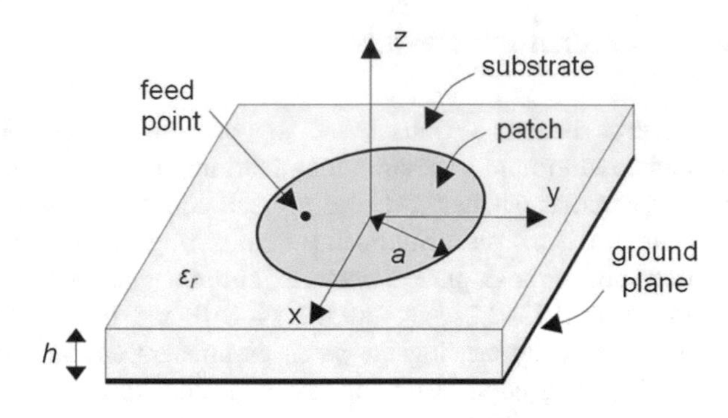

Fig. 2 Circular patch antenna

Table 1 Proposed antenna design parameters

Parameters	Units (mm)
L_1	35
W_1	30
R_1	11
R_2	6
R_3	12
L_f	7.5
W_f	1.5
h	1.6
C_1	0, 0
C_2	−3.5, 0

The width of the microstrip feed is fixed at 1.5 mm to achieve an impedance of 50 Ω; a length of 7.5 mm is chosen and can be varied. The annular patch has an external radius $R_1 = 11$ mm and an internal radius $R_2 = 6$ mm centered at $C_1 = (0, 0)$, provided by a feed of width $W_f = 1.5$ mm and length $L_f = 7.5$ mm. Another concentric ring of radius $R_3 = 12$ mm positioned at $C_2 = (-3.5, 0)$ is placed between the patch and substrate and is etched in the ground plane to form defective ground structure [3] as shown in Fig. 2. Parameters considered for the design of the antenna are as shown in Table 1.

The simulation software High-Frequency Structure Simulator (HFSS) [7] is used to optimize the dimensions of the proposed design on the basis of best performance. HFSS software is widely used in industries for analysis, simulation and design of different antennas, with highly accurate and high-speed structures to solve problems faced with using modern equipment; optimizing the size of antennas, bandwidth, minimized time and high efficiency. HFSS provides solution to both near field and far field; electric and magnetic field regions [9]. HFSS is the best tool to optimize the challenges faced by antennas. Antenna performance is characterized by several parameters like return loss, VSWR, gain, directivity, impedance and efficiency; these parameters are analyzed using HFSS. The design of the ARMSA is shown in Fig. 3.

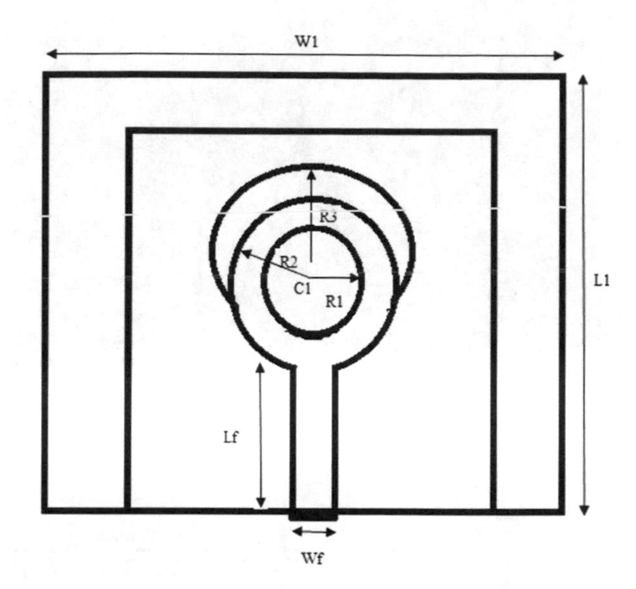

Fig. 3 Model of an annular-ring microstrip antenna

4 Performance Analysis

An ARMSA with a defective ground structure has been designed and fabricated on FR-4 substrate. The ARMSA has been fed with Wave port- feed, 50 Ω impedance is achieved by taking the width, $W_f = 1.5$ mm of the inner conductor, Length, $L_f = 7.5$ mm and the gap width, $g = 0.5$ mm between the ground plane and the inner conductor. The designed ARMSA is analyzed in terms of return loss, VSWR, directivity, gain parameters and a comparison is made by varying patch parameters.

Figure 4 shown above is the top view of the designed ARMSA with a defective ground structure obtained by simulation using HFSS. It consists of an inner annular ring with inner radius R_1 and outer radius R_2. It is defected by a bthird annular ring R_3 along the Y-axis. The entire setup is placed inside a radiation box to avoid leakage of radiation in free space.

The simulation results for return loss characteristics of the ARMSA are shown in Fig. 5. The return loss versus frequency graph shows that the antenna resonates at four resonant frequencies viz. 2.96, 5.57, 7.91 and 10.57 GHz, showing multiband behavior with a maximum return loss of −15.4 dB. It can be inferred that the designed antenna can be suitably used in S/C/X bands, with smooth curves moving from one band to another.

Simulation results for the ARMSA are shown in Fig. 6. The VSWR versus frequency graph shows that the antenna resonates at four resonant frequencies viz. 2.96, 5.57, 7.91 and 10.57 GHz, showing multiband behavior with a maximum VSWR of 1.256 dB. It has good agreement with the theoretical acceptance value.

Further simulation results for the ARMSA are shown in Fig. 7. The gain versus frequency graph shows that the antenna resonates at four resonant frequencies viz. 2.96, 5.57, 7.91 and 10.57 GHz, showing multiband behavior with a maximum

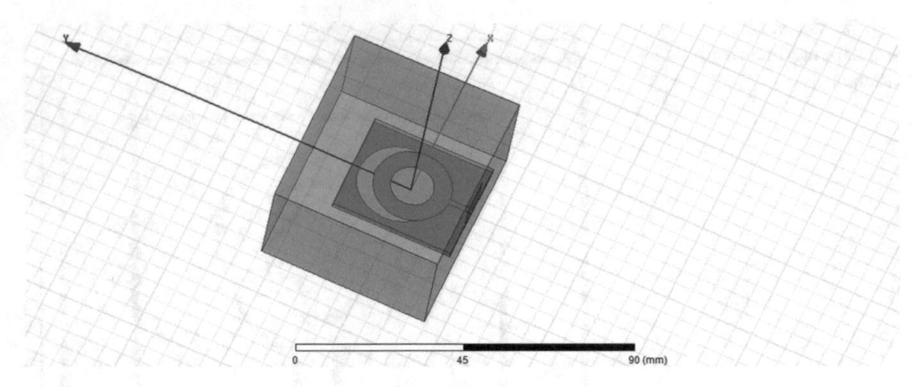

Fig. 4 Microstrip antenna with a defective ground

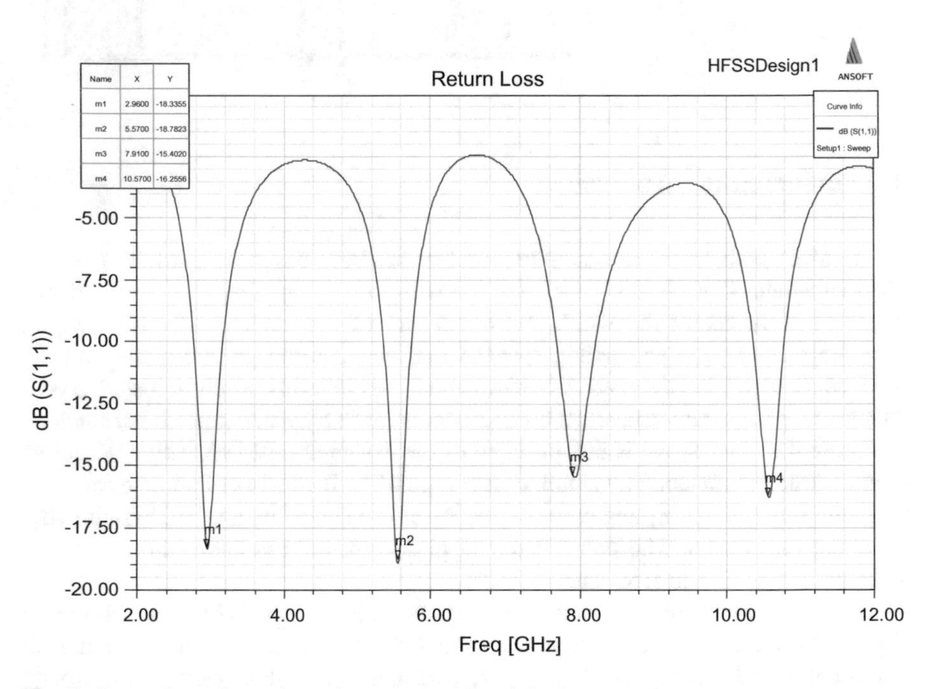

Fig. 5 Return loss of an microstrip antenna with a defective ground

gain of 4.1 dB at 3.7 GHz. Another peak gain can be obtained at 12 GHz by further optimization; a smooth variation of the graph makes it best suited in all bands.

Additional simulation results for the ARMSA are shown in Fig. 8. The directivity versus frequency graph shows that the antenna resonates at four resonant frequencies viz. 2.96, 5.57, 7.91 and 10.57 GHz, showing multiband behavior with a maximum directivity of 4.7 dB at 3.7 GHz. Peak directivity can be obtained at 12 GHz by further optimization.

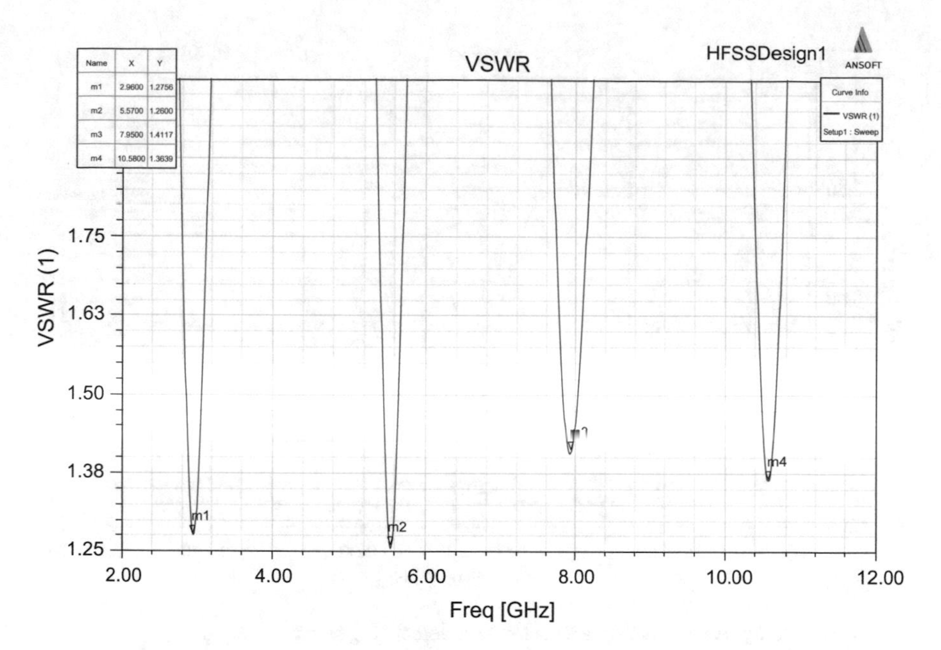

Fig. 6 VSWR of a microstrip antenna with a defective ground

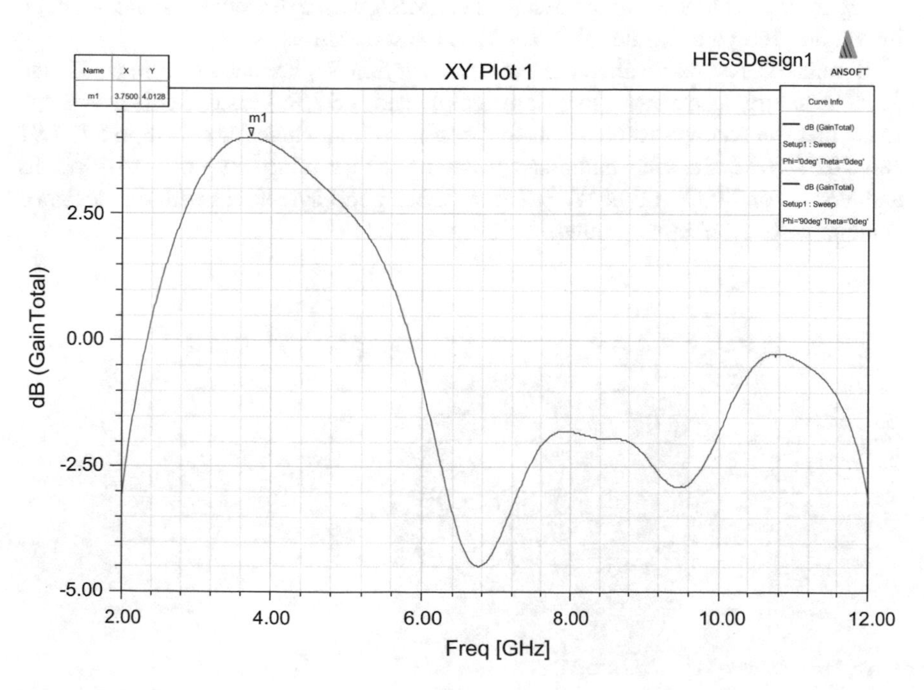

Fig. 7 Gain of a microstrip antenna with a defective ground

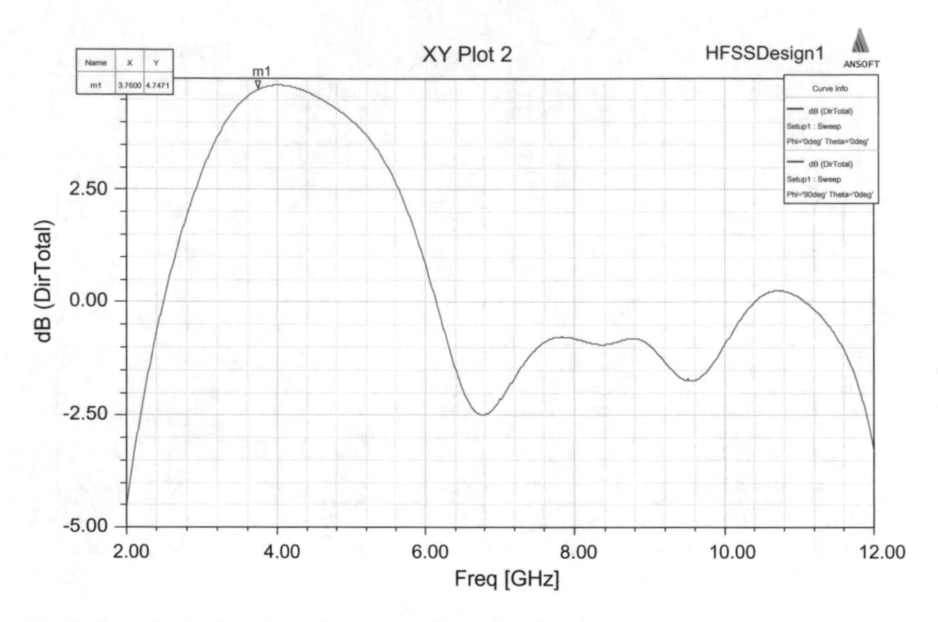

Fig. 8 Directivity of a microstrip antenna with a defective ground

Figure 9 is a top view of the designed ARMSA with a defective ground structure by varying the patch width (W_f), i.e. 1, 1.5 and 2 mm.

Simulation results for the ARMSA with varying W_f are shown in Figs. 10 and 11. The return loss versus frequency graph and VSWR versus frequency graph show that the antenna resonates at four resonant frequencies viz. 2.96, 5.57, 7.91 and 10.57 GHz, showing multiband behavior with a return loss of −14.1069 dB and VSWR of 1.0824 dB at $W_f = 2$ mm. It has good agreement with the designed antenna with a defective ground.

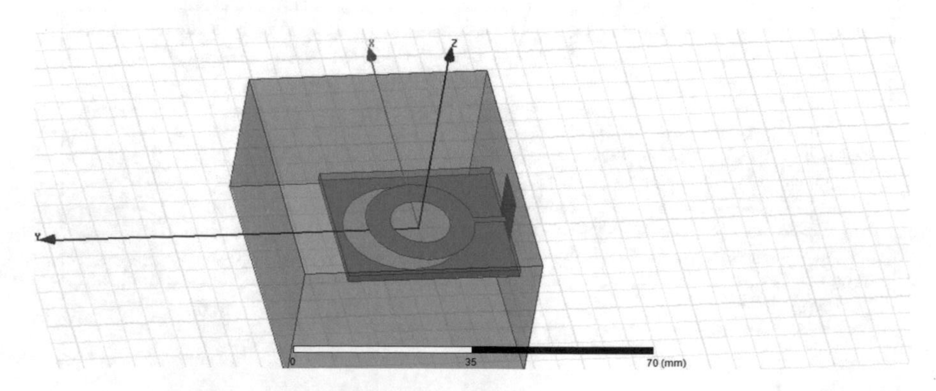

Fig. 9 Microstrip antenna with varying patch width

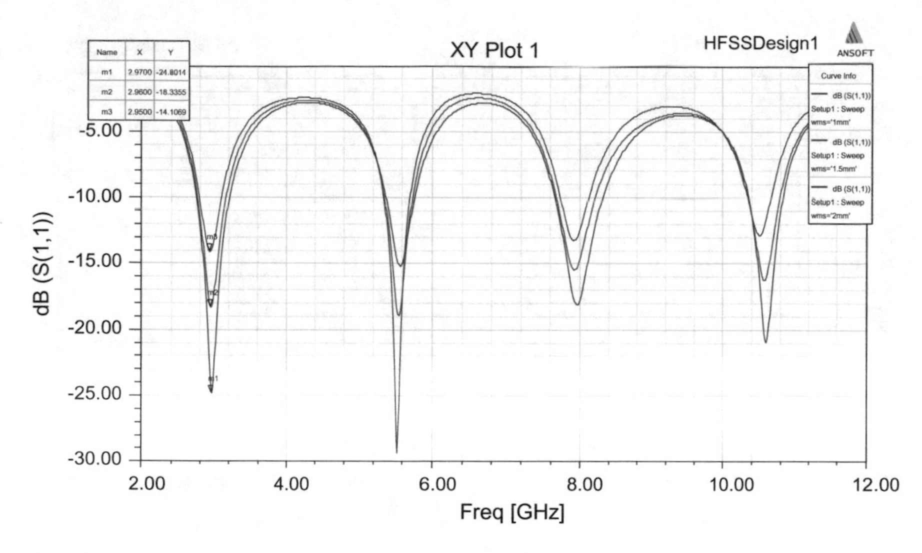

Fig. 10 Return loss of a microstrip antenna with varying W_f

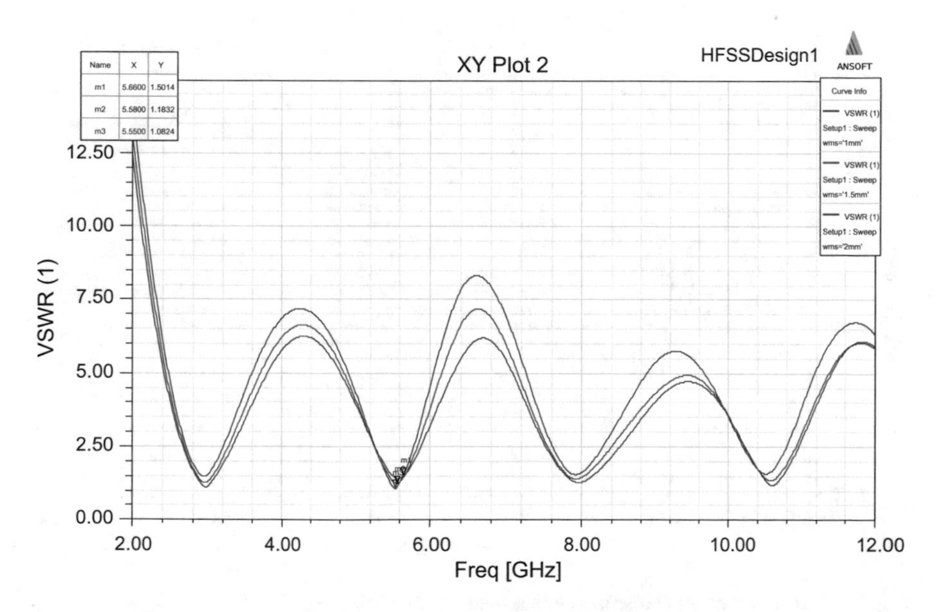

Fig. 11 VSWR of a microstrip antenna with varying W_f

The simulation results for the ARMSA with varying W_f are shown in Figs. 12 and 13. The gain and directivity versus frequency graph shows that the antenna resonates at three bands and has no smooth curves. The random variation in curves

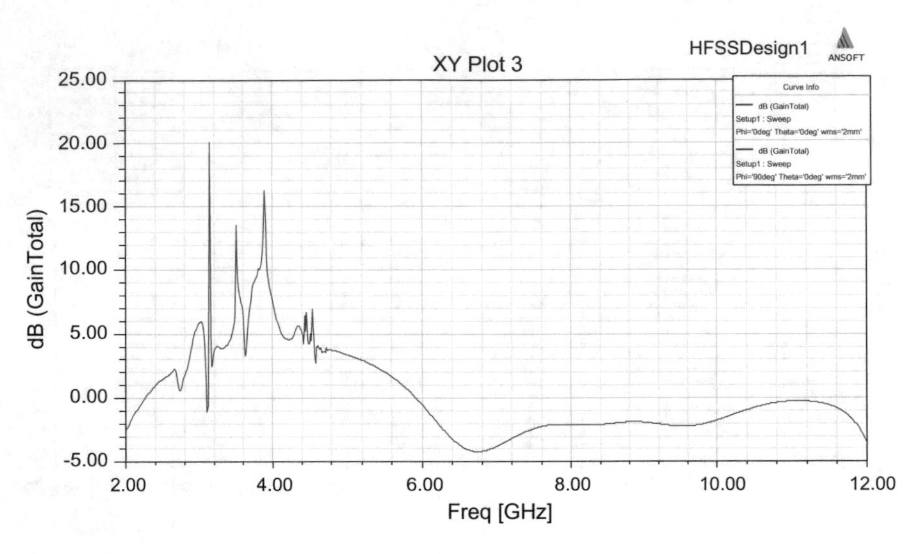

Fig. 12 Gain of a microstrip antenna with varying W_f

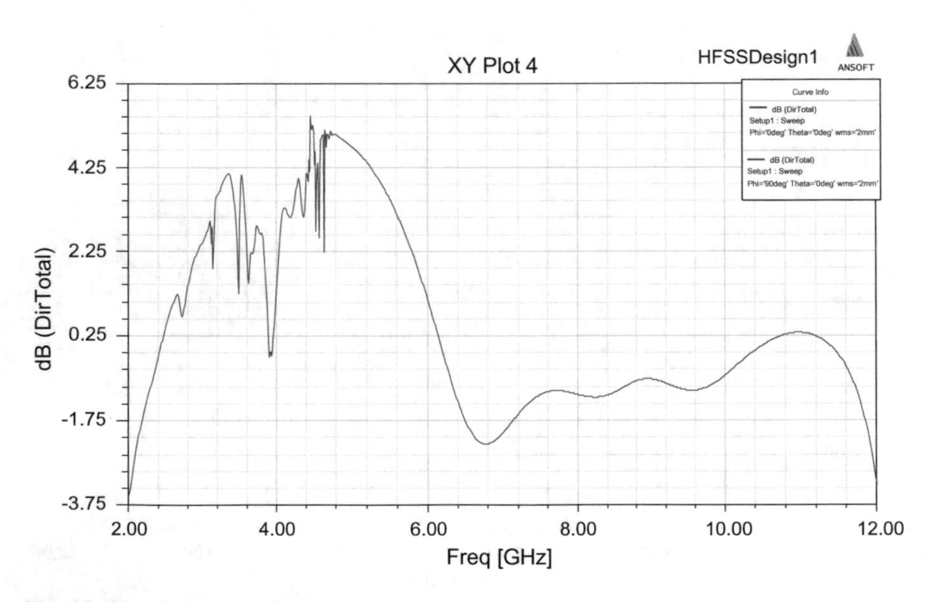

Fig. 13 Directivity of a microstrip antenna with varying W_f

shows the impairments in communication; hence further investigation must be done.

The simulation outputs are compared with previously simulated results; there is improvement in the performance. The return loss of microstrip antenna without an internal ring is observed to be −11.7 dB, and with an internal ring, is −18.06 dB.

Table 2 Comparison of performance parameters

MicroStrip Antenna	Return loss (dB)	VSWR	Gain (dB)	Directivity (dB)
Without internal ring	−11.7697	1.7606	8.8500	11.0320
With internal Ring	−18.0656	1.2856	10.1635	10.4718
With ARMSA without DGS	−27.6009	1.5760	7.2451	5.5546
With ARMSA with DGS	−15.4020	1.2600	4.0128	4.7471

Table 3 Comparison of performance parameters

	Length (mm)	Return loss (dB)	VSWR
Varying patch width	1	−24.8014	1.5014
	1.5	−18.3355	1.1832
	2	−14.1069	1.0824

Without a defective ground, the return loss is −27.6 dB, and for an antenna with a defective ground, it is −15.4 dB. The VSWR of a microstrip antenna without an internal ring is observed to be 1.76 dB and with an internal ring, it is 1.28 dB. Without a defective ground, the VSWR is 1.56 dB, and for an antenna with a defective ground, the VSWR is 1.26 dB, which is an improvement over an antenna without an annular ring and without a defective ground, observed to be −18.33 dB with a VSWR of 1.2564 dB. This is desirable for multiband operations with omni-directional radiation.

The gain of a microstrip antenna without an internal ring is observed to be 8.85 dB; with an internal ring, it is 10.1 dB. Without a defective ground, the gain is 7.2 dB, and for an antenna with a defective ground, it is 4 dB. The gain of the microstrip antenna without an internal ring is observed to be 11 dB; with an internal ring, it is 10.4 dB. Without a defective ground, it is 5.5 dB; for antenna with a defective ground, it is 4.7 dB.

The comparison made for different parameters with different conditions are shown in Table 2. From Table 2, it can be shown that the designed antenna is optimized and is capable of multiband operation, giving the ARMSA a maximum efficiency of 85%. Further implementation includes varying parameter like length of patch, which does not alter the design and varying width of patch have good return loss and VSWR; but gain and directivity curves are not smooth enough to be compatible with communication applications as shown in Table 3. Further optimizing can be done for the designed antenna that can be operated by varying patch width; to be best suited for multiband applications with improved gain and efficiency.

5 Conclusion

The design was carried out using the structural simulator Ansoft HFSS tool. An ARMSA with a defective ground structure for multiband operation is proposed. It operates at three different frequency bands, S/C/X bands. Return loss, VSWR, gain and directivity with a defective ground and varying patch width were observed. The results were compared with existing literature contributions and it was found that the performance of the proposed system is better and best suited for multiband applications.

References

1. Devashree S., Marotkar, and Prasanna Zade. 2016. Bandwidth enhancement of microstrip patch antenna using DGS. In *International conference on electrical and optimization techniques*, pp. 9939–9943.
2. Mythili, P., P. Cherian, S. Mridula, and B. Paul. 2009. Design of a compact multiband microstrip antenna. *Microwave and Optical Technology Letters* 9: 4244–4859.
3. Ado Muhammad, Nuraddeen, and Yusuf Idris. 2016. Design and analysis of rectangular microstrip antenna for global WLAN applications. *International Journal of Engineering Science and computing* 6 (4): 3732–3737.
4. Ming Mak, K.A., Hau Wah Lai, Kwai Man Luk, and Chi Hou Chan. 2015. Circular polarised patch antennas for future 5G mobile phones. *IEEE Publications* 2 (1): 1521–1529.
5. Puneet, Khanna, Kshitji Shinghal, and Arun Kumar. 2016. Multiband annular ring microstrip antenna with defected ground structure for wireless communication. *International Journal of Computer Applications* 135 (2): 19–25.
6. Paras, and Gangwar. 2016. Design of compact GA-optimized annular slot multiband circular microstrip antenna. *International Journal of Advances in Microwave Technology* 1 (1): 30–35.
7. Ado Muhammad, Nuraddeen, and Yusuf Idris. 2015. An annular microstrip patch antenna for multiband applications. *International Journal of Engineering Research and Technology* 4 (3): 546–549.
8. Ammann, M.J., and X.L. Bao. 2007. Miniaturized annular ring loaded patch antennas. *IEEE Transactions on Antennas and Wireless Propagation* 1 (7): 912–915.
9. Wu, C.M., Y.L. Chen, and W.C. Liu. 2012. A compact UWB slotted patch antenna for wireless USB dongle applications. In *IEEE antennas and wireless propagation letters*, pp. 596–599.
10. Huang, C.Y., and W.C. Hsia. 2005. Planar elliptical antenna for ultra wide band communications. *Electronics Letters* 41 (6): 21–25.
11. Raj, Kumar, and K.K. Sawant. 2010. Rectangular patch microstrip antenna: A survey. *International Journal of Microwave and Optical Technology* 5 (6): 320–327.

Authentication Using Dynamic Question Generation

Irfan Siddavatam, Dhrumin Khatri, Pranay Ashar, Vineet Parekh and Toshkumar Sharma

Abstract Privacy and security today are extremely important as we enter the digital age. Authenticating users via passwords has proven to be secure, but flaws still exist. Due to the vast availability and flexibility of password authentication, it is now the most accepted form of user authentication. Intrusion systems, using various techniques, can find ways to break into passwords. Alternate solutions have been proposed, varying from expensive hardware-based authentication to two-phase authentication protocols. We offer an alternative system to authenticate users by extracting social user data available on Facebook and other social media, then generating questions using the collected data and user personality insights. Our mobile application and website are developed using Express Framework, NodeJS, and MongoDB.

Keywords Authentication · Captcha · Challenge questions
Natural language processing · NLP · Security

1 Introduction

Corporations spend a considerable amount of money in creating secure access portals. Costly authentication techniques, with regard to acquisition and maintenance of equipment, or techniques requiring colossal exertion from the client, are inefficient and unjustified. Authentication using passwords needs the least resources but has a variety of issues, including memorability, dictionary attacks and SQL injections, to name a few. Because of memorization difficulties, some work focuses not on memo-

I. Siddavatam (✉) · D. Khatri · P. Ashar · V. Parekh · T. Sharma
K.J. Somaiya College of Engineering, Mumbai, India
e-mail: irfansiddavatam@somaiya.edu

D. Khatri
e-mail: dhrumin.k@somaiya.edu

P. Ashar
e-mail: pranay.ashar@somaiya.edu

V. Parekh
e-mail: vineet.p@somaiya.edu

A. N. Krishna et al. (eds.), *Integrated Intelligent Computing, Communication and Security*, Studies in Computational Intelligence 771,
https://doi.org/10.1007/978-981-10-8797-4_31

rized passwords but on passwords based on an entrenched knowledge that is elicited through questions. This approach is often criticized, as the traditional use of this type of authentication has proven to be flawed due to the vulnerability of social engineering. The goal in the paper [1] is to retain the approach's inherent convenience and to improve upon its weak security. Various approaches have been proposed in this case. These solutions are often either insecure or expensive with regard to data collection and maintenance. While developing an application of this type, one must weigh the advantages and disadvantages of the time spent and the cost involved in creating and maintaining it. For instance, applications which have to be lightweight or do not put much weight on data integrity do not need complex authentication systems to be in place. Over the years, many efforts have been made to authenticate users in unique ways [2–6]. In this paper, we propose an alternative method of authenticating users by generating dynamic questions based on data extracted from different social media. A well-accepted theory of psychology and human understanding is that language reflects an individual's personality, thinking style, social connections and emotional states. Therefore, we collect data that the user has uploaded over the years on social media and use it to map his personality traits, his likes and dislikes, then us that material to generate questions of various degrees of difficulty.

2 Approach

This section has the flowchart including the explanation of every block involved.

2.1 Sign-Up

Sign-up is a one-time process for registering the user. The user signs up on Facebook allowing the application to access this data to create a dataset for user authentication. The generated response with a user identity and access token is fed to data to the graph API, which replies with the information requested by the application (Fig. 1).

The next step involves asking the user a set of questions to be answered in brief, choosing questions for which the responses vary from person to person. These answers are stored in JSON format so that later, during log-in to the application, we can compare the user input using Latent Semantic Indexing and understand the similarity between the sentences. After all questions have been answered, the sign-up process is complete, and the user is granted application access to use the smart assistant.

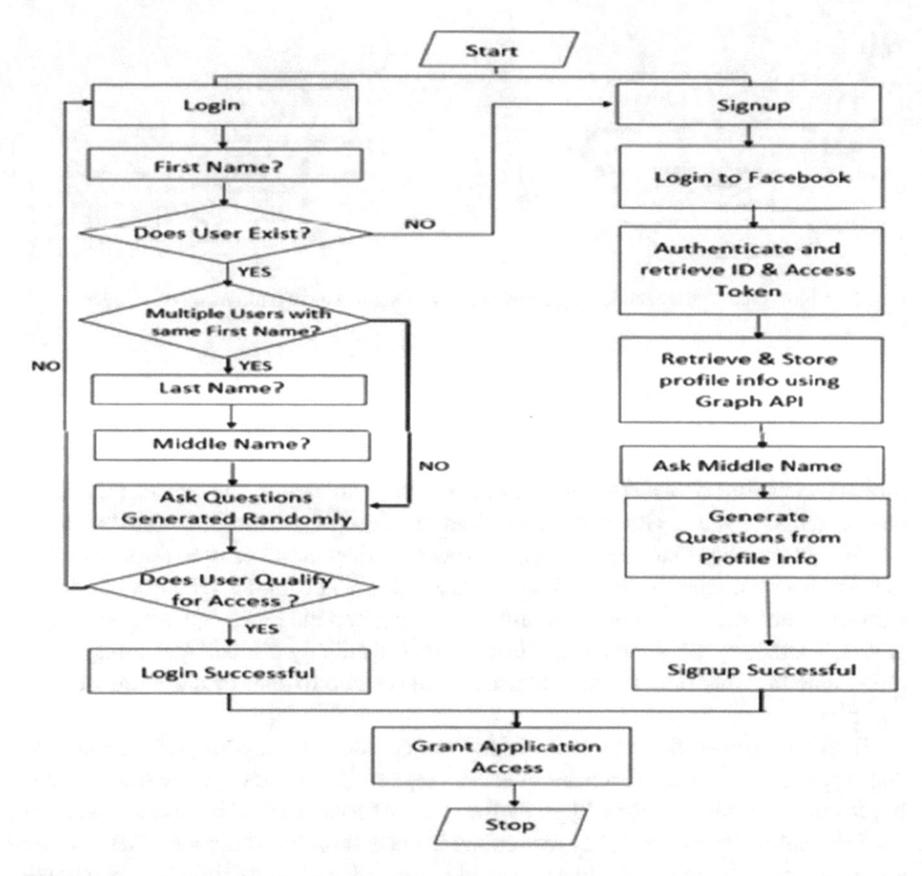

Fig. 1 Flow of application for sign-up and log-in

2.2 *Login*

The user enters his first name. The process reiterates until the application can pinpoint the particular user. If the user has not signed up, he is redirected to the sign-up page. After the system pinpoints the user, the verification process begins, and the system asks various questions of the user to authenticate his identity. The authentication is based on asking questions with varying levels of difficulty depending on the user's responses; once the user has qualified, he is allowed to log-in.

The sensitivity and specificity of the questions vary depending on the grade of the question and the answers given by the user. We have specified three grades: easy or grade 1, medium or grade 2, and difficult or grade 3. Question grade 1 has a sensitivity level of 1, grade 2 has 75% sensitivity, and grade 3 has 50% sensitivity. Easy grade questions are such that the inability to answer them implies that the person trying to authenticate is not who he says he is. Questions like, "Is Person A your brother?" or "Is Taurus your sun-sign?" are questions that most users would never fail to answer.

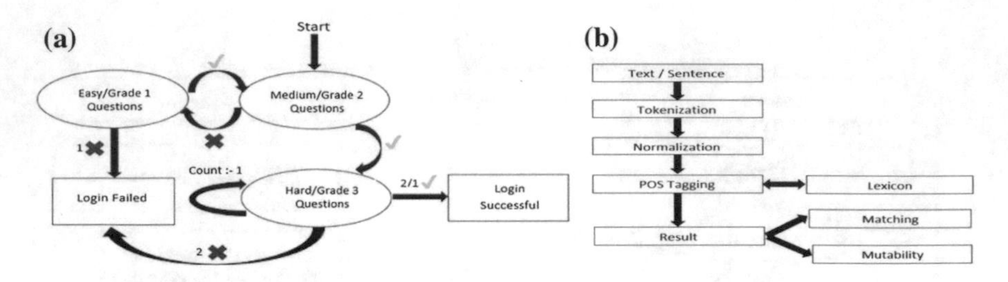

Fig. 2 **a** Flow chart for authenticating users. **b** Flow of sentences in parts of speech tagger

They have a low error acceptance rate. Grade 2 questions have a moderate level of error acceptance. Questions like, "While playing chess, which color pieces would you choose?" or pet preferences, preference of style over comfort, places visited by the user, questions of his likes and dislikes, which do not readily change over time, are considered as grade 2. Grade 3 questions are the type of questions that are based on a user's personality that rarely changes. However, they aren't easy to document due to their sheer complexity and the linguistics involved. The hard level questions allow leeway for error. We use a bag of words model to grasp the less frequently occurring words. Therefore, users can be qualified for access only by difficult questions due to the simple fact that only the user himself shall be able to answer questions of high complexity.

To extract keywords and words of importance like names, verbs, adjectives used and stop-words, we require a parts of speech tagger. For instance, if we require finding the name in the sentence, "I go by the name of John Smith" or "My friends call me John Smith," a named-entity recognizer is necessary. For these purposes, we use a parts-of-speech tagger. We have devised a flow chart for qualifying users, shown in Fig. 2.

Parts of Speech Tagging: We use a parts-of-speech tagger to extract names, understand different parts of sentences and their implementation, convert them into robot understandable format and reframe declarative sentences in an interrogative or a negative format and use them in the correct tense. We wish to accomplish tagging of different terms of the sentences. To do so, we require tokenizing the various words in terms lumped individually or in groups. Once this is accomplished, the data must be normalized to robot understandable form. Some of the various actions required are ensuring sane whitespace, removal of punctuation, rooting of words to their infinitive forms and grouping phrasal verbs like "sleep in," "beef up," or "brighten up" together as single tokens. After the sentences are converted to robot understandable format, we can pass our words to a lexicon for mapping. A lexicon is a file consisting of all our words, each assigned to its correct utility in a sentence. To achieve the fastest mapping, we keep verbs only in their infinitive form in our lexicon, along with adverbs, adjectives and stop words. Nouns are used as a fallback option to achieve higher accuracy. To keep the processing snappy and highly responsive, the lexicon is restricted in size to 110 kb, and then we add suffix/regex patterns and sentence level post-processing on each sentence for better processing. We use a toolkit called NLP

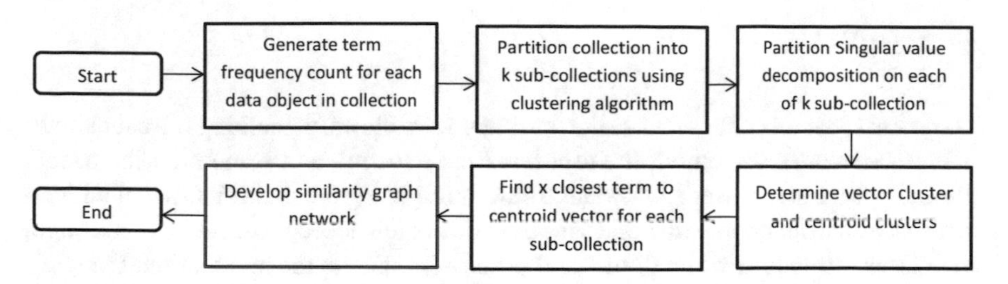

Fig. 3 Flow of data in Latent Semantic Indexing for mapping cosine similarity between two sentences

Compromise available in the Node Packet Manager to achieve this task. It requires 8–10 ms to respond to an average sentence. NLP Compromise runs with an accuracy of 86% on the Penn treebank. This implies that a 15% error rate is introduced into the application from the development end that requires compensation later (Figs. 3 and 4).

Latent Semantic Indexing: Latent semantic indexing is used in natural language processing in the category of distributional semantics. To understand how close in meaning two documents or sentences are to each other, we use the Jaccard index and cosine similarity. This is achieved by using singular-value decomposition (SVD), using Java, and principal component analysis (PCA). A vector space model, mapping all the different words in space by using matrices, first constructs a matrix consisting of word counts per sentence or paragraph while preserving the structural similarity in column attributes. This is followed by term frequency-inverse document frequency. This is followed by calculating the cosine similarity of the centroid of the vector clusters of the two paragraphs.

The results are further improved by promoting rare word occurrences and giving slack to ignore frequently occurring words. The results are generated in a range of 1 to −1, where 1 implies complete co-relation and −1 signifies no co-relation at all. We set our similarity index at 0.3 for allowing leniency in input by users. That is, the angle between the two sentences accepted by the system is approximately 72° in space.


```
{"nlp":{"str":"My name is Steve Apple Jobs",
"terms":[{"whitespace":{"preceding":"","trai
ling":" "},"text":"My","normal":"my","expans
ion":null,"reasoning":["lexicon_pass"],"pos"
:{"Possessive":true},"tag":"Possessive"},{"w
hitespace":{"preceding":"","trailing":" "},"
text":"name","normal":"name","expansion":nul
l,"reasoning":["lexicon_pass"],"pos":{"Verb"
:true,"Infinitive":true},"tag":"Infinitive"}
,{"whitespace":{"preceding":"","trailing":"
"},"text":"is","normal":"is","expansion":nul
l,"reasoning":["lexicon_pass"],"pos":{"Verb"
:true,"Copula":true},"tag":"Copula"},{"white
space":{"preceding":"","trailing":""},"text"
:"Steve Apple Jobs","normal":"steve apple jo
bs","expansion":null,"reasoning":["capital_s
ignal, lexicon_pass, capital_signal,
chunked
person-titleCase","capital_signal","chunked
person-titleCase"],"pos":{"Noun":true,"Plura
l":true,"Person":true,"MalePerson":true},"ta
g":"Person","honourific":null,"firstName":"S
teve","middleName":"Apple","lastName":"Jobs"
}],"contractions":{}}}
```


Fig. 4 Mapping results of NLP compromise

3 Results

The primary factors of authentication are stated as follows: something the user knows (like a password), something the user has (like a token), and something the user is (biometrics). Using a user's knowledge base is following the first principal of authentication. Since our application takes input via Google's speech-to-text for conversion, an error is introduced while giving and processing data to the application. Google's speech recognition APIs are said to be 92% accurate, which implies that there is an 8% chance of an error being introduced into the system. This is dealt with by allowing the user to correct any mistakes in the sentence manually. The input given is, "My name is Steve Apple Jobs." The output received generates tags, normalized output and maps the different parts of speech of the sentence as well as different named entities. First we calculate the results of compromise toolkit. We are able to bench test the accuracy at 85%. Errors are formed in complex sentences due to the limited lexicon size. However, for our usage, we find the toolkit to be more than accurate enough. The inputs and outputs are shown in Fig. 5a, b, respectively. We tabulate the errors introduced into the system using various tools used. The induced error is 18.6%. Once we have understood the various errors in the system and how they affect the authentication protocol, it is possible to understand the varying true positives and false negatives. A true positive implies that a correct user gets access to the system. A false negative implies that a user is denied access into his system due to incorrect sensitivity levels. Figure 6a shows the graph of the number of questions asked to the user versus the probability of the user answering correctly.

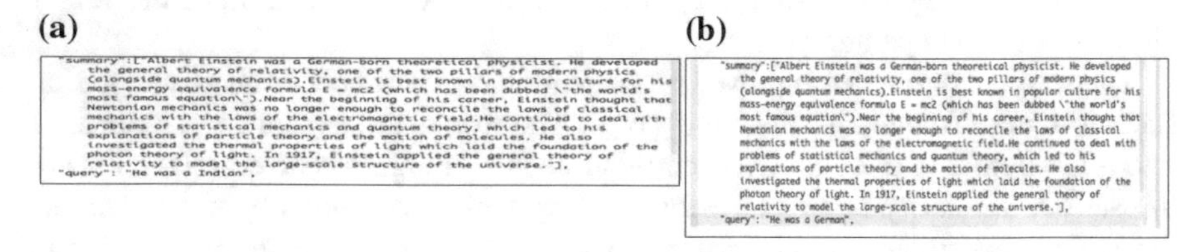

Fig. 5 **a** Two input sentences that are not similar. **b** Comparing input sentences that are similar

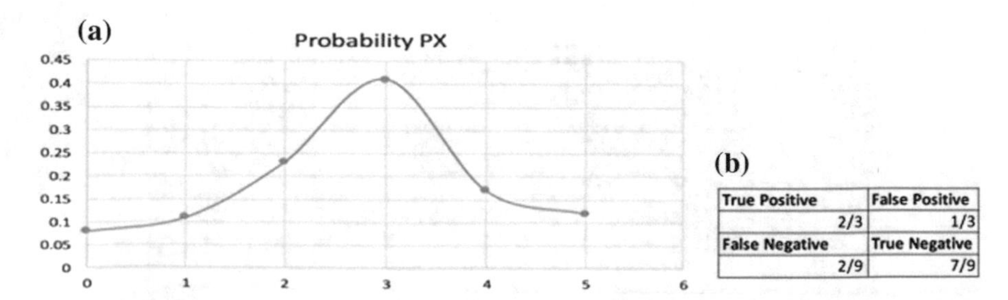

True Positive	False Positive
2/3	1/3
False Negative	**True Negative**
2/9	7/9

Fig. 6 **a** Bell curve of number of questions answered versus probability of user giving correct answers, i.e. true positives. **b** Confluence Matrix

Thus, as shown in Fig. 6b we calculate the varying values of true and false positives and negatives depending on multiple factors, including error induced in the various blocks, errors introduced by users, complexity of generated questions and shortcomings in algorithms. In order to overcome these flaws, the system has to become more accepting towards false negatives. To improve the level of security provided by the system, improvements can be made in the latent semantic indexing model by increasing the cosine similarity. However, this adversely affects the system's capacity to identify true positives, increasing the amount of false negatives.

Thus, further research needs to be done to make the model stricter to reduce chances of false positives, while not reducing the identification capability of true positives. The LSI module could be made more efficient, it being most vulnerable to error, if made more foolproof; then, the total induced error in the system would decrease significantly. This helps in two ways, by reducing the values of false positives generated as well as reducing the chances of false negatives.

4 Conclusion

In this paper, we have presented an alternative way to perform identity verification without the use of passwords. A cognitive way to identify a user is to use his personality quirks, characteristics and activities to verify his identity. The goal here is to generate a dynamic set of non-repetitive questions, so that replay attacks are deterred. Better questions can always be generated by improving the data set and understanding how to interpret different psychological traits of users. We realize that the system is time-consuming to log into in comparison to other systems; however, this system doesn't place any burden on the users, such remembering complex alphanumeric passwords or the like. The end goal is to create a general purpose system or algorithm that can generate questions to authenticate users by gathering data from appropriate, reliable sources. We can justly say that we have accomplished the same. For future research, there lies a scope of increasing the security level of the system by reducing the probability of false positives and false negatives.

References

1. OGorman, Lawrence, Amit Bagga, and Jon Bentley. 2004. Financial cryptography. In *Call center customer verification by query-directed passwords*. Springer: Berlin. https://doi.org/10.1007/978-3-540-27809-2_6
2. Furnell, S.M., I. Papadopoulos, and Paul Dowland. 2004. A long-term trial of alternative user authentication technologies. *Information Management and Computer Security* 12 (2): 178–190. https://doi.org/10.1108/09685220410530816.
3. Rabkin, A. 2008. Personal knowledge questions for fallback authentication: Security questions in the era of Facebook. In *SOUPS 2008 - Proceedings of the 4th symposium on usable privacy and security*, 13–23.

4. Brown, Alex, and Myriam Abramson. 2015. Twitter fingerprints as active authenticators. In *2015 IEEE international conference on data mining workshop (ICDMW)*, 58-63, Atlantic City, NJ.
5. Nosseir, Ann, Richard Connor, and Karen Renaud. 2006. Question-based group authentication. In *Proceedings of the 18th australia conference on computer-human interaction: design: activities, artefacts and environments*. ACM.
6. Bachmann, Michael. 2014. Passwords are dead: alternative authentication methods. In *2014 IEEE Joint intelli-gence and security informatics conference (JISIC)*, IEEE.

A Novel Key Predistribution Scheme for Minimum Angle Routing in Wireless Sensor Network

Monjul Saikia and Md. Anwar Hussain

Abstract The main objective of a sensor network is to forward the information regarding any event detected by a sensor node. For efficient routing of the packets through the network, effective and portable routing is essential. We find that a minimum angle-based routing algorithm is highly efficient with low computational cost. In addition to a fast routing mechanism, security is also a major concern of any sensor network. Symmetric key encryption is widely used for this purpose, with secret keys loaded prior to deployment. In this chapter, we propose a novel key predistribution scheme especially applicable for minimum angle-based routing.

Keywords Routing · Min angle routing · Security · KPS · Connectivity Resilience

1 Introduction

Secure routing of information in wireless sensor networks has been broadly studied by researchers in recent years. Protection of the sensed information over a vulnerable environment is essential. Many routing algorithms have been proposed to date. During deployment of any sensor network, the mechanism for routing is also bootstrapped in advance. Therefore, appropriate security measures need to be loaded prior to deployment. Instead of a single route in a base station (BS), multi-path routing gives advantages in various aspects such as congestion control, node failure, etc. Stavrou et al. [1] discussed the advantages of secure multi-path routing protocols in wireless sensor networks (WSNs). In [2, 3] various routing algorithms designed

M. Saikia (✉) · Md. Anwar Hussain
Department of Computer Science and Technology, North Eastern Regional Institute of Science and Technology, Nirjuli 791109, Arunachal Pradesh, India
e-mail: monjuls@gmail.com

A. N. Krishna et al. (eds.), *Integrated Intelligent Computing, Communication and Security*, Studies in Computational Intelligence 771,
https://doi.org/10.1007/978-981-10-8797-4_32

especially for wireless sensor networks were discussed. Saikia et al. [5] proposed a secure, energy-aware, multi-path routing scheme with key management for wireless sensor networks. Key predistribution is not a new technique for security in WSNs; it is used widely, and many schemes have been proposed to enhance its performance. A key predistribition system (KPS) can be made more efficient and robust when post-deployment knowledge is known prior to deployment. If the routing mechanism that will be used for the sensor network is known earlier, it provides advantages in designing for a special KPS. Chakraborty et al. [4] proposed a simple joint routing and scheduling algorithm for a multi-hop wireless network using minimum angle, with a reference line to the BS. The Minimum Angle-Centralized Server (MA-CS) is one of the most efficient routing algorithms in wireless sensor networks. In this approach data are routed through intermediate nodes by computing minimum angle to the base station with respect to a reference line to the BS. In our proposed work we try to develop a key predistribution scheme that is exclusively applicable for the minimum angle centralized scheduling routing algorithm, given the known coordinates of sensor nodes. Blom [6] discussed an optimal class of symmetric key generation systems that can be effectively applied in wireless sensor networks. Combinatorial design-based KPS performs well in various location independent topologies [7, 8]. Liu et al. proposed a scheme for a location-based, pair-wise key management scheme, static wireless sensor networks [9]. Chan et al. designed a KPS that is based on probability of attack on sensor nodes and gives a very good resilience to the network [10]. Du et al. [11] proposed a new technique for a key management scheme using deployment knowledge.

1.1 Organization of the Chapter

In Sect. 1 we give a basic introduction and explain the usefulness of efficient routing and KPS for the purpose of security. In Sect. 2 we discuss the fundamental working of the minimum angle-based routing protocol. Section 3 gives our proposed idea for key distribution, namely KPS: MA-CS. In Sect. 4 we give the basic algorithm for implementing the proposed KPS. Section 5 demonstrates some experimental results and concludes the chapter.

2 Minimum Angle-Based Routing Protocol for WSN

Minimum angle routing in a WSN is a routing protocol that is based on minimum angle with the base station to any neighboring sensor node. The major computation needed to find the shortest path to a given node is to find the minimum angle among all the nodes in its communication range. This can be done with the help of a vector *dot* product. The *dot* product of a vector is a useful and efficient way for computing

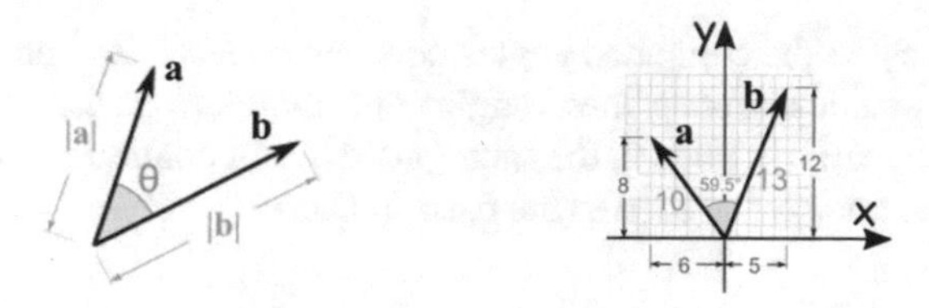

(a) Vector $\vec{a}$ and $\vec{b}$ with angle θ (b) An example of dot product

Fig. 1 Dot product explanation for finding the angle between two vectors

the angle between two vectors. If $\vec{a}$ and $\vec{b}$ are two vectors, then the dot product of $\vec{a}$ and $\vec{b}$ is $\vec{a} \cdot \vec{b} = |\vec{a}| \times |\vec{b}| \times \cos\theta$, where $|\vec{a}|$ is the magnitude of vector $\vec{a}$, $|\vec{b}|$ is the magnitude of vector $\vec{b}$ and θ is the angle between $\vec{a}$ and $\vec{b}$.

In Fig. 1a shows two vectors $\vec{a}$ and $\vec{b}$ with angle θ. The dot product facilitates finding the value of θ. An example is shown in Fig. 1b, where coordinates of the vector end points are given. From the given coordinates $\vec{a} \cdot \vec{b}$ can be evaluated as $-6 \times 5 + 8 \times 12 = 66$ and the magnitude of vector $\vec{a}$ and $\vec{b}$ are $\sqrt{(-6)^2 + 8^2} = 10$ and $\sqrt{5^2 + 12^2} = 13$. Therefore, θ can be computed as $\theta = \cos^{-1}(\frac{\vec{a} \cdot \vec{b}}{|\vec{a}| \times |\vec{b}|}) = \cos^{-1}(\frac{66}{10 \times 13}) = 59.5°$.

Figure 2 shows the minimum angle routing algorithm. The algorithm is simple and yet efficient. It computes the angle with a reference line to the base station with every node within its coverage range. Then, the path with the minimum angle is chosen in every step. The figure shows the node S trying to find a path to the base station. It first computed angle with every neighboring nodes i.e. 1, 2, 3 and 4. Let

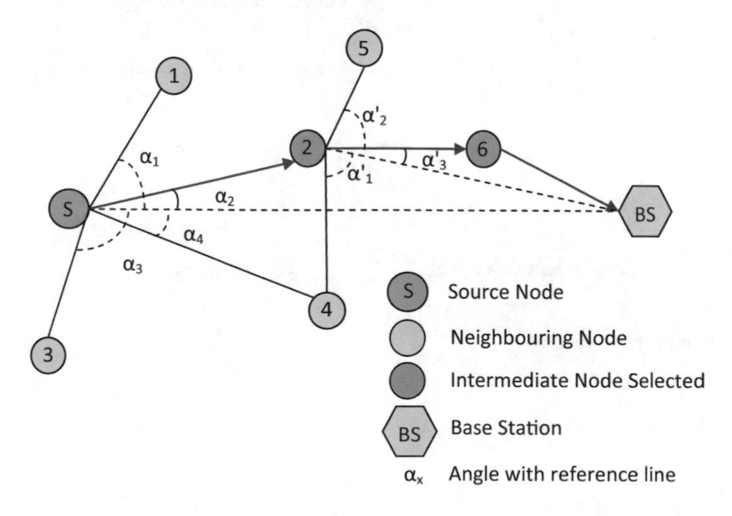

Fig. 2 Minimum angle routing

α_1, α_2, α_3 and α_4 be the computed angles between nodes 1, 2, 3 and 4 respectively. Let α_2 be the smallest among these angles and therefore packet will be routed to node 2. Consequently, in node 2, the same process will continue and the packet will be forwarded to node 6 and then to the base station.

3 Proposed Idea for KPS with Minimum Angle Routing

Our key distribution idea is to preload the minimum required keys to the sensor nodes prior to deployment, keeping in mind that the routing algorithm to be followed is minimum angle-based routing. At this moment of time we have only the expected coordinates of each sensor node. Therefore, with the help of these coordinates and the BS static coordinate we need to perform the key distribution algorithm. The basic idea of the algorithm is to assign shared keys to a sensor node which is within its communication range, and makes a minimum angle with the reference line to the base station. With the known coordinates, the algorithm finds all the neighboring nodes, finds the angle with the reference line to the BS, and then assigns a shared to the node which makes a minimum angle. In Fig. 3, node C and node 2 share a common key, key#1, as it makes a minimum angle with the base station reference line. With further enhancement, instead of keeping a key shared with only one node, we prioritized a set of k keys in ascending order of angles they make with the reference line.

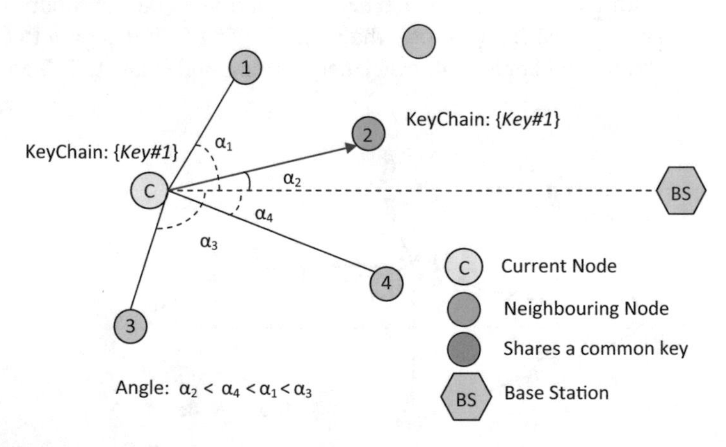

Fig. 3 Minimum angle routing

4 Design of the Algorithm

The algorithm takes the total number of sensor nodes, coordinates for each sensor node, BS coordinates, area of deployment and coverage range as input and produces a key chain as output for each of the sensor nodes. The procedure *findMinAngle()* first finds the list of nodes that makes a minimum angle with the reference line to BS. The *minAngleList* is then passed to a procedure *minAngleKPS()* that assigns keys to each of the nodes. The pseudo-code for the procedure *minAngleKPS* is given in Algorithm 1.

Algorithm 1 minAngleKPS

Require: $X, Y, BSx, BSy, Coverage$
Ensure: $keyChain$
1: *minAngleList* ← call procedure*findMinAngle(X,Y,BSx,BSy,Coverage)*
2: *listlen* ← *len(minAngleList)*
3: *ringSize* ← *maxOccuranceNode(minAngleList)*
4: *keyID* ← 1
5: $i \leftarrow 1$
6: **for** $i < listlen$ **do**
7: $l \leftarrow 1$
8: **while** $keyChain_i(l)$ NOT NULL **do**
9: $l \leftarrow l + 1$.
10: **end while**
11: $keyChain_i(l) \leftarrow keyID$
12: $m \leftarrow 1$
13: $k \leftarrow nextindex(minAngleList)$
14: **while** $keyChain_k(m)$ NOT NULL **do**
15: $m \leftarrow m + 1$.
16: **end while**
17: $keyChain_k(m) \leftarrow keyID$
18: $i \leftarrow i + 1$
19: **end for**
20: **return** *KeyChain*

5 Experimental Results

Let us start with a simple example with 30 nodes, each with their given expected location in an area of 100×100 sq. units (randomly generated coordinate values are as given in Table 1), and let us assume that the base station is fixed at coordinates (50, 50), i.e., the middle of the target area. Consider the coverage range of each sensor node to be 30 units.

Table 1 Randomly generated coordinates of sensor nodes

Node	X	Y	Node	X	Y	Node	X	Y
1	42	10	11	42	99	21	81	2
2	73	43	12	69	75	22	97	68
3	1	96	13	21	29	23	32	22
4	31	54	14	88	79	24	70	27
5	15	70	15	3	11	25	88	50
6	10	32	16	68	45	26	90	6
7	19	69	17	42	91	27	9	58
8	35	84	18	56	30	28	4	15
9	40	2	19	15	29	29	17	59
10	54	76	20	20	14	30	88	70

Table 2 Key chain for each of the sensor nodes

Node	Key chain	Node	Key chain	Node	Key chain
BS	{26, 27, 28, 29, 30}	11	{8}	22	{17}
1	{1, 7}	12	{9, 11, 25}	23	{15, 18}
2	{26}	13	{4, 10, 12, 23}	24	{16, 19, 21}
3	{2}	14	{11}	25	{20}
4	{3, 5, 10, 14, 22, 24, 27}	15	{12}	26	{21}
5	{2, 3}	16	{19, 20, 29}	27	{22}
6	{4}	17	{8, 13}	28	{23}
7	{5}	18	{1, 18, 30}	29	{24}
8	{6}	19	{14}	30	{17, 25}
9	{7}	20	{15}		
10	{6, 9, 13, 28}	21	{16}		

After successful execution of the algorithm, the key chain for each sensor node along with the BS is found as given in Table 2. Equivalently, a key graph with the assigned keys is drawn, as shown in Fig. 4.

Similarly, various experiments were performed, and their key graphs are shown in Fig. 5.

For performance evaluation of the scheme, we evaluated the average key chain length in different experiments. The Table 3 shows the average key chain length in the range of a sensor network. The scheme is found to be highly efficient, considering storage overhead of sensor nodes.

Resilience is the measure for network resistance against attack. It reflects the fraction of the network still operating when a fraction of the network is compromised. This scheme is found to be highly resilient to adversaries, which is shown

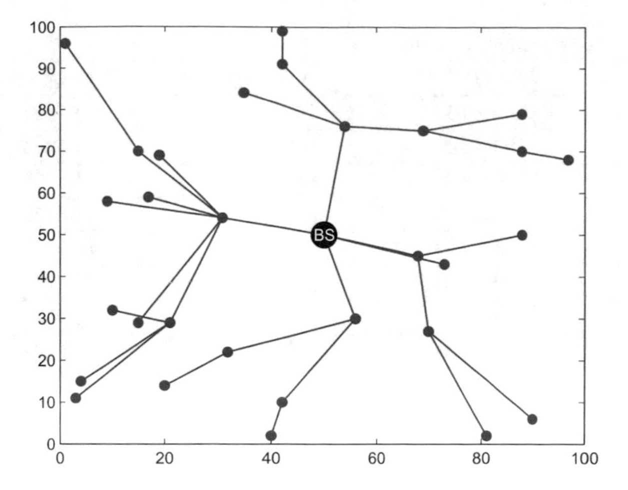

Fig. 4 Key graph with KPS with minimum angle routing for N = 30 nodes

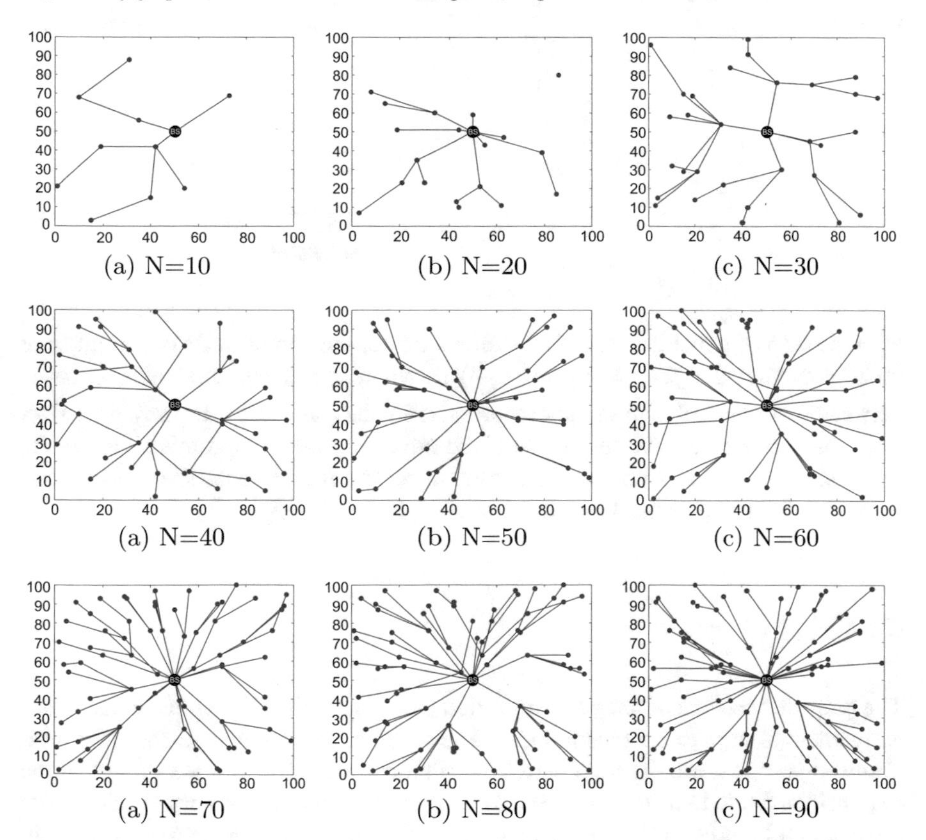

Fig. 5 Key graph for various experiments [Area = 100 × 100; Coverage = 30]

Table 3 The average key chain length

N	Area (sq. m)	Coverage (m)	Avg. keychain size
10	100×100	30	1.7000
20	100×100	30	1.5000
30	100×100	30	1.8333
40	100×100	30	1.8250
50	100×100	30	1.7000
60	100×100	30	1.7667
70	100×100	30	1.7571
80	100×100	30	1.8375
90	100×100	30	1.7111
100	100×100	30	1.8000

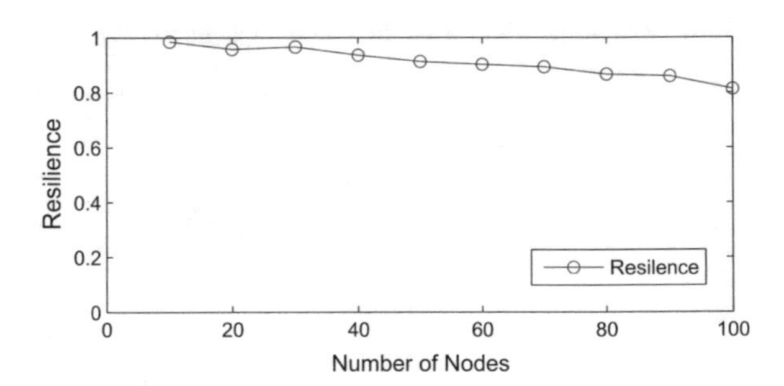

Fig. 6 Resilience of the network

with the help of a plot in Fig. 6. The resilience after the compromise of s number of nodes is evaluated as $fail_s = (1-(1-\frac{k}{|K|})^s)^q$, where, q is the number of keys shared between two randomly-chosen uncompromised nodes, and $|K|$ is the key-pool size. This leads to the intuitive result that if the adversary has compromised one node, learning k keys, and two randomly-chosen uncompromised nodes share $q = 1$ key, then the probability of the adversary knowing the key is $fail_l = \frac{k}{|K|}$.

6 Conclusion

The proposed scheme of key predistribution performs well, as it avoids unnecessary assignment of keys to the sensor nodes. The scheme is perfectly applicable for minimum angle-based routing for a wireless sensor network. The average storage of the key chain is found to be minimal, and resilience of the network is very efficient. The scheme can be effectively implemented for 3D wireless sensor networks, as well. Various experiments were performed and results were shown here.

References

1. Stavrou, Eliana, and Andreas Pitsillides. 2010. A survey on secure multipath routing protocols in WSNs. *Computer Networks* 54 (13): 2215–2238.
2. Akkaya, K., and M. Younis. 2005. A survey of routing protocols in wireless sensor networks. *Ad hoc Network, Elsevier Journal* 3 (3): 325–349.
3. Abd-El-Barr, M.I., M.M. Al-Otaibi, and M.A. Youssef. 2005. Wireless sensor networks. part II: routing protocols and security issues. In *Canadian conference on electrical and computer engineering*, 69–72, May 14.
4. Chakraborty, Ishita, and Md. Anwar Hussain. 2012. A simple joint routing and scheduling algorithm for a multi-hop wireless network. In *2012 International conference on computer systems and industrial informatics (ICCSII)*, 1–5. IEEE.
5. Saikia, Monjul Uddipta, K. Das, and Md. Anwar Hussain. 2017. Secure Energy Aware Multipath Routing With Key Management in Wireless Sensor Network. *Published in 4th International Conference on Signal Processing and Integrated Networks (SPIN 2017)*, February 2–3, Amity University Noida, India.
6. Blom, R. 1984. An optimal class of symmetric key generation systems. In *Theory and application of cryptographic techniques (EUROCRYPT 84)*.
7. Saikia, Monjul, and Md. Anwar Hussain. 2017. Combinatorial group based approach for key predistribution scheme in wireless sensor network. In *International conference on computing, communication and automation (ICCCA2017)*, 502–506, May 5–6, Galgotias University Noida, India.
8. Camtepe, S.A., and B. Yener. 2007. Combinatorial design of key distribution mechanisms for wireless sensor networks. *IEEE-ACM Transactions on Networking* 15: 346–358.
9. Liu, D., and P. Ning. 2013. Location-based pairwise key establishments for static sensor networks. In *Proceedings of the first ACM workshop security ad hoc and sensor networks.*
10. Chan S., R. Poovendran and M. Sun. 2005. A key management scheme in distributed sensor networks using attack probabilities. In *IEEE global communications conference, exhibition industry forum (Globecom).*
11. Du W., J. Deng, Y.S. Han, S., Chen and P.K. Varshney. 2004. A key management scheme for wireless sensor networks using deployment knowledge. In *Proceedings of the 24th IEEE conference on computer communications (INFOCOM).*

An Intelligent Algorithm for Joint Routing and Link Scheduling in AMI with a Wireless Mesh Network

A. Robert singh, D. Devaraj and R. Narmatha Banu

Abstract Advanced metering infrastructure (AMI) of smart grids is a network that integrates the consumer directly with the smart grid communication infrastructure. AMI generates a meter that reads the datagram periodically. This unpredictable data flow needs link scheduling between the smart meter and the data collector. The AMI network is scalable because any number of smart meters can be added, as well as removed. So, the routing method should ensure reliability even in node/link loss in the path. An intelligent link scheduling algorithm is discussed in this chapter. This method addresses the scheduling of multiple links in a single slot with assurance for immediate reception of the packets. This method is applied on an AMI network that is deployed using a wireless mesh network (WMN). The paths between the smart meters and data aggregator are identified using a hybrid wireless mesh routing protocol (HWMP). The results show that the proposed link scheduling method can ensure faster packet delivery with a desirable time slot length according to the available traffic.

A. Robert singh (✉)
Department of Computer Science and Engineering,
Kalasalingam University, Krishnankoil, India
e-mail: robertsinghbe@gmail.com

D. Devaraj
Department of Electrical Engineering, Kalasalingam University, Krishnankoil, India
e-mail: deva230@yahoo.com

R. Narmatha Banu
Department of Electrical Engineering, Velammal College of Engineering and Technology,
Madurai, India
e-mail: buvanriya@gmail.com

A. N. Krishna et al. (eds.), *Integrated Intelligent Computing, Communication and Security*, Studies in Computational Intelligence 771,
https://doi.org/10.1007/978-981-10-8797-4_33

1 Introduction

The smart grid [1] is a new generation of power system that has two types of infrastructure: the electric grid and the communication infrastructure. The entire communication infrastructure consists of three parts [2]: the home area network (HAN), which interconnects households or devices within industry in short-range coverage; the neighborhood area network (NAN), which acts as an interface between HAN and a central utility center; and the wide area network (WAN), which connects two or more NANs from different localities. Advanced metering infrastructure (AMI) [3] is a subsystem of the smart grid which automates reading and the billing processes of power consumption.

The AMI needs a two-way communication setup between a group of smart meters and the utility center through a data aggregator. Per American National Standards Institute (ANSI) Standard C12.19 [4], AMI traffic should be delay-tolerant, periodic in most cases, or event-based on a few cases with small burst size. The critical aspects [5] that influence these network characteristics are the large number of devices, high overheads and huge ups and downs in loads. Any routing method used to deliver AMI packets has to apply fair traffic scheduling to avoid packet loss due to unavailability of intermediate node(s)/links or by bandwidth unavailability or due to ineffective path assignment. A wireless mesh network (WMN) [6] is a communication network that contains a number of radio nodes with maximum connectivity. The features of WMN are described by the Institute of Electrical and Electronics Engineers (IEEE) Standard 802.11s [7]. WMN is used to design an AMI network because of the characteristics [8, 17] such as ease and low cost maintenance, ability of self-configure, scalability, flexibility and robust security.

Gabale et al. [9] classify scheduling methods for a WMN based on scheduling control, channel access type, antenna type and type of topology. Ganjali et al. [10] compare single path and multipath routing in ad hoc networks in terms of load balancing. The algorithm using WMN topology in [11] describes link selection, routing and scheduling of activation of multiple paths concurrently. Akyildiz et al. [12] analyze the requirement to use multipath routing for load balancing in WMN. In our previous work [16], a load balanced routing scheme was introduced between a smart meter and data collector. This is an enhanced HWMP algorithm with load balancing on high demand links.

Gharavi et al. [13, 15] have proposed a multi gateway routing and scheduling for AMI communication using WMN. The paper proposes a combination of routing through multi-gate, real-time traffic scheduling, and multichannel (MC)-aided wireless mesh routing. Xiaoheng et al. [19] summarize the functional requirements to deploy various smart grid applications using WMN with IEEE 802.11s standard.

In this chapter, a novel link scheduling algorithm is proposed to schedule the links of the paths identified by the HWMP routing protocol in an AMI network. The proposed method shows that more number of links can be simultaneously activated to manage concurrent transmission and reception processes. The chapter is organized as follows: Sect. 2 briefly explains the architecture of the AMI network using WMN.

The scheduling method is discussed in detail in Sect. 3. Section 4 analyzes the performance of the proposed scheduling method with HWMP. The conclusion of the work is drafted in Sect. 5.

2 AMI Architecture

The architecture of AMI using WMN topology is given in Fig. 1. The AMI network contains a number of smart meters that are the interface between HAN and NAN. In this network, smart meters are mesh points and data collector are the gateways connected by mesh topology with random connectivity. The data collectors are connected to the utility center through a wired network. HWMP [14], the default routing protocol for WMN that finds the path between mesh clients and gateway through mesh routers based on the airtime metric of each link as given in Eq. (1).

$$C_a = \left[O + \frac{B_t}{r}\right] \frac{1}{1 - e_f} \tag{1}$$

where C_a is airtime metric, O is overhead latency, B_t is test packet size, r is the data rate in Mb/sec, and e_f is the frame error rate.

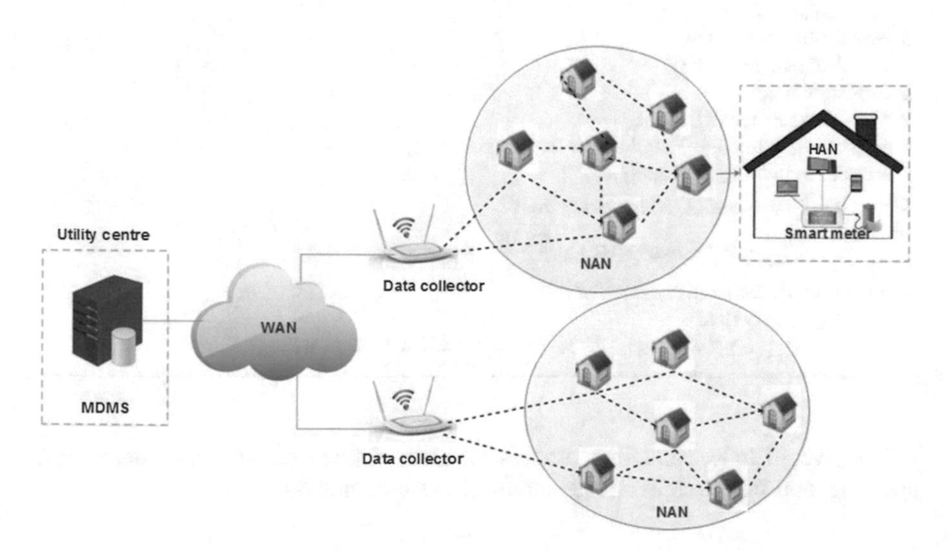

Fig. 1 AMI architecture

3 Proposed Link Scheduler for AMI Traffic

Let the AMI network be a graph $G = (V, E)$ that contains a set of vertices V and set a of edges E. G can be assumed as an adjacency matrix that represents the link between the smart meters. Let $n_i \geq 1$ be the number of links that a smart meter i can handle at the same time. Let N_i be the set of neighbors of smart meter i. Let $e_{ij} \in E$ be the link between the pair of smart meters i and j, where $i,j \in V$. In practice, the antenna of each smart meter cannot handle N_i links simultaneously. To manage this, a unique link scheduling algorithm is necessary. There are two assumptions in this problem: all smart meters are using the same channel, and a smart meter cannot send/forward a packet while a receiving process is going on and vice versa. The objective of the proposed link scheduling algorithm is to place the maximum number of links in each slot. The link scheduler algorithm is explained in Algorithm 1.

Algorithm 1: AMI link scheduler

Input : Set of paths P between various smart meters and data collectors, M and G
Output: Scheduler table for links
1 Initialization: Set $W_0 = 1$
2 Update smart meter status (M, t_i): $\forall m \in M$ update $W^m_{t_i}$
3 **Weight adjustment**$(M, t_i + 1)$**:**
4 **if** $Slots_{wasted}! = 0$ **then**
  Multiplicative_decrease();
5 **else if** $(Avg_q > 0)$ **then**
  Additive_increase();
6 **Schedule links:**
7 $\forall m \in M$ schedule to frame t_{i+1}
8 Repeat step 2 through 7 until all the queues are empty
9 **Function** *Additive_increase()*
10 **if** $\frac{Avg_q}{Max_q} \neq 0$ *and* Q_n *is not empty* **then**
   $W^n_{t+1} = W^n_t + max(1, |log_{10}(\frac{Avg^t_q \times 100}{Max^t_q})|)$
11 **Function** *Multiplicative_decrease()*
12 **if** Q_n *empty* **then**
   $W^n_{t+1} = W^n_t - max(1, |W^n_t \times (1 - \frac{Amount\ of\ unused\ slots}{2 \times Amount\ of\ allocated\ slots})|)$

The overall link scheduling process consists of three parts: smart meter state updating, real-time weight update and local election method.

3.1 Smart Meter State Update

Every scheduler frame needs the status of the smart meter's queue to manage the scheduling process and the slot usage information that is to decide the number of slots to be assigned in the future. The slot usage information will maintain the statistics of used slots and wasted slots of the scheduler frame by the current smart meter. HWMP uses a minimum airtime metric as the parameter for next hop selection. This property allows the link with the minimal airtime metric to participate in more paths. So, the queue will always have more packets to forward. According to the packets waiting in the queue, the weight can be increased or decreased. The increase is done in terms of addition, and the decrease is done exponentially based on the additive-increase/multiplicative-decrease (AIMD) concept [18]. The initial weight for all the nodes is assigned as zero. Based on the amount of traffic waiting in the queue, the weight is assigned to the node. The packets are transmitted on a first-in/first-out basis. Thus, the corresponding link will be assigned in that slot. There is a maximum limit S for each slot that is the maximum number of links allowed in the slot.

3.2 *Real-Time Weight Update*

The weight value is assigned as one for the first slot assignment. Then all the nodes are evenly allocated with slots based on their unique identifier. Then the queue status of each node is examined and they are divided into two groups:

(i) The nodes using the assigned slots completely and with more packets in the queue. These nodes need an additive increment in their weight as given in Eq. (2).

$$W_{t+1}^{n} = W_{t}^{n} + max\left(1, \left|log_{10}(\frac{Avg_{q}^{t} \times 100}{Max_{q}^{t}})\right|\right) \tag{2}$$

Avg_q^t is the average number of packets per slot entered into the queue q, and Max_q^t is the maximum number of packets until the time slot t. Thus, based on the new weight the number of slots will be assigned. The weight is incremented based on the queue occupancy percentage until the previous schedule frame.

(ii) The nodes that have no packet to send within the given slot, and part of or the whole slot is wasted, meaning that the weight can be decreased exponentially as given in Eq. (3).

$$W_{t+1}^{n} = W_{t}^{n} - max\left(1, \left|W_{t}^{n} \times (1 - \frac{Amount\ of\ unused\ slots}{2 \times Amount\ of\ allocated\ slots})\right|\right) \tag{3}$$

The amount of the slot for both wasted and allocated is calculated as the rate of slot between the range (0,1).

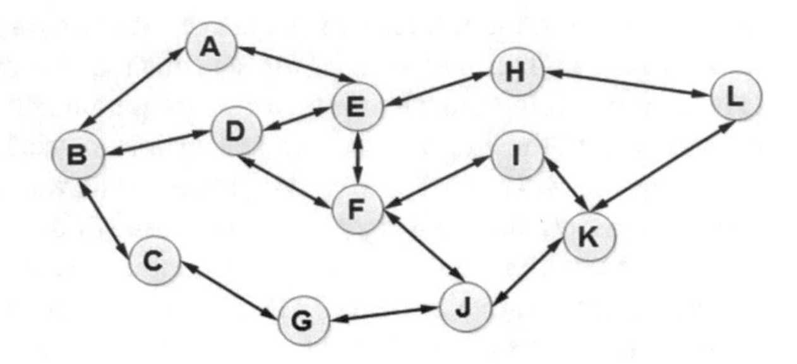

Fig. 2 AMI-link scheduler example

3.3 *Local Election Method*

The link allocation is done based on the local election method using the weights of the nodes. All the nodes that have a non-empty queue participate in the election process. The nodes with maximum weight are assigned to the first slot of the frame. If there is only one node with maximum weight, it is given higher priority, and the others in the order are scheduled. The entire slot can be allotted to a single node if the queue size is greater than or equal to the maximum limit S.

Consider an AMI network with 12 nodes as shown in Fig. 2, with random mesh topology. Each pair of neighbors is connected with a bidirectional link which is considered as a combination of two unidirectional links in opposite directions. Consider three paths waiting for link scheduling, say B-D-E-H-L, A-E-F-J-K, B-A-E-F-I and C-G-J-K. The limit for the slot S is 2. The length of the scheduler frame is 3. Initially, the weight of each node is assigned as 1. Now, nodes B,A,C have packets in their queues. There is no need for an exponential decrease in the weight, as it is the first

Table 1 Weight update for smart meters

Smart meter	Weight			
	Round 1	Round 2	Round 3	Round 4
A	1	1	1	0
B	2	2	1	0
C	1	1	0	0
D	1	1	0	0
E	1	1	2	2
F	1	0	1	1
G	1	1	1	0
H	1	0	0	1
I	1	0	0	0
J	1	0	0	1
K	1	0	0	0
L	1	0	0	0

Frame 1			Frame 2			Frame 3			Frame 4		
Slot 1	Slot 2	Slot 3	Slot 1	Slot 2	Slot 3	Slot 1	Slot 2	Slot 3	Slot 1	Slot 2	Slot 3
B-D	A-E		A-E	E-F		E-H	F-J		H-L	F-I	
B-A	C-G		D-E	G-J		E-F	J-K		J-K		

Fig. 3 Example schedule

iteration. But node B has two packets to transmit in its queue. This situation is handled by updating their weight as given in Table 1. The entire scheduling process for all four paths is given in Fig. 3. Each frame is assigned with three slots, and each slot is assigned with a maximum of two nodes. This method ensures minimum queuing delay for each packet in the nodes which have high demand due to their low airtime links.

4 Performance Analysis

The AMI network is simulated with an OPNET 14.5 modeler in Proto-C language for implementing the link scheduler algorithm. The AMI communication network is tested with 48 smart meters and four data collectors. Each data collector acts as the destination for a group of 12 smart meters. There is no restriction to establish communication between groups, but the destination belongs to the same group. The network parameters to deploy AMI are listed in Table 2. The time slot is assigned with a span of 5 ms and a maximum limit of five links per slot. Each scheduler frame can accommodate four slots. The proposed method is tested with different input data rates that are recommended by the previous studies [3]. The allowed input

Table 2 Network parameters

Parameter	Value
Number of smart meters	48
Number of data collectors	4
Propagation model	Free space path loss model
Routing protocol	HWMP
Packet size	512 bytes
Transmission energy	17.4 mA
Receiving energy	19.7 mA
Simulation time	3600 s
Packet generation interval	900 s
Traffic type	CBR
Transmission range	300 m

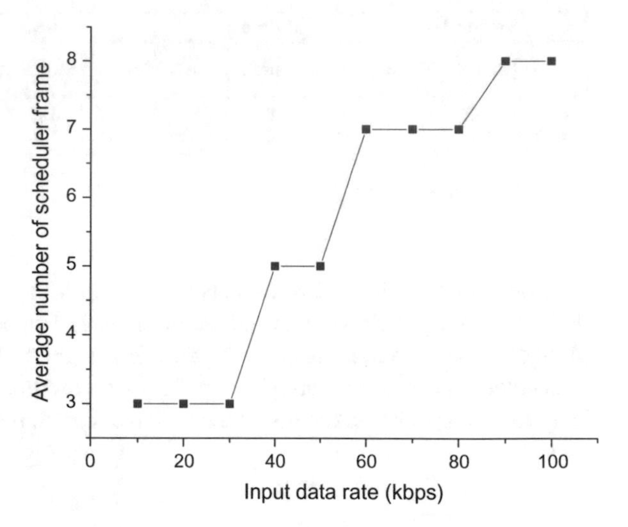

Fig. 4 Input data rate versus average number of scheduler frames

data rate is 10–100 kbps/node. With reference to this range, the average number of scheduler frame is examined for different input data rates as given in Fig. 4. As the input data rate increases, the number of frames required to place all the slots for link scheduling increases. When more meter reading information is in transit, the network needs higher bandwidth support for better transmission.

Per the standards, the path can take 2–15 s to complete successful AMI reading information. If a reasonable time slot is assigned, then a link that needs activation for two consecutive transmit and receive cycles may be completed within a single slot by leaving the packets in the recipient's queue. Different time slots in the range of 4 ms to 20 ms are considered with an interval of 4 ms against average delay. The result given in Fig. 5 shows that the delay is gradually reduced up to 16 ms. This shows that longer time slots have some unused slots even after completing the transmission. The amount of unused time slots for various slot length is shown in Fig. 6. In larger AMI networks, longer time slots can be used to reduce the amount of an unused time slot to ensure that the scalability of an AMI network can be managed by choosing the appropriate time slot length.

Slot degree is the maximum number of allowed links in a given time slot. The AMI network generates packets in a uniform interval that causes increasing traffic. In such cases, to accommodate more links in the slot, slot degree must be gradually increased. Figure 7 compares the average end-to-end delay of the transmission under various input data rates and different degrees of slot. While increasing the input data rate, more reading information will be in transit.

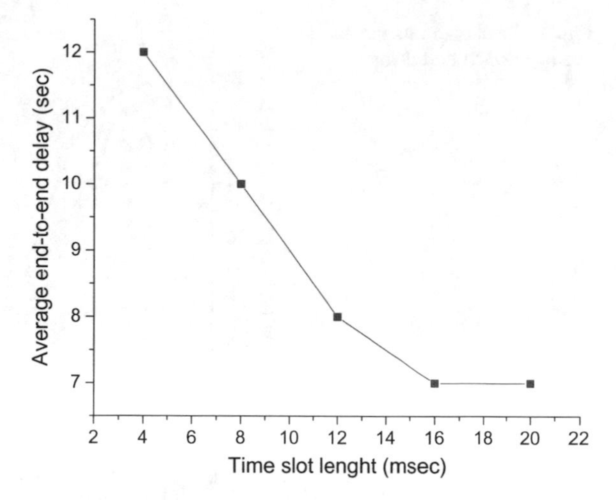

Fig. 5 Time slot length versus delay

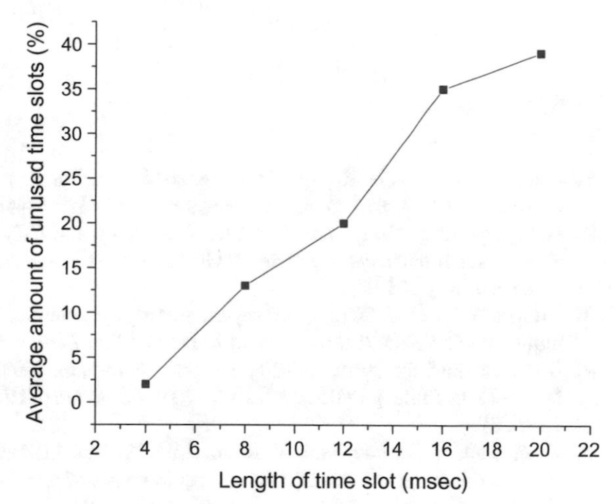

Fig. 6 Length of time slot versus unused time slot

5 Conclusion

In this chapter, a novel link scheduling algorithm has been proposed for scheduling the links of the paths identified by the HWMP routing protocol in an AMI network. The proposed method shows that more links can be activated simultaneously to manage concurrent transmission and reception processes. The unpredictable nature of AMI traffic is managed by choosing the appropriate time slot length and slot degree. The input data rate influences the unpredictability of the AMI traffic. A standard record of input data rate against the desirable time slot length and slot degree can be used in practice to maintain uninterruptible AMI meter reading packet delivery.

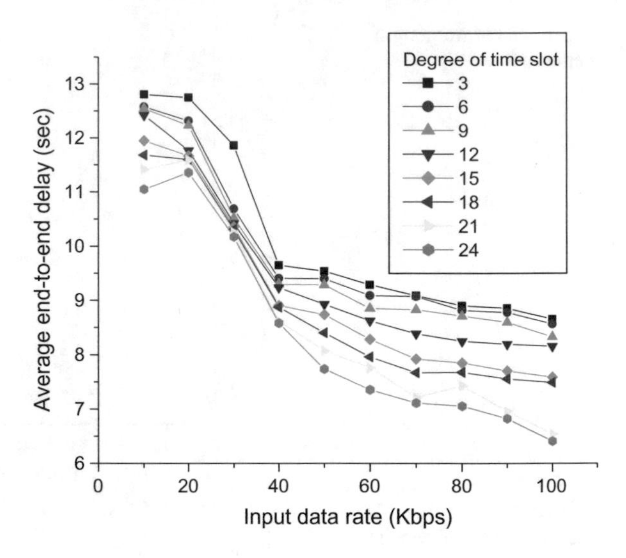

Fig. 7 Input data rate versus average end-to-end delay

References

1. Fang, X., S. Misra, G. Xue and D. Yang. 2012. Smart grid - the new and improved power grid: a survey, *IEEE Communications Surveys and Tutorials* 14(4): 944–980. (Fourth Quarter)
2. Gao, Jingcheng, Yang Xiao, Jing Liu, Wei Liang, and C.L. Philip Chen. 2012. A survey of communication/networking in Smart Grids. *Future Generation Computer Systems* 28 (2): 391–404. (February 2012)
3. Gungor, V.C., et al. 2013. A survey on smart grid potential applications and communication requirements. *IEEE Transactions on Industrial Informatics* 9 (1): 28–42.
4. IEEE standard for utility industry metering communication protocol application layer (End Device Data Tables), In IEEE Std 1377–2012 (Revision of IEEE Std 1377–1997), 1–576. (Aug 10 2012)
5. Khan, Reduan H., and Jamil Y. Khan. 2013. Review article: a comprehensive review of the application characteristics and traffic requirements of a smart grid communications network. *Computer Networks* 57 (3): 825–845. (February 2013)
6. Wang, Xudong, and Azman O. Lim. 2008. IEEE 802.11s wireless mesh networks: framework and challenges. *Ad Hoc Networks* 6 (6): 970–984. (August 2008)
7. Hiertz, G.R., et al. 2010. IEEE 802.11s: The WLAN Mesh Standard. *IEEE Wireless Communications* 17 (1): 104–111.
8. Kulkarni, P., S. Gormus, Z. Fan, and F. Ramos. 2012. AMI mesh networks - a practical solution and its performance evaluation. *IEEE Transactions on Smart Grid* 3 (3): 1469–1481.
9. Gabale, V., B. Raman, P. Dutta, and S. Kalyanraman. 2013. A classification framework for scheduling algorithms in wireless mesh networks. *IEEE Communications Surveys and Tutorials* 15 (1): 199–222. (First Quarter)
10. Ganjali, Y., and A. Keshavarzian. 2004. Load balancing in ad hoc networks: single-path routing versus multi-path routing. *IEEE INFOCOM* 2: 1120–1125.
11. Draves, R., J. Padhye, B. Zill. 2004. Routing in multi- radio, multi-hop wireless mesh networks. *MobiCom '04*, Philadelphia, PA. (Sept 2004)
12. Akyildiz, Ian F., Xudong Wang, and Weilin Wang. 2005. Wireless mesh networks: a survey. *Computer Networks and ISDN Systems* 47 (4): 445–487. (March 2005)
13. Gharavi, H., and B. Hu. 2011. Multigate communication network for smart grid. *Proceedings of the IEEE* 99 (6): 1028–1045.

14. Hiertz et al, G.R., et al. 2008. IEEE 802.11s: WLAN mesh standardization and high performance extensions. *IEEE Network* 22 (3): 12–19. (May–June 2008)
15. Gharavi, H., and C. Xu. 2012. Traffic scheduling technique for smart grid advanced metering applications. *IEEE Transactions on Communications* 60 (6): 1646–1658.
16. Robertsingh A., D. Devaraj and R. Narmathabanu. 2015. Development and analysis of wireless mesh networks with load-balancing for AM in smart grid. In *2015 international conference on computing and network communications (CoCoNet)*, 106–111, Trivandrum.
17. Jung, J.S., K.W. Lim, J.B. Kim, Y.B. Ko, Y. Kim and S.Y. Lee. 2011. Improving IEEE 802.11s wireless mesh networks for reliable routing in the smart grid infrastructure. In *IEEE international conference on communications workshops (ICC)*, 1–5, Kyoto.
18. Floyd, S. 2001. A report on recent developments in TCP congestion control. *IEEE Communications Magazine* 39 (4): 84–90.
19. Deng, X., T. He, L. He, et al. 2017. Performance analysis for IEEE 802.11s wireless mesh network in smart grid. *Wireless Personal Communication* 96: 1537.

Throughput Optimization Methods for TDMA-Based Tactical Mobile Ad Hoc Networks

G. Selvakumar, K. S. Ramesh, Shashikant Chaudhari and Manoj Jain

Abstract In today's network-centric battlefield, mobile ad hoc networks (MANET) play an important role in military operations to provide situation awareness. MANET is an infrastructure-less network that is automatically built as soon as devices connect or disconnect. Instead of relying on a central node to coordinate the flow of messages, the individual nodes are able to forward packets to each other. Tactical MANET is characterized by rapid changes in the network topology due to node mobility and intermittent line-of-sight (LOS) connectivity. In military operations, tactical wireless networks have high demands for robustness, responsiveness, reliability, availability and security. Also, wireless communication networks supporting tactical military operations are limited by bandwidth requirements. To achieve greater responsiveness and better system throughput for TDMA-based tactical mobile ad hoc networks, it is necessary to minimize the overhead parameters at various layers. In this chapter, we propose various optimization methods to maximize tactical wireless throughput by reducing the control plane and data plane overheads with the available system bandwidth.

Keywords TDMA · MANET · Throughput

1 Introduction

Today's modern tactical operations are evolving towards the concept of network-centric warfare [1], in which network infrastructure is an important parameter for tactical operation. Digitalization of the battlefield and widely diverse operations with modern command and control (C2) applications, sensors and real-time situation awareness leads to the need for higher data rates and mobile networks in the

G. Selvakumar (✉) · K. S. Ramesh · S. Chaudhari · M. Jain
Central Research Laboratory, Bharat Electronics Limited, Bengaluru, India
e-mail: selvakumarg@bel.co.in

A. N. Krishna et al. (eds.), *Integrated Intelligent Computing, Communication and Security*, Studies in Computational Intelligence 771,
https://doi.org/10.1007/978-981-10-8797-4_34

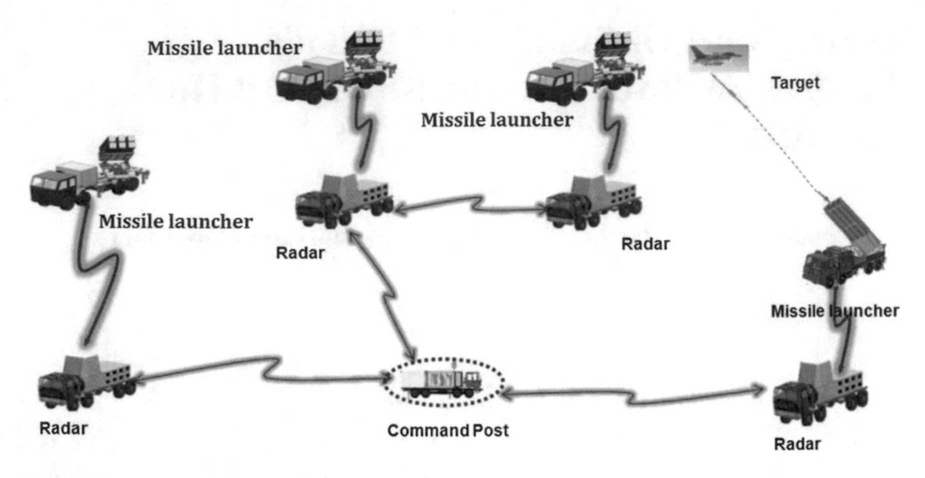

Fig. 1 Tactical MANET architecture

tactical military [2, 3]. Tactical communications are military communications in which information is conveyed from one command to another to provide situation awareness [4]. In the present stage, tactical wireless communication uses digital technology for transmission of data [5]. Figure 1 shows a tactical MANET scenario operation. In this scenario, the tactical elements such as command post vehicle unit, radar vehicle unit and missile launching unit are connected via wireless radio links [6–8]. These tactical wireless nodes are able to move in any direction. The tactical MANET is characterized by rapid changes in the network topology due to node mobility and intermittent line-of-sight (LOS) connectivity. Based on the connectivity between these tactical elements, communication can be in either a direct or indirect mode [9, 10], as shown in Fig. 1 for communication between the command post vehicle and the missile launcher vehicle. In the indirect mode of communication, the radar vehicle unit will act as a relay between these two wireless nodes.

Current tactical wireless communication is able to transmit voice and data as an IP packet. The need for data transmission over wireless channels has increased due to various tactical information such as radar data, target location, and sensor data which are required to be shared among the tactical units to provide situation awareness. As a result, a wireless networking solution has been integrated into the system to provide information sharing between the moving military vehicles/units. Information such as radar tracking data, target location and sensor data are very much required at the central military location for decision making. Based on the this information, threat priority will be analyzed and an appropriate decision will be made at a central place. Based on the threat, the appropriate commands will be sent from the central location to the missile launching location to fire on the target. Performance of the tactical wireless system is measured in terms of throughput, delay, jitter and packet error rate [6, 10]. Bandwidth is a major constraint in the tactical wireless domain. Communication networks supporting tactical operations are often constrained by wireless bandwidth. In the tactical wireless domain, throughput should be maximized, and other parameters such as delay, jitter and packet error rate should be minimized. Since current tactical wireless communication is capable of transmitting voice and data as IP packets over the network, the end-to-end packet delay should be minimized. The IP packets generated should be transmitted to the appropriate destination either directly or indirectly (relay), depending on the connectivity. To reduce latency, it is necessary to utilize the system bandwidth efficiently, and the application packet must be transmitted without queuing. In general, the control and data plane overhead will consume more bandwidth, resulting in lower throughput. The control plane and data plane overheads must be minimized in order to efficiently utilize the wireless system bandwidth [4, 9]. In this chapter, we propose various throughput enhancement methods at the control plane and data plane to increase the system throughput with the available bandwidth. The proposed method is implemented and tested in a real-time environment.

The remainder of the chapter is organized as follows: Sect. 2 describes the various throughput optimization parameters for TDMA-based mobile ad hoc networks. Section 3 describes the implementation and test results of the proposed optimization methods. Section 4 concludes the chapter.

2 Throughput Optimization Methods

2.1 Data Concatenation

In the tactical wireless scenario, the user data can be either sensor data, radar track data, short message or voice/video packets. In the modern digital tactical scenario, all these packets are transmitted as IP packets. The IP packet size can vary from a few bytes to a few hundred bytes. These packets can be generated in any order by the transmitting node. Sometimes the IP packet size will be less than the radio frame size. In TDMA-based networks, a node will transmit only in its allocated time slot. In addition, for each time slot, only one IP packet will be taken for transmission, leading to under-utilization of throughput. In these TDMA-based ad hoc networks, the incoming IP packets will be queued if more than one IP packet arrives, thus leading to latency issues. In order to maximize the throughput, instead of transmitting only one IP packet per allocated time slot, multiple IP packets must be transmitted. This can be achieved by concatenating or keeping more than one IP packet in the radio frame by adding the proper individual segment headers. With this data concatenation approach, a wireless node can transmit multiple IP packets simultaneously in its allocated time slot instead of transmitting the IP packets one at a time. The multiple data segments can be differentiated by adding the proper segment header to an individual data packet. With this approach, a node can transmit one or more IP packets that can be intended for a single destination or even for multiple destinations. The total number of data segments in a radio frame depends upon the radio frame size per time slot and the individual IP packet size.

Source Node Id	Destination Node Id	Fragment Number	Message Length

Fig. 2 Data segment header

Each data segment header can consist of a source node identifier, destination node identifier, fragment number, node stamp and message length field, as shown in Fig. 2. The source node identifier is a value of the transmitting wireless node. The destination node identifier is a value of the destination wireless node. The fragment number provides the fragmentation number in case a packet is divided into multiple fragments. The node stamp is a duplicate packet detection indicator. This will avoid the retransmission of the same packet. The message length field provides the length of the individual data segment.

The receiver node receives the entire radio frame, and decodes each and every data segment header. After decoding the received frame, if the data segment belongs to its own frame, then it is passed to higher layers; otherwise it is dropped. In the case of multi-hop routing, the receiver node also checks whether the data segment can be forwarded. The forwarding decision is based on its routing table. In such cases, the data segment is stored in a relay queue for further relaying. This data concatenation feature allows a wireless node to simultaneously transmit multiple data packets to one destination or multiple destinations on its allocated time slot, thereby improving system throughput.

2.2 *Optional Flags*

In dynamic TDMA-based ad hoc networks, the control information can be node connectivity information, time slot information or time synchronization information. This control information may not be required to transmit in its allocated time slots. Instead, the information can be transmitted at regular predefined intervals of duration. This predefined regular interval can be a user-configurable parameter. The optional flags are the flags to indicate the presence of data or control information in the transmitted frame. This flag is set by the transmitter node. The optional flag can be utilized to increase the system throughput by periodic transmission of control information.

For example, the optional flag for a TDMA-based ad hoc network can be node connectivity information flag, time slot information flag, time information flag or data flag. Depending upon the predefined duration, these flags can be set or unset by the transmitter node. Based on the flag indicator values, the receiver node decodes the frame. Sample optional flag parameters for a TDMA-based wireless network are shown in Fig. 3. This optional flag feature enhances the system throughput by

Time Information Flag	Network Connectivity Information Flag	Time Slot Information Flag	Data Flag

Fig. 3 Optional flag parameters

reducing the amount of control information transmitted in the TDMA-based ad hoc networks.

2.3 *Bit Vector-Based Representation of Control Information*

The control information is the information that is used to manage or control the wireless devices. For example, the control information can be node connectivity information, time slot information or time synchronization information in the multi-hop dynamic TDMA wireless network. In general, this information is represented in byte format. This type of representation increases the overhead size and consumes more bandwidth. To reduce the overhead, instead of representing this control information in the form of bytes, a bit vector-based method can be used. Wherever possible, this bit vector method can be utilized to minimize the control plane overhead. Since the control information is represented in the form of bit vectors rather byte vectors, the overall length to represent the control information is reduced, thereby increasing throughput. For example, in a 14-node multi-hop dynamic TDMA wireless network, if the time slot details for a 40-ms TDMA frame structure are represented in byte format, it requires 40 bytes, but if the time slot details are represented in the nibble format, only 20 bytes are required. Similarly, to represent the node connectivity, information for a table-driven proactive protocol requires $14 * 14 = 196$ bytes (each node requires 14 bytes to represent its connectivity information; overall $14 * 14$ bytes are required). Instead of representing the connectivity information in the form of bytes, if a bit vector approach is used, only $14 * 14 = 196$ bits are required (each node requires 14 bits to represent its connectivity information; overall $14 * 14$ bits are required). Figure 4 shows the bit vector-based representation of node connectivity information for an eight-node network.

In this type of proactive table-driven approach, every node maintains the global connectivity table to represent the overall network connectivity. This global connectivity table consists of all the individual node connectivity information to other nodes. Each individual direct connectivity information is represented in the form of a bit vector (0/1). Bit vector 1 shows that the individual node is directly connected with the node. The value 0 shows that no connectivity is established between the nodes. With this approach, in an "n"-node multi-hop TDMA-based wireless network, each node requires only $n * n$ bits rather than $n * n$ bytes to represent the global connectivity table if a table-driven approach is used. This bit vector-based approach increases

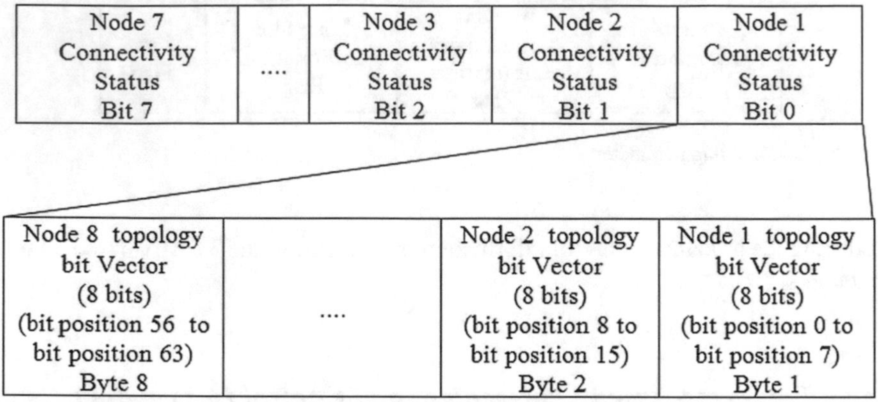

Node 7 Connectivity Status Bit 7	....	Node 3 Connectivity Status Bit 2	Node 2 Connectivity Status Bit 1	Node 1 Connectivity Status Bit 0

Node 8 topology bit Vector (8 bits) (bit position 56 to bit position 63) Byte 8	....	Node 2 topology bit Vector (8 bits) (bit position 8 to bit position 15) Byte 2	Node 1 topology bit Vector (8 bits) (bit position 0 to bit position 7) Byte 1

Fig. 4 Bit vector-based topology information representation

the system throughput by representing the control information in the form of bits rather than bytes.

2.4 *Fragmentation and Reassembly*

In the tactical wireless scenario, the user data can be either sensor data, radar track data, short message or voice/video packets. In the modern digital tactical scenario, all these packets are in the form of IP packets. The user data size varies depending on the type of application used. Sometimes the user data size will be more than the available radio frame size. In such scenarios, the user data can be divided into smaller sizes, which can be accommodated by a radio frame. In this, a larger IP packet can be divided into smaller packets with proper fragmentation number and transmitted over wireless medium. At the receiver side, the reassembly module can be used to reassemble the fragmented packets to form the original packet. With this fragmentation and reassembly approach, maximum throughput is achieved by keeping more user data in the radio frame.

3 Implementation and Test Results

The various throughput optimization methods are designed, implemented and tested in the dynamic TDMA-based mobile ad hoc network consisting of 14 wireless nodes. The system design parameters are shown in Fig. 5. The optimization methods are implemented in the Cyclone V SoC hardware platform. The Cyclone V SoC Hard Processor System (HPS)consists of a dual-core (ARM Cortex-A9) processor. We have used ARM DS-5 IDE for programming and development. The development

Fig. 5 System design parameters

System Bandwidth	10 MHz
Number of Wireless Nodes	14
TDMA Frame size	40 ms
Slot duration	1 ms
Maximum Transfer Unit/Slot	526 bytes
Modulation scheme	QPSK
Distance	8-10 KM
Number of TDMA Slots assigned per node	Dynamic
Mode	Point to Point Point to Multipoint Multipoint to Multipoint
Theoretical Throughput per TDMA slot duration	102 kbps

board consists of a baseband card along with an AD9361-based RF section. We have used a 40-ms TDMA frame structure to verify the throughput. The control plane and data plane overheads are reduced after introducing the various optimization parameters including data concatenation, fragmentation and reassembly, optional flag and bit vector-based representation of control information. Figure 6 shows the overhead comparison before and after optimization.

Optimization Parameter	Overhead Before optimization Parameter	Overhead After optimization Parameter	Throughput improvement
Bit level representation of Connectivity information, Time Slot Information	236 bytes for every time slot opportunity (46.09 kbps)	48 bytes for every timeslot opportunity (9.375 kbps)	36.71 Kbps for one slot allocation. 73.43 kbps for two slot allocation.
Bit level representation of Connectivity information, Time Slot Information + Optional Flag of 1 second duration	48 bytes for every timeslot opportunity (9.375 kbps)	48 bytes/second (0.375 kbps)	9 kbps for one slot allocation. 18 kbps for two slot allocation

Fig. 6 Overhead comparison

S. No	Mode of Operation	Theoretical Throughput per slot	Measured Throughput per slot (Before Optimization)	Measured Throughput per slot (After Optimization)	Throughput Improvement per slot
1	Point to Point	102 kbps	59.29 kbps	96 kbps	36.71 kbps
2	Multi Point to Multi Point (Multi Hop)	102 kbps	59.29 kbps	96 kbps	36.71 kbps

Fig. 7 Theoretical throughput versus measured throughput

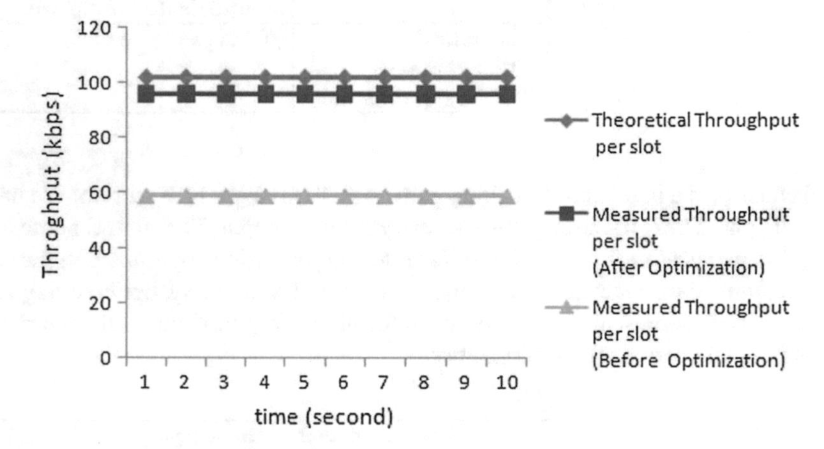

Fig. 8 Throughput comparison chart

These throughput improvement methods are tested and verified over the air for both single-hop and multi-hop scenarios. The throughput is measured using a VeEX Ethernet tester and Jperf software. End-to-end ping test and voice-over-IP (VOIP) calls are verified in both single- and multi-hop scenarios. Figure 7 shows the measured throughput considering the optimization parameters. It is observed that the end-to-end latency is reduced after introducing the optimization parameters. Figure 8 shows the comparison of measured throughput in graph format. With this combination of all the optimization approaches, the application throughput is maximized and the measured throughput is close to its theoretical baseband throughput (96 kbps vs. 102 kbps) for both single- and multi-hop scenarios.

4 Conclusion

In military operations, tactical wireless networks have high demands for robustness, responsiveness, reliability, availability and security. One key requirement for the tactical user is efficient utilization of bandwidth. To achieve greater responsiveness and better system throughput of TDMA-based tactical mobile ad hoc networks, it is necessary to minimize the overhead parameters at both control and data plane. In this chapter, we developed and tested various throughput optimization methods to reduce the overhead for the dynamic TDMA-based tactical multi-hop networks. The optimized throughput is compared with the theoretically achievable throughput and throughput without optimization parameters. With this customized optimization framework, the end-to-end latency between the packets is reduced, throughput is enhanced, and quality of service (QoS) is thereby ensured between the packets. In the future, the proposed solution will be tested for additional wireless nodes.

References

1. Pawgasame, Wichai, and Komwut Wipusitwarakun. 2015. Tactical wireless networks: A survey for issues and challenges. In *2015 Asian Conference on Defence Technology (ACDT)*. IEEE.
2. Nicholas, Paul J., Jeffrey C. Tkacheff, and Chana M. Kuhns. 2014. Analysis of throughput-constrained tactical wireless networks. In *Military Communications Conference (MILCOM), 2014 IEEE*. IEEE.
3. Suman, Bhupendra, L. Mangal, and S. Sharma. 2013. Analyzing impact of TDMA MAC framing structure on network throughput for tactical MANET waveforms. In *Conference on Advances in Commun. and Control Systems (CAC2S)*.
4. Jalaian, Brian, Venkat Dasari, and Mehul Motani. 2017. A generalized optimization framework for control plane in tactical wireless networking. In *2017 International Conference on Computing, Networking and Communications (ICNC)*. IEEE.
5. Fossa, C. E., and T. G. Macdonald. 2010. Internetworking tactical manets. In *Millitary Communication Conference, 2010-MILCOM 2010*. IEEE.
6. Elmasry, George F. 2010. A comparative review of commercial vs. tactical wireless networks. In *IEEE Communications Magazine 48.10*.
7. Aschenbruck, Nils, Elmar Gerhards-Padilla, and Peter Martini. 2008. A survey on mobility models for performance analysis in tactical mobile networks. *Journal of Telecommunications and Information Technology* 54–61.
8. Wang, Haidong. 2007. Implementing mobile ad hoc networking (MANET) over legacy tactical radio links. In *Military Communications Conference, 2007. MILCOM 2007. IEEE*. IEEE.
9. Larsen, Erlend, et al. 2010. Optimized group communication for tactical military networks. In *Military Communications Conference, 2010-MILCOM 2010*. IEEE.
10. Cheng, Bow-Nan, et al. 2014. Design considerations for next-generation airborne tactical networks. *IEEE Communications Magazine* 52.5: 138–145.

Distributed IoT and Applications: A Survey

H. S. Jennath, S. Adarsh and V. S. Anoop

Abstract The Internet of Things (IoT) has given rise to a vast network of connected smart objects, leading to the development of cyber-physical pervasive frameworks. Recent years have witnessed an exponential growth and interest in developing smart systems, enabling the identification of innovative solutions for industrial, domestic, and agricultural sectors. Various applications have been built and deployed in IoT space that leverage the increasing ubiquity of radio-frequency identification (RFID), wireless, mobile, and sensor devices. However, the current IoT paradigm with traditional cloud-based architecture is not scaled for the future, and requires all data to be transferred to a central point for analysis and action. This centralized processing model presents numerous challenges with the future need to connect hundreds of billions of devices. The major bottlenecks are associated with various dimensions such as cost of connectivity, meaningful value creation from sensor data, reliability, establishing digital trust, traceability, preventing concentration of power, unmanageable complexity, security, privacy, and latency. One solution to overcome these hurdles is to bring the concept of decentralization to the IoT space to build distributed IoT. This paper presents the current state of the research on IoT and its potential business use cases along with a case study on distributed IoT. It also discusses various challenges and opportunities in decentralized ecosystems, along with open research questions and future directions.

H. S. Jennath (✉) · S. Adarsh · V. S. Anoop
Data Engineering Lab, Indian Institute of Information Technology
and Management-Kerala (under CUSAT), Thiruvananthapuram, India
e-mail: jennath.res16@iiitmk.ac.in

S. Adarsh
e-mail: adarsh.res16@iiitmk.ac.in

V. S. Anoop
e-mail: anoop.res15@iiitmk.ac.in

A. N. Krishna et al. (eds.), *Integrated Intelligent Computing, Communication and Security*, Studies in Computational Intelligence 771,
https://doi.org/10.1007/978-981-10-8797-4_35

Keywords Internet of Things · Blockchain · Smart contracts · Decentralized computing · Supply chain

1 Introduction

The Internet of Things (IoT) is expanding exponentially, and the current IoT paradigm with a conventional centralized control model is not scaled for the future and able to handle the huge amounts of data generated from hundreds of billions of devices. Also, currently, it supports little processing or analytic power at the network edge. Traditional cloud-based architecture requires all data to be transferred to a central point for analysis and action, and hence poses the risk of connectivity disruption and latency, which makes it less suitable for real-time applications [1, 2]. In a network of this projected size, implementing trust verification, security, privacy, and anonymity will be complex, expensive, and nearly impossible with centralized systems [3]. The centralized processing model has to deal with various other challenges such as the cost of connectivity, meaningful value creation, reliability, traceability, preventing concentration of power, manageable complexity while scaling to future requirements. A solution for overcoming these challenges is in the concept of decentralized or distributed intelligence for the IoT world, where the power or control in the IoT shifts from the center to the edge. With the decentralization approach, it is possible to build autonomous IoT systems with scalability and improved latency, along with reduced infrastructure and maintenance costs, enabling the development of critical real-time applications for monitoring and control, especially in remote or hard-to-reach areas for industrial or military purposes. This approach also eliminates the risk of a single point of failure [3–5], and trust verification and device coordination can be addressed by employing trust-free device coordination techniques in the system.

Enabling technologies for distributed IoT as demonstrated by IBM, Samsung and Filament in their proof of concept project are Telehash [6], BitTorrent [7] and Blockchain [8, 9]. Telehash is a peer-to-peer data distribution and communication protocol that is decentralized and secure. IBM's ADEPT [3, 4], which is a proof of concept in decentralized IoT, uses Telehash for secure peer-to-peer messaging. There may be situations that demand the exchange of electronic files in distributed networks and BitTorrent can be used for secure file transfer. Blockchain, a universal shared distributed digital ledger that functions at the heart of decentralized system, finds application in transactional applications that maintain a continuously growing list of encrypted, secured, tamper-proof, and immutable data records. This enforces trust, accountability and transparency while ensuring smooth business processes.

Contributions of this chapter: This chapter presents a comprehensive survey on distributed IoT that discusses state-of-the-art and current development in distributed IoT space. It also discusses potential application areas of distributed IoT in the supply chain domain and then outlines challenges and potential research opportunities in distributed IoT.

Organization of the chapter: The remainder of this chapter is organized as follows: Sect. 2 discusses some game-changing innovations in IoT, and Sect. 3 outlines new market initiatives in distributed IoT considering the supply chain as a case study. In Sect. 4, the authors discuss challenges and some potential research opportunities in distributed IoT, and concluding remarks are presented in Sect. 5.

2 Game-Changing Innovations in IoT Space

This section outlines state-of-the-art systems in IoT and distributed IoT areas. Distributed intelligence in IoT is an emerging topic that is still in its infancy. Literature available in this area is sparse, and no fully decentralized commercial implementation currently exists. However, organizations such as IBM, Samsung, and Cisco are interested in this area of distributed intelligence in IoT for harnessing the huge amount of data in the industrial Internet [3–5]. IBM, together with Samsung, worked on a project called Autonomous Decentralized Peer-to-Peer Telemetry (ADEPT), which is a proof of concept for distributed or decentralized implementation of IoT in a washer with basic functionalities, along with the ability to communicate and negotiate with peers for efficient operation [3, 4].

An IBM white paper [3] proposes decentralization as a solution to address challenges such as cost, privacy, and durability to scale the IoT to connect to hundreds of billions of devices. According to the new paradigm of device democracy as stated in the paper, the system design should focus on leveraging user-centric experiences and the value of a connected device to its user, while designing a decentralized, autonomous IoT system. In other words, the devices in the network should be empowered to collaborate and communicate with peers in the network and respond in the best interests of the user, rather than employing third parties such as manufacturers or service providers to create intelligence out of any central repository. IBM [4] validates the future vision for decentralized systems to supplement current centralized solutions. It has demonstrated performing basic IoT tasks without any centralized control and empowered devices to take part autonomously in marketplace transactions in their proof of concept (PoC). The ADEPT PoC established how to enable a washer to be an autonomous device capable of executing its own maintenance, managing its consumable supply, negotiating with other devices as well as with the grid, and optimizing energy consumption, all without using a central controller mediating the devices. The practicability of ADEPT paves the way for modifying current centralized IoT solutions with more decentralized capabilities [4]. However, challenges associated with commercializing distributed systems, such as scalability, security, coordination, intellectual property management, and identity and privacy issues, are not addressed in this work.

A Cisco and Bit Stew white paper, "Distributed Intelligence: Unlocking the Benefits of the Industrial Internet" [5], proposes employing distributed intelligence in IoT to harness the data explosion in the industrial Internet. The paper emphasizes the notion that networks that blend centralized with distributed processing methods are much better able to handle the tremendous data volume. By managing and controlling the data locally, information processing takes place at the source. This avoids the problems associated with network latency and processing delays. Normalizing data at its source using utilities will enable orchestration, contextualization, and standardization of information, making data integration simpler and optimizing resource utilization.

Filament [10] introduces an open technology stack that has the integrated capabilities of cutting-edge communication and security features that enable complete autonomy of devices, such that discovery, communication, and interaction with other peer devices occurs in an absolutely distributed fashion. The open stack technology of the Filament is built on five key principles, namely, security, privacy, autonomy, decentralization, and exchange (SPADE). They developed Filament Taps, which can be freely deployed on any transportation system, such as a rail network, in which locomotives, freight and passenger cars, switch motors, and other pieces of infrastructure are networked through this inexpensive, surface-mount device. Upon deployment of Filament Taps on the transportation network, the devices connect dynamically and collect real-time data from the network autonomously and run preventive analytics. Filament technologies provide a secure foundation for decentralized interaction, which proactively manages preventive maintenance and mitigates the risk of hazardous and costly derailments.

The "robochef" by Moley Robotics [11] is a truly game-changing invention for automated cooking. It is a fully functional, aesthetically designed robot integrated into an executive kitchen, which is capable of cooking dishes with the talents of a master chef. The autonomous cook is able to retrieve the recipes and mimic cooking activities from the pre-learned sessions. The technology behind it is simply a pair of artificial hands with multiple joints, with a number of degrees of freedom, tangible sensors, and complicated control systems. Nest Thermostat [12, 13] was the first learning thermostat developed by Nest Labs, which was later acquired by Google. The main advantage of the Nest is its "learning" ability, which means that the user does not need to program it. Previously, only programmable thermostats were available on the market. Here the user simply adjusts the temperature as they go, and over time, Nest learns what temperatures suits the environment and prepares the day-to-day schedule for the home. Nest supports a Wi-Fi connection and can add any number of thermostats that can be controlled using a mobile app. The new initiatives discussed here are indications of the market trends in IoT space. Unlike traditional centralized IoT, the current trend is to impart cooperative autonomy and cognitive thinking into the world of connected things. The next section discusses some potential application areas for IoT in our day-to-day lives that can disrupt conventional automation methods.

3 Distributed IoT: New Market Initiatives and a Case Study on Supply Chain

Blockchain technology [8, 9], together with IoT, opens up a new dimension of opportunities to explore in the direction of decentralization. The distributed ecosystem built using decentralized enablers can be leveraged to address potential problems in various domains of centralized IoT. Open-standards-based distributed IoT networks have the ability to solve major issues related to security, scalability, and cost present in the current centralized cloud-based IoT solution. Blockchain provides a mechanism for a secured and immutable distributed framework that facilitates the interaction of connected edge devices with peers. Smart contracts deployed in Blockchain validate and enforce actionable decisions using device-generated data. The shared, trust-free, and immutable environment provided by Blockchain, along with distributed messaging such as MQTT (Message Queuing Telemetry Transport), reduces the requirement for manual intervention in monitoring and control, and leverages more space for device democracy. Decentralized Blockchain networks offer improved security for IoT solutions with respect to the immutable data records. In a nutshell, the self-executing smart contracts on Blockchain can embed specific consensus mechanisms that identify and eliminate fraudulent devices from the network. An architecture diagram of a distributed IoT framework is shown in Fig. 1. Here, the end devices are connected to the distributed ledger to implement distributed consensus and global collaboration. A secured shared messaging protocol is required for communication and collaboration between peer-to-peer edge devices.

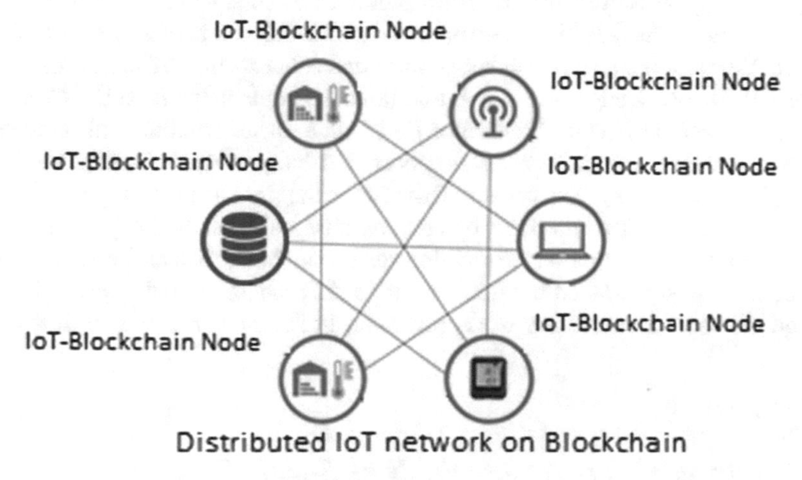

Fig. 1 Distributed IoT powered by Blockchain

3.1 Distributed IoT: New Market Initiatives

A basic application architecture of distributed IoT with Blockchain for the supply chain is the enrollment of goods transfer on the shared distributed ledger (Blockchain) as transactions that capture a variety of information regarding the various stakeholders involved, pricing information, date and time, location details, product quality and state, and any other information that would be relevant to running the supply chain. The shared nature of Blockchain makes it possible to track every product right from the point of origin to the market with clarity on the minutest details of the product along the supply chain. The cryptographically secured tamper-proof distributed ledger does not allow anyone to manipulate the data to their advantage. IBM [14] stepped into this game-changing business by rolling out a customizable service that employs Blockchain to track high-value items through complex supply chains, called Hyperledger. Everledger [15] is a firm that uses IBM's Hyperledger service to implement transparency in the supply chain use cases of expensive items such as wine, diamonds, aircraft engines. It is possible to perform various fraudulent activities along the chain from the manufacture to the customer end. The immutable shared ledger offers a trust-free platform for diamond identification, transaction verification, and so on, for all stakeholders, including insurance companies, claimants, and law enforcement agencies in the diamond pipeline.

ChainLink [16] is another initiative that uses Blockchain technology as the backbone for verifying and validating the authenticity and title of real-world items. Blockverify [17] is an anti-counterfeit solution based on Blockchain in the medical field, which provides verification of medicine authenticity. This is achieved by creating a unique identity on Blockchain for every product that comes to market. This facilitates tracking of changes in ownership or origin, which can be easily accessed and audited. Chimera-Inc [18] is another venture that connects the IoT to real-time analytics at the network edge. It is a hub that links the home network to the cloud and electrical devices around it. Provenance [19] built a trusted solution which creates a transparent and trusted platform for supply chains that provides a digital identity for all tangible goods using Blockchain. This aids in the fight against the sale of counterfeit or stolen goods or products by creating a traceable, auditable profile for any physical goods. The traceable information can be anything, such as the environmental impact or who produced the item or where the production took place. The next section discusses a case study of supply chain in Blockchain-powered distributed IoT.

3.2 Distributed IoT: A Case Study on Supply Chain

IoT can be leveraged to enable improvement and revamping of the existing supply chain process. The IoT-powered supply chain enhances the transparency of transit and can easily connect the digital and physical chain. This is beneficial for checking the quality of commodities in transit. Everledger [15] uses Blockchain technology to trace unique features such as diamond cut, color, clarity, and quality, as well as to help monitor the diamond's ownership from mine to jeweler. A unique ID is given to each diamond transacted over Everledger [15], thus preventing fraud and fake products. Customers can easily check the provenance, and even in cases of theft, the owners of each diamond can be easily traced.

Blockchain offers a distribution platform that replicates the system data among the distributed peers, where all transaction validation is mandated through smart contracts. This helps the geographically separated stakeholders work towards a unified goal. It is possible to enforce the roles and responsibilities of each of the participating users in the supply chain using smart contracts over Blockchain. Blockchain-powered organic produce is another potential application in the supply chain. Procured organic products will be continuously monitored from farm to consumer. Blockchain captures various sensor data such as acetylene, temperature, and humidity of the products along various lines of transit. It is possible to distribute the risk of quality delivery of organic products among the various stakeholders across the supply chain using Blockchain and smart contracts. The roles and responsibilities of each stakeholder along the chain are defined in the system using smart contracts. Thus, the immutable data captured in Blockchain can be used as a fingerprint for tracking and monitoring the quality of the perishable goods in terms of the various mentioned parameters. Any degradation in the quality of the product can be easily traced, and the stakeholder held responsible for the loss can be identified from Blockchain.

Freight management is a complicated supply chain process involving various transport companies. IoT-enabled Blockchain offers a transparent solution for monitoring and controlling the system to ensure the timely delivery of goods and products. The immutable shared ledger offers a trust-free platform for the geographically separated stakeholders to work on a common goal and to identify and penalize the defaulters in the game. Various data including temperature, humidity, location information, time of arrival and departure, condition of the shipping containers, and timing information are captured into the IoT powered by Blockchain. Tamper-proof Blockchain transactions guarantee that all parties in the chain can trust the data and take necessary action for quick and efficient product movement. In short, using smart contracts over Blockchain-powered IoT, the solution suite mandates and confirms that all contractual obligations are met at all points along the chain.

Another use-case for distributed IoT application is in tracking various components of vehicles, including spacecraft, aircraft, and automobiles. This distributed IoT system with Blockchain maintains an accurate and indelible history of critical components for vehicles. This is a very important issue from both a safety and regulatory compliance perspective. The replicated shared ledger offered by Blockchain stores the distributed data in a secured transparent manner that participating agents in the system will be able to audit and see component provenance throughout the vehicle's life. This provides a cost-effective, secure way of sharing this information with regulatory bodies, transporters, and manufacturers. Distributed IoT also finds application in tracking the safety status of critical machines and their maintenance operations. Blockchain provides for an immutable ledger of operational and maintenance data for various machinery including aircraft engines and elevators. This offers a means for third-party repair associates to monitor and audit Blockchain for preventive maintenance and other record-keeping. Regulatory compliance verification is also feasible using these tamper-proof traceable and auditable operational records.

4 Distributed IoT: Challenges and Potential Research Opportunities

Though the concept of distributed computing in IoT is highly acclaimed in the industry, it is still in its infancy. Major IT giants such as IBM, Samsung, Microsoft, and Intel have stepped into this vast arena to explore its potentials, but there is not yet a single operational fully distributed IoT system available in market. This opens up many opportunities for research and innovation in dealing with distributed data, behavior differentiation, decision making, business models for swarm robotic systems, autonomous mobile agents, edge intelligence, security, and trust management [20]. The design of a distributed peer-to-peer lightweight framework capable of implementing cooperative autonomy among peers in the network and efficient distributed consensus is an essential mandate. The framework should be able to handle seamless integration of the participating peer-to-peer devices with proper admission control and trust management. Lightweight protocols are essential for constraint IoT devices for enabling proper communication among peers. There should be provisions for secure and distributed data sharing among peers or participating agents. Effective mechanisms for robust device coordination and negotiation among the peers should be instantiated.

5 Conclusions

In this chapter, we have discussed innovations and current market initiatives in IoT space. Moving towards distributed intelligence, the new paradigm introduces decentralized consensus, replacing traditional centralized decision-making processes. With the built-in trusted, traceable, tamper-proof, immutable capabilities of the distributed shared ledger, it is possible to implement embedded rules for trust, identity and ownership management. The chapter also explored the potential for IoT in building smart systems in areas such as agriculture, healthcare, and transportation. We also highlighted supply chain use-cases for IoT leveraging the potential of Blockchain. Major challenges and research opportunities in distributed IoT were discussed. A case study on distributed IoT in supply chain management was also included.

Acknowledgements This work is supported by the research fellowship (Ref:Order No.1281/2016/ KSCSTE) from Women Scientists Division (WSD), Kerala State Council for Science, Technology and Environment (KSCSTE).

References

1. Heylighen, Francis. 2002. The Global Superorganism: An evolutionary-cybernetic model of the emerging network society. *Journal of Social and Evolutionary Systems* 6: 57–117.
2. Heylighen, Francis. 2013. *Self-organization in communicating groups: The emergence of coordination, shared references and collective intelligence*. Berlin: Springer.
3. Brody, Paul, and Veena Pureswaran. 2014. Device democracy: Saving the future of the Internet of Things. IBM.
4. Veena, Pureswaran, et al. Empowering the Edge-Practical Insights on a Decentralized Internet of Things.
5. Cisco and Bit Stew Point of View: Distributed Intelligence: Unlocking the Benefits of the Industrial Internet. 2014. BitStew Systems.
6. Telehash: Encrypted mesh protocol. http://telehash.org. Accessed 15 Nov 2017.
7. Pouwelse, Johan, et al. 2005. The bittorrent p2p file-sharing system: Measurements and analysis. *Peer-to-Peer Systems IV*, 205–216. Berlin: Springer.
8. Ethereum: A platform for decentralized applications. https://www.ethereum.org. Accessed 15 Nov 2017.
9. Asharaf, S., and S. Adarsh (eds.). 2017. *Decentralized Computing Using Blockchain Technologies and Smart Contracts: Emerging Research and Opportunities: Emerging Research and Opportunities*. IGI Global.
10. Filament. 2016. Foundations for the next economic revolution: Distributed exchange and the Internet of Things. https://filament.com/assets/downloads/Filament%20Foundations.pdf. Accessed 15 Nov 2017.
11. Moley. 2017. *The world's first robotic kitchen*. http://moley.com. Accessed 15 Nov 2017.
12. https://nest.com/thermostat/meet-nest-thermostat. Accessed 15 Nov 2017.
13. http://www.androidcentral.com/what-nest-thermostat-and-why-would-you-want-it. Accessed 15 Nov 2017.
14. https://www.ibm.com/internet-of-things/platform/private-blockchain. Accessed 15 Nov 2017.
15. https://www.everledger.io. Accessed 15 Nov 2017.
16. www.chainlinkresearch.com/IoT/index.cfm. Accessed 15 Nov 2017.
17. www.blockverify.io. Accessed 15 Nov 2017.
18. www.chimera-inc.io. Accessed 15 Nov 2017.
19. https://www.provenance.org/whitepaper. Accessed 15 Nov 2017.
20. Ferrer, Eduardo Castell. 2016. The blockchain: A new framework for robotic swarm systems. arXiv:1608.00695.

DNA-Based XOR Operation (DNAX) for Data Security Using DNA as a Storage Medium

V. Siddaramappa and K. B. Ramesh

Abstract Recent advances in genetic engineering have enabled the insertion, deletion and modification of original genome sequences of the living cells of organisms. Demand for data storage increases rapidly because of structured and unstructured data from the Internet of Things (IOT), sensors and Big Data. Information security has become a crucial need for almost all information transaction applications due to the large variety of hackers and attacks. Traditional techniques such as cryptography, watermarking and data hiding are basic notions and play an important role in developing information security algorithms and solutions. This chapter presents DNA XOR operation-based security and storage of encrypted data in a DNA format with long life, while maintaining artificially created DNA in specific laboratory conditions. We propose new encryption and decryption algorithms for data security and storage in DNA sequences with genes as keys. We can store 10^8 TB per 1 g of created cipher data with a life span of 4000 years. We demonstrate how to overcome the existing key length of 64, 128, 192, 256 bits and cryptanalysis of letters.

Keywords DNAX · DNA storage medium · Genes · Exon and intron

1 Introduction

DNA, is a molecule that carries genetic information from one living cell to another and has definite structures in particular organisms.

In 1953, the American biologist J. D. Watson and British physicist F. H. C. Crick proposed the three-dimensional model of physiological DNA (i.e. B-DNA) on the basis of X-ray diffraction data of DNA obtained by Franklin and Wilkins.

V. Siddaramappa (✉)
Department of CS&E, R.V. College of Engineering, VTU, Bangalore, India
e-mail: siddavmk@gmail.com

K. B. Ramesh
Department of IT, R.V. College of Engineering, VTU, Bangalore, India

A. N. Krishna et al. (eds.), *Integrated Intelligent Computing, Communication and Security*, Studies in Computational Intelligence 771,
https://doi.org/10.1007/978-981-10-8797-4_36

For this epoch-making discovery, Watson, Crick and Wilkins received the Nobel Prize in Medicine in 1962 [1].

In living cells, production of proteins from DNA molecules is a very complex problem, as it involves different components, and the production of proteins plays a very important role in living cells. DNA is located in the core or nucleus of human cells in a double-stranded spiral model of a double-helix model. Each strand is made up of millions of chemical bases such as adenine, thymine, cytosine and guanine. The order of bases appears as combinations and permutations contained in each strand. There are 23 chromosomes present in the human genome.

A gene consists of coding sequences known as exons and non-coding sequences known as intons which are involved in the transcription and translation process. Transcription is a process to produce messenger RNA (mRNA) from a DNA template, while translation produces proteins from mRNA. Genes are basic molecular units of DNA which produce proteins. The Human Genome Project has an estimated 20,000–25,000 human genes.

Alleles are one or two forms of the same gene from parental genes with small differences in their DNA sequences. These small differences make persons physically unique in world.

Today data security hiding from unauthorized persons very crucial problems exists in society. There have been many algorithms proposed and implemented to ensure security of data transmission. Public key cryptography was proposed by Diffie and Hellman [2]. Roughly 43% of enterprises and 31% of mid-market firms are investing in IoT to meet the demands of the new data economy, according to a recent study by Dell and the International Institute for Analytics. Yet, most firms are still exploring how they can process and use all this data, and a study by AT&T Labs Research has shown that poorly managing data costs U.S. companies around $600 billion per year.

A 120-PB drive could hold 24 billion typical 5-MB MP3 files or comfortably swallow 60 copies of the biggest backup of the Web, the 150 billion pages that make up the Internet Archive's WayBack Machine. The data storage group at IBM Almaden is developing the record-breaking storage system for an unnamed client that needs a new supercomputer for detailed simulations of real-world phenomena. However, the new technologies developed to build such a large repository could enable similar systems for more conventional commercial computing, says Bruce Hillsberg, director of storage research at IBM and leader of the project [3].

Digital data security is very important from a number of perspectives. Data can be converted into binary and decimal format for processing any kind of information [4].

2 Methods

The Human Genome Project has massive amounts of information stored in the form of DNA and protein sequences. Each generation consists of a set of similar functions and explores functionality based on external factors such as diet and environment, as well as lifestyle. But today the Human Genome Project and the 1000 Genomes Project have identified genes of human beings functionality same although the location and sequence of DNA sequences slightly varies.

We can retrieve a complete human genome from a single DNA sample, and identify relatives and where they are likely living. We can secure information by storage in living organisms and later extract the stored information from wet and dry labs.

We can genetically modify DNA without affecting the original DNA sequences by inserting data in terms of DNA sequences into living cells. The National Center for Biotechnology Information (NCBI) provides huge genomics and proteomics databases for all of the earth's living organisms [5]. The generally encryption done binary numbers, some problems of cryptography third party easily can predict letters based on cryptanalysis of generally used in communication between them.

In this proposed algorithm, we are using a reverse string before the encryption is performed. This reduces the possibility of finding text in cryptanalysis. The genetic DNA sequence will play a major role in the future for data security and storage methods. We can generate a complete genome of living things within 2–4 days at low cost due to advanced technology in biochemistry.

Next-generation sequencing is capable of rapid sequencing of the DNA of all living organisms. A number of platforms are available, including the Roche GS-FLX 454 Genome Sequencer, Illumina/Solexa Genome Analyzer, ABI SOLiD platform, Danaher/Dover/Azco Polonator G.007, and Helicos HeliScope. Dystrophin is a rod-shaped cytoplasmic protein and a vital part of a protein complex that connects the cytoskeleton of a muscle fiber to the surrounding extracellular matrix through the cell membrane. The dystrophin gene is one of the longest human genes known, covering 2.5 megabases (0.08% of the human genome) [6]. DNA as a storage medium has different properties compared to traditional disk-based storage [7].

These days, every aspect of life seems to rely on a digital-based architecture.

In this work, we first convert digital information into DNA-based sequence information shown in Table 1.

Table 1 Combination of DNA mapping into binary values

DNA	1	2	3	4	5	6	7	8
A	00	00	01	01	10	10	11	11
T	11	11	10	10	01	01	00	00
G	01	10	00	11	00	11	01	10
C	10	01	11	00	11	00	10	01

Table 2 DNA XOR operation

XOR	A	G	C	T
A	A	G	C	T
G	G	A	T	C
C	C	T	A	G
T	T	C	G	A

Table 3 Genes of the haematobium mitochondrial genome

Sl. no.	Gene	Location	Gene ID
1	COX3	1 … 666	4097444
2	CYTB	843 … 1946	4097445
3	ND4L	1986 … 2246	4097438
4	ND4	2219 … 3484	4097446
5	ND3	3769 … 4137	4097439
6	ND1	4311 … 5192	4097440

Table 2 differentiates binary and DNA XOR operations and shows different combinations of A, T, G and C letters, compared to 0/1 combinations. Eight different combinations are possible.

XOR of DNA has uniqueness and reflexive properties. Table 3 contains the genome of haematobium mitochondria with a length of 1503 base pairs with 12 genes. Each gene location is different in genome to consider for gene as secrete key for encryption process. Table 4 contains gene IDs with DNA sequences of genes of the haematobium genome.

Figure 1 shows how DNA is organized in human cells. Each cell of a living organism contains different pairs of DNA. Humans have 23 chromosomes that came from parents and having specific functions.

Figure 2 illustrates the steps of the encryption algorithm and information storage in the DNA format. XOR operation between key values of genes and DNA sequences of original data.

3 Results and Discussion

Compared to traditional algorithms, DNA-based algorithms for data encryption are faster and more secure because the performance of the proposed algorithm is unbreakable and it is difficult to find key values and length. Data retention in a DNA sequence format is long-term via storage in living organism. In next-generation of sequencing, we will be able to retrieve original data in the form of DNA sequences.

Digital information is converted into cipher text in the form of DNA sequences stored in specific genome, or cipher text is synthesized in laboratory conditions.

Table 4 Details of key generation from genes

Sl. no.	Gene ID	Sequence key values (keys for encryption of data)
1	4097444	atgagatatt tttctgtagt aaatctgata ttggtctttt ttttacttcc atgtatattc ttttatcatc cttggtgatt gttaggattt gttgcgatat gattagtttt atattttcgt tatattaggt gagaatttgt gttattagtt cacagtcaat gtaagatagg attttggttg tttattatga gtgaagtagt aatatttttg actttaatgt ttaaatgttt ctggtttaac gatatagaga atatttctat ctcttattcg ttttcagtgc ctataataga aacttatttg ttagtgatgt catcattaat gatttcaatg tttcatagtg gcgtagcttc tgatcatagt agtaagtata tttatcttgc tttattgttt tcattagttt ttatttgctt tgctgtagat gaatttttta atagttcatt taattcatta tgtagacctt attatgctag ttgtttcatg ttagttggtc tacatttaag acatgttgtg gtgggtagat taggtttaat tgaattaata cattttaatg aatccgagtt ggttcgatct aagagtgaga tgatagtaat atattgacat tttgttgatt ttatatgatt actcgttttt atagttgtgt atatatctaa aaataagatt ttttag
2	4097445	atgtgaaa gttagtatgt aataatttaa ttgatcttcc aactagattg gggttaaggt atttatggtg tatagggttt gttttaaggg tgtttatgat tattcaagta attactggta taatactttc tatgttttat aattcattta ataaatttg ttttatagtt tgggcggatg ataaatttat ttactgaatg gttcgttact tacatatttg aggtgttaga gtaatttttg tattattata tattcatata tgtcgtggtt tatactatag aagatacaga aaggttttaa tttgaagtag tgggtttata ttatatctgc ttgtaatggt agaggctttt ttaggttatg tcttgccttg acatcagatg tcatattgag ckgcgactgt rttaacatct atagtgttaa gtgttcctat atttggttct tctatttata gttatatagt cggtggatac tcagttacag taaatgagac attagtgcgt tttttttctg ttcatataat tctaggtata ggtttattag ggttaataat gttgcattta ttttatttgc ataagacagg ttcaagggtt ccattgtttt tatatagtag gtacagtgac aatgtgtatt ttcataagta ttatactatt aaggattttt ttgtttttat gtttatacta atgattttat ttatgtttgt gataagaagt cctaaattgg tccttgactg tgaggcattt acagaagcga atttattggt tactccttca aaaattaagc ctgagtgata ttttttggga tattatgcta tacttcgttc aattaattct aagttaggtg gtttggtatt tgttttagta atattgtttg ttatatgagt tcctaaaaat aattataggt gcatttattc attttctcgt cagttagttt tttgatttat agtttcttta ggtgttcagt tgagttatat aggtgcttgt catgctgaat atccttatgg tttagttagc caggtaagta gtattatatt tttatgtttt ttagtaattt ataagatgtt ttggttagtt ccatttagat gtgataaggt agctgaaata ttttaa
3	4097438	atgat tagtatacta atattaggcg taggtttatt attaataggg ttatttttat gtaattatag tttatttaaa tatctaataa ttttagaaaa atataaagtt atagtgttgc taataagatt aagtatatta gaaagtgggt gtcgtatgat gtttatttgt atgatgtcta tgtttgtagt agaagcttcc ttaatgttaa tgactatagg tgttagaata aagagaggat gtatacgtat accgttaggt ttataa
4	4097446	at gtatacgtat accgttaggt ttataaaggg ttgtagttac aggtttttgt ttagtttagt gatatgaaga tttcaggcat gttttgatgc ttgatgttat tgtgttaaat caagtttagt aaattggtta tttatacgtg actatttaag atttgtgatg atattattaa ctgttattat aatgtgacta tttatagtta taggtctgaa cagtatttct atatatataa gtatgtttag tgctatattg atttacgtaa ttaataaatc gttaattttt tgattttttt acgaattatc aattatttgt gctctatata tgttacttaa ggagagttta tatccagaac gttataatgc tagatgatat atgggtggat atatcttact tagtagtgtt ccattattag tatgtatctt gataatagga ttagccgagg gtagatttaa tgttttagtt tggggaagta gtaaaaatat gaacttaaat ttaatgtata taataatgat tatagtgttt tgtacaaaga ttccattagt tccatttcat agatgattgc cagtagttca tgcagaagcg agtagttcta caagtatcat attaagagga tacataatga agttaggttt agtaggtgtt gtccgtttat gtagtaagtg gatatgaaga tcaggtataa gcagagtttt agagtttata ttttattaa gactaggatt

(continued)

Table 4 (continued)

Sl. no.	Gene ID	Sequence key values (keys for encryption of data)
		tttaatttgg gcttattatg aaattgactc aaagcgttga ttgggtgtat taagtttatc tcatataatg gttagagttg ttttattagt ttatggatgt tatgatgtga aattattagc ttatattttt tgtcttggcc atggtttttc agtgtgtggg atgtttacac taatatgatg agggtataac tatgtagaat ctcgaaggtg attaatttta gctaggtttt atagttttgt acctgtaatt caagtgttgt gtgttcttat ttttataagt gtttctggtt ttccgcctac tctacaattt tttagtgaag taatattagt atgatgtagt ttaattgtaa gtaacgatat attaacattt attatatgaa tatatctgtt tgggagaagt ttagtggggt tactagtatt gggattagta ttagtttatt tatttgataa tagtagaatt attaggtgta ataataataa aagcttaagt gtaggtataa tttatattgt aatattaatg ttatctattt ttcttataat ttag
5	4097439	at gtttaaatta agttcagcat tattaacagt tatattttta gtattattac cagttatatt aactcattat aatgttttttg gagttaagag tttaaggggt aatcattata gtaatatatc agaatgatat tcaaggtttg aatgtgggtt tttagggcat gggttaaatg aaaaattttt tagcttttct tacttaaatc tattaatatt atttgtagtc tttgatttag aaatatcgtt attattaaat atagtttacg atggaatatg atattataca ttctggtgtt attttttttt ttttttttg ttgtttttttg gttacatggt agaacttaag ttaggttatg taaggtgaat taaataa
6	4097440	atgtatatgt tagagttagt tttaagaatc gagagtttaa taatgacttt attattagta gctttttata taatgagtga gcgtaaggtc ttgtcatata tacaattacg caagggacct aaaaaggttg gtattatggg actattacaa aggtttgctg attttataaa gttaataaat aagagtaagt ttaaaaattt tagatttcgc agatgatttt cgttgttagg ttgtatagtt ttaatttcat gttcagtgtc gatggtaata gtctatagat tagtaaatag gaatatatgg tgtaattgaa tattacttta tttttagtg gttggttcta tagtgagtta cagaacttta ttaataggct gatgttcgtg atcaaagtac agtctaatta gttctattcg tgtttctttt tctagtgtaa tgttcgagat gatactaatg tgcattatga tttgtttgg cttaatgtat ggtggttata ggaaacccag agttagaaga gtaatgttat tcatagcacc tttagctttt attgcatggt taatagtgtt attaagagag agtaatcgta caccatgtga ctattccgaa tctgaaagtg agctagtaag aggaattaga gtggaatata ggagagtatt atttttagtt atatttgctt gcgagtattt aataatgttt atatttagct gagtgagatt tatagtattt tgaagcataa atgaaatatt gatagtaata aacttgatgt tatttgtagt aatgcgtggt tcattttctc gacttcgatt tgatatattt gtttcagtag tttgaaaata ttgcgttgta gtgatactag tatatttaat atgtttattt agattgttat ag

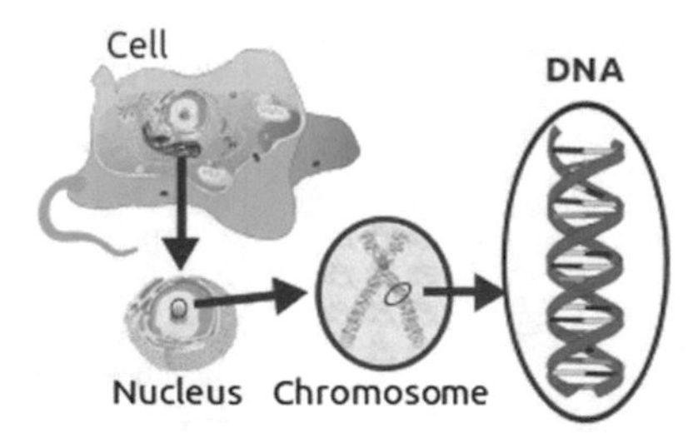

Fig. 1 DNA structure within a cell

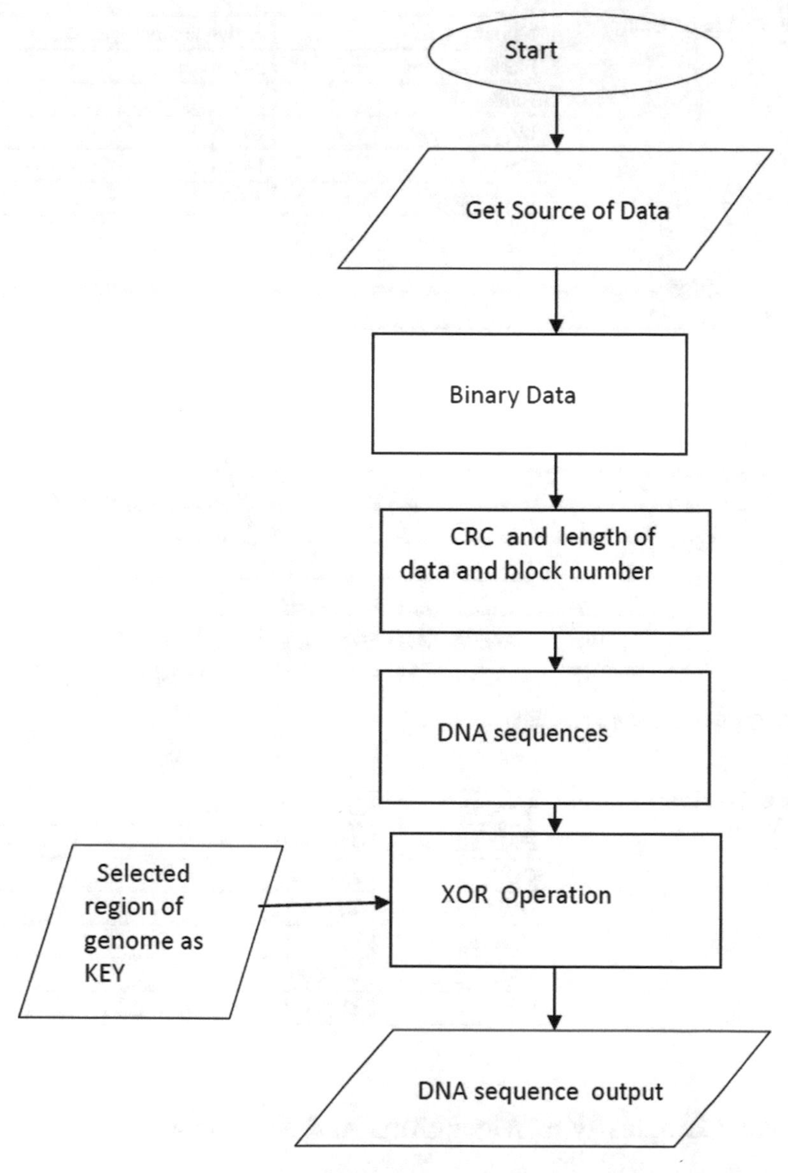

Fig. 2 Flow chart of the encryption process using an encryption algorithm to generate DNA sequence output

As shown in Table 5, data stored in a DNA format with specific preservation and storage of DNA.

In lab proved at −164 °C or dried for storage in centuries years with primer PCR reactions to reproduce DNA sequences for reading original data in terms of DNA letters (Fig. 3 and Table 6).

Table 5 Data durability

Memory type	Active archive (years)
Flash memory	2–3
Hard disk	3–5
Magnetic tape	10–15
Optical	15–50
DNA storage	4000

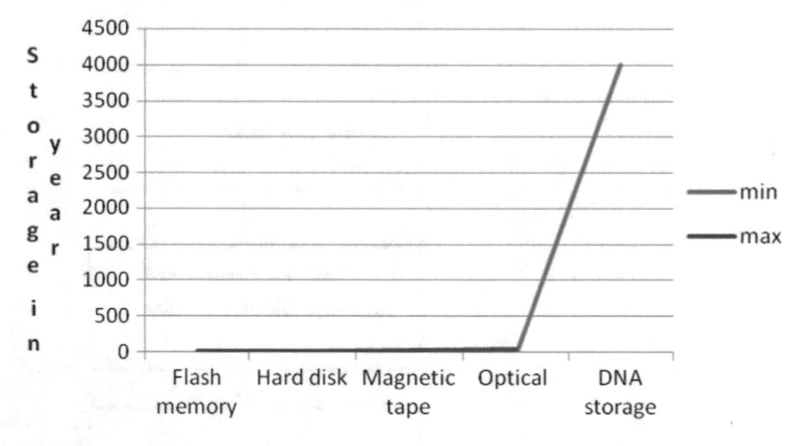

Fig. 3 Comparison of data durability

Table 6 Data storage capacity

Memory	Data storage (GB)
Flash memory	0.00391–0.0625
CD	0.68359
DVD	4.7
Blue ray	100
Tapes	122,880
DNA	22 PB/10 g of DNA or 2.2 EB/kg of DNA

4 The Complexity of Algorithms and Storage Media

The DNA of humans contains three billion base pairs such that the probability of a single letter being correct is one in three billion. The NCBI and European Molecular Biology Laboratory (EMBL) databases contain billions of both genomics and proteomics data sets. The proposed algorithm uses different genes from Table 1 and key values from Table 2. The likelihood of G3PD finding a correct letter is $1/1503 * 4 * 4 * 4 * 1103$. DNA based algorithms to encryption data compare to traditional algorithms fast and secure because performance of proposed algorithms is unbreakable and difficult to find key values and length. Data is retained long terms in DNA sequence format and stored in living organism and next generation we can retrieve original data in the form of DNA sequences. The proposed encryption and decryption algorithms for data security and storage in DNA sequences can store 10^8 TB of data per 1 g of created cipher data with a life span of 4000 years. In coming days, DNA as storage media for Big Data storage and the IOT.

5 Conclusion

DNA XOR operations have combinations that differ from the existing binary format of 0s and 1s, in that they are reflexive and unique. The proposed encryption and decryption algorithms for data security and storage in DNA sequences are very secure because genes sequences are used as key values for the DNA XOR operation. 10^8 TB per 1 g of created cipher data life span 4000 years. In coming days DNA as storage media for Big Data storage and IOT.

References

1. Watson, J.D., and F.H.C. Crick. 1953. *Nature* 171: 737.
2. Diffie, W., and M. Hellman. 1976. New directions in cryptography. *IEEE Transactions* on *Information Theory IT-22* 6: 644–654.
3. https://www.technologyreview.com/s/425237/ibm-builds-biggest-data-drive-ever/. Accessed 25 Aug 2011.
4. Siddaramappa, V., and K.B. Ramesh. 2015. *IEEE international conference on applied and theoretical computing and communication technology (iCATccT).*
5. National Center for Biotechnology Information. http://www.ncbi.nlm.nih.gov.
6. Online Mendelian Inheritance in Man. http://asia.ensembl.org/.
7. Nick, Goldman, Paul Bertone, Siyuan Chen, Christophe Dessimoz, Emily M. LeProust, Botond Sipos, and Ewan Birney. 2013. Nature, towards practical, high-capacity, low-maintenance information storage in synthesized DNA. *Nature* 494: 77–80.

Modeling a Data Rate-Based Peripheral Security System in GNU Radio

Urvish Prajapati, B. Siva Kumar Reddy and Abhishek Rawat

Abstract This chapter proposes a model for a peripheral security system based on data rate and specific signal format, which can also be used for authorization and recognition functions. This arrangement can be used for security management and monitoring of the inside or outside entry in a large peripheral area. Generally, conventional peripheral security systems detect every intruder and report authorization regardless of checking for authenticity. In the proposed method, identification of authorized persons depends upon exchange of certain fixed data rates between the transmitter and receiver modules. Exchange of data is initiated by the first layer of security provided by the IR sensor, and therefore this system can provide better protection from jamming devices or hacking. The proposed model of data rate-based peripheral security is tested and analyzed using an open-source software, GNU Radio.

1 Introduction

Perception of security is a state-of-mind concept, yet the magnitude of demands and the rising costs of security systems require new and improved capabilities and innovative approaches. A safe and secure infrastructure is an attractive and prosperous place where the respective authorities feel safe to work and live. Security is critical in areas such as military units, prisons, and nuclear power plants, where rapid response

U. Prajapati (✉) · B. S. K. Reddy · A. Rawat
Institute of Infrastructure Technology Research and Management, Ahmedabad, India
e-mail: urvishkumar.prajapati.16me@iitram.ac.in

B. S. K. Reddy
e-mail: bskreddy@iitram.ac.in

A. Rawat
e-mail: arawat@iitram.ac.in

A. N. Krishna et al. (eds.), *Integrated Intelligent Computing, Communication and Security*, Studies in Computational Intelligence 771,
https://doi.org/10.1007/978-981-10-8797-4_37

is essential. Among existing security systems, video surveillance-based systems are the most widely used today. The global video surveillance industry is expected to grow at a compound annual growth rate (CAGR) of 11.87% over the period from 2015 to 2021, reaching a total market size of US$48.69 billion by 2021 [1].

The proposed peripheral security system based on data rate was designed in GNU Radio, which is an open-source software development toolkit for building any type of real-time signal processing application. It offers methods for manipulating data flow between the blocks. Users can create new signal processing blocks, and generally they are written in C++ and put together in Python flow-graphs.

2 Literature Survey

Among the models described in the literature, Liu et al. reported work on a video surveillance system to improve tracking accuracy in complex scenes involving changing speeds and directions, and proposed advance tracking systems that monitor the moving human across many cameras using an improved color texture feature fusion (ICTFF) algorithm [2]. Such systems are expensive for monitoring large areas. Wei et al. developed an intelligent anti-theft alarm system for a large area of substation using a laser light approach to transmit the accurate location of intrusion using Zigbee devices [3]. Allwood et al. reviewed optical fiber-based techniques used in physical intrusion detection, in which the optical fiber itself can act as a transmission medium and sensor [4].

Assaf et al. designed a monitoring system which facilitates web-based user interaction and alerts on an email [5]. It is a challenging task to handle the heterogeneous data and to provide features such as availability, security, and reliability in IOT systems [6]. Chilipirea et al. discussed different CPU frequencies and different amounts of RAM in IOT devices, which creates difficulties in the interaction between devices and in applying standard fault tolerance techniques along with features such as energy efficiency and robustness [7]. Kaur et al. proposed the concept of a wireless multi-functional robot that can be a stand-in for a soldier at some border areas [8].

Previously proposed security systems have been expensive, with various limitations, and sound the alarm for every interruption without authenticity information. Therefore, in our contribution, a data rate-based peripheral security system is proposed to overcome such limitations and gives the authenticity status of an intruder by detecting the false entry. It also updates the security feature from time to time with real-time monitoring. The proposed system has the ability to secure and monitor large remote areas using the local command unit.

3 Proposed Design

The main objective of the data rate-based peripheral security system is to provide security for a large peripheral area with authenticity information and to communicate the exact location of intrusion. There are six subsystems, namely the transmitter module (T), receiver module (R), sensor transmitter (ST), sensor receiver (SR), local command (LC) unit and central command (CC) unit. ST–SR modules are planted in the probable unauthorized entry points, and each pair has a wired or wireless connection to the receiver module, which is activated when someone passes from this area, as shown in Fig. 1. The entire geographic area is classified into predefined sub-sectors, and each has one receiver module (R), which sends authenticity status to the LC unit through a transceiver. The LC unit communicates with the receiver module (R) and transmitter module (T), and also with the CC unit via satellite.

In the proposed system, each authorized person is equipped with a transmitter module which continuously transmits an RF signal with a certain predefined data rate and specific signal format in its active mode. It is then detected and cross-verified to identify the authorized entry. This module can be designed in such a way that it can be secretly planted in a shoe or mobile device. The number of transmitter modules (T) varies with the number of authorized persons and associated devices that are to be allowed, but the number of receiver modules (R) is fixed based on various factors of geographic locations. Each transmitter module consists of a transmitter unit, micro-controller and a separate receiver unit (to communicate with the LC unit) with a battery for power supply. Transmitter modules use the amplitude-shift keying (ASK) modulation technique to transmit a signal, and receiver modules demodulate these signals.

Whenever an intruder or object enters a secured area, the sensor module detects the interruption and initializes the receiver module of that sub-sector. A micro-controller which is part of the receiver module searches for the specific data rate. This data rate will be in a certain fixed limit in a certain area of a transmitter based on its range and analysis using GNU radio software, which will be discussed in Sect. 4. An unauthorized intruder will not be able to communicate due to lack of a predefined data rate and a specific signal, and the receiver module immediately reports to the LC with location information. At the LC, an alarm is triggered or notified on LCD, and it stores the status and location information of the intruder in the registry of the LC and further reports to the CC about unauthorized entry in a particular area.

When an authorized person enters a secured area, the micro-controller identifies that the received signal data rate matches the predefined data rate limit, and the receiver module then sends information to the LC about authorized entry. The LC stores this information in its registry. Other features help the system to be energy-efficient and robust in nature. The receiver module is activated only when the sensor receiver is interrupted, because it is important that predictive applications are not active at all times, in order to conserve energy. The receiver modules and LC unit are both self-checking the overall operations, so in case of any physical damage, fail-

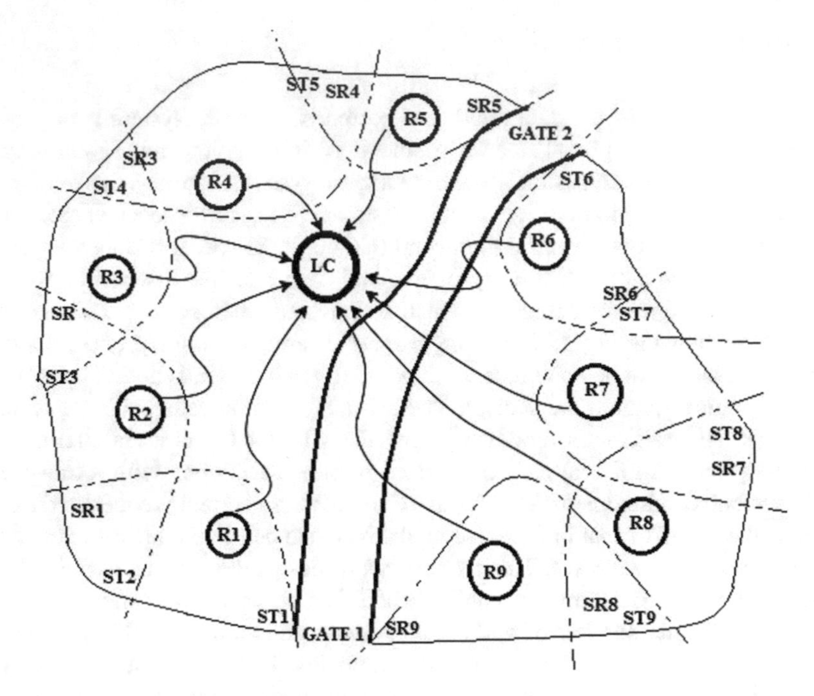

Fig. 1 Proposed design for peripheral security using nine IR sensors

ure of operational discipline or any type of attack, a warning is issued to investigate the entire system. A published patent describing this method is included [9].

4 Results and Discussion

A model of the 433 Mhz ASK transmitter module and a receiver module is developed in GNU radio as shown in Fig. 2. The signal source block generates the information signal of 100 Hz; square waves and other signal source blocks generate the sinusoidal carrier waveform of 433 MHz. Amplitude levels of a signal can be changed by reconfiguring and adding constant value.

The transmission channel model is used to reproduce the behavior of the signal during transmission in the form of an electromagnetic wave. There are different types of channels available to analyze different types of interference and noise. Additive white Gaussian noise (AWGN) is a channel model that has constant spectral density (expressed as watts per hertz of bandwidth) and a Gaussian distribution of amplitude. This model is represented by the component "Channel Model", which allows the user to adjust the intensity of AWGN noise with different values. A multiple input single output (MISO) model can be developed in GNU radio to analyze a case when more than one intruder enters simultaneously and all need to be authenticated.

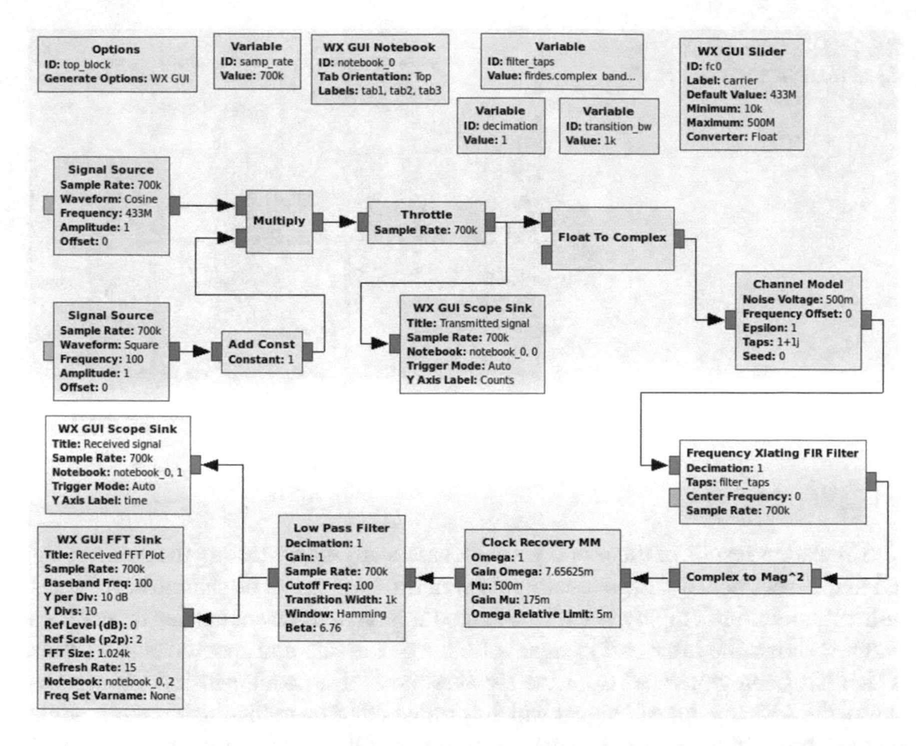

Fig. 2 Developed model of 433 MHz ASK transmitter module and receiver module in GNU radio

Fig. 3 Information signal of a specific data rate and signal format

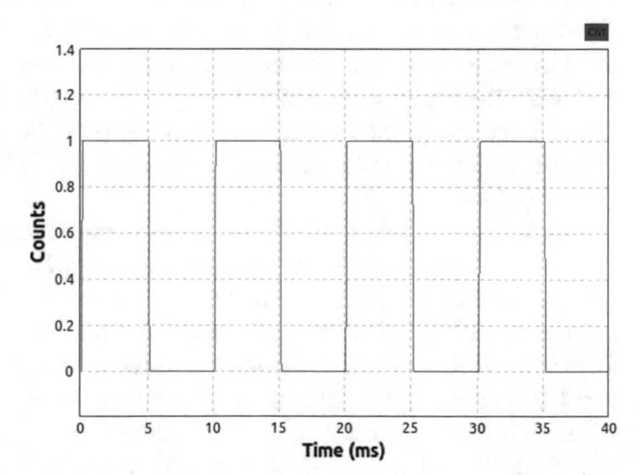

Fig. 4 ASK transmitted signal from the transmitter circuit

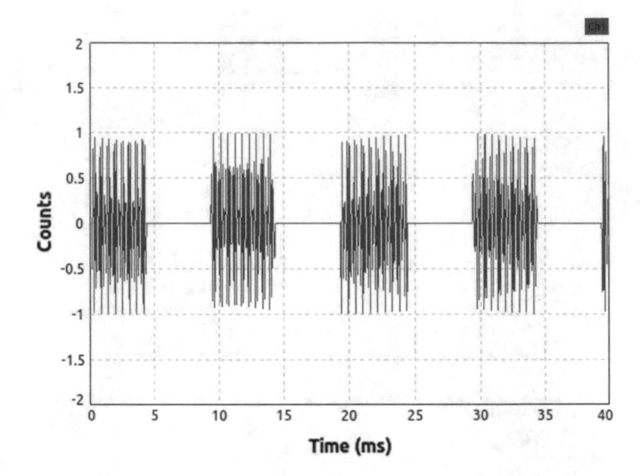

Simulation results of the proposed model are analyzed by testing the receiver signal at different noise voltage levels and using different types of channels. All transmitter modules are employed with ASK modulation at different carrier frequencies. Figure 3 shows the information signal of a fixed data rate and specific signal format which has been generated from the signal source of a developed model. Figure 4 shows the ASK modulated signal which is transmitted from the transmitter module. Figure 5 shows the recovered signal at the output of the receiver module in WX GUI scope plot, and Fig. 6 shows the corresponding signal's FFT plot.

The recovered signal is slightly destroyed by the information signal because of noise in the channel between the transmitter and receiver. As the distance between them increases, the signal power value decreases. Hence, more noise in the channel will affect the signal when moving away from the transmitter. In wireless communication, the transmitted signal follows many different paths before arriving at the receiver, which is due to scattering, reflections and refraction from buildings and other objects.

In addition to multi-path effects, noise is introduced to the transmitted signal. Most of the source noises are wideband and additive, coming from many natural sources such as the thermal vibrations of atoms, electrical noise in the receiver amplifiers, shot noise, black body radiation from the earth and other interference. Rayleigh and Rician fading channels are used to model narrowband and wideband wireless channels in a number of realistic wireless communication environments. In a Rayleigh fading model, objects in the environment scatter and reflect the transmitted signal so that there is no line-of-sight (LOS) path, and the signal becomes constructive and destructive [10]. In a Rician channel, there is a line-of-sight path between the transmitter and the receiver so that the received signal comprises both the directional and scattered (or reflected) paths [10].

The received signal is demodulated, and the bit error rate (BER) of the recovered signal is analyzed to check the permissible noise level. Differences in the (BER) are analyzed with different noise levels and sinusoids, as shown in Tables 1 and 2. From

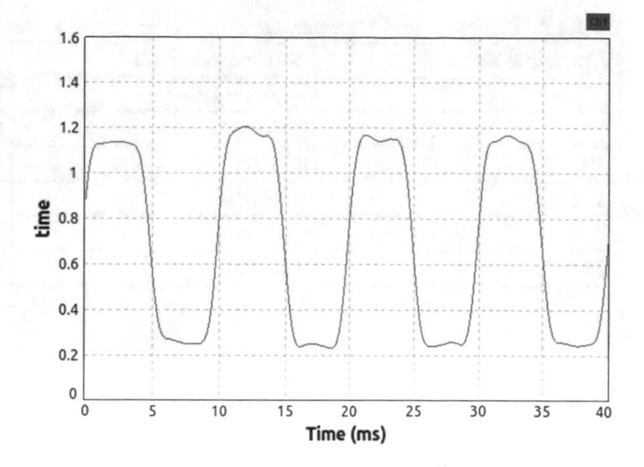

Fig. 5 Recovered signal at receiver output

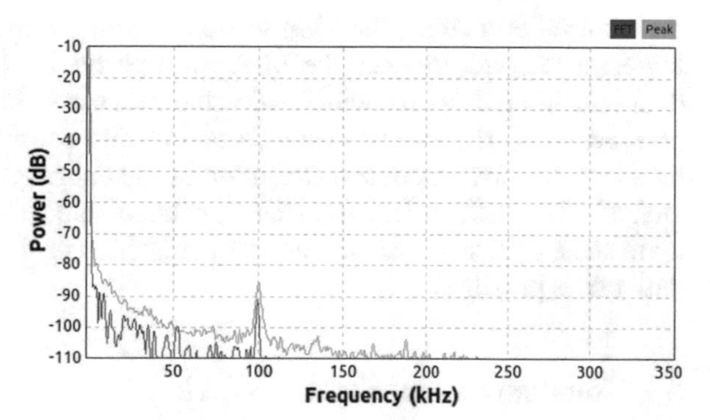

Fig. 6 FFT plot of recovered signal

Table 1 Bit error rate at different AWGN noise levels

Noise voltage (mv)	Bit error rate (units)
5	0.467327
10	0.475538
15	0.482994
20	0.488470
25	0.488565
100	0.489134
200	0.490182
300	0.492629
400	0.497369
700	0.525828
800	0.543855
900	0.676202

Table 2 Bit error rate at frequency selective fading model in Rayleigh and Rician fading

Number of sinusoids	Bit error rate (units)	
	Rayleigh fading	Rician fading
5	0.458564	0.466219
10	0.510496	0.514296
15	0.565850	0.551947
20	0.632114	0.582658
25	0.681203	0.611168

these readings, we can observe that the BER of the corresponding signal increases gradually as the noise level or number of sinusoids increases. The transmitted signal path loss depends on the noise level of the channel model, so moving far away from the transmitter module, the additive noise at the receiver module increases, which increases the path loss and the BER. At a particular distance, the received signal data rate differs from its predefined value because of high BER, so data rate-based peripheral security can be designed taking into consideration the received signal data rate. Receiver module distribution in the entire geographical area follows this analysis. The BER and the received signal data rate will be in some predefined limit up to some particular distance, and authentication on the basis of the received signal data rate is possible.

5 Conclusion and Future Work

In this chapter we have proposed a data rate-based peripheral security system and developed a model in GNU radio. The results from this model also cross-verify the proposed approach. At a certain distance from the transmitter, the BER and received signal data rate are within a predefined limit, and variations can be considered to authenticate the intruder. Data rate-based peripheral security is a unique way of providing security, which also facilitate authentication of every intruder. The proposed system is cost-effective and is also well suited for securing large peripheral areas at remote locations.

In the future, the proposed model will be tested experimentally by employing USRP N210 hardware. We will extend the work for more users with a larger coverage area.

References

1. Mali, D. 2017. Role of surveillance in securing cities, Smart Cities Council India.
2. Liu, J., K. Hao, Y. Ding, S. Yang, and L. Gao. 2017. Moving human tracking across multi-camera based on artificial immune random forest and improved color-texture feature fusion. *The Imaging Science Journal* 65 (4): 239–251.

3. Wei, C., J. Yang, W. Zhu, and J. Lv, 2010. A design of alarm system for substation perimeter based on laser fence and wireless communication. In *2010 International Conference on Computer Application and System Modeling*, V3-543–V3-546, Taiyuan.
4. Allwood, G., G. Wild, and S. Hinckley. 2016. Optical fiber sensors in physical intrusion detection systems: a review. *IEEE Sensors Journal* 16 (14): 5497–5509. 15 July 2016.
5. Assaf, M., R. Mootoo, S. Das, E. Petriu, V. Groza, and S. Biswas. 2014. Designing home security and monitoring system based on field programmable gate array*. *IETE Technical Review* 31 (2): 168–176.
6. Silva, B., M. Khan, and K. Han. 2017. Internet of things: A comprehensive review of enabling technologies, architecture, and challenges. *IETE Technical Review*.
7. Chilipirea, C., A. Ursache, D. Popa, and F. Pop. 2016. Energy efficiency and robustness for IoT: Building a smart home security system. In *2016 IEEE 12th International Conference on Intelligent Computer Communication and Processing (ICCP)*, 43–48, Cluj-Napoca.
8. Kaur, T., and D. Kumar. 2015. Wireless multifunctional robot for military applications. In *2015 2nd International Conference on Recent Advances in Engineering & Computational Sciences (RAECS)*, 1–5, Chandigarh.
9. Rawat, A., D. Deb, V. Rawat, and D. Joshi. 2017. Methods and Systems for Data Rate Based Peripheral Security, India Patent 201721005324A, 24 February 2017.
10. Nguyen, D. 2013. *Implementation of OFDM systems using GNU Radio and USRP*. Master by Research - Engineering: University of Wollongong.

Adiabatic SRAM Cell and Array

Ayon Manna and V. S. Kanchana Bhaaskaran

Abstract A new SRAM cell with better leakage control and enhanced memory retention capability is proposed. The memory array configured using the proposed SRAM cell has all its word lines and bit lines driven adiabatically using differential cascode and pre-resolved adiabatic logic (DCPAL), and it operates as a buffer for the memory array. This paper demonstrates how one of major concerns, namely, the VT variation can be controlled by modifying the ground-line and power-line voltage of the SRAM cell. The designs are implemented using 45-nm technology models operating at a supply voltage of 0.8 V.

Keywords SRAM cell design · Adiabatic logic circuit · Leakage current in memory · Adiabatic buffer circuits

1 Introduction

With power dissipation being a prime concern of very-large-scale integration (VLSI) design, researchers have presented various conventional low-power design solutions, with the major approaches being a reduced supply voltage, scaling the device size, etc. However, they incur limitations beyond a certain level, leading to newer and non-conventional techniques, such as the adiabatic or energy recovery circuits. The adiabatic circuits found in the literature, viz. 2N-2P, 2N-2N2P and positive-feedback adiabatic logic (PFAL), possess better power performance characteristics and drive than the conventional low-power circuits [1]. This prompts the use of these adiabatic circuits in driving the bit and word lines of a static random-access memory (SRAM) cell. In any nanoscale SRAM circuit, the fluctuation in the quantity of dopant atoms and resulting threshold voltage variations pose

A. Manna · V. S. Kanchana Bhaaskaran (✉)
School of Electronics Engineering, VIT University Chennai, Chennai 600 127, India
e-mail: vskanchana@gmail.com

A. Manna
e-mail: ayon.manna19@gmail.com

A. N. Krishna et al. (eds.), *Integrated Intelligent Computing, Communication and Security*, Studies in Computational Intelligence 771,
https://doi.org/10.1007/978-981-10-8797-4_38

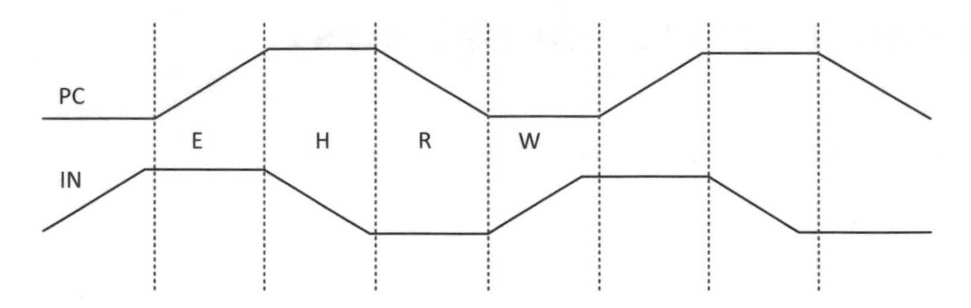

Fig. 1 Power clock and input signal

serious issues to the designer [2]. This concern can be addressed by changing the power/ground-line voltage, as discussed later in this paper.

The pass transistor logic is a popularly used alternative to the complementary metal-oxide semiconductor (CMOS) logic. While the CMOS possesses great advantage in configuring any logic function, the number of transistors required is quite large. This fact leads to the choice of pass transistor logic, employing the bare minimum quantity of transistors for implementing a logic function. Furthermore, while the CMOS logic experiences increased logic effort proportional to the primary inputs, the pass transistor logic allows the primary inputs to drive the gate and source terminals, thus incurring reduced logic efforts.

The adiabatic logic circuit works on the principle of reclaiming the expended energy from the charged output nodes through recovery paths. These circuits are more efficient in handling larger load capacitance than any other low power circuits. The novel adiabatic logic circuit arrangement proposed in this paper exhibits lower non-adiabatic power dissipation than some of the other adiabatic circuits such as the 2N2N-2P and PFAL [1, 3]. The power source supplying the adiabatic circuit is called the power clock (PC) as shown in Fig. 1. The adiabatic circuits employing the four-phase PC have their operating phases referred to as evaluation (E), hold (H), recovery (R) and wait phases (W) [1]. Evaluation occurs when the PC slowly rises from zero to the peak voltage, which is held during the hold phase. Then, the PC slowly falls from its peak to zero voltage, during the recovery phase. Before the PC rises again, 0 V is maintained during the wait phase, and this feature allows adiabatic cascading.

2 Adiabatic Logic Circuits

The 2N-2P circuit comprises two cross-coupled positive-channel metal-oxide semiconductor (pMOS) transistors, which act as a latch. The 2N-2N2P circuit has its functional block and its complement in the pull-down network, while PFAL comprises a negative-channel metal-oxide semiconductor (nMOS) functional block and its complement in the pull-up network along with the 2N-2P latch. The input signals lead the PC by 90° [1, 3]. For these adiabatic circuits, the energy loss per cycle is given by the following equations:

$$E_{2N2P} = \left(\frac{2R_P C_L}{T}\right) C_L V_{DD}^2 + C_L V_{TP}^2 \quad (1)$$

$$E_{2P2P2N} = \left(\frac{2R_N C_L}{T}\right) C_L V_{DD}^2 + C_L V_{TN}^2 \quad (2)$$

Here, C_L is the load capacitance and R_P and R_N are the turn-on device resistances of the pMOS and nMOS devices, respectively. V_{DD} is the peak PC voltage with T being the transition time and V_{TP} and V_{TN} being the threshold voltages of the devices. For any adiabatic circuit, the non-adiabatic energy loss should be minimal to derive the operational benefit.

In Eqs. 1 and 2, the first and second terms represent the adiabatic and non-adiabatic energy loss, respectively. The equations also depict the fact that adiabatic energy can be lowered by operating the circuit at very low frequency ranges. It can also be inferred that the non-adiabatic energy loss is determined by the load capacitance. Hence, it can lead to the conclusion that the adiabatic circuit cannot be used in driving load capacitances of large values, such as the bit line and word line capacitances of SRAM cells [4–7]. However, the DCPAL adiabatic logic circuit shown in Fig. 2a overcomes the difficulty of driving high load capacitance, with reduced nodal capacitance values and circuit architecture. The output nodes (out or out/) charges and discharges through P1 and P2 based on the inputs IN and IN/. During recovery, the charge stored on the output nodes gets fed back to the PC through the feed-back paths formed through P1 and P2. The comparative power energy dissipation characteristics of the adiabatic circuits while driving 200-pF load capacitance are depicted in Fig. 3a, b. In DCPAL, with the nMOS device present as a footer, the leakage current is reduced, thus reducing leakage power dissipation.

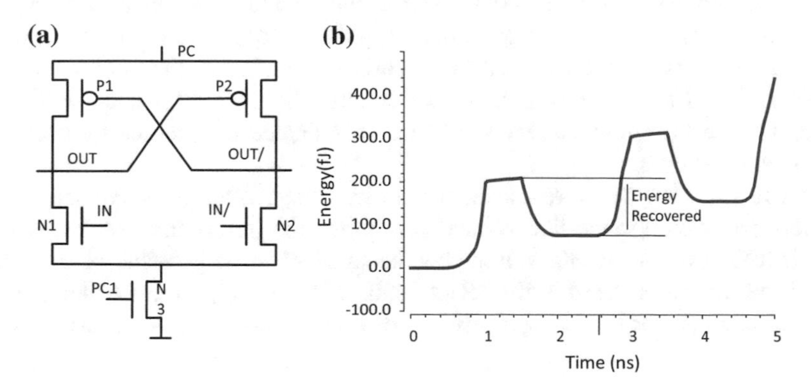

Fig. 2 DCPAL adiabatic circuit: **a** circuit diagram, **b** energy recovery behavior

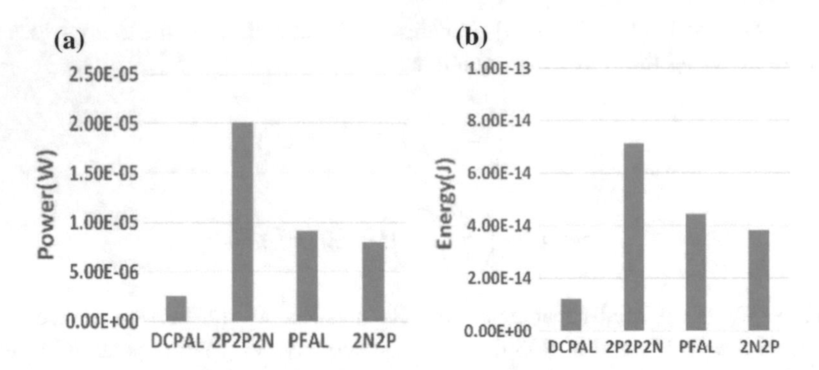

Fig. 3 **a** Power and **b** energy while driving a 200-pF load capacitance

Another advantage of DCPAL logic is its enhanced energy recovery capability as compared to the adiabatic circuit counterparts, as validated in Fig. 2b. Ensuing sections describe how the adiabatic circuit can be employed in SRAM circuits.

3 Design of an Adiabatic SRAM Cell and SRAM Array

Figure 4a depicts the proposed SRAM cell structure. This cell consists of four pMOS devices marked P1, P2, P3 and P4 and four nMOS devices N1, N2, N3 and N4, forming the storage or latch. Since the lower-technology-node devices are employed in recent VLSI circuits, problems related to the threshold voltage (V_{TH}) variation arise, because of the extended agility and statistical fluctuation of the dopant atoms [8]. This variation in threshold voltage results in increased leakage in SRAM cells. To overcome this disadvantage, the proposed SRAM cell uses two devices, N1 and N2, in series, used as a divided footer in one terminal, and a second set of N2 and N4 devices in series, used in the second terminal. In this way, the resistance of the path connecting the output node to the ground is increased. In any SRAM cell, it is important to maintain constant voltage in the output node, for successful read and write operations. To achieve this aim, P1–P2 and P3–P4 are used as headers as shown in Fig. 4a. Figure 4b, c depicts the transients during write and read operations.

An alternate way to overcome the threshold voltage variation is by changing the power-line or the ground-line voltage [2]. When the power line (MCPL)/ground line (MCGL) voltage changes, it affects the threshold voltage value of the header and footer transistors used in the SRAM cell. With the help of this technique, one can actually counter the leakage current, resulting in lower power dissipation.

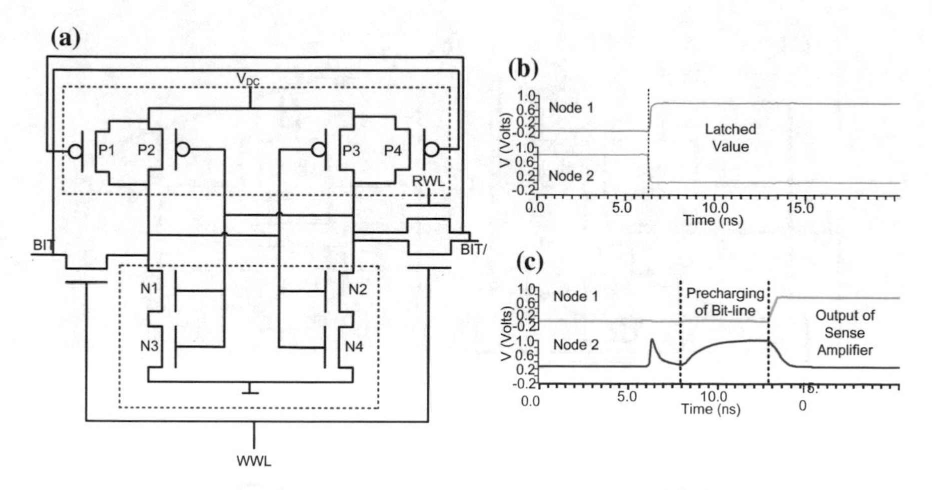

Fig. 4 **a** Proposed SRAM cell, **b** write operation and **c** read operation

3.1 Address Decoder

An 8 × 8 memory array using the proposed SRAM cell is shown in Fig. 5, with individual decoders for addressing, during read and write. Here, two 3 × 8 decoder circuits for row addressing and a single 3 × 8 decoder are used for column addressing. Since two 3 × 8 decoders are used for row addressing, two level address decoding is done using 6-bit address. The option of row and column addressing is controlled by the write enable (WE) and read enable (RE) signals. The write/read word line selection signals or the write/read address decoding signals are

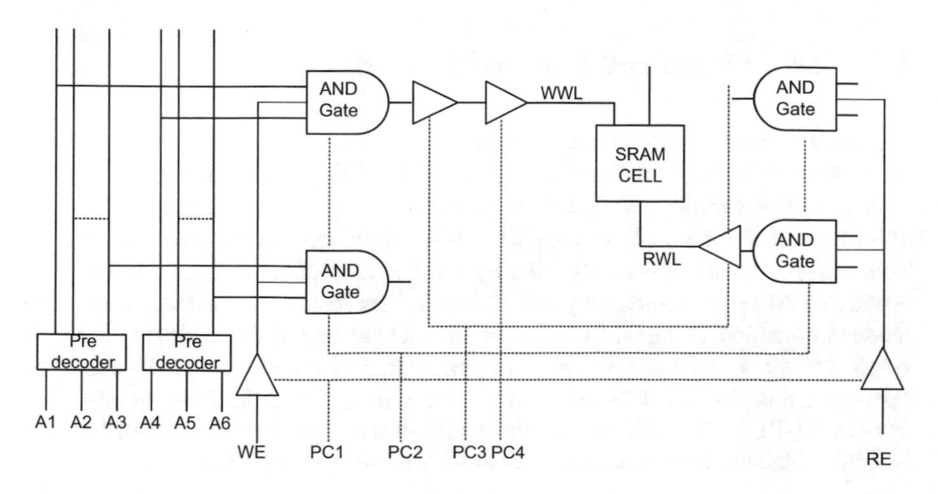

Fig. 5 SRAM cell array with a row and column address decoder

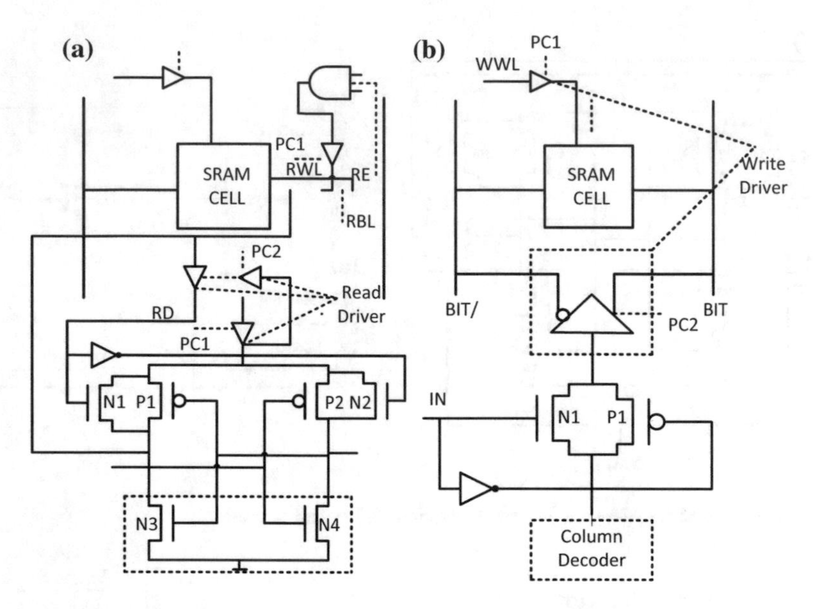

Fig. 6 **a** SRAM cell with read driver, **b** SRAM cell with write driver

produced using a three-input AND gate, with the write/read enable signal acting as one of the inputs and the remaining inputs taken from the two 3 × 8 row address decoders. Because of the ability of driving larger load capacitances, all the bit lines, write word lines (WWLs) and read word lines are effectively driven by the DCPAL circuit, as shown in Fig. 6a, b. Successful read/write operations depend on the timing of the PC applied to each of the drivers in the array.

3.2 *Sense Circuit and Read/Write Driver*

Figure 6a shows the sense amplifier circuit closely resembling the dual-transmission gate adiabatic logic circuit (DTGAL) [5]. Figure 6b depicts the write operation circuit with the write driver. After write, because of the pre-charging of bit lines, the output node voltage of the SRAM cell changes by an optimal value. This change is sensed and magnified by a sense amplifier. As shown in the proposed SRAM cell structure, only one of the bit lines (read bit line, RBL) is used for the read operation. A signal from this bit line acts as one of the inputs to the sense amplifier and a reference voltage is applied to the other. The sense amplifier operates using the PC. Charge recovery takes place from the RBL through the devices N1-P1 (or N2-P2). This whole charge recovery process is controlled by the DCPAL adiabatic circuit, whose input is also taken from the RBL.

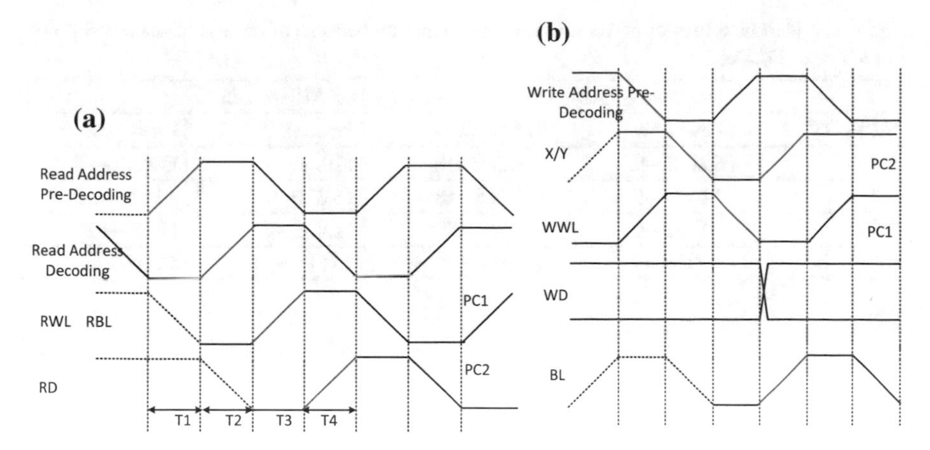

Fig. 7 **a** Read and **b** write operation transients

3.3 *Read and Write Operation Timing*

The use of an adiabatic circuit as an integral part of adiabatic SRAM array makes the timing quite pertinent. The waveform transients for read are shown in Fig. 7a. The row address decoding (or the pre-decoder) the signal and RE signal are processed during T1. The outputs of the AND gates are processed during the T2, for generation of the read word line (RWL) selection signal. During the time T3, the output signal of the adiabatic buffer is processed and the required read word line is selected.

During T4, the signal passing through the RBL is sensed by the sense amplifier. Thus, the read operation is completed within one clock cycle. Figure 7b portrays the timing of the write operation. During time T1, the row address decoding (pre-decoder) signal is processed. During T2, the outputs of AND gates are processed for generating the WWL selection signal. During T3, WWL is selected. During T4, input is applied to bit-lines, driven by the proposed adiabatic logic circuit. During T5, the write operation is completed by changing bit line voltage within one clock cycle.

3.4 *Simulation*

All the adiabatic logic circuits employed in the array structure of the proposed SRAM cell are simulated using ideal four-phase PC signals. Figure 4b–c shows read and write operations of the SRAM cell. Table 1 validates the use of DCPAL-based cell for its reduced power and energy values against the 6T SRAM cell for different operating voltages. Table 2 demonstrates the power dissipation control of SRAM cell through ground- and power-line voltage (MCGL). The

Table 1 Different values of power and energy dissipation because of change of main cell power line voltage (MCPL)

	Proposed SRAM cell		6T SRAM cell	
MCPL (V)	Power (W)	Energy (J)	Power (W)	Energy (J)
0.8	39.62E−9	318.28E−18	110.6E−9	888.5E−18
0.7	14.70E−9	117.21E−18	16.88E−9	134.6E−18
0.65	10.28E−9	79.37E−18	11.43E−9	88.26E−18
0.6	5.71E−9	42.98E−18	7.88E−9	59.32E−18

Table 2 Different values of power and energy dissipation because of change of main cell ground line voltage (MCGL)

MCGL (V)	Power (W)	Energy (J)
0.2	63.27E−9	535.3E−18
0.3	70.40E−9	598.7E−18
0.4	80.66E−9	687.6E−18
0.5	84.76E−9	729.1E−18

Table 3 Average flow of current through footer transistors of the proposed and 6T SRAM cells

	Proposed SRAM cell	6T SRAM cell
Average current (A)	0.008970	0.009635

current flowing through the footer transistors of the recommended SRAM cell is found to be lower than the conventionally structured 6T SRAM cell as demonstrated in Table 3.

4 Conclusion

A low-power SRAM cell and memory array using DCPAL logic are presented. The proposed circuits consume significantly less power due to its structure and the use of adiabatic read/write drivers than 6T SRAM-based circuits. The proposed cell incurs 39.62 nW of power consumption as compared to 110.6 nW of the 6T SRAM cell, which is 64.17% lower. The proposed read/write drivers of the memory array help in recovering charge from large nodal capacitance of SRAM memory structures, as validated through the experimental results.

References

1. Kanchana Bhaaskaran, V.S. 2011. Energy recovery performance of quasi adiabatic circuits using lower technology nodes. In *India international conference on power electronics 2010 (IICPE2010)*, 1–7, New Delhi.
2. Nakata, S., T. Kusumuto, M. Miyama, and Y. Matsuda. 2009. Adiabatic SRAM with a large margin of VT variation by controlling the cell power line and word line voltage. In *Proceedings ISCAS digest*, 393–396.
3. Kanchana Bhaaskaran, V.S., S. Salivahanan, and D.S. Emmanuel. 2006. Semi-custom design of adiabatic adder circuits. In *Proceedings of 19th international conference on VLSI design and embedded systems design*, 745–748.
4. Hu, J., H. Li, and H. Dong. 2005. A low-power adiabatic register file with two types of energy-efficient line drivers. In *48th Midwest symposium on circuits and systems*, 1753–1756.
5. Hu, J.P., X.Y. Feng, J.J. Yu, and Y.S. Xia. 2004. Low power dual transmission gate adiabatic logic circuits and design of SRAM. In *47th Midwest symposium on circuits and systems*, 565–568.
6. Sudarshan, Patil, and V.S. Kanchana Bhaaskaran. 2017. Optimization of power and energy in FinFET based SRAM cell using adiabatic logic. In *IEEE International conference on Nextgen Electronic Technologies: Silicon to software (ICNETS2)*. Chennai, 23–25 March 2017.
7. Dinesh Kumar, S., S.K. Noor Mahammad. 2015. A novel adiabatic SRAM cell implementation using split level charge recovery logic. In *IEEE 19th international symposium on VLSI design and test (VDAT)*, 1–2.
8. Nakata, S., H. Suzuki, T. Kusumuto, S.I. Mutoh, H. Makino, M. Miyama, and Y. Matsuda. 2010. Adiabatic SRAM with a shared access port using a controlled ground line and step-voltage circuit. In *Proceedings of 2010 IEEE international symposium on circuits and systems*, 2474–2477.

A Comparative Study on Prominent Strategies of Cluster Head Selection in Wireless Sensor Networks

Priti Maratha and Kapil

Abstract Wireless sensor networks (WSNs) have an imprint in every aspect of human life from very small to large-scale application. On one hand, energy-constrained sensor nodes are expected to run for a long duration. On the other hand, designing and maintaining sustainable WSNs is a very major issue nowadays. It may be very costly either to replace expired batteries or even impossible in hostile environments. So, there is a necessity to conserve the energy of the nodes so as to extend the network lifetime. After a deep review of the literature, we found that network lifetime can be extended by dividing it into groups (clusters). Decisions regarding all the operations in all groups are made by respective cluster heads (CHs). But selecting the best CHs is a critical issue to be resolved so as to utilize energy consumption most efficiently. This study can be a recommendation for researchers while optimally selecting WSN CHs.

Keywords Wireless sensor networks · Network lifetime · Wireless node Cluster · Clustering schemes

1 Introduction

Over the past few years, as micro-electro mechanical system (MEMS) technology has advanced, significant developments have also emerged in wireless sensor networks. A wireless sensor network (WSN) is a self-organized network of small sensor nodes (motes) which are deployed in large quantities to sense, monitor and understand the physical world. The nature of the surrounding environment can be dynamic, so the deployed nodes can be stationary or mobile. However, the batteries in these systems generally have very limited energy, and battery replenishment may be impossible in hostile environments. In addition, some applications require networks with longer lifetimes (e.g. months, years) [1].

P. Maratha (✉) · Kapil
National Institute of Technology, Kurukshetra, India
e-mail: niki.maratha19@gmail.com

A. N. Krishna et al. (eds.), *Integrated Intelligent Computing, Communication and Security*, Studies in Computational Intelligence 771,
https://doi.org/10.1007/978-981-10-8797-4_39

In some applications, renewable energy sources such as solar or wind have been recently examined and integrated into WSNs for longer operation. However, the intermittent nature of these sources still has a significant effect on network performance, and additional battery power is needed as well. Among resources, energy is critical and must be used with great efficiency [2, 3]. Therefore, limiting energy consumption is a critical consideration in the design of systems based on WSNs. One promising technology for conserving energy that has recently emerged is clustering [4–8].

The research community has widely adopted the idea of making groups or clusters of these wireless devices in order to enhance the network lifetime and scalability. Corresponding to each cluster, there is a leader, a cluster head (CH), which fetches data from other cluster members. The close-packed sensed data is relayed to a sink node after being aggregated. By using this principle, the volume of data communication within a network can be reduced and energy conserved. Many clustering methods [5, 9–19] have been introduced, but they have not focused on the crucial requirements of WSNs such as enhancing their lifetime and connectivity. Our work differs from other literature surveys as follows:

- This work presents a new taxonomy of novel clustering protocols that includes fuzzy logic (e.g. CHEF, EAUCF) for the selection of CHs.
- It is the first detailed review of cluster head selection by prominent clustering methods developed for WSNs.

2 Classification of Clustering Attributes

2.1 Cluster Properties

In this sub-section, we classify the WSN clustering algorithm by exploiting various sets of attributes, which will describe how intra- and inter-cluster communication will take place. In the following subsections, we specify some significant attributes:

- **Stability**: There are times when the total CH count is not predefined manually by the network operator and the node membership status evolves (changes) with time such that the resulting clustering protocol is said to be adaptive. Otherwise, the total count is assumed to be fixed because there is no switching of the sensors among different clusters, and hence the CH count remains fixed during the entire network lifetime.
- **Intra-cluster topology**: In some algorithms based on clustering, a sensor and its respective CH communicate directly. But it observed that when the communication between the sensors and CHs is done on a multi-hop basis, it is found to be more beneficial, especially when either the CH count is fixed, i.e. defined manually, or the communication range of the nodes is small.

- **Inter-CH connectivity**: Under circumstances when CHs cannot establish a long-range connection with a sink node, clustering schemes need to ensure connectivity between each CH and sink node.

2.2 CH Capabilities

In this sub-section, we discuss different attributes related to the various capabilities of CHs which vary from application to application:

- **Mobility**: Since a sensor's communication range changes dynamically when CHs are mobile, the size of the clusters needs to be continuously adjusted. In contrast, well-balanced clusters are formed if the CHs are stationary. Due to balanced cluster formation, the process of intra- and inter-cluster network adjustment is quite easy.
- **Node types**: As discussed earlier, when sensor nodes are deployed in some applications, some of them are declared CHs and the rest are just members of their respective CH group. In some protocols, deployed nodes can be heterogeneous by being equipped with more communication and computation resources.
- **Role**: It is the duty of the designated CHs to aggregate data once they collect it from their cluster members. This aggregated information is transferred to a base station (BS) by the CH. Generally, a CH is treated as a local master station. So, depending on the target event, CHs make all the decisions in their cluster.

2.3 Clustering Process

In this subsection, we discuss the attributes considered significant when cluster formation takes place and CHs are selected:

- **Methodology**: Selection of CHs can be done via two approaches. Either a centralized authority, i.e. network operator, separates the wireless nodes offline when different types of nodes are deployed and takes control of the cluster membership, or a distributed concept can be utilized when CHs are simple sensors nodes. When the CHs have abundant resources, hybrid algorithms are suggested. In the latter case, communication between the CHs is done using a distributed approach; here, each individual CH handles the duty of forming its own cluster.
- **Algorithm complexity**: Numerous clustering methods have been suggested in the literature by keeping in mind several objectives and methodology. The time and space complexity are based on the number of CHs and/or sensors, or it can be a constant value.

- **Cluster head selection**: We can select CHs either randomly from the network area where nodes are deployed, or nodes can be predesignated as CH by the network authority.

3 Protocols Related to CH Selection

A general overview of various clustering protocols is presented in Table 1. This section is segregated into two parts based on the methodology of selecting CHs, i.e. probabilistic and fuzzy logic-based.

Table 1 Comparative analysis of probabilistic and fuzzy approach based algorithms

Algorithm name	CH selection techniques	Inter-cluster routing	Network parameters used	Clustering nature	Characteristics of the algorithm
LEACH	Pure probabilistic	Single hop	–	Equal	Randomized rotation of CHs, non-uniform energy consumption
HEED	Probabilistic	Single/Multi-hop	Residual energy, Intra-cluster communication cost	Equal	Assumption: sensor having different power levels. More efficient than LEACH
DEEC	Probabilistic	Single/Multi-hop	Residual energy, average energy of the network	Equal	Multiple levels of heterogeneity, network lifetime enhanced, but hotspot problem is present
CHEF	Probabilistic and fuzzy	Single/Multi-hop	Residual energy, local distance	Equal	More efficient than LEACH [5]
EAUCF	Probabilistic and fuzzy	Single/Multi - hop	Residual energy, distance to base station	Unequal	Hotspot and energy hole problem resolved for static networks. Better than LEACH or CHEF
MOFCA	Probabilistic and fuzzy	Single/Multi-hop	Remaining energy, distance to base station	Unequal	Hotspot and energy hole problem resolved for static and evolving networks

3.1 Probabilistic Approach-Based Methods

- **Low-energy adaptive clustering hierarchy (LEACH)**: Heinzelman et al. [5] proposed a self-organizing, adaptive and pure probabilistic-based algorithm for the selection of final local BSs, i.e. CHs. By this algorithm, motes randomly recognize themselves as CHs by using a predefined probability, and via using a carrier sense multiple access (CSMA) media access control (MAC) protocol, they publicize their decision to all other nodes. After that, each and every sensor node elects the cluster it wants to be a part of. Data transmission is done using a time division multiple access (TDMA) schedule and code division multiple access (CDMA) by each cluster. It ensures that every sensor node becomes a CH at once within 1/p rounds, where p is the desired percentage of CHs. Hence, they die irregularly throughout the network. The whole performance of the system will vary according to the random number generation. In each cluster, local compressing of data is done. LEACH has two phases: set-up phase and steady-state phase.
  Set-up phase: In the set-up phase, CHs are selected. Each node generates its own random number between 0 and 1. If the random number is less than the threshold value, that sensor node becomes a CH, otherwise not. The threshold value is calculated as:

$$T(n) = \begin{cases} \frac{p}{1 - p * rmod\left(\frac{1}{p}\right)} & if\ n \varepsilon S \\ 0 & otherwise \end{cases} \quad (1)$$

  S is the set of nodes that were not CHs in the last 1/p rounds.

  Steady-state phase: During the steady-state phase, nodes send their data to the CH using a TDMA schedule. This schedule allots time slots to every node. The CH collects the data and sends it to the BS.
  After comparison with the earlier protocol, i.e. direct transmission and minimum transmit energy (MTE) routing, it is determined (see Table 2) that the LEACH protocol is eight times much more efficient in terms of first node death (FND) and three times more efficient in terms of last node death (LND).
  Various drawbacks are associated with the LEACH protocol in that it is not applicable over large regions and has not mentioned how the predetermined CHs are uniformly distributed through the network. Also, the optimal percentage of CHs is not taken into account; i.e. CHs are selected randomly.

- **Hybrid energy-efficient distributed clustering (HEED)**: To avoid the random selection of CHs [5], a new distributed clustering algorithm was developed by Ossama et al. [9]. It selects the CHs based on the hybrid combination of residual energy and intra-cluster communication cost. So, it does not support random number-based CH selection; rather, it emphasizes the use of remaining energy.

For probabilistic selection of the CHs, residual battery power is utilized as the primary parameter. The second parameter is a function of the *node degree* or *1/(node degree)*. This ensures that each node either joins the CH having a minimum degree or the one with maximum degree to form dense clusters.

Working of the HEED protocol is categorized in three subsequent modules:

Initialization module: In this module, the initial percentage of CHs is set out of all nodes, C_{prob}. So, C_{prob} basically limits the number of initial CHs announcements. Each sensor node computes its probability, i.e. CH_{prob} to become CH by following equation:

$$CH_{prob} = C_{prob} \times \frac{E_{residual}}{E_{max}} \tag{2}$$

Here, $E_{residual}$ refers to the current residual energy level of the concerned node; E_{max} corresponds to maximum battery level of a node, i.e. when the node is fully charged. Since HEED supports heterogeneous sensor nodes, E_{max} may vary for different nodes according to its functionality and capacity. CH_{prob} is not allowed to fall below a certain threshold, which is selected to be inversely proportional to E_{max}.

Repetition module: This module is iterated continuously until it finds the CH that it can transmit to, with the least transmission power (cost). If it does not find any such CH, the wireless node elects itself to be a CH and sends a *declarative message* to neighbors in close proximity, informing them about the change of level from tentative to permanent. Finally, each sensor doubles its CH_{prob} value and goes to the next iteration of this phase. It halts executing this phase when its CH_{prob} reaches 1. The time complexity of this module is O(1).

Table 2 Comparison of LEACH with Direct, MTE protocol in terms of FND and LND [5]

Energy (J/node)	Protocol	Round first node dies	Round last node dies
0.25	Direct	55	117
	MTE	5	221
	Static clustering	41	67
	LEACH	394	665
0.5	Direct	109	234
	MTE	8	429
	Static clustering	80	110
	LEACH	932	1312
1	Direct	217	468
	MTE	15	843
	Static clustering	106	240
	LEACH	1848	2608

Finalization module: CH selection is finalized here on the basis of communication cost. The tentative CH now becomes the final CH. Each node joins their respective CH.
In HEED, the problem of random node selection is eliminated and it proves to be better than the previously developed algorithm as it utilizes the availability of multiple transmission power levels at sensor nodes, minimizes the communication cost and prolongs the lifetime. However, in the HEED protocol, the hotspot problem persists and, hence, there is unbalanced energy drain off in the network.

- **Distributed energy-efficient clustering algorithm (DEEC)**: In the previously designed hierarchical clustering technique, a crucial drawback was that those were homogeneous clustering algorithms. In HEED [9], only up to two levels of heterogeneity is possible, while Qing et al. [10] presented an idea which can handle two and multiple levels of heterogeneity. Now, let us have a look at how this protocol works:

  CH selection: In this method, due to the introduction of heterogeneity, the count of each node to become the CH varies for all nodes. Selection of CHs is based on the probability (actually, the ratio) of current residual energy of the node-to-average energy of the network after each round. Average energy [$E_{Avg}(r)$] for each round is given by the following equation:

$$E_{Avg}(r) = \frac{1}{N} * E_{Total}\left(1 - \frac{r}{R}\right) \tag{3}$$

  where N refers to total nodes deployed, r is the current round value and R is total number of rounds during the network lifetime. E_{Total} is total energy. On the basis of initial and residual energy, the total number of epochs selected as CHs will vary. So, the nodes with high initial and residual energy are more prone to being selected as CHs.
  The average energy of each node is used as a reference. For the network to remain alive, each alive node is supposed to have an ideal amount of energy in the current round. By using these criteria, uniform distribution of the nodes and energy of the network is achieved, and it is found that all the nodes die at the same time.
  It is determined that the DEEC achieves an enhanced network lifetime as compared to LEACH (as shown in Fig. 1) in terms of number of nodes alive vs. number of rounds. The reason behind is that it uses the average energy of the network for the purpose of reference energy.

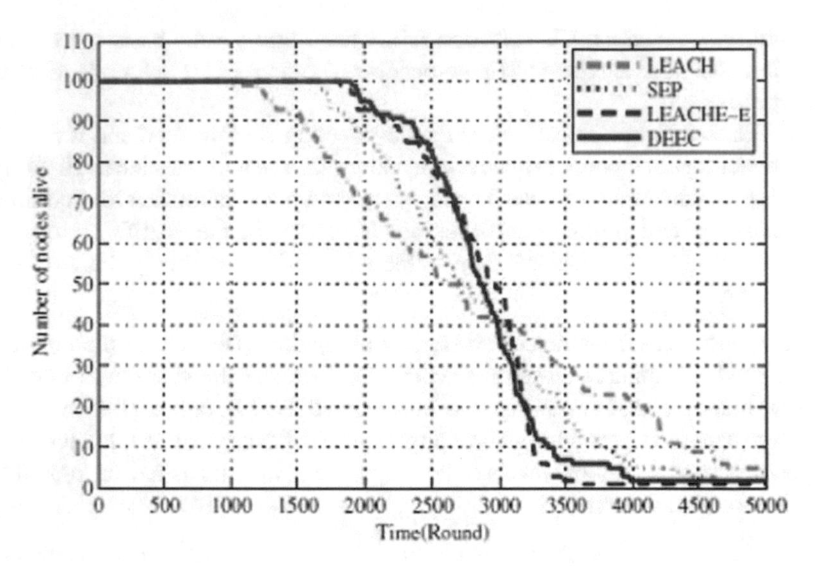

Fig. 1 Comparison of DEEC protocol with LEACH in terms of alive number of nodes [10]

3.2 *Fuzzy Logic-Based Clustering Techniques*

Past researchers wisely put their efforts into organizing clusters to extend the network lifetime. To avoid the drawbacks of the protocol suggested by Heinzelman et al. [5], Gupta et al. [8] introduced a protocol which exploits fuzzy logic for the selection of CHs. All the conditions in the previous techniques are considered too idealistic while doing the simulation. Such a scenario is not feasible as some parameters cannot be treated as crisp; so fuzziness is introduced to achieve accuracy.

- **Cluster head election mechanism using fuzzy logic (CHEF)**: In [19], a major loophole is that all the deployed nodes have to send the information regarding energy and their distance from respective CHs to the sink node. So, Kim et al. [15] proposed a distributed scheme based on fuzzy logic. Each wireless node chooses a random number between 0 and 1. If the random value corresponding to any node is lower than the predefined threshold, then those nodes select themselves as tentative CHs. In total, nine fuzzy rules are utilized to find the chance value in order to be selected as the CH. Two fuzzy linguistic variables used in the protocol are local distance and remaining energy. Here, all tentative CHs estimates their *chance* value to designate itself as an actual CH for this round. Local distance is estimated by taking the sum of the distances of all the nodes to the CH in a particular radius r. On the basis of higher chance value in a particular proximity, a sensor node selects itself as final CH by broadcasting a message consisting of information *id and chance* value. After that, other nodes join closest CHs by sending a message.

From the analysis, it is observed that CHEF is 22.7% more efficient [15] than LEACH [5] in terms of network lifetime. There is a slight possibility that some nodes remain orphans due to probabilistic selection. CHEF has somewhat overcome the problem of centralized [19] and efficient selection of CHs.

- **Energy-aware fuzzy unequal clustering (EAUCF)**: Bagci et al. [12] suggested an approach in which homogeneous nodes are deployed, i.e. all the nodes have the same amount of battery power, and their functionality is the same. In the previous protocols supporting equal clustering, the hotspot problem is associated with and due to CHs in close proximity to the sink losing their energy in a rapid and unequal manner. So, in order obviate this issue, EAUCF is suggested. The nature of clustering is unequal and distributed. Tentative CHs are selected with a probabilistic approach using the procedure proposed by Heinzelman et al. [5]. Here, a competition radius concept is used. The competition radius concept basically supports uniform energy consumption. In order to estimate the competition radius, fuzzy logic is utilized. Estimation of the competition radius is done using two parameters of the network, distance between the wireless node and sink, and current residual energy of the node. With the passage of time, each sensor node loses its energy, i.e. less residual energy as compared to earlier. So, it is quite obvious to decrease the service range of the CH with time as remaining energy decreases gradually. If this action is not taken, these sensor nodes will drain off their battery storage unequally. As the battery power of each tentative CH decreases, the competition radius is also shortened. Already-defined fuzzy logic mapping rules are used to take care of the uncertainty in the competition radius. The most popular Mamdani method is exploited to analyze the fuzzy rules. In order to defuzzify the competition radius, the center of area method is used.
  Once the temporary CHs and their corresponding radius are estimated, every CH circulates a *CandidateCH Message* in order to compete with other CHs lies in its periphery. This message disseminates the information about the sender node and it basically includes node ID, residual energy buffer and competition radius. If each tentative CH receives this information, then the CH which has more battery power will designate itself as the final CH. And resting, tentative CHs will quit by broadcasting a *QuitElection Message*. After CH election is completed, each ordinary sensor node joins the closest cluster, as in LEACH [5], CHEF [15] and EEUC [16].
  It was determined that EAUCF is 93.4%, 26.5% and 4.1% more efficient in terms of FND [12] than LEACH, CHEF and EEUC, respectively. If the half-node death metric is considered, EAUCF [12] is 19.4%, 6.8% and 8.5% better than LEACH, CHEF and EEUC, respectively. EAUCF has somewhat tackled the hotspot problem related to static networks.
- **Multi-objective fuzzy clustering algorithm (MOFCA)**: The main drawback associated with EAUCF is that it is able to resolve the hotspot problem only for static networks, but this shortcoming still exists if the nature of the network is

Table 3 Comparison of MOFCA protocol with LEACH, CHEF and EAUCF in terms of FND, HNA and TRE [13]

Algorithm	FND	HNA	TRE (J)
LEACH	88	189	43.50
CHEF	25	206	58.32
EEUC	84	254	67.47
EAUCF	83	261	68.50
MOFCA	90	268	70.34

evolving. Sert et al. [13] proposed a multi-objective approach to address the hotspot problem for evolving networks, and the energy hole problem for both type of networks. Here, CH designation is similar to the approach previously proposed by [12]. Here, tentative CHs are selected using the same pure probabilistic approach. MOFCA takes into account three input parameters for fuzzy logic, such as node residual energy, how far the node is from sink, and the density of the node. These factors help find the communication range for tentative CHs. The density of the node is crucial in resolving the hotspot problem when a network is evolving. Using all of the above-mentioned parameters, uncertainties in estimating the competition radius are greatly reduced. If the density is large, the competition radius will be enhanced correspondingly.
Here also, once the provisional CHs and their respective radius are determined, competition for the designation of final CH starts. All the tentative CHs broadcast the *CandidateCH Message* in order to establish a competition with their siblings. All the tentative CHs within the node's competition radius receive this message. This includes node id, competition radius, remaining energy level, and density of the generating node. All the tentative CHs which receive this message compare their remaining energy with each other. The final CH is selected on the basis of whichever node has higher residual energy while the CHs which have comparably less energy quit the election by sending a message. In case two or more CHs have the same amount of energy, this tie is broken using estimated node densities.
After performing the analysis, it is found that MOFCA is much better than LEACH [5], CHEF [15], EEUC [16] and EAUCF [12] in terms of FND by 41%, 51%, 19% and 12%, respectively (see Table 3). Hotspot and energy hole problems for both stationary and evolving networks are rectified using a MOFCA approach.

4 Conclusion

In the past decade, MEMS technology has gained considerable attention. Day by day, real-life applications are coming into light, especially in hostile locations such as the military, combat fields, agriculture and hospitals. While writing this paper, we have tried our best to survey the literature and segregate the different algorithms on the basis of important and relevant attributes. Clustering has emerged as a blessing both for the lifetime and scalability of a network. So, the whole paper revolves around the different ideas related to the selection of CHs. From the analysis and above table, it is quite clear that fuzzy logic with unequal clustering serves as a better option in the selection of CHs. Problems like hotspots and energy holes are reduced to a great extent by MOFCA (the best among all algorithms). In the future, better and novel algorithms can be designed by retaining the ideas from previous algorithms so as to improve network lifetime and scalability.

References

1. Akyildiz, I.F., W. Su, Y. Sankarasubramaniam, and E. Cayirci. 2002. Wireless sensor networks: A survey. *Computer Networks* 38 (4): 393–422.
2. Anastasi, Giuseppe, et al. 2009. Energy conservation in wireless sensor networks: A survey. *Ad hoc ne*tworks 7 (3): 537–568.
3. Rault, Tifenn, Abdelmadjid Bouabdallah, and Yacine Challal. 2014. Energy efficiency in wireless sensor networks: A top-down survey. *Computer Networks* 67: 104–122.
4. Afsar, M. Mehdi, and H. Mohammad, N. Tayarani. 2014. Clustering in sensor networks: A literature survey. *Journal of Network and Computer Applications* 46: 198–226.
5. Heinzelman,W.R., A. Chandrakasan, and H. Balakrishnan. (2000, January). Energy-efficient communication protocol for wireless microsensor networks. In *Proceedings of the 33rd annual Hawaii international conference on system sciences, 2000*, 10-pp. IEEE.
6. Abbasi, A.A., and M. Younis. 2007. A survey on clustering algorithms for wireless sensor networks. *Computer Communications* 30 (14): 2826–2841.
7. Logambigai, R., and Arputharaj Kannan. 2016. Fuzzy logic based unequal clustering for wireless sensor networks. *Wireless Networks* 22 (3): 945–957.
8. Gherbi, Chirihane, et al. 2017. A survey on clustering routing protocols in wireless sensor networks. *Sensor Review* 37 (1): 12–25.
9. Younis, O., and S. Fahmy. 2004. HEED: A hybrid, energy-efficient, distributed clustering approach for ad hoc sensor networks. *IEEE Transactions on Mobile Computing* 3 (4): 366–379.
10. Qing, L., Q. Zhu, and M. Wang. 2006. Design of a distributed energy-efficient clustering algorithm for heterogeneous wireless sensor networks. *Computer Communications* 29 (12): 2230–2237.
11. Mao, S., C. Zhao, Z. Zhou, and Y. Ye. 2013. An improved fuzzy unequal clustering algorithm for wireless sensor network. *Mobile Networks and Applications* 18 (2): 206–214.
12. Bagci, H., and A. Yazici. 2013. An energy aware fuzzy approach to unequal clustering in wireless sensor networks. *Applied Soft Computing* 13 (4): 1741–1749.
13. Sert, S.A., H. Bagci, and A. Yazici. 2015. MOFCA: Multi-objective fuzzy clustering algorithm for wireless sensor networks. *Applied Soft Computing* 30: 151–165.
14. Gajjar, S., M. Sarkar, and K. Dasgupta. 2016. FAMACROW: Fuzzy and ant colony optimization based combined mac, routing, and unequal clustering cross-layer protocol for wireless sensor networks. *Applied Soft Computing* 43: 235–247.
15. Kim, J.M., S.H. Park, Y.J. Han, and T.M. Chung. (2008, February). CHEF: Cluster head election mechanism using fuzzy logic in wireless sensor networks. In *ICACT 2008. 10th international conference on Advanced communication technology*, vol. 1, 654–659. IEEE.
16. Yuan, H.Y., S.Q. Yang, and Y.Q. Yi. (2011, May). An energy-efficient unequal clustering method for wireless sensor networks. In *2011 international conference on computer and management (CAMAN)*, 1–4. IEEE.
17. Chen, G., C. Li, M. Ye, and J. Wu. 2009. An unequal cluster-based routing protocol in wireless sensor networks. *Wireless Networks* 15 (2): 193–207.

18. Zhao, X., and N. Wang. (2010, May). An unequal layered clustering approach for large scale wireless sensor networks. In *2010 2nd international conference on future computer and communication (ICFCC),* vol. 1, V1-750. IEEE.
19. Gupta, I., D. Riordan, and S. Sampalli. (2005, May). Cluster-head election using fuzzy logic for wireless sensor networks. In *Communication networks and services research conference, 2005. Proceedings of the 3rd annual*, 255–260. IEEE.

Empirical Evaluation of a Real-Time Task Priority Procedure on Small/Medium-Scale Multi-core Systems

P. Pavan Kumar, Ch. Satyanarayana, A. Ananda Rao and P. Radhika Raju

Abstract Multi-core environments have become a substantial focus of research in recent years. However, many undermined hypothetical problems and pragmatic issues must be addressed if real-time environments are to be efficiently accommodated on multicore platforms. In multicore platform, every core can utilize the available shared cache resources. Numerous works have shown that real-time scheduling algorithms can be used efficiently with shared cache memory resources present on a single chip. Task reprocessing—Task rescheduling approach on multi-core platform (which is ideal for high-performance systems). This scheme is an improvement over existing cache-aware real-time scheduling, and encourage eligible task sets to be reprocessed based on heuristics called ENCAP [ENhancing shared CAche Performance]. We also address the arguments regarding the implementation of ENCAP on a Linux testbed called LITMUS^{RT}, and an empirical evaluation of ENCAP is provided. An experimental exploration of ENCAP is presented on a unified Linux system for assessing the real-time schedulability of any task set performance on a medium/large-scale multicore environment under G-EDF.

P. Pavan Kumar (✉) · A. Ananda Rao · P. Radhika Raju
Department of CSE, JNTU Anantapur, Anantapuramu, India
e-mail: paruchuripa1@gmail.com

A. Ananda Rao
e-mail: akepogu@gmail.com

P. Radhika Raju
e-mail: radhikaraju.p@gmail.com

Ch. Satyanarayana
Department of CSE, JNTU Kakinada, Kakinada, India
e-mail: chsatyanarayana@yahoo.com

A. N. Krishna et al. (eds.), *Integrated Intelligent Computing, Communication and Security*, Studies in Computational Intelligence 771,
https://doi.org/10.1007/978-981-10-8797-4_40

Keywords Multi-core processing · Shared cache memory performance ENCAP · Reprocessing · Small/medium-scale multi-core systems · Load set

1 Introduction

In multi-core systems, all processing units present on a single chip can exploit the available shared cache memory. Unlike single processor systems, multiprocessor technology is considered a cost-effective and viable solution for substantially increased usage of real-time (soft) quality-of-service (QoS)-directed and high-performance computing applications. The processing rate for a job is determined by its utilization. Similar rates are guaranteed by splitting the individual job into quantum-space sub-jobs, which will have intermediate cutoffs, defined as a pseudo-cutoff. With most known cache-aware and proportionate-fair (Pfair) schemes, sub-jobs are processed on an earliest-deadline-first (EDF) and earliest-pseudo-deadline-first basis. However, among subtasks to eliminate cutoff misses, ties with a similar cutoff should be determined with caution. In fact, for optimal scheduling algorithms, tie-breaking must be resolved heuristically [13, 24]. To ensure reprocessing, ENCAP imposes more robust restraints on the opportuneness of loads than is the case with ordered scheduling.

The problem. The issue examined here is motivated by the work of Calandrino et al. and Fedorova et al., in which reprocessing the discouraged real-time system jobs that produce constrained shared cache memory can be discouraged from being co-processed provided that real-time constraints are guaranteed. In our earlier work, we defined a heuristic set called ENCAP (ENhancing the shared CAche Performance) to reschedule loads that were ignored due to encouragement of lower-priority loads [12]. As shared memory, traffic can be co-scheduled while safeguarding the real-time constraints, but this suffers from high overhead. Previous works [23–25] have shown that task migration overhead can be minimized by per-task utilization.

Brief contributions. In this chapter, a task reprocessing scheme is proposed to enhance the performance of common cache memory on soft real-time multi-core systems, which presumes a quantum-placed universal recurring co-scheduling model. The impact of EDF-EN guidelines performance on the targeted systems is empirically evaluated, and extensive experimental results are presented. The impact of the reprocessing approach is validated with extended libraries including OPENMP as well as a Linux testbed called LITMUS^{RT} (Linux Testbed for Multiprocessor Scheduling in Real-Time Systems [35–39]). Calandrino et al. proposed an approach for co-scheduling of a lower-priority task set to be processed by ignoring higher-priority task sets. Our ENCAP approach has proven that the necessary load set can determined to be processed again, which are ignored to co-schedule in earlier works [15, 16, 21, 22]. The remainder of the chapter is organized as follows: The essential fundamental background of multi-core real-time scheduling is presented in Sect. 2. A procedure for task priority, the experimental evaluation phase of task priority procedure including an exhaustive look at performance enhancement under distinct works also represented in Sect. 3. In Sect. 4, conclusions are drawn and future work is discussed.

2 Background

In maximum prospective multi-core system environments, every core shares on-chip caches. The focus of this chapter is on fully exploiting possible parallelism in an on-chip shared cache environment, with an emphasis on improving the shared cache memory throughput by designing a load-balancing scheduler. Here we focus on the reprocessing of periodic and aperiodic loads. Every load in such a job is supplicated or discharged regularly; each such supplication is called an activity of the load. A periodic load can be derived by a slot, which implies the (definite) partition between its consecutive load discharges, and reprocessing cost, which implies the largest reprocessing time of any of its activity [4, 5]. Every activity of a load has a cutoff corresponding to the discharge time of the load's coming activity. Load slots are presumed to be essential with regard to the span of the computer's scheduling slot, but reprocessing costs may be nonessential. A load-set implementation or exertion is given by the ratio of its reprocessing cost and time. Both ENCAP and repoload are investigated in this chapter. In ENCAP, activities are rescheduled in the order of those previously discouraged [3, 15, 18, 19]. As mentioned in the introduction, tasks statically assigned to the core and those on each core are reprocessed on an ENCAP basis. In ENCAP, tasks may migrate, but once an activity embarks upon reprocessing on a core, it will process to completion on the same core without preemption. Therefore, activities may not migrate.

3 Performance Evaluation

In this work we consider multiple-task models in which all tasks contain a sequence of computational segments [1, 2, 7, 14]. All task models are synchronized and processed simultaneously [9, 10]. A common construct *for* is used to generate parallel tasks. Each task T_i (where $T_i = 0$) consists of sequential computational segments S_i, so each task consists of K segments, $1 \leqslant K \leqslant S_i$, and also contains $P_{i,j}$ concurrent threads. Every task is synchronized with periodic implicit deadlines.

Algorithm 1: Task Priorities (Max_Core, Uti_Cache, Cu_pri)

```
1        Cu_pri← NULL;
2        Uti_Cache← NULL;
3        Max_Core← 16;
4  CacheAware_EDF( );
5  while (Avail_Core ←TRUE) do
6      repeat
7          if (Neglected = TRUE) then
8              C: = max(0,Max_Cache – Uti_Cache);
9              M: = Max_Core - i+1;
10             else if ((Uti_Cache / Max_Cache) ⩾ (Max_Uti_Cache))∧(Eligible==TRUE)∧(Avail_Core≠0) then
11                 Task_Encap( );
12                 Task Partition ();
13                 Assignnewpriorities( );
14
15     until (Max_Core⩽ 16);
16     else
17         Eligible==False
```

3.1 Experimental Setup

Our empirical environment consists of one eight-core i7 processor (Intel Core, processing speed 3.8 GHz; available on the open market since May 2016). The Intel i7 currently represents the state of the art among multi-core chips. We assess ENCAP in terms of both profiler efficiency and its performance compared with the cache-aware real-time scheduler proposed by Calandrino [6, 18, 19]. We have produced load sets irregularly with different allocations identical to the Calandrino, Bastoni and Anderson [13, 21, 24] to scope the system overhead and the schedulability ratio of ENCAP. The worst-case execution time (WECT) for all loads is determined based on load span and application. We used a consistent span of load dissemination identical to Saifullah et al. [8] and Calandrino et al. [19]. In our ENCAP approach, load sets are statically assigned to processing cores and processed on a cache-aware real-time EDF basis. There are no constraints on load set migration or preemption, provided all the load sets are guaranteed with bounded tardiness in ENCAP. In our experiments, the scope of each autonomous core is the scheduling of a taskset. We consider an approach for load-partitioning for reprocessing in which selected tasks are statically split, but a selected number (at most one per processing core) are acknowledged to migrate from one core to any another while processing.

3.2 Experimental Results

We now assess our ENCAP heuristic set in terms of task set reprocessing and compare its performance with that of cache-aware scheduling. We analytically distinguish the ENCAP heuristic on the basis of schedulability. The results of our experi-

Table 1 The range of determined scheduling access time and cycle time for various cores with 600 KB storage region set (SRS) in ns

	2	4	8	16	32
Minimum	4.486	3.845	2.124	2.588	2.358
Average	14.419	13.010	11.970	11.364	9.535
Maximum	29.250	25.360	26.350	24.125	20.001
Minimum	6.756	5.565	4.895	4.289	3.825
Average	17.069	16.721	15.996	15.573	15.163
Maximum	34.125	34.186	33.455	32.955	32.214

Table 2 The range of determined scheduling access time and cycle time for 2, 4, 8, 16, and 32 processing core elements with 900 KB storage region set (SRS) in ns

	2	4	8	16	32
Minimum	3.962	4.248	4.126	3.886	3.198
Average	13.826	13.010	14.154	12.699	12.128
Maximum	29.255	29.136	28.135	27.725	27.001
Minimum	5.386	4.74565	3.9359	3.4996	2.9
Average	16.001	14.778	15.043	14.487	12.741
Maximum	32.250	29.860	30.550	29.325	24.142

mental task sets are presented in this section. In the schedulability assessment, arbitrary task sets are produced, and their schedulability under each heuristic is analyzed [18, 19, 34]. The results for access time and cycle time for various cores with 600 and 900 KB storage region set (SRS) of parallel threaded task (PTT) in ns are illustrated in Tables 1 and 2. We have used a lightweight event-tracing tool called Feather-Trace to measure the accuracy of our random task sets. Tables 1 and 2 summarize the range of determined scheduling access time and for various cores with 600 and 900 KB storage region set (SRS) in ns.

4 Conclusion

In this chapter, we have presented and experimentally evaluated a task reprocessing approach for cache-aware real-time scheduling load sets in which eligible and high-priority co-scheduling loads are discouraged (due to thrashing) and low-priority tasks are encouraged to task accomplishment. The objective is to enhance the shared cache performance and reduce overhead in real-time scheduling. We have shown that with the use of the ENCAP approach, by encouraging the reprocessing of neglected task sets, every core run, with the eligible reprocessed tasks modifying their affinities after an established time to deadline after task set release. The initial chunk of any reprocessed task should have its cutoff similar to its processing time. In effect, initial processor processes non-preemptively. This will supply the largest span to the subsequent cores for the load set to be accomplished. Our analysis demonstrates that our randomly generated task sets indicate that the scheme indeed achieves superior performance.

References

1. Saifullah, Abusayeed, Jing Li, Kunal Agrawal, Chenyang Lu, and Christopher Gill. 2012. Multi-core real-time scheduling for generalized parallel task models. *Journal of Supercomputing* 49: 404–435.
2. Lelli, Juri, Dario Faggioli, Tommaso Cucinotta, and Giuseppe Lipari. 2016. An experimental comparison of different real-time schedulers on multicore systems. *The Journal of Systems and Software* 85: 2405–2416.
3. Nie, Pengcheng, and Zhenhua Duan. 2016. An Efficient and scalable scheduling for performance heterogeneous multicore systems. *Journal of Parallel Distributed Computing* 72: 353–361.
4. Wang, Hongya, LihChyun Shu, Wei Yin, Yingyuan Xiao, and Jiao Cao. 2014. Hyperbolic utilization bounds for rate monotonic scheduling on homogeneous multiprocessors. *IEEE Transactions on Parallel and Distributed Systems* 29 (6): 1510–1521.
5. Xiang, X., Bin Bao, Chen Ding, and K. Shen. 2012. Cache conscious task regrouping on multicore processors. In *12th IEEE/ACM International Symposium on Cluster*, 603–611. Cloud and Grid Computing China: IEEE.
6. Lee, J., H.S. Chwa, J. Lee, and I. Shin. 2016. Thread-level priority assignment in global multiprocessor scheduling for DAG tasks. *The Journal of Systems and Softwares* 113: 246–256.
7. Lee, J., and K.G. Shin. 2014. Improvement of real-time multi-core schedulability with forced non-preemption. *IEEE Transactions on Parallel and Distributed Systems* 25 (5): 1233–1243.
8. Saifullah, A., D. Ferry, J. Li, K. Agrawal, C. Lu, and C. Gill. 2013, 2014. Parallel real-time scheduling of DAGs. *IEEE Transactions on Parallel and Distributed Systems* 25 (12): 3242–3252.
9. Burns, G., R.I. Davis, P. Wang, and F. Zhang. 2012. Partitioned EDF scheduling for multiprocessors using a C = D task splitting scheme. *Real-Time Systems* 49 (1): 3–33.
10. Buttazzo, G.C., M. Bertogna, and G. Yao. 2013. Limited preemptive scheduling for real-time systems. A Survey. *IEEE Tranctions on Industrial Informatics* 9 (1): 3–33.
11. Lee, J., and K.G. Shin. 2012. Controlling preemption for better schedulability in multi-core systems. In *IEEE 33rd Real-Time Systems Symposium (RTSS-12)*, 29–38 (San Juan).
12. Back, H., H.S. Chwa, and I. Shin. 2012. Schedulability analysis and priority assignment for global job-level fixed-priority multiprocessor scheduling. In *IEEE 18th Real-Time and Embedded Technology and Applications Symposium (RTAS-12)*, 297– 306 (China).
13. Anderson, J.H., J.M. Calandrino, and U.C. Devi. 2006. Real-time scheduling on multicore platforms, In *Proceedings of the 12th IEEE Real-Time and Embedded Technology and Applications Symposium (RTAS-06)*, 179–190.
14. Jin, Shiyuan, Guy Schiavone, and Damla Turgut. 2008. A performance study of multiprocessor task scheduling algorithms. *Journal of Supercomputing* 43 (1): 77–97.
15. Pavan, K.P., C.H. Satyanarayana, and A. Anandarao. 2016. Generalized approach to enhance the shared cache performance in multicore platform. In *First International Conference on Computational Intelligence and Informatics, Advances in Intelligent Systems and Computing (ICCII-16)*, 75–83.
16. Pavan, K.P., C.H. Satyanarayana, A. Anandarao, and P. Radhikaraju. 2016. Task reprocessing on real-time multicore systems. In *First International Conference on Informatics and Analytics (ICIA-16)*, Pondicherry.

17. Pavan, K.P., C.H. Satyanarayana, A. Anandarao, and P. Radhikaraju. 2017. Design and implementation of task reprocessing on medium-large multi-core architectures. *Application and Theory of Computer Technology* 2 (3): 25–34.
18. Calandrino, J., and J. Anderson. 2008. Cache-aware real-time scheduling on multicore platforms: heuristics and a case study. In *20th Euromicro Conference on Real-Time Systems*, 299–308. USA: IEEE.
19. Calandrino, J., and J. Anderson. 2009. On the design and implementation of a cache-aware multicore real-time schedule. In *21st Euromicro Conference on Real-Time Systems*, 194–204. Londan: IEEE.
20. Ramasubramaniam, N., V.V. Srinivas, and P. Pavan Kumar. 2011. Understanding the impact of cache performance on multicore architectures. In *4th International Conference On Advances in Information Technology and Mobile Communication (AIM-2011)*, 403–406. Nagapur: Springer-LNCS.
21. Bastoni, A., B. Brandenburg, and J. Anderson. 2010. An empirical comparison of global, partitioned, and clustered multiprocessor EDF schedulers. In. *31st Real-Time Systems Symposium (RTSS)*, 14–24. Italy: IEEE.
22. Block, A., B. Brandenburg, J. Anderson, and S. Quint. 2008. An adaptive framework for multiprocessor real-time syatems. In *Euromicro Conference on Real-Time Systems (ECRTS)*, 23–33. China: IEEE.
23. Anderson, J., J. Calandrino, and U. Devi. 2008. Real-time scheduling on multicore platforms. In *Euromicro Conference on Real-Time Systems*, 299–308. Paris: IEEE.
24. Anderson, J., Anand Srinivasan. 2000. Pfair scheduling: beyond periodic task systems. In *7th International Conference on Real-Time Computing Systems and Applications*, 297–306. London: IEEE.
25. Anderson, J., and Anand Srinivasan. 2004. Mixed Pfair/ERfair scheduling of asynchronous periodic tasks. *Journal of Computer and System Sciences* 68 (1): 157–204.
26. Burchard, A., J. Liebeherr, Y. Oh, and S.H. Son. 1995. New strategies for assigning real-time tasks to multimocessor systems. *IEEE Transactions on Computers* 44 (12).
27. Xu, Jia. 1993. Multiprocessor scheduling of processes with release times, deadlines, precedence, and exclusion relations. *IEEE Transactions on Computers*, 19 (2).
28. UNC Real-Time Group. LITMUS RT project. http://www.cs.unc.edu/anderson/litmus-rt/.
29. Extremetech. http://www.extremetech.com/computing/.
30. AMD. http://www.amd.com.
31. Andersson, B. 2006. Static-priority scheduling on multiprocessors. University of North Carolina, Chapel Hill, NC.
32. Devi, U. Ph.D. 2003. Soft real-time scheduling on multiprocessors. Chalmers University of Technology, Goteborg, Sweden.
33. Andersson, B., T. Abdelzaher, and J. Jonsson. 2003. Partitioned aperiodic scheduling on multiprocessors. In *International Parallel and Distributed Processing Symposium. Nice, France*, 22–26.
34. Calandrino, J. Ph.D. 2009. On the design and implementation of a cache-aware soft real-time scheduling for multicore platforms. University of North Carolina, Chapel Hill, NC.
35. Calandrino, J., H. Leontyev, A. Block, U. Devi, and J. Anderson. 2006. *LITMUS*^{RT}: A testbed for empirically comparing real-time multiprocessor schedulers. In *27th IEEE Real-Time Systems Symposium, China*, 111–123.
36. UNC Real-Time Group. *LITMUS*^{RT} homepage. http://www.cs.unc.edu/anderson/LITMUS.
37. INTEL. http://www.intel.com/pressroom/archive/releases/2010/20100531comp.html, 2010.
38. CACTI6.0, CACTI-6.0 (Cache Access Cycle Time Indicator). http://www.cs.utah.edu/~rajeev/cacti6/.
39. SESC, SESC Simulator. SESC is a microprocessor architectural simulator. http://iacoma.cs.uiuc.edu/~paulsack/sescdoc/.

Application of Swarm Intelligence in Autonomous Cars for Obstacle Avoidance

Adil Hashim, Tanya Saini, Hemant Bhardwaj, Adityan Jothi and Ammannagari Vinay Kumar

Abstract Obstacle detection is a major challenge which must be addressed for optimal implementation of self-driving cars. Various approaches have been postulated regarding the same. However, the acquisition of data by the various sensors in a car is shortsighted and constrained due the physical limitations in the scope of the sensors. In this chapter, we propose a model for obstacle avoidance in self-driving cars by integrating swarm intelligence with pre-existing conventional technologies. By establishing bi-directional communication of sensory data between the various cars which may form a network we can overcome the limitations faced by the receptors of a self-driving car.

Keywords Self-driving cars · Computer vision · Convolutional neural networks · Swarm robotics · Obstacle avoidance

1 Introduction

Obstacle detection is a crucial part of self-driving cars, and computer vision in general. The past decade has seen considerable research and advances in the technologies used for detecting obstacles. Different sensors and cameras on cars

A. Hashim (✉) · T. Saini · H. Bhardwaj · A. Jothi · A. Vinay Kumar
Amity University, Noida, India
e-mail: adil96hashim@gmail.com

T. Saini
e-mail: tanyasaini4@gmail.com

H. Bhardwaj
e-mail: bhardwajhemant84@gmail.com

A. Jothi
e-mail: research.adityan@gmail.com

A. Vinay Kumar
e-mail: vinayvinu9716@gmail.com

A. N. Krishna et al. (eds.), *Integrated Intelligent Computing, Communication and Security*, Studies in Computational Intelligence 771,
https://doi.org/10.1007/978-981-10-8797-4_41

obtain the information required. It is important to detect obstacles in the images accurately and make decisions based on that. Current technologies like light detection and ranging (LIDAR) sensors are very useful in detecting obstacles and determining distances between cars and obstacles, but the range of these sensors is limited to up to 40 m. This is where communication between cars can be very useful, as it will increase the efficiency of the cars. Cars can share information such as obstacles around the area and reduce the processing load on each car by simply sharing the information already processed by the cars around that area. Car-to-car communication will also mean fewer accidents and reduced traffic jams. Therefore, it is important to find efficient ways of communication between autonomous vehicles. In recent years, there have been a few ways of communication tried by researchers and car companies. Some auto manufacturers like Audi, Mercedes-Benz, BMW and Nissan have already explored communication via laser and light-emitting diode (LED) lights, and projections on the road.

Swarm robotics is a field gaining a lot of attention for cooperation between machines. It is capable of performing complex behaviors at the micro-level with high level of fault tolerance and scalability. The system usually consists of a large number of simple and low-cost robots which cooperate and interact locally such that complex macro-level behaviors emerge from simple local interaction between the individuals and between them and their environment [1]. The scenario we focus on in this chapter is the deployment of a robotic swarm to make self-driving cars interact with each other and share data with the other cars to help detect obstacles on the road more precisely and accurately. Some ways of communication are Bluetooth, wireless local-area network (LAN), communication via the environment (stigmergy) or infrared LEDs.

In this chapter, we have proposed Bluetooth 5.0 as a way of sharing information between cars. It has many advantages over the previous versions of Bluetooth and other ways of communication (wired or wireless). It has twice the data transfer speed over the previous version and can transfer information up to 800 feet away. This will be very important when sharing data between the cars as the size of the data will be huge and it needs to be transferred as quickly as possible to the nearest cars to be efficient and relevant. Instead of trying it on actual cars we tried our methodology on laptops. Each laptop would act as a car and share the data it has processed to other laptops.

2 Literature Survey

A self-driving vehicle is the combination of new technologies and different levels of automation to create a new human-automobile hybrid [2]. In-vehicle sensing technology is used for real-time collection of data about the vehicle and its environment. This data includes details like the car's speed, direction, acceleration, obstacles encountered and geographical coordinates among others. Vehicles store this enormous amount of data with the aid of increased processing speed and power.

Vehicle-to-vehicle and vehicle-to-infrastructure communications technology enable broadcasting the data collected in short range of the vehicle by transferring periodic messages to surrounding vehicles and infrastructure. Switched digital videos (SDVs) allow greater standardization of travel time, improved data collection, and powerful management of fleets of cars. At present, numerous security and computing technologies, along with various SDV technologies, occupy a favorable position in the field of research and have garnered popularity amongst researchers; this can be observed in the amount of research being undertaken in this area, which is rapidly increasing day by day. One such paper discusses the necessity in micro-aerial vehicles (MAVs). These MAVs or self-driving cars detect obstacles in real time in order to take timely actions [3]. Solutions to object detection are dependent on the availability of resources and the type of vehicle under action.

For instance, MAVs have limited computational capacity. An obstacle detection framework has three primary steps. A depth map for a 3D representation of the surrounding environment is constructed from each camera mounted on the car by multi-view stereo matching on the series of captures made by the vehicle in continuous motion. In addition, the wheel odometry is also a source of the required camera captures. The next step deals with extracting information on the obstacles in 2D and free space from each of the depth maps. The wheel odometry is not completely reliable in the aspect of the real travelled distance, thereby indicating some errors. The last step involves combining all the obstacle detections from the numerous cameras to provide an accurate estimation of the occupied space around the vehicle. This combining can be done either for each camera separately or all of them as a whole and this would give an overall perspective of the surrounding area.

One paper has been proposed to detect lines and signs on the road by a robot [4]. The methodology involves a central robot taking the images captured in red-green-blue (RGB) color through the camera attached to it and then converting it to hue-saturation-value (HSV) format and processing it to detect lines. Any pixel value converted into white color and anything above that is converted into black color. So, it converts the black lines that the camera detects to white lines and white background into black. The trajectory of the robot is then converted on the basis of the data obtained and the processed frame. All the processing can be done on a central computer and all the robots are connected remotely through Wi-Fi. The central computer processes the image, determines the required trajectory for the robots to follow and sends parallel virtual machine (PVM) values to each robot. Since the central computer is tracking all the robots, they can also be programmed to avoid each other when they are close by. This can be a great application for avoiding traffic by detecting blockage and calculating alternative routes.

Another paper proposes a navigation system that is loosely based on routing algorithms used in mobile ad hoc networks (MANETs) [5]. The general idea is that all robots in the swarm maintain a table with navigation information about all known target robots in the environment, similar to how nodes in a MANET maintain routing tables. Each robot periodically broadcasts the content of this table to its neighbors, so that the information spreads throughout the swarm. It allows robots of a swarm to guide a single robot to a target, without the need to adapt their own movements. This system can be used for collective swarm navigation between two targets. It was shown cooperation improves navigation performance, and that when enough robots are present, the swarm self-organizes into a dynamic structure that supports efficient navigation and is robust to disturbances and robot failures. The swarm communication used for navigation in this paper can be extended to carry more information—for example, for self-driving cars.

3 Proposed Methodology

The proposed system is composed of two primary components, namely the learning framework and the communications module. The learning framework's objective is to learn from the training data to determine the behavior of the vehicle in real-time situations. The communications module performs the task of transmitting a digest of one vehicle to another and making use of collaborative learning algorithms in order to aid vehicles to adapt better to situations.

3.1 Communication System

The proposed system utilizes communication between various cars in the system for optimizing the process of obstacle detection in self-driving cars, as shown in Fig. 1. Our proposed system involves fitting each car with a Bluetooth 5.0 module for communication between networked cars. Setting up a Bluetooth connection between two devices is comparatively easier and safer when compared to other forms of communications which can be established between vehicles. The initial phase of the communication system involves one of the cars sending out a Bluetooth broadcast signal requesting connection with other autonomous cars in range around it. On successfully connecting to cars in the vicinity, data can be shared as long as the connection is not disconnected or lost. If the initial connection attempt fails, the process is repeated until successful.

On making connection with vehicles around it, the autonomous vehicle checks its database for the 'ID tag' of the vehicles it has connected with. This connection with the database helps in the retrieval of communication protocol data and proximity hardware data, as well as other pertinent data required for inter-vehicle data transfer. The data retrieved from the database could be used in further optimizing the process of communication between the cars in the network. Each of the cars in the network also has a specialized driver inbuilt into its system which would help in neutralizing the differences in architectures and hardware of the various sensor-based systems present in the autonomous vehicle.

A car would hence be a part of a multiple-node network where each car exchanges data with the other. This connection created using Bluetooth would be dependent on the distance of the various cars from each other. A car would

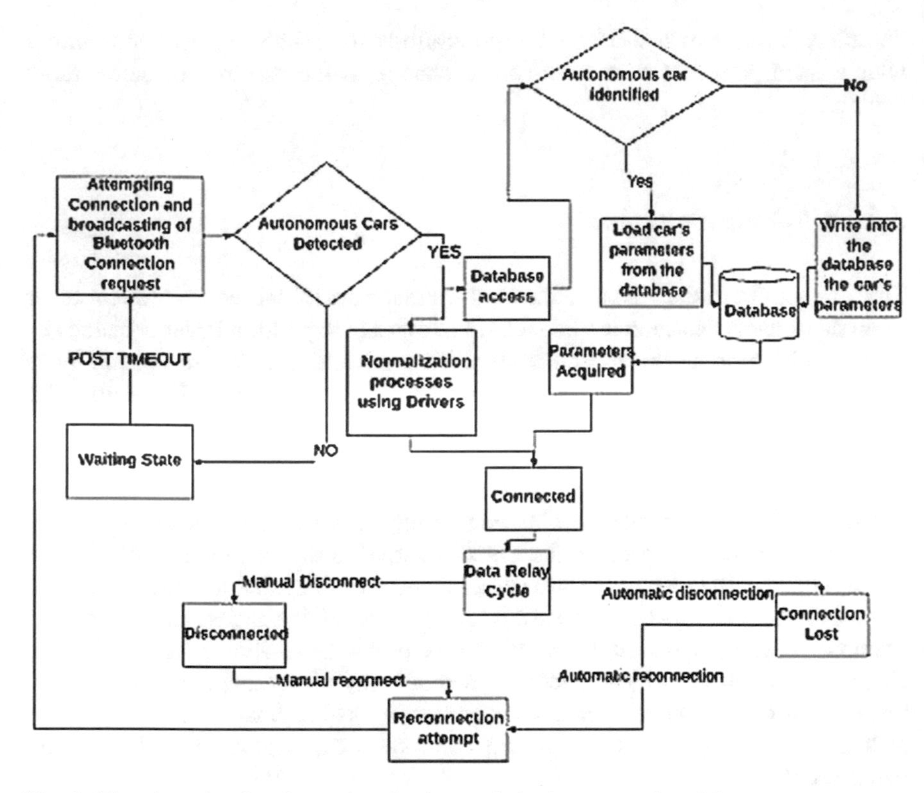

Fig. 1 Flowchart showing the communication cycle in the proposed model

automatically disconnect completely or partially from a network if it overshoots the connection radius. It may as well simultaneously connect to other new vehicles nearby, hence creating a new network in reference to the network from which it disconnected. Disconnection of a car from a network can also be a manual event. It may also happen due to various reasons and circumstances such as the breakdown of the car or sometimes a glitch in the system which leads to restart of the autonomous driving system, and consequently restart of the whole Bluetooth connection establishment process. As mentioned, in the event of connections of an autonomous car being lost, the car would try to establish the connections yet again. This event can roll out in two manners. In one approach, the car may try to establish a connection for an indefinite time until it manages to do so. In another, the car would retry to create the connections until it hits a timeout. The latter is the method chosen for the process of reestablishment of connection in our methodology as it is more suitable for the problem at hand. If no connection is reestablished after the timeout, it would be assumed by the car that there are no other autonomous cars in its vicinity at present which are in a position to communicate with it. Upon reaching this state, the car would check for any possibilities for establishment of connections after a fixed period of time to discern a change in the presence of autonomous vehicles.

3.2 Obstacle Detection

The obstacle detection module forms a crucial part of the entire system as an accurate obstacle detection module would allow cars to perform better decisions by sharing this information with other cars. This module is comprised primarily of pedestrian detection and blob detection as parameters for learning. By sharing this information between cars along with GPS coordinates, cars would be able to predict obstacle occurrence by learning from data obtained from the car in front. The module generates a feature vector comprised of the proximity of the obstacles to the car, optical flow to indicate the direction of movement of the obstacle and also a rough estimate of the speed of the object relative to the car; the coordinates are based on GPS. The problem of pedestrian detection is one that has been subject to extensive research [6–8]. This system makes use of the histogram of oriented gradients (HOG) algorithm for solving this problem, as shown in Fig. 2. This algorithm provides an object detection methodology by making use of number of instances of gradient orientation in localized portions of an image. It is computed on a dense grid of equidistant blocks and common local contrast normalization for better accuracy.

Fig. 2 Pedestrian detection using histogram of oriented gradients

Another phase of the obstacle detection module is blob detection. Blob detection is useful in identifying objects of interest that differ significantly from their background. We could use these object boundaries and optical flow calculations to estimate the speed of the object relative to the camera/car. This model makes use of the determinant of Hessian blob detector. A multivariate function's local curvature can be described by a Hessian matrix effectively and thus would be useful in blob detection.

$$\begin{array}{c} L_{xx}L_{yy} - L_{xy}^{2} \\ \det H_{\mathrm{norm}}\mathrm{L} = \mathrm{t}^{2} \end{array} \tag{1}$$

This system also makes use of a cascade classifier in order to extract cars present in the video sequence shown in Fig. 3. This method begins with training a cascade classifier using many positive samples and several negative samples as well. It follows an iterative approach where a sample is evaluated in multiple levels until it is repudiated at one level or clears all levels. This is followed by the classifier scanning each image or frame in the sequence of images for a match and highlighting any matches through this window.

Fig. 3 Cars identified using cascade classifier technique

3.3 Swarm Intelligence Implementation

The usage of swarm intelligence helps in the optimization of the process of obstacle detection in the proposed methodology. The system being proposed involves cars with multi-sensor modules attached to them. Every car would be mounted with a multi-sensor module with an accelerometer, gyroscope, camera and GPS. The diagram below depicts the structure of the multi-sensor module. The multi-sensor module would provide valuable information regarding the current state of the vehicle along with a vision system for detecting obstacles near it and thereby providing a rough estimate of the proximity. The various sensors collect data from the surroundings of the car it is attached to. This data is further processed via the Raspberry Pi module through a convolutional neural network to infer essential control parameters and is propagated to the effectors of the car. The multi-sensor module is linked to a touchscreen display that features various control parameters that provide insight to the driver regarding the car's trajectory, location and also obstacle-avoidance suggestions based on intelligence gathered from connected cars shown in Figs. 4 and 5. The multi-sensor module makes use of the learning framework's outputs to indicate whether there is an obstacle approaching which may not be visible to the current car's field of view and, correspondingly, it suggests the user to accelerate or decelerate.

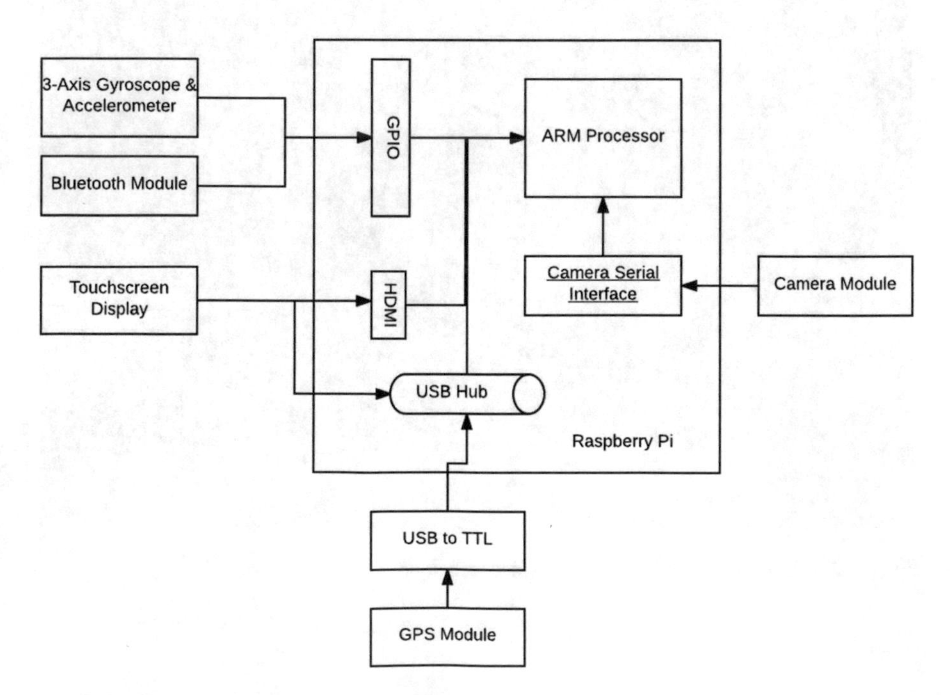

Fig. 4 Multi-sensor tracking module circuitry

Fig. 5 Real-time suggestions based on swarm intelligence shown on interactive touchscreen run using Raspberry Pi

This approach could be adapted for a self-driving car as the learning framework provides a quantified value for the acceleration/deceleration which could be applied directly to a self-driving car to travel at a velocity that would be safe. The quantified value would allow the learning framework to infer more details, make better decisions and improve obstacle avoidance methods as a whole. The additional compatibility factor described in the previous section would allow self-driving cars with different hardware specifications to communicate effectively and learn from each other through compatibility bridges. Each of the cars would be integrated with a Bluetooth module which would facilitate the process of interconnection and data relay between the various cars. The learning framework in each of the cars would receive input in the form of readings from the multi-sensor modules of various other cars in the network or from the car itself. The data which would be internally generated would help in initial decision making. Any data which is received from the external connected agents would be used to ameliorate the precision of results being computed by a car. A higher priority would be given to data which is internally generated by a car rather than data which is acquired from another car in the same network. This is to make sure that a car's system is fully functional even in the case when there are no active connections and would not be in jeopardy of failing to detect obstacles which are to be dealt with.

Swarm intelligence makes use of collective strength in knowledge about a system's surrounding. The swarm approach to dealing with obstacles helps in making up for the shortcomings of conventional obstacle detection technologies widely used today in self-driving cars. This would imply an enhanced field of view, improved decision making and overall increase in reliability can be achieved by the integration of this model into the existing self-driving cars.

4 Results

The proposed system was tested and evaluated on two cars fitted with the multi-sensor modules. The data transferred between the cars were logged and analyzed after multiple trips and the results showed improved decision making by the car using data it received from its peer. The acceleration, in the range of 0.11–0.21 m/s^2, was multiplied by a factor of 10,000 to scale up to the proximity data obtained as the output of a function of the obstacle area occupied as identified by the vision system. The correlation between the obstacle proximity and the acceleration computed can be observed in the graph below. As we can see, the learning framework upon receiving continuous high values for the proximity indicator decides to decelerate and returns to acceleration gradually as the obstacle size reduces significantly.

Figure 7 shows graphically how car acceleration changed during trips in the experiment when the car came across obstacles in its path. It is found that the driver of each car is shown a notification to decelerate and accelerate upon coming across an obstacle and upon finding no obstacles in the car's path, respectively. This notification is shown to the driver via an interactive touch screen panel operated using a Raspberry Pi module which is in turn is connected to all the other sensory modules in the car. Successful acceleration and deceleration suggestion notification was shown, according to which the driver accelerated and decelerated as shown in Fig. 7.

Figure 6 illustrates a complete trip in which car 1 is closely followed by car 2. Both cars were fitted with multiple sensors which generated location, acceleration and obstacle proximity data. Upon observing an obstacle, car 1 relayed data over to car 2, and hence car 2 learned from the relayed data and suggested a deceleration upon reaching the estimated position of the obstacle identified by car 1. Car 2

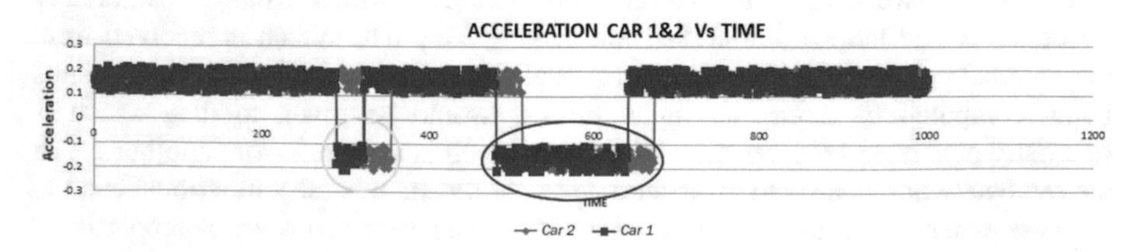

Fig. 6 Acceleration variation of the two cars in the experiment

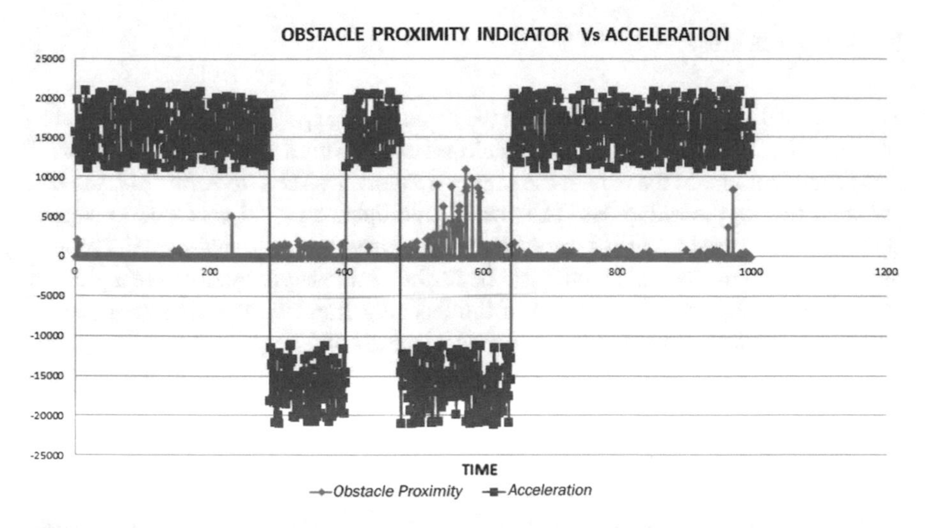

Fig. 7 Graph depicting the acceleration of a car with respect to proximity of obstacles

closely trailed car 1, and data relay continued during the entire trip, which lasted as long as car 2 was connected to car 1 via Bluetooth connection and stayed in range. This result shows how car 2 learns from the data obtained from car 1 and how obstacle avoidance is successfully implemented (Fig. 7).

We can observe the transmission of data between two cars in real time in Fig. 8. The data transmission works on the communication architecture proposed and transmits various attributes read by the multi-sensor module. We can see the frame-by-frame analysis of obstacles and the corresponding data values transmitted between the cars.

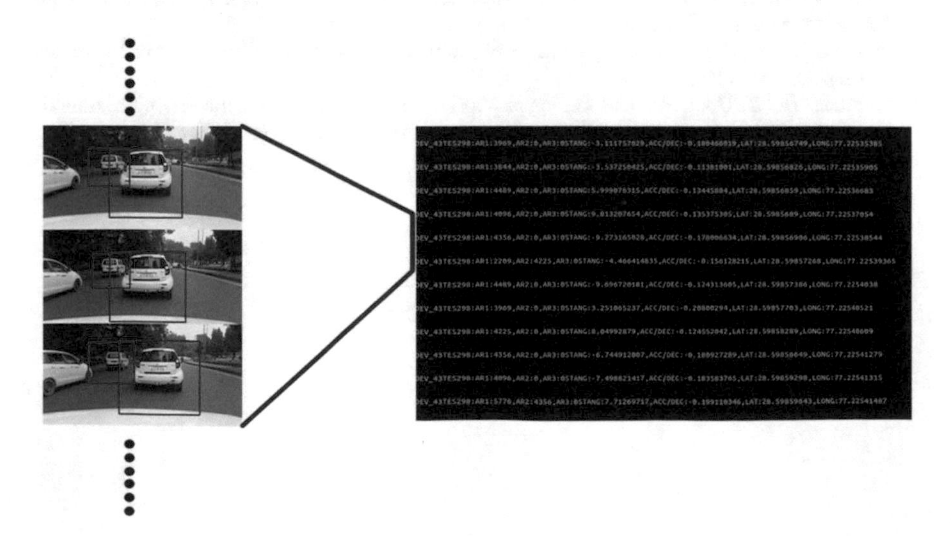

Fig. 8 Graph depicting the acceleration of a car with respect to proximity of obstacles

5 Conclusion

The proposed model was evaluated on the basis of an experiment involving a binary network of cars which were fit with multi-sensor modules to quantify car status. It was found that one of the cars in the binary network was able to learn from the data obtained from a preceding car. The swarm intelligence model was able to suggest changes in the present state of the system to the driver so as to optimize the process of obstacle avoidance. This model can be further extended to incorporate a plethora of sensors to expand the capability of the learning framework and to augment the quality of predictions made by the learning framework.

References

1. Khaluf, Y., E. Mathews, and F.J. Rammig. 2011. Self-organized cooperation in swarm robotics. in *CAIEEE international symposium on object/component/service-oriented real-time distributed computing workshops*, 217–226.
2. Blyth, Pascale-L., M.N. Mladenovic, B.A. Nardi, N.M. Su, and H.R. Ekbia. 2015. Driving the self-driving vehicle. In *International symposium on technology and society (ISTAS)*, Dublin, 1–6.
3. Häne, C., T. Sattler, and M. Pollefeys. 2015. Obstacle detection for self-driving cars using only monocular cameras and wheel odometry. In *International conference on intelligent robots and systems (IROS)*, Hamburg, 5101–5108.
4. Haque, I.U., A.A. Arabi, S. Hossain, T. Proma, N. Uzzaman, and M.A. Amin. 2017. Vision based trajectory following robot and swarm. In *International conference on control and robotics engineering (ICCRE)*, Bangkok, 35–38.
5. Ducatelle, F., G.A. Di Caro, C. Pinciroli, F. Mondada, and L. Gambardella. 2011. Communication assisted navigation in robotic swarms: Self-organization and cooperation. In *International conference on intelligent robots and systems*, San Francisco, 4981–4988.
6. Dollár, P., B. Babenko, S. Belongie, P. Perona, and Z. Tu. 2008. Multiple component learning for object detection. In *ECCV*.
7. Ess, A., B. Leibe, K. Schindler, and L. van Gool. 2008. A mobile vision system for robust multi-person tracking. In *CVPR*.
8. Shashua, A., Y. Gdalyahu, and G. Hayun. 2004. Pedestrian detection for driving assistance systems: Single-frame classification and system level performance. In *Intelligent vehicles symposium*.

Home Automation Using Hand Recognition for the Visually Challenged

Ammannagari Vinay Kumar, Adityan Jothi, Adil Hashim, Pulivarthi Narender and Mayank Kumar Goyal

Abstract Human–computer interaction has spawned much research in recent years, which has greatly reduced the complexity which existed in interactions between humans and computers. The visually challenged have benefited through systems resulting from this research. This paper aims at developing a system that allows the visually challenged to interact with the electronic systems in their vicinity. The objective of this paper is to overcome the shortcomings identified in previous systems and enhance access for the visually challenged. The paper describes two submodules whose integration allows the visually challenged to effortlessly interact with electronic devices around them.

Keywords Hand gesture recognition · Human computer interaction
Home automation · Arduino · Raspberry pi

1 Introduction

One of the problems faced by visually impaired people is navigating in confined places, which complicates their daily lives. Interaction with electrical appliances at home is also difficult for those with vision loss, who are often unable to see and

A. Vinay Kumar (✉) · A. Jothi · A. Hashim · P. Narender · M. K. Goyal
Amity University, Noida, India
e-mail: vinayvinu9716@gmail.com

A. Jothi
e-mail: research.adityan@gmail.com

A. Hashim
e-mail: adil96hashim@gmail.com

P. Narender
e-mail: narendervaschowdary@gmail.com

M. K. Goyal
e-mail: mayankrkgit@gmail.com

A. N. Krishna et al. (eds.), *Integrated Intelligent Computing, Communication and Security*, Studies in Computational Intelligence 771,
https://doi.org/10.1007/978-981-10-8797-4_42

control such systems using knobs, switches, levers and so on. Home automation looks to be a potential solution that would allow visually impaired people to control electrical appliances with ease. However, there is a potential problem to be addressed, the interaction between the user and the home automation system. In this paper, we attempt to address this issue through the development of trackers for home automation using gestures.

Home automation means controlling and automating home appliances. There are various types of applications in home automation such as HVAC, light control and indoor positioning systems. Home automation basically helps us to save power, thereby contributing to sustainable development. Advances in technology have simplified our life has become simpler such that one can sit at a place and control all the appliances in the house. In this current scenario, remote controllers and Android applications are widely used in the market, but most of them are faulty, unreliable and inaccurate.

Among the growing technologies in the field of gesture recognition there has been considerable research, of which hand gesture recognition is an important component. Hand gestures will be used as input data in human machine interaction for controlling electric appliances. In this project, we have proposed a technology in which we use two wearable trackers to control in-home appliances. Each tracker consists of accelerometers, ATmega328P microcontrollers and a Bluetooth module for communication. These components track hand movements in three directions and this input will be supplied to Raspberry Pi from where all the electronic appliances can be controlled.

We have created a module based on Raspberry Pi that facilitates the high processing power required to control electronic appliances and interprets the data from Arduino. This would work as the main hub of the system. All the data collected by Arduino would be transferred to the Raspberry Pi using a Bluetooth HC-05 module which is connected to the Arduino, and used to control all the electronic appliances connected to it.

2 Literature Survey

The tremendous increase in research in home automation systems demonstrates positive growth in more effective use of home appliances [1–13]. Especially for elderly and disabled people the advantage of doing their own daily tasks would become effortless. Various approaches proposed to achieve this task include a vision-based approach, where a camera captures the movements of the operator's hand. This approach, however, is very expensive; for cost-effective home automation system, a haptic-based approach is an alternative. This economical approach utilizes gloves fixed with different measurement devices. Various have been discussed on haptic based approach implementation and its accuracy in dealing with the home appliances.

One such paper based on hand gestures has been proposed which would help people with visual disabilities, by making use of a wearable device for the operation of home automation systems. The controlling methodology used in the pendant, which has the ability to analyze pathological tremors, is the standard X10 network which processes the image into digital form and then subsequently converts it into a signal. To understand the hand gestures very accurately with less computational complexity and less color segmentation, infrared illumination is used. Black and white CCD cameras pick infrared images properly and this is attributed to the wide 160-degree input range of the camera. The hidden Markov model for user-defined gesture recognition provides discrete output for a single gesture where images are recognized by different discrete outputs. Control gestures provide continuous control of a device. The gesture pendant can recognize control gestures and user-defined gestures with accuracy of 95% and 97%, respectively.

Other research proposed the construction of an intelligent room where an individual can operate the home appliances with intuitive gestures. The room uses three-dimensional (3D) measurement for the detection of the gestures. Charge-coupled device (CCD) cameras with pan, tilt and zoom functions are used for the 3D measurements. The cameras are distributed systems which are interconnected to collect data from the hand gestures. Monocular cameras are used for the recognition of fingers in 3D measurement. Color information is used for recognition and extraction of hand gestures. Based on the gesture that has been recognized, a control signal is propagated to that particular appliance by the infrared remote controller.

Another paper proposed a low-cost wireless intelligent hand gesture system based on the haptic approach which makes use of gloves fixed with measurement devices for tracking hand movements. There is classification of subsystems into primary and secondary types, interconnected through a pair of Bluetooth modules, each of which consists of a sensor. The data from the sensors on the gloves is passed onto an Atmega328 microcontroller in each sub-system. A dual-mode intelligent agent (DIA) is used for recognizing the gestures for the system, where the identification mode (IM) is the basic mode for the recognition and an automatic switch to enhanced identification mode (EIM) takes place when the gesture recognition fails via the IM. A bit stream error elimination (BSEE) algorithm is used for EIM for accuracy in gesture recognition. The gesture recognition process takes place in the primary system's microcontroller with the help of a DIA, and the identified gesture is displayed on a liquid crystal display (LCD). System efficiency of 80.06% was found in IM, which increased to 93.16% in EIM.

3 Proposed Methodology

The proposed system utilizes various technologies including Bluetooth communication, accelerometers, ATmega328P and Raspberry Pi to enable the visually challenged to control electronic appliances with ease. The core of the system is a Raspberry Pi 3 board that has an ARM processor with 1 GB RAM, allowing us to achieve our goal by providing low-cost yet effective computational power. The data sources are provided by the circuits made using the ATmega microcontroller.

The system can be divided into two main parts: the data sources or trackers, and the control system. The trackers are wearable gloves that are fit with various sensors and connected to an ATmega328P microcontroller, as shown in Figs. 1 and 2. The trackers consist of accelerometers which allow us to accurately identify movements in the palm and hand as a whole. Thus we are able to infer gestures and send control signals to the control system, which would in turn make use of an association table to correlate the gesture with an action to be performed in controlling an electronic appliance. The tracker systems are powered by a rechargeable lightweight battery that functions for 4–5 h without the need for recharging. The accelerometer allows us to track the movement of the hand as a whole, and conveys that information to the control system. The tracker system establishes a communication channel with the control system using a Bluetooth HC-05 module, after which the readings are transmitted.

The control system performs various tasks simultaneously to provide home automation using the gesture signals received from the tracking system. The presence of a built-in Bluetooth module allows us to build a cost-effective control system without having to depend on many modules. The control system is connected to electrical circuitry using control circuits, and when a signal is received and interpreted, the control system performs the corresponding action by manipulating these control signals to the electronic circuitry. The device was made to be reprogrammable, where the specific actions or gestures can be reprogrammed to indicate different functions, providing enhanced convenience for the user (Fig. 3).

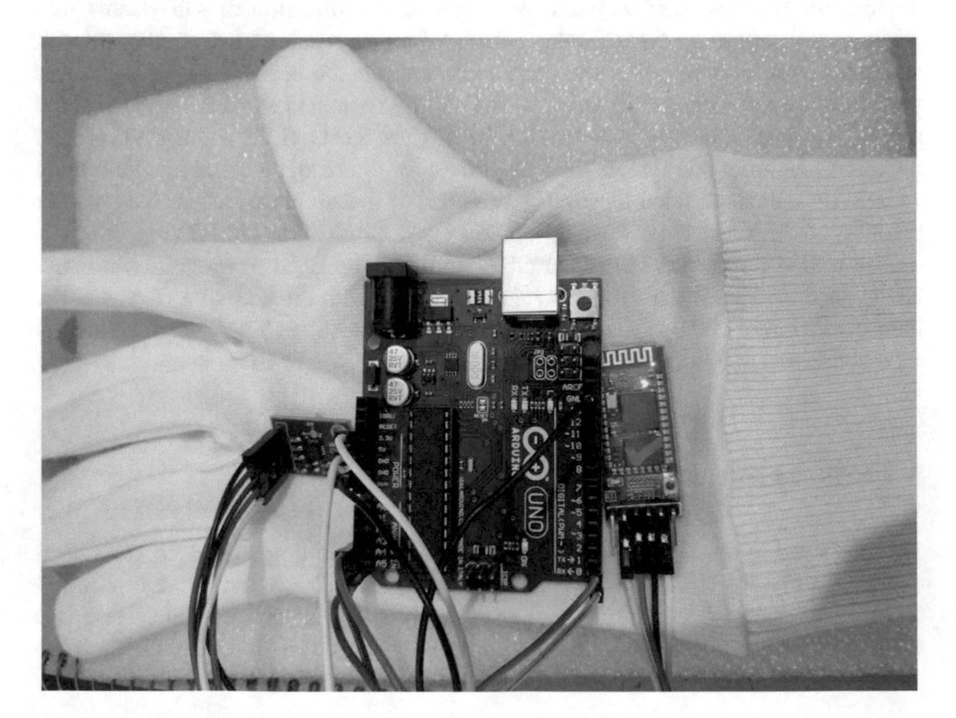

Fig. 1 Tracker system fitted onto glove

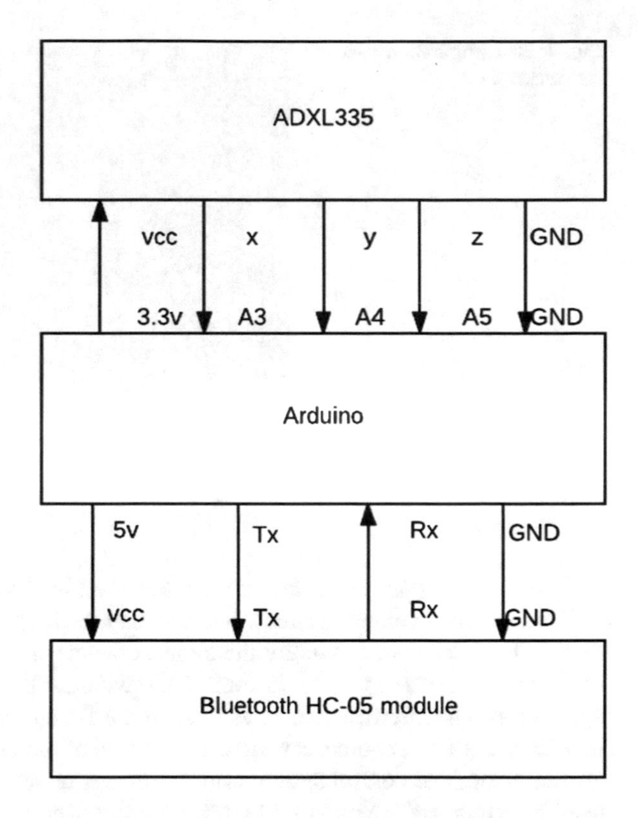

Fig. 2 Circuit of the tracking system

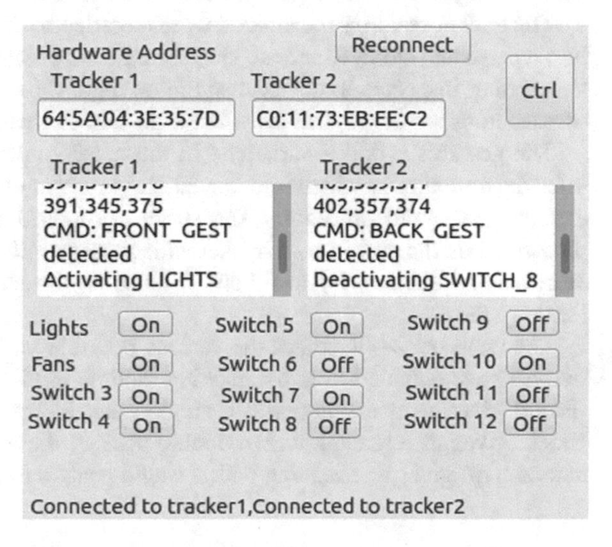

Fig. 3 Control system running on Raspberry Pi

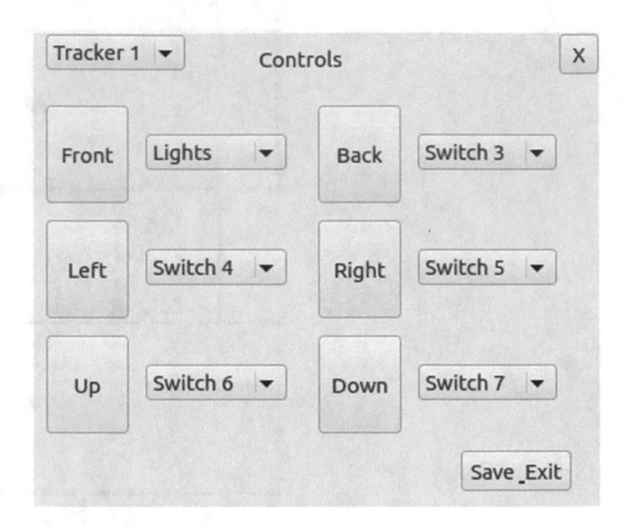

Fig. 4 Reconfiguration panel for tracker 1

The control system presents an interface used in the development phase to test the trackers and evaluate gestures. It shows that the Raspberry Pi system was able to discern the gestures and control the circuit connected to it. It supports 12 gestures overall and has the capability to control 12 switches that are connected to electrical appliances. The interface allows us to save the Bluetooth addresses of the trackers in a file at installation time, saving us the trouble of having to reconfigure the device at each boot. The control system also presents a reconfiguration panel that can be used by engineers to configure gestures for the user's comfort (Fig. 4).

The tracker readings transmitted were recorded by the control system and written into a persistent storage for analyzing the patterns created for each gesture. The data was further inspected to devise recognition algorithms that would simplify the task of executing stored procedures when a gesture is performed.

The gestures produce variations in the accelerometer readings, allowing us to infer the direction in which it is moved and to respond to the gesture performed by executing a stored procedure. The system proposed is able to detect 6 gestures performed by the user in each tracker and, consequently, facilitates the control of 12 electronic appliances. Figures 5 and 6 illustrate the readings obtained for two different gestures.

The table below illustrates the various possible hand gestures and their respective readings obtained from the tracking module in Fig. 7. The hand gesture named 'Front' gives us a reading with a single peak in the Y axis and the one named 'Back' gives us a reading with a double peak in the Y axis. Similarly, the gesture named 'Up' gives us a reading with a single peak and the one named 'Down' gives

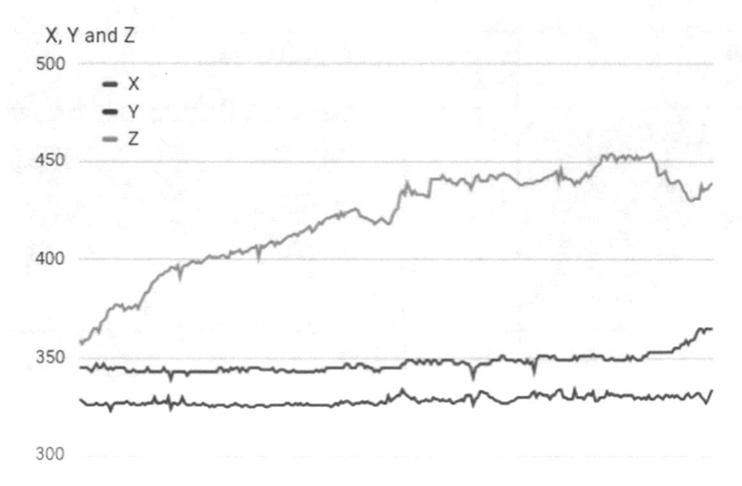

Fig. 5 Accelerometer readings for 'Up' gesture

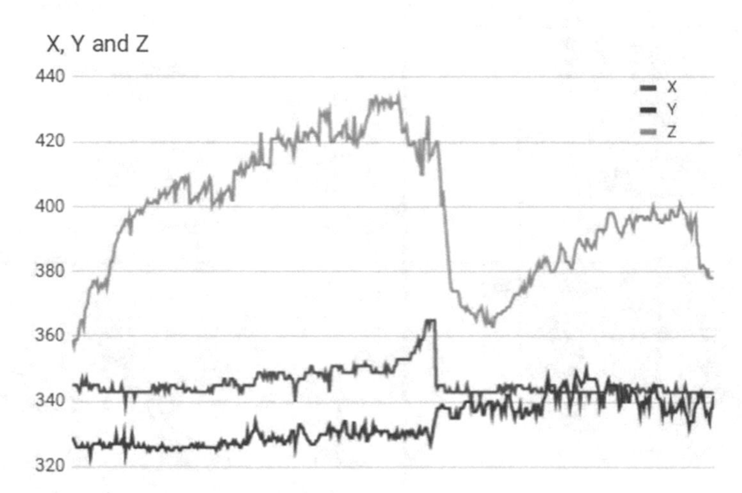

Fig. 6 Accelerometer readings for 'Down' gesture

us a reading with a double peak in the X axis. The table below shows the variation in the accelerometer readings with the gestures being performed. Various hand gestures were performed during the experiment and the accelerometer readings were obtained as follows.

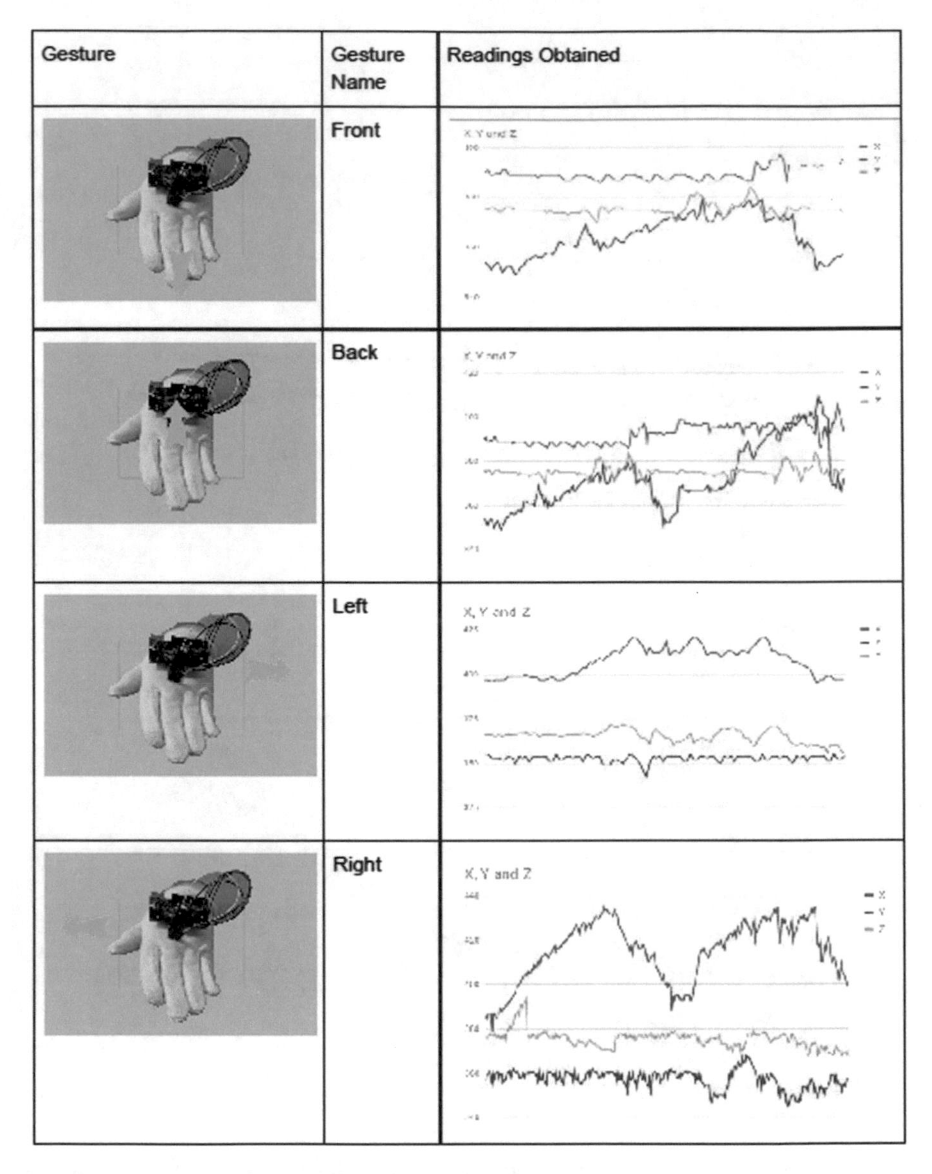

Fig. 7 Accelerometer readings along with associated gestures

4 Conclusion

The proposed system was constructed and two trackers were used to communicate with the control system to evaluate the effectiveness and ease of use in controlling home appliances. The control system was able to identify gestures and control the circuitry easily with less delay. The system was constructed keeping cost as an important aspect. However, the proposed system can be extended to incorporate a wide variety of gestures by adding gyroscopes to keep track of directions, and simultaneous data processing of the two trackers would facilitate inferring more complex gestures which would allow us to control more components.

References

1. Sulfayanti, Dewiani and Armin Lawi. 2016. A real time alphabets sign language recognition system using hands tracking. In *2016 international conference on computational intelligence and cybernetics*, 69–72, Makassar.
2. Islam, M.M., S. Siddiqua and J. Afnan. 2017. Real time Hand Gesture Recognition using different algorithms based on American Sign Language. In *2017 IEEE international conference on imaging, vision & pattern recognition (icIVPR)*, 1–6, Dhaka.
3. Kartika,. D.R., R. Sigit and Setiawardhana. 2016. Sign language interpreter hand using optical-flow. In *2016 international seminar on application for technology of information and communication (ISemantic)*, 197–201, Semarang.
4. Kumarage, D., S. Fernando, P. Fernando, D. Madushanka and R. Samarasinghe. 2011. Real-time sign language gesture recognition using still-image comparison & motion recognition. In *2011 6th International Conference on Industrial and Information Systems*, 169–174, Kandy.
5. Hatami, N., P. Prinetto and G. Tiotto. 2010. Sign language synthesis using hand motion acquisition. In *2010 East-West Design & Test Symposium (EWDTS)*, 226–229, St. Petersburg.
6. Natesh, A., G. Rajan, B. Thiagarajan and V. Vijayaraghavan. 2017. Low-cost wireless intelligent two hand gesture recognition system. In *2017 annual IEEE international systems conference (SysCon)*, 1–6, Montreal, QC.
7. Fan, Y.C., and H. K. Liu. 2015. Three-dimensional gesture interactive system design of home automation for physically handicapped people. In *2015 IEEE international symposium on medical measurements and applications (MeMeA) proceedings*, 432–435, Turin.
8. Starner, T., J. Auxier, D. Ashbrook and M. Gandy. 2000. The gesture pendant: A self-illuminating, wearable, infrared computer vision system for home automation control and medical monitoring. Digest of Papers. In: Fourth international symposium on wearable computers, 87–94, Atlanta, GA, USA.
9. Chandra, M., and B. Lall. 2016. A novel method for low power hand gesture recognition in smart consumer applications. In *2016 international conference on computational techniques in information and communication technologies (ICCTICT)*, 326–330, New Delhi.
10. Chou, P.H., et al. 2017. Development of a smart home system based on multi-sensor data fusion technology. In *2017 international conference on applied system innovation (ICASI)*, 690–693, Sapporo.
11. Hung, C.H., Y.W. Bai and H.Y. Wu. 2016. Home outlet and LED array lamp controlled by a smartphone with a hand gesture recognition. In *2016 IEEE international conference on consumer electronics (ICCE),* 5–6, Las Vegas, NV, 2016.
12. Fang, W.P. 2012. An intelligent hand gesture extraction and recognition system for home care application. In *2012 sixth international conference on genetic and evolutionary computing*, 457–459, Kitakushu.
13. Irie, K., M. Wada and K. Umeda. 2007. 3D measurement by distributed camera system for constructing an intelligent room. In *2007 fourth international conference on networked sensing systems*, 118–121, Braunschweig.

Server Communication Reduction for GPS-Based Floating Car Data Traffic Congestion Detection Method

Abdul Wahid, A. C. S. Rao and Dipti Goel

Abstract In large urban areas, traffic congestion is a perpetual problem for vehicle travelers because of the continuous and random flow of traffic. This causes congestion at multiple places due to delays in communication between server and vehicles. To reduce this communication delay and the associated costs, we have developed a server communication reduction policy using GPS-based floating car data (FCD), a traffic congestion detection method in which it is assumed that all vehicles act as sensor nodes that transmit their data to the server, and the server uses the data to calculate traffic congestion on that road segment and then broadcasts the updated real-time traffic data to the user. Using this updated data, vehicles can determine the optimal route for reaching their destination in the shortest amount of time. In this chapter, we analyze this reduction policy applied to traffic data for an Australian road network, consisting of approximately 300,000 samples from 11 different types of vehicles. We then present the results based on graphs and tables showing our improved outcomes.

Keywords Floating car data (FCD) · Server communication reduction policy
Traffic congestion · Intelligent transport system (ITS)

A. Wahid (✉) · A. C. S. Rao
Department of Computer Engineering, Indian Institute of Technology
(Indian School of Mines) Dhanbad, Dhanbad 826004, Jharkhand, India
e-mail: abdul.cspg14@nitp.ac.in

A. C. S. Rao
e-mail: acrao232@yahoo.co.in

D. Goel
Department of Computer Engineering, National Institute of Technology Kurukshetra,
Kurukshetra 136119, Haryana, India
e-mail: diptigoel12@gmail.com

A. N. Krishna et al. (eds.), *Integrated Intelligent Computing, Communication and Security*, Studies in Computational Intelligence 771,
https://doi.org/10.1007/978-981-10-8797-4_43

1 Introduction

Together, rapid growth in the number of vehicles, increasing population and urbanization of the cities raise many challenges for transport systems with increasing numbers of road accidents, more air pollution and more traffic congestion. Conventional traffic data systems utilize fixed sensors to collect real-time traffic data. Traditional fixed detectors, including magnetic loop detectors, inductive loops, video camera and radar, are subject to distribution restrictions [1] because it does not make sense to fix a huge number of fixed sensors at every street. Moreover, traditional loop detectors are subject to errors in measurement. For instance, according to the Performance Measurement System in California, 30% of traffic detectors are subject to daily breakdown [2]. In addition, fixed traffic detector devices have limited coverage, providing data collection on only a limited portion of the entire road facility [3]. Due to high costs of installation and maintenance, these technologies cannot be deployed on all road networks, especially in developing countries [4]. Table 1 shows some congestion detection techniques with their lifetime operational costs and capital costs.

Nowadays, large numbers of vehicles in major cities are equipped with global positioning system (GPS) devices. In this regard, vehicles can be considered as probes that produce a considerable number of high resolution GPS traces. A GPS device in a vehicle provides accurate information of the vehicle's velocity and location [6]. This floating car data (FCD) can largely assist in overcoming the problems associated with conventional traffic data collection methods [7]. For example, a floating car equipped with a GPS device can function as a traffic agent on the transportation network to collect real-time traffic data. Compared to a traditional fixed sensor, a mobile GPS device provides the advantages of more accurate, more comprehensive and larger scale data collecting [8]. Furthermore, information about the position of the car can be gathered at any time during traffic flow. An array of mobile GPS devices thus can be considered a viable source of traffic operating data, especially in case of urban road networks.

Table 1 Equipment cost of some detectors [5]

Cost per unit	Period (in years)	Operational cost (in $000)	Capital cost (in $000)	Total price date
Inductive loop surveillance corridor	5	0.4–0.6	3–8	2001
Inductive loop at intersection	5	0.9–1.4	8.6–15.3	2005
Machine vision sensor at intersection	10	0.2–1	16–25.5	2005
Passive acoustic sensor at intersection		0.2–0.4	5–15	2001
Passive acoustic sensor on corridor		0.2–0.4	3.7–8	2002
CCTV video camera	10	1–2.3	9–19	2005

Information can be collected and sent to a server at regular intervals to be used for traffic flow analysis. The server manages a database that is spatial in nature to store real-time traffic data of an entire road network. After receiving those data from vehicles, the server calculates road congestion, average travel speed and average travel time of each road segment, then updates the database and broadcasts updated data to the vehicles.

A significant amount of communication between the server and the vehicles is needed to update data stored at the server. In large urban areas, the large number of vehicles moving on the road networks creates a problem of data load on the server and increases communication cost between the server and the vehicles, as well as increasing the server load to process all the data collected from vehicles. These problems make it difficult for an FCD system to expand in larger areas. To overcome these problems, we can reduce the number of transmissions to the server. But reducing the number of transmission, which (happily) reduces the server load and also reduces the transmission cost between server and vehicles, also reduces the accuracy of the traffic data. In this paper we propose a server communication reduction policy that reduces transmission cost and increases the accuracy of the traffic data updated by the server compared with other server update policies.

The remainder of this chapter is organized as follows. Section 2 describes related work, while Sect. 3 describes a server communication reduction policy that reduces transmission cost between server and vehicles, and also proposes techniques that reduce transmission cost. Section 4 describes simulation and discussion on these results. Section 5 presents our conclusion and some future work.

2 Related Work

Various technologies have been applied to FCD. Traffic congestion management techniques based on FCD have assumed that every vehicle is equipped with a GPS device and a wireless communication link such as GPRS or UMTS. Each GPS device sends its current location and speed to a server at a specified interval of time [9], generally 1–4 positions per minute. The main advantage of FCD based on GPS is its accuracy in terms of location (about 10–20 m) and its wide range of geographical coverage. The main disadvantage is the cost of GPS devices, and communication cost between the server and the vehicles.

A traffic information sharing system developed by Kerner et al. [10] used a client server architecture in which vehicles act as clients that send their data to the server to measure the traffic speed of the road segments. This client server architecture reduces data transmission using a deterministic policy with some threshold value, where vehicles send their data to the server if the difference between the server's broadcast speed and its current speed exceeds a given threshold value.

A randomized update policy introduced by Tanizaki and Wolfson [11] reduces transmission cost by using a randomizing function and a threshold value. In the randomized policy, vehicles send traffic data (speed) to the server if the probability of sending those data is less than 1, which reduces the server communication cost by using a threshold value. It decides whether to send their data to the server by tossing a coin, where the probability of getting heads is "*p*", then it transmits those data to the server only if heads is tossed. But due to this, the server collects multiple values and most of the data are skewed, which is why it is not suitable for a server communication reduction policy.

Another approach is to consider a mobile peer-to-peer (MP2P) architecture, in which vehicles share their speed with other vehicles on the road without any server facilities. Goel et al. [12] consider such a system, based on MP2P architecture, where each vehicle sends its measured velocity to other vehicles and updates a wide range using P2P communication methods.

A traffic information sharing system considered by Shinkawa et al. [13] is based on mobile peer-to-peer expanded by using buses moving along static routes to send traffic related data to detached clients. Other methods also exist that decrease the communication cost by reducing the data capacity or lessening the transmission frequency. A compression method of trajectory developed by Shinya et al. [14], which is based on the breakdown of spatial and temporal constituents and distinct wavelet conversion of the temporal constituent.

3 Server Communication Reduction Policy

A server communication reduction policy is a technique used by each vehicle to decide when we have to send a data transmission to the server and how to reduce communication costs between server and vehicles without affecting the efficiency of traffic data. In this unit we present the deterministic policy, the randomized policy and our proposed policy.

A. **Deterministic Policy**: The traffic information sharing system developed by Kerner et al. [10] uses a client server architecture that reduces data transmission using a velocity threshold, in which vehicles send their data to the server if it exceeds from a threshold value. In other words, we can say that vehicle send its current speed "V_c" to the server if the difference between its current speed "V_c" and its broadcasted speed "V_b" is greater than or equal to some threshold value "γ". If γ is the threshold value, then according to this transmission rule

$$|Vc - Vb| \geq \gamma \tag{1}$$

B. **Randomized Policy**: The randomized update policy introduced by Tanizaki and Wolfson [11] reduces transmission cost by using a randomizing function and a threshold value. In the randomized policy introduced by Tanizaki and Wolfson, vehicles sends traffic data (speed) to the server if the probability of sending those data is less than 1, which reduces the server communication cost by using a threshold value as described previously.

C. **Proposed Server Update Policy**: We propose a server update policy that does not use randomization or information-cost criteria. Under this system, every vehicle has an equal chance of transmitting its data to the server. Figure 1 shows our proposed server communication reduction policy, and Fig. 2 shows a flow chart for the proposed system.

In this proposed update policy, all the vehicles send their data (related to location and speed) to a control room. After receiving data from all the vehicles, the control room uses the following server communication reduction policy: If the difference between the current speed of a vehicle and the previous speed of a vehicle is greater than or equal to some threshold value at some interval of time, then it sends further data to the server, otherwise it discards those data. In other words, we can say that a vehicle forwards its data to the server if

$$|Vc - Vp| \geq \gamma \tag{2}$$

where Vc is the current speed of the vehicle and Vp is its previous speed. After applying this server communication reduction policy, the control room dispatches those data for storage in the FCD database. After that, the server processes those data by using some steps (data preprocessing, map matching, traffic speed estimation, etc.) to calculate real-time traffic data for any part of the road network. After calculating and updating those real-time data, the server broadcasts its updated data to the vehicles with the help of GPRS, GPS, etc.

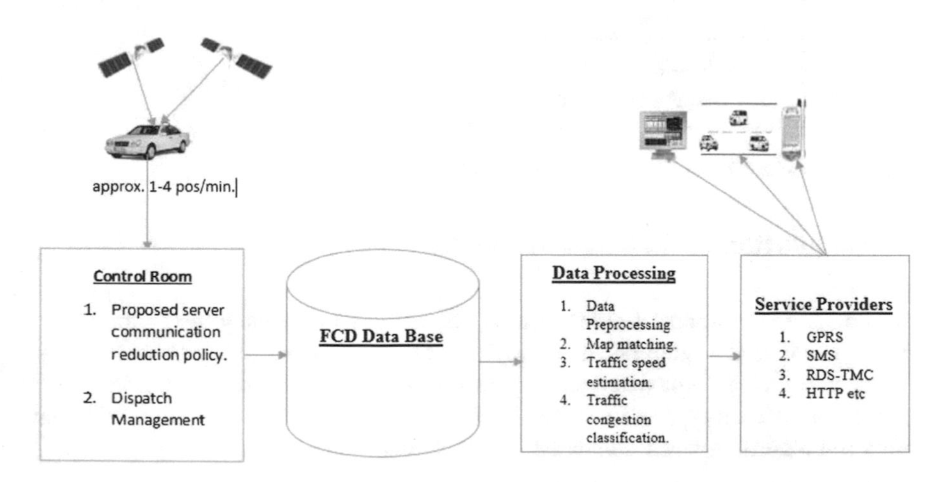

Fig. 1 Proposed architecture of real-time traffic data with server communication reduction policy

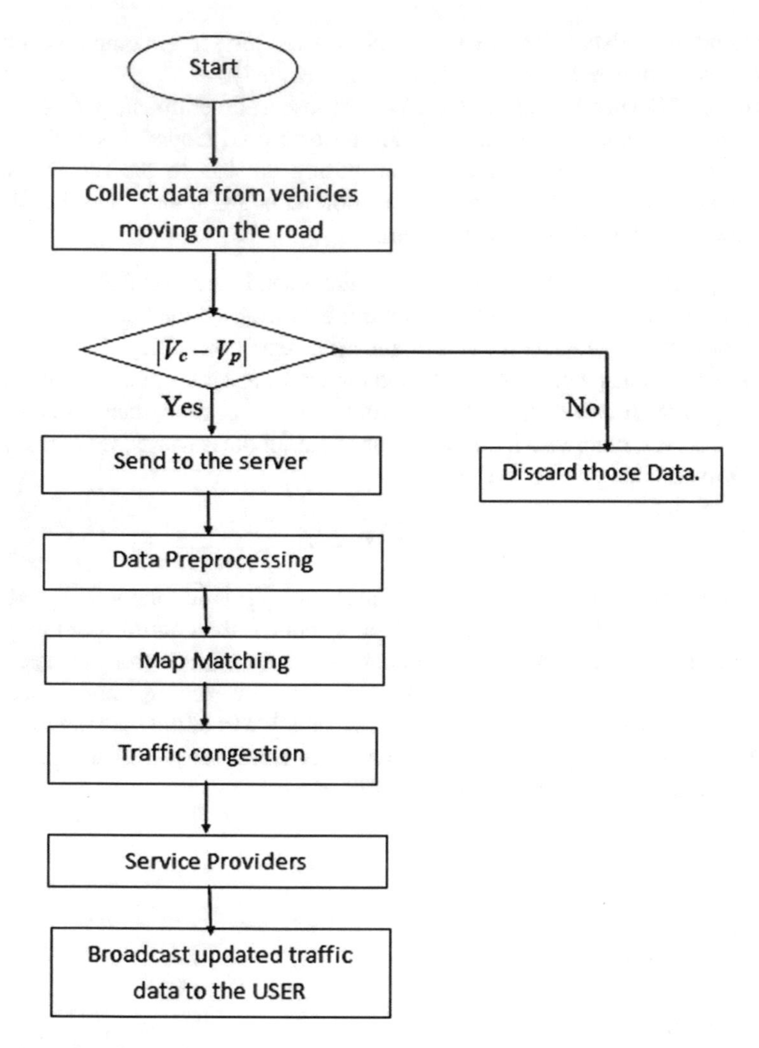

Fig. 2 Flow chart for proposed server update policy

4 Simulation and Discussion of Results

To simulate our proposed algorithm and compare its efficiency with other algorithms (such as the deterministic policy and the randomized policy described previously), we have developed a simulation system that takes real-time traffic related data from different types of vehicles. In this section, we discuss the results of our proposed algorithm when compared with existing systems.

A. **Input Data for Simulation**: To simulate our proposed algorithm, we used real-time data provided by the Australia Transportation department, which includes real-time traffic data, congestion data, and average speed of road

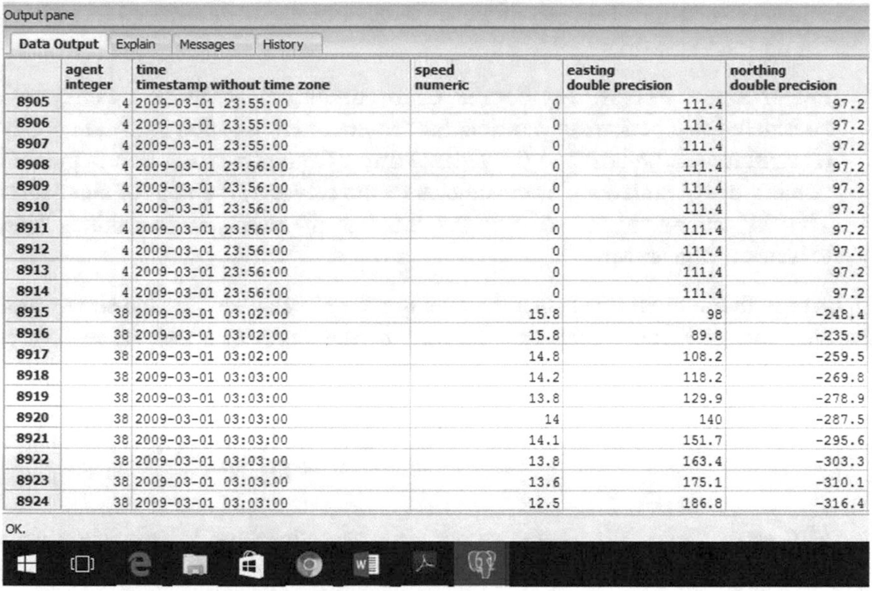

Output pane

Data Output | Explain | Messages | History

	agent integer	time timestamp without time zone	speed numeric	easting double precision	northing double precision
8905	4	2009-03-01 23:55:00	0	111.4	97.2
8906	4	2009-03-01 23:55:00	0	111.4	97.2
8907	4	2009-03-01 23:55:00	0	111.4	97.2
8908	4	2009-03-01 23:56:00	0	111.4	97.2
8909	4	2009-03-01 23:56:00	0	111.4	97.2
8910	4	2009-03-01 23:56:00	0	111.4	97.2
8911	4	2009-03-01 23:56:00	0	111.4	97.2
8912	4	2009-03-01 23:56:00	0	111.4	97.2
8913	4	2009-03-01 23:56:00	0	111.4	97.2
8914	4	2009-03-01 23:56:00	0	111.4	97.2
8915	38	2009-03-01 03:02:00	15.8	98	-248.4
8916	38	2009-03-01 03:02:00	15.8	89.8	-235.5
8917	38	2009-03-01 03:02:00	14.8	108.2	-259.5
8918	38	2009-03-01 03:03:00	14.2	118.2	-269.8
8919	38	2009-03-01 03:03:00	13.8	129.9	-278.9
8920	38	2009-03-01 03:03:00	14	140	-287.5
8921	38	2009-03-01 03:03:00	14.1	151.7	-295.6
8922	38	2009-03-01 03:03:00	13.8	163.4	-303.3
8923	38	2009-03-01 03:03:00	13.6	175.1	-310.1
8924	38	2009-03-01 03:03:00	12.5	186.8	-316.4

OK.

Fig. 3 Description of GPS floating car data (FCD)

segments. Data gathered by FCD techniques at road segments in Australia were downloaded for the period of "2009-02-28 23:59:00" to "2009-03-01 21:13:00". The dataset gathered during this collection period contains about 297351 data items from 15 different types of vehicles. These data contain vehicle type, timestamps of vehicles with speeds at that time, and location of vehicles, in intervals of 1 s, as is shown in Fig. 3.

B. **Simulation Procedure**: Inputs to the system include:

1. Vehicle data with vehicle type, timestamp, speed, longitude and latitude
2. A threshold value "γ" (difference between current velocity and previous velocity), used in the transmission of data from vehicles to the server
3. A collection period "ζ" with a length of 300 ms, which is fixed for the entire simulation
4. An updated server policy.

The simulation is executed on both server and client threads. The server receives traffic data (velocity) from the control room (if it is greater than the threshold value γ) via the clients. After receiving those data from clients, the server updates its traffic speed "Vp" at the end of each collection period "ζ". Client threads

Table 2 Parameters for simulation

Parameter	Symbol	Unit	Range
Threshold value	γ	Km/h	1, 2, 3, 4, 5
Collection period	ζ	Min	5

continuously send travel speeds "Vc" to the control room to estimate traffic congestion on road networks (Table 2).

C. **Performance Factors for Server Communication Reduction Policy**: We considered two performance factors for server communication reduction policy: (i) communication cost and (ii) average error. Communication cost is the total number of transmissions from vehicles to the server at the time of simulation. That is, communication cost is the number of messages transmitted from vehicles to the server.

Error is the difference between the average speed calculated by the server from the data gathered from vehicles at the time of the simulation and the real speed throughout the collection period. Average error can be calculated as

$$Average\,Error = \frac{\text{sum of error values}}{Total\,no.\,of\,collection\,period}\,\text{Km/h} \tag{3}$$

D. **Efficiency Metrics**: To compare our proposed algorithm with existing algorithms, we use calculate an efficiency metric by using the average error and communication cost simultaneously. In this simulation we consider communication cost (number of messages transmitted from vehicles to the server) as the input parameter to the system, and average error as the output of the system. Our aim is to reduce average error as well as communication cost

$$Efficiency = \frac{1}{Average\,error * communication\,Cost} \tag{4}$$

From this equation we can say that if average error values of the compared systems are approximately equal, then this efficiency metric will favor the policy that has the lowest communication cost. Conversely, if communication costs are almost equal across systems, then this efficiency metric will favor that policy that has the lowest average error.

Figure 4 presents a graph of communication cost and collection period showing the number of messages between servers and the vehicles at each 300 s interval. The graph shows that communication cost decreases as we increase threshold value from 1 to 5.

$$Communication\,cost \propto \frac{1}{Threshold\,value\,(\gamma)} \tag{5}$$

However, this does not fulfill our aim, because when we decrease communication cost between server and vehicles, the average error increases. From the given graph:

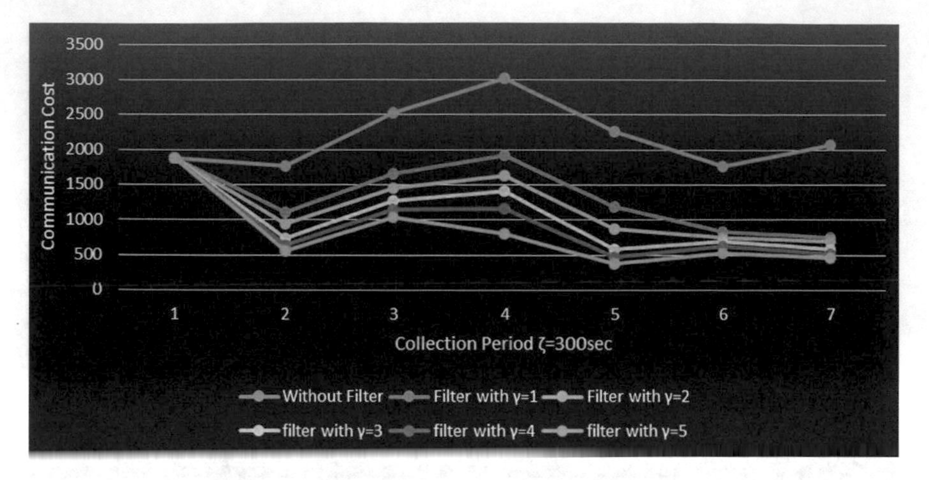

Fig. 4 Communication cost versus threshold values at 300 s interval

$$\mathit{Communication\ cost} \propto \frac{1}{\mathit{Average\ Error}} \tag{6}$$

In Fig. 4, we can see that our proposed algorithm gives better results without a data filtering method. From the graph, it gives better results at threshold value 5 km/h. Figure 5 presents a graph of average error and threshold value at a collection period of 300 s. This graph shows how average error varies with threshold value. From the given figure, we can say that

$$\mathit{Average\ Error} \propto \mathit{Threshold\ value} \tag{7}$$

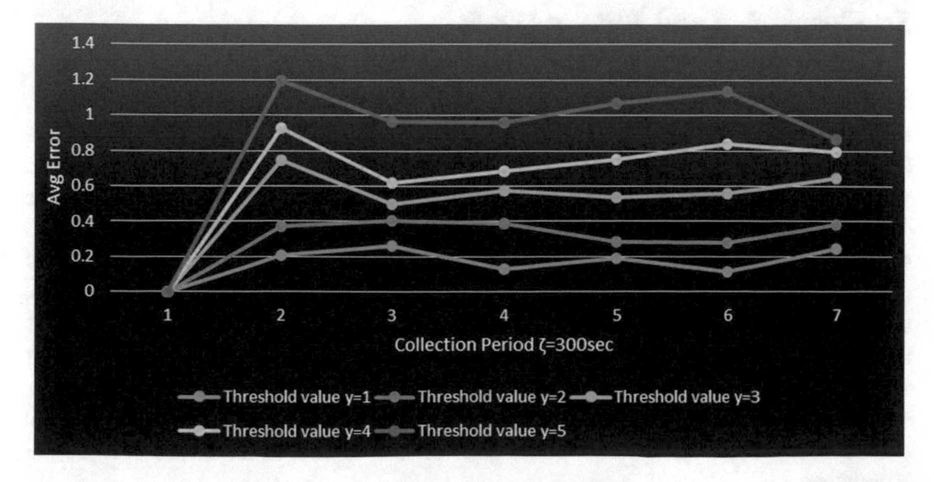

Fig. 5 Average error versus threshold values at 300 s interval

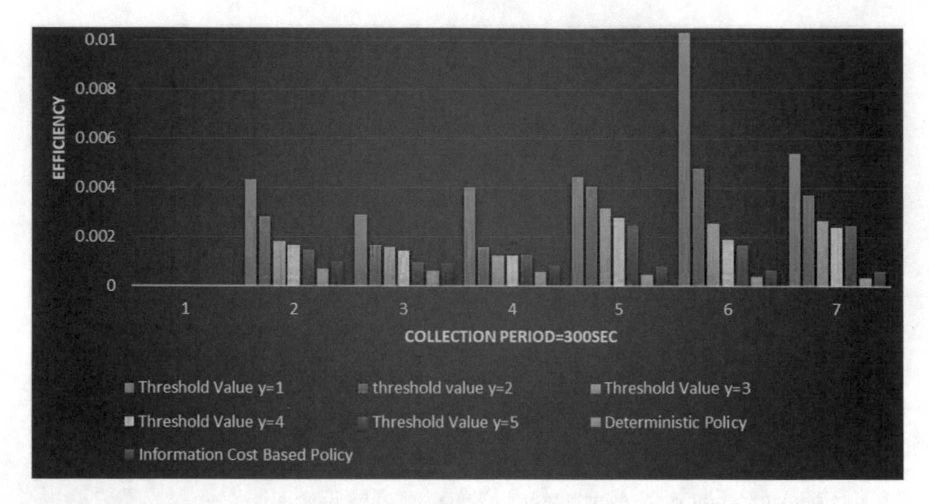

Fig. 6 Efficiency and threshold value at 300 s interval

as we increase the threshold value from 1 to 5, then average error also increases. This also does not fulfill our aim, which is to decrease average error and communication cost while simultaneously increasing the efficiency of our proposed system. From the graph, we can see that it gives a lower average error at threshold value 1 km/h. Figure 6 shows how our proposed server update policy is better than other policies in terms of efficiency. This graph shows the efficiency metric of our proposed algorithm with the other two server update policies. In some cases, average error is low as compared to other techniques but there are high communication costs.

5 Conclusion and Future Work

This chapter discussed the problem of how to reduce communication costs between server and vehicles in a real-time traffic information sharing system based on GPS floating car data (FCD) to increase the efficiency of the system. In this system all the vehicles are equipped with GPS devices and act as sensor nodes to transmit their data to a server at some interval. After collecting these data from vehicles, the server transmits updated real-time traffic data to the vehicles. After getting this real-time traffic data, vehicles can find an optimal path. We also described a server update policy that reduces the transmission cost with allowable average error and increases the efficiency of the system. Simulations and graphs were presented showing that the proposed algorithm works better than other policies in terms of efficiency. Future work in this direction would be to propose an approach for consideration of minor road networks and reduce the cost of hardware used in the simulation.

References

1. Zhou, et al. 2013. Traffic flow analysis and prediction based on GPS data of floating cars. In *Proceedings of the international conference on information technology and software engineering*. Springer Berlin Heidelberg.
2. Bar-Gera, H. 2007. Evaluation of a cellular phone-based system for measurements of traffic speeds and travel times: A case study from Israel. *Transportation Research C* 15 (6): 380–391.
3. Xiaohui, S., X. Jianping, Z. Jun, Z. Lei, and L. Weiye. 2006. Application of dynamic traffic flow map by using real time GPS data equipped vehicles. In *ITS telecommunications proceedings*, 1191–1194.
4. Herrera, Juan C., et al. 2010. Evaluation of traffic data obtained via GPS-enabled mobile phones: The mobile century field experiment. *Transportation Research C: Emerging Technologies* 18 (4): 568–583.
5. http://www.itscosts.its.dot.gov.
6. Lee, W.H., S.-S. Tseng, S.-H. Tsai. 2009. Knowledge based real-time travel time prediction system for urban network. *Expert Systems with Applications* 4239–4247.
7. Byon, Y.J., Amer Shalaby, and Baher Abdulhai. 2006. Travel time collection and traffic monitoring via GPS technologies. In *Intelligent transportation systems conference*.
8. Yong-chuan, Zhang, Zuo Xiao-qing, and Chen Zhen-ting. 2011. Traffic congestion detection based on GPS floating-car data. *Procedia Engineering* 15: 5541–5546.
9. Shi, W., and Y. Liu. 2010. Real-time urban traffic monitoring with global positioning system-equipped vehicles. *IET Intelligent Transport Systems* 4 (2): 113–120.
10. Kerner, B., et al. 2005. Traffic state detection with floating car data in road networks. In *IEEE proceedings on itelligent transportation systems,* 44–49.
11. Tanizaki, M., and O. Wolfson. 2007. Randomization in traffic information sharing systems. In *GIS '07: Proceedings of the 15th annual ACM international symposium on advances in geographic information systems*.
12. Goel, S., T. Imielinski, K. Ozbay, and B. Nath. 2003. Grassroots: A scalable and robust information architecture. DCS-TR-523.
13. Shinkawa, T., T. Terauchi, T. Kitani, N. Shibata, K. Yasumoto, M. Ito, and T. Higashino. 2006. A technique for information sharing using inter-vehicle communication with message ferrying. In *Proceedings of the 7th international conference on mobile data management*.
14. Shinya, A., N. Satoshi, and T. Teruyuki. 2005. Research of compression method for probe data-a lossy compression algorithm for probe data. *IEIC Technical Report* 104 (762): 13–18.

Random Seeding LFSR-Based TRNG for Hardware Security Applications

R. Shiva Prasad, Anirudh Siripagada, Santthosh Selvaraj and N. Mohankumar

Abstract Rapid developments in the field of cryptography and hardware security have increased the need for random number generators which are not only of low-complexity but are also secure to the point of being undeterminable. A random number generator is a part of most security systems, so it should be simple and area efficient. Many modern-day pseudorandom number generators (PRNGs) make use of linear feedback shift registers (LFSRs). Though these PRNGs are of low complexity, they fall short when it comes to being secure since they are not truly random in nature. Thus, in this chapter we propose a random seeding LFSR-based truly random number generator (TRNG) which is not only of low complexity, like the aforementioned PRNGs, but is also 'truly random' in nature. Our proposed design generates an *n*-bit truly random number sequence that can be used for a variety of hardware security based applications. Based on our proposed *n*-bit TRNG design, we illustrate an example which generates 16-bit truly random sequences, and a detailed analysis is shown based on National Institute of Standards and Technology (NIST) tests to highlight its randomness.

Keywords TRNG · PRNG · Hardware security · Cryptography NIST · Random numbers

R. Shiva Prasad · A. Siripagada · N. Mohankumar (✉)
Department of Electronics and Communication Engineering,
Amrita School of Engineering, Amrita Vishwa Vidyapeetham,
Coimbatore, India
e-mail: n_mohankumar@cb.amrita.edu

R. Shiva Prasad
e-mail: shivapuzzler@gmail.com

S. Selvaraj
Cognizant Technology Solutions, Chennai, India

A. N. Krishna et al. (eds.), *Integrated Intelligent Computing, Communication and Security*, Studies in Computational Intelligence 771,
https://doi.org/10.1007/978-981-10-8797-4_44

1 Introduction

Security issues are a major threat to modern electronic hardware. Designing secure hardware is a major challenge. Random number generators, being hardware which play a significant role in cryptographic functions such as encryption, authentication and identification, should therefore be designed to be secure, to contribute to the overall security of the hardware in which they are used. Random numbers can be broadly classified into deterministic, pseudorandom numbers and non-deterministic, truly random numbers. A typical true random number generator (TRNG) is a system that acts as an interface which converts uncertainty in observed phenomena to random numbers. Modern day TRNGs make use of natural phenomena, mainly noise such as thermal noise over a resistor and shot noise in vacuum tubes. Pseudorandom numbers are numbers which are computed from an initialized vector called a seed. The seed is the only presenter of uncertainty in pseudorandom number generation. The property of periodicity is common among all pseudorandom number generators (PRNGs), but to be truly random, a bit or number sequence must never have this property. The main reason for the existence of this property in the PRNG algorithm is its deterministic nature. However, it is possible to improve periodicity in larger values such that it is almost no longer an issue of concern. This improvement in periodicity can be attained by simply changing the seed value just before the sequence begins to repeat.

Random numbers play a prominent role in hardware security. Modern-day cryptographic algorithms make use of keys which are known only by the designers of the cryptographic algorithms. Thus, the output of the cryptographic algorithm is deemed to be unbreakable, mainly because of the unpredictability of the key used in the algorithm. Generally, a stream of random bits is presented as the key to these cryptographic algorithms and the designers make sure that this stream of random bits (used as the key) is highly random in nature to ensure overall security of the algorithm. Random number generators are also widely used in the generation of digital signatures, which are strings of random bits unique to a given circuit or system. These random strings of bits are similar to a fingerprint and ensure security of the electronic system.

2 Literature Survey

A ring oscillator based TRNG and a physical unclonable function (PUF) combination is discussed in [1]. Manoj Reddy et al. proposed an IC Authentication technique for embedding unique signatures with minimalistic hardware and area overhead, while preserving the intended functionality in [2]. In addition to verifying the reliability of the IC, it also detects any malicious activity during runtime if any hardware Trojan exists in the design by using a linear feedback shift register (LFSR) based PRNG. Threats due to hardware Trojans (HTs) are increasing day by day.

Several methods have been proposed to counter HT attacks. Methods involving current, power [3], delay [4] and multiple modular redundancy methods, such as an IC authenticity check using voting methods or virtual intelligence based methods. Aishwarya et al. [5], Bharath et al. [6] and Kamala Nandhini et al. [7], are attracting more attention. The idea of designing tamper proof hardware and designing for secure systems has become a focus. In this context of designing secured hardware, TRNGs play a pivotal role. Blum and Shub [8] is a popular pseudorandom number generator suggested by Lenore Blum, Manuel Blum and Michael Shub in 1986 which is still being used in some cryptographic applications. However, of late, there is proof that the security of B-B-S pseudorandom number generators can be reduced to the computational difficulty of solving the quadratic residuosity problem. Thus, security is an important aspect which needs to be considered when opting for a PRNG instead of a TRNG. However, as we see in [1], as security increases hardware requirements also increase [9]. Thus PRNGs which are good in both security and hardware requirements are difficult to design without going for a trade-off between the two [10, 11]. Thus TRNGs, though more expensive than PRNGs, are more secure and reliable for security based applications.

In this chapter, we propose an LFSR based TRNG which is of low complexity and operates within quality requirements in terms of random number uniqueness and reliability. In the following section, we discuss the basis for our proposed TRNG, the seed generator for our proposed design which exploits the meta-stability phenomena of the SR-latch to generate truly random seed values, thus enabling our proposed random number generator to generate truly random bit sequences.

3 Background

An LFSR, or linear feedback shift register, uses as its input a linear combination of previous states of the LFSR. The output of the LFSR depends on the location of the tap-points and the linear combination logic used. In general, the linear combination logic used is XOR logic. In our proposed design we have used 8-bit LFSRs. The 8-bit LFSR consists of 8 D-type flip-flops, with an XOR gate as the linear combination logic. Each LFSR has its own characteristic polynomial. Depending upon the tap-points in the LFSR, a characteristic polynomial is computed. The 8-bit LFSR tap-point configuration used in our design is shown in Fig. 1.

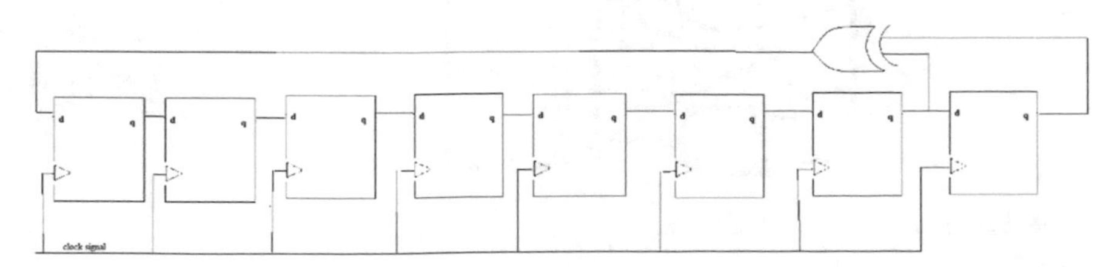

Fig. 1 8-Bit LFSR with characteristic polynomial $1 + x^7 + x^8$

The periodicity of sequences generated by an n-bit LFSR is given by the expression $2^n - 1$. Thus an 8-bit LFSR has a periodicity of $2^8 - 1 = 255$ sequences, after which the sequences begin to repeat.

3.1 Seed Generator

To ensure a low-complexity design, we included a simple seed generator, which uses the meta-stability phenomena of a set/reset (SR) latch, in our proposed TRNG module. When the input is given a ‘0’ the output is a stable ‘1’, but when the input is changed from ‘0’ to ‘1’, the output temporarily enters a metastable state and then settles to a stable ‘1’ or ‘0’. This phenomenon can be exploited to generate seed values for our proposed TRNG; the seed generator for an 8-bit LFSR is shown in Fig. 2. In this way, seeds can be generated for any n-bit LFSR by using ‘n/2’ SR-latches with their corresponding ‘S’ and ‘R’ inputs as ‘1’.

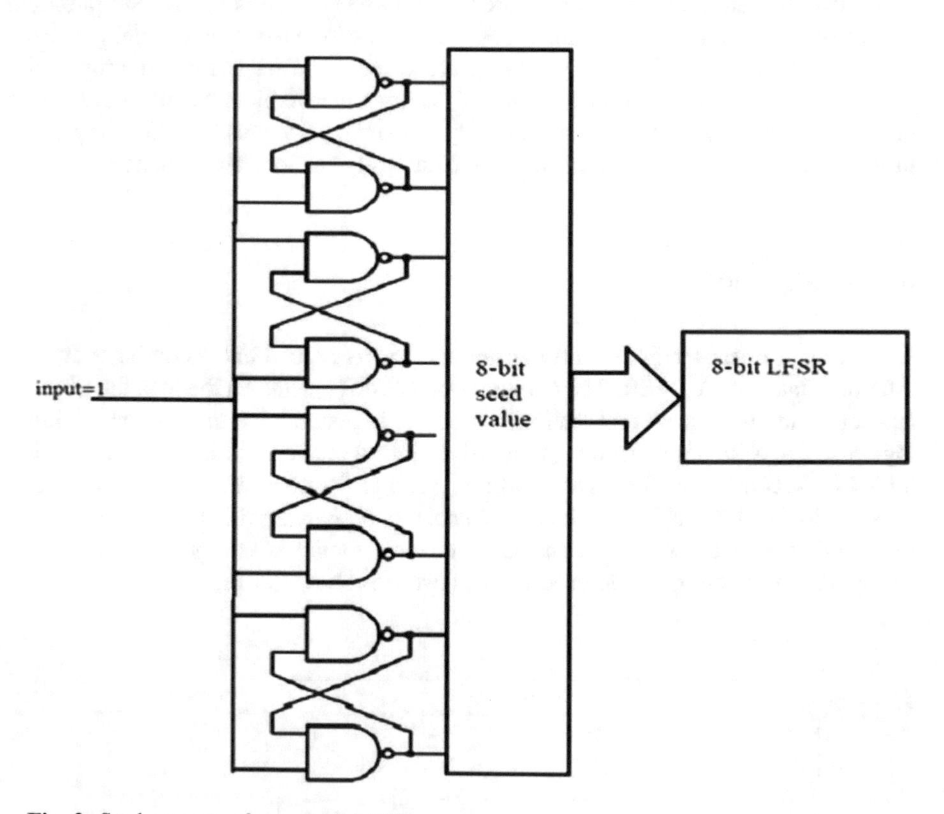

Fig. 2 Seed generator for an 8-bit LFSR

4 Proposed TRNG

In this section, we propose our design of an *n*-bit true random number generator. The block diagram shown in Fig. 3 illustrates the proposed randomly seeded LFSR-based TRNG module. The proposed 8-bit LFSR module contains *n*/2 8-bit LFSRs in parallel, a clock divider module consisting of four D-type flip-flops and a D-type flip-flop at the end of each LFSR to latch the output of the LFSR at the positive edge of the input clock signal. This design facilitates throughput generation of *n*-bit random numbers by utilizing *n*/2 8-bit LFSRs in parallel. Though the 8-bit LFSR by itself has a periodicity of only 255, the random seeding provided by the proposed seed generator makes this a true random number generator. The 'true random' nature of our proposed random number generator mainly arises from the use of our proposed seed generator module, which exploits the nature of the SR-latch to go into meta-stability when both the S and R inputs are '1'. Thus, each time the TRNG is invoked, different '*n*' bit sequences are generated, mainly because of the random seeding by the SR-latch based seed generator.

The schematic for the proposed 8-bit LFSR-based TRNG which utilizes two 8-bit LFSRs in parallel, a clock divider module and branch of D-type flip-flops to latch the output is shown in Fig. 4.

5 Results and Analysis

To justify the use of any design for random number generation, it is important to test the randomness of the numbers generated by suitable tests. We tested the randomness of the sequences generated by our designs using National Institute of Standards and Technology (NIST) tests such as frequency mono-bit test, Runs test, Frequency test within a block, Longest run of ones in a block test,

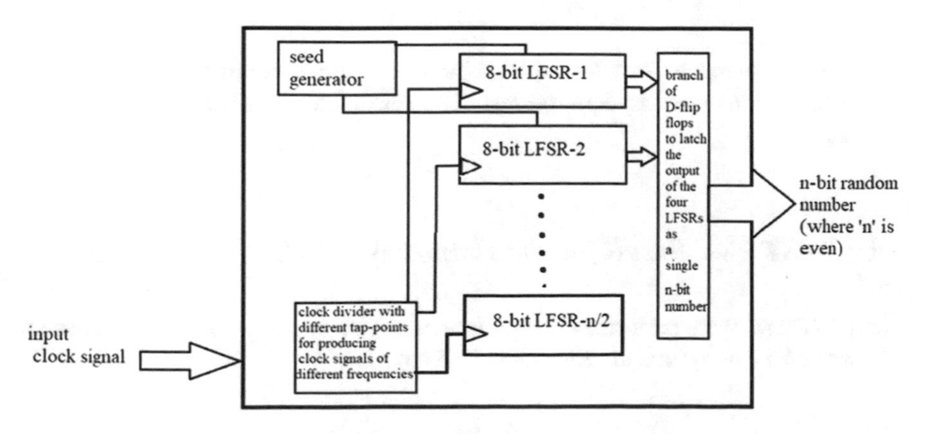

Fig. 3 Proposed 8-bit LFSR-based TRNG module

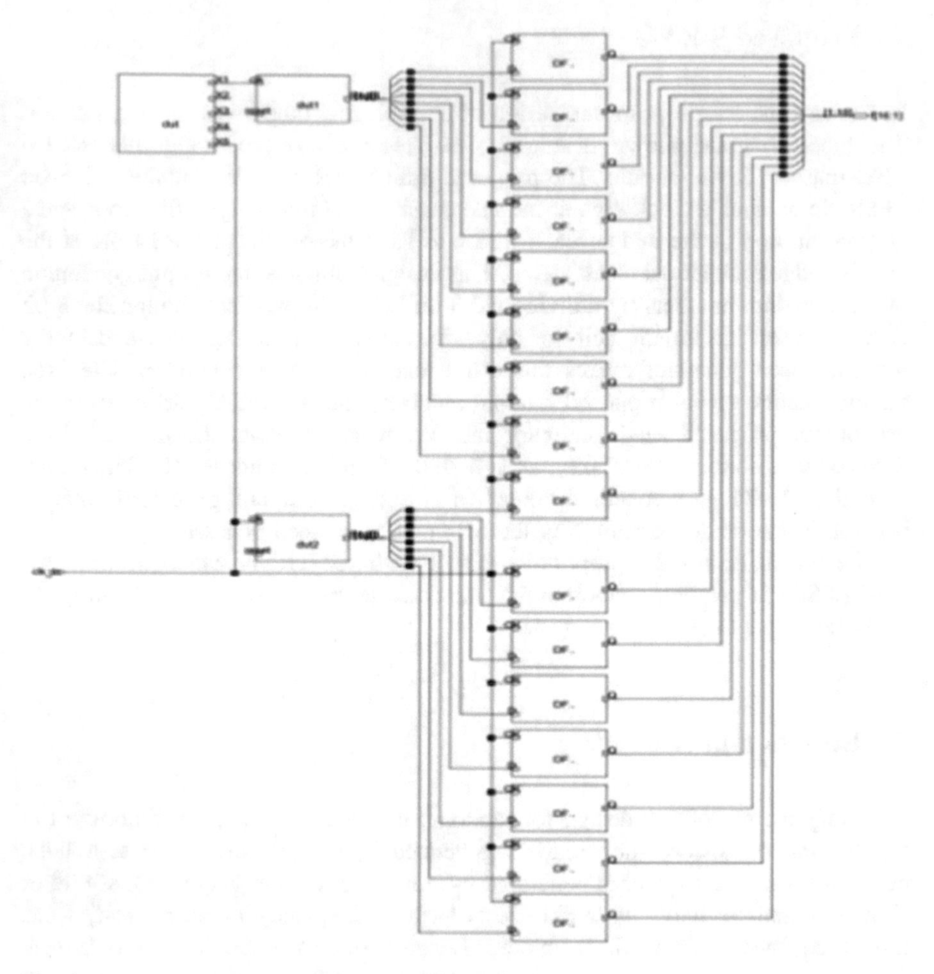

Fig. 4 Schematic of the proposed 8-bit LFSR-based TRNG

Overlapping template matching test, Non-overlapping template matching test, Cumulative sums test and Spectral test. The results are detailed in the following section.

5.1 NIST Test Result for the Proposed TRNG

The NIST tests were performed using sequences generated by our proposed design for five runs. The results are illustrated in Table 1.

Table 1 NIST test results for randomly seeded LFSR-based TRNG

Test names	Sequence 1 Seed 1	Sequence 2 Seed 2	Sequence 3 Seed 3	Sequence 4 Seed 4	Complete sequence
Frequency mono-bit test	0.3017	0.3017	0.3017	0.3017	0.0389
Frequency test within a block	1.0000	1.0000	1.0000	1.0000	1.0000
Runs test	0.0000	0.0000	0.0000	0.0000	0.0000
Longest run of ones in a block test	0.0000	0.0000	0.0000	0.0000	0.0000
Overlapping template matching	0.99050	0.9189	0.9189	0.9189	0.9905
Non-overlapping template matching test	0.0000	0.0000	0.0000	0.0000	0.0000
Cumulative sums test	1.4005	1.4005	1.4005	1.4005	1.9223
Spectral test	0.0939	0.0939	0.0939	0.0939	0.6753

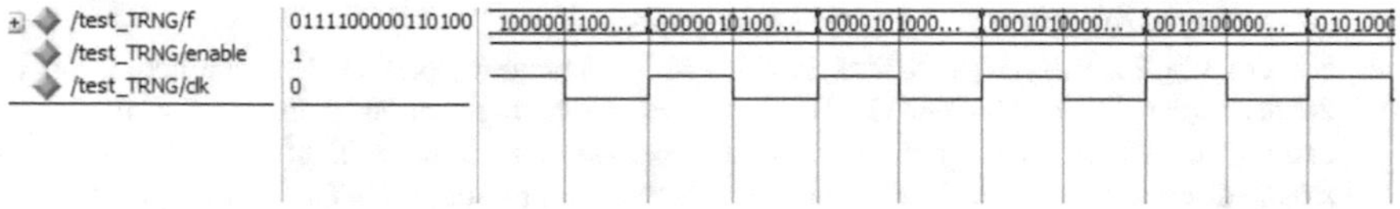

Fig. 5 Simulation result of 16-bit random seeding LFSR-based TRNG

5.2 *Functional Simulation*

The n-bit random seeding LFSR-based TRNG was coded in Verilog and simulated using suitable tools. For simulation purposes, '*n*' was considered to be 16; the result of simulation for the 16-bit random seeding LFSR-based TRNG is illustrated in Fig. 5. Functional Simulation is carried out for several clock pulses, which shows the randomness of generated sequences.

6 Conclusion

In this chapter, we have proposed an n-bit random seeding LFSR-based TRNG. The seed of the LFSR in our design was changed randomly by our proposed seed generator. This makes the resulting LFSR sequence highly random and non-predictable. We evaluated the randomness of numbers generated by our designs using NIST tests, and were able to extrapolate that they offered true randomness and are suitable for a wide range of hardware security based applications such as key generation for handshake authentication. Moreover, the simple design allows the random seeding LFSR to be an in-built module in any hardware and caters to the needs of applications related to hardware security and the Internet of Things (IoT). Thus, we have come to the conclusion that our proposed design is of low-complexity and can be used in every application that requires the use of true random number generators.

References

1. Maiti, Abhranil, Raghunandan Nagesh, Anand Reddy and Patrick Schaumont. 2009. Physical Unclonable Function and True Random Number Generator: a Compact and Scalable implementation. In *The 19th ACM great lakes symposium on VLSI*, May 2009.
2. Reddy D.M., K.P. Akshay, R. Giridhar, S.D. Karan and N. Mohankumar 2017. BHARKS: Built-in Hardware Authentication using Random Key Sequence. In *4th International Conference on Signal Processing, Computing and Control (ISPCC)*, Solan 2017, on pp. 200–204. https://doi.org/10.1109/ISPCC.2017.8269675.
3. Karunakaran D.K., and N. Mohankumar 2014. Malicious combinational hardware trojan detection by gate level characterization in 90 nm technology. In *5th International conference on computing, communication and networking technologies* (ICCCNT), China. pp. 1–7. https://doi.org/10.1109/ICCCNT.2014.6963036.
4. Koneru V.R.R., B.K. Teja, K.D.B. Reddy, M.V. GnanaSwaroop, B. Ramanidharan and N. Mohankumar 2017. HAPMAD: Hardware based authentication platform for malicious activity detection in digital circuits. In *Information System Design and Intelligent Applications*, Advances in Intelligent Systems and Computing, vol 672. Springer, Singapore. https://doi.org/10.1007/978-981-10-7512-4_60.
5. Aishwarya G., S. Hitha Revalla, S. Shruthi, V.P. Ananth and N. Mohankumar. 2017. Virtual instrumentation based malicious circuit detection using weighted average voting. In *International conference on micro-electronics, electromagnetics and telecommunications (ICMEET-2017)*, India.
6. Bharath R., et al. 2015. Malicious circuit detection for improved hardware security. In: *Security in computing and communications (SSCC'15)*, Communications in Computer and Information Science, vol 536. https://doi.org/10.1007/978-3-319-22915-7_42.
7. Kamala Nandhini S., Vallinayagam S., Harshitha H., Chandra Shekhar Azad V., Mohankumar N. 2018. Delay-based reference free hardware trojan detection using virtual intelligence. In *Information system design and intelligent applications*, Advances in Intelligent Systems and Computing, vol 672. https://doi.org/10.1007/978-981-10-7512-4_50.
8. Blum, Lenore and Milke Shub. 1986. A simple unpredictable pseudo random number generator. *SIAM Journal on Computing*.
9. Zalivako S.S., and Ivanuik A.A. The use of physical unclonable functions for true random number sequences generation.
10. Gupta, Sanjay Janusz Rajski, Jerzy Tyszer. 1996. Arithmetic additive generators of pseudo-exhaustive test patterns. *IEEE Transactions on Computers* 45 (8): 939–949.
11. Srinivasan, R., S.K. Gupta, and M.A. Breuer. 2000. Novel test pattern generators for pseudoexhaustive testing. *IEEE Transactions on Computer* 49 (11): 1228–1240.

Metaheuristic-Based Virtual Machine Task Migration Technique for Load Balancing in the Cloud

Geetha Megharaj and Mohan G. Kabadi

Abstract Cloud service centers require an efficient load balancing strategy to reduce the excessive workload on some virtual machines (VM). The VM migration technique, which achieves migration of overloaded VM from one physical machine to another, is quite popular. This technique consumes excessive time and monetary cost. Instead of migrating actual VM, migrating the extra tasks of overloaded VM has been found to be more beneficial with respect to time and cost (Ramezani et al. in International Journal of Parallel Programming 42:739–754, 2014, [1]). The task migration technique does not pause the overloaded VM, and the VM pre-copy process is not involved. This technique also provides other advantages such as elimination of VM downtime, no loss of customers' recorded activities, and better quality of service (QoS) to the customer. The VM task migration technique presented in (Ramezani et al. International Journal of Parallel Programming 42:739–754, 2014, [1]), utilizes an ineffective discriminant function to identify overloaded VM. This discriminant function may falsely identify non-overloaded VM as overloaded VM. In this chapter, a new discriminant function is designed to identify actual overloaded VM. Cost functions are designed to model the actual cost of performing task migration. A particle swarm optimization (PSO) technique is proposed to search for efficient task migration strategies. The proposed technique PSOVM is simulated in MATLAB, and the results are compared with a contemporary technique (Ramezani et al. in International Journal of Parallel Programming 42:739–754, 2014, [1]). The simulated results exhibit greater effectiveness of the proposed technique in identifying actual overloaded VM for task migration.

G. Megharaj (✉)
CSE, Sri Krishna Institute of Technology, Bangalore, India
e-mail: geethagvit@yahoo.com

M. G. Kabadi
CSE, Presidency University, Bangalore, India
e-mail: kabadimohan60@gmail.com

A. N. Krishna et al. (eds.), *Integrated Intelligent Computing, Communication and Security*, Studies in Computational Intelligence 771,
https://doi.org/10.1007/978-981-10-8797-4_45

1 Introduction

Cloud computing is basically divided into three classes: platform as a service (PaaS), infrastructure as a service (IaaS) and software as a service (SaaS). This new computing paradigm has provided new business opportunities for cloud service providers and service requesters. The cloud center provides all three services by abstracting its physical resources. Most of the cloud centers provide their services through a virtual machine (VM), which is a virtual system that provides computing services through the abstraction of physical resources.

Cloud service centers can be distributed in different geographical locations. An existing cloud center can be built by a federation of other cloud centers. This scenario creates an opportunity to provide different types of services.

Load balancing inside cloud centers is extremely important due to the huge workload and computational resource demands. Various load balancing techniques have been discussed in [2]. One popular load balancing strategy is VM migration. Here, overloaded VM are migrated from one physical machine to another in order to achieve favorable load balancing. Many efficient VM migration strategies have been proposed [3–6]. There are significant drawbacks to the use of the VM migration technique to achieve load balancing:

1. Significant memory consumption is incurred in the physical machine from where the VM is migrated and the physical machine to which the migration takes place.
2. Excessive VM downtime occurs due to pausing the migrating VM.
3. Some recent customer activities can be lost.
4. It can result in excessive cost expenditure.
5. There will be increased dirty memory after VM migration.

1.1 Motivation

It was shown in [1] that, instead of migrating the entire VM, migrating the extra or new tasks that arrive would help in overcoming all the mentioned drawbacks. This initial work proposed a PSO-based search technique for finding a favorable strategy for migration of VM tasks. In [7], the work carried out in [1] is enhanced by including fuzzy c-means (FCM) clustering to group VM with similar characteristics to minimize candidate VM for PSO, and it also includes energy consumption of the task in the PSO fitness function to make the load balancing algorithm energy efficient. The overloaded VM are identified based on the number of tasks executed inside those VM. However, this classification procedure is not effective in identifying actual overloaded VM which require task migration, because overloaded VM should be identified on the consumption of computational resources and power. A large number of tasks inside a VM does not necessarily correlate with heavy consumption of computational resources and power.

1.2 Contributions

In this chapter, the following contributions are made:

(i) The actual overloaded VM are identified through the design of a discriminant function which identifies the overloaded VM based on consumption of computational resources and power usage. (ii) The cost of migrating a single task or group of tasks is modeled through a cost function which takes into account the probable execution time of the migrated task and the cost of transmitting data belonging to the migrated task. (iii) The PSO technique is designed to seek out a favorable solution to achieve efficient task migration. Simulated results exhibit superior effectiveness of the proposed technique in identifying the actual overloaded VM and similar efficiency in searching for solutions, compared with the contemporary technique presented in [1].

2 Related Work

Distributed systems are a union of different computing components which are used by an application to execute its tasks. Different models are used for designing distributed systems which are based on sharing methodology for the computing resources to execute different tasks. Task scheduling is an important aspect of distributed systems. This procedure involves migration of task load from an overloaded processor to other lightly loaded processors so that utilization of all processors can be increased, and waiting time of the tasks can be reduced [8]. In [9], logical constraints were defined to achieve load balancing through task scheduling. The optimal task scheduling problem identifies the best strategy for task migration so that the most balanced task resource mapping can be achieved. It was proven that this problem is NP-hard [10].

Task resource mapping is also critical in achieving load balancing in primary cloud providers (PCP). The initial work for this problem was proposed in [11]. To achieve optimal resource utilization in those PCP which offer infrastructure as a service (IaaS), two task scheduling techniques were proposed in [12]. Some of the recent task scheduling approaches [13–16] have improved upon the results obtained in [11].

Load balancing in a grid computing environment through bee colony optimization and data replication techniques was proposed in [17]. The bee colony optimization technique was again used in [18] to achieve load balancing in cloud computing environments. The tasks were prioritized for migration, and those tasks that were migrated were considered honey bees which carried updated information. Another task scheduling approach for cloud computing environments using ant colony optimization was proposed in [19].

Computational grids were utilized to model task scheduling problem in [20], and game theoretic solutions were designed to address this problem. Similarly, a genetic algorithm was proposed in [9] to address task scheduling problems. In [10], multi-objective task scheduling techniques were proposed.

In [21, 22], it was reported that the PSO technique provides the best result for the task scheduling problem, outperforming the results obtained through a genetic algorithm [10]. All the described techniques address the issue of migrating tasks from overloaded physical machines.

The initial work on task migration in VM was proposed in [1, 23]. Parameters such as task execution time, task transfer cost and task power consumption values were utilized to achieve task migration. A PSO algorithm was proposed to determine a near-optimal/optimal task migration strategy.

3 VM Task Migration Technique

3.1 Overloaded VM Selection Criteria

The first task is to identify those VM that are genuinely overloaded. Consider the *y*th VM indicated by VM_y, where c_y is the number of CPU nodes present in VM_y, m_y is the memory capacity of VM_y, t_{iy} is the *i*th task inside VM_y, and c_{iy} is the CPU utilization ratio of t_{iy}. Then, if t_{iy} is executed on multiple CPUs, c_{iy} is the sum of the CPU utilization ratio for every CPU on which t_{iy} is executed, m_{iy} is the memory utilization of t_{iy}, and p_{iy} is the consumed power of t_{iy}. Since CPU utilization and memory utilization of a task are considered important factors in VM power consumption [24], higher values of these factors result in higher VM power consumption [24], so p_{iy} is approximated as the product of c_{iy} and m_{iy} in Eq. 1.

$$p_{iy} = c_{iy} \times m_{iy} \tag{1}$$

The variable p_y shown in Eq. 2 is the total consumed power of all the tasks executing in VM_y. Here, n_y is the total number of tasks executing in VM_y.

$$p_y = \frac{\sum_{j=1}^{n_y} p_{jy}}{c_y} \tag{2}$$

Two thresholds, T_{cy} and T_{py}, are defined on CPU utilization and power consumption, respectively. The function *overloaded*$(VM_y) = 1$ indicates that VM_y is overloaded; otherwise, *overloaded*$(VM_y) = 0$. This decision is made based on threshold violations. This scenario is represented in Eq. 3.

$$\mathit{overloaded}(VM_y) = \begin{cases} 1, & \text{if } (T_{cy} \le \frac{\sum_{j=1}^{n_y} c_{jy}}{c_y}) || (T_{py} \le p_y) \\ 0, & \mathit{otherwise.} \end{cases} \tag{3}$$

3.2 Task Migration Technique

The overloaded VM will stop receiving new tasks, and these extra tasks will be migrated to other lightly loaded VM. To achieve this migration, two metrics are defined. Suppose t_{iy} needs to be migrated to VM_z. Generally, an increase in the data size of a task results in increased execution time. Also, if a task has rich resources in its VM for execution, it influences the decrease in the task execution time. This intuition is utilized in defining the estimated execution time of t_{iy} in the migrated VM, which is represented in Eq. 4. Here, d_{iy} indicates the size of data used by t_{iy}. The transfer cost of t_{iy} to VM_z is shown in Eq. 5. Here, bw_{yz} is the available bandwidth between VM_y and VM_z. The migration score to perform migration of t_{iy} to VM_z is shown in Eq. 6.

$$exe_{iz} = \frac{d_{iy}}{c_z \times m_z} \tag{4}$$

$$transfer(t_{iy}, VM_z) = \frac{d_{iy}}{bw_{yz}} \tag{5}$$

$$score(t_{iy}, VM_z) = exe_{iz} + transfer(t_{iy}, VM_z) \tag{6}$$

Consider the situation with a set of tasks $[t_{i_1y_1}, t_{i_2y_2} .... t_{i_sy_s}]$ which need to be migrated. One of the candidate solutions is the VM set $[VM_{z_1}, VM_{z_2} .... VM_{z_s}]$, such that $t_{i_1y_1}$ will be migrated to VM_{z_1}, $t_{i_2y_2}$ will be migrated to VM_{z_2}, and $t_{i_sy_s}$ will be migrated to VM_{z_s}. There is no restriction that the VM in the candidate solution set should be distinct. The migration score for this candidate solution is represented in Eq. 7. Here, $t_{i_jy_j} \rightarrow VM_{z_j} (1 \le j \le s)$ indicates that the task $t_{i_jy_j}$ has already been assigned to VM_{z_j} and is executed inside it. The CPU and memory utilization ratio of $t_{i_jy_j}$ in VM_{z_j} is assumed to be same as observed when $t_{i_jy_j}$ was executed inside VM_{y_j}. The operator | is interpreted as *such that.*

$$\begin{aligned}
&migration_score((t_{i_1y_1}, VM_{z_1}), (t_{i_2y_2}, VM_{z_2}), ...(t_{i_sy_s}, VM_{z_s})) = \\
&score(t_{i_1y_1}, VM_{z_1} | t_{i_2y_2} \rightarrow VM_{z_2}, t_{i_3y_3} \rightarrow VM_{z_3}, ... t_{i_sy_s} \rightarrow VM_{z_s}) + \\
&score(t_{i_2y_2}, VM_{z_2} | t_{i_1y_1} \rightarrow VM_{z_1}, t_{i_3y_3} \rightarrow VM_{z_3}, ... t_{i_sy_s} \rightarrow VM_{z_s}) + \\
&....score(t_{i_sy_s}, VM_{z_s} | t_{i_1y_1} \rightarrow VM_{z_1}, t_{i_2y_2} \rightarrow VM_{z_2}, ... t_{i_{s-1}y_{s-1}} \rightarrow VM_{z_{s-1}})
\end{aligned} \tag{7}$$

The most beneficial candidate solution is the one that satisfies the optimization condition, which is represented in Eq. 8.

$$optimization\ condition = \arg \min_{(VM_{z_1}, VM_{z_2}, \ldots VM_{z_s})} migration\_score((t_{i_1y_1}, VM_{z_1}), (t_{i_2y_2}, VM_{z_2}), \ldots (t_{i_sy_s}, VM_{z_s})) \quad (8)$$

3.3 PSO Algorithm for Efficient VM Task Migration

The PSO algorithm [25] is a function optimizing technique which is inspired by the social behavior of birds. It was initially developed for weight balancing inside neural networks, but PSO achieved wide popularity, especially in the area of real value decision variables [26, 27]. The search elements in PSO are called particles, and they move in the defined hyper-dimensional search space. The position of each particle changes based on a certain decision function which models the tendency of an individual to emulate the success achieved by other individuals, so a particle changes its position based on its local experience and experience of other particles. Let $\overrightarrow{X_i}(t)$ indicate the position of particle i defined at iteration t. The next position is defined as represented in Eq. 9. Here, $\overrightarrow{V_i}(t+1)$ indicates the velocity of this particle at iteration $t+1$.

$$\boxed{\overrightarrow{X_i}(t+1) = \overrightarrow{X_i}(t) + \overrightarrow{V_i}(t+1)} \quad (9)$$

The velocity function is defined as shown in Eq. 10. Here, C_1 represents the degree of attraction a particle has towards its individual success, C_2 represents the degree of attraction a particle has towards the entire group success, W is the control variable that controls the impact of previous velocity on the current particle velocity, $\vec{x}_{pbest_i}$ is the individual best position of a particle, $\vec{x}_{gbest_i}$ is the position of the best particle in the entire group, and $r_1, r_2 \in [0, 1]$ are the factors that take random values.

$$\boxed{\overrightarrow{V_i}(t+1) = W\overrightarrow{V_i}(t) + C_1r_1(\vec{x}_{pbest_i} - \overrightarrow{X_i}(t)) + C_2r_2(\vec{x}_{gbest_i} - \overrightarrow{X_i}(t))} \quad (10)$$

The task migration technique considers n tasks $\hat{T} = [t_{i_1y_1}, t_{i_2y_2}, \ldots, t_{i_ny_n}]$ which need to be assigned to m VM, $\hat{VM} = [VM_{z_1}, VM_{z_2}, \ldots VM_{z_m}]$. The position of each particle is selected from one of the candidate solutions represented by $\overrightarrow{X_i} = (x_1, x_2, \ldots x_n)$. Here, the m candidate VM are indexed by real numbers $1, 2 \ldots m$, so $x_j = r(1 \le j \le n)(1 \le r \le m)$ indicates that task $t_{i_jy_j}$ is migrated to VM_{z_r}.

Algorithm 1 describes the PSO algorithm for VM task migration. Initially, the overloaded VM are selected using the selection procedure represented in Eq. 3. To execute the PSO search technique, k particles are created, which are initialized to arbitrary solutions and their corresponding positions. In every iteration, the local best score of each particle and the global best score of all the k particles will be updated, which then will be used to calculate the next possible position of every particle and the corresponding position score. After an acceptable solution is found, the calculated best global score will be the output result of the PSO algorithm.

Algorithm 1 PSO algorithm for VM task migration

for each VM inside the cloud center **do**
  Identify whether the VM is overloaded by using Eq. 3.
  if the selected VM is overloaded **then**
    Append that VM to the overloaded list $\hat{T}$.
  end if
end for
Create k particles and initialize them to arbitrary candidate solutions, which also represent the initial position of these particles.
while acceptable solution is found **do**
  for each particle **do**
    Update the local best score and global best score for the selected particle.
    Calculate the current iteration position score of the selected particle by using Eq. 7.
    Evaluate the next iteration velocity of each particle by using Eq. 10.
    Compute the next iteration position of each particle by using Eq. 9.
  end for
end while
Output the best candidate solution obtained.

4 Experimental Results and Discussion

The proposed PSO technique for VM task migration is simulated on MATLAB. This proposed technique is compared with the PSO-based VM task migration technique described in [1]. The simulation parameter settings are presented in Table 1. Each PSO search particle is assumed to be executing on an exclusive computing node, so that maximum parallelism is exploited from both these techniques. For ease of reference, the proposed PSO technique for VM task migration will be denoted as *PSO-new*, and the PSO-based VM task migration technique described in [1] will be denoted as *PSO-old*.

To conduct performance analysis on these two techniques, two metrics are defined. The first metric is denoted as *Average CPU Utilization Ratio*, which is represented in Eq. 11. Here, *overloaded_set* denotes the set of overloaded VM. This metric calculates the average CPU utilization ratio of all those overloaded VM from which tasks were migrated. Both techniques will have their corresponding overloaded VM set. For ease of reference, this metric will be denoted as *ACPUR*. The second metric is denoted as *Average Power Utilization Ratio*, which is represented in Eq. 12.

Table 1 Simulation parameter settings

Simulation parameter	Values
Number of computing nodes/CPUs in each VM	Varied between 5 and 20
Main memory capacity for each VM	Varied between 3 and 5 GB
Number of tasks executing in each VM	Varied between 10 and 50 tasks
Bandwidth between any two VM	Varied between 100 and 500 mbps
CPU utilization ratio for any task	Varied between 0.2 and 0.8
Memory utilization of each task	Varied between 0.2 and 0.8
Number of PSO search particles	10
Threshold T_{cy} $\forall$ VM	0.7
Threshold T_{py} $\forall$ VM	0.6
Size of task data	Varied between 1 and 10 GB
No. of VM considered	Varied between 1000 and 5000

This metric calculates the average power utilization ratio of all those overloaded VM from which tasks were migrated. For ease of reference, this metric will be denoted as $APUR$.

$$\boxed{ACPUR = \frac{\sum_{VM_y \in overloaded\_set}^{|overloaded\_set|} cpur(VM_y)}{|overloaded\_set|} : \text{where, } cpur(VM_y) = \frac{\sum_{j=1}^{n_y} c_{jy}}{c_y}} \quad (11)$$

$$\boxed{APUR = \frac{\sum_{VM_y \in overloaded\_set}^{|overloaded\_set|} p_y}{|overloaded\_set|}} \quad (12)$$

The first experiment analyzes the performance of PSO-new and PSO-old when the number of available VM in the cloud center varies. The result of this experiment with regard to $ACPUR$ is illustrated in Fig. 1. PSO-old decides whether the VM is overloaded based on the number of tasks being executed inside that VM. Clearly, this choice of VM selection does not identify those VM that have high CPU utilization ratio, even when the number of available VM increase, which creates a tendency to have more overloaded VM.

The execution time of PSO-old and PSO-new for the first experiment is illustrated in Fig. 2. Since both techniques are based on PSO philosophy and use a similar search method, the execution times exhibit a strong correlation. Also, in general, as the number of available VM increases, the number of overloaded VM also increases. This can result in a larger solution search space, which results in increasing execution time for both these techniques.

The result of the first experiment with regard to the $APUR$ metric is illustrated in Fig. 3. Again, PSO-old ignores those VM that have a large power utilization ratio.

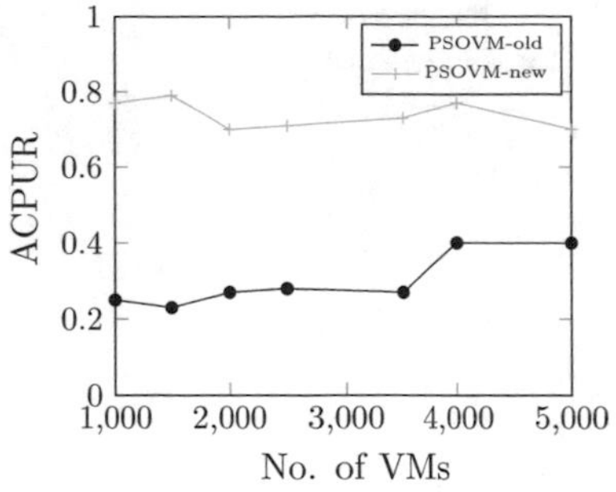

Fig. 1 No. of VM versus ACPUR

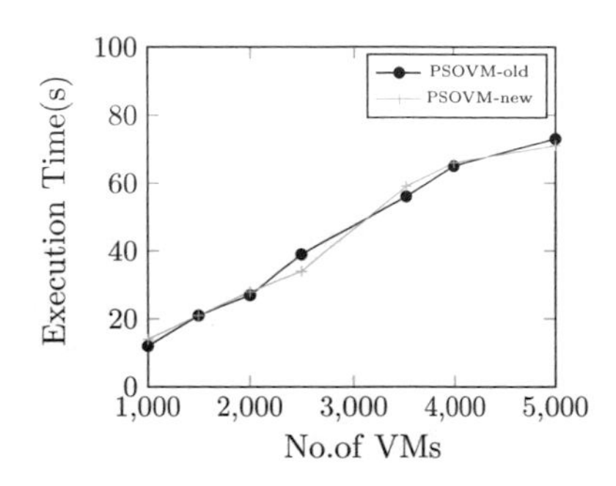

Fig. 2 No. of VM versus execution time

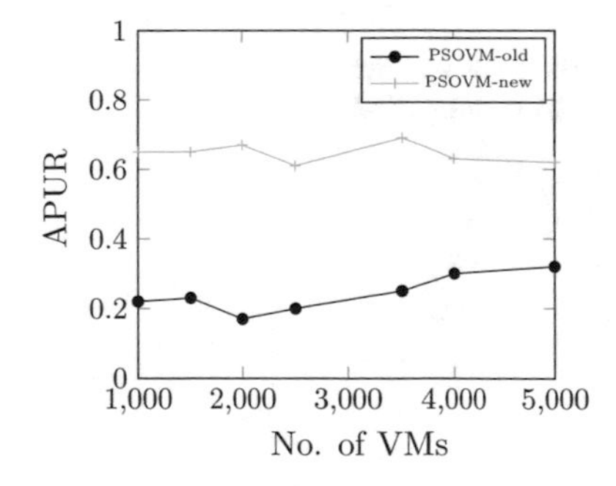

Fig. 3 No. of VM versus APUR

Many tasks can be neither data-intensive nor CPU-intensive and do not require vigorous disk access or extreme CPU utilization, and the corresponding VM with a large number of such tasks might not necessarily consume excessive power. The overloaded VM selection criteria of PSO-old suffers due to the selection of those VM with more of such tasks.

The second experiment analyzes the performance of PSO-old and PSO-new when the number of overloaded VM varies. The increase in overloaded VM creates a larger solution search space. The result of the second experiment with regard to *ACPUR* is illustrated in Fig. 4. PSO-old under-performs for the same reasons explained before.

The result of the second experiment as regards execution time is illustrated in Fig. 5. Because of the increase in the size of solution space as the number of

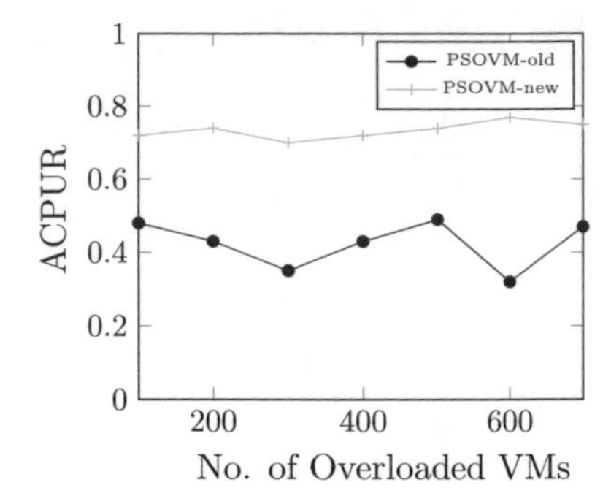

Fig. 4 No. of overloaded VM versus ACPUR

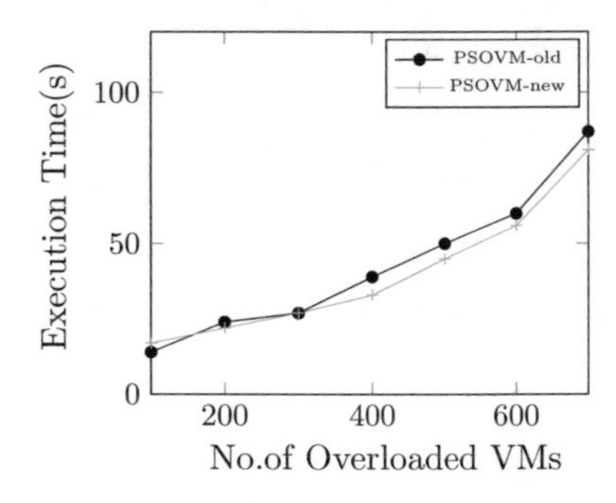

Fig. 5 No. of overloaded VM versus execution time

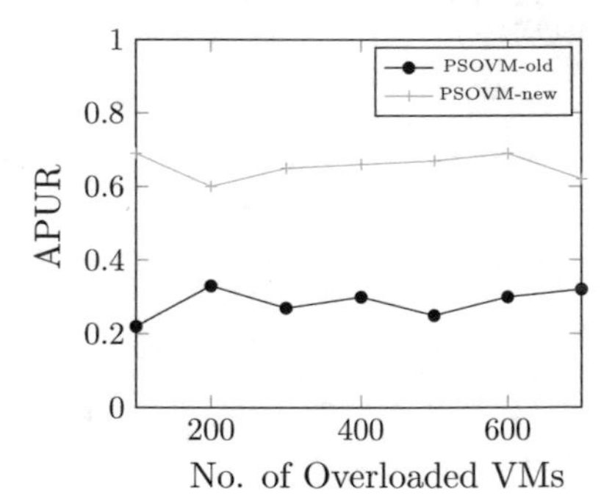

Fig. 6 No. of overloaded VM versus APUR

overloaded VM increases, the execution time of both these search techniques also increases. But the execution time of both techniques exhibits a tight correlation for the same reasons explained before.

The result of the second experiment with regard to *APUR* is illustrated in Fig. 6. PSO-old under-performs for the same reason explained before.

5 Conclusion

In this work, a new discriminant function was designed to identify overloaded VM, and scoring functions were designed to model task migration costs. A PSO-based search technique was proposed to search for ae favorable strategy for performing task migration. The simulated results exhibit superior effectiveness of the proposed technique in identifying the actual overloaded VM over the contemporary technique proposed in [1].

Future work in this area is as follows:

1. Thus far, the extra task migration problem for overloaded VM has been addressed. In some cases, it is also beneficial to migrate the existing tasks inside overloaded VM to achieve efficient load balancing. Thus, a new technique for migrating the existing tasks inside overloaded VM needs to be designed.
2. An integrated framework to accomplish both extra task and existing task migration in overloaded VM is also essential for increased customer satisfaction and for achieving better QoS.

References

1. Ramezani, F., J. Lu, and F.K. Hussain. 2014. Task-based system load balancing in cloud computing using particle swarm optimization. *International Journal of Parallel Programming* 42 (5): 739–754.
2. Megharaj, Geetha, and K.G. Mohan. 2016. A survey on load balancing techniques cloud computing. *IOSR Journal of Computer Engineering (IOSR-JCE)* 18 (2): 55–61, 12–23. e-ISSN: 2278–0661, p-ISSN: 2278–8727, Version I.
3. Jain, N., I. Menache, J. Naor, and F. Shepherd. 2012. Topology-aware VM migration in bandwidth oversubscribed datacenter networks. In *39th International Colloquium*, 586–597.
4. Kozuch, M., and M. Satyanarayanan. 2002. Internet suspend/resume. In *4th IEEE Workshop on Mobile Computing Systems and Applications*, 40–46.
5. Sapuntzakis, C.P., R. Chandra, B. Pfaff, J. Chow, M.S. Lam, and M. Rosenblum. 2002. Optimizing the migration of virtual computers. *ACM SIGOPS Operating System Review* 36 (SI): 377–390.
6. Whitaker, A., R.S. Cox, M. Shaw, and S.D. Gribble. 2004. Constructing services with interposable virtual hardware. In *1st Symposium on Networked Systems Design and Implementation (NSDI)*, 169–182.
7. Megharaj, Geetha, and K.G. Mohan. 2016. FCM-BPSO energy efficient task based load balancing in cloud computing. *Journal of Theoretical and Applied Information Technology* 94 (2): 257–264. E-ISSN 1817–3195, ISSN 1992–8645.
8. Zomaya, A.Y., and T. Yee-Hwei. 2001. Observations on using genetic algorithms for dynamic load balancing. *IEEE Transactions on Parallel and Distributed System* 12 (9): 899–911.
9. Zhao, C., S. Zhang, Q. Liu, J. Xie, and J. Hu. 2009. Independent tasks scheduling based on genetic algorithm in cloud computing. In *5th International Conference on Wireless Communications, Networking and Mobile Computing*, 1–4.
10. Juhnke, E., T. Dornemann, D. Bock, and B. Freisleben. 2011. Multi objective scheduling of BPEL workflows in geographically distributed clouds. In *4th IEEE International Conference on Cloud Computing*, 412–419.

11. Song, B., M.M. Hassan, and E. Huh. 2010. A novel heuristic-based task selection and allocation framework in dynamic collaborative cloud service platform. *2nd IEEE International Conference on Cloud Computing Technology and Science (CloudCom)* 360–367.
12. Li, J., Qiu M, Z. Ming, G. Quan, X. Qin, and Z. Gu. 2012. Online optimization for scheduling preemptable tasks on IaaS cloud systems. *Journal of Parallel and Distributed Computing* 72 (5): 666–677.
13. Milani, Alireza Sadeghi, and Nima Jafari Navimipour. 2016. Load balancing mechanisms and techniques in the cloud environments. *Journal of Network Computer Applications* 71: 86–98.
14. Abdullahi, Mohammed, Md Asri Ngadi, and Shafi'i Muhammad Abdulhamid. 2016. Symbiotic organism search optimization based task scheduling in cloud computing environment. *Future Generation Computer Systems* 56: 640–650.
15. Singh, Poonam, Maitreyee Dutta, and Naveen Aggarwal. 2017. A review of task scheduling based on meta-heuristics approach in cloud computing. *Knowledge and Information System* 1–51.
16. Razzaghzadeh, Shiva, Ahmad Habibizad Navin, Amir Masoud Rahmani, and Mehdi Hosseinzadeh. 2017. Probabilistic modeling to achieve load balancing in expert clouds. *Ad Hoc Network* 12–23.
17. Ld, D.B., and P.V. Krishna. 2013. Honey bee behavior inspired load balancing of tasks in cloud computing environments. *Applied Soft Computing* 13 (5): 2292–2303.
18. Taheri, J., Y. Choon Lee, A.Y. Zomaya, and H.J. Siegel. 2013. A bee colony based optimization approach for simultaneous job scheduling and data replication in grid environments. *Computers and Operations Research* 40 (6): 1564–1578.
19. Li, J., J. Peng, X. Cao, and H.-Y. Li. 2011. A task scheduling algorithm based on improved ant colony optimization in cloud computing environment. *Energy procedia* 13: 6833–6840.
20. Kolodziej, J., and F. Xhafa. 2011. Modern approaches to modeling user requirements on resource and task allocation in hierarchical computational grids. *International Journal of Applied Mathematics and Computer Science* 21 (2): 243–257.
21. Lei, Z., C. Yuehui, S. Runyuan, J. Shan, and Y. Bo. 2008. A task scheduling algorithm based on PSO for grid computing. *International Journal of Computational Intelligence Research* 4 (1): 37–43.
22. Liu, H., A. Abrahan, V. Snasel, and S. McLoone. 2012. Swarm scheduling approaches for workflow applications with security constraints in distributed data-intensive computing environments. *Information Sciences* 192: 228–243.
23. Ramezani, F., J. Lu, and F. Hussain. 2014. Task based system load balancing approach in cloud environments. *Knowledge Engineering and Management*, 31–42.
24. Jain Kansal, Nidhi, and Inderveer Chana. 2014. Artificial bee colony based energy-aware resource utilization technique for cloud computing. *Concurrency Computation Practice Experience*.
25. Kennedy, J., and R. Eberhart. 1995. Particle swarm optimization. In *IEEE International Conference on Neural Networks*, 1942–1948.
26. Engelbrecht, A.P. 2005. *Fundamentals of Computational Swarm Intelligence*. Hoboken: Wiley.
27. Engelbrecht, A.P. 2007. *Computational Intelligence: An introduction*. Hoboken: Wiley.

Analysis of Student Engagement and Course Completion in Massive Open Online Courses

S. Suresh Kumar and P. M. Mallikarjuna Shastry

Abstract The growth of the internet has given rise to a number of open online learning platforms, enabling access to learning materials by millions of individuals regardless of age or educational background (Zawilska et al., Social collective intelligence. Springer International Publishing, pp. 187–202, 2014) [1]. These massive open online courses (MOOC) have replaced traditional institutional learning environments, such as physical attendance of lectures, with globally accessible learning via the Web. Despite the large number of students who enrol in MOOCs, however, the percentage of student completion these course remains low. The current work investigates methods for predicting learner behaviour and analysis of overall student performance, will help service providers identify students at risk of dropout so that they can take preventive measures. Results show that students who attended lectures and engaged and interacted with course material achieved the highest grades. The galaxy schema is used to show the recommended queries for the students based primarily on their seek.

Keywords Galaxy schema · Learner prediction · Massive open online courses

1 Introduction

The growth of the internet has witnessed transformational changes in access to information and education, and the emergence of massive open online courses (MOOC) has given rise to a new learning paradigm [2]. Access to online learning

S. Suresh Kumar (✉)
Research Scholar, Department of Computer Science & Engineering,
Visvesvaraya Technological University (VTU)-RRC, Belgaum, Karnataka, India
e-mail: sureshkatte89@gmail.com

P. M. Mallikarjuna Shastry
Department of Computer Science & Engineering, Reva Institute of Technology
& Management (VTU), Bengaluru, Karnataka, India
e-mail: pmmshastry@yahoo.com

A. N. Krishna et al. (eds.), *Integrated Intelligent Computing, Communication and Security*, Studies in Computational Intelligence 771,
https://doi.org/10.1007/978-981-10-8797-4_46

permits students to participate in university courses via platforms such as weblogs and YouTube. Interactive online tools and internet resources afford participants greater convenience, and institutions provide learners with access to educational content at any time and with numerous devices. Because participation is unlimited, and users can take part regardless of age or academic level, huge numbers of learners worldwide are able to take advantage of these resources. Students who may not have access to traditional institutional learning are now able to pursue an education through online learning [3]. Despite the massive enrolment numbers, however, effectively only 7.5% of students are completing courses. Thus providers of MOOC must develop strategies to maintain student engagement in order to ensure the sustainability of the MOOC model.

To this end, a wide array of data can be collected to provide researchers with information on student engagement with course material, including student interactions with text, audio and video content. Such data can be used to profile student behaviour and to predict engagement patterns within the MOOC setting. For large records repositories, data mining and analysis tools can be used to study engagement patterns of students successfully completing the coursework, and this information can help providers developing methods to encourage greater active participation by less engaged students. This studies also gives the consumer to find the answer for the queries this is received by means of the online analytical processing (OLAP) top-k set of rules [4]. The galaxy schema is a combination of the star schema and snowflake schema, which affords a powerful approach for dealing with queries in the system. The end result shows the importance of the student involvement to finish the guides and technique for the MOOC organizer to motivate the students.

2 Literature Review

A number of works have investigated learner performance in MOOCs. Chu et al. [5] described a model for a learning assessment in which students' trust in the system and perceived experience of fun with the course were assessed as factors in user adoption of online learning. A total of 212 survey response samples from a Chinese population were collected and used for analysis of the learning method. Learner willingness and enthusiasm was assessed based on attendance and compared with the participant responses. The results showed that trust and perceived usefulness were key factors for MOOC completion, although the measurement was based on a limited number of samples.

Shapiro et al. [6] performed text analysis of interview transcripts of participants in an introductory chemistry MOOC and a data analytics and statistical analysis MOOC in order to understand students' reasons for taking the courses and obstacles to learning based on students' experiences. The authors explored factors both within and outside the course setting. Lack of time was found to be the most common barrier to course completion. Further studies are needed to apply this research to a larger MOOC context.

Phan et al. [7] assessed the relationship between learners' patterns and motives for engagement and their prior subject knowledge with regard to their performance in a digital storytelling MOOC. The results reveal that motives such as earning a continuing professional development certificate and gain competency correlated with performance superior to that of beginners with no such motivation. Students with greater subject expertise also achieved higher grades. However, the study examined only completion data, and grades were not considered. Thus a deeper learning assessment is warranted.

Chapman et al. [8] developed a two-pronged approach aimed at assessing both performance and developmental aspects of student progress, with tracking of overall performance including participation, quality and student achievement. The method offers improved evaluation questions and assessment methodology, and also develops performance monitors for individual learners within the MOOC. The statistics show that learning success is largely independent of the underlying complex adaptive system, and thus evaluation techniques must shift from individual learning to the assessment of successful MOOC strategies overall.

Hughes et al. [9] provided a method for predicting learner performance in MOOC which included identifying students at risk of dropout. In this system, student attendance data are captured analysed by means of an eRegister. The study found that the students with High attendance, interaction and engagement with the course achieved higher grades than others. This method is able to normalize the data into consistent a series so that the end result can be used by the organizer of the MOOC. This research is not done for the individual students in the course and deep learning in MOOC is required.

3 Proposed System

The proposed system involves the analysis of student participation and engagement in the MOOC and their performance. It is also provides a top-k algorithm and galaxy schema to give the best answer to the user queries. The galaxy schema is the combination of the star schema and snowflake schema, which includes fact tables to provide more efficient answers. The study is performed on the dataset from the school of computing Mathematical sciences, within the Faculty of Technology and Environment, in Liverpool John Moores University has run an attendance monitor for 7 years and another dataset is used from Coursera [10].

3.1 Student Engagement and Performance in MOOC

MOOCs attract students from all over the world, who have many different patterns of engagement, making it more difficult to monitor meaningful student participation [11]. In the traditional academic setting, students experiencing difficulty within a course can seek help from faculty, and performance problems can thus be addressed. Within a MOOC environment, no such intervention is possible. Student engagement is thus measured based on attendance and the level of engagement and interaction with the course material [12]. This information is used to predict student performance and likelihood of course completion. Some researchers have conceptualized student participation using a 'funnel of participation' methodology, which involves four stepwise processes: awareness, registration, activity and progress. The largest learner concentration is in the awareness stage, with participation progressively decreasing as students move through the four steps. Students who complete courses tend to engage more actively, similar to those in a traditional classroom, including attending lectures and completing all assignments. Students who are disengaged tend to participate actively at the beginning, but their engagement wanes as the course progresses. Measures of student engagement can be used to predict such behaviour in order to intervene in such situations before dropout occurs [13].

Figure 1 shows the MOOC architecture of with the top-k algorithm and galaxy schema. This is used to analyse student performance and engagement in the online courses, in order to encourage greater student participation.

3.2 Data Analysis in Predicting Student Performance

The prediction of student performance from the information of student engagement and interaction with course requires state-of-the-art algorithms [5]. Data mining is a famous technique that is used to expect performance of the student from the database. The statistics is stored in excessive price and its miles prepared by the data warehouse and additionally offer extra feature like cleaning, integrating and online analytical processing (OLAP).

Figure 2 shows the data warehouse architecture. The data are sent to the extract, transform, load (ETL) layer, which is used to compact the data for warehousing. The data are then stored in the warehouse and it can be used to predict the learning method. The OLAP top-k algorithm is employed for analysis the information and it's for visualized it to the user.

The information regarding the user is given as input to the OLAP, which is used to generate the user preference model and user log, and then user preference model will be built. The consumer log contains three forms of files, which is used to develop the OLAP sessions. The similarity calculation method is applied to the session by the Smith-Waterman algorithm, which is the basis of top-k recommend

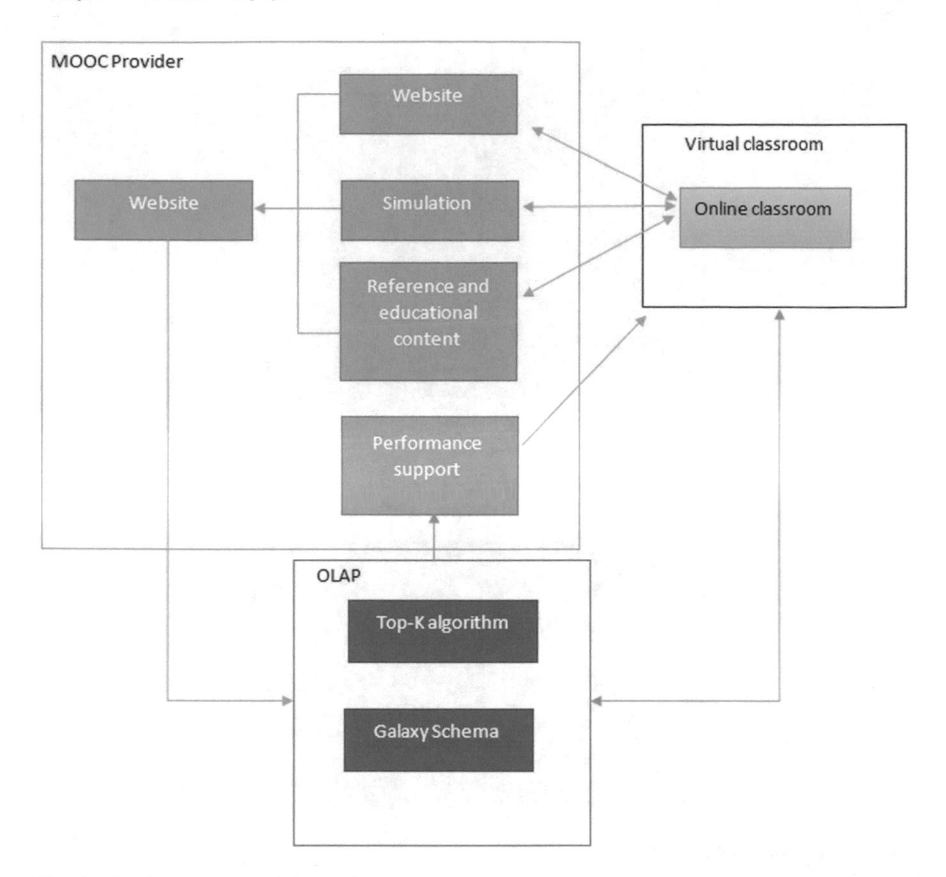

Fig. 1 Architecture of MOOC with performance support

algorithm and tag-aware method is also used for semantic understanding in the final recommendation.

The extraction is obtaining of data from the different databases and change into compactible formats such as comma-separated values (CSV) and extensible markup language (xml) files. Figure 3 shows the ETL layer and its process. Transform refers back to the wider set of techniques which includes grouping, sorting, merging and pivoting statistics. Loading of records are dependent into the galaxy schema with a purpose to solution the person queries. The galaxy schema is the combination of star schema and snow flake schema from [9].

The galaxy schema has a single fact table, which includes all numerical information; any number of dimensions is described in the fact table. Figure 4 represents the fact table and dimension table in the galaxy schema.

The suitable data has been taken to predict the student performance and transform into the compactible format. Then queries have been identified by means of

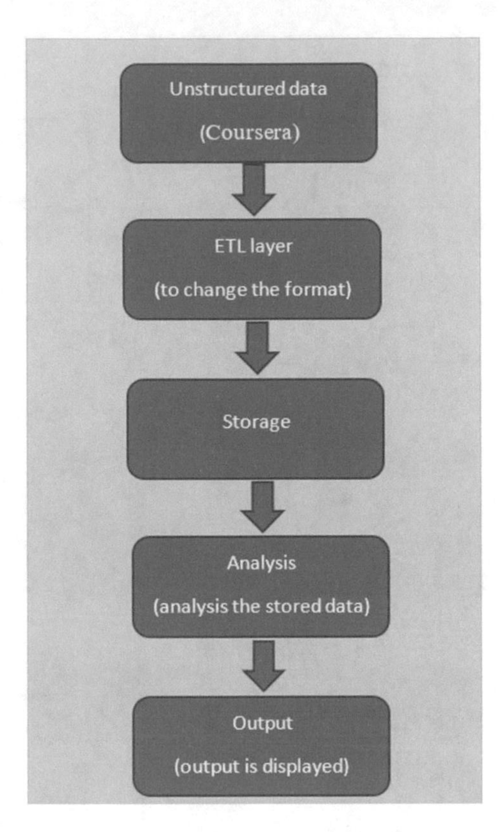

Fig. 2 Data warehouse

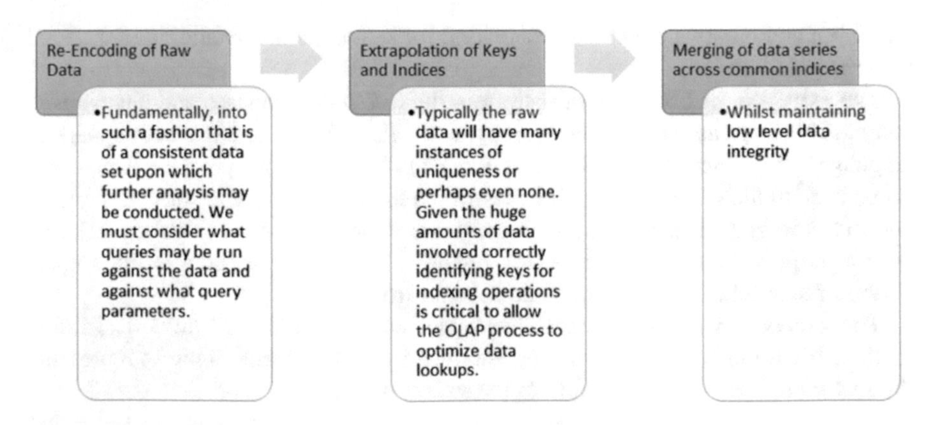

Fig. 3 ETL layer

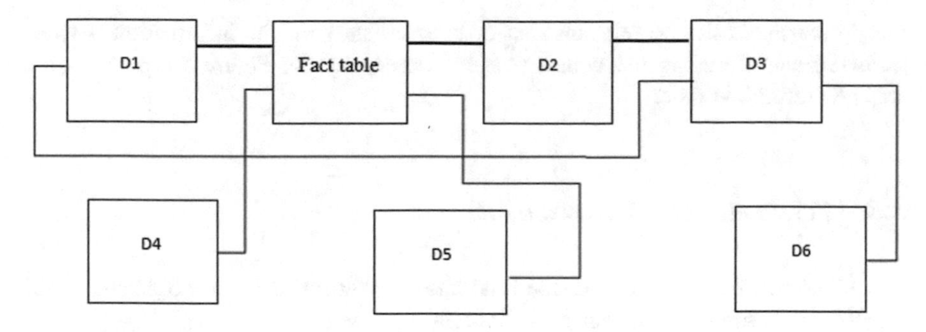

Fig. 4 Galaxy schema

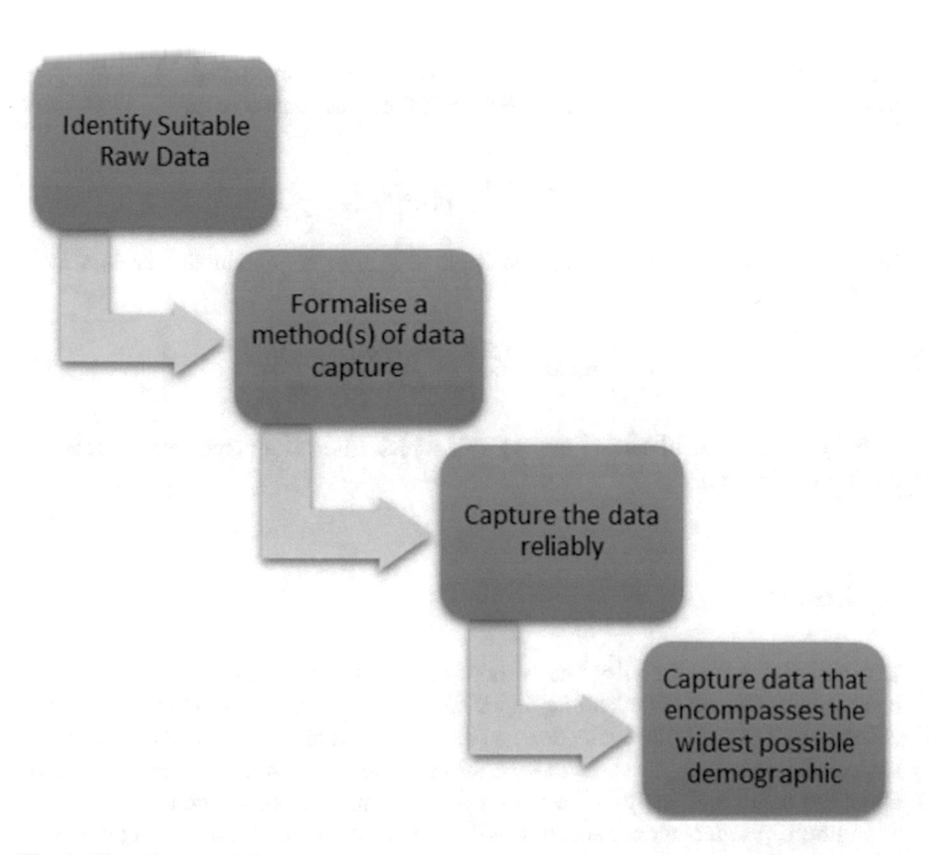

Fig. 5 Formulate data information steps

the proper methods and relevant answer is to be shown. The information is to be prepared and stored in the widest possible demographic. Figure 5 represents the steps to formulate data.

3.3 Performance Measurement

The measure of precision gives the total answers retrieved from the database and recall r gives the positive images in the total retrieved.

$$precision = \frac{|TP|}{|TP| + |FP|} \tag{1}$$

To evaluate the performance of the method precision and recall are measured. *RIR* represents the number of relevant answers is retrieved and *TP* is the total number of applicable answer retrieved. Precision P is measured in Eq. (1).

$$recall = \frac{|TP|}{|TP| + |FN|} \tag{2}$$

where *FN* is the total number of non-applicable answers obtained. Recall is measured in Eq. (2).

$$F - \text{measure} = \frac{2 \times precision \times recall}{percision + recall} \tag{3}$$

F-measure is calculated by the Eq. (3), which gives the percentage of relevant answer is obtained.

4 Results

The students can learn in online without any background check like educational background, age, etc. The learning in MOOC is increased and completion of course is less than 7.5%. So, the evaluation has been made to discover the obstacles for the students to complete the direction. In this section, the discussion has been made about the result and analysis the reason for incompletion of the course.

Figure 6 presents a comparison of attendance and final grades among students, which clearly show that students with high attendance scores also attained the highest grades, reflecting their greater level of engagement. Students who participated in lectures and higher interaction and utilization of tutorial material tended to achieve higher marks. This also helps to analysis about the student who are at the risk of dropout and alert them before they are low in attendance. This also alerts

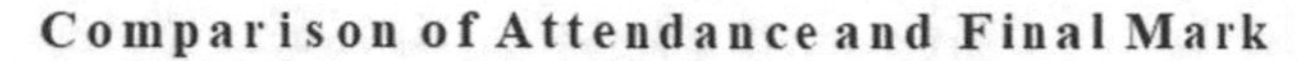

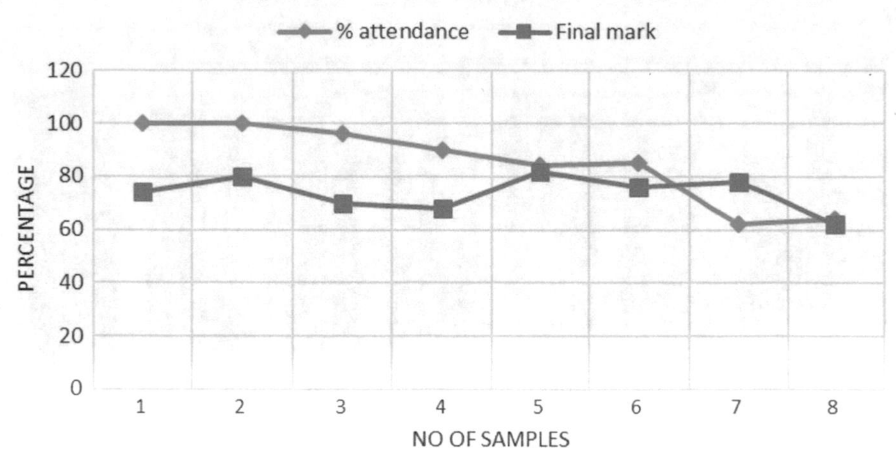

Fig. 6 Comparison of attendance and final marks among students

Table 1 F-measures for the OLAP top-k algorithm

Sets (S)	Precision	Recall	F-measure
S1	0.77	0.44	0.56
S2	0.85	0.65	0.736667
S3	0.74	0.72	0.729863
S4	0.86	0.75	0.801242
S5	0.72	0.77	0.744161
S6	0.74	0.78	0.759474
S7	0.77	0.85	0.808025

MOOC providers to inspire their students who're all on the threat [14]. By means of providing the learner's prediction, allows to motivate the students to finish the course and achieve high marks [15].

In Table 1, S represents the queries sets and precision, recall and F-measure for the queries set is determined. Precision gives the number of correct answer obtained and recall gives the number of answer obtained. The F-measure gives the percentage of applicable answer obtained by the OLAP top-k algorithm. The high F-measure are obtained for the proposed algorithm, indicates that more relevant answer is obtained for the user queries.

The user's queries answered by the algorithm of OLAP top-k, which gives relevant answer for the queries. This is more efficient than the existing system and provide quicker and more relevant answer for the queries. This answer for the queries are collected by the galaxy schema, which contains the fact tables with dimensional table. Figure 7 represents the measure of the precision, recall and F-measure for the queries set S1, S2, etc. This OLAP top-k algorithm gives high F-measure than the existing system and helps the user to obtain the answer for the

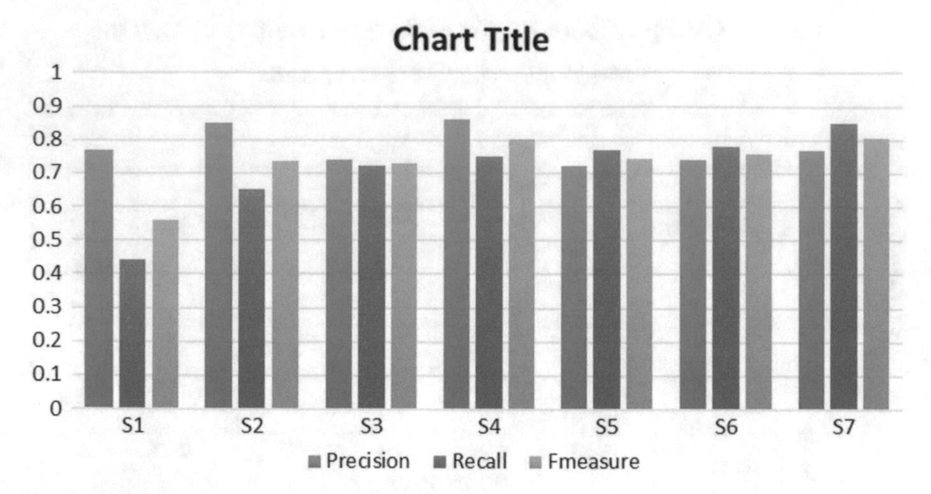

Fig. 7 F-measure of query sets

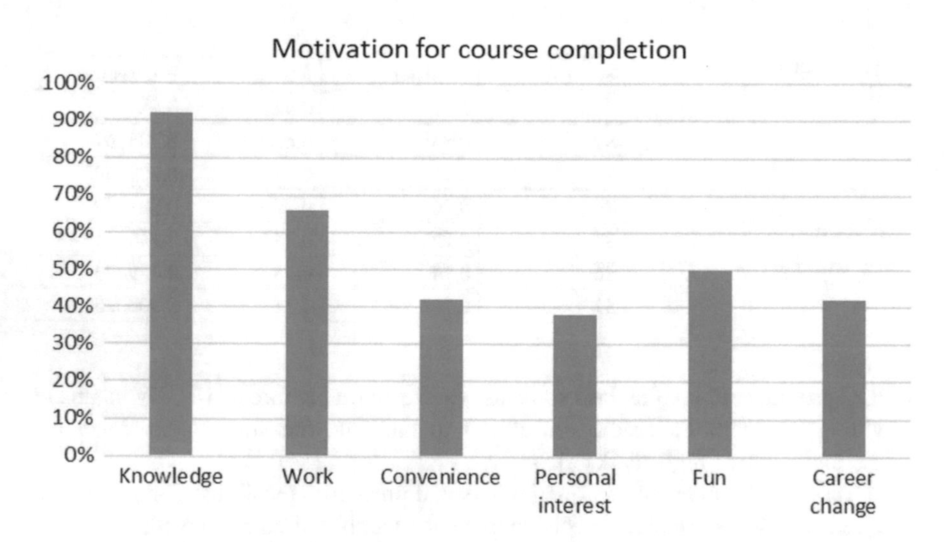

Fig. 8 Motivation for course completion

queries. The negative time control is the main obstacles for the students to finish the path and this reduces the time for the student to trying to find the answer.

Students who completed the course were interviewed and asked the survey for the motivation for finishing the path [16]. The maximum of the student offers the solution as knowledge is the main reason for complete the course and second most answer for completion of course is paintings. Other common reasons included convenience, personal interest, fun and career change. Figure 8 shows the statistical measure of the answer provide by the students.

5 Conclusion

The growth of the internet has given rise to increase the online courses over the years and many educational institutions conduct the online courses. The participants of the MOOC are also rapidly increased, but completion of the courses are very less. The students are not completing their courses because of various reasons. On this research, the analysis has been made to identify the cause of the student for not completing their course and also found the motivation for the student to finish the course. The OLAP top-k algorithm and galaxy schema are proposed to give the relevant answer to the user's queries concerning the publications. The end result actually indicates the student who has high attendance achieved higher grade when in comparison to others. This look at indicates that knowledge is the important motivation for the student to finish the course. The OLAP top-k algorithm offers the efficient answer to the user queries is measured in the system. The study provides the deep analysis for the motivation and obstacles for the completion of the course.

References

1. Zawilska, A., M. Jirotka and M. Hartswood. 2014. Surfacing collective intelligence with implications for interface design in massive open online courses. In *Social collective intelligence*, 187–202. Springer International Publishing.
2. Maté, A., E. de Gregorio, J. Cámara and J. Trujillo. 2013. Improving massive open online courses analysis by applying modeling and text mining: A case study. In *International conference on conceptual modeling*, 29–38. Cham: Springer.
3. Anderson, A., D. Huttenlocher, J. Kleinberg and J. Leskovec. 2014. Engaging with massive online courses. In *23rd international conference on World wide web*, 687–698. ACM.
4. Yuan, Y., W. Chen G. Han and G. Jia. 2017. OLAP4R: A top-k recommendation system for OLAP Sessions. *KSII Transactions on Internet and Information Systems*, 11 (6).
5. Chu, R., E. Ma, Y. Feng and I.K. Lai. 2015. Understanding learners' intension toward massive open online courses. In *International conference on hybrid learning and continuing education*, 302–312. Cham: Springer.
6. Shapiro, H.B., C.H. Lee, N.E.W. Roth, K. Li, M. Çetinkaya-Rundel, and D.A. Canelas. 2017. Understanding the massive open online course (MOOC) student experience: An examination of attitudes, motivations, and barriers. *Computers and Education* 110: 35–50.
7. Phan, T., S.G. McNeil, and B.R. Robin. 2016. Students' patterns of engagement and course performance in a massive open online course. *Computers and Education* 95: 36–44.
8. Chapman, S.A., S. Goodman, J. Jawitz, and A. Deacon. 2016. A strategy for monitoring and evaluating massive open online courses. *Evaluation and program planning* 57: 55–63.
9. Hughes, G., and C. Dobbins. 2015. The utilization of data analysis techniques in predicting student performance in massive open online courses (MOOCs). *Research and Practice in Technology Enhanced Learning* 10 (1): 10.
10. Al-Shabandar, R., A. Hussain, A. Laws, R. Keight J. Lunn, N. Radi. 2017. Machine learning approaches to predict learning outcomes in Massive open online courses. In *International joint conference on neural Networks (IJCNN)*, 713–720.
11. Ferguson, R., and D. Clow. 2015. Examining engagement: Analysing learner subpopulations in massive open online courses (MOOCs). In *5th international conference on learning analytics and knowledge*, 51–58. ACM.

12. Barak, M., A. Watted, and H. Haick. 2016. Motivation to learn in massive open online courses: Examining aspects of language and social engagement. *Computers and Education* 94: 49–60.
13. Chen, Y., Q. Chen, M. Zhao, S. Boyer, K. Veeramachaneni and H. Qu. 2016. DropoutSeer: Visualizing learning patterns in massive open online courses for dropout reasoning and prediction. In *IEEE conference on visual analytics science and technology (VAST)*, 111–120.
14. He, J., J. Bailey, B.I. Rubinstein, and R. Zhang. 2015. Identifying at-risk students in massive open online courses. In *AAAI*, 1749–1755.
15. Gallén, R.C., and E.T. Caro. 2017. An exploratory analysis of why a person enrolls in a massive open online course within MOOCKnowledge data collection. In *Global engineering education conference (EDUCON)*, 1600–1605.
16. Stich, A.E., and T.D. Reeves. 2017. Massive open online courses and underserved students in the United States. *The Internet and Higher Education* 32: 58–71.

Extended Conflict-Based Search with Awareness

Shyni Thomas, Dipti Deodhare and M. Narasimha Murty

Abstract Extended Conflict-Based Search (XCBS) is a distributed agent-based approach which has been used for path finding and scheduling of spatially extended agents on a traversable network. The algorithm arrives at an optimal schedule while resolving conflicts between pairs of agents one at a time. In this chapter, we propose XCBS with Awareness wherein a conflicting agent plan is resolved with respect to the proposed route plan of other agents. The approach allows multiple conflicts to be resolved simultaneously, avoids cascading conflicts in the new plans and shows an improved efficiency in terms of nodes explored and time taken to arrive at the solution.

Keywords Planning and scheduling · Multi-agent · Path planning
Conflict resolution · Search algorithm

1 Introduction

The problem of path finding and scheduling of agents on geographical transportation networks like roads, rails, etc. has been explored in various forms like multi-agent path finding (MAPF) [1], automated guided vehicle (AGV) routing [2] and vehicle routing [3]. Some of these problems are specifically targeted towards point objects whose size is significantly smaller than the network on which it is traversing. In some scenarios, like convoy scheduling, train scheduling, snake/thread motions in games, etc., the length of the object cannot be ignored. Such spatially extended objects have the following distinct characteristics:

S. Thomas (✉) · D. Deodhare
Centre for AI & Robotics (CAIR), DRDO, Bengaluru, India
e-mail: thomas_shyni@yahoo.com

S. Thomas · M. Narasimha Murty
Department of Computer Science and Automation, Indian Institute of Science (IISc), Bengaluru, India

A. N. Krishna et al. (eds.), *Integrated Intelligent Computing, Communication and Security*, Studies in Computational Intelligence 771,
https://doi.org/10.1007/978-981-10-8797-4_47

i. At a given time instance, they may occupy multiple edges of a road network as opposed to point objects which occupy only a single edge.
ii. Point objects transition instantaneously from one edge to another while edge transitions are durative to allow the complete length of the spatially extended object to cross over.

In this chapter, we adopt the eXtended Conflict-Based Search (XCBS) [4] which allows optimal planning and scheduling for Spatially Extended Agents (SEAs). XCBS arrives at a solution by resolving a single conflict, between a pair of agents, at a time using a search tree called the Constraint Tree (CT). As each conflict generates at least two CT nodes, for problems with multiple or cascading conflicts, the CT grows in depth and the time taken to arrive at the solution increases accordingly. In our approach, we propose a new concept called a Temporal Occupancy Graph for identifying mutually exclusive sets of conflicting agents which can be resolved simultaneously. The second novelty is in conflict resolution, wherein each agent attempts to resolve conflicts by recognizing the plans of all the other agents. This conflict-resolution strategy based on awareness of other agents' plans leads to our approach being called XCBS with Awareness (XCBS-A). We assume that the agents willingly share their plans with others to arrive at a globally optimal solution.

There have been limited attempts at handling path planning for extended-length objects. Dinh et al. [5] described a decoupled approach to route planning for multiple agents with defined length, using a delegate Multi-Agent System (MAS). The decoupled approach uses a resource agent to randomly choose an order to assign resources from among simultaneously arriving intention ants and which could lead to a globally suboptimal solution. A practical example of an extended-length object is a convoy of vehicles, and several centralized and distributed approaches have been attempted for the Convoy Movement Problem (CMP). In [6], Kumar et al. presented a Planning Domain Definition Language (PDDL) formulation of the CMP and demonstrated it with existing planners. Thomas et al. in [7] arrive at a centralized solution for CMP using A* which arrived at optimal plans but could not be scaled up due to the large search space.

2 Model

The planning and scheduling of set A of SEAs is required on a road network G with J as the set of road junctions and R as the set of roads. Each road $r_i \in$ R is D_i kms long with a speed limit S_i kmph. Further, each agent $a_j \in$ A has length L_j and moves from initial location I_j to final location F_j with an average speed of S_j kmph. Therefore, each agent a_j is a SEA and can execute two actions:

- *move*(r_i, *t*): The head of the agent enters road r_i at time t
- *wait*(t, t_d): The agent stops moving and enters a wait state at time t for t_d time units, at its current location.

Definition 1 A plan *P* of agent a_j is the sequence of actions to be executed by a_j to reach from I_j to F_j. It is of the form {*move*(r_1, t_1), *move*(r_2, t_2), *wait*(t_3, *d*)…}, where, $t_1 < t_2 < t_3$. It is consistent, i.e. the plan satisfies all constraints imposed on a_j. The cost (P_j) is the total plan time taken by the agent a_j and is a cumulative sum of the travel time and wait times, given by:

$$Cost(P_j) = \sum_{n=1}^{q} D_n/\min(S_j, S_n) + \sum_{r=0}^{l} wait_r + L_j/S_J \quad (1)$$

The travel time along each road r_j on the path depends on the minimum agent speed S_j and the road speed constraint S_n. The tail-crossing time at the destination is computed based on the length of the agent L_j and its speed S_j and is constant throughout the plan.

Definition 2 The solution set S = {P_1 …, P_k}, where P_j is the consistent plan of agent a_j and S is valid; i.e. plans are conflict-free and the cost (S) = $\sum_{j=1}^{k} cost(P_j)$ is minimal.

Definition 3 The temporal occupancy (t.o.) node γ, expressed as a tuple $<r, \tau, Z>$, describes the occupancy details of a road *r* during a given time period τ. The exact time duration of each SEA a_j spent on *r* is given by agent details Z. There exists at least one t.o. node for each road $r \in R$, which is in any plan P_j. For any t.o. node γ_r, the time duration τ_r is the maximal covering time interval of the time durations spent by agents on *r* as defined in agent details Z.

Definition 4 The t.o. graph is a directed graph G(Γ, Δ) with Δ, the set of edges, which connects the t.o. nodes Γ. An edge $\delta_{ij} \in \Delta$ connects t.o. nodes γ_i, γ_j if they are consecutive roads in a plan P_j of agent a_j. The edge property δ_{ij}*.agent* indicates the agent whose plan makes that edge valid.

Definition 5 The search space for the plans is maintained in a CT. Each CT node *N* of the CT comprises: (i) an *N.Id,* a unique identifier for the CT node; (ii) *N. Solution*: {P_1, …, P_k} a set of consistent optimal plans P_i of each agent; and (iii) *N. Cost*: CT node cost as defined in Definition 2.

3 Algorithm

The solution is arrived at by the interaction between two types of agents.

- **SEAs**: Each SEA models the spatially extended object which moves from I_j to final location F_j. The SEA performs a low-level search to generate a consistent plan with respect to its own spatial and temporal constraints/preferences. If the agent is aware of the plans of the other agents (when doing conflict resolution), then consistency is ensured with respect to the other plans as well.

- **Central agent (CA)**: The CA coordinates the search among the SEAs to achieve a valid solution. This is accomplished by a high-level search on the CT to identify valid plans leading to a solution.

The behavior of the above-defined agents is now described in detail.

3.1 High-Level Search

The high-level search shown in Algorithm 1 is executed by the CA for achieving a valid solution. The CT is initialized with the root CT node N_r with cost $N_r.Cost$. The CT node N_r is initialized with the initial plan P_i identified by the SEAs assuming that the entire road network is freely available. The least-cost CT node, N_e, is chosen for exploration from an open set O which maintains all unexplored CT nodes. The CA generates a t.o. graph for N_e to validate and identify mutually exclusive sets of conflicting agents. If conflicts are detected, CA initiates the SEAs to execute low-level search for conflict resolution. The revised plans are combined into child CT nodes which are added to O for further exploration. The search is continued till a conflict-free CT node is found.

Algorithm 1 High Level Search

1. Define root CT-Node N_r with consistent plans from each SEA.
2. Add N_r to Openset O.
3. *Result = false.*
4. While (*!O.empty* or *!Result*) do
   a. Get Least Cost N_r from O
   b. Partition P = *CreatePartitionOnTOGraph*(N_r)
   c. If P has non-singleton block
      i. *ResolveConflict*(N_r, *P*)
      ii. Add resolved child CTnodes to O
   d. Else
      i. *Result = true*
5. End While

Algorithm 2 CreatePartitionOnTOGraph

1. For each plan P_i in *Nr.Solution* do
   a. *prevTONode* , *currTONode* = null
   b. For each road r_k in P_i do
      i. t_k = time by a_i on road r_k
      ii. If $\exists \gamma \in \Gamma, \gamma.r = r_k \wedge overlaps(\gamma.\tau, t_k)$
         - Add a_i to $\gamma.Z$
         - $\gamma.\tau$ = covering($\gamma.\tau, t_k$)
      iii. *Else*
         - Create new t.o.node γ with agent values
         - Add γ to G(Γ, Δ)
      iv. *currTONode* = γ
      v. If *prevTONode*!= null
         - Connect *prevTONode* and *currTONode* by edge δ
         - δ.agent = a_i
         - *prevTONode* = *currTONode*; *currTONode* = null
   c. EndFor
2. EndFor
3. For each t.o.node γ with degree(γ) >1 or $\gamma.Z > 1$
   a. Add all agents a_k in γ.Z to block *b* of the partition.
   b. If any agent a_k exists in another block b', merge b and b'
4. Add singleton blocks for agents not part of any block, to the partition

Conflict detection: The CT node N_e is first validated to identify presence of conflicts between the plans of the agents. In point objects, the presence of multiple agents on a given road at the same time confirms a conflict. Spatially extended objects could share the same road without conflict if the inter-object distance is maintained throughout that road. So, conflict detection in spatially extended objects involves identifying whether the time at which the agents meet each other in a head-on or a rear-end collision on the road lies within the shared time interval on the road.

In XCBS, for conflict detection, the entire plan time was divided into equal time intervals. The spatial location of every pair of agent is compared to check for conflict in each interval. Hence, in case of *l* time intervals for *k* agents, where an agent could occupy *r* roads during that interval, the time taken for detection was $O(l \cdot k^2 r)$.

In XCBS-A, for conflict detection in the CT node N_e, we create a t.o. graph G(Γ, Δ) based on the agent plans defined in the N_e.*Solution*. As given in Algorithm 2, for every road r_i, occupied by agent a_j during time interval t_k, Γ is checked to get t.o. node γ, such that $\gamma \cdot r = r_i$. If $\gamma \cdot \tau$ overlaps with t_k, then $\gamma \cdot Z$ is updated to include the agent a_j and the occupancy time-duration $t_k \cdot \gamma \cdot \tau$ is updated to the covering time-period of all the agent times in $\gamma \cdot Z$. When a match in terms of road r_i or time interval t_k is not found, a new t.o. node with the relevant values is added to the graph. Every t.o. node is then linked to the previous t.o. node generated for the plan, through a t.o. edge δ. Given *k* agents with *l* edges in each plan, the creation time of the t.o. graph is $O(k \cdot l)$. The set of t.o. nodes with degree > 1 or $\gamma \cdot Z > 1$ indicate possible conflict and hence need to be evaluated for checking the conflict time. If there are *n* such conflicting t.o. nodes, the overall time for conflict detection (including creation time) is $O(k \cdot l + n \cdot k^2)$, which is an improvement over XCBS.

Generally, agents can be divided into mutually exclusive conflicts sets, and by simultaneously resolving conflicts of these sets, we can accelerate the search to

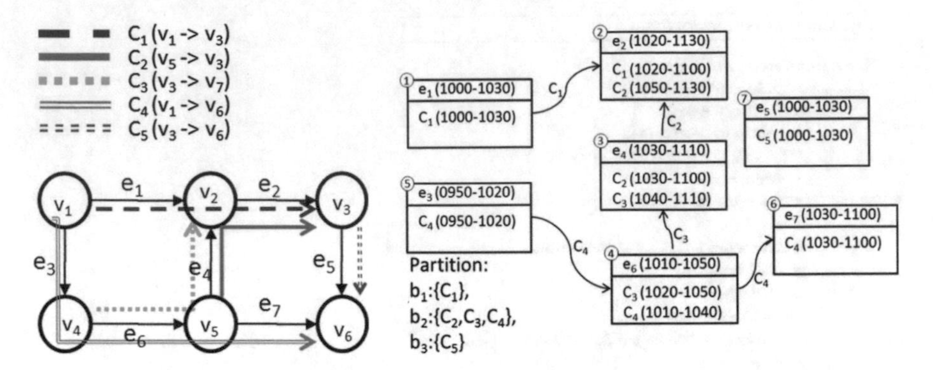

Fig. 1 Agent plans corresponding to the t.o. graph and partition set

solution. In XCBS-A, the agents are partitioned into blocks such that mutually conflicting agents, identified through the t.o. graph, belong to the same block. From the list of the t.o. nodes with degree >1, overlapping pairs of agents are added to the same block of the partition. If none of the agents preexisted in any block, then a new block is created with both agents as members. If either one of the agents is a member of some pre-existing block of the partition, the other agent is also added to the same block. When both agents belong to different blocks, then the two blocks are merged into a single block. The partition set could also contain singleton blocks comprising an agent which does not conflict with any other agents.

Figure 1 shows the t.o. graph created for five agents, with their initial and final locations shown in the grid network. The partition set comprises of block b_2 with agents C_2, C_3, C_4 and two singleton blocks with C_1 and C_5 each.

Conflict resolution: For the CT node N_e with conflicts, the CA simultaneously resolves conflicts of an agent of each block of the partition created for N_e. It passes all agents' plans to each of the SEAs belonging to any of the non-singleton partitions. When revised plans are received from the SEAs, the child CT node is created by having a revised plan of an agent from each non-singleton block; all other agents' plans remain unchanged. The child CT nodes so generated are added to the open set O to be further explored. If no conflict is detected in the CT node being explored, it means that each of the entities is moving along its shortest possible path optimally. A generic CT is shown in Fig. 2a.

For N_e with j non-singleton blocks, each child CT node N'_e will have j revised plans. The number of child CT nodes so generated will be $(j \times i)$, where i is number of agents in each of the j blocks. As an agent's revised plan is consistent with all other agents' unrevised plans, a conflict can only arise in any child CT node between the j revised plans. Hence, for j non-singleton blocks of size i identified in the root CT node, the maximum depth at which the solution is guaranteed to be found is $(i - 1) * j + 1$.

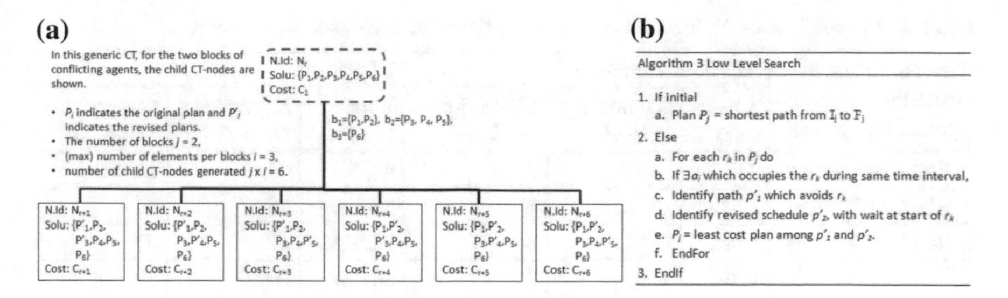

Fig. 2 **a** Generic CT created by CA **b** Algorithm for low-level search

3.2 *Low-Level Search*

The low-level search is executed by each SEA to identify consistent plans for itself, as shown in Algorithm 3. During the first run of the algorithm (no other agent plans are available), each SEA generates its optimal path using Djikstra's algorithm assuming full availability of road space. The path plan so obtained is the best plan for each agent.

In subsequent iterations of the algorithm, the CA passes the plans of all the other participating SEAs as per the CT node N_e. To identify a revised, non-conflicting, optimal plan P' for itself, each agent a_j sequentially checks each road in its current plan P_i with respect to the plans of the other agents. Sequential checking of the edges facilitates identification of conflicts as they occur. For each conflict so identified, the agent a_i, resolves its plan by avoiding the conflicting road or waiting out the conflict period at the start of the road. The plan eventually identified by the SEA is consistent with respect to its own constraints and also avoids conflicts with the current plans of all other agents.

4 Experimental Results

The performance of the proposed XCBS-A was compared with the XCBS, the A*-based solution defined in [5] and the distributed greedy-based approach described in [4]. The test suite comprising a grid network of varying number of uniform-length roads was defined in a Sparksee graph database and run on an Intel Xeon (4 Core) processor @ 3.26 GHz with 12 GB of RAM.

When XCBS-A was compared with XCBS using test cases having varying numbers of mutually conflicting agents, as Table 1 shows, XCBS-A outperformed XCBS not only in terms of time taken for evaluation but also in terms of the CT nodes explored for arrival at the solution. However, the number of nodes in the search is very large in case of XCBS-A, making scalability an issue. Further, the impact of the number of agents, the size and branching of the road network and the path length (number of roads) from initial to final location on the performance of

Table 1 Comparison of XCBS and XCBS-A in terms of time and search space

Test case (sets of mutually conflicting agents)		XCBS-A (proposed)			XCBS		
		Time (ms)	CT nodes created	CT nodes explored	Time (ms)	CT nodes created	CT nodes explored
1 Block of 5 agents		245	15	5	40,757	41	11
1 Block of 7 agents		300	21	6	85,147	85	22
1 Block of 9 agents		554	36	8	91,565	17	27
15 Agents	2 Agents/ block	490	129	2	30,337	29	8
	3 Agents/ block	649	179	3	62,493	61	16
	4 Agents/ block	708	164	4	86,632	85	22
	5 Agents/ block	785	171	5	123,041	121	31
12 Agents	6 Agents/ block	602	71	6	122,442	121	31
	7 Agents/ block	633	66	7	126,414	125	32
	8 Agents/ block	666	59	8	138,487	137	35

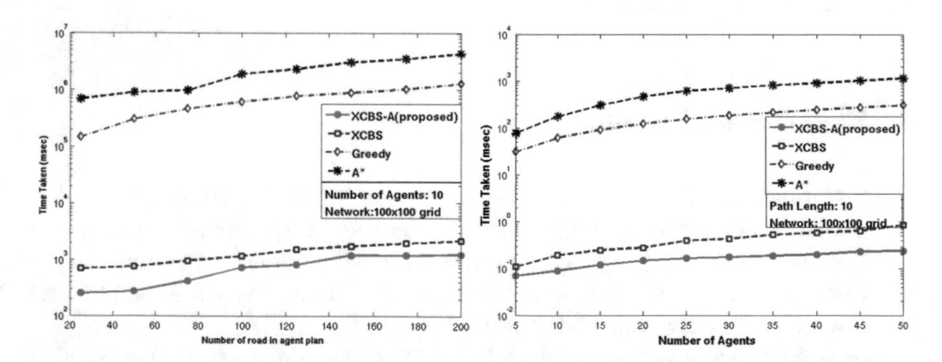

Fig. 3 Comparison of XCBS-A with greedy, A* and XCBS approaches

XCBS-A is shown in Fig. 3. As depicted, XCBS-A performed significantly better than the A* and greedy-based approaches and also surpassed XCBS in terms of time taken to arrive at the solution.

5 Conclusion

In this chapter, XCBS-A is proposed for optimal path finding of spatially extended objects. A comparative analysis of XCBS-A with other approaches in terms of road density, network size, number of agents and path lengths indicates that XCBS-A performs significantly better than the other algorithms, even with a large number of agents. Further, the number of search nodes explored by XCBS-A was significantly less than XCBS. We need to further work on pruning the search space to reduce the number of CT nodes generated. Further, detailed theoretical analysis on the optimality of the solution needs to be accomplished. A study on the performance of the algorithm when running on distributed machines over a cluster is also to be undertaken.

References

1. Standley, T., and R. Korf. 2011. Complete algorithms for cooperative pathfinding problems. In *IJCAI*, 668–673.
2. Qiuy Ling, Wen-Jing Hsuy, Shell-Ying Hunagy and Han Wang. 2002. Scheduling and routing algorithms for AGVs: A survey. *International Journal of Production Research*, 40 (3): 745–760.
3. Bodin, L.D.,B.L. Golden, A. Assad, and M. Ball. 1983. Routing and scheduling of vehicles and crews: The state of the art. *Computers and Operation Research*, 10: 63–211.
4. Thomas, S., D. Deodhare, and M.N. Murty. 2015. Extended Conflict Based Search for Convoy Movement Problem. *IEEE Intelligent Systems* 60 (3): 67–76.
5. Dinh, H.T., R.S. Rinde, T. Holvoet. 2016. Multi-agent route planning using delegate MAS. In *ICAPS proceedings of the 4th workshop on distributed and multi-agent planning*, 24–32.
6. Kumar A., I. Murugeswari, D. Khemani and N. Narayanaswamy. Planning for convoy movement problem. In *ICAART 2012-Proceedings of the 4th International Conference on Agents and Artificial Intelligence*, 495–498.
7. Thomas, S., N. Dhiman, P. Tikkas, A. Sharma and D. Deodhare. 2008. Towards faster execution of OODA loop using dynamic Decision Support System. In *3rd international conference on information warfare and security (ICIW)*.

Parametric and Functional-Based Analysis of Object-Oriented Dynamic Coupling Metrics

Navneet Kaur and Hardeep Singh

Abstract Software coupling is considered one of the most crucial issues in the software industry. The measurement of coupling gives a quantitative indication of internal features that are correlated to external quality attributes such as reliability, maintainability, extendibility, functionality and usability of the software. Various static and dynamic metrics have been proposed by several academicians and industry experts to evaluate software coupling. Although the values of static metrics can be easily extracted from the source code without its execution, researchers have emphasized the significance of dynamic metrics over static metrics in terms of program comprehensibility, change proneness, maintainability, etc. This chapter presents some of the dynamic coupling measures and their comparative analysis: (a) parametric-based analysis of six parameters, i.e. *quality focus, measured entity, theoretical validation, dynamic analysis approach*, *environment,* and *statistical analysis technique*; (b) functional-based analysis. Only those metrics which possess the required amount of information have been considered in order to make a meaningful comparison possible. Based on the comparison, the observed results reveal that empirical validation conducted for the evaluation of dynamic coupling metrics is not adequate to make them applicable to industries.

Keywords Dynamic coupling · Metrics · Software maintenance
Static measures

N. Kaur (✉) · H. Singh
Guru Nanak Dev University, Amritsar, India
e-mail: navneetsandhu02@gmail.com

H. Singh
e-mail: hardeep.dcse@gndu.ac.in

A. N. Krishna et al. (eds.), *Integrated Intelligent Computing, Communication and Security*, Studies in Computational Intelligence 771,
https://doi.org/10.1007/978-981-10-8797-4_48

1 Introduction

The key motivation behind the measurement of software factors such as coupling, cohesion and complexity is to ensure the quality attributes of the Software. The attributes like functionality, maintainability, extendibility, usability, etc. are vital to make any software system a high-quality product. For example, in evolutionary development, higher-quality software must comprise extendibility and maintainability for the allowance of changing technological needs and future requirements. These attributes are potentiating only when there is an appropriate level of coupling, cohesion and complexity. Hence, it becomes imperative to identify some system of measurement for controlling the software factors.

As the level of interdependent software modules rapidly intensifies, low coupling is becoming a challenging task for the software evolutionists. Sometimes, the rewarding characteristics of an object-oriented system (OOS) contribute to high coupling, e.g. the reusability facilitates the software development process, but at the same time, it also increases the coupling that ultimately complicates software maintenance. High coupling is detriment to the structural quality and increases the required maintenance. According to Arisholm and Sjùberg [1], high coupling is an indicator of changeability decay where changeability decay means the augmentation of changeability efforts for modification between two versions of software. Eder et al. [2] discussed three types of coupling: (a) interaction coupling, (b) component coupling and (c) inheritance coupling. To quantify these different types of software coupling, many researchers have made their contribution by formulating various coupling metrics. Chidamber and Kemerer [3] introduced a suite of six metrics, one of which is *coupling between object classes (CBO)* which finds coupling at the class level by counting the number of other classes accessed by a given class. The subsequent improvements in measuring the coupling were directed by Li and Henry [4], Abreu and Carapuça [5] and many other researchers. All of the measures mentioned above are static coupling metrics, which can be easily acquired without executing the source code.

This chapter presents a comparative analysis of dynamic coupling measures. It is necessary to mention that Geetika and Singh [6] categorized different dynamic coupling metrics on the basis of their characteristics. Although some of the parameters chosen for the current work are similar to the categories suggested by the authors, the motive of the current study is perceptively different, which is to compare selected coupling measures; this comparison has been accomplished at parametric and functional level. The remainder of the chapter is organized as follows: Sect. 2 presents various existing dynamic coupling metrics, Sect. 3 compares dynamic coupling metrics, and Sect. 4 concludes the chapter.

2 Object-Oriented Dynamic Coupling Metrics (OODCM)

The values of dynamic metrics can only be collected during the runtime. Sometimes, a program proved as highly coupled by static metrics is actually found low coupled at runtime because of two reasons: (a) during execution, only some part of the program is active, (b) presence of a high amount of dead code. Also, object oriented programming (OOP) properties like polymorphism and dynamic binding remain undetected by static metrics. For instance, if a class A has more than two methods of the same name, the actual method that will be called by the object can only be known at runtime. In the subsequent subsection, the major contribution of the authors in developing dynamic coupling measures is briefly described.

2.1 Yacoub et al. [7]

Yacoub et al. [7] identified a set of four dynamic coupling measures; i.e. *export object coupling* (*EOC*), *import object coupling* (*IOC*), *object request for service* (*OQFS*) and *object response for service* (*OPFS*), which are capable of evaluating the executable design models created for complex real-time applications. The objective of measurement was to assess the dynamic behavior of the system at an early development phase, which has been accomplished by finding the number of runtime messages exchanged between the objects in the design model. The authors considered that higher values of *EOC* and *IOC* can affect the quality attributes, such as maintainability, understandability and reusability and also increase the error flow from faulty objects to other objects. No empirical validation has been presented in the subsequent papers to prove their conception. Though these metrics are competent in finding values at an early development phase, it is pertinent to mention that the values collected from the design model are not necessarily the same as that from the final software.

2.2 Arisholm et al. [8, 9]

Arisholm et al. [8, 9] examined the idea that runtime module interdependence is an improved measure of change proneness (changes more likely to happen in the future), rather than size and static coupling. The authors introduced 12 metrics defined along the three pronged dimensions: (a) mapping: object (*O*) and class (*C*); (b) direction: import coupling (*IC*) or export coupling (*EC*); and (c) strength: number of distinct methods invocation (*M*), distinct classes (*C*) and dynamic messages (*D*). The metrics listed by the author include: *IC_OD, IC_OM, IC_OC, IC_CD, IC_CM, IC_CC, EC_OD, EC_OM, EC_OC, EC_CD, EC_CM,* and *EC_CC*. In each metric, the first two letters represent the direction of coupling, i.e.

IC or EC, the third letter indicates the mapping level, and the last letter denotes the strength of coupling. The author has shown that the high value of EC, for a given class A of system *X*, indicates the high dependability of other classes of system *X* on the functionality of the class A; therefore, EC is highly correlated to changes that can occur in subsequent software releases [8]. The author has not considered the coupling to/from the library and frame classes.

2.3 Hassoun et al. [10–12]

Hassoun et al. [10–12] proposed a metric supporting meta-level architecture, capable of altering its own execution state. The authors determined that during the execution of the software, various objects couple to other objects (each of different complexity) over a specific period of time and presented a *dynamic coupling* metric (*DCM*). According to this measure, Eq. (1) has been used to compute the coupling for object P.

$$P|\Delta t = \sum_j \sum_i [f_i(t_j)] * complexity\ of\ ith\ coupled\ object \tag{1}$$

In Eq. (1), Δt is an ordered series of program implementation step and ∑j is the total number of execution steps. ∑i is a summation set of objects coupled to P and the value of fi(tj) is either 0 or 1 depending on the association with another object.

2.4 Mitchell and Power [13–15]

Mitchell and Power conducted extensive studies on dynamic coupling measures and results with several new metrics for quality attributes evaluation. The concept of CBO outlined by Chidamber and Kemerer [3] has been extended by the authors in the field of dynamic execution and they introduced three measures: *dynamic coupling between objects* (*DCBO*), *degree of dynamic coupling between two classes* (*DDCBO*) and *degree of dynamic coupling within a given set of classes* (*DDCC*) [13]. Subsequently, the authors continued their previous work, leading to the creation of four new metrics: *runtime import coupling between objects* (R_I), *runtime export coupling between objects* (R_E), *runtime import degree of coupling* (RD_I) and *runtime export degree of coupling* (RD_E) [14]. Later, Mitchell and Power studied two metrics to prove the hypothesis that objects of the same class behave differently at runtime from the point of view of coupling [15]. The metrics used in their study include: (i) *runtime coupling between objects* (*RCBO*) which measures the number of classes accessed by a class at execution time. The metric *RCBO* is basically used to inspect whether objects of the same class type couple to the same classes at execution time; (ii) The second metric measures the number of unique accesses to other classes by each object of the class.

2.5 Zaidman and Demeyer [16, 17]

Maintainability requires the adequate understanding of existing software by the programmers and literature has witnessed that 30–60% of development time is spent in this phase, where the developer has to study an existing code, documentations and other design artifacts. To mitigate this problem, Zaidman and Demeyer proposed a metric, *class request for service* (*CRFS*), that gives a clue to the programmer as to where to start the program comprehension [16]. Basically, *CRFS* is an extension of the *OQFS* metric, introduced by Yacoub et al. [7], and works by counting every method that a certain class calls during the execution of the program. Subsequently, the authors introduced the *IC_CC* metric to accomplish the program comprehensibility process. The *IC_CC* metric is a variation of the *IC_CC* metric (proposed by Arisholm) and counts the number of called methods instead of calling methods [17].

2.6 Beszedes et al. [18]

Beszedes et al. [18] introduced the metric *dynamic function coupling* (*DFC*), which is defined as "minimum level of indirection between all possible occurrences of the two functions in the traces in order to compute the impact set for impact analysis". Impact analysis is important for evolving software to inquire as to the potential consequences of a change. The study is motivated by the work of Apiwattanapong et al. [19], and is based on the idea that if a function f1's execution is closer to the execution of function f2, then the chances of their dependency are high. The authors have also developed an algorithm that uses an execute round (ER) relation, an extension of the 'execute after' relation developed by Apiwattanapong et al. [19], to compute the set of impacted functions for a changed function set.

2.7 Sandhu and Singh Metrics [20]

Sandhu and Singh [20] introduced two set of metrics to measure dynamic coupling and dynamic inheritance. Dynamic coupling metrics outlined by the authors are *dynamic afferent coupling* (*DCa*), *dynamic key server class* (*DKSC*), *dynamic key client class* (*DKCC*), *dynamic key class* (*DKC*) and *percentage active class* (*PAC*).

In all the included papers, the authors claimed that dynamic coupling metrics are better than static coupling measures as OODCM provides improved results in terms of program comprehension, detecting change proneness, change impact analysis, testability and maintainability. But the evidences provided by the authors are inadequate to generalize their outcomes.

3 Comparative Analysis of OODCM

This section contains the comparative analysis of all metrics discussed in the previous section and this comparison has been conducted at two levels—(1) Parametric Analysis, and (2) Functional based Analysis. In the former case the comparison has been executed on the methods and techniques followed by the authors in order to formulate different metrics. And at second level, comparison has been realized on each metric on the basis of their functionality.

3.1 Parametric Analysis of OODCM

Based on the existing work, the current paper has analyzed the implicit development process of the software metric. By combining the adopted approach from literature review and our observation, the authors of this paper have figured out an explicit software metric development process (SMDP) in Fig. 1. The SMDP consists of a sequence of operations that are essential for the creation of any successful measurement system. The SMDP serves as a base for the parametric analysis of OODCM where all the parameters are opted from its sequential actions (bold text in Fig. 1). The selected parameters for comparison are: *quality focus* (P_1), *measured entity* (P_2), *theoretical validation* (P_3), *dynamic analysis approach* (P_4), *environment* (P_5) and *statistical analysis technique* (P_6) as presented in Table 1. In this chapter, the term 'environment' is used for the set of software, systems and executable models that have been used for empirical validation of the dynamic coupling measures.

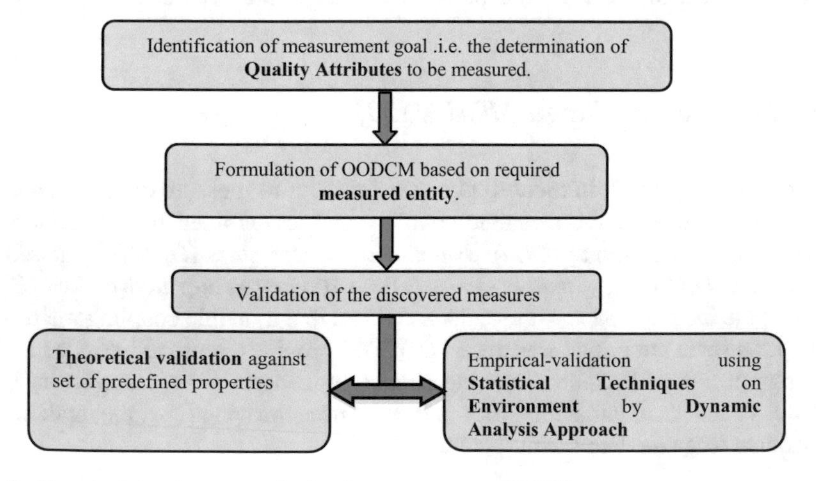

Fig. 1 Steps in the software metric development process (SMDP)

Table 1 Parametric analysis

Ref no.	P_1	P_2	P_3	P_4	P_5	P_6
[7]	Maintainability, understandability, reusability, error propagation	Object	–	Dynamic modeling diagrams	Pacemaker device	–
[8, 9]	Change proneness	Class, object, message	Braind et al. [21]	Profile-based tracing	Velocity	PCA, descriptive statistic and linear regression
[10–12]	Reusability, maintainability	Object	Braind et al., Weyuker et al. [22]	Aspect oriented based technology	Static system, meta system	Comparative study
[13–15]	Maintainability, reusability, error propagation, testing efforts	Class	–	Profile-based tracing	Java grande forum Benchmark suite, SPECjvm98 benchmark suite Jolden benchmark suite.	Principal component analysis (PCA), coefficient of variation, cluster analysis, descriptive analysis, hypothesis testing
[16, 17]	Understandability	Class	–	Profiler-based tracing for OOS and aspect-oriented based tracing for C system.	Two object-oriented systems and an industry legacy C system	Hypothesis testing
[18]	Change impact analysis	Method	–	Virtual machine instrumentation	JSubtitles, NanoXML, Java2HTML (open source Java programs)	Precision and recall
[20]	Maintainability, testability, reusability, portability	Class	–	Profile-based tracing	Educational institution project (java application	Descriptive statistics, correlation study

Table 2 List of entities and relations

Notation	Description	Notation	Description
E1	Class	R3	Object receives message from object
E2	Object	R4	Class accessed/used by another class
E3	Method	R5	Class uses instance of another class
R1	Class sends message to (or accesses) another class	R6	Method calls method
R2	Object sends message (or couples) to object/method	R7	Class receives message or call request from another class

3.2 *Functional-Based Analysis of OODCM*

In this section, the authors present a data model for object-oriented dynamic coupling metrics based on the idea outlined by Abounder and Lamb [23]. The motive behind the incorporation of a data model in this study is to compare the metrics at a more detailed level, based on their functionality. The list of entities and relationship that have been used for comparing the metrics are presented in Table 2, where entities represent the mapping level and relationship denotes the functionality performed by each metric. In Table 3, the presence of a particular mapping level and functionality of each metric is marked by '*' (asterisk) symbol. Although the formulae to calculate each metric are distinct, analysis showed that many of the measures capture similar relations, thus reflecting the fact that many of them share similar functionality. Metrics like *EOC*, *OPFS*, *IC_OD*, *IC_OM*, *IC_OC*, *IC_OD*, *IC_CM*, *IC_CC*, *EC_OC*, *EC_CC* and *DCM* possess a similar relation R2; i.e., relation R2 will be active for any metric where an object sends messages to other objects or methods. Likewise, for *IOC* (O_i, O_j), import coupling for object O_i is the number of messages received by an object O_i from object O_j; for this metric relation, R3 is appropriate. Similarly, for all other metrics, relations have been chosen on the basis of functionality owned by a particular measure.

4 Conclusions

In this chapter, the authors have discussed different dynamic coupling measures for object-oriented languages. The outcome of these metrics heavily depends on the environment input pattern. The researchers have claimed that the dynamic metrics have advantages over static metrics, as they provide accurate assessment of the software by considering the impact of design features like dead code, polymorphism and dynamic binding. The main focus of the study is to present the detailed comparison of different dynamic coupling measures under study. During parametric

Table 3 Functional-based analysis

Metrics	E1	E2	E3	R1	R2	R3	R4	R5	R6	R7
IOC		*				*				
EOC		*			*					
OQFS		*				*				
OPFS		*			*					
IC_OD		*			*					
IC_OM		*	*		*				*	
IC_OC		*			*					
IC_CD	*				*				*	
IC_CM	*		*		*				*	
IC_CC	*				*				*	
EC_OD		*				*				
EC_OM		*	*			*				
EC_OC		*			*					
EC_CD	*					*			*	
EC_CM	*		*			*			*	
EC_CC	*				*					
DFC									*	
DCBO	*			*				*		
DDCBO	*			*				*		
DDBC	*			*				*		
RD_I	*			*						
RD_E	*						*			
R_I	*			*				*		
R_E	*			*				*		
DCM		*			*					
CRFS	*			*						
DCa	*			*					*	
DKSC	*			*						
DKCC	*			*						
DKC	*			*						*
PAC	*			*					*	*

analysis, some limitations of existing dynamic coupling measures have been encountered. As both the theoretical and empirical validation of metrics are crucial for software applicability, the study clearly reveals the lack of theoretical validation of some metrics. Although empirical validations and experiments have been conducted to prove the validity of dynamic measures, as compared to static measures, their approach was confined to just a few systems for testing the relationship between measures and quality attributes. One of the main reasons for the lack of this empirical investigation is the high cost of metric data collection. In order to realize the full potential of OODCM, the identification of an advanced metric data collection tool is required. The authors have also presented the functionality-based analysis and concluded that many metrics share similar functionality. Consequently, before formulating new measures, there should be some verification approach to confirm the inadequacy of existing measures in solving the problem at hand. It is observed that the incorporated papers have not followed this approach.

References

1. Arisholm, E., and Sjùberg, D.I. 2000. Towards a framework for empirical assessment of changeability decay. *Journal of Systems and Software,* 53 (1): 3–14.
2. Eder, J., G. Kappel, and M. Schrefl. 1994. Coupling and cohesion in object oriented systems.
3. Chidamber, S., and C. Kemerer. 1994. A metrics suite for object oriented design. *Transactions on Software Engineering* 20 (6): 476–493.
4. Li, W., and S. Henry. 1993. Object-oriented metrics that predict maintainability. *Journal of Systems and Software* 23 (2): 111–122.
5. Abreu, F.B., and R. Carapuça. 1994. Object-oriented software engineering: Measuring and controlling the development process. In *4th international conference on software quality*, McLean, VA, USA.
6. Geetika, R., and P. Singh. 2014. Dynamic coupling metrics for object oriented software systems: A survey. *ACM SIGSOFT Software Engineering Notes* 39 (2): 1–8.
7. Yacoub, S., H. Ammar and T. Robinson. Dynamic metrics for object oriented designs. In *Sixth international software metrics symposium*, Boca Raton, FL, USA.
8. E. Arisholm. 2002. Dynamic coupling measures for object-oriented software. In *Proceedings of the eighth IEEE symposium on software metrics*, Ottawa, Ontario, Canada.
9. Arisholm, E., L. Briand, and A. Foyen. 2004. Dynamic coupling measurement for object-oriented software. *IEEE Transactions on Software Engineering* 30 (8): 491–506.
10. Hassoun, Y., R. Johnson and S. Counsell. 2004. A dynamic runtime coupling metric for meta-level architectures. In *Proceedings of eighth European conference on software maintenance and reengineering (CSMR 2004)*, Tampere, Finland.
11. Hassoun, Y., R. Johnson and S. Counsell. 2004. Empirical validation of a dynamic coupling metric.
12. Hassoun, Y., S. Counsell, and R. Johnson. 2005. Dynamic coupling metric: proof of concept. *IEE Proceedings—Software* 152 (6): 273–279.
13. Mitchell, A., and J.F.Power. Run-time coupling metrics for the analysis of java programs-preliminary results from the SPEC and grande suites.
14. Mitchell, A., and J.F. Power. 2004. An empirical investigation into the dimensions of run-time coupling in Java programs. In *Proceedings of the 3rd international symposium on principles and practice of programming in java,* Las Vegas, Nevada, USA.
15. Mitchell, A., and J.F. Power. 2005. Using object-level run-time metrics to study coupling between objects. In *Proceedings of ACM symposium on applied computing*, Santa Fe, New Mexico.
16. Zaidman, A., and S. Demeyer. 2004. Analyzing large event traces with the help of coupling metrics. In *Proceedings of the fourth international workshop on OO reengineering*, University Antwerpen.
17. A. Zaidman. 2006. Scalability solutions for program comprehension through dynamic analysis. PhD dissertation, University of Antwerp.

18. Beszedes, A., T. Gergely, S. Farago, T. Gyimothy and F. Fischer. 2007. The dynamic function coupling metric and its use in software evolution. In *11th European conference on software maintenance and reengineering*, Amsterdam, Netherlands.
19. Apiwattanapong, T., A. Orso and M.J. Harrold. 2005. Efficient and precise dynamic impact analysis using execute-after sequences. In *Proceedings of the 27th international conference on software engineering*, St. Louis, MO, USA.
20. Singh, P., and H. Singh. 2010. Class-level dynamic coupling metrics for static and dynamic analysis of object-oriented systems. *International Journal of Information and Telecommunication Technology* 1 (1): 16–28.
21. Briand, L., J. Daly and J. Wust. 1999. A unified framework for coupling measurement in object-oriented systems. *IEEE Transactions on Software Engineering* 25 (1): 91–121.
22. Weyuker, E. 1998. Evaluating software complexity measures. *IEEE Transactions on Software Engineering* 14 (1): 1357–1365.
23. Raymond, J.A., and D.L. Alex. 1997. A data model for object oriented design metrics. Kingston, Ontario, Canada.

Unraveling the Challenges for the Application of Fog Computing in Different Realms: A Multifaceted Study

John Paul Martin, A. Kandasamy and K. Chandrasekaran

Abstract Fog computing is an emerging paradigm that deals with distributing data and computation at intermediate layers between the cloud and the edge. Cloud computing was introduced to support the increasing computing requirements of users. Later, it was observed that end users experienced a delay involved in uploading the large amounts of data to the cloud for processing. Such a seemingly centralized approach did not provide a good user experience. To overcome this limitation, processing capability was incorporated in devices at the edge. This led to the rise of edge computing. This paradigm suffered because edge devices had limited capability in terms of computing resources and storage requirements. Relying on these edge devices alone was not sufficient. Thus, a paradigm was needed without the delay in uploading to the cloud and without the resource availability constraints. This is where fog computing came into existence. This abstract paradigm involves the establishment of fog nodes at different levels between the edge and the cloud. Fog nodes can be different entities, such as personal computers (PCs). There are different realms where fog computing may be applied, such as vehicular networks and the Internet of Things. In all realms, resource management decisions will vary based on the environmental conditions. This chapter attempts to classify the various approaches for managing resources in the fog environment based on their application realm, and to identify future research directions.

J. P. Martin (✉) · A. Kandasamy
Department of Mathematical and Computational Sciences,
National Institute of Technology, Mangalore, Karnataka, India
e-mail: johnpm12@gmail.com

A. Kandasamy
e-mail: kandy@nitk.ac.in

K. Chandrasekaran
Department of Computer Science and Engineering,
National Institute of Technology, Mangalore, Karnataka, India
e-mail: kch@nitk.ac.in

A. N. Krishna et al. (eds.), *Integrated Intelligent Computing, Communication and Security*, Studies in Computational Intelligence 771,
https://doi.org/10.1007/978-981-10-8797-4_49

1 Introduction

The latest techniques in pervasive and ubiquitous computing aim to make the world totally connected. The constant availability of computing devices empowers mobile users by providing them access to a range of cost-effective and high-performance services in different environments, including smart homes, smart cities and smart vehicles. These smart devices have the potential to interconnect distinct physical business worlds, and their usage in real-time complex applications helps to generate efficient, comfortable, simple solutions [18]. The number of devices connected to the Internet is estimated to increase to 50 billion by 2020 [16]. A large number of these devices generate huge volumes of heterogeneous data, but the limited computing and processing capabilities of these edge devices makes them unsuitable for processing these data and deriving useful results [13]. To overcome these issues, smart devices collaborate with cloud computing [11]. Use of the cloud provides efficient processing power and unlimited storage for a heterogeneous volume of data [23]. Even though much effort has been focused on efficient integration of cloud and Internet of Things (IoT), the rapid explosion in the number of services and edge devices makes the traditional cloud, where data and processing is done through a few centralized servers [3], insufficient for all applications. There are many IoT applications that require low latency (real-time applications), mobility support and location-aware processing [20]. Using existing network configurations and low bandwidth, relying on the cloud environment alone to carry out analytics operations has proved to be inadequate, especially for applications that are not delay-tolerant [2]. Researchers aim to carry out processing activities nearer the end user where data are generated, rather than the distant cloud data centers. The technology that incorporates data processing capabilities at the edge of the network, rather than holding that processing power solely in a cloud or a central data warehouse, is called edge computing [26]. The major drawback to edge computing is the lack of scalability [1]. Limitations in edge computing and cloud computing for performing real-time applications led researchers to propose a new computing model called fog computing, which can overcome these issues and handle all scenarios efficiently. Fog computing brings the computation closer to the user devices rather than deploying them on the user devices alone. Because fog computing is an emerging technology, there are still ambiguities as to what can constitute a fog node, the scenarios that can be enhanced with the introduction of fog nodes, how resource management challenges vary across domains, and so on. This chapter investigates the concepts used in fog computing and challenges faced while implementing in different application scenarios. The main contributions of the chapter are as follows:

- Identifying the expected capabilities of the fog computing system and analyzing the domains that can be enhanced with the application of fog computing.
- Identifying the limitations of the existing models and requirements of applications, which all point to the need for a fog computing system.
- A detailed review on the orchestration issues of the fog computing environment in different application domains.

- Analyzing the challenges involved and future research directions in fog computing.

The remainder of the chapter is organized as follows. Section 2 summarizes the definition and architecture of fog computing and characteristics of fog nodes. Section 3 surveys the major orchestration issues in fog computing in different application domains. Section 4 presents the major challenges and possible research directions, and Sect. 5 concludes the chapter.

2 Fog Computing

Requirements for real-time applications have led to the concept of bringing computation closer to user devices rather than on user devices, a concept which is widely known as fog computing [4, 7]. Fog provides decentralized processing power and distributed intelligence for fast processing of heterogeneous, high-volume data generated by various real-time applications. Processing of real-time data involves making complex decisions such as choosing the processing location at the right place and right time [8]. Figure 1 gives a model architecture of the fog environment. There are various types of edge devices present closest to users. Fog nodes are generally placed at a one-hop distance from edge devices. Streaming data collected by edge

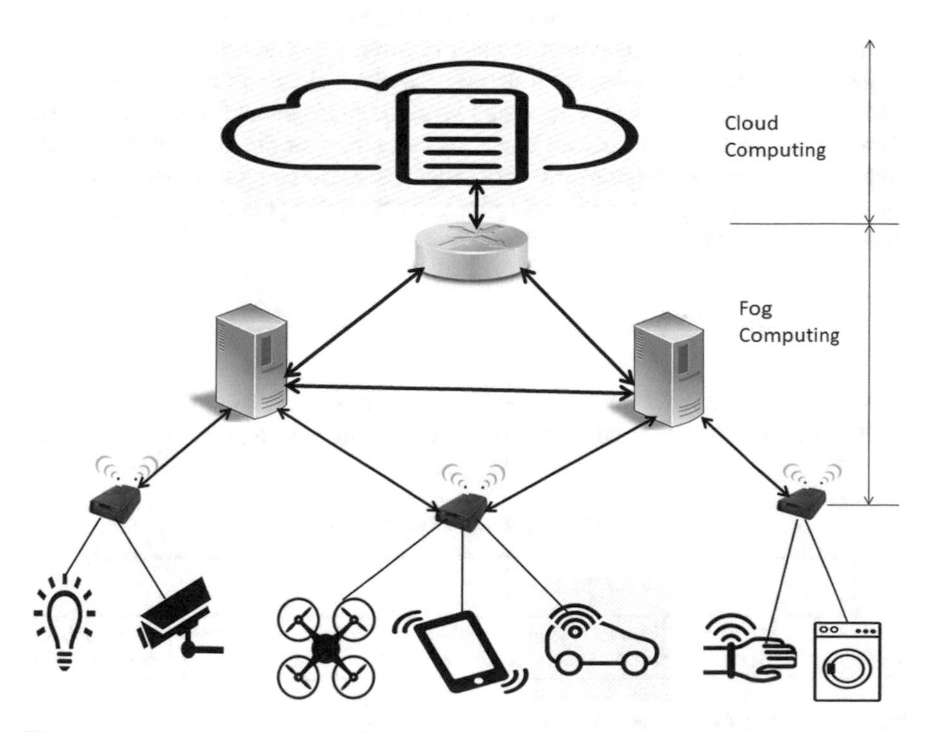

Fig. 1 Fog computing-reference architecture

devices are then transferred to the fog nodes, rather than transporting in bulk volume to the cloud. This can lead to a significant reduction in delays and latency. Fog nodes receive and process these data and make decisions that are communicated to the edge devices. Data that are required for future analysis are transported from the fog node to the cloud, which provides persistent storage mechanisms.

Fog nodes, or fog servers, are placed at different levels between the cloud and the edge of the network to enable efficient data storage, processing and analysis of data, and they reduce latency by limiting the amount of data transported to the cloud. Location, number and hierarchy of fog nodes are all decided on a case-by-case basis. The research in this field is still in its infancy, which is evident from the article by Tordera et al. [21], where the authors attempt to reach a consensus on what can be considered a fog node. However, the fog computing concept can be adopted by different application domains to better serve user requests. In the next section, we review a subset of applications that may be made more efficient by employing fog computing techniques.

3 Applicability of Fog in Different Domains and Challenges Involved

Fog computing plays a significant role in IoT. However, in recent research works, the applicability of fog computing in other networking systems (mobile networks, content distribution networks, radio access networks, vehicular networks, etc.) have also been highlighted. Orchestration and resource management issues in different scenarios depend on context. Our aim is to categorize works in the literature and obtain an overview of the management challenges in fog computing across the various domains. The proposed taxonomy for classifying the research works has been illustrated in Fig. 2.

3.1 *Internet of Things*

Wen et al. [24] explored the challenges involved in the creation of an orchestrator in IoT-based fog environments. The uncertain working environment of IoT applica-

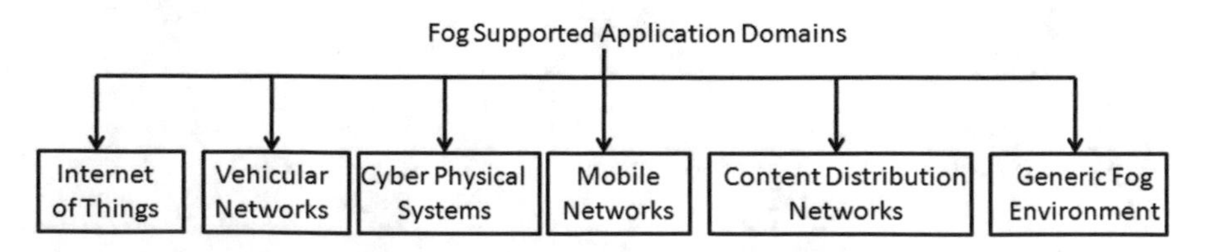

Fig. 2 Taxonomy for application domains of fog computing

tions may create some internal transformations and corresponding dynamic changes in work flow components, so a dynamic orchestrator that has the ability to cater to efficient services is required. They proposed a framework based on a parallel genetic algorithm in Spark. The experiments were made on servers in the Amazon Web Service. Wang et al. [22] proposed a resource allocation framework for big multimedia data generated by IoT devices at the cloud, edge and fog. They also proposed a multimedia communication paradigm for the application layer which is based on premium prioritization for big volumes of data in wireless multimedia communications. Their framework can incorporate diversity of multimedia data at the application layer. Thcy analyzed dependence in spatial, frequency and temporal domains. They also investigated how resource allocation strategies ensure energy efficiency in a quality-of-service (QoS)-driven environment.

3.2 Vehicular Networks

Chen et al. [5] explored the feasibility of fog computing in the area of vehicular networks and proposed two dynamic data scheduling algorithms for fog-enabled vehicular networks. They proposed an efficient three-layer vehicular cloud network (VCN) architecture for real-time exchange of data through the network. Resource management is controlled in a road-side cloud layer. The first dynamic scheduling algorithm is based on the classical concept called Join the Shortest Queue (JSQ), and the next is a complete dynamic algorithm based on predicting the time required for each incoming task. Performance of their proposed algorithm is evaluated by modeling the case scenarios through a compositional approach called Performance Evaluation Process Algebra (PEPA).

3.3 Cyber-Physical Systems

Gu et al. [12] analyzed the integration of fog computing in medical cyber-physical systems (MCPS). The major limitations of cloud computing in cyber-physical systems are the instability and high delay in the links between cloud providers and end devices. The authors claim that the integration of fog computing in MCPS improves QoS and other non-functional properties(NFP). They proposed a solution for the resource management problem in a cost-efficient manner with the required QoS. The cost minimization problem is formulated as a mixed integer non-linear programming (MINLP) problem with constraints such as base station association, task distribution, Virtual Machine (VM) placement and QoS. Zeng et al. [25] explored the features of fog computing, which supports the software-defined embedded system, and developed a model for embedded systems supported by a fog computing environment. They also proposed an effective resource management and task scheduling mechanism in such an environment with the objective of task completion time minimization. The task completion and time minimization problem can be treated as a mixed integer non-linear programming (MINLP) problem. They formulated constraints for storage and task completeness and propose a three-stage heuristic algorithm for solving highly computational complex MINLP problems. The algorithm uses the concept of “partition and join” and with the first two stages tries to minimize I/O time and computation time independently. Joint optimization is performed in the last stage.

3.4 Mobile Network/Radio Access Network

Dao et al. [6] proposed an adaptive resource balancing scheme for improving the serviceability in fog radio access networks. They used a back-pressure algorithm for balancing resource block utilization through service migration among the remote radio heads (RRH) on a time-varying network topology. Mungen et al. [19] surveyed the recent developments in the field of socially aware mobile networking in fog radio access networks(F-RANS). Their main scope of study was in radio resource allocation and performance analysis in an F-RAN environment. The presence of local cache and adaptive model selection makes F-RAN more complex and advanced. The authors examined how this local caching and adaptive model selection affects spectral efficiency, energy efficiency and latency.

3.5 Content Distribution Network

A content distribution network (CDN) is composed of distributed proxy servers that provide content to end users, ensuring high performance and availability. Jiang et al. [27] proposed a method for optimizing the performance of web services by taking advantage of fog computing architecture. Fog server caches contain all recently accessed details, and this content is used to improve performance. Cuong et al. [10] explored the usage of fog computing in content delivery systems. The usage of fog nodes at the edge of the network for streaming services reduces latency and improves QoS. They proposed a distributed solution for joint resource allocation in a distributed fog environment and to minimize the carbon footprint. The problem can be formed as a general convex optimization, and the proposed solution method is built on proximal algorithms rather than other conventional methods.

3.6 Conventional Fog Environments

Deng et al. [9] proposed a framework for optimal workload allocation among cloud and fog computing environments. Minimal power consumption and service delay are two major constraints considered when formulating the primal problem. They developed an approximate approach for solving the primal problem by dividing it

Table 1 Summary of the methodologies adopted by existing research works for management of the fog environment

Research Work	Authors	Objective	Methodology	Limitations
[25]	Deze Zeng et al.	Resource management and task scheduling in fog-enabled embedded systems	Formulated into mixed-integer non-linear programming problem and solved	Memory management not considered
[12]	Lin gu et al.	Resource management in cyber-physical systems supported by fog computing	Two-phase heuristic algorithm used for solving resource management problems	Only limited QoS parameters are considered
[9]	Ruilong Deng et al.	Optimal workload allocation among cloud and fog computing environments	The primal problem is divided into three sub-problems and solved	Only considered power consumption and delay for allocation
[22]	Wei Wang et al.	Resource allocation framework for big multimedia data generated by IoT devices	Framework is based on premium prioritization for big volumes of data	Feasibility in a real-time scenario not considered.
[5]	Xiao Chen et al.	Dynamic data scheduling algorithms for fog-enabled vehicular networks	Three-layer vehicular cloud network architecture used for real-time exchange of data through the vehicular network	Performance evaluation is done only through formal approach
[24]	Zhenyu Wen et al.	Creation of an orchestrator in an IoT-based fog environment	Proposed a solution based on parallel genetic algorithms	Does not consider a massive-scale system
[17]	Lina Ni et al.	Resource allocation and scheduling strategy for fog computing environments	Resource allocation strategy is developed based on priced timed petri net (PTPN)	Not all performance metrics are considered

(continued)

Table 1 (continued)

Research Work	Authors	Objective	Methodology	Limitations
[14]	Jingtao et al.	Developed an efficient method to share resources in a fog cluster	Steiner tree in graph theory is used for developing caching method	Does not perform well with topologies with fewer fog nodes
[27]	Jiang et al.	Web optimization using fog computing	Recently accessed details in the fog server cache are used to improve the performance.	Only developed proof-of-concept system.
[6]	Ngoc dao et al.	Implement resource balancing scheme for improving the serviceability in fog radio access networks	Back pressure algorithm is used to implement the scheme.	Number of service migrations is not considered with changing network

in to three sub-problems of corresponding subsystems. Ni et al. [17] proposed a resource allocation and scheduling strategy for fog computing environments. The objectives for the proposed schemes were to maximize resource utilization, fulfill all the QoS requirements and maximize profit for both providers and users. Their dynamic resource allocation strategy is based on a priced timed petri net (PTPN), by which the user is allowed to pick resources that satisfy requirements from a pool of pre-allocated resources, considering price cost and time cost to complete a task. Jingtao et al. [14] proposed an efficient method for sharing resources in a fog cluster to provide local service to mobile users in a cost-efficient manner. The fog cluster is composed of many functional specific servers. They applied a Steiner tree in graph theory to develop this caching method in a fog environment. A graph G is composed, where vertices represent the servers and the edge represents the connection between them. The Steiner tree is used to find the minimum cost of the subgraph, which is generated from graph G. A summary of the research works considered in this table has been provided in Table 1.

4 Discussion

Fog computing has a wide range of applicability across different domains. Usage of fog computing in various domains makes it necessary to deal with different aspects of constrained environments. The dynamic changes inherent in the runtime environment impose an additional challenge for fog management and contribute to the

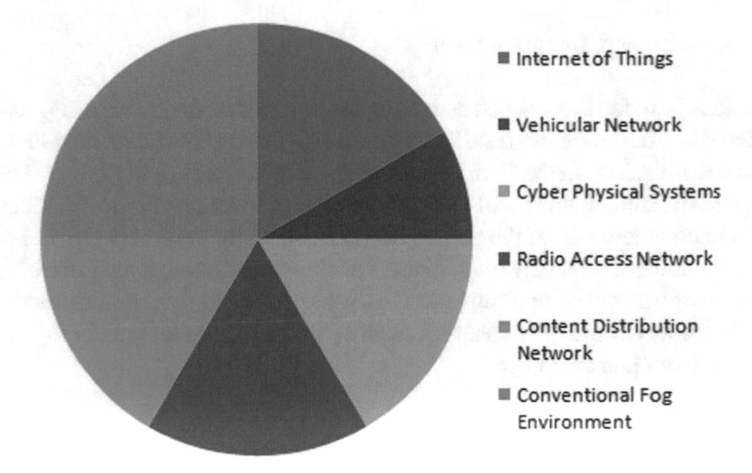

Fig. 3 Application domains harnessing the power of fog computing systems

complexity of coordination among the fog resources. Though there are a few research works attempting to solve many of the management issues in the fog environment, there are many unresolved issues, offering great opportunity for future researchers. We list a few such challenges in the next subsection. An analysis of the proportion of research contributions to the different application domains of Fog has been shown in Fig. 3.

4.1 Challenges and Future Research Directions

4.1.1 Limited Capacity and Heterogeneous Nature of Resources

Fog nodes have limited capacity, and they are heterogeneous in nature. Unlike cloud computing resources, these behaviors create an additional burden on scheduling and managing methods. Fog nodes have different Random Access Memory (RAM) capacity, Central Processing Unit (CPU) computing performance, storage space and network bandwidth. Fog management systems should be able to handle all these customized hardware and personalized user requests. Some applications may run only on hardware with specific characteristics, and many times the fog nodes have limited capacity. Development of a fog managing component in such a complex environment that meets all user needs presents many challenges.

4.1.2 Security and Integrity Issues

End devices transfer the collected data to the fog nodes for processing, and they make decisions based on the result obtained from the fog. Intruders may be present between transmissions in the form of fake networking devices or fog nodes. They can intercept confidential data of real-time applications and make appropriate decisions. Attacks aimed at any one of the nodes pose data integrity problems [15]. Malicious codes in the nodes are usually identified using signature-based detection techniques, but because of the low computing power, execution of such detection methods is not feasible in IoT environments. Ensuring security and integrity for data in fog environments is still an open challenge.

4.1.3 Fault Diagnosis and Tolerance

Fog resources are geographically distributed in nature. Scaling of these systems increases complexity and the likelihood of failures. Some minor errors can be detected that were not identifiable on a small scale, and testing may cause a significant reduction in performance and reliability in the scaled-up systems. There should be proper mechanisms to incorporate these failures without affecting the performance of the system. Developers should incorporate redundancy techniques and policies for handling failures with minimum impact on performance results.

4.1.4 Cost Models for Fog Computing Environment

The recent developments in fog computing allows multiple providers to deliver computational resources based on demand as a service. Similar to the cloud pay-as-you-go model, these services should have a charge associated with them. Research in the field of cost models for the fog computing environment is still in its infancy. Creating models that can be implemented by users to find optimal providers and by providers to achieve their profit margins is of significant importance.

4.1.5 Development of Interactive Interfaces for Fog Nodes

Fog computing systems are distributed in nature and consist of a wide variety of nodes. Fog computing allows dynamic reallocation of tasks among fog nodes and between the cloud and the edge. Efficient management of the resources requires flexible interfaces from fog to cloud, fog to edge and among fog nodes. Efficient communication through the interface allows a collaborative effort to jointly support execution of an application. Such interfaces that can enable management tasks must be designed and implemented.

4.1.6 Programming Models

The heterogeneous and widely dispersed nature of the resources in the fog environment demands the right abstraction so that programmers need not handle these complex issues. Flexible application program interfaces (APIs) and programming models are needed that hide the complexity of infrastructure and allow users to run their applications seamlessly.

5 Conclusion

Fog computing is an emerging trend which has been the recipient of increasing interest from industrialists and academia alike. Fog can be perceived as an extension of cloud services near the edge of the network. Fog brings cloud services to end users and improves QoS. Fog consists of a group of near-end user devices called fog nodes. Fog nodes or fog servers are placed at different levels between the cloud and the edge of the network to enable efficient data storage, processing and analysis of data, and they reduce latency by limiting the amount of data transported to the cloud. We compared different analytic platforms including cloud, edge and fog, and also investigated the challenges in deploying fog computing environments in different application scenarios.

References

1. Ahmed, A., and E. Ahmed. 2016. A survey on mobile edge computing. In *IEEE 10th International Conference on Intelligent Systems and Control (ISCO)*, 1–8.
2. Arkian, H.R., A. Diyanat, and A. Pourkhalili. 2017. Mist: Fog-based data analytics scheme with cost-efficient resource provisioning for iot crowdsensing applications. *Journal of Network and Computer Applications* 82: 152–165.
3. Barcelo, M., A. Correa, J. Llorca, A.M. Tulino, J.L. Vicario, and A. Morell. 2016. Iot-cloud service optimization in next generation smart environments. *IEEE Journal on Selected Areas in Communications* 34 (12): 4077–4090.
4. Byers, C.C. 2015. Fog computing distributing data and intelligence for resiliency and scale necessary for iot the internet of things.
5. Chen, X., and L. Wang. 2017. Exploring fog computing-based adaptive vehicular data scheduling policies through a compositional formal methodpepa. *IEEE Communications Letters* 21 (4): 745–748.
6. Dao, N.N., J. Lee, D.N. Vu, J. Paek, J. Kim, S. Cho, K.S. Chung, and C. Keum. 2017. Adaptive resource balancing for serviceability maximization in fog radio access networks. *IEEE Access* 5: 14548–14559.
7. Dastjerdi, A.V., and R. Buyya. 2016. Fog computing: helping the internet of things realize its potential. *Computer* 49 (8): 112–116.
8. Datta, S.K., C. Bonnet, and J. Haerri. 2015. Fog computing architecture to enable consumer centric internet of things services. In *IEEE International Symposium on Consumer Electronics (ISCE)*, 1–2.

9. Deng, R., R. Lu, C. Lai, T.H. Luan, and H. Liang. 2016. Optimal workload allocation in fog-cloud computing toward balanced delay and power consumption. *IEEE Internet of Things Journal* 3 (6): 1171–1181.
10. Do, C.T., N.H. Tran, C. Pham, M.G.R. Alam, J.H. Son, and C.S. Hong. 2015. A proximal algorithm for joint resource allocation and minimizing carbon footprint in geo-distributed fog computing. In *IEEE International Conference on Information Networking (ICOIN)*, 324–329.
11. Fernando, N., S.W. Loke, and W. Rahayu. 2013. Mobile cloud computing: a survey. *Future Generation Computer Systems* 29 (1): 84–106.
12. Gu, L., D. Zeng, S. Guo, A. Barnawi, and Y. Xiang. 2017. Cost efficient resource management in fog computing supported medical cyber-physical system. *IEEE Transactions on Emerging Topics in Computing* 5 (1): 108–119.
13. He, Z., Z. Cai, J. Yu, X. Wang, Y. Sun, and Y. Li. 2017. Cost-efficient strategies for restraining rumor spreading in mobile social networks. *IEEE Transactions on Vehicular Technology* 66 (3): 2789–2800.
14. Jingtao, S., L. Fuhong, Z. Xianwei, and L. Xing. 2015. Steiner tree based optimal resource caching scheme in fog computing. *China Communications* 12 (8): 161–168.
15. Khan, S., S. Parkinson, and Y. Qin. 2017. Fog computing security: a review of current applications and security solutions. *Journal of Cloud Computing* 6 (1): 19.
16. Networking, C.V. 2017. Ciscoglobal cloud index: forecast and methodology, 2015–2020. white paper
17. Ni, L., J. Zhang, C. Jiang, C. Yan, and K. Yu. 2017. Resource allocation strategy in fog computing based on priced timed petri nets. *IEEE Internet of Things Journal.*
18. Ning, H., H. Liu, J. Ma, L.T. Yang, and R. Huang. 2016. Cybermatics: cyber-physical-social-thinking hyperspace based science and technology. *Future Generation Computer Systems* 56: 504–522.
19. Peng, M., and K. Zhang. 2016. Recent advances in fog radio access networks: performance analysis and radio resource allocation. *IEEE Access* 4: 5003–5009.
20. Qiu, T., R. Qiao, and D. Wu. 2017. Eabs: An event-aware backpressure scheduling scheme for emergency internet of things. *IEEE Transactions on Mobile Computing.*
21. Tordera, E.M., X. Masip-Bruin, J. García-Almiñana, A. Jukan, G.J. Ren, and J. Zhu. 2017. Do we all really know what a fog node is? current trends towards an open definition. *Computer Communications.*
22. Wang, W., Q. Wang, and K. Sohraby. 2017. Multimedia sensing as a service (msaas): Exploring resource saving potentials of at cloud-edge iot and fogs. *IEEE Internet of Things Journal* 4 (2): 487–495.
23. Weinhardt, C., A. Anandasivam, B. Blau, N. Borissov, T. Meinl, W. Michalk, and J. Stößer. 2009. Cloud computing-a classification, business models, and research directions. *Business and Information Systems Engineering* 1 (5): 391–399.
24. Wen, Z., R. Yang, P. Garraghan, T. Lin, J. Xu, and M. Rovatsos. 2017. Fog orchestration for internet of things services. *IEEE Internet Computing* 21 (2): 16–24.
25. Zeng, D., L. Gu, S. Guo, Z. Cheng, and S. Yu. 2016. Joint optimization of task scheduling and image placement in fog computing supported software-defined embedded system. *IEEE Transactions on Computers* 65 (12): 3702–3712.
26. Zhang, Y., D. Niyato, and P. Wang. 2015. Offloading in mobile cloudlet systems with intermittent connectivity. *IEEE Transactions on Mobile Computing* 14 (12): 2516–2529.
27. Zhu, J., D.S. Chan, M.S. Prabhu, P. Natarajan, H. Hu, and F. Bonomi. 2013. Improving web sites performance using edge servers in fog computing architecture. *IEEE 7th International Symposium on Service Oriented System Engineering (SOSE)*, 320–323.

A Probe into the Technological Enablers of Microservice Architectures

Christina Terese Joseph and K. Chandrasekaran

Abstract Microservice architectures (MSA), composed of loosely coupled and autonomous units called microservices, are gaining wide adoption in the software field. With characteristics that are loyal to the requirements of the Cloud environment, such as inherent support for continuous integration/continuous deployment (CI/CD), MSA are actively embraced by the Cloud computing community. Containers employing lightweight virtualization have also been increasingly adopted in the Cloud environment. The containers wrap applications along with their dependencies into self-contained units, which can be deployed independently. These features make it the unanimously accepted technology to enable seamless execution of microservices in the Cloud. With this outlook, this chapter undertakes a study on how containers may be used to support the execution of microservices. The study also includes other technologies that, in collaboration with container technologies, provide the support required for running microservices in the Cloud. An interesting concern for applications running on containers is resource management. Nevertheless, this is a significant aspect for supporting microservices as well. Such issues have been identified and research works addressing all or some of these issues, have been considered. The various relevant studies have been classified into different categories and the future directions have been identified, which can be used by researchers aiming to enhance the technological support for microservices in Cloud.

C. T. Joseph (✉) · K. Chandrasekaran
Department of Computer Science and Engineering, National Institute of Technology, Karnataka 575025, Surathkal, India
e-mail: xtina_1232@hotmail.com

K. Chandrasekaran
e-mail: kch@nitk.ac.in

A. N. Krishna et al. (eds.), *Integrated Intelligent Computing, Communication and Security*, Studies in Computational Intelligence 771,
https://doi.org/10.1007/978-981-10-8797-4_50

1 Introduction

Cloud computing has been comprehensively accepted as the distributed computing paradigm that is most appropriate for serving the current generation user requirements. There is a plethora of research work being carried out in this domain.

An emerging trend in the domain of software architecture is the microservice architecture (MSA). An architectural pattern that has its roots in service-oriented architecture (SOA) [1, 2], the MSA was introduced by software engineers in 2011. Though the practice of using "micro" services is not entirely novel, the term has been in wide use only since 2014, and it during this time that the architectural pattern has been gaining wide popularity. In MSA, each application is divided into small, atomic components, which may be independently deployed [3]. Monolithic applications contained all functionalities encapsulated in a single unit as shown in Fig. 1a. On the other hand, in microservices, the functionalities are divided among the different units such that each unit handles just one functionality. These components are self-contained and communicate with the other components to provide the functionalities of the application. These small, independent components are called microservices. The major characteristics of microservices are given in Fig. 1b.

Microservices are functional units that may be deployed independently. They expose their functionalities using well-defined interfaces and communicate through lightweight protocols usually based on HTTP [3]. The independent units are generally divided across the functional boundaries of the application. This design aspect improves the scalability by enabling the components providing functionalities with higher service requests to have a larger number of instances, rather than having to scale the entire application. Another interesting feature of the microservices is their inherent support for continuous integration/continuous deployment. Different parts of the application may be modified and deployed independently with no or fewer impacts on the other parts of the application. All these features make the architectural approach well suited for Cloud-based applications.

The microservice architecture offers several advantages over conventional monolithic architectures. These advantages may be fully exploited only when it is deployed in environments that provide adequate support for the architecture. Cloud-based applications have been typically run on virtual machines (VMs). When microservices are deployed on VMs, limitations inherent in VMs, such as virtualization overhead and degraded performance, will prevent the microservices from providing the expected benefits. Based on this observation, several researchers have explored the use of lightweight virtualization techniques, including operating system virtualization, in deploying microservices [4–7]. The contributions of this study are as follows:

- Characterize the major technologies enabling the microservice architectures and place them in the context of Cloud environments.
- Review and categorize the various studies dealing with resource management in environments composed of lightweight virtualized entities, the containers.

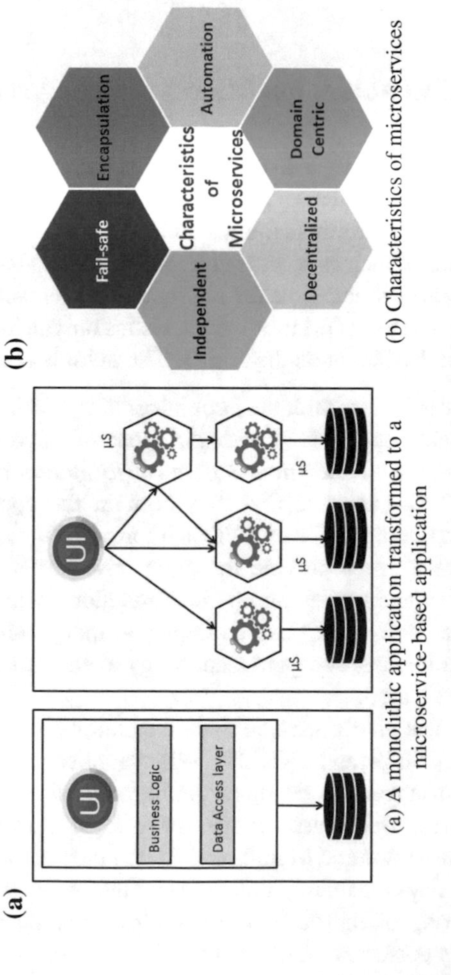

Fig. 1 **a** A monolithic application transformed to a microservice-based application. **b** Characteristics of microservices

- Identify the challenges currently existing in deriving all the promised benefits of microservices in Cloud environments which can be adopted by future researchers to enhance the support in Cloud environments for MSA.

This study focuses on technological aids for running microservices in the Cloud environment, as discussed in Sect. 2. In addition, an in-depth review of resource management issues involved with containers is presented in Sect. 3. Section 4 provides future directions for researchers, and Sect. 5 concludes the chapter.

2 Technological Enablers for Microservices Architectures

This section discusses technological developments that support the execution of microservices in the Cloud. Microservices are similar to services in terms of their operation. The difference is mainly in the size of the units. Owing to the similar functionalities, several aspects such as service discovery and load balancing are involved in providing the technical foundation for microservices as well. In addition, other technologies are adopted to support the characteristics unique to microservices, such as autonomicity and independent deployability. The notable enablers are as follows:

- Lightweight virtualization techniques, containers: Operating system containers provide a more viable option for the deployment of microservices. They have *shorter startup times* and offer *near-native performance* in several scenarios. Another feature of containers is that they contain the application and all the libraries required to run the application. This improves the *portability* of the application. These characteristics provide the agile environment required to deploy microservices. The containers wrap applications along with their dependencies into a self-contained unit, which can be deployed independently. These features make them the unanimously accepted technology to enable seamless execution of microservices in the Cloud.
- Container management platforms: The current technologies for containers include Docker [8], CoreOS Rkt [9] and LXC [10]. When applications are deployed in containers, container management becomes another significant problem. Container management systems widely used today are Google Kubernetes [11], Docker Swarm[12], Apache Mesos and Marathon [13] and the Amazon EC2 service [14]. These management systems carry out the scheduling, load balancing, configuration and monitoring of the different containers running the applications. To enhance the activity performance of such platforms, several researchers have proposed improved techniques. Section 3 classifies the research works in this domain.
- Infrastructure as code (IaC): This approach makes use of coding languages to specify the hardware and network resource configurations. The approach allows the provision of infrastructure resources in a dynamic setting. It enhances the support for continuous delivery by automating the entire provisioning process. The approach also supports various versions to record alterations made to the configuration environment. This approach provides better control of the resource provi-

sioning process, while eliminating unwanted delays involved in manually setting up the running environment.

- API gateway: The microservices are small, independent units that provide different functionalities required for an application. Clients generally submit application requests, which will in turn require the invocation of several microservice instances. The entry point for user requests will be an API gateway. The gateway will be responsible for forwarding the requests to the various constituent microservices. This provides a layer of abstraction and hides the internal details of implementation from the user.
- Service discovery: Service discovery aids the different components in locating their counterparts. In scenarios that involve dynamicity, the location of the components will not be known beforehand. When the need for communication between components (microservices) arises, the information as to where the communication must be addressed is provided through service discovery. Various technologies exist for service discovery, including the HashiCorp Consul [15], Apache ZooKeeper [16], etcd [17] and Netflix Eureka [18].
- Load balancing: A major advantage of the microservice architecture is that the functional components can be scaled without affecting the other modules of the application. This leads to scenarios where there are multiple instances of the same microservice. The details of all the instances and their network locations will be maintained in a registry, and can be extracted in the service discovery process. Once the details of all the instances have been obtained, the next step is to decide which instance can handle the current request. It is at this point that the load balancer comes into the picture. The load balancing service selects the microservice instance that can handle the incoming request while maintaining uniform distributed load among all the microservice instances. There are two types of load balancing, depending on the entity that initiates the load balancing operation: server-side load balancing and client-side load balancing. Open source technologies that provide this functionality include the Netflix Ribbon [19], Amazon elb [20] and Nginx Plus [21].
- Circuit breaker: Remote communication with distributed components can lead to situations involving unresponsiveness. These situations must be aptly handled to ensure that the effects are not exacerbated. Generally, when one component fails to receive/respond to the requests, there are possibilities of a domino effect. In order to eliminate the risk of a domino effect, the pattern of interaction between the components is structured like an electric circuit breaker, where the connections are tripped when the number of unresponsive cases exceeds a particular threshold value. Further requests to the component are returned.

As microservices are smaller, the complexity of managing several distributed units increases. Techniques that can manage and coordinate the various instances are necessary and essential [22]. A major factor that impacts the performance and implementation of the microservice architectures is the infrastructure that hosts the microservice components: the containers. Accordingly, the major issue in container technology, i.e., the resource management technique, is of significant importance in the context of microservices as well. Thus, in this study, we focus on the existing techniques for management and coordination of containers.

3 Prevailing Research Approaches for the Management of Lightweight Virtualized Entities in the Cloud

Running microservices in the Cloud has inherent challenges that may be resolved, to some extent, by the use of containers. In such cases, the orchestration of the containers to ensure successful application delivery is of prime importance. In this context, this chapter attempts to review the existing approaches for the management of various containers. The overall management includes various activities such as allocation/provisioning of containers, scheduling of containers and migrating containers among other activities. The activities can be broadly classified as shown in Fig. 2. This section focuses on the relevant works discussing approaches to coordinate the containers.

3.1 Container Deployment Approaches

Schiavoni et al. [23] propose a scheduler that can efficiently allocate containers to physical machines. The solution proposed is unique in the sense that it does not assume any prior knowledge of the workload and the resource requirements. The scheduler is inspired by the working of the garbage collector. Nardelli et al. [25] attempt to solve the container deployment problem. When compared to the existing heuristics to derive a deployment schedule for containers, Nardelli et al. propose to use integer linear programming techniques to solve the deployment problem. Since there are multiple objectives, transformation to a single optimization problem is done using the simple additive weighing (SAW) technique. The authors have validated their approach using the Csolver technique to solve the ILP. In an extended version of their work, Narddelli et al. [32] explain how the QoS conditions are met while optimizing the other requirements using their approach. Awada et al. [26] pro-

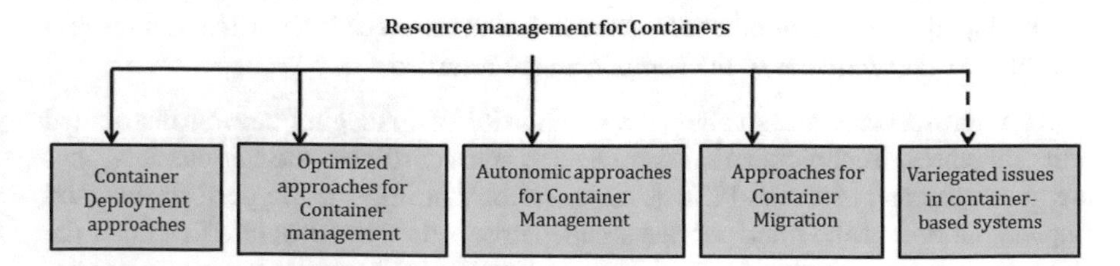

Fig. 2 Taxonomy for the works in container management

pose a deployment strategy to ameliorate the resource utilization of the nodes in the Cloud. The existing scheduling techniques do not consider the global view. Rather they schedule each incoming application to different entities. The authors propose a joint-optimization technique to make the scheduling process intuitive.

3.2 *Optimized Approaches for Container Management*

Sun et al. [24] establish the difficulties in directly applying the algorithms proposed for VM allocation in the case of containers. The main challenge is the large varieties of resource configurations available for the containers, when compared to virtual machines. The authors consider the case of web applications to be deployed onto containers running within VMs in such a manner so as to ensure that all the QoS conditions are satisfied.

3.3 *Autonomic Approaches for Container Management*

Sangpetch et al. [27] propose a solution to dynamically adjust to the unpredictable workloads requirements. They have adopted HAProxy for monitoring the requests for different metrics such as the CPU, memory utilization and the time taken to service the requests. Based on experimentation results, the authors observe that Q-Learning is best suited to capture the dynamics of the environment. Based on the study conducted by Toffetti et al. [29] it is observed that the monitoring of the applications is best done from within the applications itself. Toffetti et al. propose that monitoring may be done as part of the application. Monitoring involves the collection of metrics related to the resource level, as well as metrics specific to the application.

3.4 *Approaches for Container Migration*

Nadgowda et al. [28] identified the challenges specific to migration of containers. Despite the existence of several established techniques for migration of virtual machines, there needs to be focus on new techniques tailored for the requirements of containers, owing to the characteristics of containers such as being more agile and having lesser life spans. The commonly employed technique to migrate containers is the Checkpoint/Restore In Userspace (CRIU) technique, introduced by Virtuozzo [www.criu.org/]. The authors remark that, in addition to the in-memory data space, several other data points must be considered during the process of migration. According to their proposed approach, the in-memory data transfer is done using the page-server flavor of the CRIU, during which the container is frozen. This is the major step in the migration process impacting the application downtime in the migration

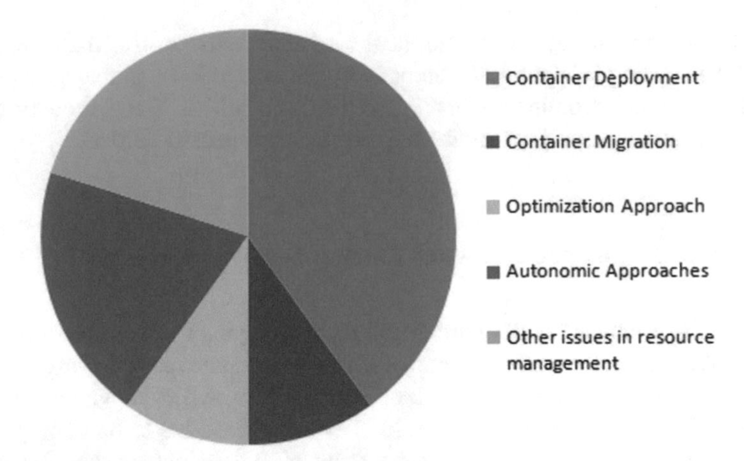

Fig. 3 Distribution of works in container management

phase. Later, local file-system is shared in a two-way transfer process employing data federation and then followed by lazy replication. In the next step, all the network file systems are unmounted and mounted at the target host.

3.5 Variegated Issues in Container-Based Systems

de Alfonso et al. [30] demonstrate the viability of containers to run scientific workloads, which have high throughput requirements. Earlier physical machines were combined together to form clusters to provide the resource requirements of scientific applications. Owing to the high start-up time of virtual machines and the performance overhead, the authors suggest the use of containers to deploy the applications. The elastic nature is incorporated with the use of CLUES manager. Kratzke attempts to enable multi-cloud deployments of cloud-native applications [31]. The inherent nature of cloud-native applications is that it is developed for a specific Cloud provider. The author argues that the applications may be migrated across various Cloud providers when required.

Table 1 summarizes the research works considered in this chapter. The research works attempt to solve various challenges such as deployment, optimization and migration. These works are very recent and there are very few journal articles, implying a low level of maturity. Many approaches have been proposed for deployment (or allocation), but the other challenges are relatively less considered, as is observable from Fig. 3. Based on these works, it is evident that the management of containers is still in its initial phase of maturity. There is a need to devise approaches or frameworks for the efficient management of containers.

Table 1 Summary of prevailing approaches for resource management

Paper title	Problem statement	Focused aspect	Solution methodology	Limitations	Year of publication
GenPack: A generational scheduler for Cloud data centers [23]	Scheduling framework for containers placement and migration in Cloud data centers	Container scheduling/re-scheduling	Monitoring framework used to derive workload resource profile+partition servers into different generations	The proposed work may not offer much savings in cases where the majority of applications are short-lived. Considers few classes of container configurations	2017
Automated QoS-oriented Cloud resource optimization using containers [24]	Automated tool to generate the optimized resource configuration based on the given static set of applications and their QoS requirements	Application deployment	Bin-pack driven benchmarking; divisible bin packing; co-location interference calculated empirically	The proposed work does not support dynamic allocation	2017
Elastic provisioning of virtual machines for container deployment [25]	Determine the container deployment on virtual machines, while optimizing quality of service (QoS) metrics	Container scheduling	Formulate problem as an ILP problem, solved periodically; multi-objective transformed to single objective using SAW	QoS parameters considered are very less, microservice specific objectives have not been considered; the time complexity to get the optimal answer may be high	2017

(continued)

Table 1 (continued)

Paper title	Problem statement	Focused aspect	Solution methodology	Limitations	Year of publication
Improving resource efficiency of container-instance clusters on Clouds [26]	A scheduling algorithm to solve optimal deployment of multi-container units on best fit container-instances across distributed Clouds, to maximize all available resources, speed up completion time and maximize throughput	Container scheduling	Formulate problem as a joint optimization problem;	While deploying, interference among multiple containers allocated on same host is not considered	2017
Thoth: Automatic resource management with machine learning for container-based Cloud platform [27]	A dynamic resource management system for PaaS, which automatically monitors resource usage and dynamically adjusts appropriate amount of resources for each application	Autoscaling	Three machine learning algorithms used: NNs, Q-Learning and Rule-based; log collector collects application request statistics and service time in addition to physical resource utilization	The frameworks performance has not been compared with existing systems not employing machine learning; no trade-off analysis has been performed, centralized approach with a single point of failure	2017
Voyager: Complete container state migration [28]	Just-in-time live container migration service, designed in accordance with the open container initiative (OCI) principles that provides consistent full-system migration	Container Migration	Three data points identified: in memory transferred using page-server mode of CRIU, local file-system transferred using data federation and lazy replication, network file system transferred using mount and unmount	Impact on network has not been considered	2017

(continued)

Table 1 (continued)

Paper title	Problem statement	Focused aspect	Solution methodology	Limitations	Year of publication
Self-managing cloud-native applications: Design, implementation, and experience [29]	High-level distributed architecture that can be used to implement self-managing cloud-native applications	Monitoring	Applications modeled using type graphs and instance graphs for global view; endpoint obtained from etcd	May not generate optimal solution; etcd may not be dependable,	2017
Container-based virtual elastic clusters [30]	Explore viability of containers to run scientific workloads which have high throughput requirements	Container-based cluster system	CLUES manager integrated; Case study approach for validation	The approach has not been tested for scalability	2017
Smuggling Multi-cloud support into Cloud-native applications using elastic container platforms [31]	enable multi-Cloud deployments of Cloud-native applications	Container-based multi-Cloud	Control process combining project-based and region-based scaling approach of container platforms	Though control process results in operational state. Optimal configuration may not be obtained	2017

4 Discussion and Future Research Directions

The domain considered provides much scope for future research. This is clearly evident from the large number of conference/workshop publications available on the topic, with very few journal articles. There is undoubtedly a need for researchers to focus on several aspects in the domain. The research works discussed here are still in the evolving stage. Though these works are on the path to solve the challenges, there is still much scope for future research. Based on the authors' understanding, the major issues that need to be resolved are as follows:

- **Energy and power consumption models for container-based systems**: The various parameters that impact the energy efficiency of the system will have to be considered. Researchers will have to analyze the effect of the parameters on the overall power consumption of the system. Based on these observations, the relationship between the parameters will have to be modeled, which will help in deriving the energy efficiency of the system.
- **Security considerations for containers**: A major concern surrounding containers is the security level. In general, the research on security for containers is in its infancy. In addition to this, there are various security issues that are dependent on the container technology employed. The underlying hosts will also have to be secured. Resource level isolation has to be ensured. Since the isolation level is less for containers when compared to virtual machines, the attack surface and the threat models will vary.
- **Financial models for systems adopting containers**: In commercial environments such as the Cloud environments, all stakeholders will aim to reap the maximum profit. The optimum level where all stakeholders suffer a minimum loss will have to be obtained. Cooperative or non-cooperative approaches, game theoretic approaches, financial option-based approaches or auction-based approaches may be used to derive financial models. All resources will have to be charged adequately as per the usage of the resources for each customer. The cost efficiency of the different offerings may also be calculated using the model.
- ***-Aware Resource Management Frameworks**: The approaches for resource management will have to consider various constraints. The temporal constraints for the activities to be carried out, the expected QoS behavior, the energy-efficiency and the overhead on the network bandwidth, are all parameters that must be considered while deciding on a resource allocation strategy.

5 Conclusion

Microservices architectural pattern has been attracting interest from software engineers. Applications are now designed as small, independent units that expose their functionalities through well-defined interfaces. Communication among the various microservices in the system is also through interfaces. A lot of research work has been carried out on designing microservices, where the software architectural aspects have been considered. This study attempts to review a lesser focused aspect: the technological support for the execution of microservices. A characterization study of the various technological tools required for supporting microservices has been carried out. Based on the observation that lightweight virtualization is most suitable for hosting microservices, a review of works dealing with resource management for lightweight virtualized entities has also been performed. The domain has many unresolved issues and offers much possibility for future research. This study also highlights a few major research directions in the domain, which will help future researchers to contribute effectively for enhancing the support for microservices.

References

1. Richards, M. 2015. *Microservices vs. service-oriented architecture*. O'Reilly Media.
2. Richardson, C. 2016. *Microservices-from design to deployment*. NGINX, Inc.
3. Newman, S. 2015. *Building microservices: designing fine-grained systems*. O'Reilly Media, Inc.
4. Amaral, M., J. Polo, D. Carrera, I. Mohomed, M. Unuvar, and M. Steinder. 2015. Performance evaluation of microservices architectures using containers. In *IEEE 14th International Symposium on Network Computing and Applications (NCA)*, 27–34. IEEE.
5. Kang, H., M. Le, and S. Tao. 2016. Container and microservice driven design for cloud infrastructure devops. In *IEEE International Conference on Cloud Engineering (IC2E)*, 202–211. IEEE.
6. Peinl, R., F. Holzschuher, and F. Pfitzer. 2016. Docker cluster management for the cloud-survey results and own solution. *Journal of Grid Computing* 14 (2): 265–282.
7. Stubbs, J., W. Moreira, and R. Dooley. 2015. Distributed systems of microservices using docker and serfnode. In *7th International Workshop on Science Gateways (IWSG)*, 34–39. IEEE.
8. Docker. 2017. https://hub.docker.com/.
9. Rkt, C. 2017. https://coreos.com/rkt.
10. Containers, L. 2017. https://linuxcontainers.org/.
11. Kubernetes. https://kubernetes.io/.
12. Swarm, D. 2017. https://github.com/docker/swarm.
13. Mesos, A. www.mesos.apache.org/.
14. ECS, A. 2017. https://aws.amazon.com/ec2/.
15. Yu, H.E., and W. Huang. 2015. Building a virtual hpc cluster with auto scaling by the docker. arXiv:1509.08231.
16. Zookeeper, A. https://cloud.spring.io/spring-cloud-zookeeper/.
17. CoreOS/etcd: (Oct 2017), https://github.com/coreos/etcd.
18. Registration, M., D. with Spring Cloud, and N. Eureka. 2015. https://spring.io/blog/2015/01/20/microservice-registration-and-discovery-with-spring-cloud-and-netflix-s-eureka.
19. Netflix: Netflix ribbon. 2017. https://github.com/Netflix/ribbon.
20. Amazon elb. https://aws.amazon.com/elasticloadbalancing/.
21. https://www.nginx.com/products/nginx/load-balancing/.
22. Olliffe, G. 2017. *Microservices: Building services with the guts on the outside*. Microservices: Building Services with the Guts on the Outside. https://blogs.gartner.com/gary-olliffe/2015/01/30/microservices-guts-on-the-outside/.
23. Havet, A., V. Schiavoni, P. Felber, M. Colmant, R. Rouvoy, and C. Fetzer. 2017. Genpack: A generational scheduler for cloud data centers. In *IEEE International Conference on Cloud Engineering (IC2E)*, 95–104. IEEE.

24. Sun, Y., J. White, B. Li, M. Walker, and H. Turner. 2017. Automated qos-oriented cloud resource optimization using containers. *Automated Software Engineering* 24 (1): 101–137.
25. Nardelli, M., C. Hochreiner, and S. Schulte. 2017. Elastic provisioning of virtual machines for container deployment. In *Proceedings of the 8th ACM/SPEC on International Conference on Performance Engineering Companion*, 5–10. ACM.
26. Awada, U., and A. Barker. 2017. Improving resource efficiency of container-instance clusters on clouds. In *Proceedings of the 17th IEEE/ACM International Symposium on Cluster, Cloud and Grid Computing*, 929–934. IEEE Press.
27. Sangpetch, A., O. Sangpetch, N. Juangmarisakul, and S. Warodom. 2017. Thoth: Automatic resource management with machine learning for container-based cloud platform. In *CLOSER 2017 - Proceedings of the 7th International Conference on Cloud Computing And Services Science*, April 24–26, 75–83. Portugal: Porto.
28. Nadgowda, S., S. Suneja, N. Bila, and C. Isci. 2017. Voyager: Complete container state migration. In *IEEE 37th International Conference on Distributed Computing Systems (ICDCS)*, 2137–2142. IEEE.
29. Toffetti, G., S. Brunner, M. Blöchlinger, J. Spillner, and T.M. Bohnert. 2017. Self-managing cloud-native applications: Design, implementation, and experience. *Future Generation Computer Systems* 72: 165–179.
30. de Alfonso, C., A. Calatrava, and G. Moltó. 2017. Container-based virtual elastic clusters. *Journal of Systems and Software* 127: 1–11.
31. Kratzke, N. 2017. Smuggling multi-cloud support into cloud-native applications using elastic container platforms. In *7th International Conference on Cloud Computing and Services Science (CLOSER 2017)*, 29–42.
32. Nardelli, M. 2017. Elastic allocation of docker containers in cloud environments. In *ZEUS*, 59–66.

Smart Poster for Tourism Promotion Through NFC Technology

Satya Naraparaju, Pranavi Jalapati and Kalyani Nara

Abstract Near-field communication (NFC) technology has been widely used in various applications over many years. The technology has been embedded in a series of devices and has not yet been completely taken advantage of. One such application is smart posters. With the growth of NFC-equipped smart phones, smart posters can be used as a low-cost marketing strategy in the field of tourism. This work proposes to utilize NFC tags to develop intuitive advertisements. The chapter further describes the feasibility with an architectural layout.

Keywords Tourism · Posters · NFC · Cloud · No-SQL · Firebase

1 Introduction

Near-field communication (NFC) is the next big disruptive technology after Bluetooth communication. It has proved its efficiency factors when blended with the existing technologies. This technology utilizes ambient intelligence to seamlessly interact with computing devices. Smart posters are the future technology for advertisements employed by tourism departments. These posters can be used in providing information, marketing campaigns, monument directions, as well as elderly services. *NFC World* has reported that an estimated 64% of cellular handsets will be NFC-enabled by 2018 [1]. This fact is represented in Fig. 1.

Also, this being an eco-friendly solution, it is a widely accepted technology that saves costs spent on stationery such as receipts, paper tickets, etc. Further, any anticipated technology recently launched in the market adds interest, especially in

S. Naraparaju (✉) · P. Jalapati · K. Nara
G. Narayanamma Institute of Technology and Science, Hyderabad, India
e-mail: satyanaraparaju@gmail.com

P. Jalapati
e-mail: pranavi.jalapati@gmail.com

K. Nara
e-mail: narakalyani3@gmail.com

A. N. Krishna et al. (eds.), *Integrated Intelligent Computing, Communication and Security*, Studies in Computational Intelligence 771,
https://doi.org/10.1007/978-981-10-8797-4_51

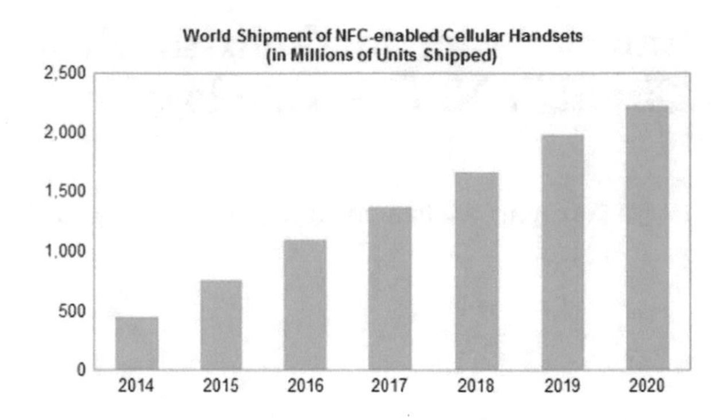

Fig. 1 Estimation of NFC-enabled cellular handset growth

the tourism industry. In short, the technology has the potential to succeed owing to its ease of use.

Data stickers attached to any object allows the users to access information simply by touching the poster with their NFC-enabled device. Smart posters can be as simple as a poster to as large as a billboard. It is a representational medium for the transfer of information. The NFC stickers are attached to objects to allow the users to interact with the service by touching the tag with their phone. Hence, simple posters, billboards and the like can be digitalized with such technology. With the increasing number of NFC-enabled devices, smart posters can attract more users [2]. For the transition period, quick response (QR) codes can be an identification technique that could be used along with NFC technology. QR code functions just as the NFC tags and use built-in cameras to identify the barcode pattern and retrieve information.

2 Existing System

Tourism marketing is more of an art that involves scenic projection. It requires officials to spread information about the locality and, most importantly, the specialty of their location. They must be keen on capturing the interest of tourists with what they have to offer to the right audience. Therefore, marketing involves the process of planning, pricing and, most significantly, promotion and distribution of ideas. Hence, overall economic conditions can have a significant impact on tourism markets. Sometimes, though a marketing strategy may be effective, it might not be feasible to implement it due to the required investment. Unfavorable economic conditions may arise due to low employment rates and local community growth. As a result, local business will not be able to promote their tourism.

With the advances in technology, new recreational products are being introduced. These technologies offer tourism business ways to significantly reduce cost and improve service quality. For instance, advances in telecommunications have led to new promotional opportunities. Further, with the advent of the internet, social media has helped promote the cultural heritage across the world. Therefore, these technological innovations have created jobs and provided a better economy. However, promotion is still a major issue for lower-income states. These states cannot afford to telecast commercials or launch campaigns to spread information. Hence, NFC technology is the latest technology that could help increase competitiveness. The objective of such technology helps states express their strengths with their intuitive technology at cheaper investment [3]. Many tourist locations cannot provide complete information about their resources through their media to attract the tourists. Without complete information, it is difficult to make decisions in the planning process. Hence, it is their responsibility to consistently lure tourists with the help of smart posters. Not only smart posters, the sites of major tourist attractions can be equipped with NFC-enabled posters to guide tourists across the city. This helps in bridging the language barrier as well.

3 NFC Technology

NFC is a short-range, wireless technology. It is primarily based on radio frequency identification (RFID) technology. NFC enables two-way interaction between electronic devices which allows users to perform contactless data transmissions. This technology can be used in three different modes including card emulation mode, reader-writer mode and peer-to-peer mode. The major advantage of NFC is that it offers an easy way to interact with consumers. Moreover, the average data retrieval time is faster when compared to RFID technology. The radio frequency (RF) signal transmission in the case of NFC depends on magnetic/electrostatic coupling between the devices. These devices operate on low electric or magnetic fields due to the short range of communication. NFC stickers are an important application of NFC devices and contactless cards. The stickers are self-adhesive and easily manageable. They can reduce the cost of production as well. Further, existing RFID technology can be incorporated as it and NFC technology are compatible. NFC technology can also play a role in student tracking, elderly aids, etc. Currently, NFC applications are mostly driven towards business environments such as payments, e-tickets, etc. Current technology has matured enough to handle bidirectional communication between NFC-enabled devices. This enables implementing various operation modes permitting payments using standard protocols [4, 5]. Lee et al. [6] proposed an NFC touch and control platform (NTCP) system for easy integration of NFC technology.

4 Proposed Model

4.1 Ideology

NFC technology has a great potential in tourism industry. These tags can be used as posters which can serve a variety of purposes. The integration of NFC tags with paper media enhances the information delivered by each marketing attempt. They tend to be more intuitive leading to more curiosity. The digitalization of information leads to new opportunities for the customer and the service provider as well [7, 8]. Smart posters could be considered an eco-friendly alternative as the content can be dynamically updated to avoid the reprinting of material. Further, these are low in cost as compared to dynamic LED displays. The existing research on NFC application is mainly focused on service-providing applications. Boes et al. [1] emphasizes the broad reach of smart posters, citing the Unified Theory of Acceptance. Lamsfus et al. [9] discusses the role of information technology in tourism industries, noting that smart posters will increase mobile marketing and information transfer. In addition to advertisements, NFC technology has been implemented from bulletin boards. Smart posters can be developed for similar implementation. Audy and Hansun [10] tested the smart poster technique in a university setting. Also, mobile bulletin systems have been implemented using a salt tokenization method to represent categorical information [11, 12]. The flow of implementation is illustrated in Fig. 2.

Fig. 2 Build procedures

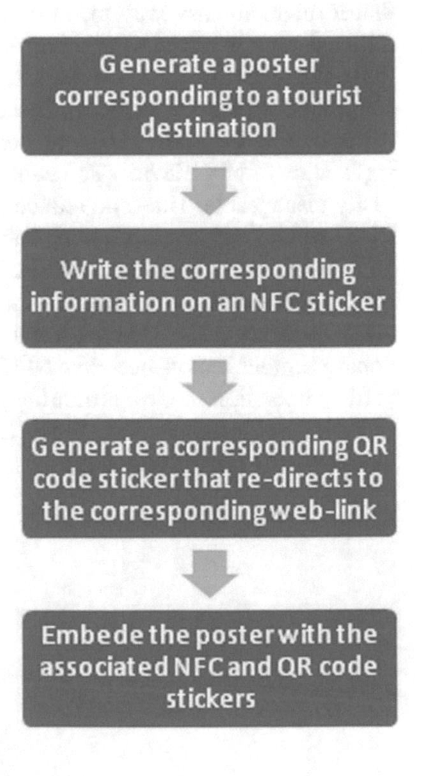

4.2 Basic Layout

The proposed system has an elegant architecture support at its root. The design has a definite set of requirements, as follows:

- The designed interface must be easy to use and navigate.
- The module must support offline data communication.
- As NFC is yet to gain significant publicity, an alternative must be available.
- The response latencies must be low.
- The application must be rich enough in containing all the information required for knowledge and navigation.

The model of the designed system is composed of the following modules as shown in Fig. 3:

- An app interface
- NFC tags
- Cloud storage

Not all phones are NFC-enabled. Hence, the application developed must consist of both NFC and QR code facilities to satisfy the current use cases. Initial prototyping of the app is based upon the end customer's readability criteria.

NFC application must be supported by a back-end database to support content management. The back end for the smart posters requires two major components including an interface for the administrators to update the changes as well as an application program interface (API). The content of smart posters can be modified by authorized staff.

There are various types of posters. For instance, the user can tap on an image to explore more about the image. This is a descriptive poster in which the user can be forwarded to a web link. On the other hand, there is a poll poster in which the

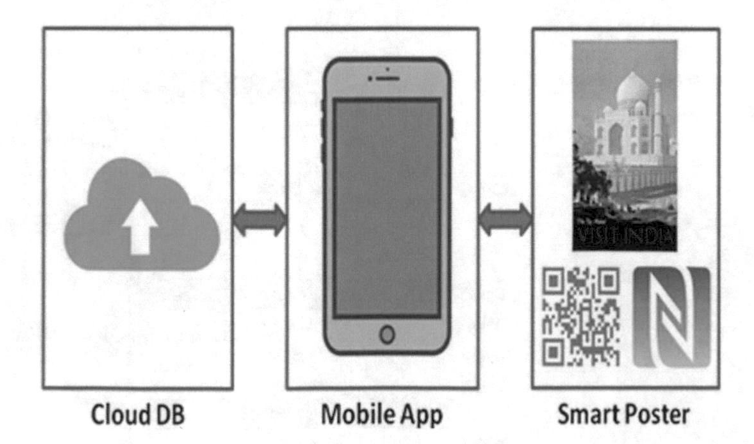

Fig. 3 Model architecture

general tendency of the crowd can be gathered. Hence, the administrator is responsible for gathering the responses, as well as setting the validity. The user can also custom search the places based on general or specific preferences.

4.3 Architecture

The key features that stand as the backbone of the project are the cloud database and NFC tags. The interface serves as the bridge between the cloud and the tags. The functional specifications of the three modules can be classified into the following three modules.

NFC Tags on the Poster: The NFC stickers corresponding to each poster are encoded with a unique identifier, name of the destination and its location. Each attribute is stored in a separate block of the second sector. The size of each block is 16 bytes. The structure of the data stored in the tag is represented in Fig. 4.

The unique ID serves as the primary key for poster identification and corresponding search in the database. The key is a combination of the city name, pin code and the allocated poster number.

AGRA28200121

Example 1: Sample poster ID

Example 1 illustrates a sample poster ID. The first four letters denote the place (AGRA). The next six digits indicate the pin code (282001) and the last two digits are the poster ID (21). This gives the crucial information about the location and is also easy to sort the posters.

Cloud Architecture: The cloud solution proposed for the architecture is non-sequential owing to its potential to satisfy the given constraints. The interface has a set of key constraints as follows:

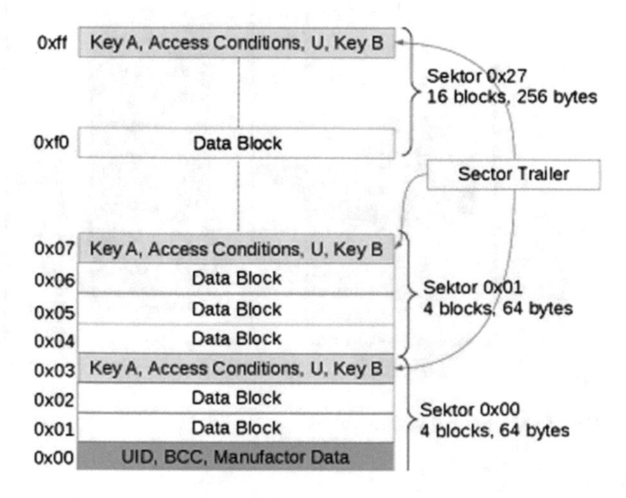

Fig. 4 Data sectors in an NFC tag

Fig. 5 Structure of data in cloud storage

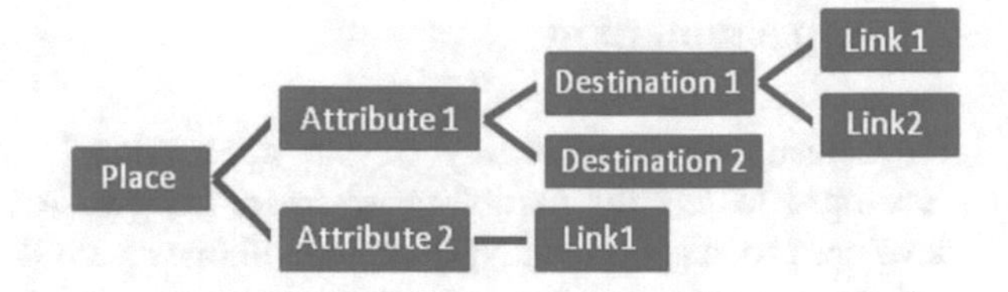

- The cloud access speed and the data retrieval latencies must be as low as possible.
- The attribute data is deeply classified.
- The architecture must be scalable.
- Load balancing is crucial.

In order to meet the imposed constraints and restrictions, data stored as nested objects have an advantage over relational data. They are highly scalable and offer low latencies. The architecture of the data is illustrated in Fig. 5.

The access to cloud is first triggered when a user scans an NFC card. The application establishes connection with the cloud and further search is driven by user preferences nested under categories or attributed. The poster ID serves as the primary key for search constraint. A secondary hash table is maintained in which every ID has a corresponding endpoint for redirection. This model is described in Fig. 6.

Mobile Interface: The mobile interface is the communication line between the NFC tags and the cloud database. On tapping the card, the primary details stored on the sector 2 blocks are retrieved. The name and location are meant for the for poster description in the case of offline usage. The poster ID is used as the primary key for navigating through the database. On connecting with the end point, the interface projects the various categories that are relevant to the poster. The pages are dynamically generated based on the user-chosen options. The poster ID is temporarily stored in the local cache of the phone for further reference in cases such as offline search.

Poster Id →

Poster Id	End-Point
XXX15...21	gov-bc15.firebaseio.com/XXX
XXX24...13	gov-bc51.firebaseio.com/XXX
⋮	⋮
XXX97...62	gov-bc51.firebaseio.com/XXX

Fig. 6 Secondary hash table

5 Implementation

To estimate the functionality of the application, a mock-up application was developed to test the feasibility of smart posters. An Android application was developed to support both NFC-enabled phones and QR code detection. The other features supported by the application are custom search and navigation history.

Fig. 7 Navigation interface

5.1 Navigation Interface

The first screen of the application provides various options to scan, custom search and navigate through previous searches. Figure 7 illustrates the navigation interface.

5.2 Scan Interface

The functionality of the scan option is to interact with the smart poster. It is equipped with options for QR code scanning and NFC tapping. This ensures that

Fig. 8 Scanning interface

the application can also serve non-NFC-enabled devices. The interface for scanning is illustrated in Fig. 8.

5.3 Search Interface

The search option comes in handy when the user wants to custom search any place and navigate to that destination. On searching for a destination, the historic details of the place are retrieved from the database. The navigation history of the user too is recorded in the cache for quick reference. Figure 9 illustrates the search option.

Fig. 9 Search interface

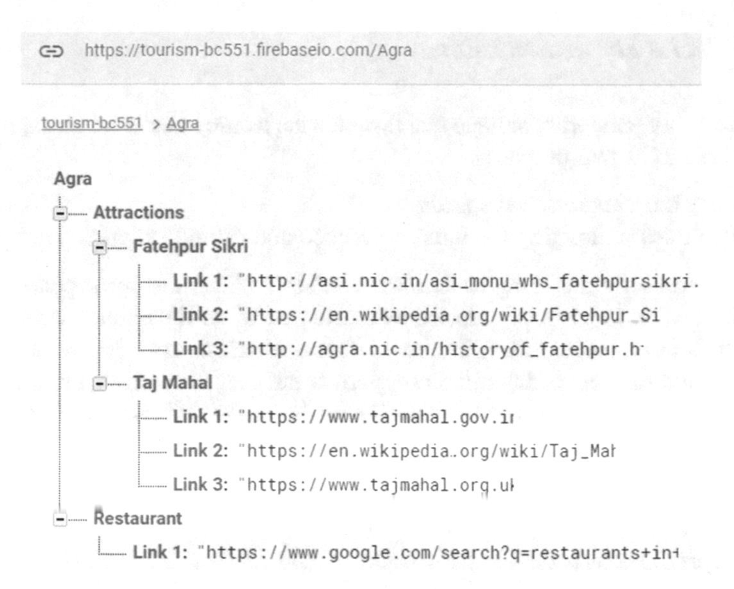

Fig. 10 Cloud layout

5.4 *Cloud Layout*

The structure of the mock cloud replicates the cloud architecture discussed above. It was implemented through the open-source Firebase cloud adhering to the performance constraints. Figure 10 is an exemplification of the cloud structure demonstrated using a Firebase console.

6 Performance Enhancers

6.1 *Database Architecture*

The structure of the database plays a key role in defining the efficiency of the entire module. The structuring of the database is aimed at optimizing the read latencies. Another important feature is load balancing as at any point of time, multiple scans need to be handled. Firebase has an inbuilt support feature to address this problem. Another key point is addressing the data redundancy. It is very likely that multiple posters would bank on similar kind of information and, as a consequence, are led to a common end point. The secondary hash table is meant for these common points and, in turn, optimizes the data storage. The nature of data is gathered from web sources and not physically maintained. This has multiple advantages like freshness of data, low maintenance cost and easy access.

6.2 Cache Implementation

A cache is any temporary storage that throttles the latency peaks. In this application, the cache serves two purposes:

- Reduce the data retrieval speeds
- Support search logging for future reference and offline support.

Once the user scans a tag or custom searches a place, the corresponding poster ID or the search key is logged in the temporary cache of the phone. In cases where the user is disconnected from the internet, the search or scan is stored for future reference and thus providing offline support. If the user is connected to the internet, the search results are logged temporarily in the cache for faster retrieval in consequent searches. Also, this significantly reduces the load on the server.

6.3 App Analytics

Recording the application statistics is an effective way to analyze the feasibility of the application. Firebase has an inbuilt analysis tool that operates on user interaction statistics and generates corresponding reports. The results generated by analyzing all the concerned factors help in building a better prediction model and thus contribute to the revenue generated by tourism. The data logged can be used to generate a genre of results as follows:

- Poster efficiency analysis
- Tourism rate in a given region
- Pulse of the tourists
- Revenue estimation
- Potential for enhancement
- Tourist plan analysis

7 Conclusion

NFC smart posters are an effective means of communication in the field of tourism at a lower investment. Therefore, they must be supported by local governments to promote their tourist attraction. They provide quality information that can be easily uploaded. The module is backed up by open-source technologies that add to its feasibility. The scalability of the solution has been ensured through the use of trending methods like object storage and cloud storage. The proposed project has a wide scope that can be explored sequentially by deploying it as a real-time system.

References

1. Boes, K., L. Borde, R. Egger, I. Tussyadiah, and A. Inversini. 2015. *The acceptance of NFC smart posters in tourism*. Springer, Cham: ICTT.
2. Yeo, Hexi, and Phooi Yee Lau. 2016. Mobile *mBus* system using near field communication. CIIS, Springer, Cham. https://doi.org/10.1007/978-3-319-48517-1_7.
3. Karmazín, J., and P. Očenášek. 2016. The state of near-field communication (NFC) on the android platform. In *International conference on human aspects of information security, privacy, and trust*, Springer, Cham. https://doi.org/10.1007/978-3-319-39381-0_22.
4. Shobha, Nahar Sunny Suresh, Kajarekar Sunit Pravin Aruna and Manjrekar Devesh Prag Bhagyashree. NFC and NFC payments: A Review. In *ICTBIG 2017*, IEEE. https://doi.org/10.1109/ictbig.2016.7892683.
5. Lopez, Ana M., Gloria Femandez and Francisco Burillo. Oct 2016. Mobile computing with near field communication (NFC) smart objects applied to archaeological research. IEEE Xplore, ISSN 2472–7571. https://doi.org/10.1109/ie2016.22.
6. Lee, Wei Hsun, Chien Ming Chou and Wei Cheng Chen. 2017. Design and implementation of an NFC-based universal touch and control platform. IEEE Xplore.
7. Coskun, Vedat, Kerem Ok and Busra Ozdenizci. 2012. Developing NFC applications. IEEE Xplore. https://doi.org/10.1002/9781119965794.ch5.
8. Vollan, Dirk, Kay Noyen, Onur Kayikci, Lukas Ackermann and Florian Michahelles. 2012. Switching the role of NFC tag and reader for the implementation of smart posters. IEEE Xplore. ISBN:988-1-4673-1209-7. https://doi.org/10.1109/nfc.2012.11.
9. Lamsfus, C., D. Martín, A. Alzua-Sorzabal, E. Torres-Manzanera, I. Tussyadiah, and A. Inversini. 2015. *Smart tourism destinations: An extended conception of smart cities focusing on human mobility*. Springer, Cham: ICTT.
10. Audy, Marcel Bonar Kristanda and Seng Hansun. Smart poster implementation on mobile bulletin system using NFC tags and salt tokenization. https://doi.org/10.1109/icsitech.2016.7852625.
11. Ayu, Media A., Teddy Mantoro, Siti Aisyah Ismail and Nurul Syafiqah Zulkifli. 2012. Rich information service delivery to mobile users using smart posters. In *DICTAP*. https://doi.org/10.1109/dictap.2012.621534.
12. Biader Ceipidor, U., C.M. Medaglia, V. Volpi1, A. Moroni1, S. Sposato, M. Carboni and A. Caridi. 2013. NFC technology applied to touristic-cultural field: A case study on an Italian museum. IEEE Xplore. https://doi.org/10.1109/nfc.2013.6482445.

Examination of Sense Significance in Semantic Web Services Discovery

Aradhana Negi and Parminder Kaur

Abstract This chapter presents the work in progress on a hybrid approach-based generic framework for semantic web services (SWSs) discovery. The novelty of this generic framework is its pertinence and coverage for diverse semantic formalisms, natural language processing of service descriptions and user queries, classification of services, and deposition of classified services to a repository known as a concept-sense knowledge base (CSKb). This manuscript investigates the significance of senses, which are extracted either from SWS concepts or user query concepts by means of natural language processing techniques. The examination of sense significance is based upon a set of three experiments on OWLS-TC V_4. The experimental evaluation signifies that the senses together with concepts substantially improve the ultimate semantic similarity score in the match-making process.

Keywords Concept-sense knowledge base · Natural language processing
Semantic web services · SWS discovery

1 Introduction

Semantic web services (SWSs) are loosely coupled, reusable, annotated software components that semantically encapsulate the discrete functionality of an application that is distributionally and programmatically accessible over the Internet. The ultimate goal of these services is the automation of Web services by the way of machine-readable annotation of service description. These services are securing approbation due to their technical lineaments for inter-organization and large environments [1]. On the contrary, the rapidly rising level of services in quantity is making the discovery of the worthy services critical, especially when one requires

A. Negi (✉) · P. Kaur
Guru Nanak Dev University, Amritsar, India
e-mail: aradhana.csersh@gndu.ac.in

P. Kaur
e-mail: parminder.dcse@gndu.ac.in

A. N. Krishna et al. (eds.), *Integrated Intelligent Computing, Communication and Security*, Studies in Computational Intelligence 771,
https://doi.org/10.1007/978-981-10-8797-4_52

refined results with minimal manual attempts and diminutive discovery time. This service discovery task has the decisive lead in the whole Web service publish-find-bind process [2], where it retrieves the specific class of potential services from a service provider in response to the capability restraint intended by the service requester.

The prime aim of our research is the betterment of SWS discovery; so, with that stated aim, three of the most analytical background discovery processes are retrieved from the existing work: (1) how to improve the match-making process; (2) how to classify new SWSs; and (3) how to create an authentic, classified, and reusable knowledge base. The advanced strategies for these analytical processes can enhance SWS discovery. By keeping these key discovery processes in mind, the outline of a hybrid approach-based generic model for SWS discovery is proposed in Fig. 1.

The existing natural language processing (NLP)-based work on SWS discovery discusses pre-processing, disambiguation or sense selection only, while none of them stressed the practice of senses beyond sense selection, i.e. sense utilization in a knowledge base. However, this chapter accentuates the utilization of senses in the match-making process through a set of three designed experiments. The experimentation examines the following hypothesis:

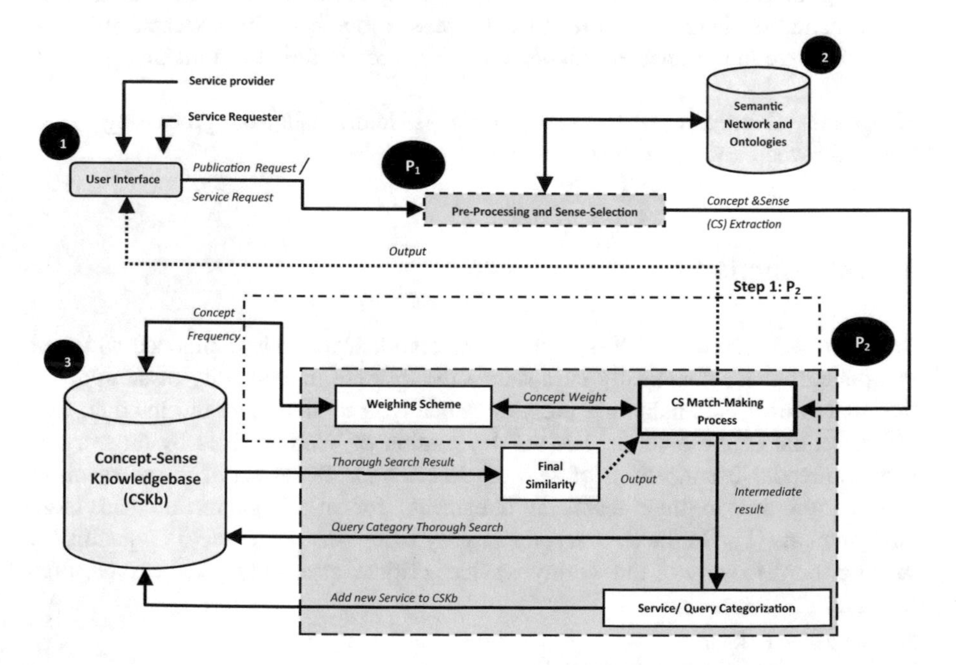

Fig. 1 Broad outline of the proposed generic SWSs discovery framework

Whether the inclusion of senses in the match-making process makes any improvement in the ultimate semantic similarity score of SWS discovery?

The chapter contains five sections. Section 2 briefly describes the related work. In Sect. 3, the hybrid approach-based generic framework for SWS discovery is proposed. The experimental evaluation and results for the hypothesis are presented in Sect. 4. Finally, the concluding remarks and outline of future work are given in Sect. 5.

2 Related Work

This section tabulates the earlier work on SWS discovery in Table 1, where the aspects like adopted *approaches*, used *semantic formalism*, *potential* of approaches and *possible extension* of reviewed frameworks have been presented. Table 1 clearly shows that not even a single work is explicitly proposed and implemented as a generic platform for available semantic formalisms, i.e. for Web service modelling ontology (WSMO), Web ontology language semantic (OWL-S), Web service description language semantic (WSDL-S), and semantic annotation for WSDL (SAWSDL). This research gap inspires us to design and implement a generic framework for SWS discovery along with the three most analytical discovery processes (as discussed in Sect. 1).

3 Proposed Work

The proposed hybrid approach-based generic framework for SWS discovery makes use of text mining, NLP and machine learning techniques. The key qualities of the proposed framework are: (a) It is not restricted to a specific semantic formalism, (b) creation of an authentic and reusable knowledge base, i.e. Concept-sense knowledge base (CSKb), (c) inclusion of senses in service classification and query service match making, and (d) to enhance query processing speed by diminishing query search space.

The numerical order as given in Fig. 1 depicts the main involved entities, whereas the dash-lined rectangle depicts the multi-level processes. Here, the discovery system takes input either as a user query from the service requester or an unclassified service/service publication request from the service provider via a common user interface. Semantic network and ontologies represent the repository of semantic relations and service terminology, respectively. The CSKb is in charge for smartly storing the categorized services and their updates.

Both the inputs use process P_1 (see Fig. 1) for their reading, parsing, and final concept-sense extraction. The extracted concepts and senses are searched in CSKb using a top-level concept matching strategy (explained in Sect. 4.2) with a CS

Table 1 Summary of related work

Ref. no.	*Approach*	*Semantic formalism*	*Potential*	*Possible extension*
[3]	Ontology mapping, service selection using QoS	OWL-S	Handling manual interoperation constraints, user query to domain specific query conversion	Experimentation for user-specific threshold values
[4]	I-Wanderer (P2P environment)	WSDL	Clustering of services, packet wandering	Computational complexity analysis
[5]	SWS clustering and discovery	General proposal (conceptual)	Clustering by capacity, complete graph construction of WS	Performance evaluation
[6]	External concept indexing for WSDL and ontological concepts	WSDL	Utilization of domain-specific ontologies, concepts and service index tables	Consideration of naive users and flexibility of user input parameters
[7]	Automatic semantic web services classifier	OWL-S	Classification and ranking of Web services	NLQP and senses inclusion
[8]	Semantics and NLP-based conceptual framework for SWS discovery	General proposal (conceptual)	Natural language query for keyword-based search and mapping of English words to ontological concepts	Pragmatic proofs for service format independence
[9]	SWS discovery using NLP techniques	WSMO	Context-based similarity measure using NLP techniques	Implementation with other semantic languages
[10]	Double-folded matching algorithm	OWL-S	Fast query processing by similarity degree calculation	Classification of services in publishing phase

match-making process as step 1 of P_2. Afterwards, the user query searches thoroughly the category which achieved the highest score in step 1, while the unclassified SWSs description is categorized to a CSKb category that scored the highest similarity in step 1. The categorized service descriptions are stored in octuple format, a designed storage format for CSKb, irrespective of their language format restrictions, while the discovered potential services in response to the user query are sent to the user interface.

The sense of a concept plays an important role in the discovery of SWSs [11]; for context-oriented language, this idea has motivated us to use senses with concept similarity. In this study, the well-organized semantic relations of WordNet [12] knowledge source are exploited. Based on WordNet semiotic relations, our match-making approach uses semantic similarity measure for nouns, verbs, adjectives, and senses for calculating the final similarity score between two set of extracted concepts. The semantic similarity measure for nouns, verbs, and adjectives utilizes a variant of semantic similarity integrating multiple conceptual relationships (SIMCR) introduced by Chen et al. [13], while the extended gloss overlapping method [14] is used as a base of measurement for sense similarity score. An overall semantic similarity (δδ) function for input (user query or unclassified service) $I_{q/s}$ and CSKb categories cat_{CSKb} is defined as:

$$\delta\delta\,(I_{q/s}, cat_{CSKb}) = \Sigma[\alpha * \acute{n}\delta\,(R_{Nc_i}, S_{Nc_j}) + \beta * \acute{v}\delta\,(R_{Vc_i}, S_{Vc_j}) + \gamma * \acute{a}\delta\,(R_{Ac_i}, S_{Ac_j}) + \acute{s}\delta\,(S_{R_{c_i}}, S_{S_{c_j}})] \quad (1)$$

where

- α, β, γ are the corresponding weights of noun, verb, and adjective. Their sum always equals 1.
- ńδ, úδ, áδ, śδ signify the similarity of nouns, verbs, adjectives, and senses, respectively.
- (R_{Nc_i}, S_{Nc_j}), (R_{Vc_i}, S_{Vc_j}), (R_{Ac_i}, S_{Ac_j}), $(S_{R_{c_i}}, S_{S_{c_j}})$ denote the noun, verb, adjective, and sense similarity inputs from $I_{q/s}$ and cat_{CSKb} respectively. The noun and verb similarity are computed using the same formula

The noun or verb similarity are calculated with Eq. (2). The notion of negative exponential semantic distance $(-S^d)$ in Eq. (2) has been employed with the fact that the lesser semantic distance leads to the greater semantic similarity. Equations (3), (4), and (5) are the complementary equations for Eq. (2).

$$\acute{n}\delta / \acute{u}\delta\,(R_{c_i}, S_{c_j}) = e^{-S^d} \quad (2)$$

$$S^d(R_{c_i}, S_{c_j}) = \sum W_{IsA}(e_i) + \sum W_{HasA}(e_j) \quad (3)$$

$$W_{IsA}(e) = \frac{1}{2 + logC(p)} * e^{-\log(D(p)+1)} \quad (4)$$

$$W_{HasA}(e) = 1 - \left[\frac{1}{2 + logC(p)}\right] * e^{-\log(D(p)+1)} \quad (5)$$

W_{IsA}, W_{HasA} in Eq. (3) denotes the weight of an “Is A” relation edge and “Has A” relation edge, respectively, between two concepts. In Eqs. (4) and (5), $C_{(p)}$ refers to the number of children nodes and $D_{(p)}$ refers to the depth of the parent node in the hierarchy.

The adjective similarity is computed using cosine similarity for five adjective relations (*rel*) of WordNet. The semantic relations *similar to*, *see also*, *derivationally used form*, and *attribute* makes one group, i.e. similarity (*sim*), whereas the relation antonym (*ant*) is a stand-alone relation which represent two opposite concepts. The similarity between two given adjectives (A_1, A_2) is calculated using Eq. (6). In Eq. (6), *rel* is antonym when $0 <= rel <= 1/(pf + 1)$ and *rel* is similar when $0 \leq rel \leq 1$. The penalty factor (pf) has the same technical implication as given in the SIMRC approach.

$$\acute{a}\delta\left(R_{Ac_i}, S_{Ac_j}\right) = \begin{cases} \frac{\cos(A_1, A_2)}{\mathrm{pf}+1} & \text{If } rel \text{ is } ant \\ \cos(A_1, A_2) & \text{else } rel \text{ is } sim \end{cases} \quad (6)$$

The sense similarity of Eq. (1) focuses on gloss-based semantic computation and is evaluated as follows:

$$\acute{s}\delta\left(S_{R_{c_i}}, S_{S_{c_j}}\right) = score[S_{R_{c_i}} \cap S_{S_{c_j}}] \quad (7)$$

The absolute sense similarity score is normalized between 0 and 1, by dividing it with the summation of all tokens presented in the participating senses from both user request and CSKb category. Both the selected match-making approaches [13, 14] are mature in their respective fields, so our preliminarily observations inspired us to use their combinational variant in proposed research work.

4 Experimental Evaluation and Result Discussion

To validate the declared hypothesis, the experiments were conducted on OWLS-TC V4 [15]. These experiments were confined to OWLS services because of the unavailability of another SWS-formatted dataset, but in the near future we intend to explore other formalisms by designing their respective Web services. The OWLS-TC V_4 dataset consists of 9 categories and 48 ontologies. The experiments have been performed in parts as: (1) experiment 1, (2) experiment 2, and (3) experiment 3. The next three sub-sections will detail the experiments and their effects.

4.1 Experiment 1

In experiment 1, we aim for determining the sense involvement in semantic similarity calculation. For experimentation, 30 (10%) services from the education (Edu.) category have been randomly selected and the experiment was executed with a split ratio of 80:20. We have trained our data on 80% while testing on remaining 20% out of the 30 services. The 30 randomly selected services are (as per occurrence order of services in Edu. of OWLS-TC V_4): 115, 107, 89, 167, 13, 201, 199, 20, 72, 8, 271, 242, 160, 39, 57, 154, 200, 114, 88, 17, 121, 184, 226, 177, 237, 62, 181, 93, 150, and 134. The randomly selected (20%) tested services are (as per occurrence order of services in Edu. of OWL-TC V_4): encyclopedia-author-book-type-service <121th> (S_1), book-author-service <57th> (S_2), academic-book-number-search <20th> (S_3), educational-organizational-lecturer <114th> (S_4), academic-degree-government-funding-service <107th> (S_5), and book-price-type-size-service <62th> (S_6). After extracting concepts (using python 2.7.13) from all selected SWSs, the proposed semantic similarity score is calculated and compared

Table 2 Services similarity with single category (Edu.)

Service	Extracted concept	SIMCR	Similarity with senses
S_1	Encyclopedia, author, book, type ...	0.788	0.874
S_2	Book, author, name, write, title ...	0.889	0.894
S_3	Academic, book, number, search, item ...	0.803	0.926
S_4	Educational, organizational, lecturer ...	0.687	0.801
S_5	Government, academic, degree, funding ...	0.699	0.824
S_6	Book, price, type, size, afford ...	0.827	0.839

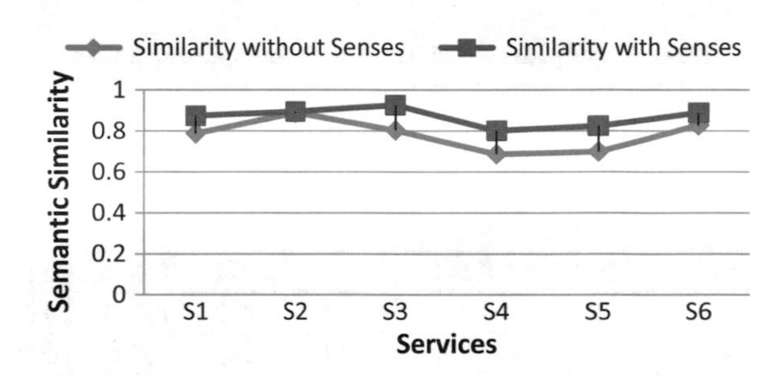

Fig. 2 Comparison of concept-sense similarity with SIMCR

with SIMCR [13] as shown in Table 2. The values for α, β, γ in Eq. (1) have been taken from the average of total extracted nouns, verbs, and adjectives from both the trained and testing sets. The overall results reveal the significance of senses in the match-making process by an average increase of 7% when compared with SIMCR. The performance of S_2 in Fig. 2 is not aligned with performance of other services. The reason for such performance can be the coverage of the trained set, e.g. the trained dataset may already have all the extracted concepts and senses of S_2, where further improvement either tends to 0 or to some very small change.

4.2 *Experiment 2*

Another test is conducted with same data as that of experiment 1. In this experiment, the semantic similarity is evaluated at different levels for one category (Edu.). Experiment 1 focused on 100% or complete trained data for the semantic similarity calculation of testing data, whereas the experiment 2 emphasized the semantic similarity as different levels of trained data. Table 3 depicts the semantic similarity from 50 to 70% of the trained set and discloses two points: (1) the proximity of *semantic similarity at level top 70% in Table* 3 with *SIMCR results in column three of* Table 2. This proximity indicates that match-making *without sense inclusio*n is close to the results of *match-making with senses using top 70% category concepts*; (2) If we do not have any concern with semantic similarity value and just want to classify the data to one category out of many categories, then instead of doing

Table 3 Services similarity with single category (Edu.) using the top-level strategy

Service	Top 50%	Top 60%	Top 70%
S_1	0.722	0.756	0.778
S_2	0.704	0.841	0.844
S_3	0.769	0.811	0.839
S_4	0.658	0.738	0.786
S_5	0.632	0.787	0.812
S_6	0.701	0.779	0.806

Table 4 Concept-sense similarity with multiple categories

Service	Edu.	Food	Comm.	Med.	Sim.	Eco.	Geo.	Weapon	Travel
S_1	0.874	0.173	0.310	0.144	0.026	0.565	0.222	0.132	0.463
S_2	0.894	0.043	0.217	0.162	0.002	0.421	0.288	0.177	0.343
S_3	0.926	0.029	0.211	0.264	0.010	0.445	0.197	0.136	0.435
S_4	0.801	0.141	0.377	0.381	0.012	0.612	0.321	0.093	0.287
S_5	0.824	0.288	0.432	0.413	0.019	0.676	0.343	0.162	0.432
S_6	0.884	0.062	0.138	0.245	0.005	0.701	0.242	0.042	0.239
Total similarity	0.8671	0.1226	0.2808	0.2681	0.0123	0.57	0.2688	0.1236	0.3665

match making with 100% data, classification at another level (let's say by considering top 50% category concepts) can be done. By doing so, we can easily categorize the service/query and can reduce the search space for classification at a higher level. This also points out that the top-ranked concept of a category can easily describe the domain specialization of that category. Therefore, by exploiting such knowledge, the query can be straightforwardly categorized first (without complete matching with all concepts of every category) and then it can be searched thoroughly from a particular matched category.

4.3 Experiment 3

In this sub-section, the proposed semantic similarity approach is calculated between testing data and all categories of OWLS-TC V_4. Table 4 shows the category-wise semantic similarity score of testing services. The selection strategy of services from all categories is the same as that for experiment 1, i.e. 10% random selection. The number of selected services from all categories are: communication (Comm.)—6, economy (Eco.)—36, food—4, geography (Geo.)—6, medical (Med.)—8, simulation (Sim.)—2, travel—17, and weapon—4.

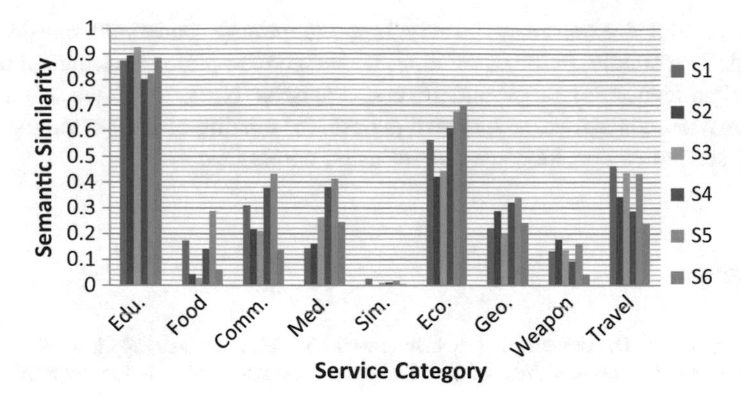

Fig. 3 Concept-sense similarity comparison of testing services with multiple categories

The result for category Eco. is slightly near category Edu. as in Fig. 3. This can be justified with the fact of *overlapped terms*, e.g. concepts like *government*, *price*, *and policy* are common in both economic and educational systems. The common terms/concepts of two or more domains can categorize such terms to more than one category, but the final similarity always includes the overall context of input not only the terms. This implies reasons for the similarity value of 0.57 for category Eco. in Fig. 3. The natural language-based queries, i.e. "*contractual jobs in teaching*" and "*CT scan lab attendant jobs on contract bases*", have the term '*contract*' in common. If the domain, let's say education and medical categories, to which these queries have to be categorized also contain the term then it might be possible that their concept-sense similarity score will remain near to each other; but, the final similarity of these queries will fall in the category that covers the overall context, i.e. *CT scan, lab, attendant,* and *job* together indicates the object and subject of medical category. In short, we can say that due to the concept similarity, a concept may have high similarity value with other categories also, but when the whole context is considered, it clearly categorizes the input and performs well.

5 Conclusion and Future Direction

This chapter addresses one of the research challenges concerning SWS discovery. We have presented an idea to improve the performance of SWS discovery by means of sense inclusion in the match-making process. This chapter focused on sense significance and examined the declared hypothesis with experimental study on an OWLS-TC V_4 dataset. A set of three experiments on pre-processed and extracted concepts have been performed by dividing the services into training data and testing data. The quantitative results obtained from experiments signify the inclusion of senses in the match-making process.

The presented experimental study is a preliminary study and considers one semantic format only. In the near future, we intend to explore the empirical study of sense significance on large and diverse datasets. Next, we have planned for (a) supervised concept-sense knowledge base, (b) ranking of the potential services, and (c) service composition with coordinated ontologies.

References

1. Nacer, H., and D. Aissani. 2014. Semantic web services: Standards, applications, challenges and solutions. *Journal of Network and Computer Applications* 44: 51–134. https://doi.org/10.1016/j.jnca.2014.04.015.
2. Ran, S. 2003. A model for web services discovery with QoS. *ACM Sigecom Exchanges* 4: 1. https://doi.org/10.1145/844357.844360.
3. Pathak, J., N. Koul, D. Caragea, and V.G. Honavar. 2005. A framework for semantic web services discovery. In *7th ACM international workshop on web information and data management*, 45–50, ACM Press, New York. https://doi.org/10.1145/1097047.1097057.
4. Zhang, C., D. Zhu, Y. Zhang, and M. Yang. 2007. A web service discovery mechanism based on immune communication. In *Convergence information technology,* 456–461, IEEE press, New York. https://doi.org/10.1109/iccit.2007.1.
5. Ying, L. 2010. Algorithm for semantic web services clustering and discovery. In *Communications and mobile computing*, 532–536, IEEE Press, New York. https://doi.org/10.1109/cmc.2010.90.
6. Liu, Y., and Z. Shao. 2010. A framework for semantic web services annotation and discovery based on ontology. In *Informatics and Computing*, 1034–1039, IEEE Press, New York. https://doi.org/10.1109/pic.2010.5688008.
7. Farrag, T.A., A.I. Saleh, H.A. Ali. 2011. ASWSC: Automatic semantic web services classifier based on semantic relations. In *Computer Engineering & Systems*, 283–288, IEEE Press, New York. https://doi.org/10.1109/icces.2011.6141057.
8. Adala, A., N. Tabbane, and S. Tabbane. 2011. A framework for automatic web service discovery based on semantics and NLP techniques. *Advances in Multimedia* 2011: 1–7. https://doi.org/10.1155/2011/238683.
9. Sangers, J., F. Frasincar, F. Hogenboom, and V. Chepegin. 2013. Semantic Web service discovery using natural language processing techniques. *Expert Systems with Applications* 40: 4660–4671. https://doi.org/10.1016/j.eswa.2013.02.011.
10. Hammami, R., H. Bellaaj, A.H. Kace. 2016. A novel approach for semantic web service discovery. In *Enabling technologies: Infrastructure for collaborative enterprises,* 250–252, IEEE Press, New York. https://doi.org/10.1109/wetice.2016.62.
11. Dietze, S., A. Gugliotta, J. Domingue. 2008. Towards context-aware semantic web service discovery through conceptual situation spaces. In *Context enabled source and service selection, integration and adaptation*, p. 6, ACM Press, New York. https://doi.org/10.1145/1361482.1361488.
12. Miller, G.A. 1995. WordNet: A lexical database for English. *Communications of the ACM* 38: 39–41. https://doi.org/10.1145/219717.219748.
13. Chen, F., C. Lu, H. Wu, and M. Li. 2017. A semantic similarity measure integrating multiple conceptual relationships for web service discovery. *Expert Systems with Applications* 67: 19–31. https://doi.org/10.1016/j.eswa.2016.09.028.

14. Banerjee, S., T. Pedersen. 2003. Extended gloss overlaps as a measure of semantic relatedness. In *18th joint conference on artificial intelligence,* 805–810, Morgan Kaufmann Publishers, San Francisco.
15. OWLS- TC V4. Retrieved June 2017, from http://projects.semwebcentral.org/frs/?group_id=89&release_id=380.

Efficient Resource Utilization by Reducing Broker Cost Using Multi-objective Optimization

B. K. Dhanalakshmi, K. C. Srikantaiah and K. R. Venugopal

Abstract Cloud computing is largely concerned with effective resource utilization and cost optimization. In the existing system, however, resources are under-utilized due to high cost. To overcome with this problem, in this chapter a new classification and merging model for reducing broker cost (CMRBC) is introduced to enable effective resource utilization and cost optimization in the cloud. CMRBC has enormous benefits. First, it has a cost-effective solution for service providers and customers. Second, for every job, virtual machine (VM) creation is avoided to reduce broker cost. Because of allocation, the creation or selection of VM resources is done based on the broker. Thus, CMRBC implements an efficient system of resource allocation that reduces resource usage cost. Our experimental results show that CMRBC achieves greater than 40% reduction in broker cost and 10% in response time.

Keywords Broker cost · Cloud computing · Classification · Cost effectiveness Merging · Multi-objective optimization · Resource utilization · Scheduling

1 Introduction

Cloud computing is an emerging technology for the storage and retrieval of data. Hence, users do not have to invest in infrastructure, which reduces cost, time and energy, and it uses a pay-as-you-go model, similar to an electricity bill [1]. Infrastructure as a service (IaaS) has a pool of hardware resources for rent or lease, including

---

B. K. Dhanalakshmi (✉) · K. C. Srikantaiah
Department of Computer Science and Engineering, SJB Institute of Technology, Bangalore, Karnataka, India
e-mail: malathi.dl@gmail.com

K. R. Venugopal
Department of Computer Science and Engineering, University Visvesvaraya College of Engineering, Bangalore, Karnataka, India

A. N. Krishna et al. (eds.), *Integrated Intelligent Computing, Communication and Security*, Studies in Computational Intelligence 771,
https://doi.org/10.1007/978-981-10-8797-4_53

memory, computing power and storage capacity. The instances of IaaS are presented by virtual machines (VMs) [2].

Computing can be modeled in two ways. A two-tier model communicates directly between the client and the infrastructure service provider. This leads to underutilized resources[3]. Another method is the three-tier model, which includes one more tier that acts as a mediator between the client and the provider. This mediator is called a broker [4]. The cloud broker plays a dual role in the context of cloud computing. When it interacts with a provider, it acts as a client, and when it interacts with a customer, it behaves as a provider. This leads to effective resource utilization, but the broker benefits more and the cost is high. To overcome this problem, we have designed a model to reduce the cost of the broker and the number of VM created, for effective resource utilization.

The cost inefficiency of existing cloud services is mainly concentrated in VM allocation and overall resource usage cost. It is restricted to a single job or single customer, and for each job a new VM is created. This process is very cost-inefficient, and the broker cost is also high. To avoid this, we have designed a new model, called CMRBC. This model classifies types of job and merges similar jobs together. By using these techniques we can reduce the number of VM created for the number of jobs. In this way, the broker cost is reduced, and on-demand resource utilization can be avoided.

The remainder of the chapter is organized as follows: Related works are described in Sect. 2, and problem definition is discussed in Sect. 3. The entire structure of the model is described in Sect. 4, experimental results are discussed in Sects. 5, and 6 concludes the chapter, highlighting future work.

2 Related Works

Kanu et al. [5] proposed a heuristic scheduling algorithm, particle swarm optimization, to reduce the total execution cost of application work flow in cloud computing. Kessaci et al. [6] explored the request type made by a user that checks the syntax, and according to the request type, the scheduler decides to which system the request is assigned. Anastasiadis et al. [7] proposed an evolutionary algorithm called Cuckoo Search Algorithm (CSA) task scheduling in cloud computing. CSA is based on optimizing the broker cost and best-fit virtual machine for a particular job. Kumar et al. [8] proposed a DLS algorithm that efficiently traces the workload in cloud computing to meet the requirements of security, time, cost and execution of tasks within the deadline.

3 Problem Definition

Given "N" number of jobs, where N = (n1,n2,n3....n) with different job size, and "m" number of VMs, where M = (V1,V2,V3.....Vn), jobs are classified into small, medium, large, or extra large based on job sizes of 50, 100,150 and 200 MB, respectively, by setting the threshold value, and merges only small jobs based on VM capacity. The profiler gives a set of requirements for each merged job where R = (Memory, CPU, RAM, Bandwidth, Start and End time); the broker interacts between user and service provider. The instances of IaaS are presented by virtual machines (VMs). The resource (VM) is denoted as a vector R = (CPU, RAM, Memory), which depicts a virtual machine that includes a number of virtual machines. The increase in cloud brokerage reduces the number of virtual CPUs (VCPU). The objective is to minimize resource utilization and broker cost in cloud computing by using multi-objective optimization.

Assumption: The simulation for every job of the execution time is assumed to be predicted execution time based on the results obtained by profiling.

4 CMRBC: Model

In this section, we present the cloud service model for CMRBC. The system model consists of the following components: user, classifier, merging, mapping phase, profiler and analyzer, cloudlet, broker, scheduler, data center. A user submits work in the form of jobs. The submitted jobs can be small, medium, large or extra large. A job consists of several tasks. Only smaller jobs are classified based on the threshold value by using priority classification. Classified jobs are passed on to a merging block.

The merged job is then passed on to a mapping phase, which ensures that service level agreement (SLA) constraints are satisfied. Next, it moves on to the profiling and analyzer phase, where it uses entire resource requirements to execute the job. These profiled jobs are converted to a cloudlet, which is said to be a task. Tasks are submitted to an intermediary broker who is responsible for scheduling the tasks to VM. The work flow scheduler performs the scheduling based on availability of VMs.

Consider "M" number of virtual machines in the system and "N" number of jobs J={ $j_1, j_2, j_3, \ldots\ldots\ldots\ldots j_n$ }, M={ VM_1, VM_2, VM_3,..........VM_n }.

The small jobs are classified based on the threshold valve. Small jobs ={ J_i | size of (J_i) < P }, where "P" is the threshold value.

Merging is done based on the VM capacity size, and merges only the smaller jobs. Once the VM capacity is full, it stops merging. These small jobs are converted into one job. Merged jobs = { $j_1 \ldots\ldots j_k$ } $\in$ small jobs | size of ($j_1 \ldots\ldots j_k$) $\leq$ size (VM_i), where $1 \leq i \leq M$ and "p" is the VM capacity. Merged jobs = { Job_1, Job_2, ...Job_p }

The cost of execution of a job "j" is given by

$$Cost_j = C_k + (M + R) \tag{1}$$

Cost of execution of job on a virtual machine: Cost(Job_j) = Number of instances of VM_j, where 1 ≤ j ≤ p,

$$Cost(Job_j) = C_k + \sum_{v=1}^{S} C_R \tag{2}$$

where "R" is the type of resource used in VM type. The overall cost of execution for all "N" jobs can be given by:

$$Overallcost = \sum_{j=1}^{P} Cost(Job_j) \tag{3}$$

The steps are detailed in Algorithm 1

Algorithm 1: Algorithm CMRBC Model

Input: D:Twitter Data Set
N:Total no of Jobs P:Threshold Value(50MB)
Output: C:Cost of Broker
Step 1 : foreach job_i ε D do.
  if *(size of(job_i)<P)* **then**
    small job ∪=job_i
end
Step 2 : Merge the small jobs based on VM Capacity
Step 3 : Profile the merged job *ProfiledJob* and assign to broker
Step 4 : Schedule jobs to VM
  foreach job_j ε *ProfiledJob* **do**
    if *(VM is available)* **then**
      assign job_j to VM
        C = $C_k + Cost_j$
    end
end
Step 4: Return C.

5 Experimental Results

The algorithm CMRBC has been implemented using Java language on the NetBeans 8.1 platform and Intel Core i7 processor environment, with 100 GB memory and 1 TB HDD. A Twitter data set is considered, the unit of which is in GB. We have considered four types of jobs: small, medium, large, extra large. The size of the job

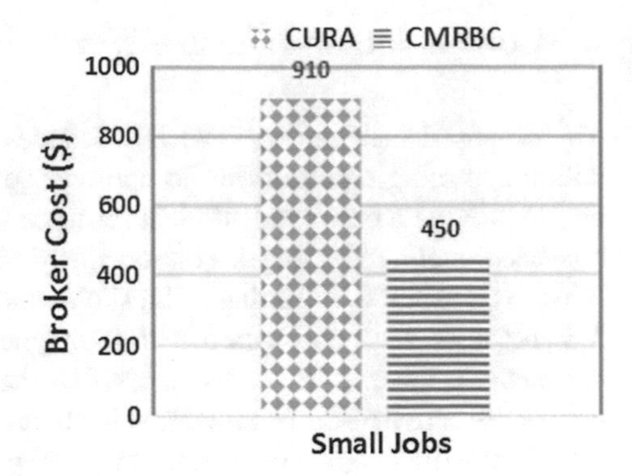

Fig. 1 Broker cost comparison for CMRBC Model

predefined as small is 50 mb, medium is 100 mb, and 150 and above is large and extra large is 200 mb. The arrival of data is based on Poisson distribution. The comparative performance is evaluated by the number of servers that calculate the cost, response time, RAM speed and bandwidth of the experiments. The experiments were run on a cloud simulator tool version 3.03; the configuration of the simulator CPU (64 bit, Intel Pentium(R) i7 CPU 2.9 GHz) with 16 GB RAM. The CMRBC implements both the scheduling with instant VM allocation and the classification and merging techniques. The results are shown in Figs. 1 and 2.

Figure 1 shows the reduction in broker cost for the number of virtual machines. Here, the cost of the broker represents the number of virtual machines that create the job. The CMRBC model shows a reduction in cost even with 500 servers obtained by the cura approach. Here, a broker is not necessary to create a VM for every job submitted, because jobs are already classified and merged to reduce the broker workload and to avoid broker charges for creating virtual machines per job.

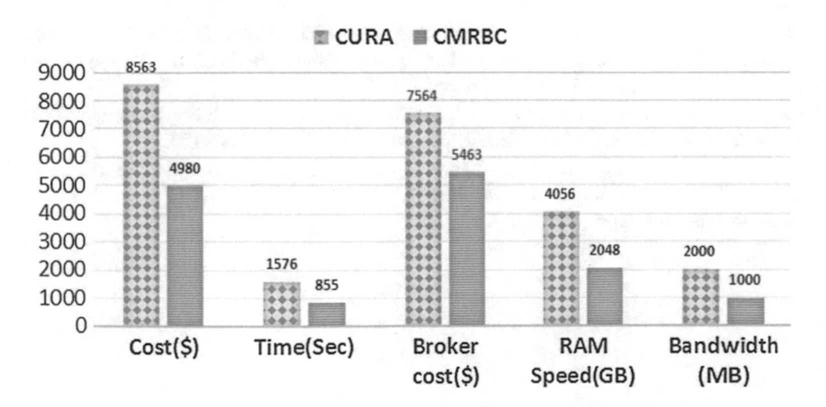

Fig. 2 Overall parameter comparison for CMRBC Model

6 Conclusion and Future Work

The proposed mechanism, CMRBC, reduces cloud brokerage cost using a classification and merging algorithm to optimize response time in order to satisfy clients and to provide a profit and efficient resource utilization to the service provider. This has been evaluated using a Poisson distribution arrival of jobs for VM requests, which is classified by setting a threshold value. The number of requests was 1000. VM requests were configured with four types of instances (small, medium, large, extra large). The results show that CMRBC helps to minimization of broker cost and achieve effective resource utilization in our algorithm, and it increases client satisfaction. This work is optimized with additional parameters, namely response time, cost, bandwidth and RAM speed, by using multi-objective optimization. Deploying our CMRBC broker approach over large jobs and extra-large jobs for effective resource utilization is not considered.

References

1. Cloud, C. 2009. Amazon elastic compute cloud.
2. Guide, D. 2010. Amazon elastic mapreduce.
3. Palanisamy, B., A. Singh, L. Liu and B. Langston. 2013. Cura: A cost-optimized model for mapreduce in a cloud. In *IEEE 27th International Symposium on Parallel & Distributed Processing (IPDPS)*, pp. 1275–1286. IEEE.
4. Singh, A., M. Korupolu, and D. Mohapatra. 2008. Server-storage virtualization: integration and load balancing in data centers. In *Proceedings of the 2008 ACM/IEEE conference on Supercomputing*, p. 53. IEEE Press.
5. Kanu, R.P., T. Shabeera, and S.M. Kumar. 2014. Dynamic cluster configuration algorithm in mapreduce cloud. *International Journal of Computer Science and Information Technologies* 5 (3): 4028–4033.
6. Kessaci, Y., N. Melab and E.-G. Talbi. 2013. A pareto-based genetic algorithm for optimized assignment of vm requests on a cloud brokering environment. In *2013. IEEE Congress on Evolutionary Computation (CEC)*, pp. 2496–2503. IEEE.
7. Anastasiadis, S.V., and K.C. Sevcik. 1997. Parallel application scheduling on networks of workstations. *Journal of Parallel and Distributed Computing* 43 (2): 109–124.
8. Kumar, B.S., and L. Parthiban. 2016. A novel approach for submission of tasks to a data center in a virtualized cloud computing environment. *International Journal of Advanced Computer Science and Applications* 7 (8): 238–242.

A Secured Steganography Algorithm for Hiding an Image in an Image

N. Gopalakrishna Kini, Vishwas G. Kini and Gautam

Abstract Recent developments in computer security have shown that steganography rather than cryptography is the better method of securing data. A commonly used technique in steganography is the least significant bits (LSB) method, which is, however, vulnerable to attacks due to its simplicity. Nonetheless, it would be difficult or impractical to discern encrypted data if a stego key is added. In this chapter, we show how a carrier image is used to hide another image and present a modified approach for embedding a stego key within it. A 24-bit color carrier image is used to hide the secret image and the same is further used to hide the stego key. We compare the peak signal-to-noise ratio (PSNR) and mean squared error (MSE) and conduct a histogram analysis to determine to what level the stego image is blurred in the carrier image.

Keywords Cryptography · Steganography · Stego key · Stego image LSB method · Histogram

1 Introduction

Data security has become a serious problem in data communication via the Internet or any other media. Cryptography and/or steganography are used to maintain the confidentiality of data, but steganography is considered to be the better method because the hidden message does not attract attention to itself for scrutiny. Steganography plays the central role in secret message communication [1].

N. G. Kini (✉) · Vishwas G. Kini · Gautam
Department of CSE, Manipal Institute of Technology,
Manipal Academy of Higher Education, Manipal, Karnataka, India
e-mail: ng.kini@manipal.edu

Vishwas G. Kini
e-mail: vgkini38@gmail.com

Gautam
e-mail: gautam00717@gmail.com

A. N. Krishna et al. (eds.), *Integrated Intelligent Computing, Communication and Security*, Studies in Computational Intelligence 771,
https://doi.org/10.1007/978-981-10-8797-4_54

In steganography, before the hiding process, the sender must select an appropriate message carrier, an effective message to be hidden as well as a stego key used as a password. The object in which the message is embedded is called the message carrier or carrier image. Varieties of entities have been used to embed messages, such as image files, video files, audio files and html pages.

Secret image refers to the image to be embedded in the carrier image, and stego image refers to the image carrying the hidden message. So, given a carrier image and a secret image, the goal of steganography is to produce a stego image which would carry the secret message. A stego key, such as a password, which is an additional secret information is required to protect the image from being decrypted by an unauthorized person.

A steganography algorithm is used to encrypt the message in more effective way. Various communication techniques can be used to send a hidden image to a receiver. After receiving the message, the receiver is able to decrypt the secret message using the extraction algorithm and a secret key [1, 2].

A digital image is a collection of computer data in the form of an array having a predetermined number of elements, wherein every element has its location and value, referred to as a pixel. This data is very large compared to the message we want to hide. Hence, digital images are preferred for hiding messages.

The objective of steganography is to make encoding more difficult and consequentially ensure that the changes to a carrier image due to injection of a secret image are visually negligible; that is to say, changes impinged on to a stego image cannot be differentiated from the carrier image. In this chapter, we show how a carrier image can be used to hide another image and present a modified approach for hiding the stego key in it.

2 Related Work

Many modifications in the LSB technique have been suggested by different authors which mostly relate to the spatial domain [1–3]. A hash-based LSB technique is discussed in [3]. Historical and more popular steganography techniques are presented in [4]. Pixel pattern-based steganography, which involves hiding messages within images using the existing red, green and blue (RGB) values, is further work contributed in [5].

To lighten the color alteration and to obtain better hiding capacity, the RGB component is encoded by mod u, mod v and mod w functions as discussed in [6]. In [7], the authors proposed an improved LSB replacement method for 24-bit color images. A study on different steganography techniques is done in [8]. Fibonacci as well as Lucas transforms are other techniques used for selecting the hiding locations. These methods can be effective when used with the frequency domain.

3 Methodology

A 24-bit color carrier image is taken to encrypt a secret image within it. Each pixel in the digital image is a color combination of RGB components. Eight bits represent each of the three color values of red, green and blue in a 24-bitmap value in each pixel, resulting in a wide variety of colors. Embedding the secret image changes the pixel intensity but does not make any noticeable change in the carrier image since the data is huge. Also, the human eye cannot notice these small changes in the pixel intensity. Maintaining picture quality is important for the protection of the secret image [5, 7].

Steganography is based on replacing the least significant bit (LSB) layer of images, the LSB technique. In this, the LSBs of the carrier image pixels are replaced by the secret image bits in order to embed the secret image into the carrier image. The secret image bits are processed before encrypting.

As far as hiding the information is concerned, the LSB technique can vary in its approach. The LSBs of pixels may be arranged by the increment/decrement of the pixel value, or randomly, or even in certain areas of the carrier image. The LSB of the bytes inside the carrier image is changed to a bit of the secret message. Hence with a 24-bit carrier image, a bit of each of the RGB components is used since they can be represented by a byte. Three bits of a pixel of a secret image can be stored in each pixel of a carrier image.

A similar technique with a modified approach is followed to encrypt the stego key into the carrier image. A password is considered as a stego key and is additional secret information in the carrier image. Before encrypting the password in the carrier image, the ASCII values are generated in binary for the password. While encrypting, the stego key selects the pixels of the carrier image in an unsystematic order.

The second LSB of the bytes inside the carrier image is changed to a bit of the stego key. The other bits of the bytes inside the carrier image can also be used to hide the password. This may possibly lead to compromising the quality of the carrier image. Hence, in order to make the altered quality of the carrier image unnoticeable to the human eye, only the first and second LSBs of the carrier image are used to encrypt the secret image and stego key. With this, only minor changes in color occur and, hence, only minor changes in the pixel intensity. This work was carried out using Visual C.

The algorithm for encrypting the stego key in the carrier image is as follows:

Step 1: Select an appropriate stego key.
Step 2: Get the ASCII value of each character in the stego key.
Step 3: Convert each ASCII value to binary.
Step 4: A random pixel is chosen by the random pixel number generator on the carrier image.
Step 5: Each bit of the ASCII value is now encrypted in the second LSB of the randomly chosen pixel.

Step 6: This pixel number is saved in a file which is later used for decrypting the stego key.
Step 7: Repeat through step 4 till all the bits of stego key are addressed

Here, the stego key such as a password is used to hide the information which is additional secret information. Security can be further enhanced by encrypting the stego key using cryptography and hiding this encryption within an image using steganography with the above algorithm.

The next phase of the work is to determine to what level the stego image containing the secret image and stego key is blurred by the carrier image. For this, we compare the calculated peak signal-to-noise ratio (PSNR) and mean squared error (MSE) and analyse histograms. Initially, MSE is calculated to find the PSNR [3, 7].

MSE is a measure of the average of the squares of the errors. It is used to measure the error difference of two images by calculating the mean error value. Let the carrier image and the stego image be of size $m \times n$; then, the pixel value (x, y) be from 0 to $m - 1$ and 0 to $n - 1$ respectively, and can be defined to each carrier image and stego image as:

$$MSE = \frac{1}{mn} \sum_{i=0}^{m-1} \cdot \sum_{j=0}^{n-1} [C(i,j) - S(i,j)]^2 \tag{1}$$

where the pixel coordinate of the carrier image is $C(i, j)$ and the pixel coordinate of the stego image is $S(i, j)$.

The PSNR is the ratio between the maximum value of an image pixel and the mean error difference of two images in consideration. This ratio helps to differentiate between the images based on their pixel values.

$$PSNR = 10 \log_{10}\left(\frac{MAX_i^2}{MSE}\right) \tag{2}$$

$$PSNR = 20 \log_{10}(MAX_i) - 10 \log_{10}(MSE) \tag{3}$$

In image steganography, the histogram analysis shows much of its importance in analyzing both carrier and stego images [1, 6, 7].

4 Results

Figure 1a–c shows our experimental cases with a carrier image size of 1080×960, secret image size of 320×240 and stego image size of 1080×960. There is absolutely no visible difference found between Fig. 1a, c.

Fig. 1 **a** Carrier image. **b** Secret image. **c** Stego image

(a)

(b)

(c)

More detail of our code run is shown in Fig. 2. During encryption, the sender is asked to enter the text message as a stego key to provide security for the hidden image embedded in the carrier image.

During decryption, only the authorized user is able to enter the right password which must be same as the stego key to initiate the extraction of the secret image. This password is compared with the stego key embedded in the stego image to authenticate the user. Now this authorized user can extract the hidden image from the stego image file to derive the secret image.

Table 1 gives the execution time for our code to encrypt and also resultant values of MSE and PSNR. A better PSNR value is obtained through our work. A larger PSNR value indicates good image quality and a negligible difference between the two images, unnoticeable to the human eye.

Figure 3a–c shows the histogram of carrier, secret and stego images, respectively. By analyzing the stego and carrier image histograms, it is observed that there are minimal changes between them. Because of this it is very difficult to conclude secret data is hidden in the stego image.

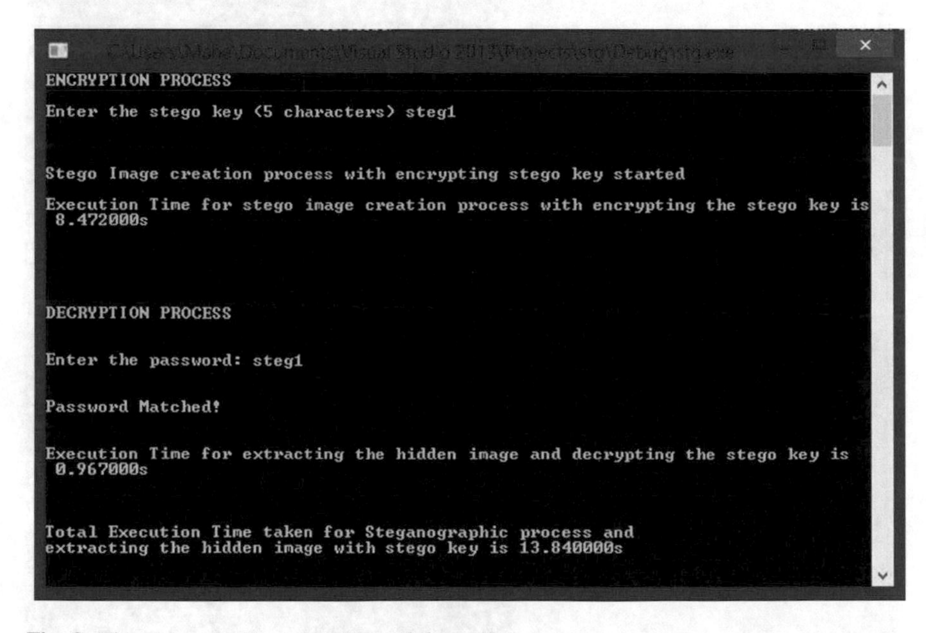

Fig. 2 Time taken for the encryption and decryption process

Table 1 MSE, PSNR and time to generate a stego image

Carrier image size	Stego image size	MSE	PSNR in dB	Execution time to generate stego image (s)
1080 × 960	1080 × 960	0.319	53.10	8.472

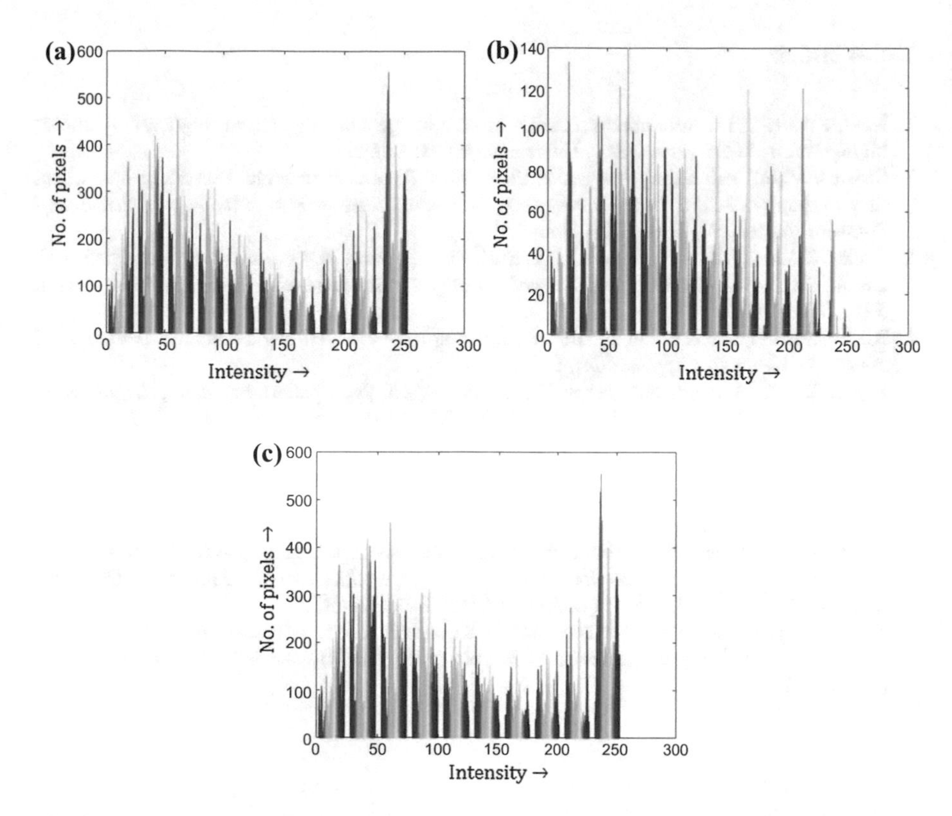

Fig. 3 **a** Histogram of carrier image. **b** Histogram of secret image. **c** Histogram of stego image

5 Conclusion

In this chapter, we describe how the LSB method is used to embed secret image bits in a 24-bit color carrier image. A modified technique to encrypt a stego key into a carrier image is also proposed. The resulting stego image looks exactly the same as the original image and it is very difficult for the human eye to identify the changes.

It is well known that the LSB method manipulates only the LSB bits, which is detectable and, hence, vulnerable to attacks. However, our proposed system provides an enhanced hiding mechanism that cannot be detected easily. We calculated the PSNR and MSE and analyzed histograms to study how stego images deviate from carrier images.

References

1. Nani, Koduri 2011. Information security through image steganography using least significant bit algorithm. M.Sc. dissertation, University of East London.
2. Sharma, Vipul, and Sunny Kumar. 2013. A New Approach to Hide Text in Images Using Steganography. *International Journal of Advanced Research in Computer Science and Software Engineering* 3 (4): 701–708.
3. Aniket, G., and Meshram, Rahul Patil. 2014. Secure secret key transfer using modified hash based LSB method. *International Journal of Computer Science and Information Technologies*, 5 (6): 7683–7685.
4. Kumar, Arvind, and Km. Pooja. 2010. Steganography-A Data Hiding Technique. *International Journal of Computer Applications*, 9 (7), 19–23.
5. Rejani, R., D. Murugan, and Deepu V. Krishnan. 2015. Pixel pattern based steganography on images. *ICTACT Journal on Image and Video Processing Communication Technology* 5 (3): 991–997. https://doi.org/10.21917/ijivp.2015.0146.
6. Nagaraj, V., V. Vijayalakshmi, and G. Zayaraz. 2011. Modulo based image steganography technique against statistical and histogram analysis. *IJCA Special Issue on Network Security and Cryptography* 4: 34–39. https://doi.org/10.5120/4347-046.
7. Rawat, Deepesh, and Vijaya Bhandari. 2013. A steganography technique for hiding image in an image using LSB method for 24 bit color image. *International Journal of Computer Applications* 64 (20): 15–19. https://doi.org/10.5120/10749-5625.
8. Tiwari, Anjali, Seema Rani Yadav, and N.K. Mittal. 2014. A review on different image steganography techniques. *International Journal of Engineering and Innovative Technology*, 3 (7): 121–124.

A Hardware Implementation of Fractal Quadtree Compression for Medical Images

S. Padmavati and Vaibhav Meshram

Abstract Medical images contain voluminous data. Transmission of medical images over the Internet using low bandwidth and fast transmission is a major challenge. Image compression techniques address this issue using cost-effective methods. Fractal image compression is a compression technique used in digital image compression to increase the transmission rate with low bandwidth. This is based on the fact that some parts of an image always resemble another part of the same image. This concept is called self-similarity. Fractal image compression has long encoding time of seconds and fast decoding time. In this paper, we propose and implement architecture for fractal image compression. This architecture has been implemented on a Xilinx Spartan-6 FPGA board and tested on medical images. The proposed implementation reduces the encoding time to nanoseconds.

Keywords Image compression · Fractal image compression · Quadtree decomposition · FPGA implementation

1 Introduction

With the growth in digital technology, image processing has become an important tool in many branches of science and engineering, including computer science engineering, electrical and electronic engineering, and robotics. Image processing presents significant challenges and is a prime area for research and development. Digital image processing is related to digital signal processing. There are various advantages to digital image processing over analog image processing. For example, digital systems use less bandwidth and are less prone to noise. Digital image processing uses digital systems such as computers to process digital images.

S. Padmavati (✉)
Department of ECE, Jain University, Bengaluru, India
e-mail: padmavj@yahoo.co.in

V. Meshram
Department of ECE, Dayananda Sagar University, Bengaluru, India

A. N. Krishna et al. (eds.), *Integrated Intelligent Computing, Communication and Security*, Studies in Computational Intelligence 771,
https://doi.org/10.1007/978-981-10-8797-4_55

The input of a digital system is a digital image, and the output is also a digital image. Image processing usually imports the image through image acquisition tools, analyzes and manipulates the image, and displays the altered image based on predefined image analysis. Digital image processing finds application in various fields such as remote sensing, medical diagnostics, agriculture, and analysis of aerial images from Google Maps. The operations involved in digital image processing include image segmentation, image enhancement, image restoration and image compression.

Image compression is a method used to reduce the number of bits required to represent an image. In lossless compression, the output image will be an exact replica of the input image, whereas in lossy compression, the output image is an approximate replica of the input image. Image compression is needed for faster transmission of data to minimize storage space, thereby making it cost effective. Existing traditional image compression methods include wavelet, Joint Photographic Experts Group (JPEG), vector quantization (VQ), and fractal. Image compression finds application in the transmission of medical images, the health industry, security, government organizations, and museums or gallery kiosks.

Compression of medical images is challenging, as the amount of data in medical images is enormous, requiring long transmit times and significant storage space. An effective image compression method [1, 2] is required to reduce storage space while maintaining fast transmission. It should also preserve the diagnostic quality of the medical image under human perception.

Fractal image compression [FIC] [3] is an effective compression method in which an entire image is divided into range blocks and domain blocks of different resolutions. Range blocks are smaller than domain blocks, which are typically selected to be twice the size of the range blocks. These blocks are mapped and searched for similar intensities. To search for the best-matching domain block, the number of operations required to search increases. Hence, FIC has a long encoding time.

A field-programmable gate array (FPGA) is a semiconductor device consisting of an array of configurable logic blocks (CLBs). Programmable interconnects are used to connect these CLBs. FPGAs are integrated circuits that can be reprogrammed by a designer for a desired application or function. FPGAs are available in different types, but dominant types are static random-access memory (SRAM)-based FPGAs. Alterdeveloped software to implement PIFSa and Xilinx are the two major FPGA manufacturing companies. An entire digital system can be embedded into a small chip such as an FPGA using very-large-scale integration (VLSI) technology. This technology reduces the required number of computations by significantly enhancing the compression ratio, storage and transmission. Because of its programmable nature, FPGA finds application in various fields including aerospace, audio, medical, video and image processing, and security. In this chapter, we implement FIC on the Xilinx Spartan-6 FPGA.

The remainder of the chapter is organized as follows: Sect. 2 briefly details the fractal image compression process. The proposed methodology and architecture of FIC is described in Sect. 3. Experimental results and analysis of the proposed methodology are discussed in Sect. 4. Finally, the chapter is concluded in Sect. 5.

2 Fractal Image Compression (FIC)

Fractal geometry was introduced in 1973 by an IBM mathematician, Benoit B. Mandelbrot. He concluded that the traditional geometry containing straight lines and smooth surfaces did not resemble the geometry of trees, clouds and mountains. Barnsley [4] came up with a new mathematical concept of iterated function systems (IFS) using fractals, popularly known as the collage theorem. The collage theorem describes how IFS should represent an image. An IFS consists of a set of contractive transformations that map a defined rectangle of the real plane to small portions of that defined rectangle. Contractive transformations make use of affine transformations to carry out translation, scaling and shearing, and for rotating point operations in the plane. This research was carried forward by one of Barnsley's Ph.D. students, who focused on how to represent an image in a partitioned iterative function system (PIFS). The transformations in PIFS will map large parts of an image to small parts of an image instead of an entire image to the parts. Qualitatively, an image will vary from one area to another. A PIFS relates the areas of an original image that are identical in appearance. The larger areas of an image are called domain blocks and the small areas are called range blocks. It is essential that each pixel of an original image belong to at least one of the range blocks. Hence, the complete pattern of range blocks of an image is known as partitioning. PIFS constructions pair each range block to a domain block that it closely resembles. These results in small PIFS encoding are compared to an original image. In 1990, Arnaud [5] developed software to implement PIFS.

The compression ratio obtained from fractal image compression typically ranges from 4:1 to 100:1, and color images are compressed to a better range than grayscale images [6]. Fractal image file size is determined from the number of transformations applied in the IFS. The bit rate scan is further reduced by using three or four levels of quadtree structure for partitioning of the range. In this way, smooth areas are represented by large range blocks where high compression is achieved, while small blocks are used to capture the fine details of an image. Fractal encoding provides better compression ratios than JPEG with respect to signal-to-noise ratio, root mean square error and mean absolute error.

Steps Involved in Fractal Image Compression:

- First, partition an image into range blocks.
- Apply affine transformations to each range block.
- The process of image decompression begins.
- Apply affine transformations repeatedly.
- After four iterations, stabilization of the attractor occurs.
- The output obtained is usually not an exact replica of the original image, but fairly close as perceived by the human eye.

2.1 Quadtree Fractal Image Compression

History of Quadtree. In 1974, Raphael Finkel and J.L. Bentley gave the name quadtree to the data structure tree. In 1985, Hannaan Samet and Robert E. Webber, along with Aluru, outlined types of quadtrees and their various applications. In the algorithm proposed by Raphael and Bentley, straightforward insertion yields O (nlogn) performance with worst case: O(n2).

The aim of quadtree partitioning is to recursively divide an image to form a quadtree. At each step of partitioning, the current node is expanded with the largest internal variance. The partitioned regions may be a square, a rectangle or any arbitrary shape. Quadtree partitioning [7] is used to achieve a high compression ratio. It finds application in binary operations on images, and in smoothening, compression and segmentation of images.

The quadtree fractal image compression algorithm is as follows:

1. Partition the given image I into four sub-images, I1–I4.
2. For each current node Ii, calculate the median color Ai and also the error Ei,
3. $Ei = \sum |I(x, y) - Ai|$.
4. Search for the sub-image with the largest error; split it further into four sub-images.
5. Repeat from step 3.
6. Finally, a clear output image is obtained.

Figure 1 shows the structure of quadtree partitioning of an image.

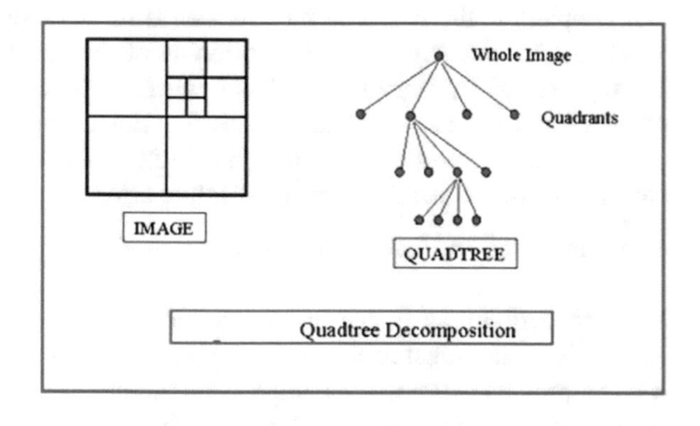

Fig. 1 General quadtree structure of an image

3 Top-Level Architecture and Implementation of FIC on Xilinx FPGA

3.1 *Salient Features of Xilinx Spartan-6 (xc6slx45t-3fgg484)*

The Spartan-6 family has significant integration capabilities. It offers low cost for a massive volume of applications. Spartan-6 is 45-nm technology-dependent. It has a dual 6 input lookup table register and built-in system blocks. Applications include digital signal processing (DSP)-oriented designs and low-cost embedded designs. Figure 2 shows the Xilinx Spartan-6.

Spartan-6 FPGAs provide great design flexibility. They load themselves automatically or load with the aid of external processors or microcontrollers onto the FPGA self-loading process. The self-loading FPGA configuration modes are called master modes. Self-loading FPGAs operate in master mode, whereas externally loaded operate in slave mode. The bit streams of the FPGA in slave mode can be stored anywhere over the entire system.

Highlights of Xilinx Spartan-6

- Useful for low-cost applications and has less static and dynamic power.
- Consists of DSP48A1 slices, four LUTs and eight flip-flops on each slice.
- An 18 × 18 multiplier, adder, and an accumulator are present on each DSP slice.
- Consists of a RAM block 18 Kb in size with fine details.
- Each RAM block can also be divided into two separate 9 Kb blocks for the applications.

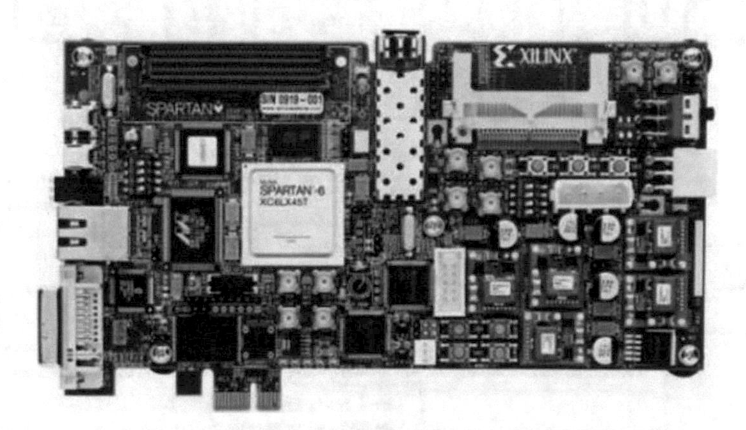

Fig. 2 Xilinx Spartan-6 (xc6slx45t-3fgg484) board

- Consists of blocks of integrated memory control.
- For the PCI Express design (LXT) applications, an integrated endpoint block is present and has high-speed connectivity for serial connections.
- For fast embedded processing applications, it consists of a cost-effective MicroBlaze soft processor, suitable industry-based IP cores and reference designs.

3.2 The Proposed Algorithm and Architecture for FIC

The proposed design method is based on the following algorithm steps:

- The input image is read.
- The resultant image is divided using fractal quadtree decomposition with a threshold value and minimum and maximum values (0.2, 2, 64), respectively.
- Mean values, x-coordinate and y-coordinate values, and block size number are noted down from quadtree decomposition.
- The encoding phase of the image is completed by reading and noting the fractal image compressed values.
- Decoding phase is applied to reconstruct the original image.

The proposed architecture of FIC with quadtree is shown in Fig. 3.

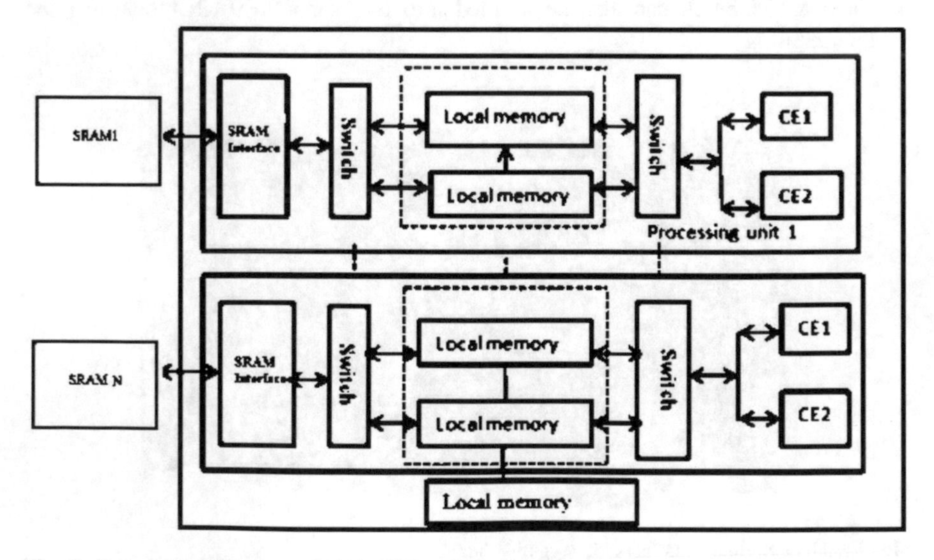

Fig. 3 Proposed architecture of FIC with quadtree

4 Results and Discussion of the Proposed Architecture

4.1 Simulation and Synthesis Results of the Designed Architecture

Figure 4 shows the output images from the proposed implementation in MATLAB. The test image for the simulation used is an eye image with dimensions of 512 × 512.

The device utilization summary of the architecture is tabulated in Table 1. From the synthesis report we find that the proposed architecture utilizes only 6% of the total slices available. The timing summary shows that the maximum frequency of operation is 212.325 MHz.

Timing Summary:

Speed grade: −2
Minimum period: 4.710 ns (maximum frequency: 212.325 MHz)
Minimum input arrival time before clock: 5.586 ns
Maximum output required time after clock: 5.897 ns
Maximum combinational path delay: 6.713 ns

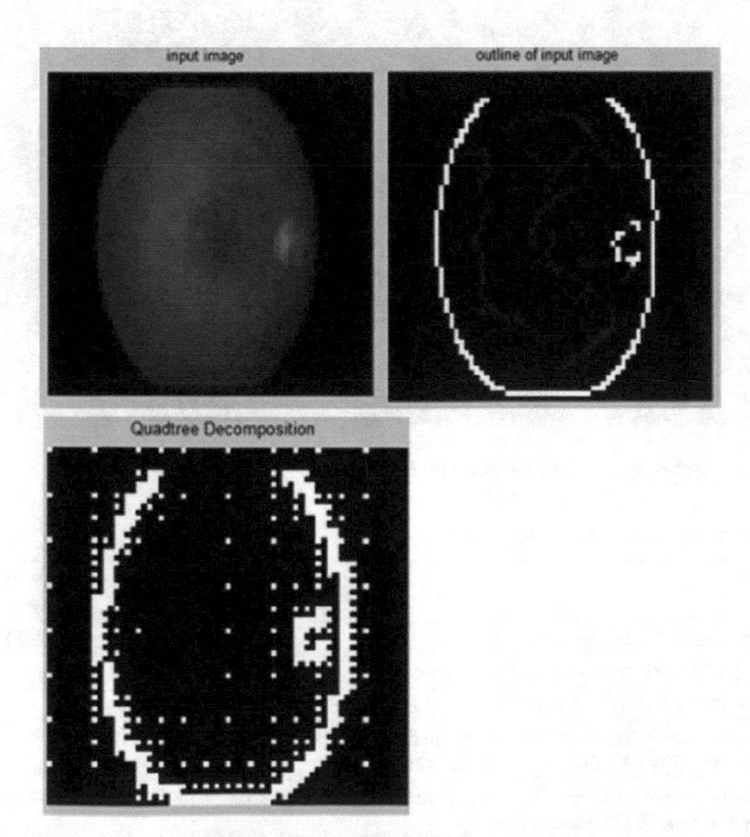

Fig. 4 Input image and fractal quadtree output image

We observed a significantly reduced encoding time of 5.897 ns. Figure 5 depicts the register transfer logic (RTL) technology schematic using Xilinx ISE version 14.2.

Table 2 shows the comparison with existing methodologies.

Table 1 FPGA resource utilization summary

Slice logic utilization	Used	Available	Utilization (%)
Number of slice registers	3,290	54,576	6
Number used as flip-flops	3,290		
Number used as latches	0		
Number used as latch	0		
Number used as AND/OR logics	0		
Number of slice LUTs	2,786	27,288	10
Number used as logic	2,430	27,288	8
Number using O6 output only	1,875		

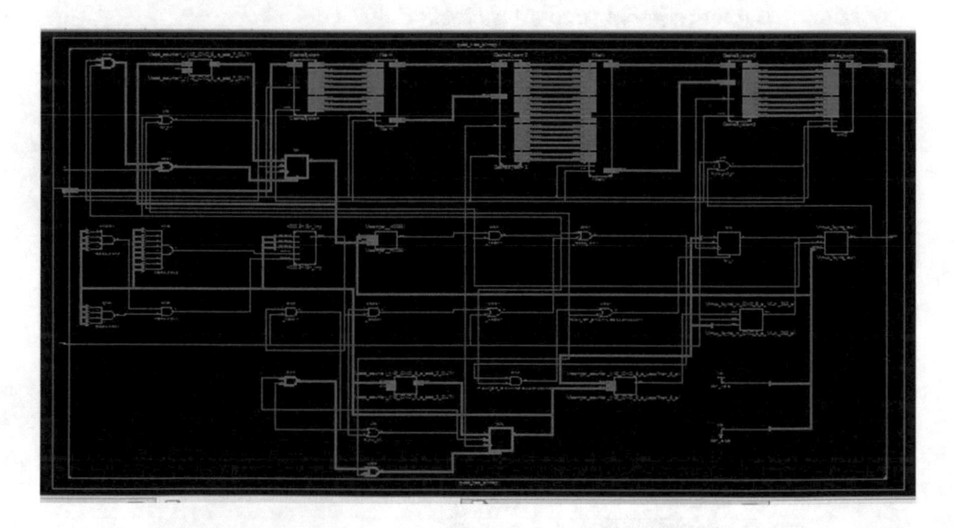

Fig. 5 RTL schematic of the proposed architecture

Table 2 Comparison with existing methods

Method	Encoding time	Resources used
Proposed method: FIC on Xilinx Spartan-6 FPGA	5.897 ns	3290
Fractal quadtree decomposition [Lena image]	1260.18 s	–
(DCT + fractal quadtree decomposition) [Lena image]	724.76 s	–
Real-time fractal image coder based on characteristic vector matching on Xilinx Virtex FPGA [Lena image]	8.3 ms	90,000
Method and architecture for FIC using multiresolution quadtree on Xilinx Virtex FPGA [medical image]	–	39,527
FIC using quadtree decomposition and Huffman coding [Lena image]	15.86 s	–

5 Conclusion

In this paper, we have proposed and designed architecture for implementing FIC with quadtree on FPGA. We have successfully implemented this architecture by designing in Verilog HDL and by MATLAB simulation. The RTL code is synthesized using Xilinx ISE version 14.2 and is targeted on a Xilinx Spartan-6 FPGA. The results of the proposed hardware implementation have been compared against existing FIC methodologies. The synthesized architecture is optimized for an encoding time of 5.897 ns, which is minimal. Our implemented architecture uses only 3,290 out of 54,576 available hardware resources. The operating frequency of the design is 212.325 MHz. The design utilizes only 6–10% of the available resources.

Acknowledgements The authors acknowledge Prof. Jayadevappa for his valuable suggestions and encouragement during the course of the work. The authors also acknowledge the sources of information used for the development of this work.

References

1. Gloria, Menegaz. 2006. Trends in medical image compression. *Current Medical Imaging Reviews* 1–21.
2. Bhavani, S., and K. Thanushkodi. 2012. A novel fractal image coding for quasi-lossless medical image compression. *European Journal of Scientific Research* 70 (1).
3. Frigaard, C., J. Gade, T. Hemmingsen, and T. Sand. 1994. *Image compression based on a fractal theory*. Institute for Electronic Systems, Aalborg University.
4. Barnsley, M.F. 1993. *Fractal everywhere,* 2nd ed. San Diego: Academic Press.
5. Arnaud, Jacquin. 1992. Image coding based on a fractal theory of iterated contractive image transformation. *IEEE Transaction on Image Processing* 1 (1): 18–30.
6. Fisher, Y. 1995. Fractal image compression. In *Theory and application.* Springer.
7. Panigrahy, M., I. Chakrabarti, and A.S. Dhar. 2016. Hardware implementation of quadtree based fractal image decoder. *IEEE Xplore.*

Multilayered Model of Soft Tissues for Surgery Simulation

K. Jayasudha, Mohan G. Kabadi and Brian Chacko

Abstract Virtual reality is an emerging technology in the area of computer graphics. In the medical field, virtual surgical training has gained increasing popularity in recent years. This is because it has aided many medical trainee surgeons to achieve proper practice and training before performing actual surgery, thus minimizing cost and time. Soft tissue modeling is an important application in surgical training simulators. It is important to study the anatomical structure of tissues so as to know the details of skin layers and their behavior, especially so that trainee surgeons can get the look and feel of skin prior to operation. This chapter focuses on the soft tissue layers present in human skin. We attempt to accomplish this through soft tissue prototype development where the skin is modeled as a triangular mesh showing the multiple layers.

Keywords Three-dimensional (3D) · Finite element method (FEM)
Functional magnetic resonance imaging (FMRI) · Soft tissue

1 Introduction

The goal in this chapter is to explain how, by using a simple frame work, one can establish the logic for soft tissue prototype development for multiple layers of skin to give a more realistic and scientifically true skin behavior. The conventional surgery simulation method involves some specific steps such as collection of the

K. Jayasudha (✉)
VTU RRC, Belgaum, India
e-mail: jayanataraja@gmail.com

M. G. Kabadi
Presidency University Bangalore, Bengaluru, India
e-mail: kabadimohan60@gmail.com

B. Chacko
Ibri College of Technology, Ibri, Sultanate of Oman
e-mail: brianchacko@gmail.com

A. N. Krishna et al. (eds.), *Integrated Intelligent Computing, Communication and Security*, Studies in Computational Intelligence 771,
https://doi.org/10.1007/978-981-10-8797-4_56

patient's data, analysis of the data and delivering a conclusion, which are helpful for the surgeon prior to performing surgery. In order to overcome these many aspects, a new emerging technology called virtual reality has come into existence, which involves visualization, modeling and training in a 3D environment. It has helped many medical students to perform mock surgery prior to performing real surgery. The researchers of Yale University have stated that surgeons who received training on computer simulations are less likely to commit fewer errors and can perform operation at about 29% faster than others.

Visualization—Visualization is the process of creating things such as diagrams, images or animations for communication. It is a way to express ideas using visual imagery. Today, visualization in the area of computer graphics has great deal of importance in many important applications such as medicine, engineering, science and education, multimedia [1]. It is hard to imagine computer graphics without visualization. Computer animation is one such advanced method of visualization. The next concept describes the previous work done related to the area of soft tissues.

2 Related Work

Skin not only protects the body, but also supports many internal organs. Many related works have been carried out for soft tissues. For example, Nebel [2] states that soft tissues are of three types, namely, epidermis, dermis and hypodermis or subcutaneous. This paper explains how to build 3D volumetric meshes using FEM. Hashim and Richens [3] use the FEM method to study the dynamic behavior of soft tissues, especially stress and strain behavior. Chacko et al. [4] use DICOM (Digital Imaging and Communications in Medicine) datasets for 3D visualization and can make incisions as part of training in a virtual environment. Holzapfel [5] uses a 3D model to describe mechanical properties such as stress and strain behavior related to all types of soft tissues such as ligaments, blood vessels, cartilage, tendons. The model uses FEM to solve complex problems related to arbitrary geometries in terms of structures. Etzel et al. [6] explains activation patterns of neurons through ROI-based hypothesis. They use MRI data for fMRI classification. Bijari et al. [7] discuss the homogeneous tissues that are present in the brain and its bifurcation in various streams. Here, the 3D coupled object is MRI scanned data and segmentation is done using tissue type information. Bouten et al. [8] describe how to re-cultivate the tissues, especially in heart valves, arteries and veins.

Jepps et al. [9] explain biomechanics and mechanobiology of growing skin, i.e. what the effect is on the skin when a drug is applied and also how it progresses deeper into the skin. This means interactions taking place between drug, product and skin. Zöllner et al. [10] use a continuum model for skin growth. This model is used to study the mechanical and biological changes occurring in the tissues. Duffy and Shuter [11] explain the viscoelastic properties, their expansion and evaluation of pig tissues. Pailler-Mattei et al. [12] describe the mechanical properties of human soft tissues and its measurements by using an indentation device. Gao and Desai

[13] describes mechanical properties of soft tissues using a tensile test. This test is conducted on liver tissues to show strain behavior. All these papers discuss the mechanical properties, tissue development, skin growth and activation patterns of soft tissues in general, but not specifically. There is a drawback in all these papers, in that they treat skin as a single layer, which is not the case. Thus, we have chosen this topic as a point of study. The next section describes the basic anatomy of skin and the simulation implementation details.

3 Implementation Details

The anatomy of soft tissues [2], shown in Fig. 1, reveals that the structure of skin is composed of three layers. So, it is evident that human skin is not just a single layer, but is of multiple layers. Hence, there is a need to focus on the properties and structure of all the layers present in the human skin. This work mainly stresses the structure of all three layers. The first layer is called the epidermis. It is thin and acts as a protective structure compared to the other two layers. The second layer is called the dermis. It is spongy, contains only flesh and supports the first layer. The third layer is called the hypodermis or subcutaneous layer. It is mainly fat and acts as a cushion, is closely connected to muscles, insulates the body and supports the upper two layers by providing essential nutrients. To simulate all these layers accurately, volumetric meshes are well suited such that navigation can be achieved in a 3D virtual environment. This is accomplished using the Visualization Tool Kit (VTK), an open source software used for visualization, image processing in a computer graphics 3D environment.

VTK allows many primitive methods that include visualization algorithms involving basic factors such as scalar, vector, texture and tensor properties. It also supports advanced modeling techniques, namely, Delaunay triangulation, implicit modeling and various techniques applied onto mesh, such as mesh formation, mesh smoothing, mesh cutting, mesh contouring, surface reconstruction, polygon reduction. This section briefly describes the methodology used in building up the prototype of soft tissues based on Delaunay 3D triangulation as explained by Maur [14]. This has the following objectives:

1. Delaunay triangulation is based on Delaunay criterion. It states that any triangulation formation n is equal to 3 for n-dimensional simplexes. Delaunay triangulation also supports formation of models from unorganized points called unstructured grids.

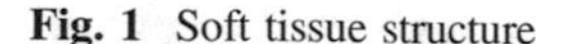

Fig. 1 Soft tissue structure

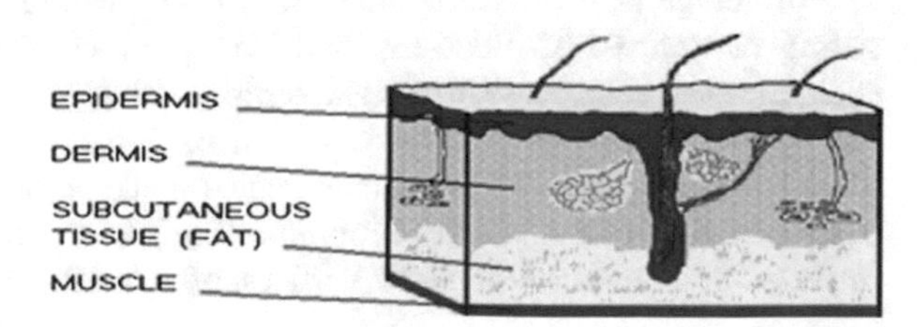

Fig. 2 Algorithm for triangulated cube

```
Let X be the input set
 While(X ≠ null) do
   Take points Vi from input set where i = 1 to 8
For (no of points == 3) do
  Connect points Vi = Vi +1 to get triangle configuration
Result = cube3D (V1 to V8)
```

2. Delaunay triangulation is used in modeling solids in a 3D environment, reconstructing the 3D objects, interpolation techniques and iso-surface extraction. It is also used in surgery simulation in a virtual reality environment.
3. Delaunay triangulation is the most popular concept for numerical computing. It is used in wide variety of areas in computer graphics such as:

- Crystallography—used to study modeling and arrangement of atoms.
- Metallurgy—used for modeling physical and chemical behavior of metallic elements.
- Cartography—used for modeling and animating maps.
- Geology—used for modeling earth science structures.

Thus, we have used Delaunay 3D triangulation to generate mesh by creating a cube data structure [15]. This method if often used for problems related to the generation of meshes. Delaunay triangulation starts mesh construction with a minimum number of points, i.e. three to form a triangle in the beginning, and if mesh is not formed, it continues the process of formation of points iteratively by subdividing the triangles until the condition is met. The following algorithm shows the formation of a triangulated cube, which continues generating triangles for eight vertices of a cube. The method is adapted to model layers of skin by placing the triangulated cube adjacently and accordingly. The algorithm for generation of the geometric multilayer soft tissue model is self-explanatory and is as shown in Fig. 2. The skin model is developed using a Pentium 4 (2 GHz) processor, 4 GB of RAM and a 512 MB graphic card. The next section briefly summarizes the simulation results obtained by developing the data structure of cube.

4 Results

The prototype of soft tissues is developed by creating a cubic data structure consisting of Vertex-List, Edge-List and Triangle-List. The basic concept is applied such that each cube consists of eight vertices, 18 edges and 12 triangles as shown in Fig. 3. The simulation concept is shown in Tables 1 and 2. Table 1 lists the 12 triangles starting from T1 to T8, constructed using vertices of cube X1–X8, with the basic concept of three vertices forming a single triangle. The data source is defined in Table 2. It consists of eight vertices of a cube, and its corresponding point

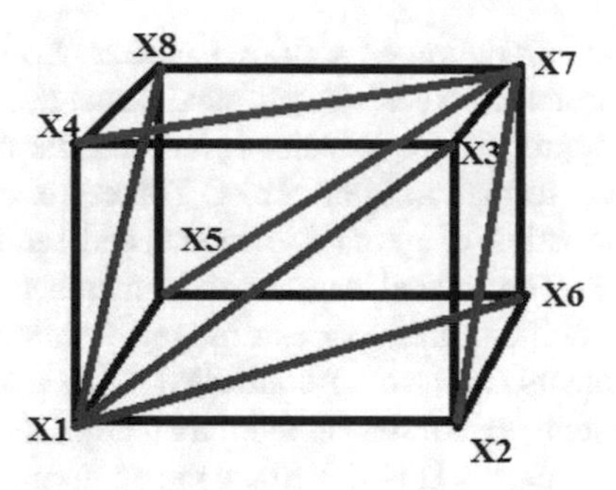

Fig. 3 Triangulated cube

Table 1 List of triangles in a cube

T1	T2	T3	T4	T5	T6	T7	T8	T9	T10	T11	T12
X1, X2, X3	X1, X3, X4	X5, X6, X7	X5, X7, X8	X2, X6, X7	X2, X7, X3	X1, X5, X8	X1, X8, X4	X1, X2, X6	X1, X6, X5	X4, X3, X7	X4, X7, X8

Table 2 Coordinates of vertices of cube

Vertices	X1	X2	X3	X4	X5	X6	X7	X8
Point Coordinates	(0, 0, 0)	(20, 0, 0)	(20, 10, 0)	(0, 10, 0)	(0, 0, 20)	(20, 0, 20)	(20, 10, 20)	(0, 10, 20)

coordinates along the x-y-z axis. Using these data, 12 triangles can be formed in a single cube. The input data are defined based on the point coordinates and is applied to generate a layered triangular mesh, and also care is taken to eliminate redundant edges in the mesh. Thus, a cube of n-dimensions is achieved along x-y-z axis. Using this data structure, the prototype of soft tissues is developed. Figure 4 shows rendering of single cube in wireframe and solid models. A wireframe model is a 3D object using lines and curves. Wireframe models consist only of points, lines, and curves that describe the edges of the object.

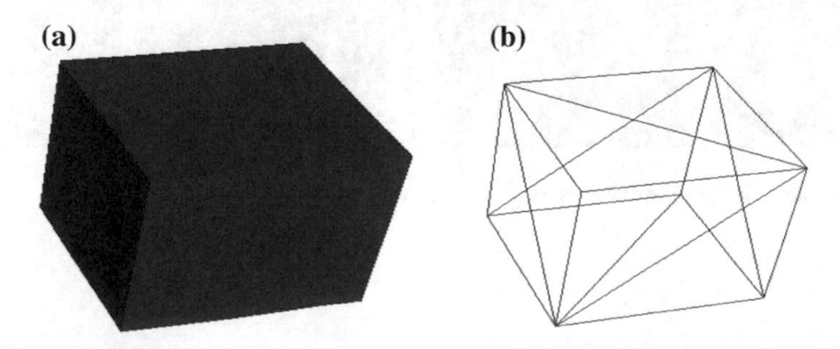

Fig. 4 Rendering the single cube **a** solid model, **b** wireframe model

Solid modeling is the modeling of solid 3D objects. Solid modeling is used in the area of computer graphics based on physical structure. In solid modeling all surfaces should meet appropriately, which proves that the object is geometrically correct, otherwise object formation is incorrect. Therefore, necessary care is taken for data collection, formation of geometric model and rendering onto the screen using a virtual camera. The virtual camera movements include elevation, yaw, pitch, azimuth, dolly, such that the user can move the camera in the virtual environment. These movements can also be controlled using a keyboard.

The current work mainly elaborates on the thickness of the layered model of soft tissues compared to other papers [11–13]. Since the epidermis is the outermost layer and is thin, it is simulated as a single-layer cubic data structure. The dermis is the next layer and is little thicker, so it is simulated as a double layer. Similarly, the subcutaneous, third layer, which is connected to muscles, is again simulated as a double-layer cubic data structure. The characteristics (attributes) of skin are realized by texture mapping that is done to all three layers to show the different properties of each layer. Texture mapping is the process of wrapping an image (jpeg, bmp, bitmap etc.) onto the model. Since the first layer is smooth, it is texture mapped with skin image. The second layer is fat and texture mapped with flesh. The third layer is thick and connected to muscles. It is texture mapped with a bone and flesh image. The outcome of simulation is shown in Fig. 5. These diagrams can be scaled to any amount and are controlled using mouse movements. The property of each layer differs from the others and is addressed in the next stage, called the deformation stage. Figure 5 shows rendering of the layered model of soft tissues in wireframe and solid models. The outcome of the development of application is that skin is modeled as triangular mesh. The three layers of skin are effectively developed based on its thickness. The performance of the proposed model is evaluated based on simulation algorithm with respect to time complexity, which is O (log N). The time taken for simulation of unit cube of skin model is 2.561 s.

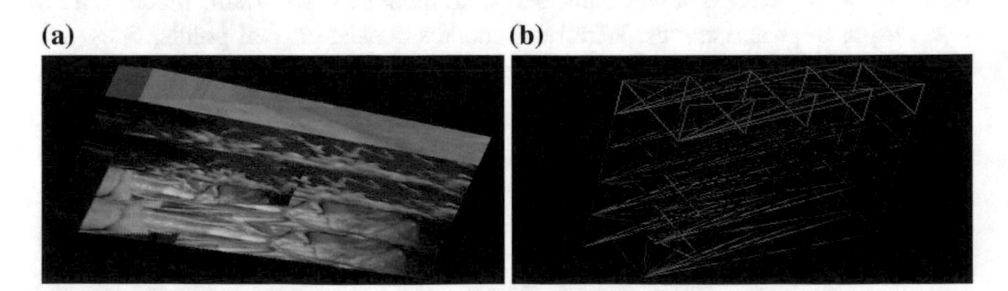

Fig. 5 Rendering the multilayer model: **a** solid model, **b** wireframe model

5 Conclusion

Real life operations require the full involvement of the surgeon on the scene. Thus, it is necessary that surgical applications be able to display complex scenes as real as possible. An attempt has been made to establish the logic of skin prototype. Different methods are available in soft tissue modeling with different criteria basically focusing on surface models, the majority of which concentrates mainly on components of skin, muscles and bones, follicle and fat. Our contribution in this paper is the development of the structure of a multilayer soft tissue prototype with three layers, namely epidermis, dermis and subcutaneous layer, using Delaunay 3D triangulation. The outcome of the simulation is the formation of a mesh model using unstructured points that mimic human skin and its layers. The current work simulates a portion of hand skin where the skin is modeled as a triangular mesh. The same logic can be implemented to any portion of the human body. We further plan to carry out deformation and the cutting process on all three layers.

References

1. Jayasudha, K., Dr. K.G. Mohan., Harshal Sawant. 2014. Review of visualization techniques in computer graphics. *International Journal of Recent Development in Engineering and Technology* 3 (1). ISSN 2347-6435 (online).
2. Nebel, Jean-Christophe. 2001. Soft tissue modeling from 3D scanned data. In *Deformable avatars*, 85–97. Springer.
3. Hashim, Shahrul, David Richens. 2006. Finite element method in cardiac surgery. *Interactive Cardiovascular and Thoracic Surgery* 5 (1): 5–8.
4. Chacko, Brian G., Harshal Sawant, S.R. Shankapal. 2008. Virtual surgery on geometric model of real human organ data. *A supplement to IEI News.*
5. Holzapfel, Gerhard A. 2001. Biomechanics of soft tissue. In *The handbook of materials behavior models*, vol. 3, no. 1, 1049–1063.
6. Etzel, Joset A., Valeria Gazzola, and Christian Keysers. 2009. An introduction to anatomical ROI-based fMRI classification analysis. *Brain Research* 1282: 114–125.
7. Bijari, Payam B., Alireza Akhondi-Asl, and Hamid Soltanian-Zadeh. 2010. Three-dimensional coupled-object segmentation using symmetry and tissue type information. *Computerized Medical Imaging and Graphics* 34 (3): 236–249.
8. Bouten, C.V.C., P.Y.W. Dankers, A. Driessen-Mol, S. Pedron, A.M.A. Brizard, F.P.T. Baaijens. 2011. Substrates for cardiovascular tissue engineering. *Advanced Drug Delivery Reviews* 63 (4): 221–241.
9. Jepps, Owen G., Yuri Dancik, Yuri G. Anissimov, Michael S. Roberts. 2013. Modeling the human skin barrier—towards a better understanding of dermal absorption. *Advanced Drug Delivery Reviews* 65 (2): 152–168.
10. Zöllner, Alexander M., Adrian Buganza Tepole, Ellen Kuhl. 2012. On the biomechanics and mechanobiology of growing skin. *Journal of Theoretical Biology* 297: 166–175.
11. Duffy, J.S., and M. Shuter. 1994. Evaluation of soft-tissue properties under controlled expansion for reconstructive surgical use. *Medical Engineering & Physics* 16 (4): 304–309.
12. Pailler-Mattei, C., S. Bec, and H. Zahouani. 2008. In vivo measurements of the elastic mechanical properties of human skin by indentation tests. *Medical Engineering & Physics* 30 (5): 599–606.

13. Gao, Zhan, Jaydev P. Desai. 2010. Estimating zero-strain states of very soft tissue under gravity loading using digital image correlation. *Medical Image Analysis* 14 (2): 126–137.
14. Maur, Pavel. 2002. Delaunay triangulation in 3D: Technical Report, Department of Computer Science and Engineering.
15. http://www.vtk.org/doc/release/6.1/html/classvtkDelaunay3D.html.

Face Recognition Based on SWT, DCT and LTP

Sunil S. Harakannanavar, C. R. Prashanth, Sapna Patil and K. B. Raja

Abstract Personal Identification based on face recognition has received extensive attention over the last few years in both research and real time applications due to increasing emphasis on security. In this paper, a face recognition methodology based on Stationary Wavelet Transform (SWT), Discrete Cosine Transform (DCT) and Local Ternary Pattern (LTP) is presented. SWT and DCT are applied on resized face images to produce features. LTP is applied on SWT features. SWT, DCT and LTP features are concatenated to get final features. Features of test and database images are compared using Euclidean distance. It is found that the total success rate of the proposed system is better than that of existing systems.

Keywords Face identification · Stationary wavelet transform
Discrete cosine transform · Local ternary pattern · Success rate

1 Introduction

Personal identification is an actively growing area of research, as it becomes clear that traditional personal authentication measures such as cards and passwords may not be enough for secure authentication of human identity. Biometric traits are efficiently used for secure authentication. Voice, face, iris, retina, fingerprint and signature are commonly used for authentication. The physiological and behavioral characteristics of human beings are measured as biometrics. These are used for uniquely recognizing humans in seamless applications such as identity access

S. S. Harakannanavar (✉)
S.G. Balekundri Institute of Technology, Belagavi, India
e-mail: sunilsh143@gmail.com

C. R. Prashanth
Dr. Ambedkar Institute of Technology, Bangalore, India
e-mail: prashanthcr.ujjani@gmail.com

S. Patil · K. B. Raja
University Visvesvaraya College of Engineering, Bangalore, India

A. N. Krishna et al. (eds.), *Integrated Intelligent Computing, Communication and Security*, Studies in Computational Intelligence 771,
https://doi.org/10.1007/978-981-10-8797-4_57

management and access control, and in surveillance applications to identify individuals in groups. Face recognition is one such biometric method which authenticates individuals by facial characteristics. Early face identification systems used simple geometric models, but the recognition process has now changed with artificial intelligence based on complex mathematical modeling, tools, techniques and matching concepts. With these major improvements and initiatives in the past ten years, the modern face recognition system is a computer application which automatically identifies or verifies a person based on a digital image acquired from any source.

Face recognition algorithms are classified in accordance by the overall techniques used; the commonly used approaches are appearance-based and geometric-based. The appearance-based approach extracts features by considering the overall image as input; in the geometric-based approach input consists of facial or physiological properties. The feature extraction process is normally rule-based, using some parameters, transformations, tools or techniques to represent high-dimensional image data. The important performance metrics used to analyze any algorithm are (i) False Rejection Rate (FRR), (ii) False Acceptance Rate (FAR), (iii) Equal Error Rate (EER), (iv) Recognition Rate (RR), and (v) Success Rate (SR).

The remainder of this chapter is organized as follows. Section 2 briefly gives an overview of related work. Section 3 presents the proposed model. Section 4 reports performance analysis, and Sect. 5 concludes the work.

2 Literature Survey

Khan and Gupta [1] proposed a face recognition system that works on two different algorithms, a sub-window algorithm for extraction of human face images and an ABC algorithm for reducing the number of sub-windows taken from the sub-window extraction algorithm. A PCA algorithm is applied on reduced sub-windows for better recognition of human face images, which are divided into various sub-windows. Patel and Shah [2] described facial feature extraction Techniques for automatic face annotation. Automatic face annotation and the different methods of comparative analysis of feature extraction techniques, such as SURF, SIFT, CNN, GABOR filter, Eigen Faces and LBP, are explained. Sharma and Gupta [3] explained a connected component-based algorithm for face recognition which is more popular for recognition, where color face images or pictures are converted into gray-level images and then a connected component-based segmentation algorithm is applied on all gray level images to extract the segments. Sobel-edge detection and a median filter are used for removing noise. A neural network is used for recognition of face images by extracting standard deviation, mean gray value, integrated density and center of mass in much less time as compared with the other algorithms like PCA, LDA and so on. Abbas et al. [4] applied various methods of classifiers with a minimum set of clusters to yield a tremendous recognition rate. The PCA method is used for feature extraction for face recognition; the technique reduces the dimensionality or variations of human poses. Three classifier methods are considered: Euclidian distance (ED), squared Euclidian distance method (SED) and city block distance method (CBD).

Kaur and Kanwal [5] compared different methods for feature extraction in face identification, including principal component analysis (PCA), linear discriminant analysis (LDA), back propagation neural networks (BPNN), genetic algorithm (GA) and support vector machine (SVM). Bhagwanrao and Kalpana [6] proposed a DCT pyramid-based human face recognition to decompose the face into two distinct components: approximation sub-band and reversed L-shaped blocks that contain very high frequency coefficients. The block-based statistical measures are computed on different sub-bands provided from the DCT Pyramid. This statistical measure helps to reduce the dimensional effect and to improve the performance of classification.

Xi et al. [7] discussed a deep learning approach for face recognition based on a local binary pattern network to extract and compare high order levels over complete features in the multi-order hierarchy. Pandey et al. [8] explained face recognition using scattering wavelets under illicit drug abuse variations. SCATNET is deep wavelet-based architecture for texture feature representation that applies Morlet wavelets at multiple scales and orientations. Chen et al. [9] proposed a face recognition system based on extended local binary patterns. Features are extracted by LBP and PCA to reduce dimensions. Kumar et al. [10] explained extraction of informative regions of a face for facial expression recognition in which the main intention is to extract useful features from informative regions of the face where LBP and IRE models estimate the importance of subregions using projection analysis of expressive images.

Padma Suresh and Anil [11] discussed face recognition using curvelet feature extraction on the JAFEE database. Patched geodesic texture transform (PGTT) experiments were conducted using BUDFE. A gradient feature matching technique using a single image is applied. Borade et al. [12] proposed face recognition using fusion of PCA and LDA for dimensionality reduction and feature extraction purposed. Zhao et al. [13] described face recognition based on LBP and Genetic algorithms. The compensation illumination algorithm, which reduces the impact of illumination on face recognition by gamma correction, histogram equalization, logarithmic transformation and exponential transform, where the system finally selects the approach based on histogram equalizations nearest neighbor (NN) and super vector machine (SVM) are used for classification.

Peng et al. [14] discussed graphical representation of heterogeneous face recognition in which weight matrices generated from Markov networks are regarded as graphical representations. The similarity between the weight matrices of heterogeneous face images is used for matching. A coupled representation similarity metric (CRSM) is explained to measure the similarities between graphical representations, and calculated similarity scores between heterogeneous face images are applied for recognition. Liu and Wu [15] proposed development of a face recognition system based on PCA and LBP for intelligent anti-theft doors.

The system uses an LBP algorithm for extracting the facial features and a PCA algorithm for reducing the dimensions of the facial features, combining these with the Euclidian distance method for future matching, which helps to realize the face recognition.

Li and Zhu [16] explained face recognition based on a mobile phone system. It consists of three parts, where the first part is to get human face images from a mobile phone camera and then to carry out some work on the obtained images, such as cropping, scaling, alignment, removing illumination effects, and so on, to get the proper preprocessed image. A convolution neural network (CNN) is used to extract the main features from face. Realization of face retrieval and matching is obtained by using cosine similarity and Euclidian distance methods. Fakhir et al. [17] proposed face recognition for images with various poses. The authors focus on identification and verification of an individual person when they are in the different poses. There are three main aspects that are used to get accurate information of face identification methods such as camera calibration, dimension measurement, face tracking and recognition. A CMT model to measure the dimensions of a human face in order to recognize it as a unique identifier is proposed. The first step of FCMT is to provide a coordinate system analysis by a Haralick algorithm.

3 Proposed Model

In this section, a proposed model of face recognition is discussed; the model is given in Fig. 1. The face images are read from a database, then preprocessing is performed to get the desired part of the face and exclude unwanted information.

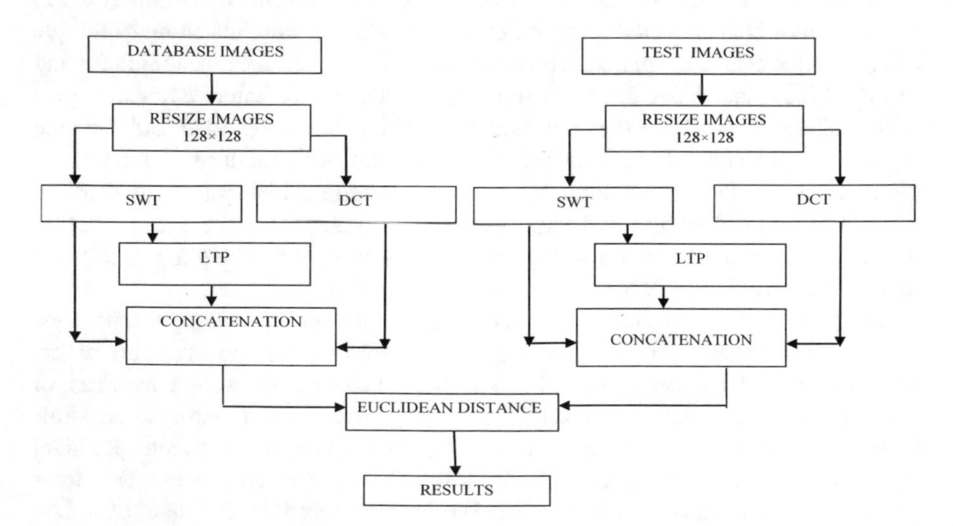

Fig. 1 Model of the proposed face recognition system

The desired features are extracted using SWT, DCT and LTP methods. Matching between the database image and test images is done using Euclidean distance.

3.1 Face Database

The ORL, Indian female and Combined databases are used as input to the system. The ORL database consists of 400 frontal images of 40 people. The size of each image is 92 * 112 and is of PGM (portable grey map) format. The Indian female database consists of 12 different images of each of 22 distinct subjects, each of size 640 * 480. The Combined database consists of 11 images of each of 120 subjects. The size of each image is 320 * 280.

3.2 Methodology

The input face images were resized to 128 * 128 for consistency of size. The stationary wavelet transform (SWT) and discrete cosine transform (DCT) algorithms were applied separately on the preprocessed images to get the corresponding coefficients. LTP was applied on the SWT coefficients. Then SWT, LTP and DCT coefficients were concatenated. Concatenated features of database image and test images were compared using Euclidean distance.

3.3 Stationary Wavelet Transform

Wavelet Transform is widely used in feature detection for MRI signal de-noising, pattern recognition and brain image classification. The discrete wavelet transform (DWT) lacks translation of variant properties, which can be nullified by using stationary wavelet transform (SWT). In SWT, even if the signal is shifted, the transformed coefficient will not change, and the method also performs better in de-noising and edge detecting. In contrast to DWT, SWT can be applied to any arbitrary size of image, rather than being limited to sizes which are multiples of powers of two, and has shown better fusion performance in medical and other applications. SWT, like DWT, is commonly known as *algorithm a trous,* which is French for *with holes,* which refers to insertion of zeros in the filter for up-sampling the filter and suppressing the down-sampling step of the DWT. As with DWT, first the filter is applied to the rows and then the columns, resulting in four images, viz., one approximation and three horizontal, vertical and diagonal detail images. Translation invariance is achieved by removing the down samples and up samples in the DWT and up-sampling the coefficients by a factor of $2j - 1$ in the jth level of the algorithm. SWT is a redundant technique, as the output of each level of the

SWT contains the same number of samples as input and improves the resolution of edge details with three groups of wavelet coefficients.

3.4 Discrete Cosine Transform

Discrete cosine transform (DCT) produces coefficients that are concentrated in the low frequency region. Therefore, it is said to have very good energy compaction properties. The 2D discrete cosine transform $X(k_1, k_2)$ of an image or 2D signal $x(n_1, n_2)$ of size $N_1 * N_2$ is defined by Eq. 1.

$$X(k_1,k_2) = \alpha(k_1)\alpha(k_2) \sum_{n_1=0}^{N_1-1} \sum_{n_2=0}^{N_2-1} x(n_1,n_2) \cos\left(\frac{\pi(2n_1+1)k_1}{2N_1}\right) \cos\left(\frac{\pi(2n_2+1)k_2}{2N_2}\right), \quad \begin{array}{l} 0 \le k_1 \le N_1 - 1 \\ 0 \le k_2 \le N_2 - 1 \end{array} \tag{1}$$

3.5 Local Ternary Pattern

LTP is a ternary or 3-valued code. In LTP the neighborhood pixel values are compared with central pixel. LTP makes use of a threshold constant ‘t’ to threshold pixel values into three different code values, that is 0, 1 and −1, and is defined by the Eq. 2.

$$\mathrm{LTP} = \mathrm{S}(P_i, i_c) = \begin{cases} 1 & P_i \ge (i_c + t) \\ 0 & |P_i - i_c| < t \\ -1 & P_i \le (i_c - t) \end{cases} \tag{2}$$

where i_c is central pixel intensity value, P_i is neighboring pixel intensity value and t is the threshold constant.

3.6 Euclidean Distance

Euclidean distance is used as a classifier for matching. It is also called Pythagorean distance. The minimum Euclidean distance gives the similarity between the unknown face image that is being tested and the images in the database. In Cartesian coordinates, if $p = (p_1, p_2, p_3, \ldots, p_n)$ and $q = (q_1, q_2, q_3, \ldots, q_n)$ are two points in Euclidean space, then the distance from p to q is given by the formula defined by Eq. 3.

$$d(p,\ q) = \sqrt{(q_1 - p_1)^2 + (q_2 - p_2)^2 + \cdots + (q_n - p_n)^2} \tag{3}$$

4 Performance Analysis

The proposed face recognition model was tested on ORL, Indian female and Combined databases for varying PID and POD combinations; the results are tabulated in Table 1. The values of performance parameters such as EER, optimum and maximum TSR of the proposed system for ORL database are computed for different combinations of PID and PODs. It is observed that the results are improved by increasing the number of persons outside the database.

Table 2 shows the values of performance parameters EER, optimum and maximum TSR of the proposed system for the Indian Female database. It is observed that the results are improved by increasing the number of persons outside the database.

Table 3 shows the values of performance parameters EER, optimum and maximum TSR of the proposed system for the Combined database. It is observed that the results are improved by increasing the number of persons outside the database.

Table 1 EER and TSR values for varying PID:POD in case of ORL database

Sl. no	PID: POD	% EER	%Optimum TSR	%Max TSR
1	10:30	15	85	100
2	20:20	13	73	85
3	30:10	11.33	72	83.33

Table 2 EER and TSR values for varying PID:POD in case of Indian Female database

Sl. no	PID:POD	%EER	%Optimum TSR	%Max TSR
1	8:14	5	95	100
2	11:11	10	90	100
3	14:8	15	85	100

Table 3 EER and TSR values for varying PID:POD in case of combined database

Sl. no	PID:POD	%EER	%Optimum TSR	%Max TSR
1	40:80	0	100	100
2	60:60	2	98	100
3	80:40	3	97	100

Table 4 Comparison of TSR value of the proposed system with existing system for ORL database

Sl. no	Techniques	%max TSR
1	LBP + ANN [18]	93.33
2	PCA + RBF NN [19]	93.6
3	SVD + Hidden Markov model [20]	97.5
4	**Proposed model** **LTP + SWT + DCT**	**100**

The percentage maximum TSR of the proposed model is compared with existing methods explained by Patil and Maka [18], Thakur et al. [19] and Bhamre and Memane [20] as shown in Table 4. It is found that the performance of the proposed method is better than the existing methods for ORL faces.

5 Conclusion

In this chapter, face recognition using stationary wavelet transform, discrete cosine transform and local ternary pattern is presented. SWT and DCT were applied to resized face images to produce features. LTP was applied on SWT features. SWT, DCT and LTP features were concatenated to get final features. Features of test and database images were compared using Euclidean distance. Total success rate of the proposed system was better than existing systems due to the invariant nature of SWT and concatenation of multiple features.

References

1. Khan, Neha, and Manish Gupta. 2016. Face recognition system using improved artificial bee colony algorithm. In *International conference on electrical, electronics and optimization techniques*, 3731–3735.
2. Patel, Twisha, and Bhumika Shah. 2017. A survey on facial feature extraction techniques for automatic face annotation. In *International conference on innovative mechanisms for industry applications*, 224–228.
3. Sharma, Shailaja, and Manish Gupta. 2016. An improved connected component based algorithm for face recognition. In *International conference on electrical, electronics and optimization techniques*, 4931–4935.
4. Abbas, Eyad I., Mohammed Safi and S.R. Khalida. 2017. Face recognition rate using different classifier methods based on PCA. In *International conference on current research in computer science and information technology*, 37–40.
5. Kaur, Gurupreet, and Navdeep Kanwal. 2016. A comparative review of various approaches for feature extraction in face recognition. In *International conference on computing for sustainable global development*, 2705–2710.
6. Bhagwanrao, Anil, and C.J. Kalpana. 2015. DCT pyramid based face recognition system. In *International conference on information processing*, 506–510.

7. Xi, Meng, Liang Chen, Desanka Polajnar, and Weiyang Tong. 2016. Local binary pattern network: A deep learning approach for face recognition. In *International conference on image processing*, 3224–3228.
8. Pandey, Prateekshit, Richa Singh and Mayank Vatsa. 2016. Face recognition using scattering wavelet under illicit drug abuse variations. In *International conference on biometrics*, 1–6.
9. Chen, Yong-Ping, Qi-Hui Chen, Kuan-Yu Chou, and Ren-Hau Wu. 2016. Low cost face recognition system based on extended local binary pattern. In *International conference on automatic control conference*, 13–18.
10. Kumar, Sunil, M.K. Bhuyan, and Bilab Ketan Chakraborty. 2016. Extraction of informative regions of a face for facial expression recognition. *Research Article on Institution of Engineering and Technology*, 10 (6): 567–576.
11. Padma Suresh, L., and J. Anil. 2016. Literature survey on face and face expression recognition. In *International conference on circuit, power and computing technologies*, 1–6.
12. Borade, Sushma Niket, Ratnadeep Deshmukh, and Shivakumar Ramu. 2016. Face recognition using fusion of PCA and LDA: BORDA count approach. In *Mediterranean conference on control and automation*, 1164–1167.
13. Zhao, L-hong, Liu Fei and Wang Yong-jun. 2016. Face recognition based on LBP and genetic algorithm. In *Chinese control and decision conference*, 1582–1587.
14. Peng, Chunlei, Xinbo Gao, Nannan Wang, and Jie Li. 2017. Graphical representation of heterogeneous face recognition. *IEEE Transactions on Pattern Analysis and Machine Intelligence* 39 (2): 301–312.
15. Liu, Zheng, and Yong Wu. 2016. Development of face recognition system based on PCA and LBP for intelligent anti-theft doors. In *International conference on computer and communications*, 341–345.
16. Li, Haifeng, and Xiaowei Zhu. 2016. Recognition technology research and implementation based on mobile phone system. In *International conference on natural computation, fuzzy systems and knowledge discovery*, 972–976.
17. Fakhir, M.M., W.L. Woo, J.A. Chambers, and S. SDlay. 2016. Novel method of face recognition from various pose. In *International conference on pattern recognition systems*, 6–11.
18. Patil, Abhilasha A., and Lakshmi Maka. 2015. User recognition based on face using local binary pattern with artificial neural network. *International Journal of Ethics in Engineering & Management Education* 2 (5): 2348–4748.
19. Thakur, S., J.K. Sing, D.K. Basu, M. Nasipuri, and M. Kundu. 2009. Face recognition using principal component analysis and RBF neural networks, vol. 10, no. 5, 7–15.
20. Bhamre, P.D., and Swati B. Memane. 2015. Face recognition using singular value decomposition and hidden Markov model. *International Journal of Modern Trends in Engineering and Research* 2 (10): 323–332.

Pseudo-Hadamard Transformation-Based Image Encryption Scheme

S. N. Prajwalasimha

Abstract In this chapter, a pseudo-Hadamard transform (PHT)-based image encryption technique has been proposed. Images are characterized by high inter-pixel redundancy. This can be varied in two phases: transformation and substitution. Correlation between adjacent pixels can be varied by the transformation phase. Pixel value variation can be made in the substitution phase. Encryption of some standard images has been done, and performance analysis is made based on correlation, entropy, mean square errors, number of pixel change rate (NPCR) and unified average changing intensity (UACI). The results obtained are comparatively better considering those of existing algorithms.

Keywords Encryption · Redundancy · Transformation · Substitution

1 Introduction

In multimedia communication, images play a significant role for information sharing. Data confidentiality can be achieved by an efficient encryption algorithm. Along with confidentiality, integrity and authentication are also important parameters with respect to text and images. Visual cryptography is a method of transforming one form of image to another. Security attacks are quite common nowadays. Among these, analytical, differential and brute force attacks are major [1–3].

Analytical attacks are also called channel attacks, which follow divide-and-conquer strategy. The different parts of the secret keys are captured and then combined analytically to get the complete secrete key.

Differential attacks basically involve the various methods of cracking the encryption algorithm. They are mainly operated on block cipher.

S. N. Prajwalasimha (✉)
Department of ECE, ATME College of Engineering, Mysore, Karnataka, India
e-mail: prajwalasimha.sn1@gmail.com

A. N. Krishna et al. (eds.), *Integrated Intelligent Computing, Communication and Security*, Studies in Computational Intelligence 771,
https://doi.org/10.1007/978-981-10-8797-4_58

Brute force attacks involve trying all combination of a secret key order to get the expected original data. Brute force attacks are also known as last resort tactics. This can be reduced by increasing the key length. An efficient algorithm should be resistant to analytical, differential and brute force attacks and utilize a small amount of execution time.

Transformation and substitution-based image encryption technique has been proposed in this chapter. Pixel dislocation is made using pseudo-Hadamard transformation. Pixel value variation is made using substitution technique.

2 Related Work

The encryption algorithm must be fast enough for the present trend. Balouch et al. have described an image encryption algorithm using Rubik's Zeta function. By this, they achieved fast and effective encryption algorithm [1].

In an image encryption scheme based on logistic maps, a considerable amount of entropy is achieved along with high NPCR [4]. The key length used is 128 bits and the key sensitivity of the algorithm is greater compared to existing algorithms. Reddy et al. have proposed an image encryption algorithm based on a residue number system. They achieved more efficiency compared to other existing algorithms [5].

Chaotic maps are more commonly adopted for encryption schemes. Hou et al. have proposed an image encryption algorithm based on switching fractional order chaotic systems. The algorithm requires less execution time compared to existing algorithms [6].

In the discrete radon transformation-based encryption algorithm, high entropy and less correlation between the adjacent pixels are observed in the encrypted image [2]. Chaotic data can be produced by many methods, among which is the Lorentz system. Butterfly effect is more clearly observed in the Lorentz system. A very low NPCR and UACI are observed in this encryption algorithm proposed by Kayhan and Kurt [7].

Zahmoul et al. have proposed a new encryption algorithm based on beta chaotic maps. High NPCR and UACI are observed by this algorithm. The key length used is 512 bits. Since the key length is greater, the algorithm is effective against brute force attacks, but takes more execution time [8]. Discrete wavelets can be used for real-time image encryption. Diffusion and confusion-based encryption algorithms are fast and efficient nowadays. Slimane et al. have described a two-stage diffusion algorithm using a nested chaotic attractor and secure hash algorithm [9].

3 Proposed Scheme

The proposed algorithm is organized with transformation and substitution. Transformation is performed using pseudo-Hadamard transform, and substitution is done using secret key and S-box (Fig. 1).

3.1 Encryption Algorithm

Step 1: The input image is subjected to pseudo-Hadamard transformation. Figure 2 shows the original images of Lena. Figure 3 shows the transformed image of Lena.

$$x = (a + b) mod\, n \qquad 1 < a, b < n \tag{1}$$

$$y = (a + 2b) mod\, n \qquad 1 < a, b < n \tag{2}$$

$$T(x, y) = I(a, b) \qquad 1 < a, b < n \tag{3}$$

where $I(a, b)$ is the input image of size n × n
$T(x, y)$ is the transformed image of size n × n.

Step 2: The 128-bit secret key is subjected to initial permutation. The initial permuted key is divided into eight groups each with 16 bits.

Step 3: S-box of size n × 1 is created using secret keys. Each row of the transformed image is subjected to XOR operation with the S-box. The process is continued for the number of rounds required to achieve better efficiency.

Step 4: The resultant image as shown in Fig. 4 is the cipher image of size n × n.

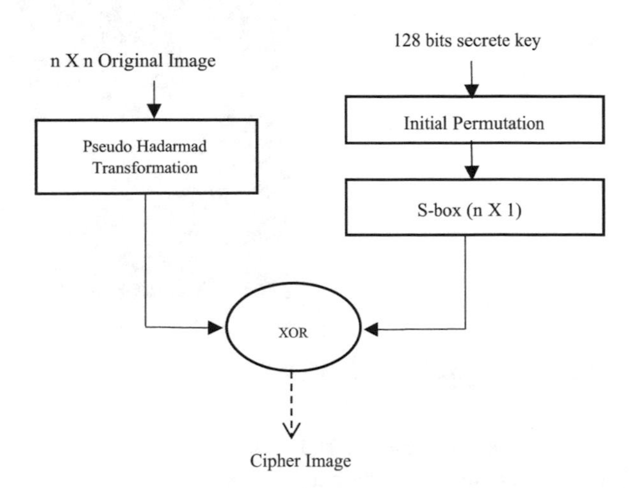

Fig. 1 Flow diagram of proposed algorithm

Fig. 2 Original image of Lena

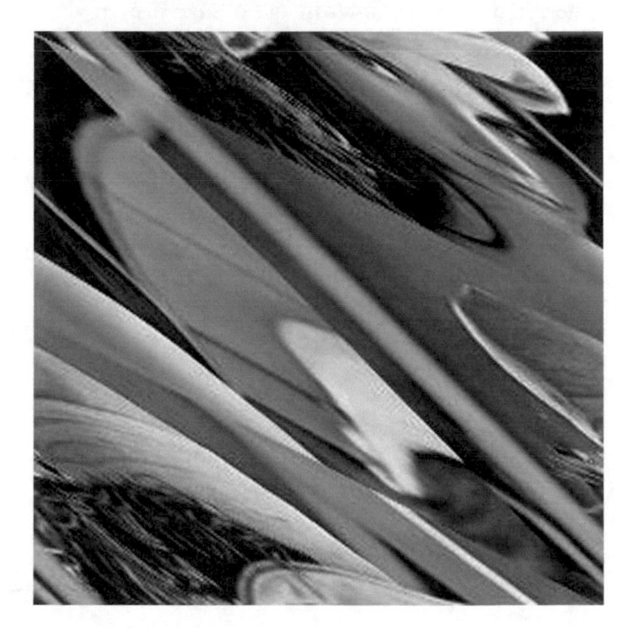

Fig. 3 Transformed image of Lena using PHT

3.2 Decryption Algorithm

Step 1: The 128-bit secret key is subjected to initial permutation as done in the encryption stage. The initial permuted key is divided into eight groups each with 16 bits.

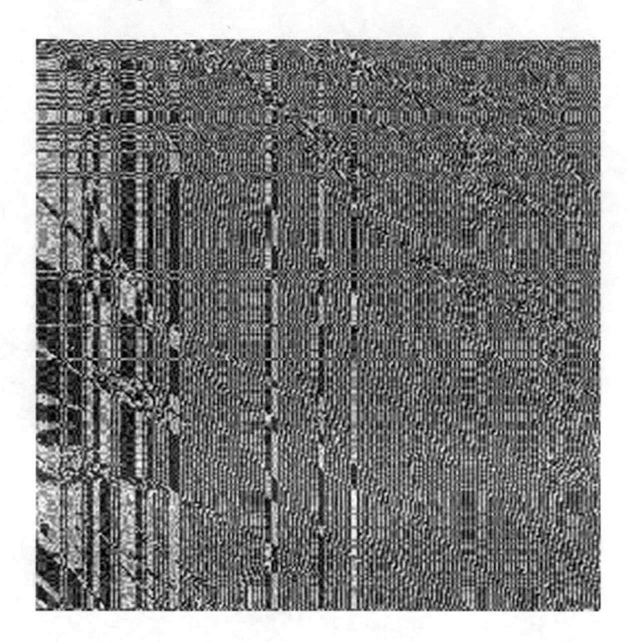

Fig. 4 Cipher image of Lena after the substitution stage

Step 2: The S-box used is same as that in the encryption stage. The cipher image is subjected to XOR operation with the S-box created using secrete keys to get the transformed image.

Step 3: The transformed image is subjected to inverse pseudo-Hadamard transformation to get the original image

$$b = (y - x) mod\, n \qquad 1 < x, y < n \tag{4}$$

$$a = (2x - y) mod\, n \qquad 1 < x, y < n \tag{5}$$

$$I(a, b) = T(x, y) \qquad 1 < x, y < n \tag{6}$$

where $I(a, b)$ is the decrypted image of size n × n
$T(x, y)$ is the transformed image of size n × n.

4 Experimental Results

From the security tests it has been observed that the correlation between the adjacent pixels has been effectively broken by PHT, but the entropy is unchanged since the pixel values are unaltered. Still, very little correlation between the adjacent pixels has been observed after the substitution phase along with a high entropy value indicating that the pixel values are altered in the encrypted image. Also, very

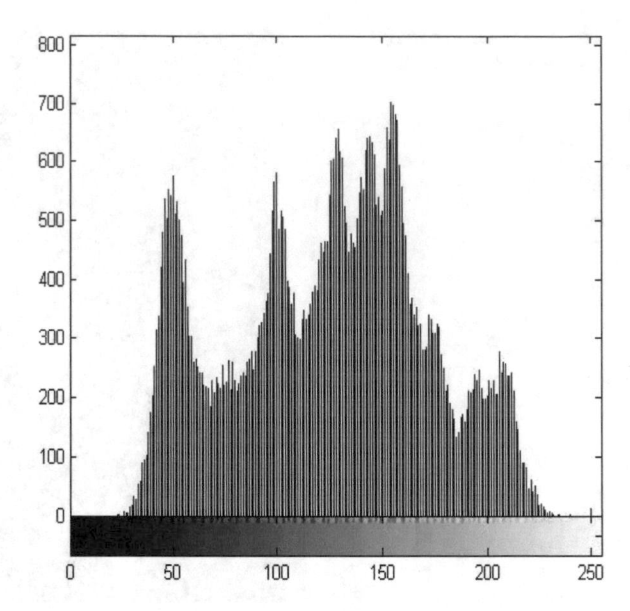

Fig. 5 Histogram of 256×256 original image of Lena

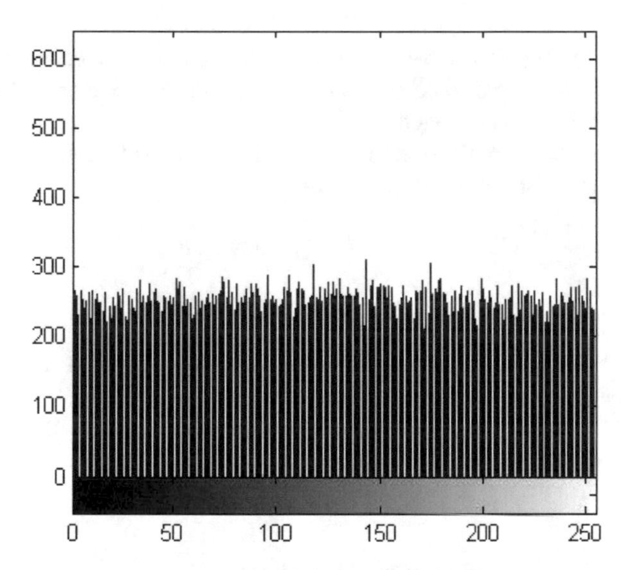

Fig. 6 Histogram of 256×256 encrypted image of Lena

small UACI and high NPCR values are noticed from the outcome of the standard images.

Figure 5 shows the histogram of original image of Lena. Figure 6 shows the histogram of final encrypted image of Lena indicating almost equal distribution of pixel values (Tables 1 and 2).

Table 1 The comparision of entropy and correlation between standard and encrypted images along with computational time for each image

Images	Entropy ≈ 8	Correlation after transformation	Correlation after substitution (Cipher Image)	Computational time in seconds
Lena	5.5407 (Blow Fish) [10]	0.0095 [2]	0.0021 [2]	1.96
	5.5438 (Two Fish) [10]			
	5.5439 (AES 256) [10]			
	5.5439 (RC 4) [10]	N/A	0.1500 [11]	
	7.5220 [11]			
	7.9950 [12]			
	7.9958 [1]	**−0.0082 (proposed scheme)**	**−0.0055 (proposed scheme)**	
	7.9970 [2]			
	7.9971 [9]			
	7.9972 (proposed scheme)			
Baboon	7.9947 [9]	**0.0228 (proposed scheme)**	**−0.0064 (proposed scheme)**	2.66
	7.9950 [12]			
	7.9954 (proposed scheme)			
Peppers	7.9954 [9]	**−0.0095 (proposed scheme)**	**−0.0095 (proposed scheme)**	2.13
	7.9960 [12]			
	7.9966 (proposed scheme)			

Table 2 Analytical test performance of the proposed algorithm for standard images

Images	UACI	NPCR
Lena	31.00 [11]	90.21 [11]
	32.01 [9]	99.60 [9]
	33.4201 [1]	99.5859 [1]
	28.3192 (proposed scheme)	**99.61 (proposed scheme)**
Baboon	30.87 [9]	99.59 [9]
	33.43 [13]	99.60 [13]
	27.078 (proposed scheme)	**99.612 (proposed scheme)**
	30.71 [9]	99.61 [9]
	29.345 (proposed scheme)	**99.541 (proposed scheme)**

5 Conclusion

In the proposed cryptosystem, pseudo-Hadamard transformation breaks the correlation between the adjacent pixels, and substitution using S-box helps to achieve high entropy in the cipher image. The algorithm uses a 128-bit secret key, which produces 2^{128} combinations in order to resist brute force attacks. Analytical tests show that the proposed algorithm gives better results compared to existing cryptosystems. Further the cryptanalysis on the algorithm can be made more complex by increasing number of rounds in the substitution stage of the algorithm.

References

1. Balouch, Zaheer Abbas, Muhammad Imran Aslam and Irfan Ahmed. 2017. Energy efficient image encryption algorithm. In *Proceedings of international conference on innovations in electrical engineering and computational technologies (ICIEECT).*
2. Prajwalasimha, S.N., and S.R. Bhagyashree. 2017. Image Encryption using Discrete Radon Transformation and Non chaotic Substitution. In *Proceedings of IEEE international conference on electrical, computer and communication technologies (ICECCT)*, vol. 2, 842–846.
3. Prajwalasimha, S.N., and Usha Surendra. Multimedia data encryption based on discrete dyadic transformation. In *Proceedings of IEEE international conference on signal processing and communication,* 492–495, July 2017, Coimbatore, Tamil Nadu, India.
4. Liu, Lingfeng, Suoxia Miao, Hanping Hu and Mengfan Cheng. N-phase logistic chaotic sequence and its application for image encryption. *IET Signal Processing* 10 (9): 1096–1104.
5. Reddy, P. Venkata Narasa, and Rajasekhar Karumuri. 2016. Image encryption and decryption in RNS domain based on moduli set. In *Proceedings of international conference on communication and electronics systems (ICCES).*
6. Hou, Jialin, Rui Xi, Ping Liu and Tianliang Liu. The switching fractional order chaotic system and its application to image encryption. *IEEE/CAA Journal of Automatica Sinica* 4 (2): 381–388.
7. Celİk,Kayhan, and Erol Kurt. 2016. A new image encryption algorithm based on lorenz system. In *Proceedings of international conference on electronics, computers and artificial intelligence (ECAI).*
8. Zahmoul, Rim, and Mourad Zaied. 2016. Toward new family beta maps for chaotic image encryption. In *Proceedings of IEEE international conference on systems, man, and cybernetics (SMC).*
9. Slimane, Nabil Ben, Kais Bouallegue, and Mohsen Machhout. 2016. Nested chaotic image encryption scheme using two-diffusion process and the Secure Hash Algorithm SHA-1. In *4th international conference on control engineering & information technology (CEIT).*
10. Cui, Delong, Lei Shu, Yuanfang Chen, and Xiaoling Wu. Image encryption using block based transformation with fractional Fourier transform. In *8th International conference on communications and networking in China (CHINACOM),* Aug 2013.
11. Hazarika, Nitumoni, Sagarika Borah and Monjul Saikia. 2014. A wavelet based partial image encryption using chaotic logistic map. In *Proceedings of IEEE international conference on advanced communication control and computing technologies (ICACCCT)*, 1–5.

12. Goel, Anis, and Kaustubh Chaudhari. 2016. Median based pixel selection for partial image encryption. In *Sixth international conference on image processing theory, tools and applications (IPTA)*.
13. Ni, Zhengchao, Xuejing Kang and Lei Wang. 2016. A novel image encryption algorithm based on bit-level improved Arnold transform and hyper chaotic map. In *IEEE international conference on signal and image processing*.

A Parallel Algorithm to Hide an Image in an Image for Secured Steganography

N. Gopalakrishna Kini, Gautam and Vishwas G. Kini

Abstract Recent developments in computer security have shown that compared to cryptography, steganography is a better way of securing messages. With the advantages offered by parallel computing platforms, a large secret image can be efficiently hidden in another image. This parallelism is achieved in steganography using the OpenCL parallel programming technique. The speed-up improvement obtained is very good with reasonably good output signal quality, even when a large amount of data is processed. The aim of this work is to analyze steganography algorithms and to show how the 24-bit color image of a carrier can be used to hide a secret image. We compare the results by calculating peak signal-to-noise ratio (PSNR), mean squared error (MSE), analysis of histogram and speedup achieved when a large amount of data is processed.

Keywords Steganography · Stego-image · LSB parallel algorithm PSNR · MSE · Histogram

1 Introduction

Data security is crucial in areas where most of the data is vulnerable to attacks. Cryptography and steganography are used to maintain the confidentiality of data. Steganography is considered to be better than cryptography because the hidden message does not attract attention to itself for scrutiny. Steganography plays a central role in secret message communication [1].

N. G. Kini (✉) · Gautam · V. G. Kini
Department of CSE, Manipal Institute of Technology, Manipal Academy of Higher Education, Manipal, Karnataka, India
e-mail: ng.kini@manipal.edu

Gautam
e-mail: gautam00717@gmail.com

V. G. Kini
e-mail: vgkini38@gmail.com

A. N. Krishna et al. (eds.), *Integrated Intelligent Computing, Communication and Security*, Studies in Computational Intelligence 771,
https://doi.org/10.1007/978-981-10-8797-4_59

In steganography, before the hiding process, the sender must select an appropriate message carrier and an effective message to be hidden. The object used to carry the message is called the message carrier or carrier image. The secret image is the image that is to be embedded into the carrier image. Stego-image refers to the image which is carrying a hidden message. So, given a carrier image and a secret image, the goal of steganography is to produce a stego-image that carries the secret message.

A steganography algorithm is used to encrypt the message in an effective way. Any communication technique can be used by the sender to send the hidden image to the receiver. After receiving the message, the receiver is able to decrypt the secret message using the extraction algorithm and a secret key [1, 2].

The work described in this chapter uses the least significant bit (LSB) algorithm to hide and extract an image. If the size of the image is colossal, then it takes a huge amount of time to encode and decode the image. So, while providing security, the algorithm must also be expeditious. We discuss how a carrier image can be used to hide another image using the parallel programming language OpenCL, and we implement a sequential program using the LSB algorithm so that we can compare the execution time of both the parallel and sequential implementations.

2 Related Work

Many modifications in the LSB technique have been suggested by different authors. In [1], the authors discuss hiding the data over an image using different steganography algorithms and a comparison of those algorithms. A steganography algorithm for hiding text files in images is described in [2]. An overview of different steganography techniques from the basic early systems to the recent approaches is discussed in [3].

History related to steganography and some of the popular steganography techniques are presented in [4]. In [5], there is a description of applying varieties of LSB methods to a 24-bit color image, comparing the results with an 8-bit color image and examining the results using PSNR (peak signal to noise ratio), MSE (mean squared error) and histogram analysis. Pixel pattern-based steganography, which involves hiding the message within an image by using the existing RGB values, is another work contributed in [6].

In [7], it has been described how to leverage the OpenCL framework to build interesting and useful applications. In [8], it has been shown that the steganography algorithms can be parallelized to hide text messages in a grey scale cover image, hence improving the execution speed. A way of implementing an image steganography algorithm in parallel on GPU systems is explained in [9].

An analysis of steganography algorithms as applied to binary images and, further, how parallel execution dramatically reduces response time for data-intensive operations has been shown in [10].

3 Methodology

In steganography, before the encryption process, the sender selects a carrier image and a secret image. The stego-image is obtained after consolidation of the secret image with the carrier image. This proposed work is a parallel LSB technique for hiding and extracting a bitmap image in and from another bitmap image, respectively.

The encryption is done using the LSB parallel algorithm. The stego-image and the header of the secret image are sent to the receiver by the sender. Here, the header of the secret image acts as a key for the receiver to decode the hidden image. This algorithm precludes any intermediary attacks by sending the header file and the encrypted image separately. The receiver then uses the LSB parallel algorithm to decrypt the secret image using the header of the secret image as a key.

The carrier image (*C*) is selected to be sufficiently large enough to be able to hide the secret image (*S*) completely in order to obtain a stego-image (*SG*). The header information (*H*) of *S* acts as a key, without which *S* cannot be extracted from *SG*. *H* is shared between the sender and receiver of *S*. The algorithm performs the encryption and decryption process in milliseconds thanks to parallelism.

Each pixel in a digital image is a color combination of RGB components. Eight bits represent each of the three color values Red (R), Green (G), and Blue (B) in a 24-bit value at each pixel. Embedding the secret image changes the pixel intensity but does not make any noticeable change in the carrier image, since the quantity of data is huge, and the human eye cannot notice these small changes in the pixel intensity. Maintaining picture quality is important for the protection of the secret image.

Two phases are defined, namely, the embedding phase, which is carried out by the sender of the secret image, and the extraction phase, which is carried out by the receiver of the secret image.

In the embedding phase, each bit of the R, G and B components of each pixel of *S* is hidden in parallel in the least significant bit (LSB) of each byte of the R, G and B components of each pixel of *C*. This process is carried out by the kernel where each work item executes in parallel. The result of the embedding phase is the stego-image *SG*. In the host, the header information of *S* is written into a text file, which is then shared with the intended receiver of *S*.

In the extraction phase, the size (*s*) of the image is obtained from the header information of *S*. The LSBs of each byte of the R, G and B components of every 8 pixels of the stego-image *SG*, taken in order, are combined in parallel to form one byte of the R, G and B components of each pixel of S. Therefore, (s times 8) pixels of *SG* are read in order to form the *s* pixels of *S*. The header information of *S*, along with the *s* pixels, are combined by the receiver to form the secret image *S′*.

3.1 Embedding Phase

```
Algorithm - I: For HostCode ()

Input: Carrier image C and Secret image S

Output: Stego-image SG, Header information H of secret
image S

Steps:

1. Read the cover image C and the secret image S.

2. Write the header information H of the secret image S
to a text file.

3. Read the number of pixels in S from H and store it in
s.

4. Store the R, G and B values of the first s x 8 pixels
of C in arrays Rc, Gc and Bc respectively.

5. Store the R, G and B values of each pixel of S in
arrays Rs, Gs and Bs respectively.

6. Call KernelCode() with Rc, Gc, Bc, Rs, Gs, Bs and (s
x 8) as the parameters, which returns the R, G and B
values of s x 8 pixels of the stego-image SG in the
arrays Rg, Gg and Bg respectively.

7. Write the header information H of C and the Rg, Gg
and Bg values to a new image file, to obtain the stego
image SG.
```

Algorithm - II: KernelCode()

Input: Arrays Rc, Gc, Bc containing the R, G, B values, respectively, of the pixels of the cover image *C*, arrays Rs, Gs, Bs containing the R, G, B values, respectively, of the pixels of the secret image *S* and number of work items *s* (which is equal to no. of pixels in S times 8)

Output: Arrays Rg, Gg, Bg containing the R, G, B values, respectively, of the pixels of the stego-image *SG*.

Steps:

1. For each work-item *i* ranging from 0 to $((s \times 8) - 1)$ executing in parallel.

    1.1 Convert $Rs_i/8$, $Gs_i/8$ and $Bs_i/8$ value into its binary equivalent $Rs_i'/8$, $Gs_i'/8$ and $Bs_i'/8$ respectively

    1.2 Obtain the bit at position (*i* mod 8) of the byte from $Rs_i'/8$, $Gs_i'/8$ and $Bs_i'/8$ of $(i/8)^{th}$ pixel of S and store them as tr, tg and tb respectively.

    1.3 Obtain the LSB of the byte from Rc_i, Gc_i and Bc_i of the ith pixel of *C* and store them as cr, cg and cb respectively.

    1.4 Compare tr and CR

    1.4.1 If cr = tr, then

    1.4.1.1 Set $Rg_i = Rc_i$

    1.4.2 Else if cr = 1 and tr = 0, then

    1.4.2.1 Set $Rg_i = Rc_i - 1$ // Subtracting 1 from decimal value of Rc_i since we want to encode 0 in the LSB of Rg_i

    1.4.3 Else if cr = 0 and tr = 1, then

    1.4.3.1 Set $Rg_i = Rc_i + 1$ // Adding 1 to decimal value of Rc_i Since we want to encode 1 in the LSB of Rg_i

    1.5 Repeat the same procedure as step 1.4 for tg, cg and tb, cb

3.2 Extraction Phase

Algorithm – III: For HostCode()

Input: Stego-image *SG*, Header information *H* of secret image *S*

Output: Secret image *S'*

Steps:

1. Read the stego-image *SG*.

2. Read the header information *H* and obtain the number of pixels of *S* and store it in *s*.

3. Store the R, G and B values of the first *s* x 8 pixels of *SG* in arrays Rg, Gg and Bg respectively.

4. Call KernelCode() with Rg, Gg and Bg as the parameters, which returns the R, G and B values of *s* pixels of the secret image *S'* in the arrays Rs, Gs and Bs respectively.

5. Write the header information *H* of *S* and the Rs, Gs and Bs values to a new image file, to obtain the secret image *S'*.

Algorithm – IV: KernelCode()

Input: Arrays Rg, Gg, Bg containing the R, G, B values, respectively, of the pixels of the stego-image *SG* and the number of work-items *s* (which is equal to number of pixels in *S*)

Output: Arrays Rs', Gs', Bs' containing the R, G, B values, respectively, of the pixels of the secret image *S'*

Steps:

1. For each work-item i ranging from 0 to *s* – 1 executing in parallel.

   1.1 Set Rs_i, Gs_i and Bs_i to 0

   1.2 For $j = i \times 8$ to $((i + 1) \times 8 - 1)$

      1.2.1 Read the LSB of the byte from Rg_j, Gg_j and Bg_j

      1.2.2 Add the corresponding LSB value to Rs_i, Gs_i and Bs_i in order to form the binary values of Rs_i, Gs_i and Bs_i of the ith pixel of *S'* respectively

   1.3 Convert the binary values of Rs_i, Gs_i and Bs_i into their decimal equivalent values Rs_i', Gs_i' and Bs_i'

The next part of the work is to determine to what level the stego-image containing the secret image is blurred by the carrier image. For this we compare the result by calculating peak signal-to-noise ratio (PSNR), mean squared error (MSE) and analyze the histogram. Initially, MSE is calculated to find the PSNR [5].

MSE is a measure of the average of the squares of the errors. It is used to measure the error difference of two images by calculating the mean error value. Let the carrier image and the stego-image be of size $m \times n$; then, the pixel value (x, y) be from 0 to $m - 1$ and 0 to $n - 1$ respectively, and can be defined to each carrier image and stego-image as:

$$MSE = \frac{1}{mn} \sum_{i=0}^{m-1} \cdot \sum_{j=0}^{n-1} [C(i,j) - SG(i,j)]^2 \tag{1}$$

where the coordinates of a pixel of the carrier image are $C\ (i, j)$, and the coordinates of a pixel of the stego-image are $SG\ (i, j)$.

PSNR is the ratio between the maximum value of an image pixel and the mean error difference of two images under consideration, as shown in Eqs. (2) and (3). This ratio helps to differentiate between the images based on their pixel values.

$$PSNR = 10 log_{10}\left(\frac{MAX_i^2}{MSE}\right) \tag{2}$$

$$PSNR = 20 \log_{10}(MAX_i) - 10 \log_{10}(MSE) \tag{3}$$

In image steganography, the histogram analysis is important in analyzing both carrier image and stego-image [1, 5].

4 Result

Figures 1, 2 and 3 show our experimental cases of a carrier image of size 1024 × 768, secret image of size 240 × 160 and the stego-image of size 1024 × 768. There is absolutely no visible difference found between the carrier image in Fig. 1 and the stego-image in Fig. 3. Similar analyses performed with secret images of larger sizes have arrived at the same conclusion.

Table 1 gives the execution time for our code to encrypt secret images of various sizes and resultant values of MSE and PSNR. Results of execution time for the sequential code are also given in order to compute the speedup achieved.

A better PSNR value is obtained through our work. A larger PSNR value is an indication of good image quality and negligible difference between carrier image and the stego-image. Speedup results show that we successfully exploited massive parallelism in encryption.

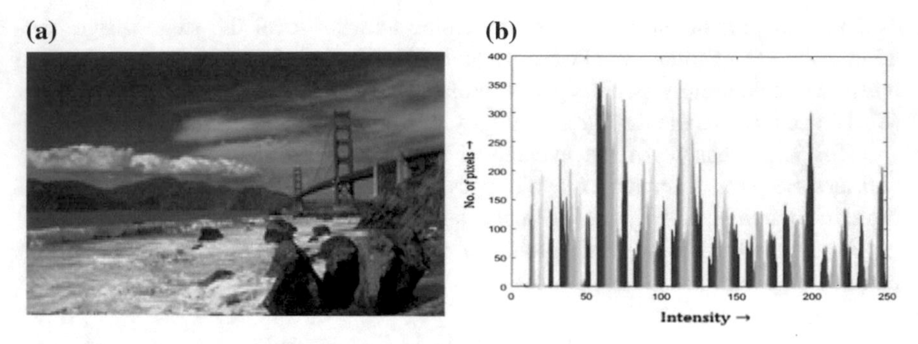

Fig. 1 **a** Carrier image. **b** Histogram of carrier image

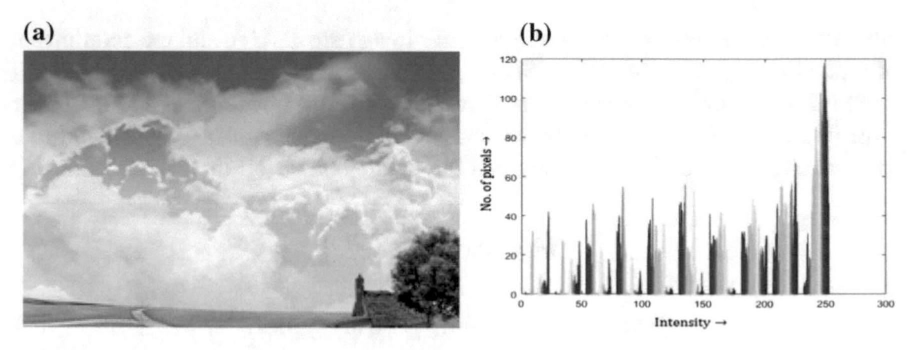

Fig. 2 **a** Secret image before embedding. **b** Histogram of secret image

Fig. 3 **a** Stego-image. **b** Histogram of stego-image

Figures 1a, 2a and 3a show the histograms of the carrier image, secret image and stego-image, respectively. By analyzing the histograms of the stego-image and the carrier image, it is observed that there are minimal changes between them. Because of this it is very difficult for an observer to perceive that secret data is hidden in the stego-image.

Table 1 MSE, PSNR, time to generate stego-image and speedup obtained in embedding phase

Carrier image size	Secret image size	MSE	PSNR in dB	Execution time in milliseconds		Speedup
				Sequential	Parallel	
1024 × 768	240 × 160	0.2452	54.2365	131454.68	17.52	7503.12
	280 × 190	0.3045	53.2955	293560.69	20.29	14468.24
	320 × 240	0.4262	51.8347	656556.21	29.21	22477.10
	360 × 270	0.6782	49.8170	872273.67	34.52	25268.65

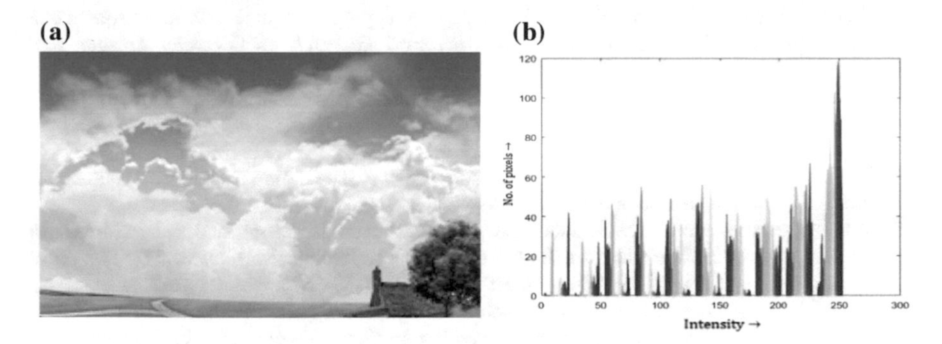

Fig. 4 **a** Secret image after decode. **b** Histogram of secret image after decode

Figure 4 gives the secret image extracted from the stego-image by the receiver. Figure 4a shows the histogram of secret image after decoding it from the stego-image. Comparing the histogram of the secret image before embedding, Fig. 2a, with the secret image after extraction from the stego-image, Fig. 4a, shows that the secret image after extraction is apparently very similar to the original secret image before embedding.

5 Conclusion

In this chapter we have described our algorithm for hiding a secret image in a 24-bit color image. We have proposed how a parallel LSB technique for hiding and extracting a bitmap image in and from another bitmap image, respectively, using an OpenCL parallel programming technique. The encryption is done using the LSB parallel algorithm. The receiver uses the LSB parallel algorithm to decrypt the secret image using the secret image header as a key.

Larger secret images can be hidden inside other large images, since the processing time with the parallel computing platform is minimal. The speed-up improvement obtained is noteworthy, with reasonably good output signal quality, when large amounts of data are processed. We compare the results by calculating peak signal-to-noise ratio (PSNR), mean squared error (MSE), histogram analysis and speedup achieved with good output signal quality when large amounts of data are processed.

References

1. Nani Koduri. 2011. Information security through image steganography using least significant bit algorithm. MSc dissertation, University of East London.
2. Sharma, Vipul, and Sunny Kumar. 2013. A new approach to hide text in images using steganography. *International Journal of Advanced Research in Computer Science and Software Engineering* 3 (4): 701–708.
3. Tiwari, Anjali, Seema Rani Yadav, and N.K. Mittal. 2014. A review on different image steganography techniques. *International Journal of Engineering and Innovative Technology* 3 (7): 121–124.
4. Kumar, Arvind, and Km. Pooja. 2010. Steganography-A data hiding technique. *International Journal of Computer Applications* 9 (7): 19–23.
5. Rawat, Deepesh, and Vijaya Bhandari. 2013. A steganography technique for hiding image in an image using LSB method for 24-bit color image. *International Journal of Computer Applications* 64 (20): 15–19. https://doi.org/10.5120/10749-5625.
6. Rejani, R., D. Murugan, and Deepu V. Krishnan. 2015. Pixel pattern based steganography on images. *ICTACT Journal on Image and Video Processing Communication Technology* 5 (3): 991–997. https://doi.org/10.21917/ijivp.2015.0146.
7. Gaster, Benedict, Lee Howes, and David R. Kaeli. 2012. *Heterogeneous computing with OpenCL*, 2nd ed. Elsevier Publications.
8. Jyothi Upadhya, K., U. Dinesh Acharya, S. Hemalatha. 2013. Speed-up improvement using parallel approach in image steganography. In *The second international conference on information technology convergence and services*, 125–136. https://doi.org/10.5121/csit.2013.3913.
9. Jyothi Upadhya, K., U. Dinesh Acharya, and S. Hemalatha. 2014. A parallel approach for image steganography using OpenCL in GPU based systems. In *International conference on graph algorithms, high performance implementations and applications*, 621–632.
10. Greca, Silvana, and Edlira Martiri. 2012. Wu-Lee steganographic algorithm on binary images processed in parallel. *International Journal of Video & Image Processing and Network Security* 12 (3): 1–4.

Medical Image Segmentation Using GA-Based Modified Spatial FCM Clustering

Amiya Halder, Avranil Maity and Ananya Das

Abstract This chapter proposes a unique method for unsupervised segmentation of medical images using genetic algorithm (GA)-based spatial fuzzy C-means (SFCM) clustering. The aim of the algorithm is to segment the medical image into an appropriate number of clusters, whereby the required number of clusters is computed automatically. SFCM takes into account the effect of neighborhood pixels on a central pixel, and thus the set of clusters obtained by SFCM forms the basis for a genetic algorithm where different genetic operators are used to further calibrate the centroids. A validity index is used to obtain the optimal number of clusters. The experimental results of the proposed method are compared with existing methods for further validation.

Keywords Image segmentation · SFCM · Genetic algorithm · Medical image

1 Introduction

Image processing is the manipulation of images using certain need-specific algorithms to enhance or obtain some information from them. It has wide application in almost every field, including remote sensing, pattern recognition, image sharpening, robotics and medical image processing [1]. Image segmentation is a pivotal aspect of image processing. It is the process of partitioning the pixels into disjoint

---

A. Halder (✉) · A. Maity · A. Das
STCET, 23, Kidderpore, Kolkata, West Bengal, India
e-mail: amiya.halder77@gmail.com

A. Maity
e-mail: avranilmaity97@gmail.com

A. Das
e-mail: ananya15.das@gmail.com

A. N. Krishna et al. (eds.), *Integrated Intelligent Computing, Communication and Security*, Studies in Computational Intelligence 771,
https://doi.org/10.1007/978-981-10-8797-4_60

homogeneous regions, based on characteristics such as intensity, color, texture and tone. A segmented image provides much information that previously could have been overlooked. A noisy image often leads to inaccurate segmentation. In medical image analysis, image segmentation is a stepping stone. In the analysis of MRI images of the brain, segmentation helps highlight anatomical structures and affected regions such as tumors, and also helps remove noise from these images to provide medical practitioners a better and clearer understanding. The K-means algorithm [2, 3] is one of the earliest segmentation algorithms. The FCM algorithm [4–6] partitions a given set of N-dimensional data into M number of subsets or clusters. FCM has previously been used to segment medical images in [7], but it still has certain disadvantages. Ahmed et al. overcame these disadvantages in [8] by introducing a new method, SFCM [4, 9]. Apart from these, Kole and Halder [10] have also introduced improved methods of image segmentation using a genetic algorithm [11, 12]. Concepts of a rough set was incorporated by Maji and Pal in [13]. A new method for segmenting medical images using Rough Kernelized Fuzzy C-means (RKFCM) is proposed by Halder and Guha [14]. This chapter presents an evolutionary method for segmenting medical images using SFCM coupled with the genetic algorithm. This is a GA-based segmentation, which utilizes the searching ability of the genetic algorithm to obtain the optimized set of cluster centers.

2 Genetic Algorithm

The genetic algorithm [10, 12, 15] is an optimization technique based on the "survival of the fittest theory". Traditional optimization techniques are too slow [16] and time-consuming compared to this heuristic search method. Five major phases are considered in a genetic algorithm: initialization, fitness function, selection, crossover and mutation.

3 Fuzzy C-means Clustering with Spatial Constraints

In order to overcome the shortcomings of FCM, Ahmed proposed a new algorithm called SFCM [8], which introduces local spatial constraints. This algorithm takes into account the effect of the immediate neighboring pixels to compensate for the inhomogeneity of the clusters. The objective function is modified to

$$J_{sfcm} = \sum_{\tau=1}^{C} \sum_{j=1}^{N} v_{\tau j}^{m} ||\rho_j - \vartheta_\tau||^2 + \frac{\alpha}{N_r} \sum_{\tau=1}^{C} \sum_{j=1}^{N} v_{\tau j}^{m} \sum_{r \in N_j} ||\rho_r - \vartheta_\tau||^2 \quad (1)$$

where $v_{\tau j}$ is the membership value of a pixel, ρ_j is a pixel or data point of an image, ϑ_τ is the cluster center of the τth cluster, N_R is the cardinality (i.e., the number of neighboring pixels) and m is the fuzziness index. α is crucial for controlling the trade-off between the original image and the mean-filtered image. The proposed algorithm is described in Algorithm 1.

Algorithm 1: GA BASED SFCM

Input: A medical image *Im* of size $N = P \times Q$

Output: Segmented image

1 **for** $n = 2......K_{max}$ **do**

2 // n is the number of cluster

3 Initializing the set of chromosomes

4 **for** $\tau = 1.....\omega$ **do**

5 **for** $j = 1.....n$ **do**

6 Initialize the cluster centers randomly

7 // Applying SFCM

8 **for** $l = 1.....\omega$ **do**

9 // l is the chromosome number

10 **for** $j = 1.....N$ **do**

11 **for** $\tau = 1.....n$ **do**

12 $v_{\tau j} = \frac{[||\rho_j - \vartheta_\tau||^2 + \frac{\alpha}{N_r} \sum_{r \in N_j} ||\rho_r - \vartheta_\tau||^2]^{\frac{-1}{m-1}}}{\sum_{k=1}^{C} [||\rho_j - \vartheta_\tau||^2 + \frac{\alpha}{N_r} \sum_{r \in N_j} ||\rho_r - \vartheta_\tau||^2]^{\frac{-1}{m-1}}}$

13 **for** $\tau = 1.....n$ **do**

14 $\vartheta_\tau = \frac{\sum_{j=1}^{N} v_{\tau j}^m (\rho_j + \frac{\alpha}{N_R} \sum_{r \in N_j} \rho_r)}{(1+\alpha) \sum_{j=1}^{N} v_{\tau j}^m}$

15 // Applying GA

16 **repeat**

17 Calculation of fitness function

18 **for** $l = 1.....\omega$ **do**

19 // l is the chromosome number

20 **for** $\tau = 1.....N$ **do**

21 **for** $j = 1.....n$ **do**

22 **if** $||\rho_\tau - \vartheta_j|| < ||\rho_\tau - \vartheta_p||, p = 1, 2, \ldots, K$ **then**

23 Assign ρ_τ to cluster ϑ_τ

24 **for** $j = 1.....n$ **do**

25 $\vartheta_j = \frac{1}{N_j} \sum_{\rho_\tau \in C_j} \rho_\tau$

26 **for** $\tau = 1.....N$ **do**

27 **for** $j = 1.....n$ **do**

28 $\Phi = \sum_{\rho_\tau \in C_j} ||\rho_\tau - \vartheta_j||$

29 $\chi = \frac{1}{\Phi}$

30 **for** $l = 1.....\omega$ **do**

31 Apply Roulette Selection

32 Crossover

33 Mutation

34 **until** $|\vartheta_{new} - \vartheta_{old}| < \delta$, $\delta \in [0, 1]$;

Table 1 Validity index of the various medical images using K-means, FCM, SFCM, RFCM, KFCM, RKFCM, GAFCM and the proposed method

Methods	K-means	FCM	SFCM	RFCM	KFCM	RKFCM	GAFCM	GASFCM
Brain 1	2.31	0.1775	0.0761	1.31	0.344	0.141	0.0665	0.0615
Brain 2	0.99	0.0977	0.0530	0.11	0.070	0.06	0.0525	0.0471
Brain 3	5.28	0.178	0.1804	0.29	1.24	0.160	0.0712	0.0693
Brain 4	2.274	0.0549	0.0522	0.36	0.26	0.05	0.0356	0.0346
Brain 5	2.18	0.080	0.0902	0.16	1.55	0.080	0.0693	0.0684
Brain 6	8.163	0.1016	0.0448	0.80	0.16	0.060	0.0383	0.0351
Brain 7	5.29	0.09	0.0454	0.50	0.182	0.083	0.0343	0.0332
Brain 8	5.96	0.1280	0.0616	1.11	0.188	0.077	0.0535	0.0406
Brain 9	4.097	0.0838	0.0614	0.09	0.160	0.069	0.0463	0.0425
Brain 10	4.59	0.1025	0.0677	0.52	0.41	0.068	0.0568	0.0506

4 Experimental Results

The proposed algorithm is applied on a dataset of 100 brain images [17] to study the effect in different situations. Table 1 records the validity indices of the proposed method in comparison to existing methods including K-means, FCM, rough FCM, kernelized FCM, rough kernelized FCM, and GAFCM of 10 images. We have used Turi's Validity Index [18], which must be minimized to obtain the optimal number of clusters. From Table 1, K-means has a large validity index, indicating that hard clustering is not a viable option for medical images, as most regions overlap in these types of images. Out of the remaining fuzzy clustering methods, it is evident that the proposed method gives us the minimum value of the validity index for each image. The optimal number of clusters of these images for GASFCM is in the range of 2–4, with a validity index much smaller than the minimum validity index of the other methods. Thus, GA-based SFCM gives a better segmented output than the existing methods. For SFCM, we have taken the value of α, as 0.2. $\alpha = 0.2$ preserves the intricate details of the brain images while also reducing the noise. Apart from the fact that this algorithm is self-starting, one of the greatest advantages of this method is that we do not need to predict the number of clusters beforehand, but instead it gives the optimal number of clusters based on the validity index. For the experiment, we choose $\omega = 10$ and $K_{max} = 8$. Figure 1 shows the segmented images using K-means, FCM, SFCM, RFCM, KFCM, rough KFCM, GAFCM and the proposed method.

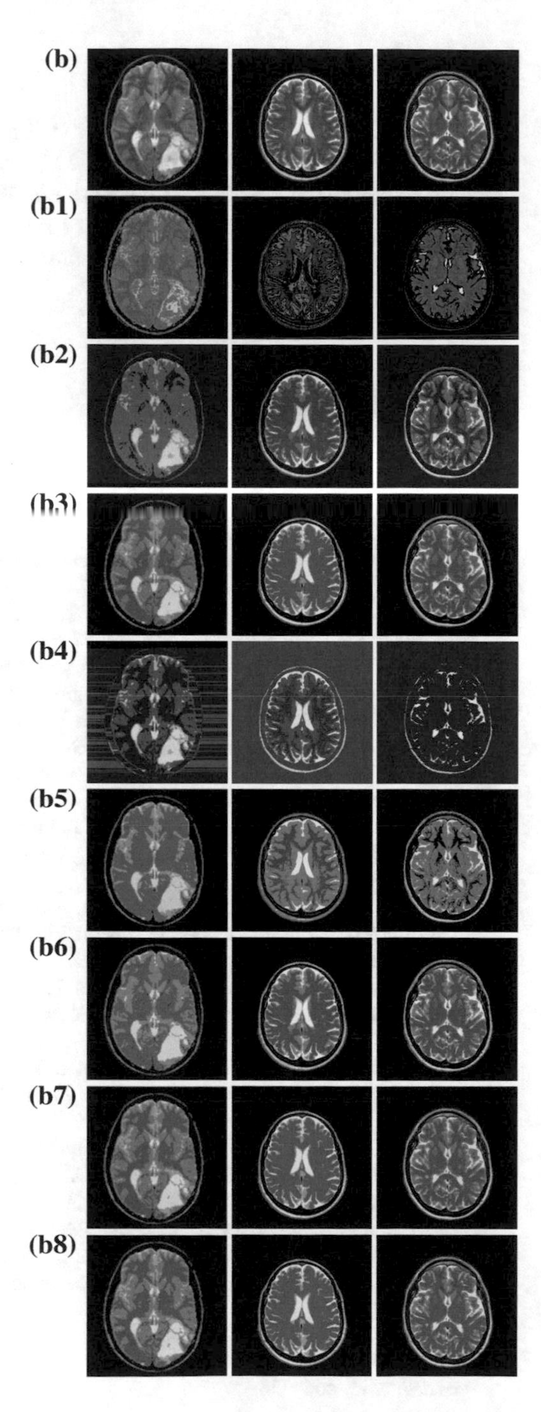

Fig. 1 (**b**) Original image and output segmented image using the methods (**b1**) K-means, (**b2**) FCM, (**b3**) SFCM, (**b4**) RFCM, (**b5**) KFCM, (**b6**) RKFCM, (**b7**) GAFCM and (**b8**) the proposed method (GASFCM)

5 Conclusions

This chapter has presented a novel method for segmenting medical images using spatial fuzzy C-means based on the genetic algorithm. GASFCM has distinct advantages over previous methods as it incorporates the search capability of GA and the noise reduction capability of SFCM. This makes it a helpful aid for medical image segmentation. However, further improvements are always possible for better segmentation.

References

1. Pham, D.L., C. Xu, and J.L. Prince. 2000. Current methods in medical image segmentation. *Annual Review of Biomedical Engineering* 2 (1): 315–337.
2. Na, S., L. Xumin, and G. Yong. 2010. Research on k-means clustering algorithm: An improved k-means clustering algorithm. In *Third International Symposium on Intelligent Information Technology and Security Informatics*, 63–67.
3. Wagstaff, K., C. Cardie, S. Rogers and S. Schrdl. 2001. Constrained k-means clustering with background knowledge. In *ICML*, vol. 1, 577–584.
4. Cai, W., S. Chen, and D. Zhang. 2007. Fast and robust fuzzy c-means clustering algorithms incorporating local information for image segmentation. *Pattern Recognition* 40 (3): 825–838.
5. Bezdek, J.C., R. Ehrlich, and W. Full. 1984. FCM: The fuzzy c-means clustering algorithm. *Computers and Geosciences* 10 (2): 191–203.
6. Gamarra, Daniel Fernando Tello. 2015. Fuzzy image segmentation using validity indexes correlation. *International Journal of Computer Science and Information Technology* 7: 15–26.
7. Kaur, P., A.K. Soni, and A. Gosain. 2013. A robust kernelized intuitionistic fuzzy c-means clustering algorithm in segmentation of noisy medical images. *Pattern Recognition Letters* 34 (2): 163–175.
8. Ahmed, M.N., S.M. Yamany, N. Mohamed, A.A. Farag, and T. Moriarty. 2002. A modified fuzzy c-means algorithm for bias field estimation and segmentation of MRI data. *IEEE Transactions on Medical Imaging* 21 (3): 193–199.
9. Chuang, K., H. Tzeng, S. Chen, J. Wu, and T. Chen. 2006. Fuzzy c-means clustering with spatial information for image segmentation. *Computerized Medical Imaging and Graphics* 30 (1): 9–15.
10. Kole, D.K., and Amiya Halder. 2010. An efficient image segmentation algorithm using dynamic GA based clustering. *International Journal of Logistics and Supply Chain Management* 2 (1): 17–20.
11. Halder, A., S. Pramanik, and A. Kar. 2011. Dynamic image segmentation using fuzzy c-means based genetic algorithm. *International Journal of Computer Applications* 28 (6): 15–20.
12. Halder, A., and N. Pathak. 2011. An evolutionary dynamic clustering based colour image segmentation. *International Journal of Image Processing* 4 (6): 549–556.
13. Maji, P., and S.K. Pal. 2007. Rough set based generalized fuzzy C-means algorithm and quantitative indices. *IEEE Transactions on Systems, Man, and Cybernetics Part B (Cybernetics)* 37 (6): 1529–1540.
14. Halder, A., and S. Guha. 2017. Rough kernelized fuzzy c-means based medical image segmentation. *International Conference on Computational Intelligence, Communications, and Business Analytics*, 466–474.

15. Bhattacharya, R.K. 2012. Introduction to genetic algorithms, *IIT Guwahati* 12.
16. Ghose. T. 2002. Optimization technique and an introduction to genetic algorithms and simulated annealing. In *Proceedings of International workshop on Soft Computing and Systems*, 1–19.
17. http://brain-development.org/ixi-dataset.
18. Turi. R.H. 2001. Clustering-based colour image segmentation. PhD thesis, Monash University.

Selective Video Encryption Based on Entropy Measure

Rohit Malladar and R. Sanjeev Kunte

Abstract The significance of securing information has reached a critical level in recent years due to digital terrorism, infringement of people's seclusion, and the obligation of communicating without intrusion and hacker attacks. Present encryption algorithms to preserve text data are not suitable for multimedia application because of the huge data volume and real time restriction. In the concept of video on demand (VOD), the proposed work emphasizes encrypting only the specific part of the frame which is of greatest interest to the user. To achieve this, a novel selective video encryption based on the entropy measure of the blocks is proposed. The technique is analyzed using different parameters including histogram, number of pixels change rate (NPCR), peak signal-to-noise ratio (PSNR), correlation coefficient, etc.

Keywords Selective video encryption · Entropy · Digital terrorism

1 Introduction

Recent progress in computers and interchanges have a huge market for sharing multimedia through the Internet. Images, audio and video data are now being used extensively in applications such as video-on-demand (VoD), video conferencing and broadcasting. However, this expansion in digital documents, multimedia handling devices and the overall Internet connectivity has also created a stage for fraud and dissemination of multimedia, bringing security to the forefront of utilizing interactive media content. Subsequently, the issue of efficient media encryption has increased its concentration in both academia and industry.

R. Malladar (✉)
Department of CS & E, Jain Institute of Technology, Davangere, India
e-mail: powerohit@gmail.com

R. Sanjeev Kunte
Department of CS & E, J. N. N. College of Engineering, Shimoga, India
e-mail: sanjeevkunte@gmail.com

A. N. Krishna et al. (eds.), *Integrated Intelligent Computing, Communication and Security*, Studies in Computational Intelligence 771,
https://doi.org/10.1007/978-981-10-8797-4_61

Multimedia encryption finds its application in numerous fields. Video encryption is essential in securing recordings used in applications and organizations such as VoD [1], video conferencing/learning and for securing privacy in medical recordings. There are two methodologies in video encryption, full and selective encryption. Full encryption is the process in which an entire content is encrypted, and selective encryption is the process in which only selected portions of the video are encrypted.

The kind of encryption basically depends upon the requirements of the application. Application requiring only partial encryption demands less computationally effective encryption. Apart from regular streaming of video content over the Internet, transmission through cable technology, VoD and pay-per-view (PPV) are gaining popularity due to there being various options available for customers.

1.1 Video on Demand (VoD)

Selective video encryption finds use in various fields including medical image processing, defense, entertainment and digital video broadcasting. In this work, we mainly focus on VoD, where customers can choose the content they are interested in watching instead of watching whatever is broadcast. VoD is used to deliver pre-recorded content such as movies and TV shows and it allows the user to watch what they want and when they want.

1.2 Pay Per View (PPV)

PPV [2] is another method of video dissemination which allows users to watch video content such as sports and wrestling. PPV has a fixed schedule to follow, unlike VoD, which provides flexibility to the user. The user is expected to subscribe for every PPV event and only pay for the same. Pre-existing cable technology is used to deliver the content to the user who has subscribed, and selective video encryption can be used here; ideally, this will generate curiosity and interest in the medium and increase the number of users.

VoD and PPV require a method capable of offering a good compression ratio, a secure encryption scheme and a robust hiding scheme for audio, text and data transmission [3]. In this work, we focus our encryption on the blocks in video with more information; i.e., blocks with entropy values higher than the threshold value. We present a novel method to find the blocks with higher information and we employ "exclusive-OR" (XOR) encryption. In the next section, we present the work done so far in this field.

2 Literature Review

In this section, we discuss the different works on selective video encryption. A sign bit encryption algorithm was combined with Advanced Encryption Standard (AES) output feedback (OFB) by Wang et al. [4], and DC and parts of AC coefficients were encrypted. This method exhibited low complexity and good security.

An encryption technique to encrypt only the I frame for every MPEG cluster per video was proposed by Spanost and Maples [5]. Encryption of only the DC coefficients with event shuffling was proposed by Liu et al. [6]. A group of codewords were identified for DC coefficients, and AC coefficients were encrypted by the event-shuffling algorithm. Visual quality was not degraded to a greater extent in a perceptual video encryption algorithm (PVEA), so the encryption of AC coefficients was proposed by Weng et al. [7]. Sign bits encryption of transform coefficients along with motion vectors was proposed by Thomas et al. [8], making way for transcoding irrespective of decryption terms. Frequently occurring patterns in the discrete cosine transform (DCT) coefficients were exploited by Raju et al. [9]. These DCT coefficients were then encrypted using the RC5 algorithm. Higher-intensity bits were selected using the one bit selection algorithm to obtain higher visual degradation by Roy et al. [10]. The bits were further encrypted by AES, since it uses less memory and increases the level of the security.

Selective encryption of context-adaptive variable-length coding (CAVLC) was proposed by Shahid et al. [11], in which the permutation of identical-length codes from a variable-length coding (VLC) table was used for converting the CAVLC into an encryption cipher. Hu and Wang [12] proposed a novel encryption technique in which every hundredth (or any other number) macro-block (MB) is selected for encryption. Further, the block is replaced by a substitute set of values which are selected according to the type of frame which followed specific rules. The set substitute values are again encrypted with AES.

The encryption in VoD involving a set of secret keys to be generated by the sender was proposed by Posch et al. [13] in 2013. The receiver obtain a small subset of these keys and one of the keys is randomly chosen for the session. Partial bit-plane encryption which encrypts the signs of the motion vector was proposed by Lian et al. [14]. In the next section, we propose a unique selective encryption algorithm based on the entropy values of the blocks.

3 Proposed Work

In this presented work, we mainly concentrate on H.264 format used in VoD which uses the bidirectional compression using both the I frame and B frame to predict the redundancy.

The region of interest in the proposed work is the part of the frame which is of more value or interest for the user.

Entropy is the measure which depicts the amount of the information embed in the media. In the path of achieving a breakthrough in the field of selective encryption, we propose a technique to encrypt the I frame by matching the entropy of the macro-blocks with a threshold value. If the entropy of the macro-block is above the threshold value, then the block undergoes a XOR operation in which the key shall be supplied by the user.

The entropy of the macro-block to make a decision of the encryption is discerned before the quantization step. In our work, we use chaotic maps [15] to generate a sequence of 0 s and 1 s in a bit stream using the key entered by the user and, in turn, the bits are grouped into units of eight. These bytes are encrypted with AC coefficients of the block using XOR encryption. The random sequence generated is expressed as the recurrent relation given as:

$$x_{n+1} = r\,x_n(1 - x_n) \tag{1}$$

where r is a parameter that belongs to the [0, 4] interval; the map behavior is also determined by this parameter. n is the number of iterations that represents a discrete quantity of time.

The block diagram shown in Fig. 1 represents the proposed work.

As shown in Fig. 1, a block undergoes encryption only according to the entropy value. Every number generated in the sequence serves as a key for XOR encryption of each of the macro-block selected for the encryption. The main aim of the selective video encryption employed here is to encrypt the data partially with very high visual degradation in computationally very effective time. The technique presented here successfully encrypts some of the critical parts of the frame and hence the aim is achieved.

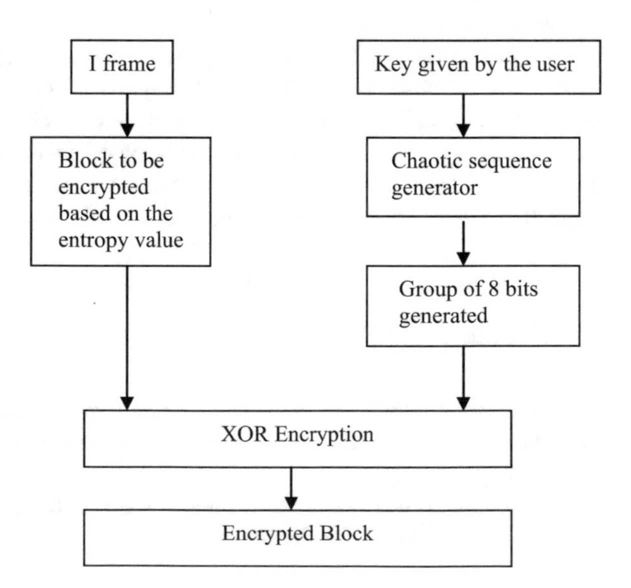

Fig. 1 Block diagram of the proposed work

Step 1: Read the threshold value (thresh) from the user. Any value above seven is treated as a considerable entropy value, so the value is chosen; *n* is taken as the total number of macro-blocks in the frame.

Step 2: Consider the key *k* from the chaotic map for XOR encryption.

Step 3: For MB = 0 to n:

Find the entropy value *e* of the MB.

If e ≥ thresh,

perform XOR operation on the MB with the key k.

Otherwise:

no change in the intensity values of the macro-block.

End

Step 4: Continue with the quantization and zig-zag ordering processes.

The user dynamically enters the preferred threshold value and thus gives the freedom to the user in deciding the visual degradation of the video. If the degradation is not enough, then the system allows the user to enter the threshold value again. The result of the encryption operation will be the selective encryption of all the macro-blocks matching the criteria shown.

This novel way of selective encryption encrypts only those blocks containing a lot of information and hence, when encrypted, it affects the critical areas of the frame.

During the decryption of the video, only the actual receiver will know the key to perform (by generating the chaotic map) the XOR and the threshold value. As the selective encryption was performed just before the quantization, the decryption process shall be started after the quantization process at the receiver end. The process of decryption involves the following steps:

Step 1: Read the threshold value (thresh) from the user; *n* is taken as the total number of macro-blocks in the frame.

Step 2: Consider the key *k* from the chaotic map for XOR encryption.

Step 3: For MB = 0 to *n*:

Find the entropy value, *e* of the MB.

If e ≥ thresh,

perform XOR operation on MB with the key *k*.

Otherwise:

no change in the intensity values of the macro block.

Step 4: Continue with the quantization step and inverse DCT (IDCT).

In the next section, a detailed analysis using various parameters including entropy, number of pixels change rate (NPCR) and correlation coefficient are presented.

Fig. 2 **a** Original I frame. **b** Encrypted I frame

4 Experimental Setup

The work has been carried out in the Windows environment using Matlab 2013a. A sample encrypted and decrypted frame from a video has been considered for the analysis.

Sample frames are shown in Fig. 2.

5 Analysis

The technique used in the chapter to selectively choose the parts of the video frame to encrypt uses the entropy measure to decide which macro-blocks are to be encrypted. The analysis starts with comparing the entropy values of the encrypted and decrypted frames. Further, we show the NPCR, correlation coefficient, etc.

5.1 Entropy Analysis

Entropy is defined as the amount of information which must be coded by the compression algorithm.

$$Entropy = -\sum_{1}^{n}(p_i \log(p_i)) \tag{2}$$

where $p_i \in [0.0, 1.0]$ and $-log\ p_i$ represents the information associated with a single occurrence of p_i. Using Eq. (1), entropy is found for the sample original, encrypted and decrypted frame of the video.

The results are shown below:

The entropy of the original frame = 7.4541.
The entropy of the encrypted frame = 7.4535.
The entropy of the decrypted frame = 7.4537.

Entropy results in randomness (statistical), and distortion results in the encrypted images.

The entropy measure of the encrypted images should be close to an ideal value from a security viewpoint. The entropy values obtained are near the ideal value of eight. This implies that the information outflow in the encrypted video is trifling and hence proves its strength against an entropy-based attack.

5.2 *Number of Pixels Change Rate (NPCR)*

It is used to measure the outcome of each pixel modification with respect to the entire image. The proportionality of different pixels is shown between the images.

$$NPCR = \sum_{i=0}^{m} \sum_{j=0}^{n} D(i,j) * \frac{100}{m * n} \tag{3}$$

where

m signifies the row count,
n signifies the column count
$D(i, j) = 0$ if $I_{original}(i, j) = I_{encrypted}(i, j)$; otherwise, it is equal to 1

Table 1 shows the different values for randomly chosen I frames. Approximately 99% is the average rate achieved in the work which specifies that the encrypted image is 99% different than the original image.

Table 1 NPCR values of three I frames

Sample no.	Frames chosen	NPCR (%)
1	1st I frame	99
2	25th I frame	99
3	50th I frame	99

Table 2 Correlation coefficients of different frames

Sample no.	Frames chosen	Correlation coefficient
1	1st I frame	0.010054
2	25th I frame	0.010053
3	50th I frame	0.010055

5.3 Correlation Coefficient

Correlation of two adjacent pixels is given by this parameter, which is calculated as:

$$Corr.Coefficient = \frac{\frac{1}{r^*c-1}\sum_{n=1}^{r^*c}\left(x_n - x_{avg}\right)\left(y_n - y_{avg}\right)}{\sqrt{\frac{1}{r^*c}\sum_{1}^{r^*c}(x_i-\mu)^2}\sqrt{\frac{1}{r^*c}\sum_{1}^{r^*c}(y_i-\mu)^2}} \quad (4)$$

A correlation coefficient of 0.010053 is achieved in this work and is found to be very near to the ideal, which is zero. Table 2 shows the coefficient values for different frames chosen. This value evidences that the original frame and the encrypted frame are absolutely dissimilar.

5.4 Peak Signal-to-Noise Ratio

PSNR is used to find the proportion linking the highest potential power of a signal and the supremacy of corrupting noise that changes the fidelity of its representation. It is given by the formula:

$$PSNR = 20 \cdot log_{10}(MAX_I) - 10 \cdot log_{10}(MSE) \quad (5)$$

where

MAX_i is the highest achievable value of a pixel,
MSE is the mean squared error given by Eq. 6.

$$MSE = \frac{1}{mn}\sum_{i=0}^{m-1}\sum_{j=0}^{n-1}|(I(i,j) - K(i,j))|^2 \quad (6)$$

where
m and n are rows and column of the images, respectively.

The PSNR value obtained for our technique is 15.74 dB, which indicates that there is much less resemblance between the encrypted frame and the original frame. Table 3 shows different PSNR values.

Table 3 PSNR values of randomly chosen frames

Sample no.	Frames chosen	PSNR (dB)
1	1st I frame	15.73
2	25th I frame	15.74
3	50th I frame	15.74

5.5 *Unified Average Change Intensity*

The average intensity of difference in pixels between the original and encrypted image is measured using the unified average changing intensity (UACI) given by the formula

$$UACI = \sum_{i=0}^{m} \sum_{j=0}^{n} \frac{\left|\left(I_{orig}(i,j) - I_{encr}(i,j)\right)\right|}{255} * \frac{100}{m * n} \quad (7)$$

where

$I_{orig}(i,j)$ refers to the original image pixel value at (*i*, *j*),
$I_{encr}(i,j)$ refers to the encrypted image pixel value at (*i*, *j*),
m and *n* are rows and column of the images, respectively.

We have recorded the UACI percentage as 69.76 with original and encrypted images.

In the next section, we present the conclusion and future enhancement to this technique.

6 Conclusion

In this chapter, we have presented a novel technique of selectively encrypting a video by choosing the macro-blocks to be encrypted by measuring the entropy value. Only the part of the video which creates more interest to the user is encrypted, to increase curiosity. The user sets the threshold value and provides the key for XOR encryption. In this way, the technique is a dual-secured technique with selective encryption. The results achieved are near ideal values and are ideal for the application of the technique in the field of VoD.

References

1. Modules, C., and S. Diego. 2015. Introduction to CDN and VOD principles CDN and VOD. *Journal of Entropy* 1–7.
2. Villegas, Á., P. Pérez, J. María Cubero, E. Estalayo, and N. García. 2012. Network assisted content protection architectures for a connected world. *Bell Labs Technical Journal* 16 (4): 85–96.
3. Bourbakis, N. 2003. SCAN-based compression—encryption—hiding for video on demand, 79–87.
4. Wang, Y., M. Cai, and F. Tang. Design of a new selective video encryption scheme based on H.264. In *2007 International conference on computational intelligence and security (CIS 2007)*, 883–882, Dec 2007.
5. Spanost, G.A., and T.B. Maples. 1995. Performance study of a selective encryption scheme for the security of networked, real-time video, 2–10.
6. Liu, G., T. Ikenaga, S. Goto, and T. Baba. 2006. A selective video encryption scheme for MPEG compression standard. *IEICE Transactions on Fundamentals of Electronics Communications and Computer Sciences* E89–A (1): 194–202.
7. Weng, L., K. Wouters, B. Preneel, and K.U. Leuven. 2006. Extending the selective MPEG encryption algorithm PVEA, 0–3.
8. Thomas, N.M., D. Lefol, D.R. Bull, and D. Redmill. 2006. A novel secure H.264 transcoder using selective encryption. In *Proceedings—International conference on image processing, ICIP*, vol. 4, 85–88.
9. Raju, C.N., G. Umadevi, K. Srinathan, and C.V. Jawahar. 2008. Fast and secure real-time video encryption. In *Proceeedings—6th Indian conference on computer vision, graphics and image processing ICVGIP 2008*, 257–264.
10. Roy, M., and C. Pradhan. 2011. Secured selective encryption algorithm for MPEG-2 video. In *ICECT 2011—2011 3rd International Conference on Electronics Computer Technology*, vol. 2, 420–423.
11. Shahid, Z., M. Chaumont, and W. Puech. 2009. Fast protection of H.264/ Avc by selective encryption of CABAC, vol. 21, no. 5, 1038–1041, UMR CNRS, M. Ii, and M. Cedex.
12. Hu, Y., and X. Wang. 2012. A novel selective encryption algorithm of MPEG-2 streams, 2315–2318.
13. Posch, D., H. Hellwagner, and P. Schartner. 2013. On-demand video streaming based on dynamic adaptive encrypted content chunks. In *Proceedings—International conference* on *network protocol, ICNP*.
14. Lian, Shiguo, Dengpen Ye, and Jinsheng Sun. 2004. Perceptual MPEG4 video encryption and its usage in video on demand. In *International symposium on consumer electronics*, vol. c, 83–86.
15. Shukla, P.K., A. Khare, M.A. Rizvi, S. Stalin, and S. Kumar. 2015. Applied cryptography using chaos function for fast digital logic-based systems in ubiquitous computing. *Entropy* 17 (3): 1387–1410.

Improved Blurred Image Splicing Localization with KNN Matting

P. S. Abhijith and Philomina Simon

Abstract Image splicing is a forgery technique where some regions are cropped or pasted from the same or different images. Splicing localization becomes challenging when post-processing techniques are used to remove the anomalies of splicing traces. In this chapter, an improved method is proposed for blurred image splicing localization based on K-nearest neighbor (KNN) matting. The proposed method minimizes computation time without compromising the quality of the result. Quantitative and qualitative results analysis show the proposed method obtains better splicing than existing systems.

Keywords Image splicing · Forgery · Out-of-focus blur · Motion blur Deblurring · K-nearest neighbor matting

1 Introduction

In today's digital world of technological advancements, different techniques are available to manipulate images and, hence, image forgery is becoming a huge challenge for current forensic techniques. Image splicing is the process of manipulating digital images by combining regions from different images with the intent of deceiving viewers. The need for image forgery detection thus arises from the necessity of preserving image authenticity and integrity. Forgery detection is requisite in fields such as publishing, criminal investigation, security issues in social media and in medical insurance. This chapter introduces a framework for blurred image splicing forgery detection and localization based on K-nearest neighbor (KNN) matting. Image blurring is a result of camera or object movement or due to

P. S. Abhijith (✉) · P. Simon
Department of Computer Science, University of Kerala, Kariavattom,
Thiruvananthapuram, India
e-mail: sreenu.abhijith@gmail.com

P. Simon
e-mail: philomina.simon@gmail.com

A. N. Krishna et al. (eds.), *Integrated Intelligent Computing, Communication and Security*, Studies in Computational Intelligence 771,
https://doi.org/10.1007/978-981-10-8797-4_62

focusing issues. KNN matting is proposed as a method for enhancing detection accuracy and thereby minimizing computation time. The basic assumption is that images under investigation contain some blur. In [1], the authors discussed a method for splicing detection in which they classified different blur types as out-of-focus blur and motion blur based on lesser directional frequency energy. Liu et al. discussed [2] a method for categorizing different blur types based on the shifted block correlation feature. Su et al. presented a work [3] for identifying different blur regions based on alpha channel feature. In other work, Khosro et al.'s [4] forgery detection method utilizes the inconsistencies in blur types to detect forgery.

The remainder of the chapter is organized as follows. Section 2 describes the features applied for digital image splicing detection. In Sect. 3, a detailed description of the proposed method is presented. In Sect. 4, the concepts behind KNN matting are discussed. Section 5 provides inferences and interpretations based on the experimental results and analysis. Section 6 concludes the chapter.

2 Feature Detection for Blurred Regions

Blur found in images such as motion or out-of-focus blur can be represented in parametric models, but only in cases of simple types of blur. Most realistic images contain complex types of blur which cannot be expressed in parametric models. Usually, they are represented as a two-dimensional (2D) matrix. Motion blur kernels are sparse whereas out-of-focus blur kernels are more common [4]. To extract features from the kernels, the pixel value distributions are fit to a generalized Gaussian distribution [4] as shown below.

$$f(\mathbf{K};\mu,\gamma,\sigma)=\left(\frac{\gamma}{2\sigma\Gamma\left(\frac{1}{\gamma}\right)\sqrt{\frac{\Gamma\left(\frac{1}{\gamma}\right)}{\Gamma\left(\frac{3}{\gamma}\right)}}}\right)e^{-\left(\frac{K-\mu}{\sigma\sqrt{\frac{\Gamma\left(\frac{1}{\gamma}\right)}{\Gamma\left(\frac{3}{\gamma}\right)}}}\right)^{\gamma}}$$

In the above expression, K represents a blur kernel obtained from a selected region, Γ(.) represents the gamma function, μ represents the mean, σ represents the standard deviation and γ (where $\gamma > 0$) represents the shape parameter of the generalized Gaussian distribution. The values of γ and σ are calculated for the blur kernel distributions. The shape parameter is estimated using the approach described by. Sharifi et al. [5]. This method is used for estimating the shape parameter for video sub-bands. With values of γ and σ, a classifier can be modeled for distinguishing the out-of-focus and motion blur kernels.

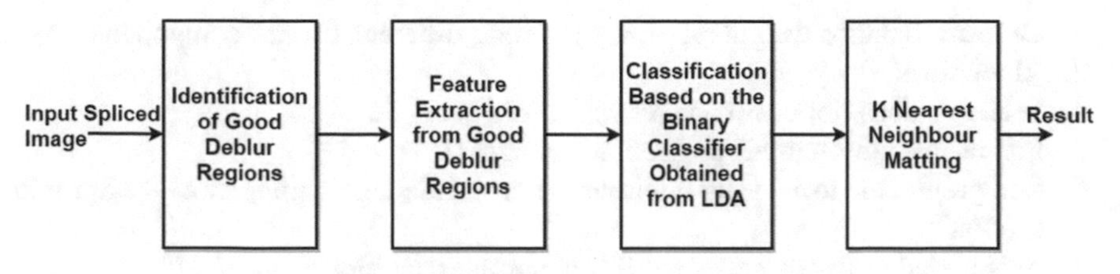

Fig. 1 Improved blurred image splicing localization with KNN matting

3 Improved Blurred Image Splicing Localization with KNN Matting—Proposed Method

The block diagram for the proposed method is shown in Fig. 1. This method is based on Khosro et al. [4] for localizing blur type inconsistencies and improving the execution time and accuracy of detection. However, there were some limitations such as long computation time and lack of a better matting method for fine boundary localization. To overcome such limitations, we used a method proposed by Hu et al. [6] for finding the suitable regions for deblurring and from that identified region, the image is divided into a grid within which a particular grid of window size is selected for estimating the blur kernel. So the grid size is increased and set as 128 × 128 and the inside-grid window size is set as 64 × 64.

Algorithm 1: Selecting good deblur regions

1. **Input:** Training image set $\mathbf{X} = \{X^m, m = 1,\ldots, M\}$; ground truth blur kernel set $\mathbf{K} = \{K^m, m = 1,\ldots, M\}$ for set $\mathbf{X}$; the set of blur kernels estimated is $\hat{K} = \left\{\hat{K}_p^m, m = 1, \ldots, M, p \in \{subwindowset\}\right\}$, from sub-windows of X; B is the input blurred image;
2. **Output:** θ is the conditional random field (CRF) Parameters; $\hat{B}$ is the subwindow.
3. **Algorithm:**
4. Obtain $\mathbf{Y} = \{Y^m, m = 1,\ldots, M\}$ as training labels; from $\mathbf{K}$ and $\hat{K}$ based on kernel similarity;
5. Create training data (**X**, **Y**) model;
6. Node and edge feature vectors are extracted from **X**;
7. Using pseudo-likelihood, learn CRF model parameters θ;
8. Create a graph for blurred image B;
9. Node and edge feature vectors are extracted from **X**;
10. Select the top-ranked sub-window $\hat{B}$ using loopy belief propagation and θ.

Algorithm 2: Learning contribution of different features

1. **Input:** The training images $\mathbf{X} = \{X^m, m = 1,\ldots, M\}$; and training labels $\mathbf{Y} = \{Y^m, m = 1,\ldots, M\}$.

2. **Output:** Balance weights $A = (\alpha, \beta)$ among different feature components.
3. **Algorithm:**
4. **While** A does not coverage (within a tolerance) do
5. **For** m goes through all the training images do
6. Run graph cuts to find the minimum energy labeling of image m:← y_*arg min $E'(y^m|x^m)$;
7. y_* is added with constraint set S^m, when the condition is $y_* \neq \hat{y}^m$.
8. Update the weights A to enforce training label has the lowest energy.
9. **End for**
10. **End while**

The first algorithm estimates the good regions for deblur, and the second algorithm is based on learning and training of features for estimating good deblur regions. The accuracy of the result increases with lower grid size. But it compromises the computation time. So, we came up with this idea of estimating good deblur regions inside a grid and then estimate the blur kernel using the method proposed by Levin et al. [7] from a grid inside that grid. The blur kernel estimation here is a non-trivial problem,

$$G_{i,j} = I_{i,j} * K_{i,j} + N_{i,j}$$

Here, $I_{i,j}$ represents a sharp image block, $K_{i,j}$ is the blur kernel, $N_{i,j}$ is the noise matrix and the '$*$' symbol is the convolution. The value for G is known while all other parameters in the equation are unknown. Thus, a blind deconvolution [7] approach is required to obtain the blur kernel.

Khosro et al. [4] proposed an efficient method for classifying motion and out-of-focus blur types. The standard deviation and shape parameter are extracted from the estimated blur kernel pixel distribution. Since two features cannot efficiently classify a two-class problem, dimensionality reduction is performed using Fisher's linear discriminant analysis (LDA) [8], by which feature [4] $v_{i,j}$ is obtained.

$$v_{i,j} = w^T x_{i,j} = \left[\omega_\gamma \omega_\sigma\right] \left[\gamma_{i,j} \sigma_{i,j}\right]^T$$

Based on the value of $v_{i,j}$, a binary classifier $B_{i,j}$ is designed [4] for classifying the motion and out-of-focus blur.

$$B_{i,j} = \begin{cases} 'M' \text{ (motion blur)}, & \text{if } v_{i,j} \geq \rho \\ 'O' \text{ (out-of-focus blur)}, & \text{otherwise} \end{cases}$$

where ρ is the threshold parameter for classifying out-of-focus or motion blur from the selected image block. The image blocks are classified into motion or out-of-focus blur. To optimize the process, smooth blocks, those without sharp edges, are not considered, and this block does not give any correct blur kernel information when we apply the blind deconvolution algorithm [7]. We estimate a smooth block as a block with less than 5% edge pixels.

When we calculate the blur kernel from the good deblur region, it gives better results. The grid size for dividing the image is set to 128 × 128. The 64 × 64 grid inside the 128 × 128 grid gives good results. When estimating the type of blur using the classifier, a matrix is used to represent the grid values and for further processing. The smooth blocks are classified into their corresponding classes based on their neighborhood pixels. After this step, the matrix is remapped to the tampered input image and then KNN matting [9] is applied.

4 K-Nearest Neighbor Matting

KNN matting is based on the non-local principle [9]. Consequently, sparse input is sufficient. It has been observed that the Laplacian matting in most cases is not good for clustering, especially whenever the local color-line model assumption fails. The non-local approach has resulted in small and localized clusters. These clusters are

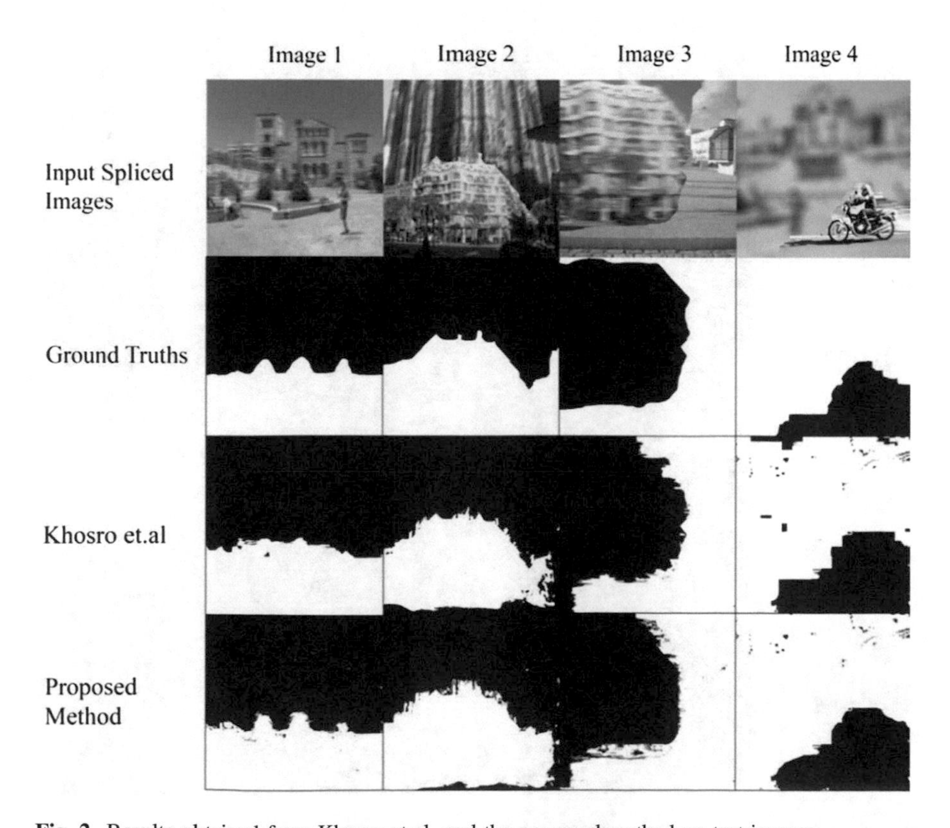

Fig. 2 Results obtained from Khosro et al. and the proposed method on test images

combined into larger ones through a nonlinear optimization scheme biased toward binary-valued alphas. This gives much more precise boundary than closed-form matting [10]. After obtaining the binary map from KNN matting, we used threshold and morphological operations to remove unwanted regions to improve the accuracy of the result. A human-based decision is required to identify the correct spliced region [11].

5 Experimental Results

To analyse the results, the Jaccard index and Dice similarity coefficient are used as evaluation measures. These evaluation measures estimate the similarity between the ground truth and the result based on overlapping regions.

From the results analysis, it is clear that the proposed method is able to produce better-quality results than the previous method. And the computation time is drastically improved since the time for executing the algorithm has been reduced to one fourth of the previous method by giving better results. The usage of a new

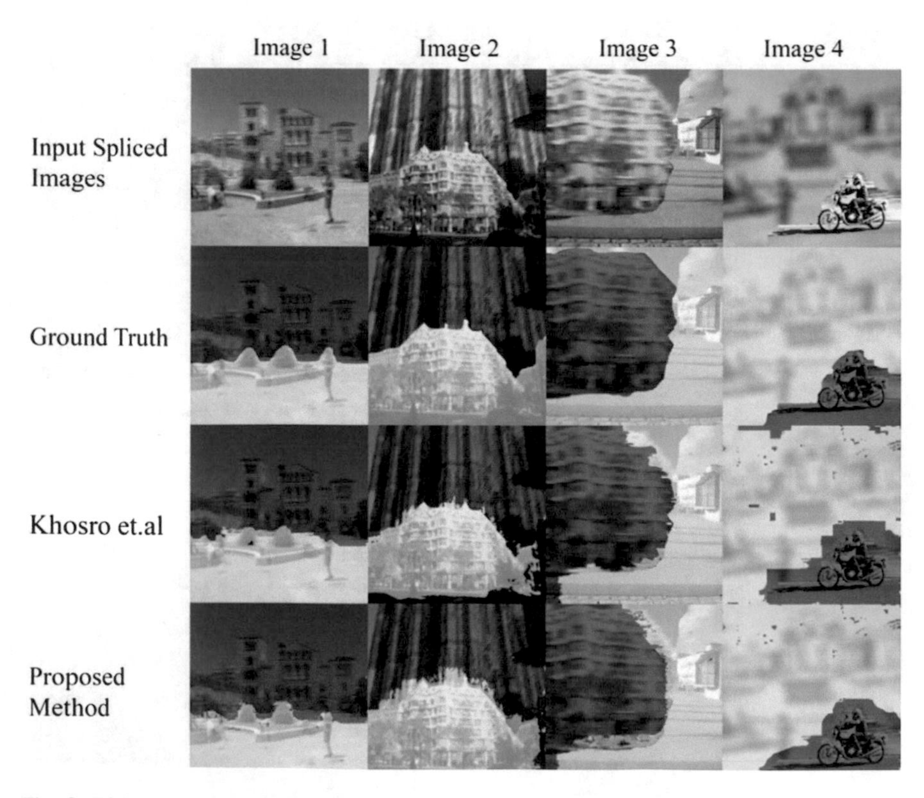

Fig. 3 Binary map overlay over the input images to show different spliced regions

Table 1 Comparison of both methods using the Jaccard index and Dice similarity coefficient

Images	Jaccard index		Dice similarity coefficient		Computation time (h)	
	Khosro et al.	Proposed	Khosro et al.	Proposed	Khosro et al.	Proposed
Image 1	0.8989	**0.9229**	0.9467	**0.9599**	2.18	**0.58**
Image 2	0.8458	**0.8483**	0.9165	**0.9179**	2.13	**0.55**
Image 3	**0.8288**	0.8122	**0.9064**	0.8963	2.21	**0.65**
Image 4	0.9120	**0.9230**	0.9539	**0.9600**	2.01	**0.48**

matting method gives better boundary precision with more accurate results (Figs. 2 and 3).

In most of the cases, the proposed method gives similar results when compared with the ground truth image and, in some cases, it gives comparable results with the existing system because of the k-NN matting. Sometimes it marks unwanted regions in the resultant image, which reduces the accuracy. A good threshold method along with morphological operations removes the unwanted artifacts and produces better results (Table 1).

6 Conclusion

A new framework for image splicing in blurred images is proposed. In this method, good blur regions are identified and then features are detected. The estimated features are used for the classification of blur types. KNN matting is applied at the end to obtain a precise boundary which distinguishes different regions. The performance of the algorithm can be improved by using some post-processing techniques such as morphological operations.

References

1. Chen, J., L. Yuan, C.-K. Tang and L. Quan. 2008. Robust dual motion deblurring. In *Proceedings of IEEE conference on computer vision and pattern recognition (CVPR)*, 18.
2. Liu, R., Z. Li and J. Jia. 2008. Image partial blur detection and classification, In *Proceedings of IEEE conference on computer vision and pattern recognition (CVPR)*, 18.
3. Su, B., S. Lu, and C. L. Tan. 2011. Blurred image region detection and classification. In *Proceedings of the 19th ACM international conference on multimedia*, 1397–1400.
4. Bahrami, Khosro, Alex C. Kot, Leida Li and Haoliang Li. 2015. Blurred image splicing localization by exposing blur type inconsistency. In *IEEE Transactions on Information Forensics and Security*, 5 (5): 999–1009.
5. Sharifi, K., and A. Leon-Garcia. 1995. Estimation of shape parameter for generalized Gaussian distributions in subband decompositions of video. *IEEE Transactions on Circuits and Systems for Video Technology* 5 (1): 52–56.

6. Hu, Zhe, and Ming-Hsuan Yang. 2012. Good Regions to Deblur. In European conference on computer vision, 59–72.
7. Levin, A., Y. Weiss, F. Durand, and W.T. Freeman. 2011. Efficient marginal likelihood optimization in blind deconvolution. In *2011* Proceedings of *IEEE conference on computer vision and pattern recognition (CVPR)*, 2657–2664.
8. McLachlan, G.J. 2004. *Discriminant analysis and statistical pattern recognition*. Hoboken, NJ, USA: Wiley.
9. Chen, Qifeng, Dingzeyu Li, and Chi-Keung Tang. 2013. KNN Matting. *IEEE Transactions on Pattern Analysis and Machine Intelligence* 35 (9): 2175–2188.
10. Levin, A., D. Lischinski, and Y. Weiss. 2008. A closed-form solution to natural image matting. *IEEE Transactions on Pattern Analysis and Machine Intelligence* 30 (2): 228–242.
11. Bahrami K., and A.C. Kot. 2014. Image tampering detection by exposing blur type inconsistency. In *Proceedings of IEEE ICASSP*, May 2014, 2654–2658.

Single-Image Super-Resolution Using Gradient and Texture Similarity

L. Rajeena and Philomina Simon

Abstract A super-resolution algorithm focuses on generating high-resolution (HR) images utilizing low-resolution (LR) images. In this proposed work, a single-image super-resolution algorithm which uses texture and gradient information is discussed. In this method, the LR texture layer is extracted from an LR input image and is used generate a corresponding HR texture layer for enhancing the texture details in the image. Finally, an optimization framework using a gradient descent algorithm is used to reconstruct the final HR image. This method is capable of creating natural high-frequency texture details as well as projecting the edges. In addition to this texture enhancement step, a gradient enhancement method is employed to suppress gradients present in noisy regions. Analysis of the results reveals that the proposed method achieves improved results even under large scaling factors and it can preserve both the texture and structure details.

Keywords Image processing · Single-image super-resolution
Texture enhancement · Gradient similarity · Gradient descent algorithm

1 Introduction

Super-resolution is an active research area in image processing. Single-image super-resolution is an efficient method requiring various priors to regulate a visually better result. Multi-image super-resolution uses different frames of the same scene to reconstruct high-resolution (HR) images, but the output image is usually blurred;

L. Rajeena (✉) · P. Simon
Department of Computer Science, University of Kerala, Kariavattom,
Thiruvananthapuram, India
e-mail: rajeenasajeer27@gmail.com

P. Simon
e-mail: philomina.simon@gmail.com

A. N. Krishna et al. (eds.), *Integrated Intelligent Computing, Communication and Security*, Studies in Computational Intelligence 771,
https://doi.org/10.1007/978-981-10-8797-4_63

so, in the final step, a deblurring algorithm is utilized to generate one fine-resolution image. Nowadays, single-image-based super-resolution has become a necessity in majority of the real-world applications such as medical imaging, remote sensing, video surveillance, and movie restoration. Some of the currently employed methods fail to preserve edges and natural texture details. Different single-image super-resolution approaches commonly used are interpolation-based, example-based and reconstruction-based methods. A majority of the conventional methods fail to preserve highly detailed information such as edge information and structural information. In this work, an improved singe-image super-resolution framework based on the texture and gradient similarity is proposed. Our method is compared with a method employed by Xian et al. for analyzing the effectiveness of the proposed algorithm. The proposed method mainly concentrates on enhancing the texture details as well as visual quality of the resultant image.

The chapter is organized as follows. Section 2 describes the different super-resolution techniques published in the literature. Section 3 provides a detailed description of the proposed super-resolution method based on texture and gradient similarity. Section 4 provides inferences and interpretations based on the experimental results and analysis, and Sect. 5 concludes the chapter.

2 Literature Review

The single-image super-resolution approach can be further divided into three categories: interpolation-based, reconstruction-based and example-based methods. Yang et al. [1] established in 1984 the concept of super-resolution. Popularly used methods based on image interpolation include nearest neighbor interpolation and bilinear- and bicubic-based interpolation. Some resultant HR images have highly blurred edges and many resultant undesired artifacts [2, 3]. Shan et al. [4] proposed a fast image or video up-sampling method; it is a simple and more effective image/video up-sampling algorithm to generate images or frames of enhanced resolution. Kanade et al. [5] proposed a method to reconstruct the super-resolution image by utilizing high-frequency data. Freeman et al. [6] discussed a method to learn the relationship between HR images and their corresponding low-resolution (LR) images through an external dataset; the generated results are noisy with abnormalities along curved edges. Reconstruction-based methods [7] generate an HR image by taking into consideration the statistical priors after up-sampling an image. Sun et al. [8] proposed a method based on using the prior information obtained from shape and sharpness from gradients. Xian and Tian [9] proposed a work based on similarity in gradient information, and a global optimization framework is used to reconstruct the resultant image which will give an HR image. In the next section, a better single-image super-resolution method is discussed which exploits the similarity of the gradient feature and texture features. This proposed method provides improved resolution, visual image perception and texture enhancement.

3 Gradient and Texture Similarity Based Super Resolution—Proposed Method

The proposed method is mainly divided into five modules such as LR texture layer extraction, construction of an HR texture layer, gradient estimation, gradient enhancement and global optimization based on a gradient descent algorithm. It uses the texture and structure layers to reconstruct an HR image. Both layers are created from the given LR image. For texture enhancement, a LR texture layer is extracted from an input image and HR texture layer is generated by applying the up-scaling method used in Xian and Tian [9]. The LR structure layer is then upscaled through bicubic interpolation to generate the HR structure layer. The constructed HR texture layer is then added with an interpolated HR structure layer to acquire the combined image. At the final step, the gradient is estimated using the first-order Gaussian derivative to extract gradient information along x and y directions (g_x and g_y) and then an optimization is performed using the gradient descent algorithm.

The framework for the proposed method is shown in Fig. 1. This method mainly uses the up-scaling method of Xian and Tian [9]. The Xian et al. method is better for preserving edge details, but the texture detail is missing. Usually, natural images lack visual quality with a realistic look since natural images contain a lot of different

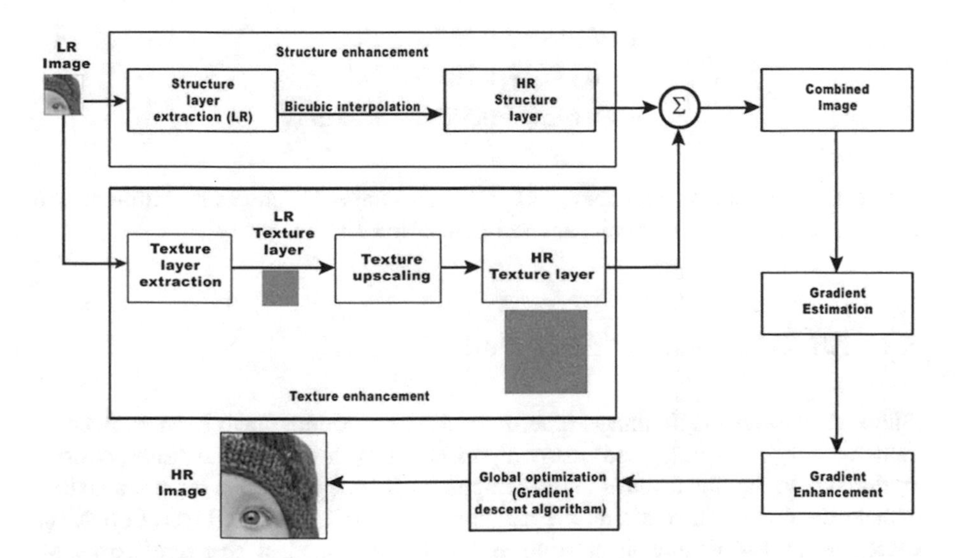

Fig. 1 Framework of the proposed method

textures. The major drawback of this method is that during the up-scaling stage, for every patch in the LR image, five patches are selected with a minimum mean squared error (MSE) value for construction of HR patches. But in some cases, we may not obtain similar patches and also the MSE will be very high. For such cases, Xian et al.'s method also chooses five patches, which is not efficient. This may lead to inconsistency between the LR image and the respective HR image. In order to deal with this issue, an efficient algorithm which extracts the structure layer is used for the decomposition of texture and structure components separately. Both the regions are processed separately and they are finally combined to obtain an HR image.

Most of the super-resolution algorithm processing is done only on the luminance channel of the input image. In an image, highly detailed information is present in the luminance components and fewer details are present in chrominance components. As such, the proposed method uses only the luminance information of the LR image. Less-detailed chrominance information of the image is upscaled using a bicubic interpolation method. In order to separate the luminance and chrominance components, the input image is converted from RGB to YUV color space for efficient representation

3.1 Conversion to YUV Color Space

The proposed single-image super-resolution method is applied in the luminance component of the LR image. Chrominance components of the image are up-scaled using linear interpolation. The equation used for this conversion is given below.

$$\begin{aligned} Y &= 0.299R + 0.587G + 0.114B \\ U &= 0.147R - 0.289G + 0.436B \\ V &= 0.615R - 0.515G - 0.100B \end{aligned}$$

where Y represents the luminance part of the LR image, U denotes the chrominance blue part and V is the chrominance red part of the LR image.

3.2 LR Texture Layer Extraction

Although various single-image super-resolution algorithms have been evolved to enhance image resolution, not many algorithms take texture features into consideration due to the fine textures in the computed LR image. For this reason, a texture enhancement concept is introduced, and applied to the LR texture layer. In order to segregate an LR texture layer from an input LR image, a structure extraction algorithm [10] is initially applied to acquire a structure layer. The texture layer can

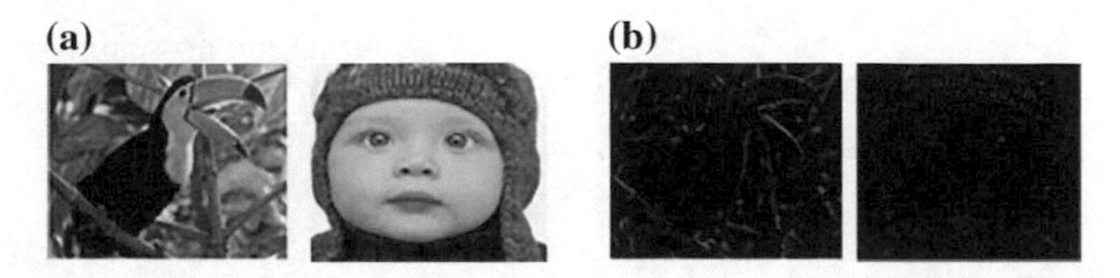

Fig. 2 Texture layer extraction. **a** LR images and **b** corresponding texture layers extracted from (**a**)

then be obtained by subtracting the structure layer from the input image. Figure 2 shows some results of texture layer extraction where the structure component is completely removed and only texture details are shown.

3.3 Construction of HR Texture Layer

In the initial step, the LR texture layer Y_{LR} is decomposed into a set of overlapping patches of size b × b (b = 5 is used in this implementation) for obtaining the HR texture layer. Then, Y_{LR} is down-sampled by factor s to acquire Y'_{LR}. A patch pool $\varnothing_x$ is created with all the patches (b × b) in Y'_{LR}, and all the patches are also normalized to have a zero mean and uniform variance to preserve structural edges accurately. For each patch p in Y_{LR}, calculate the variance and compare it with a threshold θ. If the variance is smaller than a particular threshold θ, that patch is up-scaled through bicubic interpolation. Otherwise, its top k most-similar patches are explored within the patch pool $\varnothing_x$ after the patch normalization. The MSE is calculated to find the relationship between those patches. After finding the top k most-similar patches in $\varnothing_x$ for input patch p, their output patches in Y_{LR} are selected and these selected patches are combined according to weight. Patches that are more similar to the input patch are assigned with larger weights. The weights for each patch are computed with:

$$W_i = \frac{\exp\left(-\frac{M_i}{\sum_{j=1,\ldots,K} M_j}\right)}{R}$$

W_i represents weight for patch i. M_i indicates the MSE between the input patch p and patch i. R is a normalization factor. In the final step, both layers are added to generate the combined image. Then, the gradient is computed based on the first-order Gaussian derivative. Since it is a smooth gradient operator, the amount of noise present in the image can be decreased. The amount of noise can be diminished by incrementing the value of σ_g:

$$g_i = I(x) \times \frac{\partial}{\partial_{x_i}} G(x{:}\,\sigma_g), \mathrm{i} \in (1, \ldots, N)$$

3.4 Gradient Enhancement

A gradient structure tensor is implemented for gradient enhancement, and is used in coherence estimation. It is a method to represent gradient information. The structure tensor matrix is given below.

$$S = \begin{bmatrix} I_x^2 & I_x * I_y \\ I_x * I_y & I_y^2 \end{bmatrix}$$

Eigen decomposition is performed with the structure tensor matrix 'S' to obtain eigenvalues $L_1 \,\&\, L_2$. Coherence (C) values are calculated from these estimated eigenvalues.

$$\mathrm{C} = |L_1 - L_2|$$

If the values of $L_1 \,\&\, L_2$ are equal to zero, then those regions are termed homogenous. Edges are characterized by $L_1 \gg L_2 \approx 0$, and corners by $L_1 \geq L_2 \gg 0$. Here we only enhance those regions whose coherence is larger than a particular threshold (C_{thresh}). A low-coherence region is treated as a noise region, so we suppress the gradient in those regions. The new gradient field is obtained as:

$$g_i^{'} = \begin{cases} \left(\frac{C}{\mu \cdot Avg(C)}\right)^{\rho}, & for C \geq C_{thresh} \\ \beta \cdot g_i, & for C < C_{thresh} \end{cases}$$

In the above equation, β represent the suppression factor for reducing noise, ρ and μ are constants, Avg(c) is the mean value of coherence in the structured region, and C_{thresh} represents the threshold value for gradient enhancement [11].

3.5 Reconstruction of the Final HR Image

After estimating L, G_x and G_y obtained from the above step, the target HR image H is obtained by performing a minimization operation on the following energy function:

$$A = |(H * G)\downarrow - L|^2 + \lambda |\nabla H - \nabla H_D|^2$$

where ∇H_D denotes the preferred HR gradients G_x and G_y, * is the convolution operator, G denotes the Gaussian kernel with standard variance σ, and σ varies for different scaling factors. The objective function can be minimized by applying the gradient descent algorithm as follows:

$$H^{t+1} = H^t - \delta \cdot \left(((H^t * G)\downarrow_s - L)\uparrow^s * G - \lambda \cdot (\nabla^2 H^t - \nabla^2 H_D)\right)$$

Here, H^0 is initially the bicubic interpolated version of L. H^t denotes the output after the tth iteration and δ indicates the step width. λ is a parameter used to adjust the weight between these two similarity terms. The first term keeps the image level similarity and the second term maintains the gradient level similarity.

4 Experimental Results and Analysis

To evaluate the success of the proposed method, we consider two assessment metrics, peak-to-signal noise ratio (PSNR) and structural similarity (SSIM). Both metrics measure how much reconstructed image is similar to the ground truth image. Performance analysis of the proposed method is performed by these matrices. From the experimental analysis, it can be understood that the proposed method generates visually better results when compared with the Xian et al. method and preserves the texture and structure details. The qualitative analysis of the results proved that the proposed method enhances the visual quality. A high PSNR value generally suggests that the reconstruction is of better quality and its value is taken from 0 to infinity. A larger SSIM value indicates the reconstructed image is more similar to the ground truth image and its value ranges from −1 to +1.

From Fig. 3, it is can be observed that the proposed method performs well for all the images compared to the super-resolution method of Xian et al. There is an improvement of 1 db in the achieved result images. This is evidenced by the edge maps given below when results of the proposed method are compared with the ground truth image. The usage of a new texture enhancement method preserves fine details in the generated image and the gradient enhancement method suppresses gradient in the noise region which corresponds to low coherence. Figure 3 also shows an edge map extracted from resultant HR images. From Fig. 3, it is evident that the proposed method preserves edge information relatively better compared to the Xian et al. method and requires less execution time.

From Tables 1 and 2, it is clear that the proposed method maintains relatively better accuracy than the Xian et al. method. Based on performance metrics, we can conclude that the proposed method produces quantitatively and qualitatively

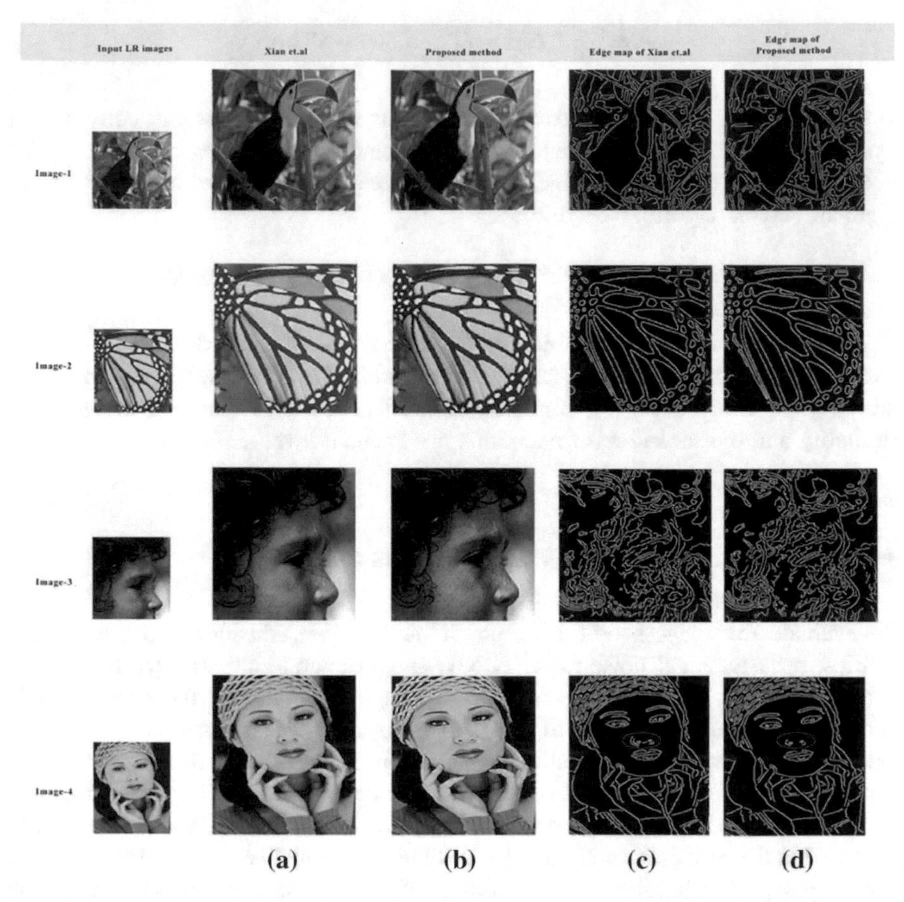

Fig. 3 Results obtained from **a** Xian et al. and **b** the proposed method, and **c**, **d**, their edge maps, respectively

Table 1 Comparison of different super-resolution methods using PSNR

Images	PSNR			
	Shan et al. [4]	Yang et al. [1]	Xian and Tian [9]	Proposed
Image 1	29.298	33.28	34.9537	**35.9902**
Image 2	23.42	25.04	26.297	**27.0954**
Image 3	29.214	30.743	31.4807	**31.7808**
Image 4	26.271	29.437	30.866	**31.7694**

Bolded entries repesent that the proposed method achieve better PSNR and SSIM value

accurate results. The efficiency of the method can be qualitatively observed from the edge map, which displays the same structures and fine details as that of the ground truth image.

Table 2 Comparison of different super-resolution methods using SSIM

Images	SSIM			
	Shan et al. [4]	Yang et al. [1]	Xian and Tian [9]	Proposed
Image 1	0.9604	0.9723	0.9909	**0.99263**
Image 2	0.8991	0.9531	0.9604	**0.96529**
Image 3	0.9011	0.9172	0.9281	**0.9299**
Image 4	0.9214	0.9528	0.9789	**0.98126**

Bolded entries repesent that the proposed method achieve better PSNR and SSIM value

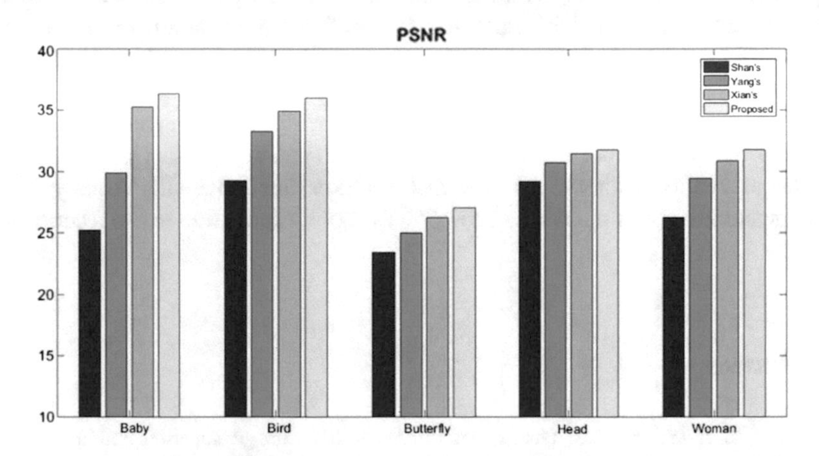

Fig. 4 PSNR of the proposed method compared to other algorithms

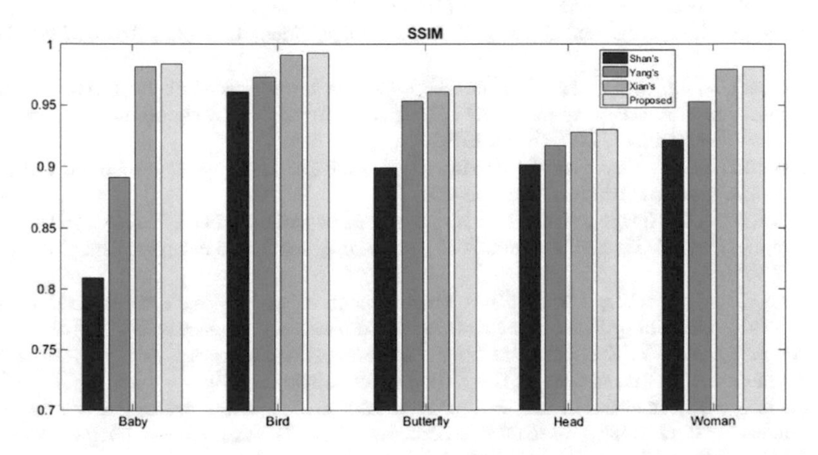

Fig. 5 SSIM of the proposed method compared to other algorithms

From the bar graphs in Figs. 4 and 5, it is evident the proposed method achieves better PSNR and SSIM values when compared with other super-resolution algorithms. Higher values indicate the reconstructed image is more similar to the original image.

5 Conclusion

The main objective of a super-resolution algorithm is to generate HR images from a single LR image or a set of LR images. In this method, we used an effective texture enhancement method and incorporated the gradient information in single-image super-resolution based on the similarity of gradient and texture information. This method sufficiently restores fine image details. The texture enhancement method highlights the superior detail and gradient-based methods to enhance the structure details uniformly without artifacts. This technique can accomplish good results for HR image creation with various scale factors. Experimental results visually display the competence of this method since it reveals more information in the reconstructed images.

References

1. Yang, J., J. Wright, T.S. Huang, and Y. Ma. 2010. Image super-resolution via sparse representation. *IEEE Transactions on Image Processing* 19: 2861–2873.
2. Su, D., and P. Willis. 2004. Image interpolation by pixel-level data-dependent triangulation. *Computer Graphics Forum* 23 (2): 189–201.
3. Muresan, D.D., and T.W. Parks. 2004. Adaptively quadratic (aqua) image interpolation. *IEEE Transactions on Image Processing* 13 (5): 690–698.
4. Shan, Q., Z. Li, J. Jia, and C. Tang. 2008. Fast image/video upsampling. In *ACM SIGGRAP Asia*.
5. Kanade, T., J.F. Cohn, and Y. Tian. 2000. Comprehensive database for facial expression analysis. In *Proceedings of the Fourth IEEE International Conference on Automatic Face an Gesture Recognition, 2000*, 46–53. IEEE.
6. Freeman, W., T. Jones, and E. Pasztor. 2002. Example-based super resolution. *Computer Graphics and Applications* 22 (2): 56–65.
7. Fattal, R. 2007. Image upsampling via imposed edge statistics. In *ACM SIGGRAPH*.
8. Sun, J., J. Sun, Z. Xu, and H. Shum. 2008. Image super-resolution using gradient profile prior. In *CVPR*.
9. Xian, Yang, and Yingli Tian. 2015. Single image super resolution via internal gradient similarity. *Journal of Visual Communication and Image Representation* 35: 91–102.
10. Xu, L., Q. Yan, Y. Xia, and J. Jia. 2012. Structure extraction from texture via relative total variation. *ACM Transactions on Graphics* 31 (6): Article ID 139.
11. Sheikh, H.R., M.F. Sabir, and A.C. Bovik. 2006. A statistical evaluation of recent full reference image quality assessment algorithms. *IEEE Transactions on Image Processing* 15 (11): 3440–3451.

An Unsupervised Approach for Moving Object Detection Using Distinction Features

G. Augusta Kani and P. Geetha

Abstract Moving object segmentation in a video is currently a most crucial task. The approach proposed in this work is used in many applications such as object detection, object tracking, surveillance monitoring and behavior analysis of moving objects in video sequences. Unsupervised moving object segmentation attempts automatic segmenting of moving objects in a video without prior understanding. The main objective of this work is to extract an object of interest from a series of successive video key frames rather than all frames. In the existing approach, object segmentation requires a tool to manually select an object area, making the process more tedious and time-consuming. To avoid this, the dynamic *k*-means algorithm for object segmentation is proposed. In this chapter, an unsupervised approach for moving foreground object segmentation is proposed by generating a saliency map from the key frames. A saliency map is computed based on global cues computed by employing the dynamic *k*-means clustering algorithm to cluster individual pixels. Finally, the moving object is extracted with the key frame based on the cluster, global cues and a saliency map. This proposed work is analyzed on a mixture of standard motion database.

Keywords Moving object (human/object) detection · Saliency map
Key frame extraction

G. Augusta Kani (✉) · P. Geetha
Department of Information Science and Technology, College of Engineering Guindy,
Anna University, Chennai, Tamil Nadu, India
e-mail: augus.jesus@gmail.com

P. Geetha
e-mail: geethap@annauniv.edu

A. N. Krishna et al. (eds.), *Integrated Intelligent Computing, Communication and Security*, Studies in Computational Intelligence 771,
https://doi.org/10.1007/978-981-10-8797-4_64

1 Introduction

The incredible growth of vision-based applications in recent years such as video surveillance, content-based video analysis, traffic monitoring and analysis of sports video, has resulted in many problems in segmenting objects of interest. Segmentation of objects of interest from a sequence of frames is one of the basic and essential operations in computer vision. Video object segmentation is the process of sub-diving a video sequence into regions with respect to object features and properties. It is useful for several computer vision tasks including video analysis, object tracking, object recognition, video retrieval and activity recognition. The approach is carried out into two ways. The supervised method of object segmentation has the user selecting an object area to be extracted. The unsupervised method of object segmentation extracts the object without having any prior information about the object. There are many approaches for supervised object segmentation, but notably fewer that focus on an unsupervised method. An unsupervised method is simpler and can be applied in many fields, such as video analysis, video indexing, video retrieval and activity understanding.

In this chapter, a saliency-based method to segment the foreground object from the unlabeled data has been proposed. First, key frames are generated by using the proposed method [1]. Then, color and texture features are extracted and given as input to dynamic *k*-means algorithm. Based on the cluster generated by the dynamic clustering algorithm, contrast and spatial cues are computed and multiplied element-wise to generate the saliency map based on the method [2]. The saliency map shows the foreground object which needs to be segmented.

1.1 Related Work

Recently, many approaches have been developed to extract foreground objects. Gu et al. [3] and Wang et al. [4] proposed a video scene segmentation approach with content coherence and contextual dissimilarity. Brox and Malik [5] proposed a method based on trajectory. A cycle of trajectories is extracted and then processed by spectral clustering. Ochs and Brox [6] extend the trajectory method to acquire a denser foreground region. Lee et al. [7] showed that these methods lack an explicit notion of the foreground object and low-level grouping of pixels, which results in over-segmentation. Visual saliency has been used to form the foreground object segmentation [8]. This method explores object-based segmentation in a static image and achieves significant progress. Those methods generate multiple object and rank hypotheses with their scores. The most contemporary research done by Xiaochun Cao et al. [9] and Zhang et al. [10] proposed a method to extract a foreground object automatically without any prior awareness about the object. Objects like regions are segmented in successive frames based on their appearance and motion. Then the shortest-path algorithm is explored to obtain the global optimum solution from the graph model. A Gaussian mixture model (GMM) is applied in an iterative manner to produce more filter results; when the GMM is matched with the results, then the iteration is stopped and returns the foreground object. Fan and Loui [11] successfully focused on video segmentation as a task of grouping pixels in successive video frames into perceptually coherent regions. This is graph-based approach in a high-dimensional feature space that sinks the inter-class connectivity between different objects. The graph model combines appearance, spatial and temporal information into semantic spatiotemporal key segments. Continuous grouping of a video sequence enables extraction of a moving object of interest from a video sequence based on its unique properties. Carsten Rother et al. [12] addressed the problem of an efficient interactive extraction of a foreground object in a difficult environment whose background cannot be subtracted easily. The grab cut approach is an extension of the graph cut approach; it has an iterative version of optimization. Ma and Latecki [13] attempted to address this video object segmentation problem by utilizing relationships between object proposals in adjacent frames. The object region candidates are selected simultaneously to construct a weighted region graph.

Zhang et al. [14] proposed a novel approach to extract primary objects in videos, in which a directed acyclic graph (DAG)-based framework is explored for detection and segmentation of the primary object in a video.

2 System Design

In surveillance monitoring, traffic monitoring systems, and moving objects in a static background, the difficulty of finding moving foreground objects in an unsupervised manner is a tedious task. Low-level segmentation is done in previous approaches, whereas our contribution in this approach is object-level segmentation. It is made by extracting key frames. All key frames are processed to extract feature vector based on color and texture. Then, a dynamic *k*-means algorithm is employed to cluster the pixel based on the feature vector. The cluster map is generated based on a dynamic clustering algorithm. Contrast and spatial cues are computed to generate a saliency map. Finally, the object of interest is segmented based on the saliency map. Figure 1 shows the proposed architecture for segmenting moving foreground objects from the given set of key frames. In this paper, the system is implemented using the Freiburg–Berkeley motion segmentation (FBMS) dataset.

2.1 Key Frame Detection

A key frame is a representative video reference frame able to reflect the abstract of video content. With video data containing several shots, it is necessary to identify individual shots for key frame extraction. Visual features and temporal information

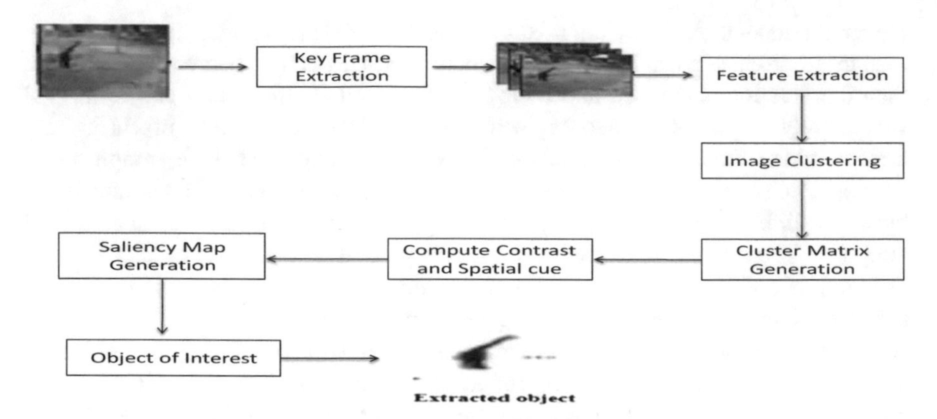

Fig. 1 Proposed system architecture

are used to determine the key frame. Key frames are extracted to reduce processing time.

Key frames are extracted based on setting a threshold value and comparing all frames with the threshold value; frames satisfying the threshold condition are considered key frames. The threshold is computed using Eq. (1):

$$T = \mu + \sigma \tag{1}$$

where μ and are the mean and standard deviation of the absolute difference, respectively. Steps for key frame extraction are as follows:

Steps:

1. Extract frames one by one
2. Compute the histogram difference between two consecutive frames
3. Calculate the mean and standard deviation of the absolute difference
4. Compute the threshold
5. Compare the mean and threshold
6. If it is greater than the threshold, set it as a key frame
7. Continue till end of the video.

2.2 *Feature Extraction*

Feature extraction reduces the amount of characters required to depict a huge set of data. A large amount of memory and computation power is needed to analyze a huge set of data. To avoid this, the process contains a feature vector containing color and texture features. A Gabor filter is employed to obtain a feature vector with texture features. A distance vector for each pixel is also calculated and the original key frame is resized for further processing; the distance vector is calculated based on the Euclidean distance. The Euclidean distance is given in Eq. (2). Texton feature extraction is an efficient feature for texture analysis. Based on their texton frequency, histogram texture classifiers classify textures by texton features. There are three main stages of building texton-based texture classifiers:

(1) Creation of a texton codebook,
(2) Calculation of a texton frequency histogram,
(3) Classification of the texture based on the texture frequency histograms.

Texture patches are created from the texture image and then a filter is applied for extracting textons. The texton codebook is quantized into a vector and compared with the codebook to classify the texture.

$$d(p,q) = \sqrt{(q1-p1)2} + \sqrt{(q1-p1)2} \cdots + \sqrt{(q1-p1)2} \tag{2}$$

where p and q are Euclidean vectors.

2.3 Image Clustering and Cluster Matrix Generation

Image clustering is the task of grouping individual pixels to the corresponding cluster based on the color and texture feature of pixels. A distance vector is also calculated to determine the centroid of the image cluster. Image clustering can be achieved using many algorithms for grouping similar pixels in an unsupervised manner. In this paper, centroid-based dynamic clustering is done to group the image pixels based on the feature vector and distance vector. A dynamic k-means algorithm is applied to cluster the image. The following are the steps of the cluster matrix generation algorithm.

Steps:

1. Select k
2. Select k points as an initial centroid
3. From the initial centroid and observation value, calculate the Euclidean distance
4. Assign the cluster based on the maximum distance
5. Update the initial centroid based on the cluster assignment
6. Again, compute the Euclidean distance for the updated centroid and next observation
7. Repeat the steps until all observations fit the k cluster.

The cluster matrix is generated using the dynamic k-means clustering algorithm. It is one of the simplest unsupervised clustering algorithm with its steps shown above. The grouping is done by determining the minimum sum of squared distance between items and the corresponding centroid. The Euclidean distance for the dynamic k-means algorithm is calculated using Eq. (2) and the cluster centroid is updated using Eq. (3):

$$cp(x1, x2, \ldots, xn) = \frac{\sum_{j=1}^{n} x1}{n}, \frac{\sum_{j=1}^{n} x2}{n}, \ldots. \tag{3}$$

2.4 Global Cues Computation

The contrast and spatial cues are called global cues. The biological vision system is responsive to contrast in visual signaling. So, the saliency of a pixel as its color contrasts with all other pixels in an image is defined in Eq. (4):

$$S_{cc}(I_k) \propto \sum_{\forall I_i \in \mathrm{I}} D(I_k, I_i) \tag{4}$$

where $D(I_k, I_i)$ is the color difference metric between pixels I_k and I_i in lab color space. Pixels of the same color have the same saliency value based on the equation defined in color space. To compute the same-color saliency pixel value, another equation is defined in Eq. (5):

$$S_{cc}(I_k) \ \alpha \ \sum_{j=1}^{n} f_i D(C_i C_j) \tag{5}$$

where c_1 is the color value of pixel I_k, n is the number of distinct pixel colors and f_j is the probability of pixel color c_j occurring in image I. The color difference metric is proposed as a Gaussian system as given in Eq. (6):

$$D(C_i C_j) = 1 - e^{\frac{-d(c_i c_j)^2}{2\sigma}} \tag{6}$$

Spatial cue information can be used to distinguish the foreground from the background in a given frame. The pixels of a salient object often lay more compactly and background pixels are often distributed more dispersedly. The colors distributing more compactly would have higher probability of belonging to the salient object. The spatial cue is defined in Eq. (7):

$$U_{sv}(I_k) \propto \frac{1}{m} \sum_{j=1}^{n} (x_{c_l}^{(i)} - x_{cl})^2 + (y_{cl}^{(i)} - y_{cl})^2 \tag{7}$$

where cl is the color value of pixel I_k, m is the number of pixels that have color value $(x_{cl}^{(i)}, y_{cl}^{(i)})$ is the coordinate of one of these pixels and x_{cl}, y_{cl} is the average coordinate of these pixels. $U_{sv}(I_k)$ represents the degree of non-saliency.

2.5 Saliency Map Generation

A saliency map is an object of interest generated in an efficient manner by computing the spatial and contrast cues, followed by element-wise multiplication. Element-wise multiplication notation is given in Eq. (8):

$$z = x * y \tag{8}$$

3 Experimental Analysis

The dataset used for the development of the proposed system is a motion segmentation dataset. Key frames are extracted based on setting a threshold value and comparing all frames with the threshold value; frames satisfying the threshold condition are considered as key frames. Figure 2 shows the key frame extraction stored in a separate folder. This process contains a feature vector containing color and texture features. A Gabor filter is employed to obtain a feature vector with texture features. A distance vector for each pixel is also calculated, and the original key frame is resized for further processing.

Image clustering is done using a dynamic *k*-means algorithm. This algorithm clusters n objects based on attributes into k partitions, where $k < n$. A cluster bin number is given to cluster the pixels. Based on the number of bins and feature vector, a cluster matrix is generated. The contrast cue is computed based on Eq. (4) and it is normalized between 0 and 1. To normalize the cue between 0 and 1, an array is given to the Gaussian normal function. The contrast cue array is [22:1743; 14:6794; 30:7943] and it is normalized as [0:5643; 0:1353; 1:0000]. The spatial cue is computed with Eq. (7) and is also normalized between 0 and 1 based on the Gaussian normal distribution. It is computed based on cluster index, bin number, distance vector and width of the frame, and it is stored in an array. The saliency

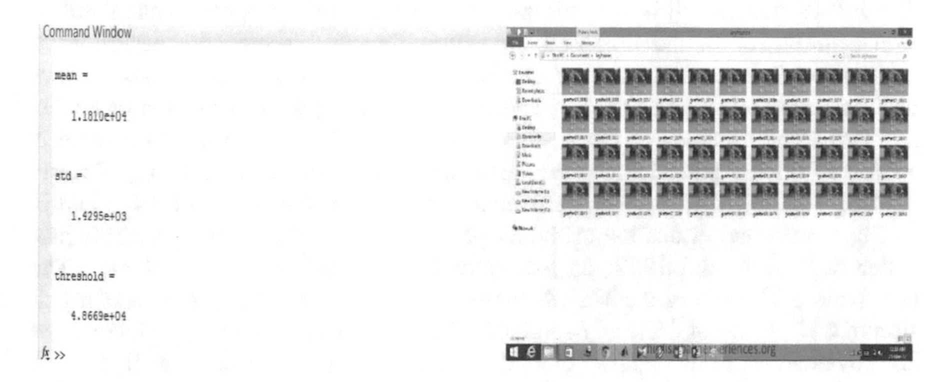

Fig. 2 Threshold calculation and key frame extraction

Fig. 3 Single- and multiple-object segmentation

map is generated based on the contrast cue and spatial cue. The global cues are combined to produce a saliency weight array for generating the saliency map. The cluster map, saliency weight array and bin number are given as an input to generate the saliency map.

Object segmentation is done based on the saliency map generated and it is also normalized to produce fine results of object segmentation. The left side of Fig. 3 shows the single-object segmentation, by computing the contrast and spatial cues. Based on the cues, a single saliency map is generated to segment the object.

The right side of Fig. 3 shows the multiple-object segmentation on sequential key frames. Contrast and spatial cues for all frames are computed and combined to generate a saliency map. Based on the saliency map, the object is segmented on sequential key frames.

4 Performance Evaluation

The FBMS dataset [15] is used for the development of the proposed system and is evaluated based on the proposed algorithm. Figure 4 shows the performance evaluation of the dataset used for developing the proposed system. It also shows that the proposed system produces average precision, recall and *f*-scores of 83, 92 and 87%. The accuracy of the system is about 95.43%. The proposed system is evaluated on the publicly available Extended Complex Scene Saliency Dataset (ECSSD) and MSRA, and their performance and results are given in Figs. 5 and 6.

The results and evaluation of the ECSSD shows that the system produces precision of 83%, recall of 95%, an *f*-score of 87% and an accuracy of 94.01%. The result and evaluation of the MSRA dataset shows that the system produces precision of 80%, recall of 85%, an *f*-score of 83% and accuracy of 91.05%. The output and evaluation results show that the proposed system works well for all state-of-the-art datasets and produces the most accuracy for all datasets.

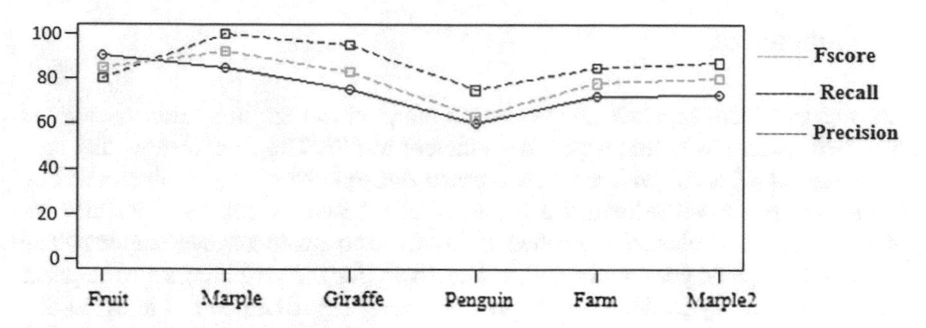

Fig. 4 Performance evaluation

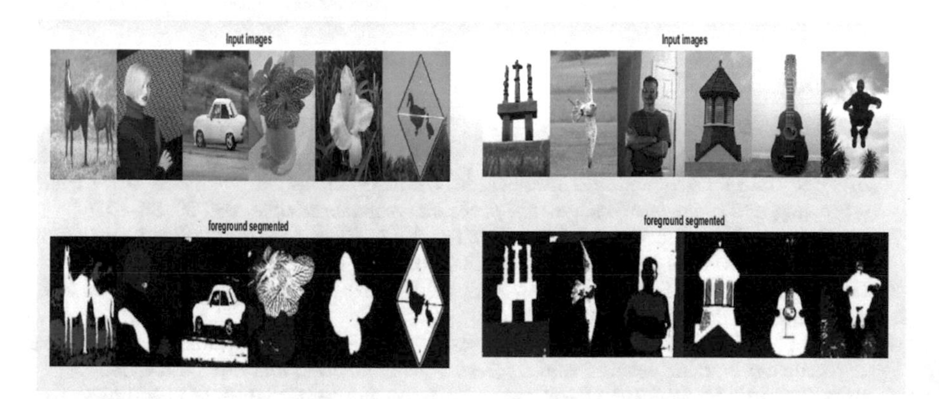

Fig. 5 Results of the ECSSD and MSRA dataset

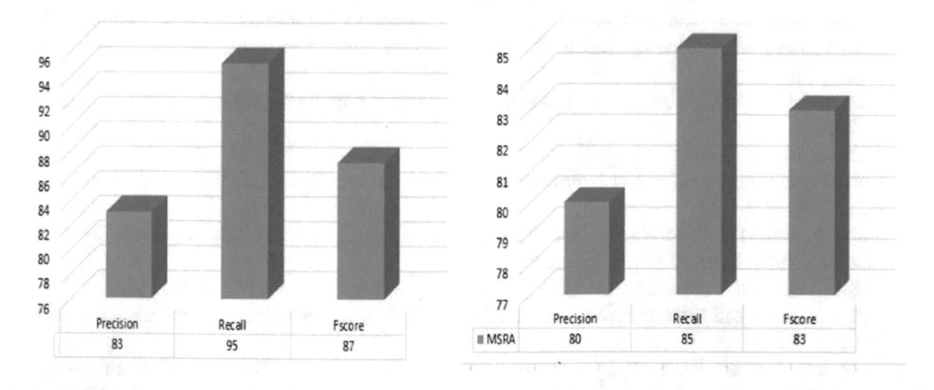

Fig. 6 Evaluation results of the ECSSD and MSRA dataset

5 Conclusion

An unsupervised approach for video moving object segmentation system is designed in such way that it provides efficient results. The results show that segmentation is achieved with a saliency-based dynamic clustering method. All key frames are processed to extract a feature vector based on color and texture. An effective dynamic clustering method is involved to avoid serious learning. The contrast and spatial cues are computed effectively for the given dataset to segment the foreground object. An effective preprocessing step is carried out to avoid difficulty. This video object segmentation approach is an effective and distinct approach and it is used for many vision applications.

References

1. Sheena, C. Va, and N.K. Narayanana. 2015. Key-frame extraction by analysis of histograms of video frames using statistical methods. In *4th international conference on Eco-friendly computing and communication systems, procedia computer science*, vol. 70, 36–40.
2. Fu, Huazhu, Xiaochun Cao, and ZhuowenTu. 2013. Cluster-based co-saliency detection. *IEEE Transactions on Image Processing*, 22 (10).
3. Gu, Z., T. Mei, X.S. Hua, X. Wu, and S. Li. 2007. Ems: Energy minimization based video scene segmentation. In *ICME*, 520–523.
4. Wang, J., X. Tian, L. Yang, Z.J. Zha, and X.S. Hua. 2008. Optimized video scene segmentation. In *IEEE international conference on multimedia and expo*, 301–304.
5. Brox, T., and J. Malik. 2010. Object segmentation by long term analysis of point trajectories. In *ECCV*, 282–295.
6. Ochs, P., and T. Brox. 2011. Object segmentation in video: a hierarchical variational approach forturning point trajectories into dense regions. In *ICCV*, 1583–1590.
7. Lee, Y.J., J. Kim, and K. Grauman. 2011. Key segments for video object segmentation. In *ICCV*, 1995–2002, Nov 6, National Center for Biotechnology Information. http://www.ncbi.nlm.nih.gov.
8. Endres, I., and D. Hoiem. 2010. Category independent object proposals. In *ECCV*, pp. 575–588.
9. Cao, Xiaochun, Feng Wang, Bao Zhang, Huazhu Fu, and Chao Li. 2015. Unsupervised pixel-level foreground object segmentation via shortest path algorithm. *Neurocomputing*, 172: 235–243.
10. Zhang, Bao, Handong Zhang, and Xiaochun Cao. 2012. Video object segmentation with shortest path. In *ACM transactions*, 801–804.
11. Fan, Lei, and Alexander C. Loui. 2015. A graph-based framework for video object segmentation and extraction in feature space. In *IEEE international conference on multimedia (ISM)*, IEEE.
12. Carsten, Rother, Vladimir Kolmogorov, and Andrew Blake. 2004. GrabCut: Interactive foreground extraction using iterated graph cuts. In *ACM transactions on graphics (TOG)—Proceedings of ACM SIGGRAPH*, vol. 23, 309–314.
13. Ma, T., and L.J. Latecki. 2012. Maximum weight cliques with mutex constraints for video object segmentation *CVPR*, 670–677.

14. Zhang, Dong, Omar Javed, and Mubarak Shah. 2013. Video object segmentation through spatially accurate and temporally dense extraction of primary object regions. *IEEE Transcations* 32: 628–635.
15. http://lmb.informatik.uni-freiburg.de/resources/.

A Fast and Efficient Method for Image Splicing Localization Using BM3D Noise Estimation

Aswathy M. Das and S. Aji

Abstract Image manipulation is the process of altering the content of an image with a particular intention or interest. By the developments in the camera industry and powerful editing tools, the image manipulation has become an easy task for even a less skilled person. Therefore, ensuring the authenticity of an image from an external source is necessary before using it. The proposed method deals with detecting an image forgery method known as splicing. It is a comparatively fast and reliable method for image splicing localization. The variation in noise level of different segments, non-overlapped segmentation using simple linear iterative clustering (SLIC), is estimated with the help of Block-Matching and 3D Collaborative Filtering (BM3D). The estimated noise variation is used to cluster the segments where the clusters with relatively higher noise level are located as suspicious or forged regions. It is found that the proposed method takes comparatively less time for execution and the false positive rate is also considerably low. Experimental results show better performance of the proposed method against existing methods of image splicing forgery detection.

Keywords Image splicing · SLIC · BM3D · Noise level · K-means clustering

1 Introduction

In today's world, digital technology makes it possible to manage, share, store and process images easily and quickly. Digital imaging technology plays an important role in almost all areas of human life, including satellite imaging, forensic inves-

A. M. Das (✉) · S. Aji
Department of Computer Science, University of Kerala, Thiruvananthapuram, India
e-mail: aswathymdasarunodayam@gmail.com

S. Aji
e-mail: aji_12345@yahoo.com

A. N. Krishna et al. (eds.), *Integrated Intelligent Computing, Communication and Security*, Studies in Computational Intelligence 771,
https://doi.org/10.1007/978-981-10-8797-4_65

tigation and medical imaging [1]. In some situations these developments are misused to make unethical modifications to the original images. Nowadays, advances in graphics editing tools make this task much easier, and even the human eye cannot perceive such modifications. There are different approaches for image forgery [2] including copy-move, image retouching and image splicing. Here we focus on a common operation known as splicing, which is the process of piecing together two or more photographic images to form a new fake image. The traces of tampering are visually deceiving unless no post-processing operations such as noise addition, contrast altering or distortions are applied to image.

Hsu et al. [3] presented an automatic method for detecting image splicing based on exploring the camera response function using geometric invariants. To classify the segments as original or spliced, CRF-based cross-fitting and local image features are computed and fed to a classifier. Zhang et al. [4] proposed a method that is based on exploring the noise variance between original and spliced images.

Lyu et al. [5] described a method based on estimating the kurtosis concentration of the natural image in the band-pass domain. Julliand et al. [6] presented an approach based on the inconsistency in noise parameters. This method uses a quadtree to decompose the spliced areas into subregions. Pun et al. [7] described a method based on multi-scale noise estimation for image splicing forgery detection. In order to detect image splicing forgery, the noise levels of all segments on multiple scales are used as evidence.

In order to decrease the false-positive rates, a reliable and robust noise estimation scheme is needed. A robust noise estimator (BM3D) [8] is introduced in the proposed method to detect the noise level of the superpixels. Experimental results show that the proposed method is superior to the existing methods for image splicing forgery detection.

This chapter is organized as follows. Section 2 provides a description of the proposed method. The experimental results and analysis are given in Sect. 3. The conclusion of the work is presented in Sect. 4.

2 Image Splicing and BM3D Noise Estimation

It will be always better to view an image, I, in smaller segments to locate forged regions correctly. As mentioned in the related works [5–7], our proposed method also considers the deviation in noise level of smaller regions to detect the spliced area.

$$I = S_1, S_2, S_3 \cdots Sm \tag{1}$$

where the image I is segmented into m segments, $S_1, S_2, S_3 \cdots Sm$ of the same size, and the noise distribution, μ, contains l levels of noise obtained by:

$$\mu = \mu_1, \mu_2, \mu_3 \cdots \mu_l \tag{2}$$

where μi represents the ith noise level, i $\in l$ and $l \leq$ m.

2.1 SLIC Superpixel

The clustering algorithm (SLIC) [9] is used to divide the input image into non-overlapping segments. The SLIC algorithm performs clustering in a combined five-dimensional space of both color and image planes $[l, a, b, x, y]$; the k cluster centers are initialized as $C_i = [l_i, a_i, b_i, x_i, y_i]^T$. The color and spatial proximities are normalized by their distance N_s and N_c. The pixels are assigned to the nearest cluster according to the distance

$$D' = \sqrt{\left(\frac{d_c}{N_c}\right)^2 + \left(\frac{d_s}{N_s}\right)^2} \tag{3}$$

where

$$d_c = \sqrt{(l_j - l_i)^2 + (a_j - a_i)^2 + (b_j - b_i)^2} \tag{4}$$

$$d_s = \sqrt{(x_j - x_i)^2 + (y_j - y_i)^2} \tag{5}$$

The cluster centers are updated to the mean $[l, a, b, x, y]$ and the clustering error calculated using L2 norm. This process will be continued until the error converges.

2.2 Noise Level Estimation

In the proposed method, the noise level of the segments is considered a feature. Spliced images may contain regions from other images; therefore, the noise level will vary at different regions. When regions from different images with different noise levels are combined or noise is intentionally added in forged regions to conceal tampering [3], the inconsistency in the noise levels in different regions of the image can be used to expose the tampered regions.

BM3D [8] is an image denoising algorithm based on enhanced sparse representation in the transform-domain. The enhancement of the sparsity is obtained by grouping similar 2D image fragments into 3D data arrays called groups. The collaborative filter is applied to obtain the 3D estimate of the group, which consists of an array of jointly filtered 2D fragments. Because of the similarity between grouped fragments, the transform can achieve a highly sparse representation of the true signal so that the noise can be separated by shrinkage. The convolution of BM3D with the original image results in the formation of a restored image, which can be given as:

$$g(x, y) = bm3d * f(x, y) \tag{6}$$

where $f(x, y)$ is the original image, $g(x, y)$ is the restored image and $*$ represents the convolution operator. The restored image will be subtracted from the original noisy image to extract the noise content as given below:

$$\eta(x,y) = f(x,y) - g(x,y) \tag{7}$$

Next, the variation in noise in each segment is estimated to identify the spliced regions, and is given in Eq. 8:

$$\eta(S_j) = \sqrt{\frac{1}{|S_j|} \sum_{\eta(x,y)\in S_j} \left(\eta(x,y) - \overline{\eta(S_j)}\right)^2} \tag{8}$$

where $\eta(S_j)$ indicates the estimated noise of the input image, $|S_j|$ is the pixel count of the corresponding segment, and $\overline{\eta(S_j)}$ represents the mean value of the estimated noise of the corresponding segment, and can be calculated using:

$$\overline{\eta(S_j)} = \frac{\sum_{\eta(x,y)\in S_j} \eta(x,y)}{|S_j|} \tag{9}$$

The variation in noise level in each segment is calculated for the next step in the method.

2.3 *Clustering the Segments*

The superpixel segments are classified into two groups according to the estimated noise level, where the clusters with higher noise levels are identified as suspicious forged regions. The K-means algorithm [10], the classical method for clustering, is used to group the segments according to the noise level.

3 Experimental Results

We have evaluated our proposed strategy with a series of experiments. We used the dataset reported by Liu et al. [7] for the experiments. Figures 1 and 2 show the results of the proposed method. Figure 1 consists of only one spliced object, and Fig. 2 consists of two spliced objects. Our method can detect image splicing forgery with multiple objects pasted into the original images. This section also provides a detailed analysis of the proposed method based on evaluation metric precision, recall and F1 score. To evaluate the efficiency of the proposed method, we have compared the results with some existing methods: Liu's model [7] and Lyu's model [5].

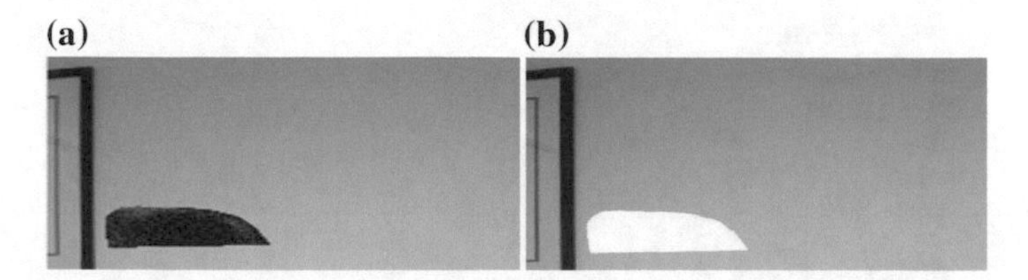

Fig. 1 **a** Original image with one object spliced; **b** result for the proposed method

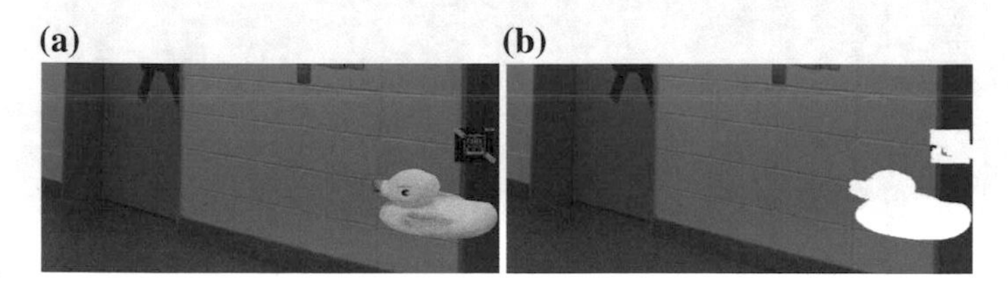

Fig. 2 **a** Original image with two object spliced; **b** result for the proposed method

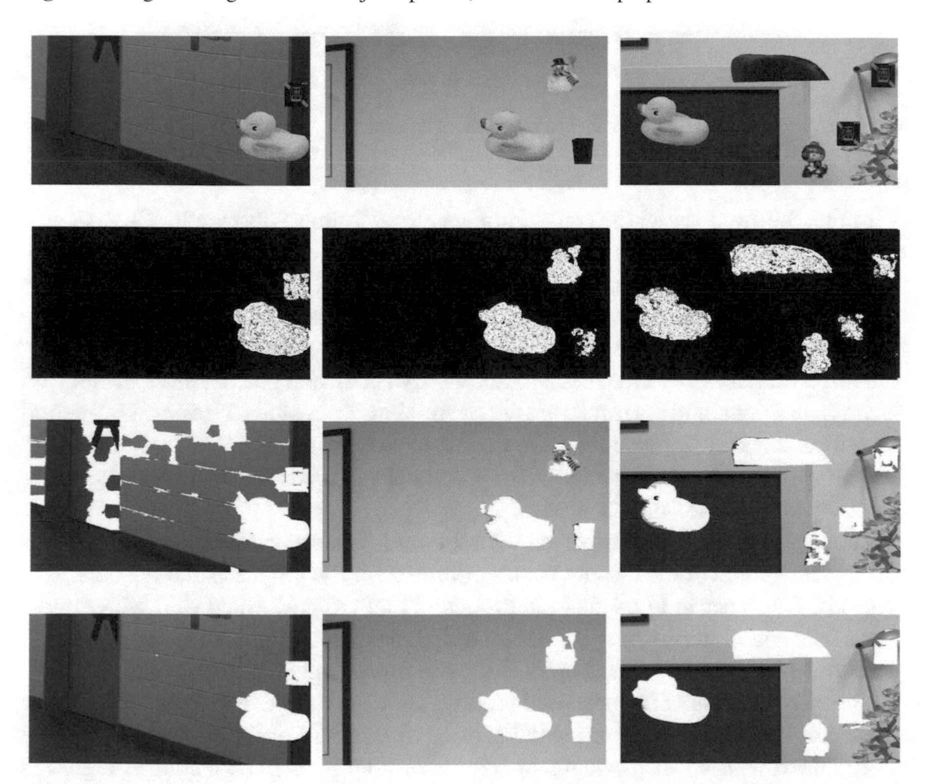

Fig. 3 Figure 3 Detection Result Row 1: Original images with forged object. Row 2: Result for Lyu et al.'s method. Row 3: Result for Liu et al. method. Row 4: Result for proposed method

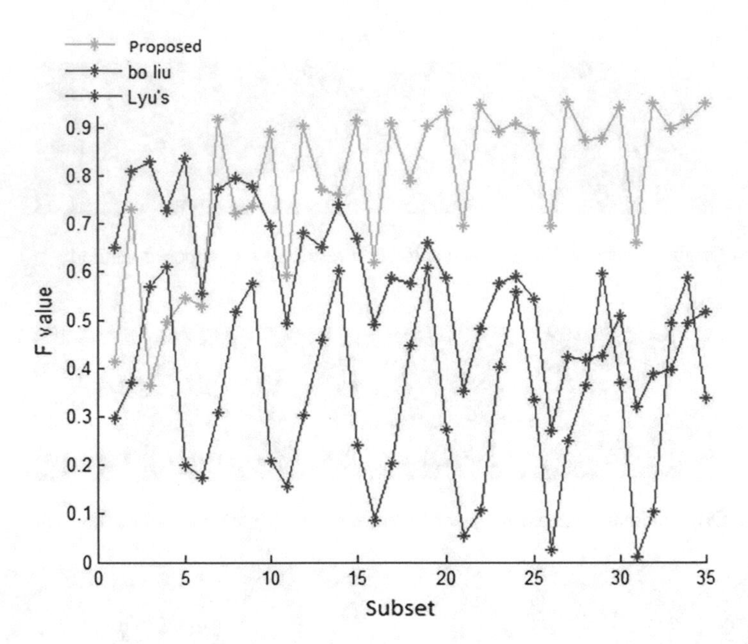

Fig. 4 Comparison of F-Score values for the entire dataset

In Fig. 3, Row 1 shows the spliced images, Row 2 shows the results of Lyu et al.'s method [5], Row 3 shows the results of Liu et al.'s method [7], and Row 4 shows the results of the proposed method.

Table 1 shows the precision recall, F-score values and time complexity for the following images using Lyu et al.'s, Liu et al.'s, and the proposed method.

The results shows that the method used by Lyu et al. and that of Liu et al. detected the forged regions, and the results are not promising. The results of Liu et al.'s method are not satisfactory because the false-positive rates are higher. The results visually demonstrate that the performance of the proposed method is satisfactory and promising comparing to the other methods.

The experiments were performed with the entire datasets; Fig. 4 gives the graphical representation of the F-score values obtained for the proposed and the other two methods. It is clear in Fig. 4 that the proposed method performed well with almost 25 subsets and obtained an F-Score value above 90. As compared to the other two methods, the proposed method outperforms in more than 30 subsets.

The effectiveness of the proposed method depends on the parameter selected for segmentation. In order to locate small spliced objects successfully, the segment size is initialized with 45. If the segment size is very large, the small spliced region is submerged and not detectable. The quantitative and qualitative analyses show that the proposed method is promising and reliable for detecting spliced areas in images.

Table 1 Precision, recall, F-score values and time complexity for the images shown in Fig. 3

	Image 1				Image 2				Image 3			
	Prec.	Recall	F-score	Time(s)	Prec.	Recall	F-score	Time(s)	Prec.	Recall	F-score	Time(s)
[5]	88.08	79.9	83.79	31.21	93.71	75.63	83.49	30.93	90.15	71.09	79.49	31.01
[7]	57.01	62.05	59.42	305.01	89.17	76.05	82.08	300.03	88.05	80.19	83.93	303.05
Proposed	**99.08**	**98.91**	**98.99**	**13.14**	**97.50**	**98.28**	**97.89**	**13.91**	**94.26**	**99.11**	**96.63**	**12.76**

4 Conclusion

Digital images have become a convenient method for representing information. Image splicing is one of the most important image tampering techniques, in which regions from one image are copied and pasted into another image. The proposed method focused on detection of image splicing and relies on the assumption that the spliced region and the original image have different intrinsic noise variances. It is a fast and efficient method for splicing localization. In order to locate the forged regions correctly, the image is segmented into non-overlapping segments. The noise level of the image areas are determined using block-matching and 3D collaborative filtering (BM3D). The deviation of noise level is used to cluster the segments into two groups. The clusters with relatively higher noise level are treated as forged regions. The use of BM3D decreases the time complexity and false-positive rates and also increases the accuracy of the proposed method. The experimental results show that the proposed method is promising compared to the other methods. The method does not address the condition in which the spliced regions have a noise distribution similar to the original image.

References

1. Pan, X., X. Zhang, and S. Lyu. 2012. Exposing image splicing with inconsistent local noise variances. In *IEEE international conference on computational photography (ICCP)*, 1–10.
2. Lin, S.D., and T. Wu. 2011. An integrated technique for splicing and copy-move forgery image detection. In *4th international congress on image and signal processing (CISP)*, 1086–1090.
3. Hsu, Y.F., and S.F. Chang. 2010. Camera response functions for image forensics: An automatic algorithm for splicing detection. *IEEE Transactions on Information Forensics and Security* 816–825.
4. Pan, Xunyu, Xing Zhang, and Siwei Lyu. 2012. Exposing image forgery with blind noise estimation. In *IEEE international conference on computational photography (ICCP)*.
5. Lyu, S., X. Pan, and X. Zhang. 2014. Exposing region splicing forgeries with blind local noise estimation. *International Journal of Computer Vision* 110 (2): 202–221.
6. Julliand, T., and V. Nozick. 2015. Automated image splicing detection from noise estimation in raw images. In *IEEE conference*, 1086–1090.
7. Pun, Chi-Man, Bo Liu, and Xiao-Chen Yuan. 2016. Multi-scale noise estimation for image splicing forgery detection. *Journal of Visual Communication and Image Representation* 195–206.
8. Dabov, Kostadin. 2007. Image denoising by sparse 3-D transform-domain collaborative filtering. *IEEE Transactions on Image Processing* 2080–2095.
9. Achanta, R., A. Shaji, K. Smith, A. Lucch, and S. Susstrunk. 2012. SLIC superpixels compared to state-of-the-art superpixel methods. *IEEE Transactions on Pattern Analysis and Machine Intelligence* 2274–82.
10. Kanungo, T., and D.M. Mount. 2002. An efficient k-means clustering algorithm: Analysis and implementation. *IEEE Transactions on Pattern Analysis and Machine Intelligence*.

A Single Image Haze Removal Method with Improved Airlight Estimation Using Gradient Thresholding

M. Shafina and S. Aji

Abstract Image dehazing is a technique to recover the intensity and quality of images captured in special climatic conditions such as fog, haze or mist. In this chapter, we present a computationally less expensive and reliable method that employs gradient thresholding airlight and weight-guided image filtering (WGIF) with a color attenuation prior approach for dehazing. Color attenuation prior is used for computing the depth of a scene. The depth information is refined with WGIF to avoid halo artifacts. By adopting the improved gradient thresholding method for airlight estimation, better results can be produced in less time.

Keywords Image processing · Dehazing · Degraded image
Image restoration · Depth restoration

1 Introduction

Generally, images captured in bad weather have low quality and contrast due to the attenuation of light and other atmospheric parameters [1]. Particles in the atmosphere reflect the airlight, which increases the brightness of the scene and thereby reduces the contrast. These kinds of natural parameters that affect the quality of the image are a major issue for many vision applications including surveillance, recognition systems, transportation systems and aerial photography.

Haze is actually an atmospheric phenomenon, where fragments existing in the atmosphere cause absorption and scattering of light. Due to haze, scene radiance changes from the original intensity to a gray or bluish color. According to the weather conditions—haze, fog, clouds, snow, etc.—the type, size and density of the particles may vary. Haze is the dispersed system of tiny particles mixed with gas, and its origin can be smoke, dust, air pollution or other sources. Haze particles are larger than air particles, and will alter the natural quality of the scene.

M. Shafina (✉) · S. Aji
Department of Computer Science, University of Kerala, Thiruvananthapuram, India
e-mail: shafina.m90@gmail.com

A. N. Krishna et al. (eds.), *Integrated Intelligent Computing, Communication and Security*, Studies in Computational Intelligence 771,
https://doi.org/10.1007/978-981-10-8797-4_66

Dehazing methods can be broadly classified into three categories—enhancement, fusion and restoration—according to the basic operations performed in the algorithm. Tan [2] proposed a dehazing method by observing the contrast and airlight of hazy images, and modeled the airlight using a Markov random field model by enhancing the local contrast of the image.

Wang et al. [3] later proposed a multiscale fusion method to remove haze from a single image. They proposed a wavelet fusion approach for eliminating blocky artifacts and preserving the scene depth. Ancuti et al. proposed another fusion-based method [4] that employs two hazy input images by performing input contrast enhancement and white balancing mechanisms. After joining the significant information of the input images, the artifacts are minimized by the multiscale method with Laplacian pyramid representation.

He et al. introduced the dark channel prior (DCP) [5] in which the intensity of one of the color channels in RGB [red/green/blue] is very small and likely to be zero. However, the DCP fails to restore the hazy image with a large sky area. Zhu et al. [6] proposed a color attenuation prior-based dehazing method where they created a linear model for the depth information of the hazy image. Using this information, the scene radiance is restored in the atmospheric scattering model. In this work, we improve the color attenuation prior-based dehazing method by employing a gradient thresholding method [7] of airlight estimation and WGIF.

The rest of the chapter is organized as follows. In Sect. 2, the atmospheric scattering model is explained. The proposed method is explained in Sect. 3. Experimental results and analysis are given in Sect. 4, and a summary of the work is provided in Sect. 5.

2 Model of Atmospheric Scattering

The atmospheric scattering model, which was proposed by McCartney in 1976 [8], is used to describe hazy image formation. It is generally used in image processing and vision applications. Nair and Narasimhan [1] derived the model as follows:

$$I(x) = J(x)t(x) + A(1 - t(x)) \tag{1}$$

$$t(x) = e^{-\beta d(x)} \tag{2}$$

In Eqs. (1) and (2), $I(x)$ is the input hazy image, $J(x)$ is the haze-free image, $t(x)$ is the transmission through the medium representing the amount of irradiance energy reaching an observer after traveling through the medium, A is the global atmospheric light, β is the scattering coefficient of the atmosphere, and d is the depth of the image. The aim of dehazing is to determine $J(x)$, $t(x)$ and A from the hazy image generation model Eq. (1).

When a beam of light propagates through a hazy environment, two mechanisms are involved, attenuation and airlight. Due to attenuation, the scene radiance diminishes as the distance from the viewer increases [6]. When light is scattered from the incident beam of light, the remaining fragment of energy is known as the direct transmission or direct attenuation [parameter $J(x)t(x)$ in Eq. 1]. The second mechanism in which the atmosphere acts as a source of light is referred to as airlight (parameter $A(1 - t(x))$ in Eq. 1). Airlight occurs due to the scattering of environmental illumination by particles present in the atmosphere. The sources of environmental illumination include diffuse skylight, light reflected from the ground, and direct sunlight. Airlight increases with increased depth or distance from the observer. Because of this feature, scene brightness increases with path length.

3 Proposed Method

Figure 1 shows the block diagram of the proposed method. The RGB color model is converted into HSV [hue/saturation/lightness] for utilizing the saturation and brightness component in the prior information. Here we are adopting the color attenuation prior for calculating the depth data. Thereafter, the depth data is refined with weighted guided image filtering [9] instead of guided image filtering (GIF) for removing the halo artifacts. The gradient thresholding approach [7] has been used for determining the airlight, which could further reduce the computation time compared to the other methods working with the brightest pixels. By employing the scattering model, we can easily calculate the transmission map and effectively restore the original haze-free image.

3.1 Depth Map Estimation

By analyzing various hazy images, it is found that the haze concentration and difference between saturation and brightness increase along with the difference or

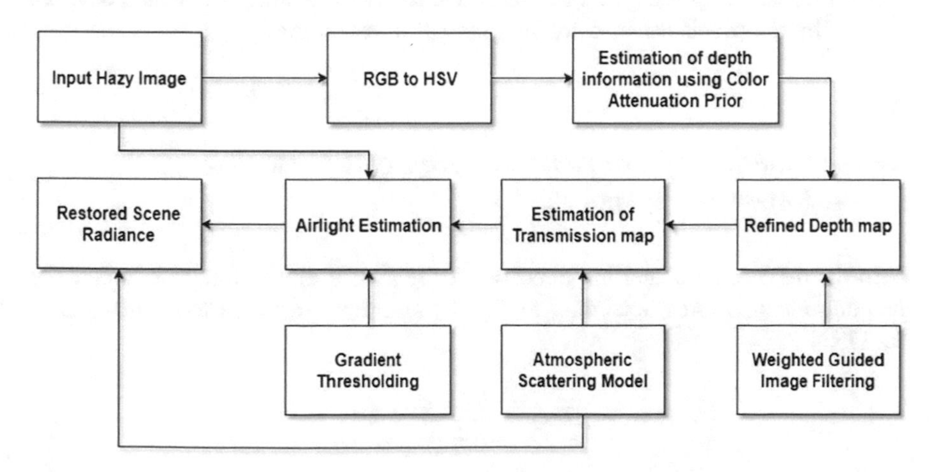

Fig. 1 Single image dehazing with improved airlight and weight-guided image filtering

change in scene depth. It can be inferred the depth data of the scene correlates positively with the concentration of the haze; the difference between the brightness and saturation is referred to as color attenuation prior. This prior linear model is exerted to compute the depth map.

$$d(x) = \theta_0 + \theta_1 v(x) + \theta_2 s(x) + \varepsilon(x) \tag{3}$$

where x is the location within the image, d is the depth, v is the brightness of the input hazy image, s is the saturation component and θ_0, θ_1 and θ_2 are 0.121779, 0.959710 and –0.780245, respectively. Here, we use a Gaussian density for ε with zero mean and variance σ^2 with $\sigma = 0.041337$.

3.2 Refinement of the Depth Map

The depth map is refined with WGIF with linear time complexity [9]. The halo artifacts are generated by the local filters, and hence the estimated depth map will be filtered by WGIF instead of GIF.

3.3 Airlight Estimation

The airlight estimation algorithm has three major steps:

Step 1: Compute the Sobel operator on the input image.
Step 2: Pixels with magnitude of gradient higher than an arbitrary threshold are selected.
Step 3: Chose airlight as the maximum intensity at the farthest distance detected by the pixels selected by the gradient thresholding.

3.4 Estimation of Transmission Map and Restoring a Haze-Free Image

With the depth data d and the atmospheric light A, it is possible to compute the medium transmission t according to Eq. (2) and recover the restored image J in Eq. (1).

$$J(x) = \frac{I(x) - A}{e^{-\beta d(x)}} + A \tag{4}$$

To reduce the amplification of noise, the value of the transmission $t(x)$ is restricted to between 0.1 and 0.9. The final function used for restoring the haze-free image J in the proposed method can be realized by:

$$J(x) = \frac{I(x) - A}{\min\left\{\max\{e^{-\beta d(x)}, 0.1\}, 0.9\right\}} + A \quad (5)$$

In the proposed approach, we are improving Zhu et al.'s dehazing method [6] by adopting the airlight method using gradient thresholding [7] as proposed by Tarel et al. The depth data of the scene is filtered with WGIF. The color attenuation prior-based approach produced better and fast dehazing results. Instead of the GIF, for refining the depth map we have used a WGIF approach. Through the gradient thresholding-based airlight estimation, we were able to further reduce the computation time.

4 Experimental Results

4.1 Qualitative and Quantitative Evaluation

We have conducted both qualitative and quantitative analysis of the results to verify the performance of the proposed method. Figures 2 and 3 show the evaluation of

Fig. 2 Qualitative evaluation of real-world images

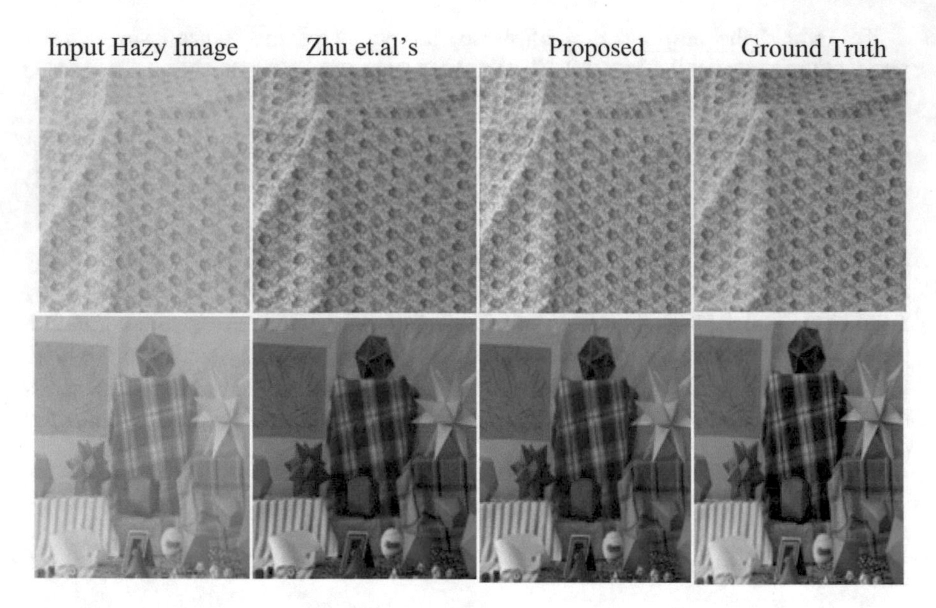

Fig. 3 Qualitative evaluation of synthetic images

real-world images and synthetic images, respectively, revealing that enhancement of the visual quality of the images is better with the proposed method than with Zhu et al.'s method [6]. For evaluation purposes, the ground truth images are taken from the Middlebury stereo datasets [11].

In the case of quantitative analysis of the algorithms, we have calculated the mean squared error (MSE) and structural similarity (SSIM) of the outputs. A low MSE value reveals the restored result is better, and a high SSIM value denotes great

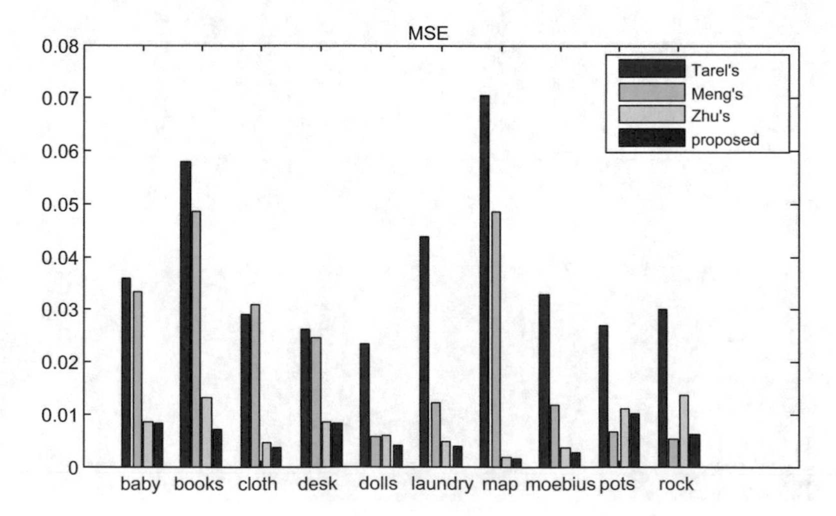

Fig. 4 MSE of the proposed method compared with other algorithms

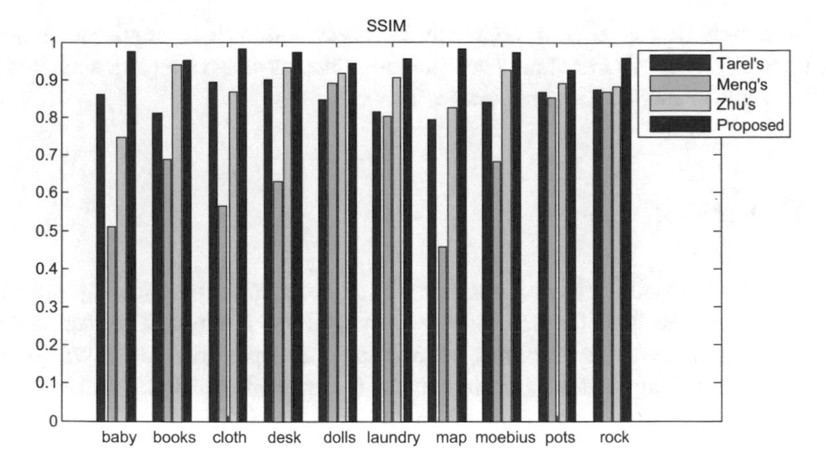

Fig. 5 SSIM of the proposed method compared with other algorithms

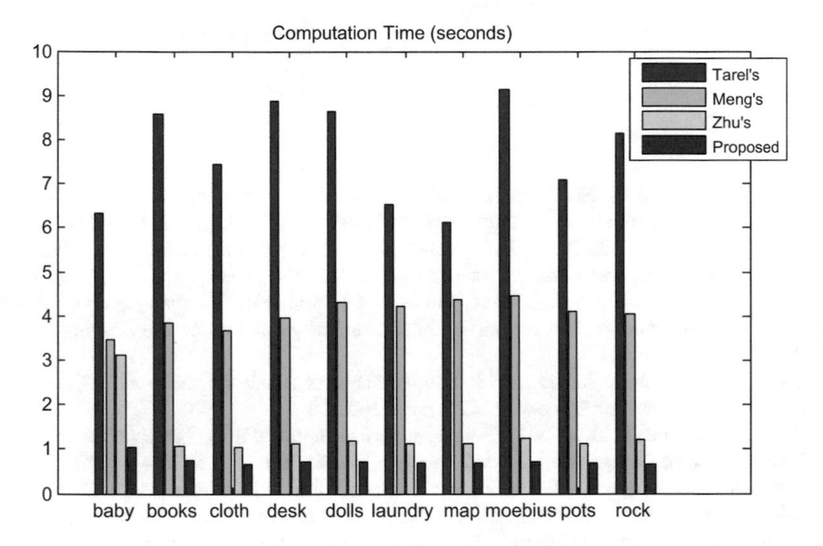

Fig. 6 Computation time of different algorithms

structural similarity between the restored haze-free image and the ground truth image. Figures 4, 5 and 6 compare the performance of our method with three other methods: Tarel et al. [12], Meng et al. [13] and Zhu et al. [6].

Figures 4 and 5 show that Tarel et al.'s method produced the highest MSE and Meng et al.'s method produced the lowest SSIM. It is clear that the proposed work achieved the best MSE and SSIM values in almost all images when compared with the other algorithms.

As shown in Fig. 6, our approach is much faster than previous dehazing methods. WGIF needs only linear computation time, as does GIF [10], such that the method can be applied to larger images as well.

5 Conclusion

Halo artifacts, amplification of noise in sky regions, and color fidelity are three major problems to be addressed for single image haze removal. We were able to effectively address these problems by refining the depth map with WGIF. The gradient thresholding method is better suited for estimating airlight, and hence we produced better results with minimal computational time. The experimental results show that the proposed approach yields a promising MSE (from 0.0014 to 0.0103) and SSIM (from 0.9247 to 0.9818), which is better than the other methods. All the parameters used for evaluation show that the proposed method is a fast and reliable approach for single image dehazing.

References

1. Nair, S.K., and S.G. Narasimhan. 1999. Vision in bad weather. In *Proceedings IEEE international conference on computer vision workshops (ICCV)*, vol. 2, 820–827.
2. Tan, R.T. 2008. Visibility in bad weather from a single image. In *Proceedings IEEE conference on computer vision pattern recognition (CVPR)*, 1–8.
3. Wang, Wei, Wenhui Li, Qingji Guan, and Miao Qi. 2015. Multiscale single image dehazing based on adaptive wavelet fusion. In *Mathematical problems in engineering*. Hindawi Publishing Corporation.
4. Ancuti, C.O., and C. Ancuti. 2013. Single image dehazing by multi-scale fusion. *IEEE Transactions on Image Processing* 22 (8): 3271–3282.
5. He, K., J. Sun, and X. Tang. 2011. Single image haze removal using dark channel prior. *IEEE Transactions on Pattern Analysis and Machine Intelligence* 33 (12): 2341–2353.
6. Zhu, Qingsong, Jiaming Mai, and Ling Shao. 2015. A fast single image Haze removal algorithm using color attenuation prior. *IEEE Transactions on Image Processing* 24 (11).
7. Caraffa, Laurent, and Jean Philippe Tarel. 2014. Daytime fog detection an density estimation with entropy minimization. In *ISPRS annals of the photogrammetry, remote sensing and spatial information sciences (PCV'14)*, II-3: 25–31.
8. McCartney, E.J. 1976. *Optics of the atmosphere: Scattering by molecules and particles*. NewYork, NY, USA: Wiley.
9. Li, Z.G., J.H. Zheng, Z.J. Zhu, W. Yao, and S.Q. Wu. 2015. Weighted guided image filtering. *IEEE Transactions on Image Processing* 24 (1): 120–129.
10. He, K., J. Sun, and X. Tang. 2013. Guided image filtering. *IEEE Transactions on Pattern Analysis and Machine Intelligence* 35 (6): 1397–1409.
11. Scharstein, D., and R. Szeliski. 2003. High-accuracy stereo depth maps using structured light. In *Proceedings IEEE conference on computer vision pattern recognition (CVPR)*, I-195–I-202.

12. Tarel, J.P., N. Hautière, L. Caraffa, A. Cord, H. Halmaoui, and D. Gruyer. 2012. Vision enhancement in homogeneous and heterogeneous fog. *IEEE Intelligent Transportation Systems Magazine* 4 (2): 6–20.
13. Meng, G.F., Y. Wang, J. Duan, S. Xiang, and C. Pan. 2013. Efficient image dehazing with boundary constraint and contextual regularization. In *Proceedings IEEE international conference on computer vision workshops (ICCV)*, 617–624.

Rough K-Means and Morphological Operation-Based Brain Tumor Extraction

Oyendrila Dobe, Apurba Sarkar and Amiya Halder

Abstract This chapter proposes a novel approach towards extraction of brain tumor images from T1-type magnetic resonance imaging (MRI) scan images. The algorithm includes segmentation of the scan image using a rough set-based K-means algorithm. It is followed by the use of global thresholding and morphological operations to extract an image of the tumor-affected region in the scan. This algorithm has been found to extract tumor images more accurately compared than existing algorithms.

Keywords Rough set · K-means clustering · Image segmentation
Morphological operation

1 Introduction

Cells are the basic building blocks of our bodies. As we grow, cells are constantly becoming old and dying. Simultaneously, new cells are being produced to replace them. The old cells undergo a process of programmed cell death, called apoptosis, that occurs in multicellular organisms. Apoptotic cells are not allowed to further divide. In old cells where the process of apoptosis breaks down, unwanted cell division occurs, leading to a mass of abnormal cells that grows out of control. A tumor

O. Dobe (✉) · A. Halder
STCET, 4, D.H. Road, Kidderpore, Kolkata 700023, West Bengal, India
e-mail: oyendrila.dobe@gmail.com

A. Halder
e-mail: amiya.halder77@gmail.com

A. Sarkar
IIEST Shibpur, Howrah 711103, India
e-mail: as.besu@gmail.com

A. N. Krishna et al. (eds.), *Integrated Intelligent Computing, Communication and Security*, Studies in Computational Intelligence 771,
https://doi.org/10.1007/978-981-10-8797-4_67

is basically an uncontrolled growth of cells. A benign tumor is not cancerous but has fatal effects on the brain. On the other hand, malignant tumors are cancerous and fatal. Brain tumors cause headaches, mental changes, problems with vision, or seizures. It is the most common cancer among those aged 0–19 (leukemia being the second). It is the second leading cause of cancer-related deaths in children (males and females) under age 20. Hence, a faster, more accurate extraction algorithm would accelerate the initiation of treatment. The beginning of the tumor treatment lies in the detection and extraction of tumors using various imaging processes and computational algorithms. Imaging processes include X-ray imaging, where X-rays are allowed to image excess depositions of calcium in bones that a tumor might have caused. Dyes are used to trace the flow of blood in the tumor region; computed tomography (CT) scans and angiography use intravenous dyes and contrast dyes to easily detect blood vessels. In a biopsy, a tissue sample is collected from the tumor-affected region to check for malignancy. However, the most efficient technology is magnetic resonance imaging (MRI), which uses gadolinium contrast media, a special dye to detect an abnormal region without the use of radiation. Also, MRI scans provide an image that digs deeper than other scans, so MRI scans are used frequently for analytical medical images. Various extraction methods have previously been proposed [1–7]. Patil et al. [8] suggested soft computing techniques for extracting brain tumors from MRIs using MATLAB functions. Halder et al. [9] proposed a combination of genetic algorithm-based fuzzy C-means algorithm (GA-FCM) and morphological operators.

2 Proposed Method

In this approach, we initially segment the scan image into a predetermined number of clusters. The segmented image then undergoes global thresholding, application of morphological operations, and finally an area-wide thresholding to accurately extract the tumor affected region of the MRI scan. However this algorithm is only applicable on T1-type MRI scan images. The MRI scan images have been collected from several databases available for research purposes. These research databases contain medical images of patients with brain tumors to facilitate the development and validation of new image processing algorithms [10]. Images produced by BrainWeb [11, 12] have also been used. This is a simulated brain database created to help validate medical image analysis algorithms. Only T1-type images have been utilized from each of the mentioned databases.

2.1 Preprocessing

The MRI images were initially converted to gray scale with intensity values ranging from 0–255. Further, a simple skull stripping algorithm was applied on each of the scan images to get rid of the unwanted portion of the brain which might interfere with the result of the segmentation step.

2.2 Segmentation of Scan Image

A unique quality of a T1-type MRI scan is that fat and abnormalities appear bright in the image produced, whereas water or fluid-filled cells appear dark. Hence, T1 type scans are better suited for tumor related analysis. The scan images are represented by a range of pixel values. Even after preprocessing, we still have 256 different pixel values that are present in the scans to represent different portions of the brain. Even the tumor segment of the brain would be represented using more than one pixel value. Under these circumstances it becomes difficult to differentiate the tumor-affected area from the rest of the brain. To solve this problem, we propose to segment the scan image into a specific number of clusters, to accurately describe the original image while reducing complexity in the extraction process. To determine the optimal number of clusters for each scan image, we utilize the value of the cluster validity index for each image, as suggested by Ray and Rose Turi [13].

2.3 Rough Set-Based K-Means Algorithm

The rough-based K-means [14] algorithm implements principles of rough set theory in the K-means segmentation algorithm to improve its accuracy in segmenting boundary pixels. In the traditional K-means algorithm, we hard cluster each pixel, making it compulsory that each pixel can belong to one cluster only. However, this may not be always true for boundary pixels. The rough set-based K-means algorithm considers the fact that boundary pixels may be associated with more than one cluster. However, a boundary pixel may have equal probability of belonging to more than one cluster. The introduction of the concepts of a rough set efficiently resolves such cases. Considering a given set of objects, $Ob = \{ob_1, ob_2, \dots, ob_n\}$, which we need to divide into k number of clusters, $Y = \{y_1, y_2, \dots, y_k\}$. Each cluster center will have a set of lower approximation pixels, $L(y_i)$ and upper approximation objects, $U(y_i)$. They will be generated using the following standard rules:

1. Any object ob_i can be in the lower approximation set of at most one cluster center. This ensures that no two lower approximation sets overlap.
2. Any object ob_i which belongs to the lower approximation set of a cluster center belongs to upper approximation by default $(ob_i \in L(y_i) \rightarrow ob_i \in U(y_i))$. This suggests that lower approximation of a cluster center is a subset of its corresponding upper approximation $(L(y_i) \subseteq U(y_i))$.
3. Any object ob_i which is not part of the lower approximation set of any cluster center belongs to the upper approximation sets of two or more cluster centers. This introduces the fact that an object cannot belong to a single boundary cluster.

Hence there is an amendment on the cluster center calculation formula of the traditional K-means. Also, extra overhead is added when we undertake to determine to which cluster a boundary object should be assigned, if it were to be included in more than one upper approximation set. Incorporating the above restrictions, the centroid calculation formula can be modified as follows [14]:

if $L(y) \neq \phi$ and $U(y) - L(y) = \phi$ **then**
$y_j = \frac{\Sigma_{v \in L(y)} v_j}{|L(y)|}$
else
if $L(y) = \phi$ and $U(y) - L(y) \neq \phi$ **then**
$y_j = \frac{\Sigma_{v \in (U(y)-L(y))} v_j}{|U(y)-L(y)|}$
else
$y_j = c_{lower} * \frac{\Sigma_{v \in L(y)} v_j}{|L(y)|} + c_{upper} * \frac{\Sigma_{v \in (U(y)-L(y))} v_j}{|U(y)-L(y)|}$
end if
end if

Where c_{lower} and c_{upper} refers to the importance given to the upper and lower approximation sets.

2.4 Morphological Operations

Morphology is a mathematical tool [15] used to extract components of an image which are useful in representing and describing the shape of a region, such as boundaries, skeletons and convex hulls. A basic morphological operation is performed on two sets: the image and the structuring element (also known as the kernel). The structuring element usually used is much smaller in comparison to the image, often a 3×3 matrix. Foreground pixels are represented by logical 1's and background pixels are represented by logical 0's in a binary image.

2.5 Proposed Algorithm

The image has to undergo a pre-processing step as mentioned earlier. The following is the algorithm for the whole extraction process:

1. Read the scan image as input.
2. Segment the image into its respective optimal number of clusters using the rough set-based k-means algorithm.
3. Find the highest intensity value among the cluster centers, say T.
4. Globally convert the image to binary using T as a threshold.
5. Perform closing operation on the resulting image.
6. Calculate the total area of foreground pixels and divide it by 3. This step enables extraction of multiple tumors, if present in the scan image.
7. Perform the opening operation considering the above area threshold. This ensures elimination of unwanted pixels that were initially considered as tumor.

3 Experimental Results

The results of the proposed method were evaluated on more than 150 tumor-affected T1-type MRI scan images. The extraction of tumors of different MRI images using the proposed method is shown in Fig. 1 (five images). Figure 1b–e shows the extracted tumor from original images using K-means, kernel-based fuzzy C-means (KFCM) [16], GA-FCM [9] and our proposed method, respectively, and Fig. 1a, f shows the original image and ground truth images. The comparison of performance of each of the algorithms is shown using measuring indices, e.g., the number of missed alarms (MA), the number of false alarms (FA) and the number of overall

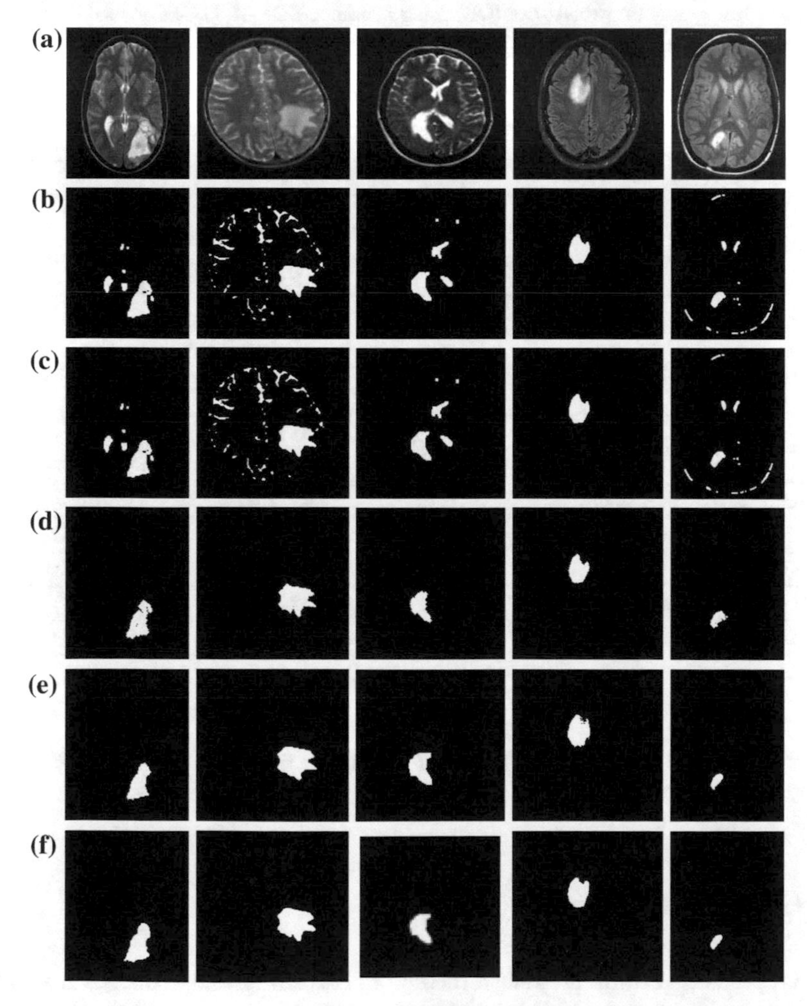

Fig. 1 Output image using (**b**) K-means (**c**) K-FCM (**d**) GA-FCM (**e**) proposed method, where (**f**) shows the ground truth images, (**a**) shows the original images

Table 1 Comparison of false alarms (FA) and missed alarms (MA) using K-means, K-FCM, GA-FCM, and our method

	FA	FA	FA	FA	MA	MA	MA	MA
	K-means	K-FCM	GA-FCM	PROP	K-means	K-FCM	GA-FCM	PROP
IMAGE1	55	103	110	2	129	248	15	122
IMAGE2	660	660	641	133	4446	4446	16	199
IMAGE3	82	144	130	32	117	230	1	25
IMAGE4	108	108	146	23	11	11	9	89
IMAGE5	0	0	1	5	445	807	152	41

Table 2 Comparison of overall error (OE) using K-means, K-FCM, GA-FCM, and the proposed method

	OE	OE	OE	OE
	K-means	K-FCM	GA-FCM	PROP
IMAGE1	184	351	125	124
IMAGE2	5106	5106	657	332
IMAGE3	199	374	131	57
IMAGE4	119	119	155	112
IMAGE5	445	807	153	46

Table 3 Comparison of accuracy (ACC) using K-means, K-FCM, GA-FCM, and the proposed method

	ACC	ACC	ACC	ACC
	K-means	K-FCM	GA-FCM	PROP
IMAGE1	98.4949	99.2110	99.4640	99.468
IMAGE2	98.8087	98.8087	99.5893	99.7925
IMAGE3	98.3377	99.1155	99.4177	99.7466
IMAGE4	99.4711	99.4711	99.3111	99.7308
IMAGE5	98.3998	99.1176	99.6966	99.9087

errors (OE) and accuracy (ACC) values [9]. The percentage of accuracy of the proposed method is better than that of the existing algorithms (shown in Tables 1, 2 and 3).

4 Conclusion

The proposed algorithm presents a method of extracting tumor images from MRI scan images using a rough set-based K-means algorithm and morphological operations. The proposed method extracts tumor images more accurately than other existing tumor image extraction techniques.

References

1. Zhou, J., K.L. Chan, V.F.H. Chong, and S.M. Krishnan. 2005. Extraction of brain tumor from MR images using one-class support vector machine. In *27th Annual conference on IEEE engineering in medicine and biology*, 6411–6414.
2. Sivaramakrishnan, A., and M. Karnan. 2013. A novel based approach for extraction of brain tumor in MRI images using soft computing techniques. *International Journal of Advanced Research in Computer and Communication Engineering* 2 (4): 1845–1848.
3. Thapaliya, K., and Goo-Rak Kwon. 2012. Extraction of brain tumor based on morphological operations. In *8th International conference on computing technology and information management*, 515–520.
4. Masroor Ahmed, M., and M. Dzulkifli Bin. Segmentation of brain MR images for tumor extraction by combining Kmeans clustering and Perona–Malik anisotropic diffusion model. *International Journal of Image Processing* 2 (1): 27–34.
5. Gordilloa, N., E. Montseny, and P. Sobrevilla. 2013. State of the art survey on MRI brain tumor segmentation. *Magnetic Resonance Imaging* 31 (8): 1426–1438.
6. Zacharaki, E.I., S. Wang, S. Chawla, D. Soo Yoo, R. Wolf, E.R. Melhem, and C. Davatzikos. 2009. Classification of brain tumor type and grade using MRI texture and shape in a machine learning scheme. *Magnetic Resonance in Medicine* 62: 1609–1618.
7. Gladis Pushpa Rathi, V.P., and S. Palani. 2012. Brain tumor MRI image classification with feature selection and extraction using linear discriminant analysis.
8. Patil, R.C., and A.S. Bhalchandra. 2012. *International Journal of Electronics, Communication and Soft Computing Science and Engineering (IJECSCSE); Amravati* 2 (1): 1–4.
9. Halder, A., A. Pradhan, S.K. Dutta, and P. Bhattacharya. 2016. Tumor extraction from MRI images using dynamic genetic algorithm based image segmentation and morphological operation. In *International conference on communication and signal processing*, 1845–1849.
10. Mercier, L., R.F. Del Maestro, K. Petrecca, D. Araujo, C. Haegelen, and D.L. Collins. 2012. On-line database of clinical MR and ultrasound images of brain tumors. *Medical Physics* 39 (6): 3253–61.
11. http://www.bic.mni.mcgill.ca/brainweb/.
12. Cocosco, C.A., V. Kollokian, R.K. Kwan, and A.C. Evans. 1997. BrainWeb: Online interface to a 3D MRI simulated brain database, neuroImage. In *Proceedings of 3-rd international conference on functional mapping of the human brain*, 5(4).
13. Ray, S., and H. Rose Turi. 2001. Clustering-based color image segmentation. Ph.D. Thesis, Monash University, Australia.
14. Lingras, P., and G. Peters. 2012. Rough sets: Selected methods and applications in Engineering and management.
15. Gonzalez, R.C., and R.E. Woods. 1992. *Digital image processing*. Boston: Addison-Wesley.
16. Selvakumar, J., A. Lakshmi, and T. Arivoli. 2012. Brain tumor segmentation and its area calculation in brain MR images using K-mean clustering and fuzzy C-mean algorithm. In *IEEE-international conference on advances in engineering, science and management*.

A Mechanism for Detection of Text in Images Using DWT and MSER

B. N. Ajay and C. Naveena

Abstract The study of video optical character readers (OCR) is an eminent field of research in image processing due to various real-time applications. Hence, in this chapter, an algorithm is proposed for text detection in images. The proposed method consists of mainly three stages in which the image is initially sharpened using a Gaussian filter. A discrete wavelet transform (DWT) is then applied to the sharpened image. After this step, the maximally stable extremal region (MSER) is detected to obtain the average of the detailed components of the wavelet images. To detect foreground components in an image, the Connected Component Analysis (CCA) algorithm is applied to localized text components. The proposed method is evaluated on standard MSER and MSRA-TD500 datasets. Experimental results are satisfactory.

Keywords DWT · MSER · Gaussian · CCA · OCR

1 Introduction

Document image analysis is an area of research in the field of digital image processing and pattern recognition. The document is considered as an image and processed to produce a machine-readable form. With the enormous growth of digital data, recognizing a text pattern in an image is a challenging task. According to official statistics of the popular video portal YouTube, more than four billion videos are viewed per day and about 60 h of video are uploaded every minute [1]. Recognizing text within an image is challenging because some of the text detection involves a content-based search, extracting text from scene images, or identifying

B. N. Ajay (✉)
Department of Computer Science and Engineering, VTU-RRC, Belagavi, India
e-mail: ajaybn30@gmail.com

C. Naveena
Department of Computer Science and Engineering,
SJB Institute of Technology, Bangalore, India

A. N. Krishna et al. (eds.), *Integrated Intelligent Computing, Communication and Security*, Studies in Computational Intelligence 771,
https://doi.org/10.1007/978-981-10-8797-4_68

vehicle number plates, sign detection and translation for helping tourists. The challenges that arise are due to large variation in text character font, color, size, texture, motion, shape and geometry [2].

A text detection and recognition framework has been proposed in [1]. An edge-based multi-scale text detector is used for identifying potential text candidates, and stroke width transform (SWT)- and support vector machine (SVM)-based verification procedures are used to remove false alarms. In [2], a different orientation of English and Chinese text is taken into consideration. Deep learning models are demonstrated to classify textual and non-textual components in [3]. In [4], an adaptive color scheme is developed by observing image color histograms. The neural network is combined with extreme machine learning to separate the textual contents from non-textual contents. Pipeline concepts are used to obtain fast and accurate text [5]. This concept is used to predict the text of arbitrary orientation. Text regions are estimated by applying the fully convolutional neural network concept. For every group of characters, a part-based tree-structured model [TSM] is constructed so as to make use of character-specific structure information; additionally, local appearance information is presented in [6]. SnooperText, a multiresolution system for text detection in complex visual scenes, is used to detect the textual information embedded in photos of building facades; it was developed for the "iTowns" urban geographic information project reported in [7]. A video text detection and tracking system is presented in [8]. The candidate text regions are identified roughly by using the stroke model and morphological operation. Stroke map and edge response are combined to localize text lines in each candidate text region. To verify the text blocks, several heuristic and SVM methods are used.

In [7], an in-vehicle real-time system able to localize text and communicate it to drivers is presented. In this chapter, the region of interest (ROI) is localized as the maximally stable extremal region (MSER) and divided into 4×4 cells; the number of edges in each cell are then counted and compared to a well-defined threshold to differentiate the text from the image. An enhanced MSER algorithm for detecting candidate characters is applied, and character candidates are then filtered by stroke width variations for removing regions with too many variations. A hybrid algorithm for detection and tracking of text in natural scenes is presented in [9] where MSER is used to detect text asynchronously, and in a separate thread, detected text objects are tracked by MSER propagation. An algorithm for detecting text in natural and complex images was presented as a two-step approach combining MSER and stroke width transformation for the detection of text in [10]. In [11], text detection in complex images of natural scenes is presented. Initially concentrated areas of the text candidate are identified based on the location. Then, fine detection of candidate characters is done by the enhanced MSER algorithm based on the luminance, followed by edge detection.

Challenges in text detection in any images include varying font, background and illumination. Applying the MSER technique enables detection of stable extremal regions of character candidates but fails to extract small candidates from blurred images and also with varying frequency of the images. This can easily be handled by a discrete wavelet transform (DWT), since it is able to extract a varying-frequency image from the original image. Motivated by these advances, in this chapter we have proposed text detection from an image using a combination of the DWT and MSER.

The rest of the chapter is organized as follows. Section 2 presents the proposed method. Section 3 discusses the experimental results obtained using standard datasets. Finally, in Sect. 4, conclusions are drawn.

2 Proposed Method

In this section, we present the proposed methodology. The proposed method consists of three stages. The image is initially sharpened using a Gaussian filter. A DWT is then applied to the sharpened image. After this step, the maximally stable extremal region (MSER) is detected to obtain the average of the detailed components of the wavelet images. To detect foreground components in an image, the Connected Component Analysis (CCA) algorithm is finally applied to localize text components. A detailed explanation is provided in the following subsections.

2.1 Gaussian Function

A Gaussian filter is a non-uniform low-pass filter used to sharpen an image using a Gaussian function, as shown in Eq. 1.

$$f\left(x|\mu,\sigma^2\right) = \frac{1}{\sqrt{2\pi\sigma^2}} e^{-\frac{(x-\mu)^2}{2\sigma^2}} \tag{1}$$

where σ represents the standard deviation of the distribution, σ^2 is the variance and μ is the mean. The filter uses two dimensions as a point-spread function. The Gaussian nature of the filter is maintained by increasing the value of standard deviation (σ); hence, Gaussian filtering is more effective for sharpening images. Figure 1 shows the sharpened image of the input image.

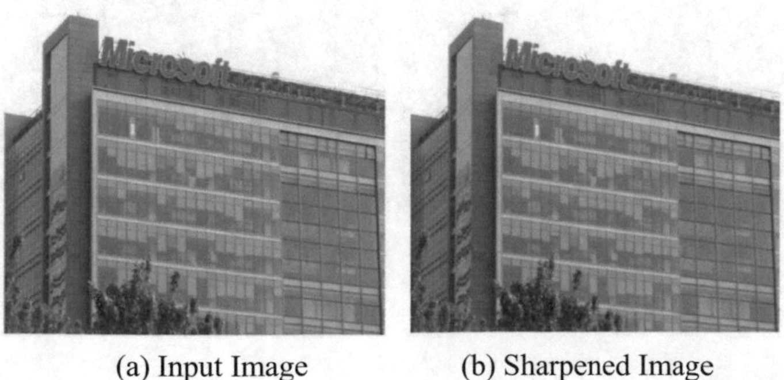

(a) Input Image (b) Sharpened Image

Fig. 1 Output obtained after applying the Gaussian function

2.2 DWT

After sharpening the image, a DWT is applied, whereby an image is decomposed into a set of mutually orthogonal wavelet basis functions. The two-dimensional DWT of an image function $s(n_1, n_2)$ of size $N_1 \times N_2$ may be expressed as Eqs. 2 and 3:

$$W_{\varphi}(j_0, k_1, k_2) = \frac{1}{\sqrt{N_1 N_2}} \sum_{n_1=0}^{N_1-1} \sum_{n_2=0}^{N_2-1} s(n_1, n_2)\varphi_{j_0, k_1, k_2}(n_1, n_2) \tag{2}$$

$$W_{\omega}^{i}(j_0, k_1, k_2) = \frac{1}{\sqrt{N_1 N_2}} \sum_{n_1=0}^{N_1-1} \sum_{n_2=0}^{N_2-1} s(n_1, n_2)\omega_{j_0, k_1, k_2}^{i}(n_1, n_2) \tag{3}$$

where $i = \{H, V, D\}$ indicates the directional index of the wavelet function. As in a one-dimensional case, j_0 represents any starting scale, which may be treated as $j_0 = 0$ (Fig. 2).

2.3 MSER

We have used combined detail components produced from the DWT for MSER to detect foreground components. The MSER is used for locating the potential character in a cropped word image as shown in Fig. 3a. It is one of the segmentation techniques to extract the candidate region. The main idea is to identify regions which remain stable over a threshold range in terms of intensity values. The connected component-based methods utilize the difference between the text and the background to extract the connected regions, as shown in Fig. 3b. Then, heuristic

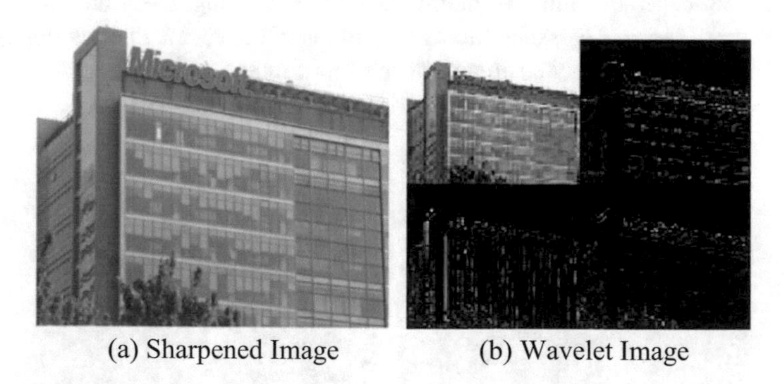

(a) Sharpened Image (b) Wavelet Image

Fig. 2 Output obtained after applying the DWT

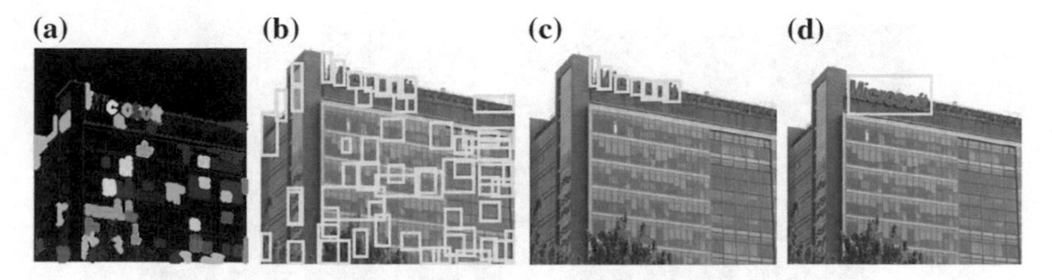

Fig. 3 Extraction of text from the input image

rules such as aspect ratio, size, geometric features and SWT [1] to filter the non-text connected regions are used, as shown in Fig. 3c. Finally, the bounding boxes are merged, as shown in Fig. 3d.

3 Experimental Results

To evaluate the performance of proposed method and performance is analyzed using Matlab 14 on 3.1 GHz with 8 GB RAM. The method was tested on 2 different datasets such as the multi-script robust reading competition (MRRC) [12] and (MSRA-TD500) [13] datasets. The performance measures used in the proposed method are recall (R), precision (P) and f-measure (F).

The MRRC dataset consists of 167 training images and 167 testing images with all kinds of challenges. The images in the dataset contain challenges like shading, slant, gloss, occlusion, shear, engrave, low resolution, depth, night vision, emboss, artistic, illumination, multi-script, multi-color, handwritten and curve, with a variety of degradation in languages like Kannada, English, Hindi and Chinese. Figure 5 shows sample results of the proposed method. Table 1 presents analysis results of the proposed method and an existing method using the MRRC dataset. The proposed method yields better results than the existing method (Fig. 4).

The MSRA-TD500 dataset is a publicly available benchmark containing 500 images collected from various locations. The images captured contain varieties like signs, doorplates, caution plates, guide boards, bill boards and many more items with complex backgrounds. The images pose various challenges like font, size, color and orientation with a diversity of text and background complexity. The dataset contains languages like Chinese, English and a mixture of both. Figure 7 shows the sample results of the proposed method. Table 2 presents the quantitative

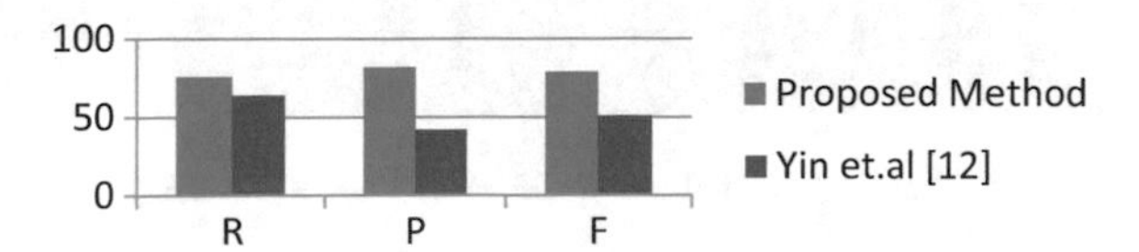

Fig. 4 Graphical representation of the results obtained (with the MRRC dataset)

Fig. 5 Results obtained with the MRRC dataset

Table 1 Comparison between the proposed method and an existing method (with the MRRC dataset)

Method	R	P	F
Proposed method	76	82	79
Yin et al. [12]	64	42	51

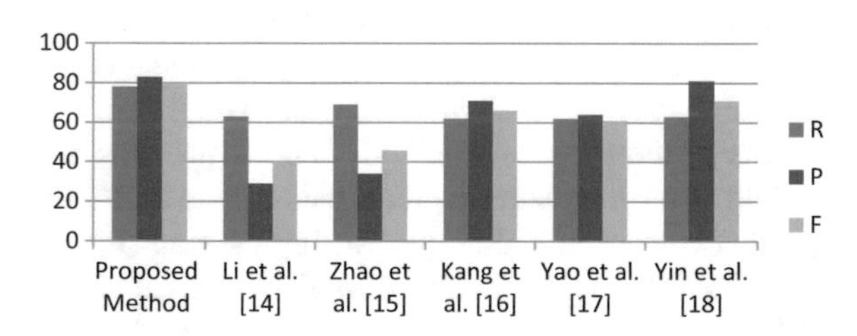

Fig. 6 Graphical representation of the results obtained (using the MSRA-TD500 dataset)

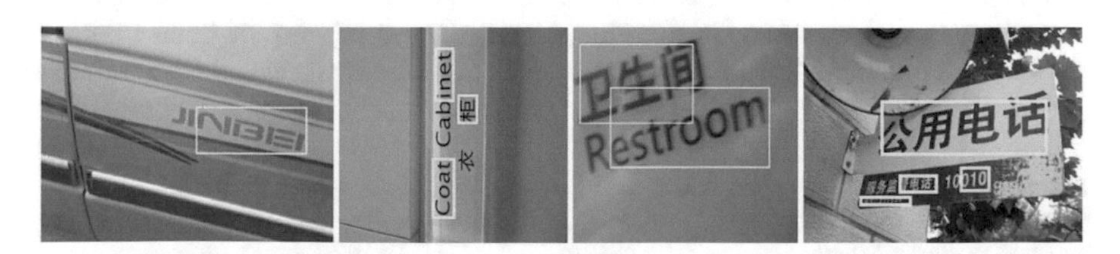

Fig. 7 Results obtained using the MSRA-TD500 dataset

Table 2 Comparison between the proposed method and existing method (using the MSRA-TD500 dataset)

Method	R	P	F
Proposed method	78	83	80
Li et al. [14]	63	29	40
Zhao et al. [15]	69	34	46
Kang et al. [16]	62	71	66
Yao et al. [17]	62	64	61
Yin et al. [18]	63	81	71

analysis of the proposed and existing method using the MSRA-TD500 dataset (Fig. 6). The proposed method yields better results compared to the existing method.

4 Conclusion

This chapter proposes a detection algorithm for images which consists of mainly three stages. Initially, the image is sharpened using a Gaussian filter. Then, a DWT is applied to the sharpened image to enhance the text recognition capability of MSER in varying frequency. Experiments conducted using standard MRRC and MSRA-TD500 datasets indicate that the proposed method performs better than existing methods. A limitation of our method is non-detection of curved text, and a few false alarms were also found, which will be resolved in the future.

References

1. Yang, H., B. Quehl, and H. Sack. 2014. A framework for improved video text detection and recognition. *Multimedia Tools and Applicattion, Springer* 69: 217–245.
2. Phan, T.Q., P. Shivakumar, S. Bhowmick, S. Li, and C. L. Tan. 2014. Semi-automatic ground truth generation for text detection and recognition in video images. *IEEE Transactions on Circuits and Systems for Video Technology.*
3. He, T., W. Huang, Y. Qiao, and J. Yao. 2016. Text-attentional convolutional neural network for scene text detection. *IEEE Transactions on Image Processing* 25 (6): 2529–2541.
4. Wu, H., B. Zou, Y.Q. Zhao, and J. Guo. 2017. Scene text detection using adaptive color reduction, adjacent character model and hybrid verification strategy. *The Visual Computer* 33 (1): 113–126.
5. Zhou, X., C. Yao, H. Wen, Y. Wang, S. Zhou, W. He, and J. Liang 2017. *EAST: an efficient and accurate scene text detector*. arXiv:1704–03155, 2017.
6. Shi, C., C. Wang, B. Xiao, S. Gao, and J. Hu. 2014. *End-to-end scene text recognition using tree-structured models*. Pattern Recognition: Elsevier.
7. Minetto, R., N. Thome, M. Cord, N.J. Leite, and J. Stolfi. 2013. SnooperText: A text detection system for automatic indexing of urban Scenes. *Computer Vision and Image Understanding, Elsevier* 122: 92–104.
8. Yusufu, T., Y. Wang, and X. Fang. 2013. *A video text detection and tracking system*, 522–529. IEEE: In the proceedings of international symposium on multimedia.

9. Gomez, L., and D. Karatzas. 2014. *MSER- based real-time text detection and tracking*. In *22nd international conference on pattern recognition*, IEEE, pp. 3110–3115.
10. Tabassum, A., and S.A. Dhondse. 2015. Text detection using MSER and stroke width transform. In *Fifth international conference on communication system and network technologies*, IEEE, 568–571.
11. Turki, H., M. B. Halima, and A. M Alimi. 2015. *Scene* text detection images with pyramid image and MSER enhanced. In *15th international conference on intelligent systems design and application* (ISDA), IEEE.
12. Yao, C., X. Bai, W. Liu, Y. Ma, and Z. Tu. 2012. *Detecting texts of arbitrary orientations in natural images*. CVPR, 1083–1090.
13. Shivakumar, P., H.T. Basavaraju, D.S. Guru, and C.L. Tan. 2013. Detection of curved text in video: quad tree based method. In 12th international conference on document analysis and recognition(ICDAR), 594–598.
14. Lu, C., C. Wang, and R. Dai. 2005. *Text detection in images based on unsupervised classification of edge-based features*. In *Proceedings of ICDAR*, 610–614.
15. Zhao, X., K.H. Lin, Y. Fu, Y. Hu, Y. Liu, and T.S. Huang. 2011. Text from corners: a novel approach to detect text and caption in videos. *IEEE Transactions on IP* 20: 790–799.
16. Kang, L., Y. Li, & D. Doermann. 2014. Orientation robust text line detection in natural images. In *Proceedings of CVPR*, 4034–4041.
17. Yao, C., X. Bai, and W. Liu. 2014. A Unified framework for multi-oriented text detection and recognition. *IEEE Transactions IP* 23: 4737–4749.
18. Yin, X.C., W.Y. Pei, J. Zuang, and H.W. Hao. 2015. Multi-orientation scene text detection with adaptive clustering. *IEEE Transactions PAMI* 37: 1930–1937.

A Novel Approach for Segmentation of Arecanut Bunches Using Active Contouring

R. Dhanesha and C. L. Shrinivasa Naika

Abstract Arecanuts are among the main commercial crops of southern India. Identifying ripeness is important for harvesting arecanut bunches and directly affects the farmer's profits. Manual identification and harvesting processes, however, are very tedious, requiring many workers for each task. Therefore, in recent years, image processing and computer vision-based techniques have been increasingly applied for fruit ripeness identification, which is important in optimizing business profits and ensuring readiness for harvesting. Thus, segmentation of arecanut bunches is required in order to determine ripeness. There are several techniques for segmenting fruits or vegetables after harvesting to identify ripeness, but there is no technique available for segmenting bunches before harvesting. In this chapter, we describe a computer vision-based approach for segmentation using active contouring, with the aim of identifying the ripeness of arecanut bunches. The experimental results confirm the effectiveness of the proposed method for future analysis.

Keywords Ripeness · Harvesting · Segmentation · Erosion · Closing · Arecanut bunches · Active contour

1 Introduction

Arecanuts play a prominent role in religious, social and cultural functions, as well as in the economic life of Indian people. In India, around 600 hectares (HA) contain different varieties of arecanut plants. Farmers can incur losses from harvesting

---

R. Dhanesha (✉) · C. L. Shrinivasa Naika
Department of Computer Science and Engineering, U.B.D.T. College of Engineering, Davanagere 577002, Karnataka, India
e-mail: dhaneshphddoc@gmail.com

C. L. Shrinivasa Naika
e-mail: naika2k6@gmail.com

A. N. Krishna et al. (eds.), *Integrated Intelligent Computing, Communication and Security*, Studies in Computational Intelligence 771,
https://doi.org/10.1007/978-981-10-8797-4_69

immature arecanuts. The price of arecanuts depends upon the maturity level, and this in turn is dependent on the arecanut content.

Rathod et al. [1] reported that each arecanut contains 50–60% sugar, 15% lipids (glycerides of lauric, myristic and oleic acids), 15% condensed tannins (phlobatannin, catechin), 15% polyphenols and 0.2–0.5% alkaloids (arecoline, arecaidine, guvacine and guvacolin). The tannin content determines the maturity level. By market standards, if the tannin content is high, the nut is immature (near mature level) and can be sold at a good price compared to matured nuts.

There are different levels of arecanuts with respect to ripeness, including Api, Bette, Minne and Gotu. The farmer needs to know the ripeness level of the arecanut bunches before harvesting. Nowadays, there are fewer laborers available for agriculture field work, so it is difficult for farmers to complete the work in time. As mentioned previously, there are different varieties of arecanuts available, and if the farmer fails to harvest the arecanut bunches in time, they may incur a huge monetary loss.

The segmentation of natural images prior to classification is an important task in computer vision applications. Arecanut images are natural images, and analysis of these images plays an important role in the Indian market. To classify the arecanut, visual descriptors of an image are used. There are several automated techniques available for other fruits and crops, but no technique is available for identifying ripeness of arecanut bunches before harvesting. Thus, such an identification system is needed, and this requires segmentation of arecanut bunches.

1.1 Literature Review

In precision agriculture, there are many methods for classifying food, crops and fruits. Mustafa et al. [2] developed a grading system for fruit images. In their work, a few artificial intelligence methods are discussed, including adaptive boosting (AdaBoost), nearest-neighbor classifier (k-NN), linear discriminant classifier (LDC), fuzzy nearest-neighbor classifier (fuzzy k-NN) and support vector machines (SVM), with AdaBoost and SVM found to be successful in defect segmentation.

Sarkate et al. [3] proposed a system for predicting the yield of flowers and flower segmentation from an image using color segmentation by setting a threshold. With the use of a hue saturation value color model and examination of the histogram, flower color was able to be differentiated.

A robust technique has been proposed for segmenting food images from a background using color images [2]. This algorithm includes (i) extraction of a high-contrast gray image using red-green-blue (RGB) components; (ii) computation of a global threshold by statistical methods; and (iii) use of morphological operation gaps to fill in the segmented binary image. Physical properties such as volume, minor axes and major axes are measured to identify the parameters of the detection method. Mery et al. presented a robust method for classifying tomato images [4], and an aspect ratio method was presented for the detection of tall, medium and short kiwi fruits in [5].

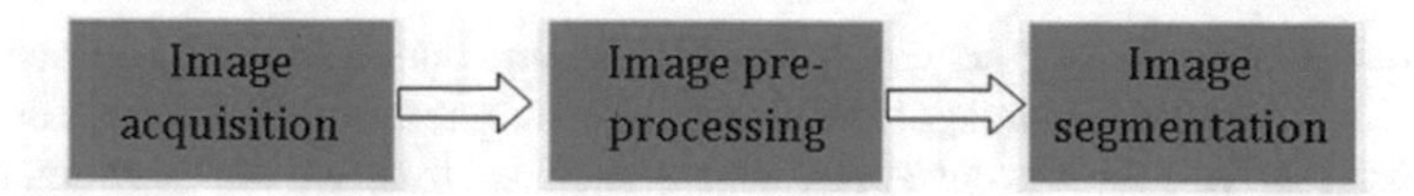

Fig. 1 System architecture

Lee et al. [6] proposed a method of rapid color grading for fruit quality evaluation using color segmentation. The paper described the segmentation of an image by considering color as the main feature for determining the grading of fruits such as tomatoes and dates. Another paper [7] described a segmentation technique for detecting defects in oranges using a graph-based approach and k-means clustering.

Danti and Suresha [8] proposed a method for classifying raw arecanuts based on variations in the upper and lower limits of RGB components in arecanuts. The two categories are classified using red and green as dominant components.

The literature survey showed that the segmentation technique is a major step in classifying agricultural products such as fruits [2, 4] and flowers [3], and can be done effectively to aid farmers in the grading of products, including the classification of arecanuts after ripening [8]. However, methods for identifying ripeness before harvesting are lacking.

As described in this chapter, in order to segment arecanut bunches, images went through several stages of preprocessing including resizing, grayscale conversion, binary thresholding and morphological techniques for region enhancement. Active contouring was then used for segmenting bunches of natural arecanut images. In this experiment, we used 20 images, which were captured using mobile cameras. Further details are given in the methodology section (Fig. 1).

2 Methodology

The first step of the considered architecture is reading the input image. This is followed by the preprocessing step, which includes converting the RGB image to a grayscale image and then to a binary image by setting the threshold. The binary image is created with morphological operations including erosion and closing to identify the region of interest. Morphology is a set of mathematical operations used in image processing that starts with considering the structuring element, applying it to an input image, and then creating an output image of the same size as the input image. Here, we use the erosion and closing morphology operations. Erosion is used to remove the pixels on the object boundaries, considering a disc as a structuring element, and the closing operation is used to fill the gaps in an image. Closing is performed by applying a dilation operation on an image, after which erosion is applied using the same disc as a structuring element.

A major step in image processing is the segmentation of the image to extract the necessary features. In the identification of the ripeness level of arecanut bunches before harvesting, the segmentation of the arecanut bunch is the main task. This consists in partitioning an image into homogeneous regions that share some common properties. Active contours have been widely used as attractive image segmentation methods because they always produce sub-regions with continuous boundaries.

In this work, we used the active contour method to segment the image. The idea behind this is quite simple, and uses the sparse-field level-set method. It segments the image into foreground and background using a mask. In this work, the state of the active contour is initially given by a mask window. The mask window size is the same as the input image size with pixel value. Later, the mask window size decreases based on the contour of the object region. The input image for the active contour method is a gray image. The output image is a binary image where the foreground is white and the background is black. The output images of active contour methods are performed with morphological operations. The final step of this work is converting the output image to the original RGB image. Thus, arecanut bunches are segmented.

3 Results and Discussion

The results of segmentation by active contouring are presented in Fig. 2 by considering two images. In our study, in-field images of arecanut bunches are captured at different ripeness levels, because there are no datasets available. Figure 2a shows that the input images are RGB, converted to grayscale, and then converted to a binary image by setting the threshold to perform fast computation. Then morphological operations are used to obtain the region of interest.

Figure 2b shows the output of active contouring used to segment the arecanut bunches from an input image. Figure 2c shows the output image of active contouring converted to an RGB color image. The steps for segmentation of the arecanut bunches are as follows:

- Read the input image.
- Convert to grayscale.
- Covert the grayscale image to a binary image by setting the threshold.
- Apply morphology erosion and closing operations.
- Apply active contouring.
- Segmentation of the arecanut bunches is obtained.

Table 1 shows the performance metric evaluation used to measure the segmentation technique. In this work, two metrics are used, volumetric overlap error (VOE) and dice similarity coefficient (DSC). VOE can be calculated using the formula given below:

$$VOE = ((|A \cap B|/|A \cup B|)1)100. \tag{1}$$

DSC can be calculated using the formula:

$$DSC = 2|A \cap B|/(|A| + |B|). \tag{2}$$

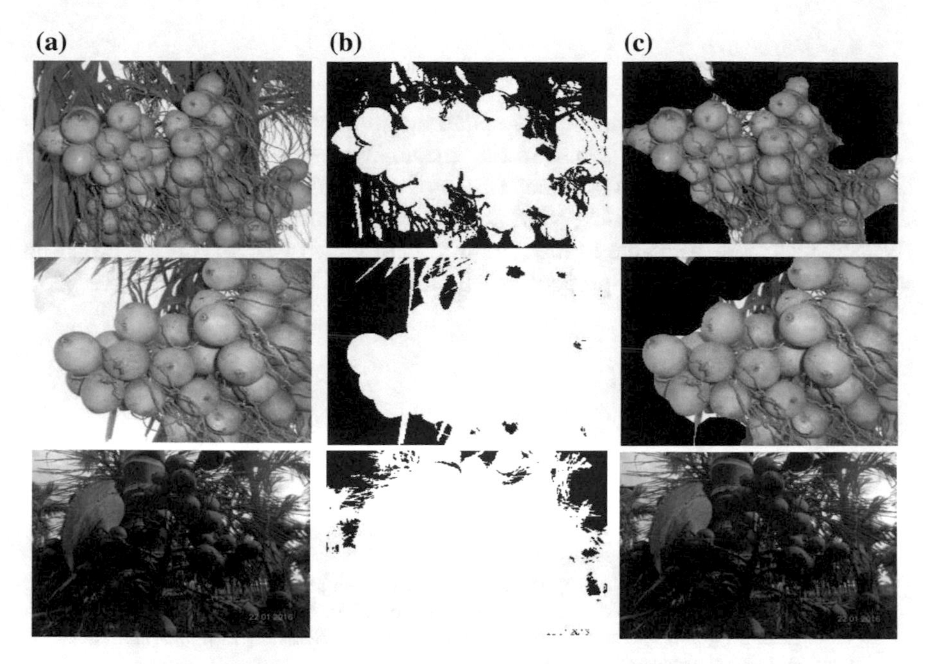

Fig. 2 Segmentation results for image 1, image 2 and image 5. **a** Input image. **b** Active contour result **c** Segmentation of an arecanut bunch

Table 1 Performance metrics used for segmentation evaluation

Input images	VOE	DSC
Image 1	−0.0352	0.0060
Image 2	−0.0156	0.0052
Image 3	−0.0391	0.0049
Image 4	−0.0010	0.0075
Image 5	−0.2852	0.0028

Table 1 shows that the obtained values of VOE and DSC are comparatively good. As we mentioned, there is no technique available for segmentation of arecanut bunches, so we compared our results with the ground truth. As shown in Table 1, image 5 does not perform as well in VOE and DSC because the captured image is of poor quality. We conducted the experiment for 20 images. The mean and standard deviation for VOE are −0.0227 and 0.0156, respectively. The mean and standard deviation for DSC are 0.0059 and 0.0265, respectively. Hence, our technique is effective, and provides a simple means for segmenting arecanut bunches from a given image.

4 Conclusion

In this chapter, we have presented segmentation using active contouring, which accurately segments arecanut bunches in the image. We tested the method on a set of 20 images, which included different stages of arecanut ripeness. The accuracy of the results was measured using VOE and DSC performance metrics. There are no benchmark results available in the literature with which to compare the proposed method for efficiency. The proposed method may not segment properly for poor-quality images.

References

1. Rathod, Kirankumar, M. Shivprasad, and Rajshekhar. 2015. Characterization and extraction of tannin from areca nut waste and using it as rust deactivator. *International Journal of science, Engineering and Technology* 366–372.
2. Mustafa, Nur Badariah Ahmad, Syed Khaleel Ahmed, Zaipatimah Ali, Wong Bing Yit, Aidil Azwin Zainul Abidin, and Zainul Abidin Md Sharrif. 2009. Agricultural produce sorting and grading using support vector machines and fuzzy logic. In *IEEE international conference on signal and image processing applications*.
3. Sarkate, Rajesh S., N.V. Kalyankar Dr, and P.B. Khanale Dr. 2013. Application of computer vision and color image segmentation for yield prediction precision. In *International conference on information systems and computer networks*.
4. Mery, Domingo, and Franco Pedreschi. 2005. Segmentation of color food images using a robust algorithm. *Journal of Food Engineering* 66: 353–360.
5. Rashidi, Majid and Keyvan Seyfi. 2007. Classification of fruit shape in kiwifruit applying the analysis of outer dimensions. *International Journal of Agriculture and Biology*; 15608530, 095759762.
6. Lee, Dah-Jye, James K. Archibald, and Guangming Xiong. 2011. Rapid color grading for fruit quality evaluation using direct color mapping. *IEEE Transactions on Automation Science and Engineering* 8 (2).
7. Pham, Van Huy, and Byung Ryong Lee. 2015. An image segmentation approach for fruit defect detection using K-means clustering and graph-based algorithm. *Vietnam Journal of Computer Science* 2 (1): 25–33. https://doi.org/10.1007/s40595-014-0028-3.
8. Danti, Ajit, and Suresha. 2012. Segmentation and classification of raw arecanuts based on three sigma control limits. *Elsevier, Procedia Technology* 4: 215–219.

Prediction of Crime Hot Spots Using Spatiotemporal Ordinary Kriging

Shilpa S. Deshmukh and Basava Annappa

Abstract Prediction can play a very important role in many types of domains, including the criminal justice system. Even a little information can be gained from proper police assignments, which can increase the efficiency of the crime patrolling system. Citizens can also be aware and alert for possible future criminal incidents. This was identified previously, but the proposed solutions use many complex features, which are difficult to collect, especially for developing and underdeveloped countries, and the maximum accuracy obtained to date using simple features is around 66%. Few of these countries have even started collecting such criminal records in digital format. Thus, there is a need to use simple and minimal required features for prediction and to improve prediction accuracy. In the proposed work, a spatiotemporal ordinary kriging model is used. This method uses not only minimal features such as location, time and crime type, but also their correlation to predict future crime locations, which helps to increase accuracy. Past crime hot spot locations are used to predict future possible crime locations. To address this, the Philadelphia dataset is used to extract features such as latitude, longitude, crime type and time of incident, and prediction can be given for every 0.36 square km per day. The city area is divided into grids of 600×600 m. According to the evaluation results, the average sensitivity and specificity obtained for these experiments is 90.52 and 88.63%, respectively.

Keywords Crime analysis · Vector point pattern · Spatiotemporal interpolation
Geostatistics · Kriging · Predictive analytics · Spatiotemporal smoothing
Covariance function

S. S. Deshmukh (✉) · B. Annappa
National Institute of Technology, Surathkal 575025, Karnataka, India
e-mail: shilpa10aug@gmail.com

B. Annappa
e-mail: annappa@ieee.org

A. N. Krishna et al. (eds.), *Integrated Intelligent Computing, Communication and Security*, Studies in Computational Intelligence 771,
https://doi.org/10.1007/978-981-10-8797-4_70

1 Introduction

Nowadays, computer statistics processes are used in most modern law enforcement techniques and strategies. This can help in tracking a particular crime and responding to it in a structured way [8]. The goal of the predictive policy is to prevent crime from happening rather than responding to these crimes faster. This can help in dropping crime risk or solving earlier crimes, deciding correct law enforcement and police assignment for a particular area. In geostatistics, kriging is used to predict the response over a spatial region given data at various locations throughout the region. A variogram first estimates the spatial dependence. Weighted averages of observed values are then used to predict the response at many locations. An advantage of ordinary kriging is that it provides a measure of the probable error associated with these estimates. The covariance models and the spatial anisotropy are similar in each year while crime number varies year to year. No personal data of any kind are used to generate the predictions. The different types of kriging include ordinary kriging, universal kriging, indicator kriging, co-kriging, simple kriging and others. The type of kriging is selected according to the characteristics of the data and type of spatiotemporal model desired. In the spatiotemporal estimation method of ordinary kriging, error variance (kriging variance) is minimized. It is based on the variogram and configuration of the data, so it is homoscedastic. It is not dependent on the data to be used for estimation.

Many developing countries have datasets with limited features such as crime location, time and type of crime. Also, crime data are highly sensitive in nature, as they include related personal data. Therefore, accessing and sharing such data is highly restricted and can even become a national security concern. Similar work has been done using the Dhaka City dataset [7], but the probabilistic model used does not utilize the correlation information available from the limited features. Many hidden patterns of crime can be considered if the correlation among attributes is considered. We have proposed an approach where we can exploit the maximum information gained by the limited features available in the dataset. Along with predicting criminal hot spots, the mapping frequency of crimes would be calculated for the respective region, depending on the *Dispatch_Date_Time* data.

2 Related Work

Prediction using a point pattern density model [4] is a multivariate method, but it uses many features such as families per square mile, per-household annual expenditure on personal care, insurance, pension, monthly income, distance from a critical border, and drug dealing centers. Thus, even though it provides higher prediction accuracy, we cannot apply this model to the dataset with limited features. Also, tools such as CRIMETRACER and approaches such as data mining use information from social media to predict crime locations, which is not the case with the Philadelphia dataset. The mathematical model approach used for crime prediction, which considers limited data features, is also a good model, but its prediction accuracy is low [12]. Similarly, there are some prediction software tools such as PredPol [8] and IBM SPSS Crime Prediction Analytic Solution (CPAS) [2]. PredPol is built using the Epidemic Type Aftershock Sequence (ETAS) model, which is a self-learning algorithm used for predicting aftershocks from earthquakes. It uses three features, namely, crime location, time and crime type. But the estimation of parameters for the ETAS models is performed by the simulated annealing (SA) algorithm. SA is a local search method invented to avoid local minima. The main shortcomings of this algorithm are its slowness and difficulty in tuning of its parameters [10]. PredPol has outperformed the kernel density model, which has been proven as a highly accurate spatial estimator [1, 13]. Also, the IBM SPSS CPAS [2] uses structured as well as unstructured historical data, which include crime events, behaviors, maps, typology and enabling factors such as weather and trigger events, holidays or paydays. Since such complex features are not available in the Philadelphia dataset, another approach is required. Random forest modeling [9] is also used to forecast crime in the city of Philadelphia, but the accuracy obtained with this method is 66%. According to the novel approach used by Dhaka City to identify spatiotemporal crime patterns [7], the sensitivity and specificity of crime prediction are 79.24 and 68.2%, respectively. The features used in this prediction methodology are limited, but instead of considering the correlation among the features, it uses the joint probability model to predict future crimes. Thus there is a need to build a prediction model which can use limited features while providing the best possible accuracy.

3 Proposed Methodology

The Philadelphia dataset from January 2011 to December 2016 is considered for the proposed work. It has 1,117,775 crimes information with 14 attributes each. Some information is missing for 8664 crimes and attributes are Dc_Dist, PSA, Dispatch_Date_Time, Dispatch_Date, Dispatch_Time, Hour, Dc_Key, Location_Block, UCR_General, Text_General_Code, Police_Districts, Month, Longitude and Latitude. During data cleaning, the missing spatial coordinates are calculated using *Location_Block* and the remaining missing values are removed. In data reduction, the unwanted features that contribute little to prediction, such as UCR_General, DC_Key and derived attributes such as Month, Dispatch_Date, Dispatch_Time and Hour of Dispatch_Date_Time, are removed. Dc_District, PSA, and Police_Dist show some significant pattern during analysis which can contribute to prediction, but the goal is to use the minimal and most contributory features. Therefore, five attributes are selected, which are Dispatch_Date_Time, Location_Block, Text_General_Code, Latitude, and Longitude. Feature engineering is performed over the selected parameters, namely, Latitude, Longitude and Dispatch_Date_Time, and features such as month, day and week of the year are created by using domain knowledge, which makes a machine learning algorithm work. These feature engineered attributes are further processed to generate input data frames for kriging experiments. Since crimes such as offenses against family and children, homicide, gambling violations, receiving stolen property, arson, liquor law violation, forgery and counterfeiting, embezzlement and public drunkenness tend to zero, only nine major crime categories are considered and are grouped into six categories: thefts, vandalism, other thefts which include thefts from vehicle and motor vehicle theft, narcotics, fraud, and burglary. The dataset dimensions after data pre-processing are $1{,}086{,}814 \times 8$. The data from January 2011 to December 2015 are used for training, while the 2016 data are used for testing. A required data transformation is performed as well. The data are analyzed to understand pattern, trend and seasonality of crime data according to month, year, day and hour. After the trend removal, according to the seasonality of data, the window size for the month, week and day are set as 6, 27 and 46, respectively. These input frames are used as observed points, and the prediction is calculated at non-observed points. The total observed points considered for the above-mentioned crimes are 159, 101, 104, 144, 145 and 132, respectively. These are selected after checking their threshold value of a number of criminal records for each crime category. Thresholds are set as 70, 20, 20, 30, 17 and 7, respectively, depending on the maximum crime count at a particular location and the total number of crime locations. These maximum crime counts are 2718, 287, 131, 442, 354 and 227, respectively. Both point and block prediction can be performed using the proposed work. In point prediction, the next day's crime locations can be predicted, while in block prediction, the next 7 days' crime locations can be predicted. The R package used is *CompRandFld* from the CRAN repository [6]. It is applied in RStudio software.

3.1 Estimate Parameters of Covariance Function and Choose the Best Fit Covariance Model

After completing the spatiotemporal structural analysis, valid covariance model fitting is performed. This gives the estimation of parameters, which are included in the spatiotemporal ordinary kriging equation. These parameters are useful in obtaining weights, which are used in the spatiotemporal predictor. These parameters are $scale_s$, $scale_t$, $power_s$, $power_t$, *nugget* and *sill*. Here, sill is the common variance of a random field, while the nugget is the local variation at the origin, which is also the white noise. $scale_s$ and $scale_t$ are spatial and temporal scales, respectively, while range represents the distance limit beyond which the data are no longer correlated. The three different ways to fit a covariance model automatically are the least squares (LS) method, maximum likelihood-based (ML) method and composite likelihood (CL) method. A parametric valid covariogram is fitted to the pseudo data derived from any of these methods. LS uses covariances of the observed data at each lag distance, while ML deals with the data directly, assuming that the random field is Gaussian. CL, which is like the LS method, uses pseudo data instead of observed data. But the advantage of working even if the data are not in a standard setting

Table 1 Mean square error (MSE) after applying corresponding covariance model

$C(h, u)$ Model name	Month (Win. Size = 6)	Week (Win. Size = 27)	Day (Win. Size = 46)
Separable covariance models			
Gneiting	20.62823	0.2335314	0.02325184
Exp_exp	3.13401	0.2162201	0.0232527
Exp_cauchy	277.885	0.3491654	0.02325269
Matern_cauchy	NA	NA	NA
Stable_stable	NA	NA	NA
Non-separable covariance models			
Gneiting	1.24173	0.2162218	0.02999429
Gneiting_GC	18.84342	0.2162389	0.0232597
Iacocesare	13.8873	0.2220265	NA
Porcu	11.6992	NA	0.02329915
Porcu2	2.69023	0.217113	0.02325078

is obtained by using the CL method [11]. Thus, maximum CL fitting of Gaussian random fields is used when fitting experimental data.

The covariance matrix is the output of the fitting phase. It is used while applying covariance models and kriging is performed. The covariance models [6] which are applied and the best models which are selected on the basis of least mean square error (MSE) value for the concerned dataset are shown in Table 1. In this experiment, since porcu2, exp_exp, and Gneiting are giving least MSE, they are selected models for the day, week and month data input frames, respectively.

Let the observations in the spatiotemporal domain be captured as $\{Z(s,t) : s \in D, t \in T\}$, with $D \subset R^2$ and $T \subset R$ [5] and correlation model be $R(h, u)$ where h represents the distance between spatial points and u represents time lags in terms of number of days, and let (s_i, t_i) and (s_j, t_j) be any two pairs of spatiotemporal locations on $R^2 \times R$ where $i = 1, 2, 3, \ldots, n$. Here, h and u are given as $h = s_i - s_j$ and $u = t_i - t_j$; then the covariance model is [6] as shown in Eq. (1):

$$\begin{aligned} C(h,u) &= sill + nugget, \quad if \quad h = 0 \quad and \quad u = 0 \\ C(h,u) &= sill * R\left(\frac{h}{scale_s}, \frac{u}{scale_t}, \frac{h}{power_s}, \frac{u}{power_t}\right), \quad if \quad h > 0 \quad or \quad u > 0 \end{aligned} \tag{1}$$

$$R(h,u) = \frac{e^{\frac{-h^{\nu}}{(1+u^{\lambda})^{0.5\gamma\nu}}}}{1+u^{\lambda}}, \quad R(h,u) = \frac{e^{-h^{\nu}(1+u^{\lambda})^{0.5\gamma\nu}}}{(1+u^{\lambda})^{1.5}}, \quad R(h,u) = R(h)R(u) \tag{2}$$

Among the spatiotemporal models shown in Eq. (2), the first two are non-separable and used for monthly (Gneiting) and daily (Porcu2) input, respectively. The range of parameters v and λ is [0,2], while parameter γ is [0,1], and for $\gamma = 0$ it represents a separable model. The third is the separable model for weekly (exp_exp) input. It is obtained as the multiplication of a spatial correlation model and a temporal correlation model as shown previously.

3.2 Train Model and Process the Data

Let $Z(s_0, t_0)$ be an unknown point value at an unobserved point (s_0, t_0) and $Z(s_0, t_0)$ be a second-order stationary random field with a constant but unknown μ and a known covariance function C(h,u). The ordinary kriging equations provide the weights needed for the kriging prediction of the random field at an unobserved point, and these can be expressed in terms of covariance function. The intermediate outputs of the model fitting process are shown in Fig. 1. The spatiotemporal ordinary kriging equations in terms of the covariance function are shown in Eq. (3) as given by [5]:

$$\sum_{j=1}^{n} \lambda_j C(s_i - s_j, t_i - t_j) - \alpha = C(s_i - s_0, t_i - t_0), \quad \forall i = 1, \ldots, n$$

$$\sum_{i=1}^{n} \lambda_i = 1, \quad \textit{otherwise} \tag{3}$$

where α is the Lagrange multiplier associated with the condition of being unbiased. The corresponding prediction variance is shown in Eq. (4):

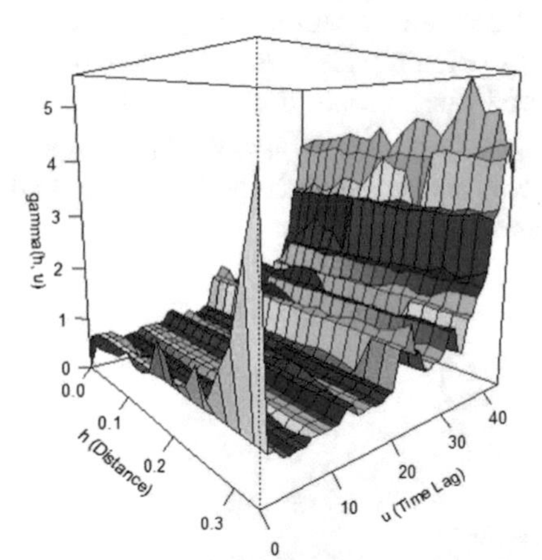

(a) Empirical spatio-temporal semivariogram corresponding to the data of crime category of thefts

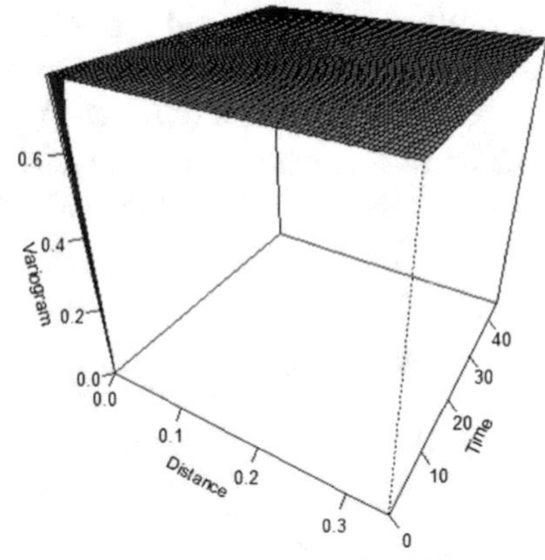

(b) Fitted semivariogram corresponding to the porcu2 covariance function for the crime category of thefts

Fig. 1 Intermediate outputs of model fitting

$$V(Z^*(s_0, t_0) - Z(s_0, t_0)) = C(0, 0) - \sum_{i=1}^{n} \lambda_i C(s_i - s_0, t_i - t_0) + \alpha \quad (4)$$

The post-processing is carried out on the predicted values of *i*th day, that is, $pred_{dy,i}$, $pred_{wk}$ and $pred_{mn}$, to obtain the improved results. The final prediction is given after the following three conditions are satisfied for the *i*th day:

$$pred_{dy,i} >\geq= \beta_1 \quad and \quad ((pred_{wk} \times P_{w,i}) >= \beta_2 \quad or \quad (pred_{mn} \times P_{m,i}) >= \beta_3)$$

where $P_{m,i}$ is the probability of a crime occurring in a month and $P_{w,i}$ is the probability of a crime occurring in a week on the *i*th day, given as follows:

$$P_{m,i} = \frac{count_i}{\sum count_{m,i}} \quad P_{w,i} = \frac{count_i}{\sum count_{w,i}}$$

The threshold values set for the day, week and month data predictions are β_1 as 0.9, β_2 as 0.11 and β_3 as 0.1, respectively. These are finalized using a heuristic approach. The sensitivity and specificity are calculated based on the crime hot spots detected per day per 0.36 square km. The total area of Philadelphia is divided into equal size grids of 600 × 600 m. These total to 1019 grids in Philadelphia.

3.3 *Results*

According to the analysis, the trend decreases, while the crime seasonality pattern is the same every year. The total number of crimes is higher in the middle period of the year, especially during March to September. The prediction of $Z(s_0, t_0)$ is carried out for each crime category using the fixed thresholds, and output for thefts on 10th August 2016 is shown in Fig. 2.

For each of the crime categories—thefts, vandalism, other thefts, narcotic, fraud and burglary—the obtained accuracy in terms of percentage is 86, 84.61538, 92, 83.33333, 82.6667 and 98.03922, respectively. The overall average prediction accuracy is 87.78%. The average sensitivity and specificity obtained is 90.52 and 88.63%, respectively.

4 Conclusion and Future Work

There is an improvement in prediction of around 21% compared to PredPol, which is around 66% [8]. The only difference is that the accuracy of PredPol is in terms of spatial locations, while kriging is in terms of the spatial grid. There is the flexibility to

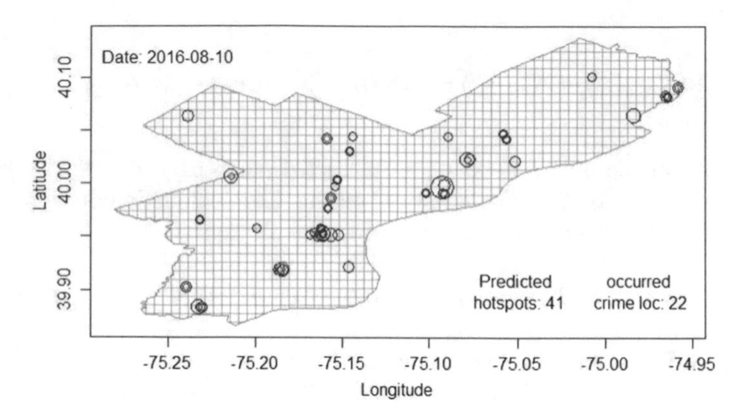

Fig. 2 Prediction performed for crime category of thefts

choose the specified locations as an input. In order to satisfy the goal of minimal generalized features and maximum accuracy of the outcome, univariate spatiotemporal ordinary kriging are used. Specifically, for Philadelphia, one can apply co-kriging using Dc_Key, PSA and Police_District attributes. If spatiotemporal co-kriging is used and filtering-based clustering [3] is applied before kriging, the results can be improved even more.

References

1. Chainey, Spencer, Lisa Tompson, and Sebastian Uhlig. 2008. The utility of hotspot mapping for predicting spatial patterns of crime. *Security Journal - Palgrave Macmillan* 21: 1–2.
2. IBM Business Analytic Software Services. 2012. IBM SPSS Crime Prediction Analytics Solution (CPAS) Service. IBM Corporation.
3. Kang, Qiao, Wei-keng Liao, Ankit Agrawal, and Alok Choudhary. 2016. A filtering-based clustering algorithm for improving spatio-temporal kriging interpolation accuracy. In *Proceedings of the 25th ACM international on conference on information and knowledge management*, 2209–2214.
4. Liu, Hua, and Donald E. Brown. 2003. Criminal incident prediction using a point-pattern-based density model. *International Journal of Forecasting - Elsevier* 19: 603–622.
5. Mateu, Jorge, and others. 2015. *Spatial and spatio-temporal geostatistical modeling and kriging*, vol. 998, 162–267. New York: Wiley.
6. Padoan, Simone and Moreno Bevilacqua. 2015. Composite-Likelihood Based Analysis of Random Fields. https://cran.r-project.org/web/packages/CompRandFld/CompRandFld.pdf
7. Parvez, Md Rizwan, Turash Mosharraf, and Mohammed Eunus Ali. 2016. A Novel approach to identify spatio-temporal crime pattern in Dhaka city. In *Proceedings of the eighth international conference on information and communication technologies and development*. ACM.
8. PredPol. 2015. How PredPol Works: We Provide Guidance on Where and When to Patrol. http://www.predpol.com/
9. Ritter, Nancy. 2013. Predicting recidivism risk: New tool in Philadelphia shows great promise. *National Institute of Justice Journal* 271: 4–13.

10. Vakil-Baghmisheh, Mohammad-Taghi, and Alireza Navarbaf. 2008. A modified very fast simulated annealing algorithm. In *IEEE international symposium on telecommunications, IST 2008*, 61–66.
11. Varin, Cristiano, Nancy Reid, and David Firth. 2011. An overview of composite likelihood methods. In *Statistica Sinica - JSTOR*, 5–42.
12. Zheng, Xifan, Yang Cao, and Zhiyu Ma. 2011. A mathematical modeling approach for geographical profiling and crime prediction. In *2011 IEEE 2nd international conference on software engineering and service science (ICSESS)*, 500–503.
13. Zhou, Zhengyi, and David S. Matteson. 2015. Predicting ambulance demand: A spatio-temporal kernel approach. In *Proceedings of the 21th ACM SIGKDD international conference on knowledge discovery and data mining*, 2297–2303.

Sentiment Analysis for Text Extracted from Twitter

G. Jalaja and C. Kavitha

Abstract In today's world, reviews and opinions available to us are a key factor shaping our perspectives and affecting the success of a brand, service or product. With the emergence and proliferation of social media, Twitter has become a popular means for individuals to express their opinions. While Twitter data is incredibly illuminating, analyzing the data presents a challenge given its sheer size and disorganized nature. This work is focused on gathering complicated information and conducting sentiment analysis of tweets related to colleges, including neutral tweets and other than pre-tagged lexicons present in dictionary. To begin with, gathering of unstructured information from Twitter, directs to preprocessing of the same leads in finding of user's sentiment. The established naïve Bayes-based algorithm is performed to classify the data, and the tweets are analyzed to determine user sentiment. The results are represented graphically.

Keywords Sentiment analysis · Assessment analysis · Probability theory
Opinion investigation · Sentiment polarity

1 Introduction

Today's world has witnessed a rapid proliferation of web services, web innovations, and various types of web-based social networking sites—for example, discussions forums, microblogs, and distributed systems—which provide massive amounts of data. In the area of education, the web allows students to post their feelings and opinions about a number of topics, creating an effective medium for them to express their thoughts and insights regarding specific learning environments, including the college experience.

G. Jalaja (✉)
Visvesvaraya Technological University, Belagavi, Karnataka, India
e-mail: jalajag.shaker@gmail.com

C. Kavitha
Department of CSE, Global Academy of Technology, Bengaluru, India

A. N. Krishna et al. (eds.), *Integrated Intelligent Computing, Communication and Security*, Studies in Computational Intelligence 771,
https://doi.org/10.1007/978-981-10-8797-4_71

Billions of clients are utilizing Twitter that is utilized as rich source of information for mining data. There is an interface for recovering the Twitter information known as Twitter corpus which gives free data as a stream. Analysis of this data has prompted an assortment of research. Illustrations of incorporate expectation of decisions and the share trading system for example information of public health, natural disasters analysis, notification of events such as earthquakes, public sentiment estimation during recession and elections. Such user-generated information can be extremely valuable in surveying the overall population's assessments, estimation, and repercussion towards colleges.

Recent survey has been uncovered such type of online surveys from public that has assumed essential part in colleges and web-based business organizations. The data posted on such platforms is an enormous resource for acquiring the opinion of the overall population. The recovery and investigation of such data is referred to as sentiment analysis or opinion mining.

In today's world, reviews and opinions available to us are a key factor shaping our perspectives and affecting the success of a brand, thing or organization. With advances in electronic person-to-person communication on a global scale, partners routinely take to conveying their decisions on noticeable online interpersonal interaction, to be particular Twitter. Although Twitter information is extremely enlightening, it shows a test for investigation in view of its humongous and confused nature. This undertaking rotates around gathering this complicated information and playing out an assumption investigation of the colleges.

Sentiment analysis allows consumers to conduct research on products so that they can make informed purchase decisions. Marketers and businesses use this analytical data to understand customer sentiment toward goods or services in order to tailor these offerings to meet customer needs. Print or textual information retrieval systems concentrate on processing, searching or analyzing factual data. Facts have a target segment be that as it may; there are some other printed substances which express subjective qualities. These substances are mostly conclusions, notions, evaluations, states of mind, and feelings, which forms the center of sentiment analysis (SA). It offers many testing opportunities to develop new applications, generally of the enormous improvement of accessible information on online sources like websites and social organizations. For instance, suggestions of things proposed by a suggestion framework can be anticipated by considering contemplations, for example, negative or positive emotions about those stuff through the use of sentiment analysis.

2 Related Work

Twitter is a constant data and smaller scale blogging administration that enables clients to post updates. The administration quickly increased overall fame that interfaces with the most recent stories, thoughts, conclusions, and news. It is an effective apparatus for continuous method for speaking with individuals by joining messages that rush to compose, simple to peruse, open and available anyplace. On Twitter anybody can read, compose and share messages or tweets. Conclusion mining is a kind of regular vernacular getting ready for following the perspective of individuals all in about a particular thing. Assessment mining, which is furthermore called data analysis, incorporates building a system to accumulate and investigate assessments about the thing made in blog passages, comments, reviews or tweets. Web-based social networking assumes a critical part in gathering the assessment of the creators. This work concentrates on tweets that will bring about breaking down the perspective of people in general and large about subjects. A tweets puller is produced that naturally gathers irregular feelings and classifier apparatus that performs arrangements on that corpus gathered from Twitter. Order depends on highlights extricated and arranged into NEGATIVE, POSITIVE and NEUTRAL. The outcomes additionally assessed and closed to gather the execution of the characterization through support vector machines (SVM) [1].

An enlarged lexicon based strategy for substance level opinion investigation is connected. In spite of the fact that strategy gives great exactness, the review may be very small. To enhance the review, do the following steps: Initially concentrate some additional obstinate indicators (e.g. tokens and words) through the Chi-square test on the outcomes of the lexicon-based procedure. With the help of the new determined pointers, additional obstinate tweets can be recognized. A brief span later, a supposition classifier is prepared to allot out assessment polarities for components in the as of late perceived tweets. The arrangement data for the classifier is the eventual outcome of the lexicon-based strategy [2]. Along these lines, the whole technique has no manual checking. The proposed approach is an unsupervised procedure beside the basic supposition vocabulary, which is freely available. The reason that our procedure works is that conclusion articulations (counting space particular assessment words, emoticons, everyday articulations, shortened forms, and so on.) relies upon the supposition setting [3].

The approach is to use distinctive ideas of machine learning classifiers and feature extractors. Machine learning classifiers include maximum entropy (MaxEnt), Naïve Bayes and SVM. The feature extractors are unigrams, bigrams and unigrams with syntactic element names [3]. Manufacture a system that regards classifiers and highlight extractors as two unmistakable parts. This structure enables us to effectively experiment with various blends of classifiers and highlight extractors [4].

Totunoglu et al. designed a Wikipedia-based semantic smoothing approach. Naïve Bayes (NB) is one of the most part algorithm for sentiment analysis. The component extractors are unigrams, bigrams, and unigrams with speech form labels. Naïve Bayes having a few favorable circumstances for lower complexity and less difficult preparing technique, it encounters sparsity [5]. Smoothing can be a response for this issue, generally Laplace smoothing is used; however in this paper proposes Wikipedia based semantic smoothing approach.

Sentiment analysis is a growing area of research zone in the field of content mining. Individuals post their opinions in the form of unstructured information so feeling extraction gives general conclusion of audits so it does best occupation for client, individuals, association and so on. The principal point of this work is to discover approaches that produce yield with great precision [5]. This work presents late updates on works identified with grouping of assessment investigation of actualized different methodologies and calculations. Contribution of research gives idea about feature selection and classification of existing gives better results.

3 System Design

3.1 System Architecture

The proposed architecture displays the design and performance of an automated asset administration framework that accomplishes a good adjusts. The proposed framework makes the following three contributions.

Web-based social networking examination helps in understanding trending points better. Inclining themes are subjects and attitudes that have a high volume of posts in online networking. Opinion analysis, also called sentiment mining, utilizes online networking tools to determine attitudes towards an item or idea and here in the usage towards different colleges [6]. Real-time Twitter trending investigation is an extraordinary case of an examination device, in light of the fact that the hash label membership show empowers to tune into particular watchwords and create assessment investigation. Therefore Twitter of micro-blogging service chooses as a data source because it serves as a major forum for online discussions and comments.

3.2 Tweet Retrieval

Gather every important tweet progressively from the whole Twitter movement by means of Tweepy, an easy-to-use python library for accessing the Twitter application program interface (API). Since our application targets diverse schools, manually develops and decides that are straightforward coherent catchphrase combinations to retrieve relevant tweets—those about a particular college and its audits.

3.3 Data Preprocessing

Data preprocessing is an information mining system that involves converting a large amount of data into a reasonable configuration. Real data is often fragmented, conflicting, or potentially ailing in specific practices or inclines, and is probably going to contain more errors. Information preprocessing is a demonstrated strategy for resolving such issues. Information preprocessing prepares raw data for additional processing.

3.4 Naïve Bayes Classification Algorithm

In machine learning, naïve Bayes classifiers are a family of direct probabilistic classifiers based on the application of Bayes' theorem, with strong (naïve) independence assumptions between features. Naïve Bayes is an essential technique for constructing classifiers, which are models that assign class labels to problem events, represented as vectors of feature values, where the class labels are drawn from some finite set of class variables.

Bayes' theorem introduces the concept of posterior probability, $p(C_k \mid x)$, which can be formulated as follows:

$$P(C_k \mid X) = \frac{(P(C_k) * P(X \mid C_k))}{P(X)}$$

In the above condition, $p(C_k)$ is known as the class prior probability, which is the likelihood of the class occurring without considering the informational index; $p(x \mid C_k)$ represents the probability, or the likelihood of the occurrence of an element for a given class; and $p(x)$ is called prove, which is the probability of the information without considering any class.

3.5 Sentiment Polarity

The result of the classification is the probability of the considered feature based on whether it belongs to a specific class. The class with the maximum probability is the outcome of the classification.

3.6 Sentiment in Graphical Representation

The outcome acquired by the classification is converted into a graphical representation for better appreciation by the user. This is finished by plotting the edge of each tweet acquired regarding a school against the time span at which the tweet was posted (Fig. 1).

The data flow diagram (DFD) shown below illustrates the plan or record the particular makeup of a framework. The last yield of the graph is an boundary showing the degree of positive and negative tweets about a college. The user gives the name of the college whose tweets are to be retrieved. This is finished by the

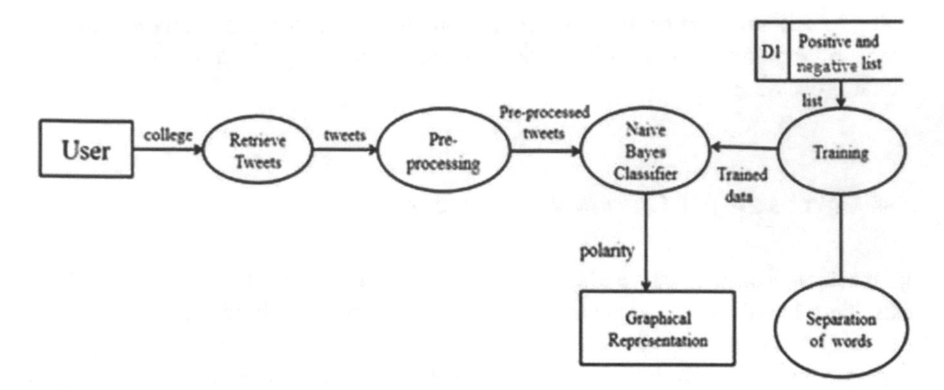

Fig. 1 System architecture

tweet retrieval process. The tweets are retrieved from Tweepy, an open source library that enables Python to communicate with the Twitter platform. The retrieved tweets are then given as input to the preprocessing stage, where they are preprocessed and converted into a readable form. The preprocessing stage is further divided into sub-stages that include the removal of stop words, removal of spam tweets, and the conversion of tweets into an intelligible configuration. These tweets are then fed to the classifier with the goal of classification. The naïve Bayes classifier gives as yield the farthest point that shows to what degree the tweets are negative or positive.

4 Results and Case Studies

The results of the execution and the deductions to be produced using the testing are recorded in the accompanying area. The assessment measurements have been recorded and the outcomes have been evaluated. A real dataset was considered program which runs on some different circumstances for various schools. The outcome is particular to the school bolstered in as the information and gives a sign with regard to the sentiment held by the general population on the individual college.

It contains the different experiments by utilizing the proposed technique of the past segments, these investigations are directed in a controlled situation where the parameters of each of the progressive iterations are kept to the past comparable from the preparation and utilized for learning inside the algorithms. This area is vital and enlightening as it gives us experimental evidence and that the technique received is superior to the current framework set up. By analyzing the effects of the results displayed in Figs. 2 and 3, can go to an educated choice on the viability of the algorithm when contrasted with others utilized as a part of a similar field.

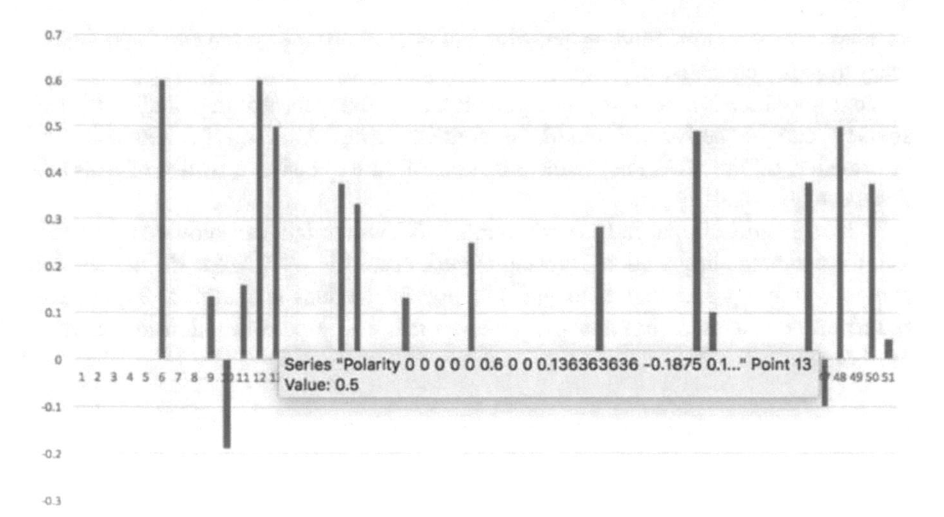

Fig. 2 Graphical representation of output 1

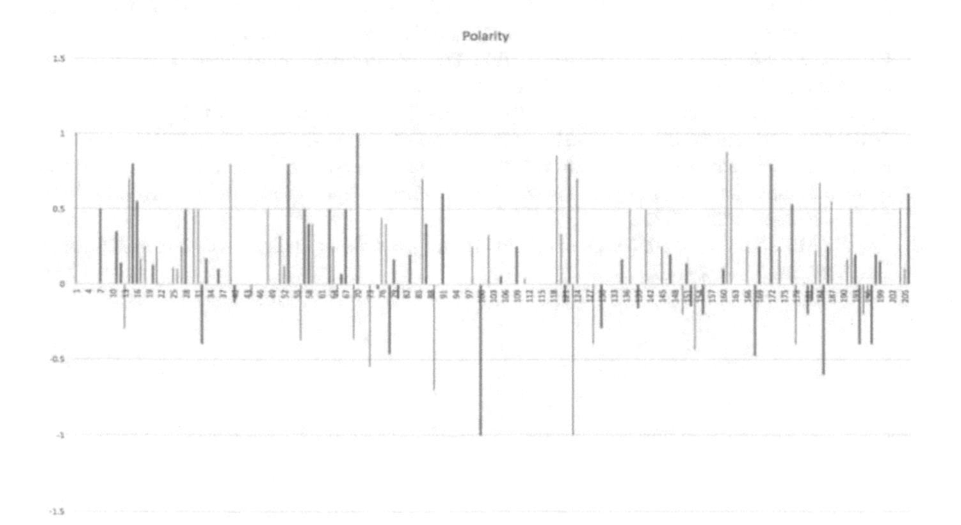

Fig. 3 Graphical representation of output 2

5 Conclusion

To conclude, this paper has presented a successful method for sentiment analysis which can be performed for colleges by collecting a sample opinions and sentiments from Twitter. Throughout the task, a wide range of data analysis was utilized to collect, clean and mine sentiment from the dataset. Such an examination could give significant input to makers and help them to detect a negative hand over watcher's observation. Finding negative patterns at an opportune time can enable them to make educated decisions.

Text classification is a key component in textual data mining, and sentiment analysis can be performed based on content grouping. With the explosion of information on the web, the characterization of an expansive quantity of information becomes challenging.

Within an educational framework, mining of Twitter data can provide a valuable tool for understanding student learning experience. This work helps the association and in a training framework to and the present student educational experience. Based on this organization and participation can easily take decision in the engineering studies.

References

1. Shrivatava, A., S. Mayor, and B. Pant. 2014. Opinion mining of real time twitter tweets. *International Journal of Computer Applications* 100 (19).
2. Zhang, L., R. Ghosh, M. Dekhil, M. Hsu, and B. Liu Combining lexicon-based and learning-based methods for twitter sentiment analysis. Technical report, HP.
3. Go, A., R. Bhayani, and L. Huang. 2014. Twitter sentiment classification using distant supervision, CS224 N Project report, 1–12. Stanford.
4. Totunoglu, Dilara, Gurkan Telseren, Ozgun Sagturk, and Murat C. Ganiz. 2013. *Wikipedia based semantic smoothing for twitter sentiment classification.* IEEE.
5. Pak, Alexander, and Patrick Paroubek. 2013. Twitter as a corpus for sentiment analysis and opinion mining. In: *Proceedings of the seventh conference on international language resources and evaluation*, 1320–1326.
6. Liu, B. 2012. Sentiment analysis and opinion mining. *Synthesis Lectures on Human Language Technologies* 5 (1): 1–167.

Printed in the United States
By Bookmasters